THE OXFORD HANDBOOK OF

PHILOSOPHY OF MATHEMATICS AND LOGIC

OXFORD HANDBOOKS IN PHILOSOPHY

PAUL K. MOSER, GENERAL EDITOR

THE OXFORD HANDBOOK OF

PHILOSOPHY OF MATHEMATICS AND LOGIC

Edited by

STEWART SHAPIRO

UNIVERSITY PRESS

2005

OXFORD
UNIVERSITY PRESS

Oxford University Press, Inc., publishes works that further
Oxford University's objective of excellence
in research, scholarship, and education.

Oxford New York
Auckland Cape Town Dar es Salaam Hong Kong Karachi
Kuala Lumpur Madrid Melbourne Mexico City Nairobi
New Delhi Shanghai Taipei Toronto

With offices in
Argentina Austria Brazil Chile Czech Republic France Greece
Guatemala Hungary Italy Japan Poland Portugal Singapore
South Korea Switzerland Thailand Turkey Ukraine Vietnam

Published by Oxford University Press, Inc.
198 Madison Avenue, New York, New York 10016

www.oup.com

Library of Congress Cataloging-in-Publication Data
The Oxford handbook of philosophy of math and logic / edited by Stewart Shapiro.
p. cm.—(Oxford handbooks in philosophy)
Includes bibliographical references and index.
ISBN-13 978-0-19-514877-0
ISBN 0-19-514877-0
1. Mathematics—Philosophy. 2. Logic, Symbolic and mathematical—Philosophy.
I. Shapiro, Stewart, 1951– II. Series.
QA8.4.O94 2004
510'.1—dc22 2004044847

1 3 5 7 9 8 6 4 2

Printed in the United States of America
on acid-free paper

Preface

This volume provides comprehensive and accessible coverage of the disciplines of philosophy of mathematics and philosophy of logic, including an overview of the major problems, positions, and battle lines. In line with the underlying theme of the series, each author was given a free hand to develop his or her distinctive viewpoint. Thus, the various chapters are not neutral. Readers see exposition and criticism, as well as substantial development of philosophical positions. I am pleased to report that each chapter breaks new ground. The volume not only presents the disciplines of philosophy of mathematics and philosophy of logic, but advances them as well.

For many of the major positions in the philosophy of mathematics and logic, the book contains at least two chapters, at least one sympathetic to the view and one critical. Of course, this does not guarantee that every major viewpoint is given a sympathetic treatment. For example, one of my own pet positions, *ante rem* structuralism, comes in for heavy criticism in two of the chapters, and is not defended anywhere (except briefly in chapter 1). In light of the depth and extent of the disciplines today, no single volume, or series of volumes, can provide extensive and sympathetic coverage of even the major positions on offer. And there would hardly be a point to such an undertaking, since the disciplines are ever evolving. New positions and new criticisms of old positions emerge with each issue of each major philosophy journal. Most of the chapters contain an extensive bibliography. In total, this volume provides a clear picture of the state of the art.

There is some overlap between the chapters. This is to be expected in a work of this scope, and it was explicitly encouraged. Authors often draw interesting, but distinctive, conclusions from the same material. There is, of course, no sharp separation between the philosophy of mathematics and the philosophy of logic. The main issues and views of either discipline permeate those of the other. Just about every chapter deals with matters mathematical and matters logical.

After the Introduction (chapter 1), the book begins with a historical section, consisting of three chapters. Chapter 2 deals with the modern period—Kant and his intellectual predecessors; chapter 3 concerns later empiricism, including John Stuart Mill and logical positivism; and chapter 4 focuses on Ludwig Wittgenstein.

The volume then turns to the "big three" views that dominated the philosophy and foundations of mathematics in the early decades of the twentieth century: logicism, formalism, and intuitionism. There are three chapters on logicism, one dealing with the emergence of the program in the work of Frege, Russell, and Dedekind (chapter 5); one on neologicism, the contemporary legacy of Fregean logicism (chapter 6); and one called "Logicism Reconsidered," which provides a technical assessment of the program in its first century (chapter 7). This is followed by a lengthy chapter on formalism, covering its historical and philosophical aspects (chapter 8). Two of the three chapters on intuitionism overlap considerably. The first (chapter 9) provides the philosophical background to intuitionism, through the work of L. E. J. Brouwer, Arend Heyting, and others. The second (chapter 10) takes a more explicitly mathematical perspective. Chapter 11, "Intuitionism Reconsidered," focuses largely on technical issues concerning the logic.

The next section of the volume deals with views that dominated in the later twentieth century and beyond. Chapter 12 provides a sympathetic reconstruction of Quinean holism and indispensability. This is followed by two chapters that focus directly on naturalism. Chapter 13 lays out the principles of some prominent naturalists, and chapter 14 is critical of the main themes of naturalism. Next up are nominalism and structuralism, which get two chapters each. One of these is sympathetic to at least one variation on the view in question, and the other "reconsiders."

Chapter 19 is a detailed and sympathetic treatment of a predicative approach to both the philosophy and the foundations of mathematics. This is followed by an extensive treatment of the application of mathematics to the sciences; chapter 20 lays out different senses in which mathematics is to be applied, and draws some surprising philosophical conclusions.

The last six chapters of the volume focus more directly on logical matters, in three pairs. There are two chapters devoted to the central notion of logical consequence. Chapter 21 presents and defends the role of semantic notions and model theory, and chapter 22 takes a more "constructive" approach, leading to proof theory. The next two chapters deal with the so-called paradoxes of relevance, chapter 23 arguing that the proper notion of logical consequence carries a notion of relevance, and chapter 24 arguing against this. The final two chapters concern higher-order logic. Chapter 25 presents higher-order logic and provides an overview of its various uses in foundational studies. Of course, chapter 26 reconsiders.

Throughout the process of assembling this book, I benefited considerably from the sage advice of my editor, Peter Ohlin, of Oxford University Press, USA, and from my colleagues and friends, at Ohio State, St. Andrews, and other institutions. Thanks especially to Penelope Maddy and Michael Detlefsen.

Contents

Notes on the Contributors

John P. Burgess, Ph.D. in Logic, Berkeley (1975), has taught since 1976 at Princeton, where he is now Director of Undergraduate Studies. His interests include logic, philosophy of mathematics, metaethics, and pataphysics. He is the author of numerous articles on mathematical and philosophical logic and philosophy of mathematics, and of *Fixing Frege* and (with Gideon Rosen) *A Subject with No Object* (Oxford University Press, 1997).

Charles Chihara is Emeritus Professor of Philosophy at the University of California, Berkeley. He is the author of *Ontology and the Vicious Circle Principle* (1973), *Constructibility and Mathematical Existence* (Oxford University Press, 1990), *The Worlds of Possibility: Model Realism and the Semantics of Modal Logic* (Oxford University Press, 1998), and *A Structural Account of Mathematics* (Oxford University Press, 2004).

Peter J. Clark is Reader in Logic and Metaphysics and Head of the School of Philosophical and Anthropological Studies in the University of St. Andrews. He works primarily in the philosophy of physical science and mathematics and is editor of the *British Journal for the Philosophy of Science.*

Roy Cook is a Visiting Professor at Villanova University and an Associate Fellow at the Arché Research Centre at the University of St. Andrews. He has published on the philosophy of logic, language, and mathematics in numerous journals including *Philosophia Mathematica, Mind, The Notre Dame Journal of Formal Logic, The Journal of Symbolic Logic*, and *Analysis.*

William Demopoulos has published articles in diverse fields in the philosophy of the exact sciences, and on the development of analytic philosphy in the twentieth century. He is a member of the Department of Logic and Philosophy of Science of the University of California, Irvine.

Michael Detlefsen is Professor of Philosophy at the University of Notre Dame. He is the author of *Hilbert's Program: An Essay on Mathematical Instrumentalism* (1986) and editor of *Notre Dame Journal of Formal Logic.*

Solomon Feferman is Professor of Mathematics and Philosophy and the Patrick Suppes Professor of Humanities and Sciences, Emeritus, at Stanford University. He is the author of numerous articles on logic and the foundations of mathematics and of *In the Light of Logic* (Oxford University Press, 1998), editor in chief of the *Collected Works of Kurt Gödel* (vols. I–V, Oxford University Press, 1986–2003), and author with Anita B. Feferman of *Truth and Consequences: The Life and Logic of Alfred Tarski* (forthcoming). Feferman received the Rolf Schock Prize for Logic and Philosophy for 2003.

Juliet Floyd is Associate Professor of Philosophy at Boston University, working primarily on the interplay between logic, mathematics, and philosophy in early twentieth-century philosophy. She has written articles on Kant, Frege, Russell, Wittgenstein, Quine, and Gödel, and (with Sanford Shieh) edited *Future Pasts: The Analytic Tradition in Twentieth Century Philosophy* (Oxford University Press, 2001).

Bob Hale is Professor of Metaphysical Philosophy at the University of Glasgow.

Geoffrey Hellman is Professor of Philosophy at the University of Minnesota. He is author of *Mathematics Without Numbers* (Oxford University Press, 1989) and edited *Quantum Measurement: Beyond Paradox* (1998) with Richard Healey. He has published numerous research papers in philosophy of mathematics, philosophy of physics, and general philosophy of science. He also has an interest in musical aesthetics and remains active as a concert pianist.

Ignacio Jané is Professor of Philosophy in the Department of Logic and the History and Philosophy of Science of the University of Barcelona. His main interests are in the foundations of mathematics, philosophy of mathematics, and philosophy of logic. He is the author of "A Critical Appraisal of Second-order Logic" (*History and Philosophy of Logic*, 1993), "The Role of Absolute Infinity in Cantor's Conception of Set" (*Erkenntnis*, 1995), and "Reflections on Skolem's Relativity of Set-Theoretical Concepts" (*Philosophia Mathematica*, 2001).

Fraser MacBride is a Reader in the School of Philosophy at Birkbeck College London. He previously taught in the Department of Logic & Metaphysics at the University of St. Andrews and was a research fellow at University College London. He has written several articles on the philosophy of mathematics, metaphysics, and the history of philosophy, and is the editor of *The Foundations of Mathematics and Logic* (special issue of *The Philosophical Quarterly*, vol. 54, no. 214 January 2004).

Penelope Maddy is Professor of Logic and Philosophy of Science at the University of California, Irvine. Her work includes "Believing the Axioms" (*Journal of Symbolic Logic*, 1988), *Realism in Mathematics* (Oxford University Press, 1990), and *Naturalism in Mathematics* (Oxford University Press, 1997).

D. C. McCarty is a member of the Logic Program at Indiana University.

Carl Posy is Professor of Philosophy at the Hebrew University of Jerusalem. His work covers philosophical logic, the philosophy of mathematics, and the history of philosophy. He is editor of *Kant's Philosophy of Mathematics: Modern Essays* (1992). A recent publication on logic and the philosophy of mathematics is "Epistemology, Ontology and the Continuum" (in *Mathematics and the Growth of Knowledge*, E. Grossholz, ed., 2001). A recent paper on the history of philosophy is "Between Leibniz and Mill: Kant's Logic and the Rhetoric of Psychologism" (in *Philosophy, Psychology, and Psychologism: Critical and Historical Readings on the Psychological Turn in Philosophy*, D. Jacquette, ed., 2003).

Dag Prawitz is Professor of Theoretical Philosophy at Stockholm University, Emeritus (as of 2001). Most of his research is in proof theory, philosophy of mathematics, and philosophy of language. Some early works include *Natural Deduction: A Proof–Theoretical Study* (1965), "Ideas and Results in Proof Theory" (*Proceedings of the Second Scandinavian Logic Symposium*, 1971), and "Philosophical Aspects of Proof Theory" (*Contemporary Philosophy, A New Survey*, 1981). Some recent ones are "Truth and Objectivity from a Verificationist Point of View" (*Truth in Mathematics*, 1998), "Meaning and Objectivity" (*Meaning and Interpretation*, 2002), and replies to critics in *Theoria* (1998) (special issue, "The Philosophy of Dag Prawitz").

Agustín Rayo received his degree from MIT in 2000, and then spent four years at the AHRB Research Centre for the Philosophy of Logic, Language, Mathematics, and Mind, at the University of St. Andrews. He is Assistant Professor of Philosophy at the University of California, San Diego, and works mainly on the philosophy of logic, mathematics, and language.

Michael D. Resnik is University Distinguished Professor of Philosophy at the University of North Carolina at Chapel Hill. He is the author of *Mathematics as a Science of Patterns* (Oxford University Press, 1997) and *Frege and the Philosophy of Mathematics* (1980), as well as a number of articles in philosophy of mathematics and philosophy of logic.

Gideon Rosen is Professor of Philosophy at Princeton University. He is the author (with John P. Burgess) of *A Subject with No Object: Strategies for Nominalistic Interpretation of Mathematics* (Oxford University Press, 1997).

Lisa Shabel is an Assistant Professor of Philosophy at The Ohio State University. Her articles include "Kant on the 'Symbolic Construction' of Mathematical Concepts" (*Studies in History of Philosophy of Science*, 1998) and "Kant's 'Argument from Geometry'" (*Journal of the History of Philosophy*, 2004). She has also published a monograph titled *Mathematics in Kant's Critical Philosophy: Reflections on Mathematical Practice* (2003).

Stewart Shapiro is the O'Donnell Professor of Philosophy at The Ohio State University and Professorial Fellow in the Research Centre Arché, at the University of St. Andrews. His publications include *Foundations Without Foundationalism: A Case for Second-order Logic* (Oxford University Press, 1991) and *Philosophy of Mathematics: Structure and Ontology* (Oxford University Press, 1997).

John Skorupski is Professor of Moral Philosophy at the University of St. Andrews. Among his publications are *John Stuart Mill* (1989), *English-Language Philosophy, 1750–1945* (1993), and *Ethical Explorations* (1999).

Mark Steiner is Professor of Philosophy at the Hebrew University of Jerusalem. He received his B.A. from Columbia in 1965, was a Fulbright Fellow at Oxford in 1966, and received his Ph.D. at Princeton in 1972 (under Paul Benacerraf). He taught at Columbia from 1970 to 1977, when he joined the Philosophy Department of the Hebrew University. He is the author of *Mathematical Knowledge* (1975) and *The Applicability of Mathematics as a Philosophical Problem* (1998).

Neil Tennant is Distinguished Humanities Professor at The Ohio State University. His publications include *Anti-realism in Logic* (Oxford University Press, 1987) and *The Taming of the True* (Oxford University Press, 1997).

Alan Weir is Senior Lecturer at Queen's University, Belfast, Northern Ireland. He has also taught at the universities of Edinburgh and Birmingham and at Balliol College, Oxford. He has published articles on logic and philosophy of mathematics in a number of journals, including *Mind*, *Philosophia Mathematica*, *Notre Dame Journal of Formal Logic*, and *Grazer Philosophische Studien*.

Crispin Wright is Bishop Wardlaw Professor at the University of St. Andrews, Global Distinguished Professor at New York University, and Director of the Research Centre, Arché. His writings in the philosophy of mathematics include *Wittgenstein on the Foundations of Mathematics* (1980); *Frege's Conception of*

Numbers as Objects (1983); and, with Bob Hale, *The Reason's Proper Study* (Oxford University Press, 2001). His most recent books, *Rails to Infinity* (2001) and *Saving the Differences* (2003), respectively collect his writings on central themes of Wittgenstein's *Philosophical Investigations* and those further developing themes of his *Truth and Objectivity* (1992).

THE OXFORD HANDBOOK OF

PHILOSOPHY OF MATHEMATICS AND LOGIC

CHAPTER 1

PHILOSOPHY OF MATHEMATICS AND ITS LOGIC: INTRODUCTION

STEWART SHAPIRO

1. Motivation, or What We Are Up to

From the beginning, Western philosophy has had a fascination with mathematics. The entrance to Plato's Academy is said to have been marked with the words "Let no one ignorant of geometry enter here." Some major historical mathematicians, such as René Descartes, Gottfried Leibniz, and Blaise Pascal, were also major philosophers. In more recent times, there are Bernard Bolzano, Alfred North Whitehead, David Hilbert, Gottlob Frege, Alonzo Church, Kurt Gödel, and Alfred Tarski. Until very recently, just about every philosopher was aware of the state of mathematics and took it seriously for philosophical attention.

Often, the relationship went beyond fascination. Impressed with the certainty and depth of mathematics, Plato made mathematical ontology the model for his Forms, and mathematical knowledge the model for knowledge generally—to the extent of downplaying or outright neglecting information gleaned from the senses. A similar theme reemerged in the dream of traditional rationalists of extending what they took to be the methodology of mathematics to all scientific and philosophical knowledge. For some rationalists, the goal was to emulate Euclid's *Elements of Geometry*, providing axioms and demonstrations of philosophical principles. Empiricists, the main opponents of rationalism, realized that their orientation to knowledge does not seem to make much sense of mathematics, and they went to some lengths to accommodate mathematics—often distorting it beyond recognition (see Parsons [1983, essay 1]).

Mathematics is a central part of our best efforts at knowledge. It plays an important role in virtually every scientific effort, no matter what part of the world it is aimed at. There is scarcely a natural or a social science that does not have substantial mathematics prerequisites. The burden on any complete philosophy of mathematics is to show how mathematics is applied to the material world, and to show how the methodology of mathematics (whatever it may be) fits into the methodology of the sciences (whatever *it* may be). (See chapter 20 in this volume.)

In addition to its role in science, mathematics itself seems to be a knowledge-gathering activity. We speak of what theorems a given person knows and does not know. Thus, the philosophy of mathematics is, at least in part, a branch of epistemology. However, mathematics is at least prima facie different from other epistemic endeavors. Basic mathematical principles, such as "$7+5=12$" or "there are infinitely many prime numbers," are sometimes held up as paradigms of necessary truths and, a priori, infallible knowledge. It is beyond question that these propositions enjoy a high degree of certainty—however this certainty is to be expounded. How can these propositions be false? How can any rational being doubt them? Indeed, mathematics seems essential to any sort of reasoning at all. Suppose, in the manner of Descartes's first Meditation, that one manages to doubt, or pretend to doubt, the basic principles of mathematics. Can he go on to think at all?

In these respects, at least, logic is like mathematics. At least some of the basic principles of logic are, or seem to be, absolutely necessary and a priori knowable. If one doubts the basic principles of logic, then, perhaps by definition, she cannot go on to think coherently at all. Prima facie, to think coherently just is to think logically.

Like mathematics, logic has also been a central focus of philosophy, almost from the very beginning. Aristotle is still listed among the four or five most influential logicians ever, and logic received attention throughout the ancient and medieval intellectual worlds. Today, of course, logic is a thriving branch of both mathematics and philosophy.

It is incumbent on any complete philosophy of mathematics and any complete philosophy of logic to account for their at least apparent necessity and apriority. Broadly speaking, there are two options. The straightforward way to show that a given discipline appears a certain way is to demonstrate that it is that way. Thus the philosopher can articulate the notions of necessity and apriority, and then show how they apply to mathematics and/or logic. Alternatively, the philosopher can argue that mathematics and/or logic does not enjoy these properties. On this option, however, the philosopher still needs to show why it *appears* that mathematics and/or logic is necessary and a priori. She cannot simply ignore the long-standing belief concerning the special status of these disciplines. There must be something about mathematics and/or logic that has led so many to hold, perhaps mistakenly, that they are necessary and a priori knowable.

The conflict between rationalism and empiricism reflects some tension in the traditional views concerning mathematics, if not logic. Mathematics seems necessary and a priori, and yet it has something to do with the physical world. How is this possible? How can we learn something important about the physical world by a priori reflection in our comfortable armchairs? As noted above, mathematics is essential to any scientific understanding of the world, and science is empirical if anything is—rationalism notwithstanding. Immanuel Kant's thesis that arithmetic and geometry are synthetic a priori was a heroic attempt to reconcile these features of mathematics. According to Kant, mathematics relates to the forms of ordinary perception in space and time. On this view, mathematics applies to the physical world because it concerns the ways that we perceive the physical world. Mathematics concerns the underlying structure and presuppositions of the natural sciences. This is how mathematics gets "applied." It is necessary because we cannot structure the physical world in any other way. Mathematical knowledge is a priori because we can uncover these presuppositions without any particular experience (chapter 2 of this volume). This set the stage for over two centuries of fruitful philosophy.

2. Global Matters

For any field of study *X*, the main purposes of the philosophy of *X* are to interpret *X* and to illuminate the place of *X* in the overall intellectual enterprise. The philosopher of mathematics immediately encounters sweeping issues, typically concerning all of mathematics. Most of these questions come from general philosophy: matters of ontology, epistemology, and logic. What, if anything, is mathematics about? How is mathematics pursued? Do we know mathematics and, if so, how do we know mathematics? What is the methodology of mathematics, and to what extent is this methodology reliable? What is the proper logic for mathematics? To what extent are the principles of mathematics objective and independent of the mind, language, and social structure of mathematicians? Some problems and issues on the agenda of contemporary philosophy have remarkably clean formulations when applied to mathematics. Examples include matters of ontology, logic, objectivity, knowledge, and mind.

The philosopher of logic encounters a similar range of issues, with perhaps less emphasis on ontology. Given the role of deduction in mathematics, the philosophy of mathematics and the philosophy of logic are intertwined, to the point that there is not much use in separating them out.

A mathematician who adopts a philosophy of mathematics should gain something by this: an orientation toward the work, some insight into the role of mathematics, and at least a tentative guide to the direction of mathematics. What sorts of problems are important? What questions should be posed? What methodologies are reasonable? What is likely to succeed? And so on?

One global issue concerns whether mathematical *objects*—numbers, points, functions, sets—exist and, if they do, whether they are independent of the mathematician, her mind, her language, and so on. Define *realism in ontology* to be the view that at least some mathematical objects exist objectively. According to ontological realism, mathematical objects are prima facie abstract, acausal, indestructible, eternal, and not part of space and time. Since mathematical objects share these properties with Platonic Forms, realism in ontology is sometimes called "Platonism."

Realism in ontology does account for, or at least recapitulate, the necessity of mathematics. If the subject matter of mathematics is as these realists say it is, then the truths of mathematics are independent of anything contingent about the physical universe and anything contingent about the human mind, the community of mathematicians, and so on. What of apriority? The connection with Plato might suggest the existence of a quasi-mystical connection between humans and the abstract and detached mathematical realm. However, such a connection is denied by most contemporary philosophers. As a philosophy of mathematics, "platonism" is often written with a lowercase "p," probably to mark some distance from the master on matters of epistemology. Without this quasi-mystical connection to the mathematical realm, the ontological realist is left with a deep epistemic problem. If mathematical objects are in fact abstract, and thus causally isolated from the mathematician, then how is it possible for this mathematician to gain knowledge of them? It is close to a piece of incorrigible data that we do have at least some mathematical knowledge. If the realist in ontology is correct, how is this possible?

Georg Kreisel is often credited with shifting attention from the existence of mathematical objects to the objectivity of mathematical truth. Define *realism in truth-value* to be the view that mathematical statements have objective truth-values independent of the minds, languages, conventions, and such of mathematicians. The opposition to this view is *anti-realism in truth-value*, the thesis that if mathematical statements have truth-values at all, these truth-values are dependent on the mathematician.

There is a prima facie alliance between realism in truth-value and realism in ontology. Realism in truth-value is an attempt to develop a view that mathematics deals with objective features of the world. Accordingly, mathematics has the objectivity of a science. Mathematical (and everyday) discourse has variables that range over numbers, and numerals are singular terms. Realism in ontology is just the view that this discourse is to be taken at face value. Singular terms denote objects, and thus numerals denote numbers. According to our two realisms, mathematicians mean what they say, and most of what they say is true. In short, realism in ontology is the default or the first guess of the realist in truth-value.

Nevertheless, a survey of the recent literature reveals that there is no consensus on the logical connections between the two realist theses or their negations. Each of the four possible positions is articulated and defended by established philosophers of mathematics. There are thorough realists (Gödel [1944, 1964], Crispin Wright [1983] and chapter 6 in this volume, Penelope Maddy [1990], Michael Resnik [1997], Shapiro [1997]); thorough anti-realists (Michael Dummett [1973, 1977]); realists in truth-value who are anti-realists in ontology (Geoffrey Hellman [1989] and chapter 17 in this volume, Charles Chihara [1990] and chapter 15 in this volume); and realists in ontology who are anti-realists in truth-value (Neil Tennant [1987, 1997]).

A closely related matter concerns the relationship between philosophy of mathematics and the practice of mathematics. In recent history, there have been disputes concerning some principles and inferences within mathematics. One example is the law of excluded middle, the principle that for every sentence, either it or its negation is true. In symbols: $A \vee \neg A$. For a second example, a definition is *impredicative* if it refers to a class that contains the object being defined. The usual definition of "the least upper bound" is impredicative because it defines a particular upper bound by referring to the set of all upper bounds. Such principles have been criticized on philosophical grounds, typically by anti-realists in ontology. For example, if mathematical objects are mental constructions or creations, then impredicative definitions are circular. One cannot create or construct an object by referring to a class of objects that already contains the item being created or constructed. Realists defended the principles. On that view, a definition does not represent a recipe for creating or constructing a mathematical object. Rather, a definition is a characterization or description of an object that already exists. For a realist in ontology, there is nothing illicit in definitions that refer to classes containing the item in question (see Gödel [1944]). Characterizing "the least upper bound" of a set is no different from defining the "elder poop" to be "the oldest member of the Faculty."

As far as contemporary mathematics is concerned, the aforementioned disputes are over, for the most part. The law of excluded middle and impredicative definitions are central items in the mathematician's toolbox—to the extent that many practitioners are not aware when these items have been invoked. But this battle was not fought and won on philosophical grounds. Mathematicians did not temporarily don philosophical hats and decide that numbers, say, really do exist independent of the mathematician and, for that reason, decide that it is acceptable to engage in the once disputed methodologies. If anything, the dialectic went in the opposite direction, from mathematics to philosophy. The practices in question were found to be conducive to the practice of mathematics, as mathematics—and thus to the sciences (but see chapters 9, 10, and 19 in this volume).

There is nevertheless a rich and growing research program to see just how much mathematics can be obtained if the restrictions are enforced (chapter 19 in this volume). The research is valuable in its own right, as a study of the logical power of the various once questionable principles. The results are also used to support the underlying philosophies of mathematics and logic.

3. Local Matters

The issues and questions mentioned above concern all of mathematics and, in some cases, all of science. The contemporary philosopher of mathematics has some more narrow foci as well. One group of issues concerns attempts to interpret specific mathematical or scientific results. Many examples come from mathematical logic, and engage issues in the philosophy of logic. The compactness theorem and the Löwenheim–Skolem theorems entail that if a first-order theory has an infinite model at all, then it has a model of every infinite cardinality. Thus, there are unintended, denumerable models of set theory and real analysis. This is despite the fact that we can prove in set theory that the "universe" is uncountable. Arithmetic, the theory of the natural numbers, has uncountable models—despite the fact that by definition a set is countable if and only if it is not larger than the set of natural numbers. What, if anything, do these results say about the human ability to characterize and communicate various concepts, such as notions of cardinality? Skolem (e.g., [1922, 1941]) himself took the results to confirm his view that virtually all mathematical notions are "relative" in some sense. No set is countable or finite *simpliciter*, but only countable or finite relative to some domain or model. Hilary Putnam [1980] espouses a similar relativity. Other philosophers resist the relativity, sometimes by insisting that first-order model theory does not capture the semantics of informal mathematical discourse. This issue may have ramifications concerning the proper logic for mathematics. Perhaps the limitative theorems are an artifact of an incorrect logic (chapters 25 and 26 in this volume).

The wealth of independence results in set theory provide another batch of issues for the philosopher. It turns out that many interesting and important mathematical questions are independent of the basic assertions of set theory. One example is Cantor's *continuum hypothesis* that there are no sets that are strictly larger than the set of natural numbers and strictly smaller than the set of real numbers. Neither the continuum hypothesis nor its negation can be proved in the standard axiomatizations of set theory. What does this independence say about mathematical concepts? Do we have another sort of relativity on offer? Can we only say that a given set is the size of a certain cardinality relative to an interpretation of set theory? Some philosophers hold that these results indicate an indeterminacy concerning mathematical *truth*. There is no fact of the matter concerning, say, the continuum hypothesis. These philosophers are thus anti-realists in truth-value. The issue here has ramifications concerning the practice of mathematics. If one holds that the continuum hypothesis has a determinate truth-value, he or she may devote effort to determining this truth-value. If, instead, someone holds that the continuum hypothesis does not have a determinate truth-value, then he is free to adopt it or not, based on what makes for the most convenient set theory. It is not clear whether the criteria that the realist might adopt to decide the continuum hypothesis are different from the criteria the anti-realist would use for determining what makes for the most convenient theory.

A third example is Gödel's incompleteness theorem that the set of arithmetic truths is not effective. Some take this result to refute mechanism, the thesis that the human mind operates like a machine. Gödel himself held that either the mind is not a machine or there are arithmetic questions that are "absolutely undecidable," questions that are unanswerable by us humans (see Gödel [1951], Shapiro [1998]). On the other hand, Webb [1980] takes the incompleteness results to support mechanism.

To some extent, some questions concerning the applications of mathematics are among this group of issues. What can a theorem of mathematics tell us about the natural world studied in science? To what extent can we *prove* things about knots, bridge stability, chess endgames, and economic trends? There are (or were) philosophers who take mathematics to be no more than a meaningless game played with symbols (chapter 8 in this volume), but everyone else holds that mathematics has some sort of meaning. What is this meaning, and how does it relate to the meaning of ordinary nonmathematical discourse? What can a theorem tell us about the physical world, about human knowability, about the abilities-in-principle of programmed computers, and so on?

Another group of issues consists of attempts to articulate and interpret particular mathematical *theories* and *concepts*. One example is the foundational work in arithmetic and analysis. Sometimes, this sort of activity has ramifications for mathematics itself, and thus challenges and blurs the boundary between mathematics and its philosophy. Interesting and powerful research techniques are often suggested by foundational work that forges connections between mathematical fields. In addition to mathematical logic, consider the embedding of the natural numbers in the complex plane, via analytic number theory. Foundational activity has spawned whole branches of mathematics.

Sometimes developments within mathematics lead to unclarity concerning what a certain concept is. The example developed in Lakatos [1976] is a case in point. A series of "proofs and refutations" left interesting and important questions over what a polyhedron is. For another example, work leading to the foundations of analysis led mathematicians to focus on just what a function is, ultimately yielding the modern notion of function as arbitrary correspondence. The questions are at least partly ontological.

This group of issues underscores the *interpretive* nature of philosophy of mathematics. We need to figure out what a given mathematical concept *is*, and what a stretch of mathematical discourse *says*. The Lakatos study, for example, begins with a "proof" consisting of a thought experiment in which one removes a face of a given polyhedron, stretches the remainder out on a flat surface, and then draws lines, cuts, and removes the various parts—keeping certain tallies along the way. It is not clear a priori how this blatantly dynamic discourse is to be understood. What is the logical form of the discourse and what is its logic? What is its ontology? Much of the subsequent mathematical/philosophical work addresses just these questions.

Similarly, can one tell from surface grammar alone that an expression like "dx" is not a singular term denoting a mathematical object, while in some circumstances, "dy/dx" does denote something—but the denoted item is a function, not a quotient? The history of analysis shows a long and tortuous task of showing just what expressions like this mean.

Of course, mathematics can often go on quite well without this interpretive work, and sometimes the interpretive work is premature and is a distraction at best. Berkeley's famous, penetrating critique of analysis was largely ignored among mathematicians—so long as they knew "how to go on," as Ludwig Wittgenstein might put it. In the present context, the question is whether the mathematician must stop mathematics until he has a semantics for his discourse fully worked out. Surely not. On occasion, however, tensions within mathematics lead to the interpretive philosophical/semantic enterprise. Sometimes, the mathematician is not sure how to "go on as before," nor is he sure just what the concepts are. Moreover, we are never certain that the interpretive project is accurate and complete, and that other problems are not lurking ahead.

4. A Potpourri of Positions

I now present sketches of some main positions in the philosophy of mathematics. The list is not exhaustive, nor does the coverage do justice to the subtle and deep work of proponents of each view. Nevertheless, I hope it serves as a useful guide to both the chapters that follow and to at least some of the literature in contemporary philosophy of mathematics. Of course, the reader should not hold the advocates of the views to the particular articulation that I give here, especially if the articulation sounds too implausible to be advocated by any sane thinker.

4.1. Logicism: a Matter of Meaning

According to Alberto Coffa [1991], a major item on the agenda of Western philosophy throughout the nineteenth century was to account for the (at least) apparent necessity and a priori nature of mathematics and logic, and to account for the applications of mathematics, without invoking anything like Kantian intuition. According to Coffa, the most fruitful development on this was the "semantic tradition," running through the work of Bolzano, Frege, the early Wittgenstein, and culminating with the Vienna Circle. The main theme—or insight, if you will—was to locate the source of necessity and a priori knowledge in the use of *language*. Philosophers thus turned their attention to linguistic matters concerning the pursuit of mathematics. What do mathematical assertions mean? What is their logical form? What is the best semantics for mathematical language? The members of the semantic tradition developed and honed many of the tools and concepts still in use today in mathematical logic, and in Western philosophy generally. Michael Dummett calls this trend in the history of philosophy the *linguistic turn*.

An important program of the semantic tradition was to show that at least some basic principles of mathematics are *analytic*, in the sense that the propositions are true in virtue of meaning. Once we understood terms like "natural number," "successor function," "addition," and "multiplication," we would thereby see that the basic principles of arithmetic, such as the Peano postulates, are true. If the program could be carried out, it would show that mathematical truth is necessary—to the extent that analytic truth, so construed, is necessary. Given what the words mean, mathematical propositions have to be true, independent of any contingencies in the material world. And mathematical knowledge is a priori—to the extent that knowledge of meanings is a priori. Presumably, speakers of the language know the meanings of words a priori, and thus we know mathematical propositions a priori.

The most articulate version of this program is *logicism*, the view that at least some mathematical propositions are true in virtue of their logical forms (chapter 5 in this volume). According to the logicist, arithmetic truth, for example, is a species of logical truth. The most detailed developments are those of Frege [1884, 1893] and Alfred North Whitehead and Bertrand Russell [1910]. Unlike Russell, Frege was a realist in ontology, in that he took the natural numbers to be objects. Thus, for Frege at least, logic has an ontology—there are "logical objects."

In a first attempt to define the general notion of cardinal number, Frege [1884, §63] proposed the following principle, which has become known as "Hume's principle":

> For any concepts *F*, *G*, the number of *F*'s is identical to the number of *G*'s if and only if *F* and *G* are equinumerous.

Two concepts are equinumerous if they can be put in one-to-one correspondence. Frege showed how to define equinumerosity without invoking natural numbers. His definition is easily cast in what is today recognized as pure second-order logic. If second-order logic is logic (chapter 25 in this volume), then Frege succeeded in reducing Hume's principle, at least, to logic.

Nevertheless, Frege balked at taking Hume's principle as the ultimate foundation for arithmetic because Hume's principle only fixes identities of the form "the number of *F*'s = the number of *G*'s." The principle does not determine the truth-value of sentences in the form "the number of *F*'s = *t*," where *t* is an arbitrary singular term. This became known as the Caesar problem. It is not that anyone would confuse a natural number with the Roman general Julius Caesar, but the underlying idea is that we have not succeeded in characterizing the natural numbers as objects unless and until we can determine how and why any given natural number is the same as or different from any object whatsoever. The distinctness of numbers and human beings should be a consequence of the theory, and not just a matter of intuition.

Frege went on to provide explicit definitions of individual natural numbers, and of the concept "natural number," in terms of *extensions* of concepts. The number 2, for example, is the extension (or collection) of all concepts that hold of exactly two elements. The inconsistency in Frege's theory of extensions, as shown by Russell's paradox, marked a tragic end to Frege's logicist program.

Russell and Whitehead [1910] traced the inconsistency in Frege's system to the impredicativity in his theory of extensions (and, for that matter, in Hume's principle). They sought to develop mathematics on a safer, predicative foundation. Their system proved to be too weak, and ad hoc adjustments were made, greatly reducing the attraction of the program. There is a thriving research program under way to see how much mathematics can be recovered on a predicative basis (chapter 19 in this volume).

Variations of Frege's original approach are vigorously pursued today in the work of Crispin Wright, beginning with [1983], and others like Bob Hale [1987] and Neil Tennant ([1987, 1997]) (chapter 6 in this volume). The idea is to bypass the treatment of extensions and to work with (fully impredicative) Hume's principle, or something like it, directly. Hume's principle is consistent with second-order logic if second-order arithmetic is consistent (see Boolos [1987] and Hodes [1984]), so at least the program will not fall apart like Frege's did. But what is the philosophical point? On the *neologicist* approach, Hume's principle is taken to be an explanation of the concept of "number." Advocates of the program argue that even if Hume's principle is not itself analytic—true in virtue of meaning—it can become known a priori, once one has acquired a grasp of the concept of cardinal number. Hume's principle is akin to an implicit definition. Frege's own technical development shows that the Peano postulates can be derived from Hume's principle in a standard, higher-order logic. Indeed, the only essential use that Frege made of extensions was to derive Hume's principle—everything else concerning numbers follows from that. Thus the basic propositions of arithmetic enjoy the same privileged epistemic status had by Hume's principle (assuming that second-order deduction preserves this status). Neologicism is a reconstructive program showing how arithmetic propositions can become known.

The neologicist (and Fregean) development makes essential use of the fact that impredicativity of Hume's principle is impredicative in the sense that the variable *F* in the locution "the number of *F*'s" is instantiated with concepts that themselves are defined in terms of numbers. Without this feature, the derivation of the Peano axioms from Hume's principle would fail. This impredicativity is consonant with the ontological realism adopted by Frege and his neologicist followers. Indeed, the neologicist holds that the left-hand side of an instance of Hume's principle has the same truth conditions as its right-hand side, but the left-hand side gives the proper logical form. Locutions like "the number of *F*'s" are genuine singular terms denoting numbers.

The neologicist project, as developed thus far, only applies basic arithmetic and the natural numbers. An important item on the agenda is to extend the treatment to cover other areas of mathematics, such as real analysis, functional analysis, geometry, and set theory. The program involves the search for abstraction principles rich enough to characterize more powerful mathematical theories (see, e.g., Hale [2000a, 2000b] and Shapiro [2000a, 2003]).

4.2. Empiricism, Naturalism, and Indispensability

Coffa [1982] provides a brief historical sketch of the semantic tradition, outlining its aims and accomplishments. Its final sentence is "And then came Quine." Despite the continued pursuit of variants of logicism (chapter 26 in this volume), the standard concepts underlying the program are in a state of ill repute in some quarters, notably much of North America. Many philosophers no longer pay serious attention to notions of meaning, analyticity, and a priori knowledge. To be precise, such notions are not given a primary role in the epistemology of mathematics, or anything else for that matter, by many contemporary philosophers. W. V. O. Quine (e.g., [1951, 1960]) is usually credited with initiating widespread skepticism concerning these erstwhile philosophical staples.

Quine, of course, does not deny that the truth-value of a given sentence is determined by both the use of language and the way the world is. To know that "Paris is in France," one must know something about the use of the words "Paris," "is," and "France," and one must know some geography. Quine's view is that the linguistic and factual components of a given sentence cannot be sharply distinguished, and thus there is no determinate notion of a sentence being true solely in virtue of language (analytic), as opposed to a sentence whose truth depends on the way the world is (synthetic).

Then how is mathematics known? Quine is a thoroughgoing empiricist, in the tradition of John Stuart Mill (chapter 3 in this volume). His positive view is that *all* of our beliefs constitute a seamless web answerable to, and only to, sensory stimulation. There is no difference in kind between mundane beliefs about material objects, the far reaches of esoteric science, mathematics, logic, and even so-called truths-by-definition (e.g., "no bachelor is married"). The word "seamless" in Quine's metaphor suggests that everything in the web is logically connected to everything else in the web, at least in principle. Moreover, no part of the web is knowable a priori.

This picture gives rise to a now common argument for realism. Quine and others, such as Putnam [1971], propose a hypothetical-deductive epistemology for mathematics. Their argument begins with the observation that virtually all of science is formulated in mathematical terms. Thus, mathematics is "confirmed" to the extent that science is. Because mathematics is indispensable for science, and science is well confirmed and (approximately) true, mathematics is well confirmed and true as well. This is sometimes called the *indispensability argument.*

Thus, Quine and Putnam are realists in truth-value, holding that some statements of mathematics have objective and nonvacuous truth-values independent of the language, mind, and form of life of the mathematician and scientist (assuming that science enjoys this objectivity). Quine, at least, is also a realist in ontology. He accepts the Fregean (and neologicist) view that "existence" is univocal. There is no ground for distinguishing terms that refer to medium-sized physical objects, terms that refer to microscopic and submicroscopic physical objects, and terms that refer to numbers. According to Quine and Putnam, *all* of the items in our ontology—apples, baseballs, electrons, and numbers—are theoretical posits. We accept the existence of all and only those items that occur in our best accounts of the material universe. Despite the fact that numbers and functions are not located in space and time, we know about numbers and functions the same way we know about physical objects—via the role of terms referring to such entities in mature, well-confirmed theories.

Indispensability arguments are anathema to those, like the logicists, logical positivists, and neologicists, who maintain the traditional views that mathematics is absolutely necessary and/or analytic and/or knowable a priori. On such views, mathematical knowledge cannot be dependent on anything as blatantly empirical and contingent as everyday discourse and natural science. The noble science of mathematics is independent of all of that. From the opposing Quinean perspective, mathematics and logic do not enjoy the necessity traditionally believed to hold of them; and mathematics and logic are not knowable a priori.

Indeed, for Quine, *nothing* is knowable a priori. The thesis is that everything in the web—the mundane beliefs about the physical world, the scientific theories, the mathematics, the logic, the connections of meaning—is up for revision if the "data" become sufficiently recalcitrant. From this perspective, mathematics is of a piece with highly confirmed scientific theories, such as the fundamental laws of gravitation. Mathematics appears to be necessary and a priori knowable (only) because it lies at the "center" of the web of belief, farthest from direct observation. Since mathematics permeates the web of belief, the scientist is least likely to suggest revisions in mathematics in light of recalcitrant "data." That is to say, because mathematics is invoked in virtually every science, its rejection is extremely unlikely, but the rejection of mathematics cannot be ruled out in principle. No belief is incorrigible. No knowledge is a priori, all knowledge is ultimately based on experience (see Colyvan [2001], and chapter 12 in this volume).

The seamless web is of a piece with Quine's *naturalism*, characterized as "the abandonment of first philosophy" and "the recognition that it is within science itself... that reality is to be identified and described" ([1981, p. 72]). The idea is to see philosophy as continuous with the sciences, not prior to them in any epistemological or foundational sense. If anything, the naturalist holds that science is prior to philosophy. *Naturalized epistemology* is the application of this theme to the study of knowledge. The philosopher sees the human knower as a thoroughly natural being within the physical universe. Any faculty that the philosopher invokes to explain knowledge must involve only natural processes amenable to ordinary scientific scrutiny.

Naturalized epistemology exacerbates the standard epistemic problems with realism in ontology. The challenge is to show how a physical being in a physical universe can come to know about *abstracta* like mathematical objects (see Field [1989, essay 7]). Since abstract objects are causally inert, we do not observe them but, nevertheless, we still (seem to) know something about them. The Quinean meets this challenge with claims about the role of mathematics in science. Articulations of the Quinean picture thus should, but usually do not, provide a careful explanation of the application of mathematics to science, rather than just noting the existence of this applicability (chapter 20 in this volume). This explanation would shed light on the abstract, non-spatiotemporal nature of mathematical objects, and the relationships between such objects and ordinary and scientific material objects. How is it that talk of numbers and functions can shed light on tables, bridge stability, and market stability? Such an analysis would go a long way toward defending the Quinean picture of a web of belief.

Once again, it is a central tenet of the naturalistically minded philosopher that there is no first philosophy that stands prior to science, ready to either justify or criticize it. Science guides philosophy, not the other way around. There is no agreement among naturalists that the same goes for mathematics. Quine himself accepts mathematics (as true) only to the extent that it is applied in the sciences. In particular, he does not accept the basic assertions of higher set theory because they do not, at present, have any empirical applications. Moreover, he advises mathematicians to conform their practice to his version of naturalism by adopting a weaker and less interesting, but better understood, set theory than the one they prefer to work with.

Mathematicians themselves do not follow the epistemology suggested by the Quinean picture. They do not look for confirmation in science before publishing their results in mathematics journals, or before claiming that their theorems are true. Thus, Quine's picture does not account for mathematics as practiced. Some philosophers, such as Burgess [1983] and Maddy [1990, 1997], apply naturalism to mathematics directly, and thereby declare that mathematics is, and ought to be, insulated from much traditional philosophical inquiry, or any other probes that are not to be resolved by mathematicians qua mathematicians. On such views, philosophy of mathematics—naturalist or otherwise—should not be in the business of either justifying or criticizing mathematics (chapters 13 and 14 in this volume).

4.3. No Mathematical Objects

The most popular way to reject realism in ontology is to flat out deny that mathematics has a subject matter. The *nominalist* argues that there are no numbers, points, functions, sets, and so on. The burden on advocates of such views is to make sense of mathematics and its applications without assuming a mathematical ontology. This is indicated in the title of Burgess and Rosen's study of nominalism, *A Subject with No Object* [1997].

A variation on this theme that played an important role in the history of our subject is *formalism*. An extreme version of this view, which is sometimes called "game formalism," holds that the essence of mathematics is the following of meaningless rules. Mathematics is likened to the play of a game like chess, where characters written on paper play the role of pieces to be moved. All that matters to the pursuit of mathematics is that the rules have been followed correctly. As far as the philosophical perspective is concerned, the formulas may as well be meaningless.

Opponents of game formalism claim that mathematics is inherently informal and perhaps even nonmechanical. Mathematical language has meaning, and it is a gross distortion to attempt to ignore this. At best, formalism focuses on a small aspect of mathematics, the fact that logical consequence is formal. It deliberately leaves aside what is essential to the enterprise.

A different formalist philosophy of mathematics was presented by Haskell Curry (e.g., [1958]). The program depends on a historical thesis that as a branch of mathematics develops, it becomes more and more rigorous in its methodology, the end result being the codification of the branch in formal deductive systems. Curry claimed that assertions of a mature mathematical theory are to be construed not so much as the results of moves *in* a particular formal deductive system (as a game formalist might say), but rather as assertions *about* a formal system. An assertion at the end of a research paper would be understood in the form "such and such is a theorem in this formal system." For Curry, then, mathematics is an objective science, and it has a subject matter—formal systems. In effect, mathematics is metamathematics. (See chapter 8 in this volume for a more developed account of formalism.)

On the contemporary scene, one prominent version of nominalism is *fictionalism*, as developed, for example, by Hartry Field [1980]. Numbers, points, and sets have the same philosophical status as the entities presented in works of fiction. According to the fictionalist, the number 6 is the same kind of thing as Dr. Watson or Miss Marple.

According to Field, mathematical language should be understood at face value. Its assertions have vacuous truth-values. For example, "all natural numbers are prime" comes out true, since there are no natural numbers. Similarly, "there is a prime number greater than 10" is false, and both Fermat's last theorem and the Goldbach conjecture are true. Of course, Field does not exhort mathematicians to settle their open questions via this vacuity. Unlike Quine, Field has no proposals for changing the methodology of mathematics. His view concerns how the results of mathematics should be interpreted, and the role of these results in the scientific enterprise. For Field, the goal of mathematics is not to assert the true. The only mathematical knowledge that matters is knowledge of logical consequences (see Field [1984]).

Field regards the Quine/Putnam indispensability argument to be the only serious consideration in favor of ontological realism. His overall orientation is thus broadly Quinean—in direct opposition to the long-standing belief that mathematical knowledge is a priori. As we have seen, more traditional philosophers—and most mathematicians—regard indispensability as irrelevant to mathematical knowledge. In contrast, for thinkers like Field, once one has undermined the indispensability argument, there is no longer any serious reason to believe in the existence of mathematical objects.

Call a scientific theory "nominalistic" if it is free of mathematical presuppositions. As Quine and Putnam pointed out, most of the theories developed in scientific practice are not nominalistic, and so begins the indispensability argument. The first aspect of Field's program is to develop nominalistic versions of various scientific theories. Of course, Field does not do this for every prominent scientific theory. To do so, he would have to understand every prominent scientific theory, a task that no human can accomplish anymore. Field gives one example—Newtonian gravitational theory—in some detail, to illustrate a technique that can supposedly be extended to other scientific work.

The second aspect of Field's program is to show that the nominalistic theories are sufficient for attaining the scientific goal of determining truths about the physical universe (i.e., accounting for observations). Let P be a nominalistic scientific theory and let S be a mathematical theory together with some "bridge principles" that connect the mathematical terminology with the physical terminology. Define S to be *conservative* over P if for any sentence Φ in the language of the nominalistic theory, if Φ is a consequence of $P + S$, then Φ is a consequence of P alone. Thus, if the mathematical theory is conservative over the nominalist one, then any physical consequence we get via the mathematics we could get from the nominalistic physics alone. This would show that mathematics is dispensable in principle, even if it is practically necessary. Field shows that standard mathematical theories and bridge principles are conservative over his nominalistic Newtonian theory, at least if the conservativeness is understood in model-theoretic terms: if Φ holds in all models of $P + S$, then Φ holds in all models of P.

The sizable philosophical literature generated by Field [1980] includes arguments that Field's technique does not generalize to more contemporary theories like quantum mechanics (Malament [1982]); arguments that Field's distinction between abstract and concrete does not stand up, or that it does not play the role needed to sustain Field's fictionalism (Resnik [1985]); and arguments that Field's nominalistic theories are not conservative in the philosophically relevant sense (Shapiro [1983]). The collection by Field [1989] contains replies to some of these objections.

Another common anti-realist proposal is to reconstrue mathematical assertions in *modal* terms. The philosopher understands mathematical assertions to be about what is possible, or about what would be the case if objects of a certain sort existed. The main innovation in Chihara [1990] is a modal primitive, a "constructibility quantifier." If Φ is a formula and x a certain type of variable, then Chihara's system contains a formula that reads "it is possible to construct an x such that Φ." According to Chihara, constructibility quantifiers do not mark what Quine calls "ontological commitment." Common sense supports this—to the extent that the notion of ontological commitment is part of common sense. If someone says that it is possible to construct a new ballpark in Boston, she is not asserting the existence of any ballpark, nor is she asserting the existence of a strange entity called a "possible ballpark." She only speaks of what it is possible to do.

The formal language developed in Chihara [1990] includes variables that range over *open sentences* (i.e., sentences with free variables), and these open-sentence variables can be bound by constructibility quantifiers. With keen attention to detail,

Chihara develops arithmetic, analysis, functional analysis, and so on in his system, following the parallel development of these mathematical fields in simple (impredicative) type theory.

Unlike Field, Chihara is a realist in truth-value. He holds that the relevant modal statements have objective and nonvacuous truth-values that hold or fail independent of the mind, language, conventions, and such of the mathematical community. Mathematics comes out objective, even if it has no ontology. Chihara's program shows initial promise on the epistemic front. Perhaps it is easier to account for how the mathematician comes to know about what is possible, or about what sentences can be constructed, than it is to account for how the mathematician knows about a Platonic realm of objects. (See chapters 15 and 16 in this volume.)

4.4. Intuitionism

Unlike fictionalists, traditional *intuitionists*, such as L. E. J. Brouwer (e.g., [1912, 1948]) and Arend Heyting (e.g., [1930, 1956]), held that mathematics has a subject matter: mathematical objects, such as numbers, do exist. However, Brouwer and Heyting insisted that these objects are mind-dependent. Natural numbers and real numbers are mental constructions or are the result of mental constructions. In mathematics, to exist is to be constructed. Thus Brouwer and Heyting are antirealists in ontology, denying the *objective* existence of mathematical objects. Some of their writing seems to imply that each person constructs his own mathematical realm. Communication between mathematicians consists in exchanging notes about their individual constructive activities. This would make mathematics subjective. It is more common, however, for these intuitionists, especially Brouwer, to hold that mathematics concerns the *forms* of mental construction as such (see Posy [1984]). This follows a Kantian theme, reviving the thesis that mathematics is synthetic a priori.

This perspective has consequences concerning the proper practice of mathematics. Most notably, the intuitionist demurs from the law of excluded middle—$(A \vee \neg A)$—and other inferences based on it. According to Brouwer and Heyting, these methodological principles are symptomatic of faith in the transcendental existence of mathematical objects and/or the transcendental truth of mathematical statements. For the intuitionist, every mathematical assertion must correspond to a construction. For example, let P be a property of natural numbers. For an intuitionist, the content of the assertion that not every number has the property P—the formula $\neg\forall x Px$—is that it is refutable that one can find a construction showing that P holds of each number. The content of the assertion that there is a number for which P fails—$\exists x \neg Px$—is that one can construct a number x and show that P does not hold of x. The latter formula cannot be inferred from the former because, clearly, it is possible to show that a property cannot hold universally without constructing a number for which it fails. In contrast, from the realist's perspective, the content of $\neg\forall x Px$ is simply that it is false that P holds universally, and $\exists x \neg Px$ means that there is a number for which P fails. Both formulas refer to numbers themselves; neither has anything to do with the knowledge-gathering abilities of mathematicians, or any other mental feature of them. From the realist's point of view, the two formulas are equivalent. The inference from $\neg\forall x Px$ to $\exists x \neg Px$ is a direct consequence of excluded middle.

Some contemporary intuitionists, such as Michael Dummett ([1973, 1977]) and Neil Tennant ([1987, 1997]), take a different route to roughly the same revisionist conclusion. Their proposed logic is similar to that of Brouwer and Heyting, but their supporting arguments and philosophy are different. Dummett begins with reflections on language acquisition and use, and the role of language in communication. One who understands a sentence must grasp its meaning, and one who learns a sentence thereby learns its meaning. As Dummett puts it, "a model of meaning is a model of understanding." This at least suggests that the meaning of a statement is somehow determined by its *use*. Someone who understands the meaning of any sentence of a language must be able to manifest that understanding in behavior. Since language is an instrument of communication, an individual cannot communicate what he cannot be observed to communicate.

Dummett argues that there is a natural route from this "manifestation requirement" to what we call here anti-realism in truth-value, and a route from there to the rejection of classical logic—and thus a demand for major revisions in mathematics.

Most semantic theories are *compositional* in the sense that the semantic content of a compound statement is analyzed in terms of the semantic content of its parts. Tarskian semantics, for example, is compositional, because the satisfaction of a complex formula is defined in terms of the satisfaction of its subformulas. Dummett's proposal is that the lessons of the manifestation requirement be incorporated into a compositional semantics. Instead of providing satisfaction conditions of each formula, Dummett proposes that the proper semantics supplies *proof* or *computation* conditions. He thus adopts what has been called "Heyting semantics." Here are three clauses:

A proof of a formula in the form $\Phi \vee \Psi$ is a proof of Φ or a proof of Ψ.
A proof of a formula in the form $\Phi \rightarrow \Psi$ is a procedure that can be proved to transform any proof of Φ into a proof of Ψ.
A proof of a formula in the form $\neg\Phi$ is a procedure that can be proved to transform any proof of Φ into a proof of absurdity; a proof of $\neg\Phi$ is a proof that there can be no proof of Φ.

Heyting and Dummett argue that on a semantics like this, the law of excluded middle is not universally upheld. A proof of a sentence of the form $\Phi \vee \neg\Phi$ consists of a proof of Φ or a proof that there can be no proof of Φ. The intuitionist claims that one cannot maintain this disjunction, in advance, for every sentence Φ.

A large body of research in mathematical logic shows how intuitionistic mathematics differs from its classical counterpart. Many mathematicians hold that the intuitionistic restrictions would cripple their discipline (see, e.g., Paul Bernays [1935]). For some philosophers of mathematics, the revision of mathematics is too high a price to pay. If a philosophy entails that there is something wrong with what the mathematicians do, then the philosophy is rejected out of hand. According to them, intuitionism can be safely ignored. A less dogmatic approach would be to take Dummett's arguments as a challenge to answer the criticisms he brings. Dummett argues that classical logic, and mathematics as practiced, do not enjoy a certain kind of justification, a justification one might think a logic and mathematics ought to have. Perhaps a defender of classical mathematics, such as a Quinean holist or a Maddy-style naturalist, can concede this, but argue that logic and mathematics do not need this kind of justification. We leave the debate at this juncture. (See chapters 9 and 10 in this volume.)

4.5. Structuralism

According to another popular philosophy of mathematics, the subject matter of arithmetic, for example, is the *pattern* common to any infinite system of objects that has a distinguished initial object, and a successor relation or operation that satisfies the induction principle. The arabic numerals exemplify this *natural number structure*, as do sequences of characters on a finite alphabet in lexical order, an infinite sequence of distinct moments of time, and so on. A natural number, such as 6, is a place in the natural number structure, the seventh place (if the structure starts with zero). Similarly, real analysis is about the real number structure, set theory is about the set-theoretic hierarchy structure, topology is about topological structures, and so on.

According to the structuralist, the application of mathematics to science occurs, in part, by discovering or postulating that certain structures are exemplified in the material world. Mathematics is to material reality as pattern is to patterned. Since a structure is a one-over-many of sorts, a structure is like a traditional universal, or property.

There are several ontological views concerning structures, corresponding roughly to traditional views concerning universals. One is that the natural number structure, for example, exists independent of whether it has instances in the physical world—or any other world, for that matter. Let us call this *ante rem* structuralism, after the analogous view concerning universals (see Shapiro [1997] and Resnik [1997]; see also Parsons [1990]). Another view is that there is no more to the natural number structure than the systems of objects that exemplify this structure. Destroy the systems, and the structure goes with them. From this perspective, either structures do not exist at all—in which case we have a version of nominalism—or the existence of structures is tied to the existence of their "instances," the systems that exemplify the structures. Views like this are sometimes dubbed *eliminative structuralism* (see Benacerraf [1965]).

According to *ante rem* structuralism, statements of mathematics are understood at face value. An apparent singular term, such as "2," is a genuine singular term, denoting a place in the natural number structure. For the eliminative structuralist, by contrast, these apparent singular terms are actually bound variables. For example, "$2 + 3 = 5$" comes to something like "in any natural number system *S*, any object in the 2-place of *S* that is *S*-added to the object in the 3-place of *S* is the object in the 5-place of *S*." Eliminative structuralism is a structuralism without structures.

Taken at face value, eliminative structuralism requires a large ontology to keep mathematics from being vacuous. For example, if there are only finitely many objects in the universe, then the natural number structure is not exemplified, and thus universally quantified statements of arithmetic are all vacuously true. Real and complex analysis and Euclidean geometry require a continuum of objects, and set theory requires a proper class (or at least an inaccessible cardinal number) of objects. For the *ante rem* structuralist, the structures themselves, and the places in the structures, provide the "ontology."

Benacerraf [1965], an early advocate of eliminative structuralism, made much of the fact that the set-theoretic hierarchy contains many exemplifications of the natural number structure. He concluded from this that numbers are not objects. This conclusion, however, depends on what it is to be an object—an interesting philosophical question in its own right. The *ante rem* structuralist readily accommodates the multiple realizability of the natural number structure: some items in the set-theoretic hierarchy, construed as objects, are organized into systems, and some of these systems exemplify the natural number structure. That is, *ante rem* structuralism accounts for the fact that mathematical structures are exemplified by other mathematical objects. Indeed, the natural number structure is exemplified by various *systems of natural numbers*, such as the even numbers and the prime numbers. From the *ante rem* perspective, this is straightforward: the natural numbers, as places in the natural number structure, exist. Some of them are organized into systems, and some of these systems exemplify the natural number structure.

On the *ante rem* view, the main epistemological question becomes: How do we know about structures? On the eliminative versions, the question is: How do we know about what holds in all systems of a certain type? Structuralists have developed several strategies for resolving the epistemic problems. The psychological mechanism of pattern recognition may be invoked for at least small, finite structures. By encountering instances of a given pattern, we obtain knowledge of the pattern itself. More sophisticated structures are apprehended via a Quine-style postulation (Resnik), and more robust forms of abstraction and implicit definition (Shapiro).

None of the structuralisms invoked so far provide for a reduction of the ontological burden of mathematics. The ontology of *ante rem* structuralism is as large and extensive as that of traditional realism in ontology. Indeed, *ante rem* structuralism *is* a realism in ontology. Only the nature of the ontology is in question. Eliminative structuralism also requires a large ontology to keep the various branches of mathematics from lapsing into vacuity. Surely there are not enough physical objects to keep structuralism from being vacuous when it comes to functional analysis or set theory. Thus, eliminative structuralism requires a large ontology of nonconcrete objects, and so it is not consistent with ontological anti-realism.

Hellman's [1989] *modal structuralism* is a variation of the underlying theme of eliminative structuralism which opts for a thorough ontological anti-realism. Instead of asserting that arithmetic is about all systems of a certain type, the modal structuralist says that arithmetic is about all *possible* systems of that type. A sum like "$2 + 3 = 5$" comes to "in any *possible* natural number system *S*, any object in the 2-place of *S* that is *S*-added to the object in the 3-place of *S* is the object in the 5-place of *S*" or "*necessarily*, in any natural number system *S*, any object in the 2-place of *S* that is *S*-added to the object in the 3-place of *S* is the object in the 5-place of *S*." The modal structuralist agrees with the eliminative structuralist that apparent singular terms, such as numerals, are disguised bound variables, but for the modal structuralist these variables occur inside the scope of a modal operator.

The modal structuralist faces an attenuated threat of vacuity similar to that of the eliminative structuralist. Instead of asserting that there are systems satisfying the natural number structure, for example, the modalist needs to affirm that such systems are possible. The key issue here is to articulate the underlying modality. (See chapters 17 and 18 in this volume.)

5. Logic

The above survey broached a number of issues concerning logic and the philosophy of logic. The debate over intuitionism invokes the general validity, within mathematics, of the law of excluded middle and other inferences based on it (chapters 9–11 in this volume), and questions concerning impredicativity emerged from a version of logicism.

There is traffic in the other direction as well, from logic to the philosophy of mathematics. Perhaps the primary issue in the philosophy of logic concerns the nature, or natures, of logical consequence. There is, first, a deductive notion of consequence: a proposition Φ follows from a set Γ of propositions if Φ can be justified fully on the basis of the members of Γ. This is often understood in terms of a chain of legitimate, gap-free inferences that leads from members of Γ to Φ. A similar, perhaps identical, idea underlies Frege's development of logic in defense of logicism, and occurs also in neologicism. To show that a given mathematical proposition is knowable a priori and independent of intuition, we have to give a gap-free proof of it. There is also a semantic, model-theoretic notion of consequence: Φ follows from Γ if Φ is true in every interpretation (or model) of the language in which the members of Γ are true. Deductive systems introduced in logic books capture, or model, the deductive notion of consequence, and model-theoretic semantics captures, or tries to capture, the semantic notion.

There are substantial philosophical issues concerning the legitimacy of the model-theoretic notion of consequence and over which, if either, of the notions is primary. Of course, the resolution of these issues depends on prior questions concerning the nature of logic and the goals of logical study (chapters 21 and 22 in this volume). If both notions of consequence are legitimate, we can ask about relations between them. Surely it must be the case that if a proposition Φ follows deductively from a set Γ, then Φ is true under every interpretation of the language in which Γ is true. If not, then there is a chain of legitimate, gap-free inferences that can take us from truth to falsehood. Perish the thought. However, the converse seems less crucial. It may well be that there is a semantically valid argument whose conclusion cannot be deduced from its premises.

Issues surrounding higher-order logic, which were also broached briefly in the foregoing survey, turn on matters relating to the nature(s) of logical consequence. Second-order logic is inherently incomplete, in the sense that there is no effective deductive system that is both sound and complete for it. Does this disqualify it as logic, or is there some role for second-order logic to play? What does this say about the underlying nature of mathematics? (See chapters 25 and 26 in this volume.)

Finally, there is a tradition, dating back to antiquity and very much alive today, that maintains that a proposition Φ cannot be a logical consequence of a set Γ unless Φ is somehow *relevant* to Γ. On the contemporary scene, the main targets of relevance logic are the so-called paradoxes of implication. One of these is the thesis that a logical truth follows from any set of premises whatsoever, and another is *ex falso quodlibet*, the thesis that any conclusion follows from a contradiction. The extent to which such inferences occur in mathematics is itself a subject of debate (chapters 23 and 24 in this volume).

Acknowledgment Some of the contents of this chapter were culled from Shapiro [2000b] and [2003b].

REFERENCES AND SELECTED BIBLIOGRAPHY

Aspray, W., and P. Kitcher (editors) [1988], *History and philosophy of modern mathematics*, Minnesota Studies in the Philosophy of Science 11, Minneapolis, University of Minnesota Press. A wide range of articles, most of which draw philosophical morals from historical studies.

Azzouni, J. [1994], *Metaphysical myths, mathematical practice*, Cambridge, Cambridge University Press. Fresh philosophical view.

Balaguer, M. [1998], *Platonism and anti-Platonism in mathematics*, Oxford, Oxford University Press. Account of realism in ontology and its rivals.

Benacerraf, P. [1965], "What numbers could not be," *Philosophical Review* 74, 47–73; reprinted in Benacerraf and Putnam [1983], 272–294. One of the most widely cited works in the field; argues that numbers are not objects, and introduces an eliminative structuralism.

Benacerraf, P., and H. Putnam (editors) [1983], *Philosophy of mathematics*, second edition, Cambridge, Cambridge University Press. A far-reaching collection containing many of the central articles.

Bernays, P. [1935], "Sur le platonisme dans les mathématiques," *L'Enseignement mathématique* 34, 52–69; translated as "Platonism in mathematics," in Benacerraf and Putnam [1983], 258–271.

Boolos, G. [1987], "The consistency of Frege's *Foundations of arithmetic*," in *On being and saying: Essays for Richard Cartwright*, edited by Judith Jarvis Thompson, Cambridge, Massachusetts, The MIT Press, 3–20; reprinted in Hart [1996], 185–202.

Boolos, G. [1997], "Is Hume's principle analytic?," in *Language, thought, and logic*, edited by Richard Heck, Jr., Oxford, Oxford University Press, 245–261. Criticisms of the claims of neologicism concerning the status of abstraction principles.

Brouwer, L.E.J. [1912], *Intuitionisme en Formalisme*, Gronigen, Noordhoof; translated as "Intuitionism and formalism," in Benacerraf and Putnam [1983], 77–89.

Brouwer, L.E.J. [1949], "Consciousness, philosophy and mathematics," in Benacerraf and Putnam [1983], 90–96.

Burgess, J. [1983], "Why I am not a nominalist," *Notre Dame Journal of Formal Logic* 24, 93–105. Early critique of nominalism.

Burgess, J., and G. Rosen [1997], *A subject with no object: Strategies for nominalistic interpretation of mathematics*, New York, Oxford University Press. Extensive articulation and criticism of nominalism.

Chihara, C. [1990], *Constructibility and mathematical existence*, Oxford, Oxford University Press. Defense of a modal view of mathematics, and sharp criticisms of several competing views.

Coffa, A. [1982], "Kant, Bolzano, and the emergence of logicism," *Journal of Philosophy* 79, 679–689.

Coffa, A. [1991], *The semantic tradition from Kant to Carnap*, Cambridge, Cambridge University Press.

Colyvan, M. [2001], *The indispensability of mathematics*, New York, Oxford University Press. Elaboration and defense of the indispensability argument for ontological realism.

Curry, H. [1958], *Outline of a formalist philosophy of mathematics*, Amsterdam, North Holland Publishing Company.

Dummett, M. [1973], "The philosophical basis of intuitionistic logic," in Dummett [1978], 215–247; reprinted in Benacerraf and Putnam [1983], 97–129, and Hart [1996], 63–94. Influential defense of intuitionism.

Dummett, M. [1977], *Elements of intuitionism*, Oxford, Clarendon Press. Detailed introduction to and defense of intuitionistic mathematics.

Dummett, M. [1978], *Truth and other enigmas*, Cambridge, Massachusetts, Harvard University Press. A collection of Dummett's central articles in metaphysics and the philosophy of language.

Field, H. [1980], *Science without numbers*, Princeton, New Jersey, Princeton University Press. A widely cited defense of fictionalism by attempting to refute the indispensability argument.

Field, H. [1984], "Is mathematical knowledge just logical knowledge?," *The Philosophical Review* 93, 509–552; reprinted (with added appendix) in Field [1989], 79–124, and in Hart [1996], 235–271.

Field, H. [1989], *Realism, mathematics and modality*, Oxford, Blackwell. Reprints of Field's articles on fictionalism.

Frege, G. [1884], *Die Grundlagen der Arithmetik*, Breslau, Koebner; *The foundations of arithmetic*, translated by J. Austin, second edition, New York, Harper, 1960. Classic articulation and defense of logicism.

Frege, G. [1893], *Grundgesetze der Arithmetik*, vol. 1, Jena, H. Pohle; reprinted Hildesheim, Olms, 1966. More technical development of Frege's logicism.

Gödel, K. [1944], "Russell's mathematical logic," in Benacerraf and Putnam [1983], 447–469. Much cited defense of realism in ontology and realism in truth-value.

Gödel, K. [1951], "Some basic theorems on the foundations of mathematics and their implications," in his *Collected Works*, vol. 3, Oxford, Oxford University Press, 1995, 304–323.

Gödel, K. [1964], "What is Cantor's continuum problem?," in Benacerraf and Putnam [1983], 470–485. Much cited defense of realism in ontology and realism in truth-value.

Hale, Bob [1987], *Abstract objects*, Oxford, Basil Blackwell. Detailed development of neologicism, to support Wright [1983].

Hale, Bob [2000a], "Reals by abstraction," *Philosophia Mathematica 3rd ser.*, 8, 100–123.

Hale, Bob [2000b], "Abstraction and set theory," *Notre Dame Journal of Formal Logic* 41, 379–398.

Hart, W.D. (editor) [1996], *The philosophy of mathematics*, Oxford, Oxford University Press. Collection of articles first published elsewhere.

Hellman, G. [1989], *Mathematics without numbers*, Oxford, Oxford University Press. Articulation and defense of modal structuralism.

Heyting, A. [1930], "Die formalen Regeln der intuitionistischen Logik," *Sitzungsberichte der Preussischen Akademie der Wissesschaften, physikalisch-mathematische Klasse*, 42–56. Develops deductive system and semantics for intuitionistic mathematics.

Heyting, A. [1956], *Intuitionism: An introduction*, Amsterdam, North Holland. Readable account of intuitionism.

Hodes, H. [1984], "Logicism and the ontological commitments of arithmetic," *Journal of Philosophy* 81 (13), 123–149. Another roughly Fregean logicism.

Kitcher, P. [1983], *The nature of mathematical knowledge*, New York, Oxford University Press. Articulation of a constructivist epistemology and a detailed attack on the thesis that mathematical knowledge is a priori.

Lakatos, I. (editor) [1967], *Problems in the philosophy of mathematics*, Amsterdam, North Holland Publishing Company. Contains many important papers by major philosophers and logicians.

Lakatos, I. [1976], *Proofs and refutations*, edited by J. Worrall and E. Zahar, Cambridge, Cambridge University Press. A much cited study that is an attack on the rationalist epistemology for mathematics.

MacLane, S. [1986], *Mathematics: Form and function*, New York, Springer-Verlag. Philosophical account by an influential mathematician.

Maddy, P. [1990], *Realism in mathematics*, Oxford, Oxford University Press. Articulation and defense of realism about sets.

Maddy, P. [1997], *Naturalism in mathematics*, Oxford, Clarendon Press. Lucid account of naturalism concerning mathematics and its relation to traditional philosophical issues.

Malament, D. [1982], Review of Field [1980], *Journal of Philosophy* 19, 523–534.

Parsons, C. [1983], *Mathematics in philosophy*, Ithaca, New York, Cornell University Press. Collection of Parsons's important papers in the philosophy of mathematics.

Parsons, C. [1990], "The structuralist view of mathematical objects," *Synthèse* 84, 303–346; reprinted in Hart [1996], 272–309.

Posy, C. [1984], "Kant's mathematical realism," *The Monist* 67, 115–134. Comparison of Kant's philosophy of mathematics with intuitionism.

Putnam, H. [1971], *Philosophy of logic*, New York, Harper Torchbooks. Source for the indispensability argument for ontological realism.

Putnam, H. [1980], "Models and reality," *Journal of Symbolic Logic* 45, 464–482; reprinted in Benacerraf and Putnam [1983], 421–444.

Quine, W.V.O. [1951], "Two dogmas of empiricism," *The Philosophical Review* 60, 20–43; reprinted in Hart [1996], 31–51. Influential source of the attack on the analytic/synthetic distinction.

Quine, W.V.O. [1960], *Word and object*, Cambridge, Massachusetts, The MIT Press.

Quine, W.V.O. [1981], *Theories and things*, Cambridge, Massachusetts, Harvard University Press.

Resnik, M. [1980], *Frege and the philosophy of mathematics*, Ithaca, New York, Cornell University Press. Exegetical study of Frege and introduction to philosophy of mathematics.

Resnik, M. [1985], "How nominalist is Hartry Field's nominalism?," *Philosophical Studies* 47, 163–181. Criticism of Field [1980].

Resnik, M. [1997], *Mathematics as a science of patterns*, Oxford, Oxford University Press. Full articulation of a realist-style structuralism.

Schirn, M. (editor) [1998], *The Philosophy of mathematics today*, Oxford, Clarendon Press. Proceedings of a conference in the philosophy of mathematics, held in Munich in 1993; coverage of most of the topical issues.

Shapiro, S. [1983], "Conservativeness and incompleteness," *Journal of Philosophy* 80, 521–531; reprinted in Hart [1996], 225–234. Criticism of Field [1980].

Shapiro, S. (editor) [1996], *Mathematical structuralism, Philosophia Mathematica 3rd ser.*, 4. Special issue devoted to structuralism; contains articles by Benacerraf, Hale, Hellman, MacLane, Resnik, and Shapiro.

Shapiro, S. [1997], *Philosophy of mathematics: Structure and ontology*, New York, Oxford University Press. Extensive articulation and defense of structuralism.

Shapiro, S. [1998], "Incompleteness, mechanism, and optimism," *Bulletin of Symbolic Logic* 4, 273–302.

Shapiro, S. [2000a], "Frege meets Dedekind: A neo-logicist treatment of real analysis," *Notre Dame Journal of Formal Logic* 41 (4), 335–364.

Shapiro, S. [2000b], *Thinking about mathematics: The philosophy of mathematics*, Oxford, Oxford University Press. Popularization and textbook in the philosophy of mathematics.

Shapiro, S. [2003a], "Prolegomenon to any future neo-logicist set theory: Abstraction and indefinite extensibility," *British Journal for the Philosophy of Science* 54, 59–91.

Shapiro, S. [2003b], "Philosophy of mathematics," in *Philosophy of science today*, edited by Peter Clark and Katherine Hawley, Oxford, Oxford University Press, 181–200.

Skolem, T. [1922], "Einige Bemerkungen zur axiomatischen Begründung der Mengenlehre," in *Matematikerkongressen i Helsingfors den 4–7 Juli 1922*, Helsinki, Akademiska Bokhandeln, 217–232; translated as "Some remarks on axiomatized set theory" in van Heijenoort [1967], 291–301.

Skolem, T. [1941], "Sur la portée du théorème de Löwenheim–Skolem," in *Les Entretiens de Zurich, 6–9 décembre 1938*, edited by F. Gonseth, Zurich, Leeman, 1941, 25–52.

Steiner, M. [1975], *Mathematical knowledge*, Ithaca, New York, Cornell University Press. Articulation of some epistemological issues; defense of realism.

Tennant, N. [1987], *Anti-realism and logic*, Oxford, Clarendon Press. Articulation of anti-realism in truth-value, realism in ontology; defends intuitionistic relevance logic against classical logic.

Tennant, N. [1997], *The taming of the true*, Oxford, Oxford University Press. Detailed defense of global semantic anti-realism.

Van Heijenoort, J. (editor) [1967], *From Frege to Gödel*, Cambridge, Massachusetts, Harvard University Press. Collection of many important articles in logic from around the turn of the twentieth century.

Webb, J. [1980], *Mechanism, mentalism and metamathematics: An essay on finitism*, Dordrecht, D. Reidel.

Whitehead, A.N., and B. Russell [1910], *Principia Mathematica*, vol. 1, Cambridge, Cambridge University Press.

Wright, C. [1983], *Frege's conception of numbers as objects*, Aberdeen, Scotland, Aberdeen University Press. Revival of Fregean logicism.

CHAPTER 2

APRIORITY AND APPLICATION: PHILOSOPHY OF MATHEMATICS IN THE MODERN PERIOD

LISA SHABEL

In the modern period[1]—which might be thought to have begun with a new conception of the natural world as uniquely quantifiable—the term "science" was used to denote a systematic body of knowledge based on a set of self-evident first principles. Mathematics was understood to be the science that systematized our knowledge of magnitude, or quantity. But the mathematical notion of magnitude, and

This material is based upon work generously supported by the National Science Foundation under grant no. SES-0135441. I am grateful to Gary Hatfield and Stewart Shapiro for helpful comments and suggestions, and to Sukjae Lee for ongoing discussion.

[1] For the purposes of this chapter, I am construing this period in the standard way, as ranging from Descartes to Kant. I do not intend to offer an encyclopedic account of the issues in the philosophy of mathematics that faced the modern philosophers, nor even a survey of each major philosopher's account of mathematical cognition. I hope, rather, to tell a story about some key issues in the philosophy of mathematics in the modern period.

the methods used to investigate it, underwent a period of radical transformation during the early modern period: Descartes's new analytic methods for solving geometric problems, Vieta's and Fermat's systems of "specious arithmetic," and Newton's and Leibniz's independent discoveries of the calculus are just a few of the developments in mathematical practice that witness this transformation. These innovations in mathematical practice inspired similarly radical innovations in the philosophy of mathematics: philosophers were confronted with a changing mathematical landscape, and their assessment of the ontological and epistemological terrain reveals that both the basic mathematical notions and the philosophers' tools for comprehending and explaining those notions were in transition.

While mathematical methods were becoming decidedly more analytical and abstract, they nevertheless remained anchored by a concrete conception of magnitude, the object of mathematics. Ultimately, mathematics was meant to provide a quantitative description of *any* quantifiable entity: the number of cups on the table, the volume and surface area of a particular cup, the amount of liquid the cup contains, the temperature of the liquid, and so on. The term "magnitude" was used to describe both the quantifiable entity and the quantity it was determined to have. That is, for the moderns, magnitudes have magnitude.[2] In the seventeenth and eighteenth centuries, modern mathematical practices had not yet become sufficiently abstract so as to fully detach the latter notion from the former: mathematics was not yet about number or shape (much less set or manifold) conceived independently from numbered or shaped *things.* This is not to suggest that the modern mathematician lacked the resources to represent such quantifiables in abstract mathematical form; indeed, the period was rich in such representational innovation. The point is, rather, that modern mathematical ontology included both abstract mathematical representations and their concrete referents. Accordingly, modern mathematical epistemology could not rest with an account of our cognitive ability to manipulate mathematical abstractions but had also to explain the way in which these abstractions made contact with the natural world.

The state of modern mathematical practice called for a modern *philosopher* of mathematics to answer two interrelated questions. Given that mathematical ontology includes quantifiable *empirical* objects, how to explain the paradigmatic features of pure mathematical reasoning: universality, certainty, necessity. And, without giving up the special status of pure mathematical reasoning, how to explain the ability of pure mathematics to come into contact with and describe the

[2] Thus, for example, the pens on my table are a discrete magnitude having, say, a magnitude of four units where the unit is one pen; the surface of my desk is a continuous magnitude having, say, a magnitude of two units where the unit is square meters. For a discussion of Newton's conception of magnitude, and of his distinction between quantum and quantity, see McGuire (1983), p. 75. For a discussion of Wolff and Kant on the same distinction, see Shabel (2003), pp. 124–126.

empirically accessible natural world.[3] The first question comes to a demand for *apriority*: a viable philosophical account of early modern mathematics must explain the apriority of mathematical reasoning. The second question comes to a demand for *applicability*: a viable philosophical account of early modern mathematics must explain the *applicability* of mathematical reasoning. Ultimately, then, the early modern philosopher of mathematics sought to provide an explanation of the relation between the mathematical features of the objects of the natural world and our paradigmatically a priori cognition thereof, thereby satisfying both demands.

At the end of the modern period, Kant attempts to meet these demands with his doctrine of Transcendental Idealism, including arguments for the synthetic a priori status of mathematical cognition. In the course of defending his own theory of how we come to cognize the mathematical world, Kant distinguishes two strains of thought in his predecessors' competing accounts of mathematical cognition, which he determines to be inadequate and misguided.[4] Taking the science of Euclidean geometry, an exemplar of pure mathematical reasoning, to provide substantive cognition of space, spatial relations, and empirically real spatial objects, Kant claims that prior attempts to account for our cognitive grasp of geometry and its objects "come into conflict with the principles of experience."[5] On Kant's view, both the "mathematical investigators of nature" (who suppose that the spatial domain of geometric investigation is an eternal and infinite subsisting real entity) and the "metaphysicians of nature" (who suppose that the spatial domain of geometric investigation comprises relations among confused representations of real entities that are themselves ultimately nonspatial) fail to provide a viable account of early modern mathematics in the sense described above. In particular, the "mathematicians" fail to meet the applicability demand, and the "metaphysicians" fail to meet the apriority demand.[6] Kant's claim, of course, is that his own theory of synthetic a priori mathematical cognition meets with greater success.

[3] The closest analog to our pure/applied distinction in the modern period is captured by a distinction between pure and "mixed" mathematics. According to Christian Wolff, the mixed mathematical disciplines are those that "consider and measure the particular magnitude of things found in nature," while the pure or unmixed mathematical disciplines "consider only the magnitude as magnitude" (Wolff (1965), pp. 866, 868).

[4] It is obviously problematic to assess Kant's early modern predecessors from Kant's own perspective. But Kant's way of articulating his own position in contrast to competing positions does provide us with a nice framework for understanding the entire period. I will do my best to present the views of Kant's predecessors objectively, with the caveat that I am telling a story in which Kant gets the final word.

[5] Kant (1998), B56.

[6] These failures are not straightforward since, as we will see below, one can explicate the view of a "mathematician" like Descartes in a way that seems clearly to satisfy both demands. The sense in which, according to Kant, the apriority and applicability demands are not met by any view prior to his own will be explained in section 3.

In what follows, I will discuss these three major attempts to provide a viable philosophical account of early modern mathematical practice. Descartes and Newton stand as examples of the "mathematicians"; Leibniz exemplifies the "metaphysicians"; and Kant sees himself as correcting the common error that, on his view, leads both "mathematician" and "metaphysician" astray. I will begin by providing a brief account of a relevant aspect of early modern mathematical practice, in order to situate our philosophers in their historical and mathematical context.

1. Representational Methods in Mathematical Practice

As noted above, the modern mathematician's task included systematizing the science of quantity. This required, first and foremost, a systematic method for representing real, quantifiable objects mathematically, as well as a systematic method for manipulating such representations. The real, quantifiable objects were conceived to include both discrete magnitudes, or those that could be represented numerically and manipulated arithmetically, and continuous magnitudes, or those that could be represented spatially and manipulated geometrically. These categories, however, were neither fully determinate nor mutually exclusive; they did not serve to demarcate two distinct sets of real, quantifiable objects because, for example, a given magnitude might be considered with respect to both shape *and* number, and thus might be treated geometrically for one purpose and arithmetically for another. Mathematical progress required a representational system that was adequate to symbolize any and all quantifiable features of the natural world in such a way that comparisons could be easily made among them. Moreover, mathematical progress required that such a representational system be as abstract and general as possible while nevertheless retaining its purchase on the concrete and particular quantifiables of the natural world.

For a simple example, consider a bag of marbles. A geometric diagram of a sphere can be taken to represent the shape of any individual marble and, indeed, any spherically shaped object. This sort of representation can be mathematically manipulated using classical techniques of Euclidean solid geometry, but its intermathematical utility is limited: while a diagram of a sphere can be useful in plainly geometric reasoning about spheres (and cross sections thereof, tangents thereto, etc.), it is impotent for the purpose of more abstract reasoning, such as might be required mathematically to compare the sphere to nonspatial magnitudes. A qualitative geometric diagram can represent a magnitude qua spatial object, but cannot aid us in abstracting from spatiality in order to represent it as a discrete quantity.

Moreover, while a number can be taken to represent the determinate ratio of all of the marbles in the bag to a single marble, it can do so only upon specification of a marble as the unit of measure. On the modern conception, a number "arises" upon consideration of a group of things of a particular kind in relation to a single thing of that kind and thus is, in a sense, context-sensitive: the use of number as a quantitative tool is inextricably tied to the choice of unit, which is not fixed. On this conception, it follows that numbers are representationally inflexible: it is difficult to see how to make quantitative comparisons among magnitudes of different *kinds*. More generally, on this conception the possibility of a notion of number that ranges beyond the positive integers appears remote.

Early modern innovations in representational flexibility began with Descartes's observation that we can represent and compare magnitudes uniformly and in the simplest possible terms by formulating ratios and proportions between and among their "dimensions." "By 'dimension,'" Descartes writes, "we mean simply a mode or aspect in respect of which some subject is considered to be measurable."[7] According to Descartes, dimensionality comprises countless quantitative features which include length, breadth, and depth, but also weight, speed, the order of parts to whole (counting), and the division of whole into parts (measuring). More importantly, a finite straight line segment is identified as the most versatile tool for representing any dimension of magnitude; any and all quantitative dimensions can be represented simply and uniformly via configurations and ratios of straight line segments.[8] Notably, in Descartes's system the unit segment, to which any other representative segment stands in relation, is problem-specific: for solution of a particular problem "we may adopt as unit either one of the magnitudes already given or any other magnitude, and this will be the common measure of all the others."[9] Once the unit magnitude is chosen and represented as a particular finite line segment, representations of all other relevant magnitudes can be constructed in relation to that unit.[10]

[7] Descartes (1985), *Rules*, AT 10:447.

[8] Descartes discusses this procedure in rules 14 through 18 of his *Rules for the Direction of the Mind*. Descartes (1985) AT 10: 438–468. For further discussion see Shabel (2003), pp. 65–67.

[9] Descartes (1985), *Rules*, AT 10:450.

[10] Thus, Descartes's unit segment cannot be identified with the real unit interval [0, 1], nor with sections of the orthogonal number lines that we use to construct what we anachronistically call the "Cartesian" coordinate system. Descartes's unit segment is a line segment of arbitrary length that stands for whatever particular magnitude functions as the unit in a particular problem. Even if the representational system were generalized and a fixed unit segment were chosen to represent the unit of magnitude functioning in any mathematical context, nevertheless the Cartesian unit segment would still serve as unit by virtue of the ratio in which it stands to the other magnitudes of the problem, as evidenced by their relative lengths, but not by virtue of its structure as a dense linear ordering.

To solve a mathematical problem, one thus begins by identifying and representing the magnitudes involved, no matter whether they be determinate or indeterminate, known or unknown, geometric or arithmetic, extended or multitudinous, continuous or discrete. One chooses a unit of measure and conceives each relevant magnitude in general terms—in abstraction from the object or group of objects that is ordered or measured but in relation to the chosen unit—by following Descartes's technical constructive procedures. Algebraic symbols can then be used as a heuristic aid for manipulating the line segments, and algebraic equations formulated to express the mathematical relations in which the segments stand to one another.[11] Famously, the Cartesian representational system liberates formerly heterogeneous magnitudes: on the Cartesian system, the multiplication of two line segments yields another line segment, rather than the area of a rectangle. It follows that products, quotients, and roots of linear magnitudes can be construed to stand in proportion to the magnitudes themselves; likewise, the degree of algebraic variables need not be taken to indicate strictly *geometric* dimensionality.

Descartes's own use of his representational system was primarily directed at solving problems in geometry,[12] though he certainly thought it adequate for other mathematical manipulations. Modern mathematicians who adopted the Cartesian system emphasized its utility across mathematical disciplines, and used straight lines to represent numbers for arithmetic and number-theoretic purposes. Once a single straight line is designated as unity, any positive integer can be straightforwardly represented as a simple concatenation of units. One advantage of the Cartesian representational system is that it allows the notion of number to expand beyond the positive integers so constructed: numbers can now be conceived as ratios between line segments of any arbitrary length and a chosen unit. Rational numbers are identified as those segments that are (geometrically) commensurable with unity, and irrational numbers those that are incommensurable. These representational innovations witness a number concept in transition: the moderns are able to treat rational and irrational magnitudes numerically, but the notion of number remains tied to the geometric concept of commensurability. Importantly, the moderns are *unable* to use the Cartesian representational system to conceive negative quantities, a disadvantage of the system. Consequently, negative quantities are variously deemed "absurd," "privative of true," "wanting reality," and "not real."

Actual mathematical progress in the modern period might be said to outpace the fundamental and foundational tools and concepts available to mathematicians

[11] For a discussion of the details of Descartes's technical constructive procedures, see Shabel (2003), pp. 65–69.

[12] His *Géometrie* opens with the claim "Any problem in geometry can easily be reduced to such terms that a knowledge of the lengths of certain straight lines is sufficient for its construction" (Descartes (1954), p. 2).

at the time. Geometric problems far more sophisticated than the ancients could have imagined are solved on ancient geometric foundations; algebraic methods are developed and fruitfully applied long before algebra is conceived to be a mathematical discipline in its own right, with its own domain of problems distinct from those of arithmetic and geometry; number-theoretic investigations and the discovery of the calculus proceed despite the foundational inadequacy of a number concept that cannot handle negative or infinitesimal quantities with logical rigor.[13] In this sense, early modern mathematical practice awaits deep philosophical response in the nineteenth century, when logical and mathematical foundations are reassessed in the face of the striking developments in mathematical practice in the seventeenth and eighteenth centuries. The early modern philosopher remains occupied by the more basic philosophical problems raised by a newly mathematized natural world.[14] It is to those problems, and their solutions, that we now turn.

2. "Mathematicians," "Metaphysicians," and the "Natural Light"

In the Fifth Meditation of his *Meditations on First Philosophy*, Descartes famously argues that we have a clear and distinct perception of the essence of material substance, and that that essence is pure extension. According to Descartes, our clear and distinct perception of the immutable and eternal natures of all material objects, which we find represented in our idea of extension, is the basis for sure and certain knowledge of those objects. This knowledge is systematized by our pure mathematics, in particular by geometry, which thus provides us with an a priori

[13] The calculus is based on a nonnumerical notion of the infinitesimally small differentials between magnitudes conceived as finite line segments, called *differences* by Leibniz and *moments* by Newton. See Struik (1986), pp. 272, 300–301.

[14] Berkeley's critique of the techniques of the calculus is an obvious exception to this generalization, though I believe a case could be made that Berkeley's criticisms prefigured and anticipated issues that did not become fully transparent until the nineteenth century. For this reason, and also due to the scope of this chapter, I have chosen not to address Berkeley's philosophy of mathematics here. For discussion of Berkeley's philosophy of mathematics, see Jesseph (1993). For discussion of the modern debates that are the precursors to the nineteenth-century debates on the foundations of the calculus, see Mancosu (1996).

science of the nature of the corporeal world. For Descartes, then, the essential nature of the material world is perfectly mathematical and perfectly knowable.[15]

Descartes begins his Fifth (and penultimate) Meditation, "The Essence of Material Things, and the Existence of God Considered a Second Time," by proposing to investigate "whether any certainty can be achieved regarding material objects."[16] Before demonstrating that we have certain knowledge of the real existence of material things, as he will do in the Sixth (and final) Meditation, Descartes must first investigate his *ideas* of material things: "But before I inquire whether any such things exist outside me, I must consider the ideas of these things, in so far as they exist in my thought, and see which of them are distinct, and which confused."[17] He proceeds to show that his clear and distinct ideas of extension provide insight into "that corporeal nature which is the subject-matter of pure mathematics."[18]

First, Descartes confirms the clarity and distinctness of his idea of quantity, in particular "the extension of the quantity (or rather of the thing which is quantified) in length, breadth, and depth." He continues, "I also enumerate various parts of the thing, and to these parts I assign various sizes, shapes, positions and local motions; and to the motions I assign various durations."[19] The clarity and distinctness of our general idea of quantity is best understood by considering the mathematically fundamental representations discussed above, in section 1. For Descartes, particular quantities are imaginable via line segments; lengths, breadths, and depths, as well as enumerations of parts, sizes, shapes, positions, motions and durations, are, qua quantities, all mathematically represented with straight line segments, relations among straight line segments, and algebraic equations expressing those relations. These precise and perspicuous mathematical symbols witness the

[15] On its surface, it should appear that Descartes's account of our mathematical cognition easily satisfies *both* the applicability and the apriority demands. Below, we will discuss the specific sense in which—according to Kant—it fails to satisfy the applicability demand.

[16] Descartes (1985), AT 7:63.

[17] Descartes (1985), AT 7:63.

[18] Descartes (1985), AT 7:71. I take it that whereas the Sixth Meditation demonstrates the *existence* of material things, the Fifth Meditation shows that *if* material things exist, *then* they have the *properties* that I clearly and distinctly perceive them to have, that is, all the properties which "viewed in general terms, are comprised within the subject-matter of pure mathematics." See passages at AT 7:65 and AT 7:80.

[19] Descartes (1985), AT 7:63. In this passage Descartes actually speaks of quantity as that which he distinctly "imagines." Throughout the Fifth Meditation he moves back and forth between his imaginings and his ideas, but his conclusions are formulated with respect to the clear and distinct deliverances of the intellect. For a discussion of the relation between the imagination and the intellect, and their roles in our knowledge of geometry, see Hatfield (1986), pp. 62–63.

quantitative properties they symbolize. But our perception of both these symbols and the real quantitative properties they symbolize depends on our having clear and distinct perception of pure extension and its modes—that is, on an idea of quantity in general that is accessible to our intellect.

Descartes claims further that his perception of particular features of quantity is in harmony with his very own nature, "like noticing for the first time things which were long present within me although I had never turned my mental gaze on them before."[20] This is a claim that is buttressed by the "wax argument" of the Second Meditation. There Descartes identifies extension, or extendedness, as that feature of the wax that makes it a material thing, its nature or essence. More important, perhaps, he identifies his own intellect as that cognitive tool that allows him to perceive the extendedness of all material things: Descartes's perception of the essential feature of material substance is due neither to sensation nor to imagination, but to a "purely mental scrutiny" which enables his clear and distinct perception of pure extension.

Descartes combines these results with his more general (and independently demonstrated) conclusions that what he clearly and distinctly perceives is *true*; that his idea of God is innate; that God exists; and that God is no deceiver. Since Descartes has claimed that he clearly and distinctly perceives the quantitative features of material objects, it follows that these quantitative properties are "true"—that is, that material objects (if they exist) really do have the quantitative properties Descartes perceives them to have. The existence of a nondeceiving God underwrites Descartes's perception in a strong sense: his intellectual capacity for clear and distinct perception is a God-given "natural light" providing him with the ability to access mathematical truths, God's own "free creations."[21]

Upon demonstrating in the Sixth Meditation that material substance exists and is really distinct from immaterial substance, Descartes completes his Meditative project. In the process of pursuing his broad metaphysical and epistemological goals, Descartes can be understood to have articulated a philosophy of mathematics according to which we have a priori knowledge of mathematical truths about real, extramental material substance. In particular, our intellect affords us clear and distinct perception of the quantitative features of mind-independent entities, which we represent to ourselves via ideas of extension. Moreover, we know that our ideas of extension perfectly describe the really extended material world—that is, that our ideas of extension are in harmony with the really extended matter that is their object.[22] Descartes's epistemological and metaphysical conclusions

[20] Descartes (1985), AT 7:64.

[21] More specifically, the existence of a good God underwrites Descartes's ability to retain mathematical knowledge that is based on clear and distinct perception. See the argument at AT 7:70.

[22] Descartes uses the harmony metaphor at AT 7:64.

vis-à-vis mathematical objects are thus entwined: he has identified a mental faculty, the rational intellect, to be a cognitive tool for accessing mind-independent material natures. Since, on Descartes's view, those natures are quantitative and mathematically describable, he conceives the "natural light of reason" to illuminate the metaphysically essential features of the mind-independent world and to provide us with "full and certain knowledge . . . concerning the whole of that corporeal nature which is the subject-matter of pure mathematics."[23]

It appears that if Descartes's arguments are accepted, his rationalist philosophy of mathematics satisfies both of the demands on a viable account of early modern mathematical practice, identified above. He accounts for the apriority of mathematics with his theory that the intellect, a faculty of mind independent of the bodily faculties of imagination and sensation, provides direct access to innate mathematical ideas of eternal and immutable natures. He accounts for the applicability of mathematical reasoning by identifying the essence of the natural material world with pure extension, the very object of our innate mathematical ideas: an explanation of how a priori mathematics applies to the natural world is easily forthcoming from a theory that directly identifies the essential features of the natural world with the subject matter of pure mathematics. Since, on Descartes's view, our a priori mathematical knowledge systematizes mathematical truths about a mind-independent and really extended natural world, he thus appears to have satisfied both the apriority and the applicability demands. It will follow from Kant's critique, however, that a theory like Descartes's goes too far in satisfying the applicability demand: the mathematical sciences as Descartes understands them apply far beyond the bounds of what Kant takes to be the limits of our "possible experience." To see why this is so, it will be helpful to consider the views of another "mathematician," and Kant's real target, Isaac Newton.

Newton did not concede all (or even many) of Descartes's metaphysical and physical conclusions; indeed, he spends much of his famous essay "On the Gravity and Equilibrium of Fluids"[24] disputing Descartes's theory of space. On Descartes's view, space is not distinct from the extended material world of spatial things:

> There is no real distinction between space, or internal place, and the corporeal substance contained in it; the only difference lies in the way in which we are accustomed to conceive of them. For in reality the extension in length, breadth and depth which constitutes a space is exactly the same as that which constitutes a body.[25]

Thus, for Descartes, any knowledge we acquire via geometrical cognition of space is, likewise, knowledge of bodily extension. By contrast, Newton conceives space

[23] Descartes (1985), AT 7:71.

[24] Newton (1962).

[25] Descartes (1985), *Principles*, AT 8:45.

to be distinct from bodies, arguing that we can clearly conceive of extension independent of bodies: "... we have an exceptionally clear idea of extension, abstracting the dispositions and properties of a body so that there remains only the uniform and unlimited stretching out of space in length, breadth and depth."[26] It follows that, for Newton, our geometrical cognition of space affords us knowledge of an entity that is infinitely extended, continuous, motionless, eternal, and immutable, but that is not itself corporeal and that can be conceived as empty of bodies. Moreover, space is a unified whole of strictly contiguous parts: the single infinite space encompasses every possible spatial figure and position that a bodily object might "materially delineate."[27] Bodies, for Newton, are the movable and impenetrable entities that *occupy* space, and thus provide us with corporeal instances of spatial parts.

Thus, Descartes and Newton disagree in a deep sense about how to understand the ontology of space and matter, about whether extension can be distinguished from corporeal reality, and thus about whether we have claims to mathematical knowledge of space that are distinct from our claims to mathematical knowledge of spatial things.[28] For Descartes, our knowledge of quantifiable things just is knowledge of the nature of quantity in general, whereas for Newton our knowledge of quantifiable things (which occupy *parts* of space) is acquired via knowledge of the general nature of extension, knowledge of how parts of space relate to space conceived as an infinite whole. But Descartes and Newton nevertheless agree in their accounts of how we come to acquire knowledge of what they both conceive to be an extramental reality that is the ultimate subject matter of pure mathematics. Newton follows Descartes in positing a faculty of understanding as the real source of mathematical cognition, a tool with which we can comprehend the eternal and immutable nature of extension, which he conceives as infinite space.[29] While sensation allows us to represent "materially delineated" bits of extension (i.e., bodies), and imagination allows us to represent *indefinitely* great extension, according to Newton only the faculty of understanding can clearly represent the true and general nature of space/extension.[30]

[26] Newton (1962), p. 132

[27] Newton (1962), p. 133.

[28] They also disagree about infinite versus indefinite extension, and about how to understand the connection between created extension and the mind of God. These are issues that I cannot pursue here.

[29] Newton (1962), p. 134.

[30] McGuire makes this point: "In this sense, then, the understanding possess [*sic*] a non-sensuous representation of infinite distance which has its ultimate ground in the real but uncreated nature of extension itself" (1983, p. 107). According to Newton, the understanding can likewise represent the finite but infinitesimally small quantities that are the basis of his calculus: "fluxions are finite quantities and real, and consequently ought to have their own symbols; and each time it can conveniently so be done, it is preferable to

Newton, like Descartes, thus accounts for our applicable a priori mathematical cognition by claiming that we have a clear perception of the mathematical features of an extramental natural world. Both Descartes and Newton conceive this clear perception of the mathematical features of an extramental natural world to be illuminated by the *natural light* of reason, a metaphor for the sense in which our faculty of understanding is, on their views, an acute mental vision bestowed by a nondeceiving God.[31] As before, it appears that Newton's account of mathematical cognition satisfies both the apriority and the applicability demands: according to Newton, we have the necessary cognitive tools to acquire a priori knowledge of the mathematical features of the natural world. As noted, however, Kant considered Newton's account of mathematical cognition—representative of what he took to be the "mathematician's" position—to be deeply inadequate. He considered Leibniz's alternative account—representative of what he took to be the "metaphysician's" position—just as problematic, as we will see in detail below.

Famously, Leibniz claims that the ultimate constituents of reality, the true substances, are monads: partless, simple, sizeless entities that have perceptual states, like minds. It follows that, for Leibniz, no extended thing is an ultimate constituent of reality and the extended natural world is merely a "phenomenon": our perceptions of a spatial material world are confused representations of a metaphysically fundamental realm of nonspatial monads. Space, time, and "the other entities of pure mathematics" are "always mere abstractions." Their perfect uniformity indicates that they are not intrinsic denominations, internal properties of things, but extrinsic denominations, orderings or relations among things: "[Space] is a relationship: an order, not only among existents, but also among possibles as though they existed. But its truth and reality are grounded in God, like all eternal truths."[32] While it might be accurate to say that, on Leibniz's view, mathematics describes relations among material entities, it is important to remember that for Leibniz material entities are exhausted by our confused phenomenal representations of a perceptually inaccessible underlying reality. This underlying monadic realm is populated by entities that are *not* mathematically describable; monads as Leibniz understands them *cannot* be identified with mathematical points, either

express them by finite lines visible to the eye rather than by infinitely small ones" (1982, p. 107). Of course, that these finite lines be literally "visible to the eye" is, on his own view, irrelevant. What matters is that the finite lines be mathematically manipulable symbols of infinitely small but nevertheless *real* quantities, the ultimate subject matter of the calculus.

[31] As mentioned above, God plays different roles in the epistemological systems of Descartes and Newton; this difference cannot be explored here. For discussion, see McGuire (1983) and Stein (2002).

[32] Leibniz (1996), II.xiii.149.

geometric or numeric. Given his monadology, Leibniz thus owes us an account of the status of mathematical claims to knowledge: Since our *metaphysical* knowledge of reality does not include *mathematical* knowledge, how do we acquire mathematical knowledge, and what is it about?

Leibniz expresses an important component of his mature philosophy of mathematics in a letter to Queen Sophie Charlotte of Prussia,[33] his student, friend, and epistolary interlocutor. There he explains his position that an internal "common sense" allows us to form those clear and distinct notions that are for Leibniz the ultimate subject matter of pure mathematics, such as number and shape. These are ideas of qualities more "manifest" than those sensible and "occult" qualities we access via the clear but confused notions of a single sense, such as colors, sounds, and odors. The notions we acquire through the "common sense" describe qualities accessible to more than one external sense: "Such is the idea of *number*, which is found equally in sounds, colors, and tactile qualities. It is in this way that we also perceive *shapes* which are common to colors and tactile qualities...."[34] Leibniz further identifies the imagination as the faculty of mind that operates on the clear and distinct notions of the common sense, the ideas that arise from the perception of qualities common to more than a single external sense, to produce our conceptions of mathematical objects:

> ...these clear and distinct ideas [of the common sense], subject to imagination, are the objects of the mathematical sciences, namely arithmetic and geometry, which are pure mathematical sciences, and the objects of these sciences as they are applied to nature, which make up applied (mixtes) mathematics.[35]

He adds, however, that sense and imagination must be augmented by understanding in order to attain a full conception of mathematical objects and "to build sciences from them." It is, according to Leibniz, the understanding or reason that assures that the mathematical sciences are demonstrative and that mathematical truths are *universally* true, for without the application of such a "higher" faculty of reasoning, the mathematical sciences would consist merely of observations and inductive generalizations therefrom. Indeed, Leibniz claims that despite the fact that our notions of mathematical objects originate in sensible experience, nevertheless a demonstrative mathematical truth is fully "independent of the truth or the existence of sensible and material things outside of us." Thus, for Leibniz, mathematical truths describe features of our sensible experience despite finding their justification in the understanding alone.

[33] Leibniz (1989), pp. 186–192. This letter is also known as "On What Is Independent of Sense and Matter." For more discussion of this letter and Leibniz's philosophy of mathematics, see McCrae (1995).

[34] Leibniz (1989), p. 187.

[35] Leibniz (1989), pp. 187–188.

Leibniz resolves this apparent tension—between the sensible source of our notions of mathematical objects and the intelligible source of our justification of mathematical truths—with recourse to the "natural light" of reason. The "natural light" of reason is solely responsible for our recognition of the necessary truth of the axioms of mathematics and for the force of demonstrations based on such axioms. Mathematical demonstration thus depends solely on "intelligible notions and truths, which alone are capable of allowing us to judge what is necessary."[36] For Leibniz, then, our pure mathematical knowledge is formal knowledge of the logic of mathematical relations, which are not directly dependent on any sensible data. This pure mathematical knowledge might be described as *verified* by our sensible experience, but ultimately our ideas of the mathematical features of sensible things conform to the mathematical necessities that we understand by the natural light:

> [Experience is] useful for verifying our reasonings as by a kind of proof.... But, to return to *necessary truths*, it is generally true that we know them only by this natural light, and not at all by the experiences of the senses.[37]

According to Leibniz, we have intelligible knowledge by the "natural light" of "what must be" and "what cannot be otherwise"—that is to say, of necessary (mathematical) truths such as the law of noncontradiction and the axioms of geometry. We can apply these laws to particular sensible qualities with our "common sense," which represents for us particular ideas of magnitude and multitude, thereby instancing universal mathematical truths. But, on this picture, we can discover necessary truths about many more things than we can possibly imagine, in Leibniz's sense: there are mathematical objects about which we can deduce mathematical truths but which our "common sense" cannot access. For example, even though "one finds ordinarily that two lines that continually approach finally intersect," the geometer nevertheless makes use of asymptotes, "extraordinary lines... that when extended to infinity approach continually and yet never intersect."[38] Likewise, the arithmetician operates with negative magnitudes, the analyst with infinitesimally small magnitudes, the algebraist with imaginary roots of magnitudes. Leibniz calls such notions "useful fictions," that is, notions which cannot be found in nature and which may seem even to contradict our sensible experience, but which nevertheless have obvious mathematical utility. He is in a unique position to account for them, despite the general inadequacies of modern symbolic systems to do so, because of another aspect of his approach to mathematics: his formalism.

[36] Leibniz (1989), p. 189.
[37] Leibniz (1989), p. 189.
[38] Leibniz (1989), p. 191.

Leibniz's formalist attitude toward mathematical reasoning allows him to conceive the symbols of arithmetic, algebra, and analysis independently from the geometric figures they were originally devised to stand for, thus allowing those symbols to be manipulated formally with only "intermittent attention" paid to the figures that are their referent.[39] The use of "fictional" mathematical notions is thus warranted rather than deemed absurd, since, on Leibniz's view, mathematical formalisms are detachable from their domain of application. In this respect, Leibniz can construe the objects of mathematical reasoning more abstractly than could either Descartes or Newton, both of whom used formal symbolic manipulation only as an aid to fundamentally geometric reasoning. Since Leibniz, by contrast, identifies the systematization of logical reasoning as the primary aim of the mathematical sciences, and separates his claims to mathematical knowledge from his claims to metaphysical knowledge, he can disregard the suggestion that metaphysical absurdities arise out of the use of effective mathematical symbolisms, thus initiating a transition to a more abstract and formal conception of the mathematical sciences.[40]

How, then, does Leibniz's account of mathematical cognition fare in the face of our two demands? His account seems to satisfy the apriority demand, since the role of the understanding in a mathematical context is specifically to provide us with knowledge of the mathematical axioms and the laws of logic. On Leibniz's view, then, our claims to mathematical knowledge are a priori because they are all founded on and deduced from universal and necessary truths known to us by the "natural light." Moreover, according to Leibniz, these claims to mathematical knowledge find application in the natural world: the clear and distinct deliverances of the "common sense" demarcate the domain of mathematical applicability. That our "common sense" notions are not notions of a metaphysically fundamental monadic realm is of no bother to Leibniz's account of mathematical cognition: despite the fact that what is metaphysically fundamental is for Leibniz not mathematically describable, we nevertheless have substantive mathematical knowledge of relations among objects of the *phenomenal* natural world.

We are now in a position to understand and appreciate Kant's critique of the philosophies articulated by the "mathematician" and the "metaphysician" who preceded him, and to explicate his alternative account of mathematical cognition.

[39] Leibniz (1996), II.xxi.186.

[40] Leibniz was skeptical that algebra could fully systematize this more formal conception of mathematics, stating in the *New Essays* that "algebra falls far short of being the art of invention, since even it needs the assistance of a more general art" (1996, IV.xvii. 489). He implies in the same passage that the "general procedures" of his infinitesimal calculus are more promising, though he speaks elsewhere of an even more general mathematical art.

3. Kant's Response

A coherent philosophy of mathematics—including an account of prevailing mathematical practice and an articulation of the epistemological and metaphysical conditions on the success of such practice—is a vital component of Kant's critical project. In an important portion of the "Transcendental Aesthetic,"[41] Kant provides the general outlines of his philosophy of mathematics, and touts its virtues, by drawing a contrast between his view and those held by his predecessors. Kant's claims in this section are that the "metaphysician's" account of mathematical cognition has failed the apriority demand; that the "mathematician's" account has failed the applicability demand; and that his own view satisfies both.

Famously, Kant's own view includes the doctrine of Transcendental Idealism, according to which space and time are the pure forms of our sensible intuition, and the sources of synthetic a priori cognitions.[42] Mathematics, geometry in particular, provides a "splendid example" of such cognitions: according to Kant, geometry is the science of our pure cognition of space and its relations, and provides us with a priori knowledge of the spatial form of the objects of our possible experience, those sensible things about which we can make objectively valid judgments.[43] As we will see in more detail below, Kant claims that his own account fully satisfies both the apriority and the applicability demands. According to Kant, mathematical cognition is cognition of our own intuitive capacities, of our own pure intuitions. Since Kant argues that our intuition of space is prior to and independent of our experience of empirical spatial objects, geometric cognition is paradigmatically a priori; thus his account satisfies the apriority demand. But, on Kant's view, mathematical cognition is also cognition of the empirical objects that we represent as having spatiotemporal form, that is, of the objects of our possible experience. Inasmuch as we have a priori geometric cognition of space, we have a priori geometric cognition of the spatial *form* of real spatial objects. The domain of our a priori mathematical cognition extends beyond the pure intuition of space to the formal conditions under which we represent empirical objects as being in space, thus allowing a priori mathematical cognition to find application in the realm of real empirical objects. Importantly, for Kant our a priori mathematical cognition extends to, but not beyond, the bounds of possible experience: a priori

[41] Kant (1998), A39/B56–A41/B58.

[42] See Shabel (2004) for a discussion of how this particular claim relates to the general doctrine of Transcendental Idealism.

[43] For the sake of simplicity, and because of the direct connection between space and mathematical (geometric) cognition, I will concentrate here on Kant's views of space and geometry, and will not discuss his account of time or other mathematical disciplines. For discussion of Kant's theory of algebra, see Shabel (1998).

mathematical cognition applies to all and only the spatiotemporal objects of a possible experience.

Kant defends his own account of mathematical cognition with positive arguments in its favor (some of which we will examine below) as well as with his critique of competing accounts. Having proclaimed the transcendental ideality of space and time, and offered an associated account of the apriority and applicability of our mathematical cognition, he objects that "Those, however, who assert the absolute reality of space and time, whether they assume it to be subsisting or merely inhering, must themselves come into conflict with the principles of experience."[44] Kant here takes both of his opponents to defend the *absolute reality* of space; on this basis, he judges that both opponents fail to solve the problem of identifying the a priori principles of our experience of the natural world.[45] But the details of Kant's critique, and the contrasting ways in which his opponents fail where he succeeds, emerge only upon discussion of what distinguishes those who "assume [space] to be subsisting," the "mathematical investigators of nature," from those who "assume [space] to be... merely inhering," the "metaphysicians of nature." We will begin by examining Kant's objections to the "metaphysicians," proceed to his objections to the "mathematicians," and, finally, rehearse his own view.

Kant claims that Leibniz, Wolff, and other "metaphysicians of nature" hold "space and time to be relations of appearances... that are abstracted from experience though confusedly represented in this abstraction...."[46] On this view, according to Kant, our notions of space and time, the alleged source of our mathematical cognition, are "only creatures of the imagination, the origin of which must really be sought in experience." The representations that lie at the basis of mathematical reasoning and cognition are abstracted by the imagination from or out of our sensible contact with appearances, and thus have an empirical origin. On this view, space and time are "only inhering" because they are relations among objects and not self-subsisting entities; our representations of space and time are constructed by abstraction from spatiotemporal relata. Note that from Kant's perspective, space and time are on this relationist view nevertheless "absolutely real": the objects or appearances that stand in such spatiotemporal relations are conceived to be the source of our notions of space and time and, thus, to be "absolutely real," that is, independent of what Kant takes to be transcendental conditions on experience. It follows from the metaphysicians' view that the subject matter of mathematics is derived from our experience of the natural world, and is not a factor in our own construction of that experience.

[44] Kant (1998), A39/B56.

[45] This should make clear the sense in which Kant sees his opponents' project from the perspective of his own. His criticism of their accounts of mathematical cognition is not wholly objective, since they would surely have formulated the "problem" quite differently.

[46] Kant (1998), A40/B57. Subsequent quotations are from this same passage.

Kant admits that on such an account, mathematical cognition does not exceed what he takes to be the limits of possible experience: if mathematical cognition derives directly *from* our engagement with the realm of appearances, then our application of mathematical cognition to all and only appearances seems guaranteed. This is a virtue of the metaphysicians' account and explains the sense in which Kant accepts it as satisfying the applicability demand. But the metaphysicians' account has, according to Kant, the fatal defect of not satisfying the apriority demand. As Kant puts it, the metaphysicians "must dispute the validity or at least the apodictic certainty of a priori mathematical doctrines in regard to real things (e.g., in space), since this certainty does not occur *a posteriori*...." Here Kant claims that because the original source of mathematical cognition is, on the metaphysicians' view, experiential, our geometric cognition of the spatial features of empirical objects must be a posteriori. This might seem not to be a problem for the metaphysician; he could reply to Kant's criticism by reiterating his account of our a priori knowledge of the laws and axioms of formal mathematics, available to us via the understanding. Thus the metaphysician would claim to satisfy the apriority demand by having given an account according to which some mathematical cognition is a priori (e.g., knowledge of laws, axioms, and theorems derivable therefrom) while some is a posteriori (e.g., knowledge of the mathematical features of real objects to which those theorems might be thought to apply).

But this sort of response misses the real thrust of Kant's criticism. Kant claims that the metaphysicians "can neither offer any ground for the possibility of a priori mathematical cognitions... nor can they bring the propositions of experience into necessary accord with those assertions." According to Kant, the metaphysicians' account serves to detach the two primary features of mathematical cognition—apriority and applicability—in such a way that there is no explanatory or meaningful harmony between the universal and a priori mathematical laws known by the understanding and the substantive but a posteriori mathematical cognition of the natural world acquired via the "common sense." Kant's charge is that formal a priori mathematical cognition as the metaphysician understands it is altogether isolated from the domain of objects taken to be mathematically and scientifically describable; universal and a priori mathematical truths can *only* be about "useful fictions" and not "real things." If our account of a priori mathematical cognition does not give us a priori knowledge of the objects of our possible experience, then our account has failed the apriority demand, at least as Kant conceives it.

Proceeding to the "mathematicians," Kant claims that "they must assume two eternal and infinite self-subsisting non-entities (space and time) which exist (yet without there being anything real) only in order to comprehend everything real within themselves." Kant here describes Newton's absolutist view that space is a container existing independently of the real spatial objects it contains, though is not itself a real empirical entity. Kant's assessment of this view is brief; he writes that "[The mathematicians] succeed in opening the field of appearances for mathematical assertions. However, they themselves become very confused through precisely these conditions if the understanding would go beyond this field." In the first sentence, Kant makes clear that he understands the strength of the mathematicians' account to be its defense of our ability to achieve a priori insight into the mathematical features of the empirical natural world. Whereas the metaphysician could only explain a posteriori knowledge of such features, the mathematician proffers apriority via our perfect understanding of extension. Thus, on the mathematicians' account we have a priori mathematical knowledge of the field of appearances, of *all* of the objects that are contained within the domain of mathematical applicability. In the second sentence, however, Kant identifies the defect in the mathematicians' account: on this view, our a priori mathematical knowledge is applicable to all—*but not only*—appearances. It attempts to extend our a priori mathematical knowledge beyond the domain of appearances without explanation or justification of that extension. That is, in positing our special knowledge of an absolutely real and self-subsisting space or extension, the mathematician makes a claim to valid mathematical knowledge of entities that cannot themselves be described as within the realm of our possible experience. Given the mathematicians' absolutist conception of space, our achieving a priori knowledge of the mathematical features of the spatiotemporal natural world (e.g., knowledge of the geometry of spatial objects) requires that we also achieve a priori knowledge of the features of the *supra*natural world (e.g., knowledge of space itself, conceived independently from both our understanding and the objects it is thought to contain). Kant considers this to be a kind of "confusion": the "mathematician" achieves apriority only by extending the domain of applicability beyond the bounds of our possible experience. For Kant, the cost of apriority cannot (and need not) be that high. Thus, Kant takes the mathematicians' account to fail the applicability demand in a special sense: on the mathematicians' account, a priori mathematical cognition is applicable, but it is applicable beyond acceptable limits, that is, beyond the limits of our possible experience and knowledge of nature.

Kant thus charges that both of his opponents "come into conflict with the principles of experience," albeit in different ways. The metaphysician conflicts with the alleged apriority of those principles as applied to experience; the mathematician with the limits of their domain of applicability. While Kant's predecessors each satisfied only one of the two demands on a successful account of mathematical cognition, Kant considers his own theory to satisfy both the apriority and the applicability demands without conflicting with the principles of experience. As noted above, his theory hinges on his claim that space and time are pure forms of sensible intuition and sources of synthetic a priori cognitions. In the case of space, our pure intuition of space is the source of our claims to geometric knowledge: we have an a priori representation of space that is the ground for geometric reasoning and cognition. Kant's defense of this notorious claim is complex and beyond the scope of this chapter;[47] it will suffice for our purposes to consider the sense in which Kant's theory of space provides him with an account of mathematical cognition that satisfies both the apriority and the applicability demands.

Kant's argument runs roughly as follows. Because space and time are forms of sensibility and cognitive sources of a priori mathematical principles, space and time "determine their own boundaries" and "apply to objects only so far as they are considered as appearances.... Those alone are the field of their validity, beyond which no further objective use of them takes place." Kant is making a claim about the connection between our way of intuiting, representing, and knowing the structure of space and our way of intuiting, representing, and knowing the features of the objects we experience to be in space: the former *determines* the latter. For this reason, Kant claims that our a priori representation of space determines its own domain of applicability, its "field of validity." Insofar as our a priori mathematical cognition has its source in our *sensible* faculty, including our a priori representation of space, such cognition can be objectively valid with respect to all and only those objects we can *sense*, the realm of appearances or objects of our "possible experience." Thus, for Kant, mathematics is a body of synthetic a priori cognition providing us with knowledge of *both* (1) the pure conditions on our sensible representations *and*, by this means, (2) the objects that appear to us under those conditions. As Kant says, geometry (which is available to us via our a priori representation of space) provides the paradigm example. In the first sense, geometry is the science of a cognitive capacity, in particular, a capacity to represent spatial relations; geometry is the unique science that describes the various ways in which that capacity is both warranted and constrained. In the second sense, as explained in Kant's transcendental philosophy, geometry is the science of the natural or empirical objects that we conceive to stand in such spatial relations as we are conditioned to represent. Thus, geometry can inform us as to the proper function of our cognitive capacity for pure spatial intuition while also providing us with knowledge of the spatial form of those objects we intuit as empirically real spatial things.

Because, for Kant, mathematics provides us with knowledge of the pure conditions on our sensible representations, mathematical cognition is a priori. Being knowledge of the cognitive conditions on our *having* sensible experience, mathematics is cognition that is necessarily acquired prior to and independent from such experience. Because, for Kant, mathematics provides us with knowledge of the objects that appear to us under such cognitive conditions, mathematical cognition is applicable to the natural empirical world. Being knowledge of the formal features of sensible spatiotemporal objects, mathematics is cognition that is necessarily about the things that inhabit the natural world, at least insofar

[47] I take these claims to be defended in the "Metaphysical Exposition" and the "Transcendental Exposition of the Concept of Space" see Shabel (2004).

as we represent them. If we are willing to accept this last caveat—that the natural world comprises all and only those things that we have the cognitive capacity to represent—then we can secure the "certainty of experiential cognition": our a priori mathematical claims have direct and complete purchase on (our experience of) the natural world. Kant thus claims to have provided an account of mathematical cognition that satisfies both the apriority and the applicability demands with which we began—and that, moreover, does not *conflict* with, but rather helps to establish, the "principles of experience."

REFERENCES

Descartes, R. (1954). *The Geometry of René Descartes* (David Eugene Smith and Marcia L. Latham, eds. and trans.). New York: Dover.

Descartes, R. (1985). *The Philosophical Writings of Descartes* (John Cottingham, Robert Stoothoff, and Dugald Murdoch, eds. and trans.). Cambridge: Cambridge University Press.

Hatfield, G. (1986). "The Senses and the Fleshless Eye," in *Essays on Descartes' Meditations* (A. Rorty, ed.). Berkeley: University of California Press.

Jesseph, D. (1993). *Berkeley's Philosophy of Mathematics.* Chicago: University of Chicago Press.

Kant, I. (1998). *Critique of Pure Reason* (Paul Guyer and Allen Wood, eds. and trans.). Cambridge: Cambridge University Press.

Leibniz, G.W. (1989). *Philosophical Essays* (Roger Ariew and Daniel Garber, eds. and trans.). Indianapolis: Hackett.

Leibniz, G.W. (1996). *New Essays in Human Understanding* (Peter Remnant and Jonathan Bennett, eds. and trans.). Cambridge: Cambridge University Press.

McCrae, R. (1995). "The Theory of Knowledge," in *The Cambridge Companion to Leibniz* (N. Jolley, ed.). Cambridge: Cambridge University Press.

McGuire, J.E. (1983). "Space, Geometrical Objects and Infinity: Newton and Descartes on Extension," in *Nature Mathematized* (W. Shea, ed.). Dordrecht: D. Reidel.

Mancosu, P. (1996). *Philosophy of Mathematics and Mathematical Practice in the Seventeenth Century.* Oxford: Oxford University Press.

Newton, I. (1962). "On the Gravity and Equilibrium of Fluids," in *Unpublished Scientific Papers of Isaac Newton* (A. Rupert Hall and Marie Boas Hall, eds. and trans.). Cambridge: Cambridge University Press.

Newton, I. (1982). "On the Quadrature of Curves," in *The Mathematical Papers of Isaac Newton* (D.T. Whiteside, ed.), vol. 8. Cambridge: Cambridge University Press.

Shabel, L. (1998). "Kant on the 'Symbolic Construction' of Mathematical Concepts." *Studies in History and Philosophy of Science*, 29 (4), 589–621.

Shabel, L. (2003). *Mathematics in Kant's Critical Philosophy: Reflections on Mathematical Practice.* New York: Routledge.

Shabel, L. (2004). "Kant's 'Argument from Geometry.'" *Journal of the History of Philosophy.* 42:2, 195–215.

Stein, H. (2002). "Newton's Metaphysics," in *The Cambridge Companion to Newton* (I.B. Cohen and G.E. Smith, eds.). Cambridge: Cambridge University Press.

Struik, D.J. (ed. and trans.) (1986). *A Source Book in Mathematics*. Princeton, N.J.: Princeton University Press.

Wolff, C. (1965). *Mathematisches Lexicon*, Abt.1, Bd.11. Hildesheim: Georg Olms Verlag.

CHAPTER 3

LATER EMPIRICISM AND LOGICAL POSITIVISM

JOHN SKORUPSKI

1. The Historical and Philosophical Context

The empiricist approaches to mathematics discussed in this chapter belong to an era of philosophy which we can begin to see as a whole. It stretches from Kant's Critiques of the 1780s to the twentieth-century analytic movements which ended, broadly speaking, in the 1950s—in and largely as a result of the work of Quine.[1]

Seeing this period historically is by no means saying that its ideas are dead; it just helps in understanding the ideas. That applies to the two versions of empiricism that were most prominent in this late modern period: the radical empiricism of Mill and the "logical" empiricism associated with Vienna Circle positivism of the late 1920s and early 1930s. Mill and the logical positivists shared the empiricist doctrine that no informative proposition is a priori. Thus they rejected two main tenets of Kantian epistemology: the claims that the possibility of knowledge requires that there be synthetic a priori propositions and that there are such

I am grateful to Agustín Rayo and Stewart Shapiro for helpful comments.

[1] See Quine (1953, 1966a, 1966b).

propositions. But as to the status of logic and mathematics, they differed sharply. Mill took logic and mathematics to consist of informative universal truths and denied that they are a priori; the positivists held logic and mathematics to be a priori and denied that they contained informative truths. Furthermore, in working out the doctrine that logic and mathematics are exact and a priori because they are—in their sense—"analytic," the logical positivists renewed vital elements of Kantian thought even as they denied Kant's doctrine that mathematics has informative content.

To bring out the contrast between these two empiricisms, and to address the element of continuity just noted between the views of Kant and those of logical positivism, we should take note of an influential conception of knowledge. Genuine knowing is of something wholly distinct from the knowing itself—something to which cognition must conform if it is to be knowledge. Cognition is receptive. It is not formative or constitutive of its objects. All this seems inherent in the very idea of knowledge; but there have been many conceptions of knowledge, of which this "mirroring" conception is only one. As far as empiricism in our late modern period is concerned, the mirroring conception may seem inherent in the naturalistic starting point that Mill and logical positivists share. They take it for granted that cognition is a natural process within the natural world; it conforms to its object by reflecting information received from it. And the information-receiving processes by which it does this—the senses—must themselves be fully characterizable within a scientific theory of the world.

Clearly the cognitive status of logic and mathematics is a central challenge for this conception of knowledge. For as Kant famously noted, on this conception a priori knowledge seems impossible:

> Up to now it has been assumed that all our cognition must conform to the objects; but all attempts to find out something about them *a priori* through concepts that would extend our cognition have, on this presupposition, come to nothing. Hence let us try whether we do not get farther with the problems of metaphysics by assuming that the objects must conform to our cognition. . . .

This attempt is Kant's "Copernican revolution":

> If intuition [sensory cognition of objects] has to conform to the constitution of objects, then I do not see how we can know anything of them *a priori*; but if the object (as an object of the senses) conforms to the constitution of our faculty of intuition, then I can very well represent this possibility to myself.[2]

Kant holds that knowing requires a framework to which every knowable object must conform. He locates it in the a priori structures of our sensibility and bases

[2] *Critique of Pure Reason*, Bxvi–xvii.

the synthetic a priori character of arithmetic and geometry on these. The resulting transcendental idealism about our cognition is inconsistent with a naturalistic view, according to which knowers and their knowledge are, straightforwardly, only a part of the world they know: it demotes the naturalistic philosopher's "empirical" view of our place in nature to just one viewpoint out of two—and not the one in which the apriority of mathematics can be elucidated.

Viennese empiricism was not as distant from the Kantian or "Critical"[3] tradition as was Mill.[4] Specifically, logical positivists agreed with the general Critical idea that knowledge is possible only within an a priori framework to which knowable objects must conform. But the way they worked it out differed fundamentally from Kant's transcendental-idealist interpretation. For them, the necessary framework was provided not by forms of intuition or (necessary) categories of understanding, but by a system of conventions. Knowledge, they agreed, required an a priori framework; there was, however, no unique framework that it required. The conventions that provided the framework could vary. Moreover, contra transcendental idealism, there was no meaningful distinction to be made between objects as they could be known and objects as they really were. There was no significant standpoint outside the empirical standpoint—only free choices of framework.[5] Things were as they could in principle be known to be within a framework stipulated a priori: that was the import of their "verificationist" conception of meaning. And this conception opened up, as they thought, a new account of the apriority of logic and mathematics. Logic and mathematics owed their apriority, their exactness, and their certainty to the fact that they belonged to the framework of conventions. By the same token, they were empty of content. Content can exist, and can be conveyed, only within a framework. This conventionalist doctrine was influential well beyond the Vienna Circle; it became a cornerstone of the twentieth-century analytic tradition as a whole, and gave that whole tradition its Critical aspect.

In contrast, no such scheme/content distinction is made by Mill (and this goes far toward explaining the low esteem in which his view was held in this tradition's heyday). "Our thoughts are true," he said, "when they are made to correspond with Phaenomena"—in other words, with what we are or could be aware of through the senses, the observable facts. And this applies to logic and mathematics, of which he takes a straightforwardly universalist view. These

[3] I capitalize "Critical" to indicate that it refers to the post-Kantian epistemological tradition. To be non-Critical is not to be uncritical.

[4] Affinities between Kant's views and those of the logical positivists and Wittgenstein have been discussed by a number of interpreters; among them are Coffa (1991), Friedman (1991), Garver (1996), and Stroud (1984).

[5] See, for example, Carnap (1956) on "internal" and "external" questions.

are, he thinks, empirical sciences which deal with the most general laws of nature.[6]

It is also worth noting that in both cases, that of Mill and that of logical positivism, their naturalism stands in apparent tension with some of their other doctrines. Mill's *System of Logic* (1843) propounds a "naturalised epistemology" which includes a thoroughly naturalistic treatment of semantics and the methodology of science; yet in his *Examination of Sir William Hamilton's Philosophy* (1865) he puts forward a phenomenalist analysis of matter as permanent possibility of sensation. He saw no inconsistency in this. But he was not sure what to say about the mind. Can the mind also be resolved into "a series of feelings, with a background of possibilities of feeling?" Mill is unwilling to accept "the common theory of Mind, as a so-called substance," yet the fact that self-consciousness is involved in memory and expectation drives him to "ascribe a reality to the Ego—to my own Mind—different from that real existence as a Permanent Possibility, which is the only reality I acknowledge in Matter."[7] This could have led him in a transcendental direction: if self-consciousness has this nonsubstantial yet irreducible kind of reality, can it impose an objectivity-constituting framework on its own experience? Yet, despite these qualms, Mill seems more poised to go in the Machian direction which followed than in the Kantian direction which preceded. Mach takes reality to consist of elements that can indifferently be construed as mental or physical, treats the self as a "thought-economical construct," and views mathematics as a universal empirical science. The *Examination* is a stage on the road to that view.

If Mill's naturalism is challenged by his uncertainty about the status of the self-conscious subject, isn't that of the logical positivists challenged by their "Critical" stance? If cognition is a purely natural process, how can it have a constitutive role in determining the objectivity of content?

To answer this question, we must distinguish between naturalism and what I will call "realism," using the term to refer to a correspondence conception of what it is for a proposition to be true—or, indeed, of what it is to *be* a proposition. The Millian view of logic and mathematics results (as we shall see) from combining naturalism and realism. In contrast, the scheme/content distinction made by the logical

[6] *An Examination of Sir William Hamilton's Philosophy, Collected Works, IX*, p. 384. By "universalism" about logic and mathematics I mean a view which takes these subjects to be on a par, ontologically, with other sciences which propound universal laws; they are distinguished by the generality of the subject matter whose laws they propound (Compare Ricketts (1996), pp. 59–60). In this sense Mill and Frege were both universalists, though they disagreed about whether logic and mathematics are a priori. Kant, Wittgenstein, and the logical positivists were not.

[7] *An Examination of Sir William Hamilton's Philosophy, Collected Works, IX*, p. 208.

positivists requires rejection of that correspondence conception. Conventionalism is in fact not just nonrealist, it is anti-realist—it requires an epistemic (in the positivists' case, verificationist) conception of meaning. It was in this sense that Wittgenstein and Carnap "offered in the early 1930s the first genuine alternative to Kant's conception of the a priori."[8] The new conception of meaning was the crucial innovation by which they were able to combine naturalism with a conception of logic and mathematics as exact and a priori—because it allowed them, as we shall see, to put forward a new doctrine of analyticity.

One can express empiricism as the view that no factual proposition is a priori. What does "factual" mean, though, if we try to refine vague phrases like "about the world?" Mill thinks of facts in the correspondence theorist's way, as truth-makers. But for the logical positivists, "fact" is elucidated in terms of "true factual statement," and factual statements exist only within a fixed logico-mathematical framework. They are the nonanalytic statements in a total scientific theory. Viennese empiricism does not come, as in the case of Mill, from a combination of naturalism and realism. So where does it come from?

In his impressive pre-logical-positivist work, The *General Theory of Knowledge*, Moritz Schlick had formulated naturalism as follows:

> According to our hypothesis the entire world is in principle open to designation by [the conceptual system of physics]. Nature is all; all that is real is natural. Mind, the life of consciousness, is not the opposite of nature, but a sector of the totality of the natural.[9]

But for logical positivism in its developed form this could not be a "hypothesis." The apparently substantive proposition that the natural facts are all the facts there are, reduces to a definitional truth: the true propositions, or contents, within a given linguistic scheme are a subset of the contents expressible within that scheme. Thus it is also a definitional truth that our knowledge is either "knowledge" of the scheme (which the positivists would not regard as genuine knowledge, knowledge of content) or factual knowledge based on sense experience.

To round off the historical picture: Quine's empiricism has affinities to both the Millian and the Viennese versions. Mill and Quine obviously differ in important ways. Mill is an inductivist and phenomenalist, Quine is an abductivist and physicalist; Mill rejects abstracta, Quine countenances them. Yet there are resemblances that are at least as important. For both of them, logic and mathematics are a part of science, known by the a posteriori methods of science. They have a similar conception of "naturalized epistemology," though Quine gives (in the opinion of most philosophers today) a better methodology of science than Mill, and obviously has a more advanced conception of the scope of logic

[8] Coffa (1991), p. 3.
[9] Schlick (1985), p. 296.

and higher mathematics. Nevertheless, Quine's view is also in important ways continuous with logical positivism. It is, one might say, logical positivism without the "two dogmas of empiricism." The scheme/content distinction is one of these dogmas, and Quine decisively rejects it.[10] Whereas Mill gets his universalism about logic and mathematics from combining naturalism and realism, Quine gets it from combining naturalism with rejection of the scheme/content distinction. But that does not mean he rejects the verificationism that made the distinction possible. A verificationist conception of language and reality was crucial to the twentieth-century version of the Critical tradition—in holistic form (the unit of meaning is the theory) it endures in Quine's thinking.

2. Analyticity in Mill and in Logical Positivism

Central to the differences between these empiricisms is the notion of analyticity. Three different conceptions of analyticity will matter to us: the Kant/Mill and the Kant/Frege conceptions, as I will call them, and the "implicit-definition conception," as I will call it, of the logical positivists.

Mill does not use the "analytic/synthetic" terminology. He distinguishes instead between "verbal"and "real" propositions and "merely apparent" and "real inferences." He notes, however, that this distinction coincides with Kant's distinction between analytic and synthetic judgments, "the former being those which can be evolved from the meaning of the terms used."[11] Thus we can use the Kantian terminology in describing his views. The crucial point in the Kant/Mill notion is that an analytic proposition conveys no information, and an analytic inference moves to no new assertion—its conclusion has been literally asserted in its premises. Mill takes this notion very strictly. In effect, on his view, an analytic inference must have the form

$$\frac{p1, \ldots, pi, \ldots, pj, \ldots, pn}{pi, \ldots, pj}.$$

[10] Quine (1953). The other dogma is reductionism.

[11] *System of Logic, Collected Works VII*, p. 116. Compare *Critique of Pure Reason*, A6–7, B10–11. Here Kant puts forward a narrower notion of analyticity than in the passage mentioned in note 13, at least on standard interpretations of the latter.

Inferences reducible to this form by explicit definitions of terms are analytic, and analytic propositions are the corresponding conditionals of such inferences.[12] With regard to propositions and inferences that are analytic in this sense, no question arises about how acceptance of them is justified, because there is no content there to justify. Hence Mill's terminology: merely verbal, merely apparent.

Compare the Kant/Frege account. This characterizes analyticity as deducibility, with explicit definitions, from logic.[13] Thus it is true by definition that logic itself is analytic. But the definition does nothing to show that logic is empty of content, or that there is no philosophical question to raise about the justification of deduction. For a philosopher who believes that logic is simply too basic for questions about its justification to arise, the point may pose no problem. But from an empiricist standpoint this belief is dogmatic. For empiricism says that no proposition that has content is a priori. Likewise, from an empiricist point of view, logicism, however successful, would cut no epistemological ice unless accompanied by a philosophical elucidation of the contentlessness of logic. This the Kant/Frege concept of analyticity does not do.

On Mill's account some propositions and inferences of logic will be analytic in the Kant/Mill sense. He holds that asserting a conjunction, A and B, is asserting A and asserting B. He defines "A or B" as "If not A, then B, and if not B, then A;" and he takes "If A, then B" to mean "The proposition B is a legitimate inference from the proposition A." He discusses generality in a way which is insightful in detail but hard to interpret overall; it fits with much of what he says to treat a universal proposition as asserting a license to infer. We can then stipulate that a universal proposition is analytic if and only if all its substitution instances are analytic. For example, "All ABs are A" says "Any proposition of the form 'X is A' is legitimately inferable from the corresponding proposition of the form 'X is A and B'." And each substitution instance of this is analytic.[14]

Mill takes "It is not the case that A" to be equivalent in meaning to "It is false that A". If we further assume the equivalence in meaning of "A" and "It is true that A," the principle of contradiction becomes the principle of exclusion—as he puts it, "the same proposition cannot at the same time be false and true." He makes analogous remarks about excluded middle, which turns—on these definitions—into

[12] The definition can be extended to make "a = a" analytic: an inference of the form "Fa, a = b, therefore Fb" becomes analytic for the case where "a" is substituted for "b". Interestingly, Mill thinks *all* name–name identities are "verbal" (because he thinks names have no "connotation"). The thought is that they assert no fact to obtain.

[13] *Critique of Pure Reason*, A151, B190–191. The sense of this passage ("The *principle of contradiction*... must be recognised as being the universal and completely sufficient *principle of all analytic knowledge*") is not entirely clear, but is often interpreted in the broad "Kant/Frege" sense. For Frege, see Frege (1950), §§12, 88–89.

[14] Obviously this stipulation raises questions. Is it within the Kant/Mill spirit to take it as analytic that a given proposition is a substitution instance of a given logical form?

the principle of bivalence: "Either it is true that P or it is false that P." He holds these principles, bivalence and exclusion, to be synthetic—"real" or "instructive"—propositions. And thus the question is how they can be a priori.

From a Kantian point of view, if Mill's analysis of logic is sound it puts logic in the same synthetic a priori class as mathematics; and the answer to how it can be a priori will then have to be the same. The apriority of logic will have to rest on the form of our intuition. Bivalence and exclusion, like other synthetic a priori truths, will be restricted to the domain of phenomena. They cannot be known to hold, as Kant himself believed, of things-in-themselves. This conclusion would undercut the epistemological aims of logicism just as much as Mill's empiricism does. Showing that mathematics is logic, and thus analytic in the Kant/Frege sense, would not show that it is analytic in the Kant/Mill sense, and thus empty of content. It would not show, without appeal to a priori intuition, how mathematics could be a priori—since logic itself would now be seen to rest on a priori intuition.

But when we turn to logical positivism, we find a conception of analyticity that is distinct from both the Kant/Mill and the Kant/Frege account. Fundamental to it is the notion of implicit definition. Its main source seems to be the thinking about the status of geometry done at the turn of the twentieth century, notably by Hilbert and Poincaré. Moritz Schlick was among the first to appreciate its full philosophical potential. In his *General Theory of Knowledge* he holds, like Mill, that "Every judgement we make is either definitional or cognitive." But he thinks that "the radical empiricism of John Stuart Mill" has no way of "saving the certainty and rigor of knowledge in the face of the fact that cognition comes about through fleeting, blurred experiences."[15] It is here that implicit definition comes to the rescue. It can show how concepts can be given a meaning in a way that does not found them directly on "immediate experience." Their sharpness can then be guaranteed independently of the degree of sharpness in our experience. "We would no longer have to be dismayed by the fact that our experiences are in eternal flux; rigorously exact thought could still exist."[16]

Hilbert[17] had provided a set of axioms for Euclidean geometry and suggested that the primitive terms used in them—"point," "straight line," "plane," "between," "outside of"—should be thought of as implicitly defined thereby. A definition of this kind is 'implicit' in that no explicit metalinguistic stipulation about meaning is made, whereas in a definition such as "'Square' means 'plane figure bounded by four equal rectilinear sides'" we have an explicit stipulation of synonymy. Treating axiom sentences as implicit definitions can be thought of as stipulating that they are to be understood as denoting, taken together, any entities which systematically satisfy them. Schlick also recognizes what he calls "concrete

[15] Schlick (1985), pp. 69, 30.

[16] See Schlick (1985), p. 31.

[17] Hilbert (1971).

definitions," as against "logical definition proper;" these are required to complete the account of the meanings of terms which appear in the axioms.[18]

In a concrete definition the meaning of a simple concept is ostensively exhibited in experience—for example, the note "A" is concretely defined by sounding a tuning fork. Pure geometry becomes physical geometry when its primitive concepts are "coordinated" by means of concrete definitions with physical facts. Schlick holds that such coordinations essentially involve conventions, in a sense which he ascribes to Poincaré.[19] The coordination is between an implicit definition and a concrete definition—for example, of "point" by ostending a grain of sand, or "straight line" by ostending a taut string. We must also determine units of measurement: crucially, Schlick holds that this determination is also a matter of convention. Thus a unit length could be defined by a particular taut string, or a rigid rod, which is then laid against other straight lines to measure their length. Schlick's example is the unit of time measurement. We can define a day as the period the earth takes to rotate about its axis (the sidereal day). We could equally have chosen "the pulse beats of the Dalai Lama" as units. But

> the rate at which the processes of nature run their course would then depend on the health of the Dalai Lama; for example, if he had a fever and a faster pulse beat, we would have to ascribe a slower pace to natural processes, and the laws of nature would take an extremely complicated form.

We choose the metric which yields the simplest laws. So, for example, the sidereal day may come to be abandoned as a unit of measurement—it may be more "practical"

> to assert that, as a consequence of friction due to the ebb and flow of the tides, the rotation of the earth gradually slows down and hence sidereal days grow longer. Were we not to accept this, we would have to ascribe a gradual acceleration to all other natural processes and the laws of nature would no longer assume the simplest form.[20]

Thus the simplicity of the total system of laws of nature taken as a whole is the only criterion for choosing units of measurement.

Given that the axioms are implicit definitions, Schlick thinks that the theorems are definitions, too:

> In the class of definitions in the wider sense, we include also those propositions that can be derived by pure logic from definitions. Epistemologically, such derived propositions are the same as definitions, since . . . they are interchangeable with them. From this standpoint purely conceptual sciences, such as

[18] Schlick (1985), p. 30.
[19] Schlick (1985), p. 71.
[20] Schlick (1985), p. 72.

> arithmetic, actually consist exclusively of definitions; they tell us nothing that is in principle new, nothing that goes beyond the axioms.[21]

Now an explicit definition is a rule of substitution, not a proposition, so it cannot play the role of premise or conclusion. But an *implicit* definition is a stipulation that a certain set of object-language sentences is to be so understood as to be *true*. (We will come back to this.) What does Schlick say, though, about the status of the "pure logic" by which the theorems of the axiom system are derived? If the truths of this logic are not contentless—because they are a posteriori or synthetic a priori—then neither are those of pure geometry. We must reapply the same treatment to logic itself, treating its axiomatized principles as implicit definitions of the primitive logical constants. It seems that Schlick has something like this in mind when he says that the principles of contradiction and excluded middle "merely determine the nature of negation" and "may be looked upon as its definition."[22]

3. The Empiricist Argument in Mill

Now that we have set out the three conceptions of analyticity that concern us, let us turn to their role in empiricist argument. We start with Mill.

The view that logic and mathematics are a priori because they are entirely empty of content is one of the rival views Mill considers. He calls it "Nominalism." Against it he argues by direct semantic analysis that logic and mathematics contain synthetic inferences. The central failure of the Nominalists, he thinks, is that they fail to make his distinction between connotation and denotation of terms—"seeking for their meaning exclusively in what they denote."[23] Thus, for example, contrary to the Nominalists, arithmetical identities such as "$2+1=3$" are real propositions because the terms flanking the equality have different connotations. But Mill also argues indirectly. If logic did not contain synthetic inferences, in the Mill/Kant sense, valid deductive reasoning could not produce new knowledge. The conclusion of any valid deduction would literally be asserted in the premises. In that case, in knowing that each proposition asserted in the premises was true, one would already know the truth of the conclusion. (Mill allows for definitional substitution, so this argument requires the principle that if sentences A and B are definitionally substitutable, knowing the truth expressed by

[21] Schlick (1985), p. 73.

[22] Schlick (1985), p. 64; compare p. 337.

[23] *System of Logic, Collected Works VII*, p. 91.

A is knowing the truth expressed by B.) He is impatient with metaphorical talk about conclusions being implicitly contained in premises:

> It is impossible to attach any serious scientific value to such a mere salvo, as the distinction drawn between being involved by implication in the premises and being directly asserted in them.[24]

Deduction plainly produces new knowledge, and therefore logic must contain synthetic ("real") inferences. Mill applies the same dual argument, direct and indirect, to the case of mathematics.

Mill also considers two other rival views: "Conceptualism" and "Realism." Realists hold that logical and mathematical knowledge is knowledge of universals existing in a mind-independent abstract domain; the terms that make up sentences are signs that stand for such universals. Conceptualists hold that the objects studied by logic are concepts and judgments conceived as psychological states and acts.

Realism is the view Mill takes least seriously, thinking it extinct (he did not know that something like it was destined for revival). But there was probably a deeper reason: although both Nominalists and Conceptualists hold that logic and mathematics can be known nonempirically, they both accept that no instructive proposition about a mind-independent world can be so known. Realism, in contrast, abandons this constraint, and in doing so, abandons one of the things Mill found most obvious.

Conceptualism had been given a new lease on life by the increasing influence of Kantian ideas. In the *System of Logic* Mill responds to it by emphasizing the distinction between judgment and the content of judgment as strongly as any "Realist" would; in the *Examination* he returns to it and engages with it more extensively.[25] Conceptualists emphasize the role of intuition as against induction in our acceptance of logical and mathematical claims. Mill takes such appeals to be appeals to what we can *imagine* experiencing (as against what we do experience); and he accepts that we are justified in basing many arithmetical or geometrical claims on such appeals. But he thinks that that justification is itself a posteriori. Experiential imagination is indeed a largely reliable guide to real possibilities—however, that is itself no more than an empirical truth. Mill also considers the Kantian point that "Experience tells us, indeed, what is, but not that it must necessarily be so, and not otherwise." So if we have insight into the necessary truth of certain propositions, that insight cannot be based on experience and must be a priori. In response, he rejects any metaphysical distinction between necessary and contingent truth; like Quine he thinks the highest kind of necessity is natural

[24] *System of Logic, Collected Works VII*, p. 185.

[25] The idea that the *Examination* propounds a psychologistic view of logic is a misconception. On the contrary, that is the "Conceptualist" view Mill wishes to attack. See Skorupski (1989), ch. 5, "Appendix."

necessity. The only other sense of "necessary truth" he is prepared to concede is "proposition the negation of which is not only false but inconceivable":

> This, therefore, is the principle asserted: that propositions, the negation of which is inconceivable, or in other words, which we cannot figure to ourselves as being false, must rest on evidence of a higher and more cogent description than any which experience can afford.[26]

In response, in the *System of Logic*, Mill largely dwells on associationist explanations of inconceivability. But in the *Examination* he makes the underlying epistemological basis of his reply clearer:

> ... even assuming that inconceivability is not solely the consequence of limited experience, but that some incapacities of conceiving are inherent in the mind, and inseparable from it; this would not entitle us to infer, that what we are thus incapable of conceiving cannot exist. Such an inference would only be warrantable, if we could know a priori that we must have been created capable of conceiving whatever is capable of existing: that the universe of thought and that of reality ... must have been framed in complete correspondence with one another. ... That this is really the case has been laid down expressly in some systems of philosophy, by implication in more ... but an assumption more destitute of evidence could scarcely be made. ...[27]

The Conceptualist needs to show, *and show a priori*, that what we are "incapable of conceiving cannot exist."

The "mirroring" idea of knowledge that I mentioned at the beginning is clearly crucial in this line of thought: genuine knowing is of something distinct from the knowing itself—to which the knowing must conform if it is indeed knowledge. Mill applies the point single-mindedly to the most fundamental synthetic truths. To prove that "a contradiction is unthinkable" is not to prove it "impossible in point of fact." It is the latter, not the former, claim which is required to vindicate "the thinking process." "Our thoughts are true when they are made to correspond with phaenomena":

> If there were any law necessitating us to think a relation between phaenomena which does not in fact exist between the phaenomena, then certainly the thinking process would be proved invalid, because we should be compelled by it to think true something which would really be false.[28]

This, then, is Mill's general epistemological framework. It has simplicity and power, and it clearly requires him to treat mathematics and logic itself as belonging to

[26] *System of Logic, Collected Works VII*, pp. 237–238.

[27] *An Examination of Sir William Hamilton's Philosophy, Collected Works IX*, p. 68.

[28] *An Examination of Sir William Hamilton's Philosophy, Collected Works IX*, pp. 382, 384, 383.

our *empirical* knowledge. Mill's more detailed accounts of geometry and arithmetic reflect more specific concerns: he wants to show how these sciences fit into an inductivist epistemology and nominalist (in our contemporary sense) ontology. These aims are prima facie distinct from the general empiricist framework we have just described. It is a matter of debate whether they follow from it; if they do not, they may seem less powerful as self-standing philosophical claims than does Mill's general epistemological framework itself. Does naturalistic empiricism force nominalism? Does it force inductivism? Neither point is obvious. It may be that these positions stem from Mill's combination of naturalism and realism, rather than from naturalism alone. But let us sketch in some details of Mill's account.

The *System of Logic* develops a rather elegant historical picture of "the inductive process." Humanity begins with "spontaneous" and "unscientific" inductions about particular, apparently unconnected, natural phenomena. Generalizations accumulate, interweave, and are found to stand the test of time or are corrected by further experience. As they accumulate and interweave, they justify the second-order inductive conclusion that all phenomena are subject to uniformity and, more specifically, that all have discoverable sufficient conditions. This conclusion in turn provides (Mill believes) the grounding assumption for a new style of reasoning about nature—eliminative induction, which he formulates in his "Methods of Empirical Inquiry." The improved scientific induction which results from this style of reasoning spills back onto the principle of uniformity on which it rests, raising its certainty. That in turn raises our confidence in the totality of particular enumerative inductions from which the principle is derived. Thus the amount of confidence with which one can rely on the "inductive process" as a whole depends on the point which has been reached in its natural history.

How does logical and mathematical knowledge fit into this? Mill thinks it comes from those earliest "spontaneous" and "unscientific" enumerative inductions which are retrospectively confirmed by the success of the inductive process. Thus he considers the principle of exclusion, that "the same proposition cannot at the same time be false and true," to be "one of our first and most familiar generalisations from experience."[29] The same goes for the parallels postulate. But this account of how one comes to believe such principles may well strike one as quite implausible. What kinds of instances could these generalizations involve?[30] Even in his own inductivist terms Mill might have found a little more wiggle room. He could have agreed that our acceptance of these principles is based on intuition, that is, "inconceivability of the negation"—but that (as in the case of geometry) the reliability of the conceivability test is itself ultimately a posteriori. This approach

[29] *System of Logic, Collected Works VII*, p. 277.

[30] "What kind of enumerative induction could lead us to the principle of mathematical induction?" (Shapiro, 2000, p. 100).

might have led him to a more pragmatist, or conservative holist, view. His argument now would be that through the progress of science, theoretical grounds remote from experience might lead us to reject a proposition supported by the conceivability test. However, to take this line he would have to find some way of defending a theory of the default authority of intuition.

Mill tries to give a nominalist analysis of both geometry and arithmetic. Geometrical objects—points, lines, planes—are ideal or "fictional" limits of ideally constructible material entities; Euclidean propositions are thus true of physical space only in the limit. The real empirical assertion underlying an axiom such as "Two straight lines cannot enclose a space" is something like "The more closely two lines approach absolute breadthlessness and straightness, the smaller the space they enclose." Likewise, the real content of the geometrical assertion that rectangles exist emerges, on a Millian view, as "The construction of a plane surface approximating to any required degree of accuracy to a Euclidean rectangle is compatible with the geometry of space." (Mill did not doubt that these Euclidean propositions of physical geometry were true.) Note the essential reference made here to laws of *space*: such constructions may be empirically impossible for us as natural beings, or incompatible with the laws of matter.

The nominalist analysis of arithmetic faces Mill with far greater difficulties, which he probably did not fully appreciate. Number terms denote "aggregates" (or "collections" or "agglomerations") and connote certain attributes of aggregates. He does not say that they denote those attributes of the aggregates, so he seems to think of them as general rather than singular. For example, *being a five* is an attribute of all aggregates composed of five elements. We are to think of these aggregates as identified by the elements that make them up, so that the aggregate denoted by "the cards in that pack," which is a 52, is not the same as the aggregate denoted by "the suits in that pack," which is a four. Yet aggregates are not sets; Mill's nominalism requires that they be concrete entities individuated by a principle of aggregation.

This account escapes some of the rather unfair criticisms Frege later made of it, but nonetheless it certainly isn't clear that it can provide an adequate ontological base for number theory. To pursue that question we would have to (1) develop a theory of what aggregates are and (2) admit clearly that number theory must allow for possible as well actual aggregates—the possibility in question being a highly idealized one. Consider terms like "the cars in the parking lot" or "the books on the shelf." I might say "The books took up all the space on the shelf, but I couldn't see how many there were." Thus here "the books" seems to refer to a perceptible entity, and its attribute of number is also perceptible in principle, though not in this case. But plainly this does not apply to physical aggregates whose elements are widely separated in space and time, such as all the utterances of a particular sound, or to aggregates which are not physical, such as the number of figures of the syllogism. Reflection on this leads to the conclusion that aggregates should be thought of as products of acts of aggregation or collection. It is the collecting that has to be concrete, not the things collected. One can think of collection as the making of a mental or physical list. The basic process with which number theory is concerned is the making and dividing of such lists, not the existence or otherwise of their intentional content. And the laws of number are the laws of listmaking and dividing. Turning to (2), however, we must obviously allow for more numbers than are given to us by *actual* lists, or by lists that *we*, even in collective relays, could produce. The idealized listmaker must be able to make uncountable lists.

Making an uncountable list sounds uncomfortably close to counting the uncountable. Still, this is an interesting line of thought which is very much in the Millian spirit, though it goes far beyond the basic ideas Mill supplies.[31] It could draw on resources provided by formalism and constructivism. But, as already noted, it is actually not evident how much should turn, for an epistemological empiricist, on the apparent ontological distinction between possible concrete aggregates and actual abstract sets; thus the prospects for a Millian ontology of number theory should not be linked too closely to the prospects for radical Millian empiricism. And finally, now that we have sketched Mill's account of geometry and arithmetic, it is worth noting that he gives no attention to the ontology of logic. If the empirical laws of geometry concern "geometrically possible" constructions, and those of arithmetic concern "arithmetically possible" collections, what do the empirical laws of logic concern? Do they concern the "logically possible" truths? What kind of concrete objects are truths? Or do they concern negation and implication? What kind of concrete objects could these be? Can there be empirical laws that do not deal with a distinctive subject matter?

4. Logical Positivism

We saw the view of mathematics and logic as a system of implicit definitions being developed by Schlick. In the 1920s this implicit-definition conception of analyticity begins to appear in the thinking of Wittgenstein and Carnap.

In its intention, Wittgenstein's idea of apriority as tautology in the *Tractatus* had been in line with the Kant/Mill notion of analyticity. That is, it was meant to show rigorously that tautologous inference is purely formal, empty of content, wholly noninstructive. True, it did so only if bivalence and exclusion are somehow purely formal or empty. A Millian could justly point out that the *Tractatus*'s doctrine of "bipolarity" as a condition of the sense of propositions takes that

[31] It is resourcefully developed by Philip Kitcher (1997). See also Kitcher (1983).

dogmatically for granted. But now, in his conversations with the Vienna Circle, Wittgenstein rejects the idea that tautology is the essence of the a priori:

> That inference is a priori means only that syntax decides whether an inference is correct or not. Tautologies are only one way of showing what is syntactical.[32]

He still holds that such inferences are noninstructive, but the way in which this is now to be explained is much simpler.[33] Analytic sentences have no declarative content because they simply articulate rules of the language. The a priori is still the analytic, but we have a new view of what is analytic.

Yet this new view is philosophically challenging, in a way that the Kant/Mill criterion of analyticity, at least when understood strictly, is not. How can syntax "decide" whether an inference is correct or not? Doesn't that depend on whether the inference is truth-preserving? The answer must be that syntax decides what is truth-preserving. But the natural objection is that we can't just stipulate that a sentence is true, or an inference rule is truth-preserving: truth is not up to us, it is not for us to stipulate it.

Let us go back to the case of geometry. What, exactly, is "up to us" here?

(1) We can freely lay down uninterpreted axioms and rules of inference, and analyze their formal consequences. The question of their truth or validity does not then arise.
(2) We can stipulate that the terms occurring in the axiom-sentences are to be understood to have (unspecified) referents of which the sentences are true.

Certainly (2) is up to us. But does this activity of ours suffice to confer meaning on the sentences? Only, according to the natural objection, if it determines a unique set of referents of which the sentences are true. For if more than one interpretation makes the axioms true, then we haven't specified what referents are intended, and if no interpretation makes them true, then we haven't found an interpretation at all. It follows that stipulation, as in (2), only gives us metalinguistic knowledge—knowledge that the axioms should be thought of as expressing something or other that's true. To know *what* the axioms say, and that *it* is true, we would need to know what objects they are about and that those objects satisfy them. That knowledge may be a priori or a posteriori. Either way, it is not the stipulation that delivers it. Thus the stipulation at most gives us a constraint on how certain words are going to be used.

[32] Wittgenstein (1979), p. 92. Note that Wittgenstein thinks they are a way. He is evidently using "syntax" broadly.

[33] Introducing ruthless simplifications that cut through supposed clutter was one of logical positivism's most modernist features.

It follows that the notion of an implicit definition cannot explain the nature of a priori knowledge. The point applies wherever we are considering an implicit-definitions approach—be it the theory of logic or the theory of phlogiston. One can stipulate that certain sentences containing the words "if" and "not," or the word "phlogiston," are to be so understood as to be true. But only if the world co-operates does that confer meaning on them. Take the sentence "Phlogiston is driven off whenever a metal or combustible material is heated." We can stipulate that either the word "phlogiston" in this sentence has a referent that makes the sentence true or it has no referent. It is then a priori that either the sentence is true or it lacks truth-value. Thus we do not know a priori that there is a true proposition (content) to the effect that phlogiston is driven off whenever a metal or combustible material is heated. Indeed what is actually the case is that (depending on one's semantic views) there either is no such proposition or, if there is, it is not true. Similarly with the stipulation that the sentence "Not-not *p* if and only if *p*" should be so understood as to be true. What is a priori, on this stipulation, is that either that sentence is true or it lacks truth-value. It does not follow from this alone that there is an priori true proposition that not-not *p* if and only if *p*. If we have a priori knowledge of that proposition, it is not the implicit-definition account of analyticity that gives it to us.

According to this objection, the only way to show an inference to be purely verbal or empty would be the Mill/Kant way. The objection seems decisive, but the logical positivists did not accept it. Their view rested on a rethinking of language, truth, and logic which was truly revolutionary and would, if sound, refute it.

It is natural to think that particular languages are merely contingent vehicles for the expression of contents of thought—concepts, propositions, epistemic norms, and so on. Thought-contents themselves are a language-independent domain about which there are language-independent truths. The Vienna School denied this. There is no such domain of thought-contents. There are no language-independent facts or norms constituting it. There are only rules of particular languages and "empirical" or "natural" facts expressible by true sentences in a particular language. That a particular set of linguistic rules exists is itself an empirical fact; on the other hand, without some set of rules there can be no expression of empirical facts. All discourse and thought is language-relative.

One can sum up this "linguistic turn" as follows:

1. Significant discourse or thought presupposes a determinate language: a framework of rules which renders it possible and at the same time fixes its "limits" (what sentences have sense).
2. These rules confer ("empirical" or "factual") content on a given sentence by determining when it is assertible within the framework.

3. The rules themselves are arbitrary or free conventions.
4. Sentences in the language which simply express or record these rules are empty of content.

The notion of a linguistic rule elucidates—or, better, just replaces—the classical philosophical notion of the a priori. This radically new Critical standpoint was expounded by Wittgenstein in his conversations with the Vienna Circle and developed in independent, divergent ways by him and by Carnap, among others, from the late 1920s on. It implied the end of philosophy, which should be replaced by the systematic activity of describing or laying down rules of "logical syntax" (Carnap) or, alternatively, kept at bay by administering unsystematic reminders of "logical grammar" wherever it threatened to break out (Wittgenstein).[34]

The meaning of a sentence, on this new view, is determined by the conditions in which it is assertible, and this determination occurs only in the framework of language-rules. "Verificationism" is the familiar name for this, but it should be noted that a *definition* of truth in terms of verifiability is not essential to it. The essential point is that the meaning, and thus the truth-value, of a sentence is internal to a framework.[35] Whether or not an assertion is licensed is determined by data of experience plus the conventions constituting the framework; these conventions are its "syntax." Epistemology reduces to syntax: the set of conventions licensing moves within the language. There is nothing beyond these conventions. There are no extralinguistic, or nonconventional, norms. "There are only empirical facts and decisions." And just as *truth* is internal to the conventions, so is *existence*.

We can now see the Viennese reply to the natural objection mentioned earlier in this section. It is a generalization of Schlick's treatment of implicit and concrete definitions. Once geometrical axioms have been set down and coordinated with experience by ostensive and metric conventions, there is no further question about the unique existence of appropriate geometrical objects. A priori knowledge is knowledge of the rules—and that can include existential knowledge, since existence is internal to the rules. Likewise, if we simply stipulate that phlogiston is driven off whenever a metal or combustible material is heated, there is no further question about whether it exists. And in the case of logic itself the same point

[34] Compare Coffa (1991), p. 263: "One could, in fact, mimic Kant's famous 'Copernican' pronouncement to state their point. If our a priori knowledge must conform to the constitution of meanings, I do not see how we could know anything of them a priori; but if meanings must conform to the a priori, I have no difficulty in conceiving such a possibility." Of course this Critical standpoint, unlike transcendental idealism, admitted no unknowable facts or necessary frameworks.

[35] A survey of these issues, including the connection between an epistemic conception of meaning and an epistemic conception of truth, can be found in Skorupski (1997a). Carnap (1963b) gives an account of his evolving views on truth.

applies. The only questions concern the pragmatic convenience of this or that language.

> Can an ostensive definition come into collision with the other rules for the use of a word? It might appear so; but rules can't collide, unless they contradict each other. That aside, it is rules that determine a meaning; there isn't a meaning that they are answerable to and could contradict.
>
> Grammar is not accountable to any reality. It is grammatical rules that determine meaning (constitute it), and so they themselves are not answerable to any meaning and to that extent are arbitrary.[36]
>
> In logic, there are no morals. Everyone is at liberty to build up his own logic (i.e., his own form of language) as he wishes. All that is required of him is that if he wishes to discuss it, he must state his methods clearly, and give syntactical rules instead of philosophical arguments.[37]

The application to mathematics now looks simple: it is simply that mathematics as a whole belongs to the framework.[38] At a stroke we have preserved empiricism while catering to Schlick's search for certainty and rigor.[39] But now we must turn to the formidable difficulties, which can be considered under three headings.

1. Can mathematics be shown to be part of the linguistic framework?
2. How do we tell whether a sentence expresses a linguistic rule or conveys an empirical assertion (i.e., whether it is analytic or synthetic)?
3. How tenable is the "linguistic turn?" Is it true that "There are only empirical facts and (linguistic) decisions?"

5. Criticisms of Logical Positivism and of Millian Empiricism

How can we show that mathematics belongs to the linguistic framework? One might think, prima facie, that the framework should be specified by explicit individual stipulation of each sentence or inference rule contained in it. But if all logic and mathematics is to be included in the framework, this cannot be done. It

[36] Wittgenstein (1974), p. 184. See the application to the meaning of "not" which follows, and the whole discussion of convention and arbitrariness in section 5.

[37] Carnap (1937), p. 52.

[38] Wittgenstein alludes to this point, "the paradox that mathematics consists of rules," in a marginal MS comment quoted in the note on p. 184 of Wittgenstein (1974).

[39] Note that in this framework, as in Mill's, logicism becomes a mathematical rather than an epistemological issue.

is not even possible to specify a procedure that effectively decides whether any given sentence belongs to the framework.

What, then, about giving an effective *axiomatization* of the framework—that is, stipulating that any theorem of the axiom system is to be counted a convention? With a fully determinate metalogic we would have a fully determinate specification of the analytic sentences in the object-language—so this would answer question (2) as well as question (1). But this approach gives rise to deep questions.

What is the status of the metalogic? Consider a sentence to the effect that an object-language sentence is derivable in the system. Is it itself part of the framework? Presumably it has to be, on pain of allowing that there really are nonconventional, and thus nonanalytic, logical truths. This sort of problem eventually led Wittgenstein to his famous rule-following considerations.[40] The point at stake is not merely that the metalogic itself must be regarded as a set of conventions. Consider the metalanguage sentence "Given the metalogic we have stipulated, sentence S is derivable in the object-language." Does *this* sentence express a stipulation? It seems that the only possible answer for the logical positivist is that it does—this is the view that Michael Dummett has called "radical conventionalism." The only alternative available within logical positivism would be the thesis that it is an empirical sentence[41]—but that idea would undermine the framework/empirical content distinction. What belongs to the framework would become an empirical matter, but empirical matters are supposed to be determinate only in the context of a determinate framework. Yet the radical-conventionalist answer also puts in question the determinacy of the framework.[42] There are, on this view, no rails to guide us; at every step, in applying the framework we make new decisions. The framework cannot be specified "in advance"; indeed, no unique framework is "given" at any time. Also noteworthy in this context is Gödel's incompleteness theorem of 1931 (of which Carnap and Wittgenstein were aware): true nonprovable propositions can be formulated within the language of any consistent, effectively specified axiomatization of arithmetic. It reinforces—in a highly unexpected way—the general point that language frameworks as envisaged by logical positivism cannot be formally specified.

Quine famously raised question (2). He asks whether threads of fact and convention can be neatly separated out of our fabric of sentences. In part, his critique consists in a far-reaching assault on the very notion of synonymy, which if effective would undermine the analytic/synthetic distinction as such and not

[40] Or so I argue in Skorupski (1997b).

[41] For example, an empirical sentence to the effect that members of the speech community would accept the derivation.

[42] It's a separate issue whether Wittgenstein's discussion of rules in his later work should be regarded as adhering to radical conventionalism or, on the contrary, as looking for a way to avoid it. See Dummett (1959, 1993) and Stroud (1965). See also Skorupski (1997b).

just the logical positivists' understanding of it. The Mill/Kant conception, for example, assumes that we can identify a conclusion sentence as saying the same thing as a premise sentence. But if Quine's arguments against synonymy are sound, they undermine that assumption (along with much else). For even when the conclusion sentence is type-identical to the premise sentence, there is no "fact of the matter" as to whether it has the same meaning.

Suppose, however, that these very radical arguments against synonymy can be set aside. Then it can look as though what Quine is saying boils down to a point about the difficulty of establishing exact meanings in naturally evolving languages. Carnap took it that way, and insisted in response that "analytic" was definable only relative to a clearly specified framework. Natural language could in principle be sharpened or just replaced by completely free language construction. The question of what language to construct is then practical, in the sense that it should simply reflect our overall aims. These aims may tend to cost-effectiveness in intellectual effort, or to nostalgic sentiment in preserving old ways of talking; but whatever they may be, it is certainly not the task of a specialist adviser from Carnap Language-Construction Industries to moralize about them.

How should Quine reply? After all, he still adheres to the language-relative view propounded by the logical positivists: experience and convention alone determine "meaning," insofar as we can talk about "meaning." He does not reject verificationism; it's just that his verificationism takes the "unit of meaning" to be the theory rather than the sentence. Thus truth and existence remain internal to the fabric of sentences, which means that the natural objection to implicit-definition conventionalism, considered above in section 2, does not arise. Thus, if Quine's all-out assault on meaning is set aside, the point that remains must be that Carnapian linguistic constructions simply won't stay rigid in use. Any fabric we use is always stretched and restretched on an inextricable mix of experience, theory, and convention—and mathematics is just a part of that fabric.

But if verificationism, and with it the language-relative view, is abandoned, then the implicit-definitions conception of analyticity will succumb not just to these Quinean points but also to the more fundamental objection just mentioned.[43] For the positivist defense against that objection crucially depended on the idea that truth and existence are themselves internal to a language framework.

In considering question (1), we noted that defending the framework idea seems to call for a radical conventionalism which accepts that the framework is always "open," or "in process of construction." But now we should also consider question (3): Is it tenable to hold that there are only empirical facts and linguistic decisions? More generally, we can ask what status any empiricist can give to epistemological principles. The options seem to reduce to treating them in the

[43] Here I disagree with Boghossian (1997). There has recently been renewed interest in the implicit-definition notion of analyticity. See, e.g., Hale and Wright (2000).

logical positivist manner as implicitly definitional of their subject matter, or treating them as statements of fact. On the latter view, for example, the principle of inference to the simplest explanation would be a factual sentence about the world: "The world is simple." Enumerative induction would be the factual sentence that the world is induction-friendly. But Mill and Quine seem to approach the matter differently. Their naturalizing approach to epistemology seems to consist in appealing to how we actually reason. As Mill put it, "The laws of our rational faculty, like those of every other natural agency, are only learnt by seeing the agent at work."[44] Is this a kind of psychologism, open to the familiar objections? Can it explain the normativity of epistemology?

If we observe rational agents at work, we find that as well as issuing rules, they make factual and *normative* claims—and in both cases discuss their truth-value. It does not seem, to prima facie observation, that the thinking agent does nothing other than make observation statements, apply rules, and express feelings. So why should one impose the rule/fact dichotomy on this observed trichotomy—factual, stipulative, and normative? Why do all declarative sentences have to be licensed by experiential data? What prevents the empiricist from accepting that as well as stipulations and statements of fact there are declarative, truth-apt, purely normative sentences? We characterized empiricism as the view that no factual sentence is a priori. But must all truth-apt sentences be factual?

It seems that what is in play here is not observation of thinkers at work, but a metaphysical commitment to the factuality of content, which then combines with the empiricist's naturalistic view of what facts there are. Where does this commitment come from? One might see it in Mill's case as coming from the correspondence conception of truth and the mirroring conception of knowledge that goes with it. But the correspondence conception is not supposed to be present in logical positivism. So where does the idea that content must be factual come from there? One interpretation is that it comes from an unargued assumption that all knowledge must be scientific knowledge. But then that just looks like scientistic prejudice.

Once we have observed that normative judgment is inherent in what the thinking agent does, we see it everywhere. In ethics and aesthetics, of course, but also in epistemology, both at the level of general principles and at the level of specific judgments of what, in the light of the data, there is sufficient reason to believe. And as Wittgenstein eventually argued, any judgment about the right way to apply a rule is neither factual nor stipulative, but normative.

So why not explore the possibility that logic and mathematics are themselves normative? In a way this would continue the Critical idea. The idea would now be that as well as factual propositions about objects and their properties, there are normative propositions which are not, at least in the same sense, about the

[44] Mill, *System of Logic, Collected Works VIII*, p. 833.

properties of objects as against norms for reasoning about objects. A normative, as against a universalist, view of logic would say that logic has no subject matter of its own. It is not a set of factual propositions about a specific domain of objects. It codifies norms governing our thinking about *all* objects. Factual propositions are seen, on this approach, as internal not to rules of language but to norms of reason. Logicism would be a way of showing that mathematics is normative, too, though it might not be the only way. This normative view would still be in an important way empiricist, in that it retained the thesis that no factual statement is a priori; but it would drop the dogmas of realism and verificationism. It might be the best way for empiricist descendants of Mill or logical positivism to go.

BIBLIOGRAPHY

Boghossian, Paul. 1997. "Analyticity." In Hale and Wright (1997), pp. 331–368.
Carnap, Rudolf. 1937 (1934). *The Logical Syntax of Language* (*Logische Syntax der Sprache*). London: Kegan Paul, Trench & Trubner.
———. 1956. "Empiricism, Semantics and Ontology." In Carnap, *Meaning and Necessity*. Chicago: University of Chicago Press.
———. 1963a. "W.V. Quine on Logical Truth." In Schilpp (1963), pp. 915–922.
———. 1963b. "Intellectual Autobiography." In Schilpp (1963), pp. 3–84.
———. 1996 (1935). *Philosophy and Logical Syntax*. Bristol, U.K.: Thoemmes.
Coffa, Alberto. 1991. *The Semantic Tradition from Kant to Carnap: To the Vienna Station*. Cambridge: Cambridge University Press.
Dummett, Michael. 1959. "Wittgenstein's Philosophy of Mathematics," *Philosophical Review*, 63: 324–348. Reprinted in Michael Dummett, *Truth and Other Enigmas*. London: Duckworth, 1978.
Dummett, Michael. 1993. "Wittgenstein on Necessity: Some Reflections." In Dummett, *The Seas of Language*. Oxford: Clarendon Press. Also in Bob Hale and Peter Clark, eds., *Reading Putnam*. Oxford: Oxford University Press, 1994.
Frege, G. 1950 (1884). *The Foundations of Arithmetic* (*Grundlagen der Arithmetik*) Oxford: Basil Blackwell.
Friedman, Michael. 1991. "The Re-evaluation of Logical Positivism," *Journal of Philosophy*, 10: 505–519.
Garver, Newton. 1996. "Philosophy as Grammar." In Sluga and Stern (1996), pp. 139–170.
Hale, Bob, and Wright, Crispin (eds). 1997. *A Companion to the Philosophy of Language*. Oxford: Blackwell.
Hale, Bob, and Wright, Crispin. 2000. "Implicit Definition and the A Priori." In Paul Boghossian and Christopher Peacocke (eds), *New Essays on the A Priori*. New York : Oxford University Press.
Hilbert, David. 1971 (1899). *Foundations of Geometry* (*Grundlagen der Geometrie*). La Salle, Ill.: Open Court.

Kant, Immanuel. 1998 (1781, 1787). *Critique of Pure Reason* (*Critik der reinen Vernunft*). Translated and edited by Paul Guyer and Allen W. Wood. Cambridge: Cambridge University Press.

Kitcher, Philip. 1983. *The Nature of Mathematical Knowledge.* New York: Oxford University Press.

———. 1997. "Mill, Mathematics, and the Naturalist Tradition." In Skorupski (1998), pp. 57–111.

Mill, John Stuart. 1963–1991. *The Collected Works of John Stuart Mill.* Gen. ed. John M. Robson. 33 vols. Toronto: University of Toronto Press.

Quine, W.V.O. 1953 (1951). "Two Dogmas of Empiricism." In Quine, *From a Logical Point of View.* Cambridge, Mass.: Harvard University Press.

———. 1966a (1935). "Truth by Convention." In Quine, *The Ways of Paradox.* New York: Random House.

———. 1966b (1962). "Carnap and Logical Truth: The Ways of Paradox." In Quine, *The Ways of Paradox.* New York: Random House.

Ricketts, Thomas. 1996. "Pictures, Logic, and the Limits of Sense in Wittgenstein's *Tractatus.*" In Sluga and Stern (1996), pp. 55–99.

Schilpp, Paul Arthur (ed). 1963. *The Philosophy of Rudolf Carnap.* La Salle, Ill.: Open Court.

Schlick, Moritz. 1985. (1918, 1925). *General Theory of Knowledge* (*Allgemeine Erkenntnislehre*). La Salle, Ill.: Open Court.

Shapiro, Stewart. 2000. *Thinking About Mathematics.* Oxford: Oxford University Press.

Skorupski, John. 1989. *John Stuart Mill.* London: Routledge.

———. 1997a. "Meaning, Verification, Use." In Hale and Wright (1997), pp. 29–59.

———. 1997b. "Logical Grammar, Transcendentalism and Normativity," *Philosophical Topics*, 25: 189–211.

———. (ed). 1998. *The Cambridge Companion to John Stuart Mill.* Cambridge: Cambridge University Press.

Sluga, Hans, and Stern, David G. (eds). 1996. *The Cambridge Companion to Wittgenstein.* Cambridge: Cambridge University Press.

Stroud, Barry. 1965. "Wittgenstein and Logical Necessity," *Philosophical Review*, 74: 504–518.

Stroud, Barry. 1984. *The Significance of Philosophical Scepticism.* Oxford: Clarendon Press.

Wittgenstein, L. 1961 (1921). *Tractatus Logico-Philosophicus* (*Logisch-Philosophische Abhandlung*). London: Routledge and Kegan Paul.

Wittgenstein, L. 1974. *Philosophical Grammar.* Oxford: Basil Blackwell.

Wittgenstein, L. 1975. *Philosophical Remarks.* Oxford: Blackwell

Wittgenstein, L. 1979. *Wittgenstein and the Vienna Circle: Conversations Recorded by Friedrich Waismann.* Oxford: Basil Blackwell.

CHAPTER 4

WITTGENSTEIN ON PHILOSOPHY OF LOGIC AND MATHEMATICS

JULIET FLOYD

LUDWIG Wittgenstein (1889–1951) wrote as much on philosophy of mathematics and logic as he did on any other topic, leaving at his death thousands of pages of manuscripts, typescripts, notebooks, and correspondence containing remarks on (among others) Brouwer, Cantor, Dedekind, Frege, Hilbert, Poincaré, Skolem, Ramsey, Russell, Gödel, and Turing. He published in his lifetime only a short review (1913) and the *Tractatus Logico-Philosophicus* (1921), a work whose impact on subsequent analytic philosophy's preoccupation with characterizing the nature of logic was formative.[1] Wittgenstein's reactions to the empiricistic reception of his early work in the Vienna Circle and in work of Russell and Ramsey led to further efforts to clarify and adapt his perspective, stimulated in significant part by developments in the foundations of mathematics of the 1920s and 1930s; these

I am indebted in this chapter to support from the Fulbright Program, Wellesley College and a sabbatical fellowship from the American Philosophical Society, as well as to stimulating conversations with members of my spring 2002 seminar on Wittgenstein at Boston University, with Burton Dreben, Warren Goldfarb, Jin Ho Kang, Montgomery Link, Thomas Ricketts, and especially Akihiro Kanamori.

[1] There were, strictly speaking, also a lecture published, though never delivered and immediately disowned by Wittgenstein (SRLF), a dictionary designed to teach spelling (1926), and a letter to the editor of *Mind* (1933).

never issued in a second work, though he drafted and redrafted writings more or less continuously for the rest of his life.

After his death large extracts of writings from his early (1908–1925), middle (1929–1934), and later (1934–1951) philosophy were published. These include his 1913 correspondence with Russell (see CL), 1914 dictations to Moore (MN), his *Notebooks 1914–1916* (NB), transcriptions of some of his conversations with the Vienna Circle (WVC), student notes of his Cambridge lectures and dictations (1930–1947), and many other typescripts and manuscripts, the most widely studied of which are *Philosophical Investigations* (PI) and *Remarks on the Foundations of Mathematics* (RFM). Not until 2000 did a scholarly edition of his middle period manuscripts (WA) and a CD-ROM of his whole *Nachlass* appear (1993– , 2000), thereby making a more complete record of his writings accessible to scholars. Since Wittgenstein often drafted multiple versions of his remarks, reinserting them into new contexts in later manuscripts, and since his later, "interlocutory" multilogue style of writing is so sensitive to context, this new presentation of the corpus has the potential to unsettle current understandings of his work, especially since a fair number of remarks on mathematics and logic have yet to be carefully scrutinized by interpreters.[2] Nevertheless, because of the wide circulation of the previously edited volumes there is a substantial and fairly settled body of later writings that has exercised a significant and continuous influence on philosophy since the 1950s.

Wittgenstein's discussions of mathematics and logic take place against the background of a complex, wide-ranging investigation of our notions of *language, logic,* and *concept possession,* and the fruitfulness of his work for the philosophy of mathematics has arisen primarily through his ability to excavate, reformulate, and critically appraise the most natural idealizing assumptions about the expression of knowledge, meaning, and thought in language on which philosophical analyses of the nature of mathematics, logic, and truth have traditionally depended. His probing questions brought to light tendencies implicit in many traditional epistemological and ontological accounts to gloss over, miscast, and/or underrate the complexity of the effects of linguistic expression on our understanding of what conceptual structure and meaning consist in. Though he held, like Frege and Russell, that a proper understanding of the logic of language would subvert the role that Kant had tried to reserve for a priori intuition in accounting for the objectivity, significance, and applicability of mathematics, his focus on the linguistic expression of thought in language ultimately led him to question the general philosophical significance of their logicist analyses of arithmetic. His resistance to logicism ultimately turned on a diverse, multifaceted exploration of the questions "What is logic?" "What is it to speak a language?" and, more generally, "What do we mean by rule- (or concept-) governed behavior?"

[2] Some recent discussions of these issues may be found in Marion (1998), Floyd (2001b), Rodych (2002), and Mancosu and Marion (2002).

Wittgenstein's overarching philosophical spirit was anti-rationalist, in sharpest contrast, among twentieth-century philosophers, to that of Gödel. For Wittgenstein, as earlier for Kant, philosophy and logic are quests for self-understanding and self-knowledge, activities of self-criticism, self-definition, and reconciliation with the imperfections of life, rather than special branches of knowledge aiming directly at the discovery of impersonal truth. These activities should thus aim, ideally, at offering improved modes of criticism, clarity, and authenticity of expression, rather than at a certain or explicit foundation in terms of general principles or an enlarged store of knowledge. Like Plato in the *Meno*, or Kant in the *Critique of Pure Reason*, when Wittgenstein discusses a particular logical result or a mathematical example, he is most often model- or picture-building: pursuing, through a kind of allegorical analogy, not only a better understanding of the epistemology of logic and mathematics, but also a more sophisticated understanding of the nature of philosophy conceived as an activity of self-expression and disentanglement from metaphysical confusion, for purposes of an improved mode of life. He is investigating how and why mathematical analyses affect and expand our self-understanding as human beings, rather than assessing their cogency as pieces of ongoing mathematical research. Rather than adopting empiricism or a clearly formulated doctrine about mathematics and logic to combat rationalism, Wittgenstein fashioned a novel appropriation of the dialectical, Augustinian side of Kant's philosophy, according to which philosophy is an autonomous discipline or activity not reducible to, though at the same time centrally concerned with, natural science. This sets his philosophy apart both from Frege's and Russell's philosophies and from logical empiricism, though his thought bears important historical relations to these and cannot be understood apart from assessing his efforts to come to terms with these rival approaches.

Because of philosophers' tendency to focus on general questions of concept possession and meaning in connection with Wittgenstein's work, his philosophy of mathematics is the least understood portion of his corpus. At present there is no settled consensus on the place of his writings on mathematics and logic within this overarching philosophy, nor is there agreement about the grounds for assessing its philosophical worth or potential for lasting significance. His post-1929 remarks have in particular met with an ambivalent reaction in contemporary philosophy of mathematics and logic, drawing both the interest and the ire of working logicians.[3] This is not surprising, in light of the controversial nature of Wittgenstein's overarching philosophical attitude and his having presented few clear indications of how to present a precise logico-mathematical articulation of his ideas. The main divide in interpretations of Wittgenstein's philosophy of mathematics is between those who take his remarks on logic and mathematics to offer a restrictive epistemic

[3] See, in particular, the critical discussions of Bernays, Goodstein, Gödel, Kreisel, and Dummett.

resource argument (constructivist, social-constructivist, anti-realist, formalist, verificationist, conventionalist, finitist, behaviorist, empiricist, or naturalist) based on the finitude of human powers of expression, and those that stress his emphasis on expressive complexity to forward criticisms of all such global epistemic positions. Most agree that the heart of his philosophy of mathematics contains a recasting of traditional conceptions of a priori knowledge, certainty, and necessary truth-an attempt, not unlike Kant's in his day, to find a way to resist naïve Platonism or rationalism without falling into pure empiricism, skepticism, causal naturalism, or fictionalism. Yet, unlike Kant, Wittgenstein rejected the idea that the objectivity of mathematical and/or logical concepts may be satisfactorily understood in terms of an overarching purpose, norm, or kind of truth to which they must do justice. Wittgenstein came to suggest that a detailed understanding of how human beings actually express themselves in the ongoing, evolving stream of life shows not the falsity, but the lack of fundamental interest, of this idea. The main interpretive debate about his philosophy concerns the basis and content of this suggestion.

One thing is clear. In his later writings Wittgenstein explicitly insisted that his own philosophical remarks were a propadeutic, an "album" of pictures, crisscrossing the same terrain in different directions from differing points of view for differing, localized purposes. On this view the philosopher's task is to investigate, construct, explore, and highlight the philosophical effects and presuppositions of many different possible models or "pictures," each having a certain naturalness given actual linguistic and mathematical practice; the point is to explore problems rather than defend or enunciate general truths. In his later writings he suggested that we conceive of human language as a variegated, polymorphous, overlapping collection of "language-games." Such a "game" he took to be a simplified structure, model, or picture (*Bild*) of a portion of our language exhibiting various kinds of distinctive patterns and regularities of human action and interaction.

Wittgenstein's resistance to defending a single, unified conception, and his many appeals to the notion of *Bild*, reflect his philosophy's lineage from the dialectical side of Kant—a side ignored and/or minimized in the mature philosophies of Frege and Russell. Kant had insisted that human claims to possess unconditional truth about reality, especially when grounded in the apparently absolute and universal truths of formal logic, would inevitably end in conceptual contradiction and paradox. The notion of *Bild* has a long philosophical history, but in the German tradition it is connected with the general nature of symbolism, with the mathematical notions of *image* and *mapping*, with the intuitive immediacy and transparency to thought of a geometrical diagram,[4] and with the notion of education or self-development (*Bildung*). Wittgenstein's distinctive uses of this

[4] In *The World as Will and Representation*, a book that influenced Wittgenstein in his teenage years, Schopenhauer emphasized the superiority of proof by diagrams in geometry, contrasting them to the axiomatic, deductive style, which he associated with Euclid.

notion were stimulated, in part, by his observation of the use of scale models in engineering, but he was also especially taken with an idea he found in Hertz's *Principles of Mechanics.* Hertz had suggested that metaphysical confusions and contradictory conceptions of fundamental notions (e.g., of force) cannot be resolved by the discovery of new knowledge, since they are caused in the first place by the accumulation of a rich store of relations and associations around these notions in the context of present-day knowledge. On this view, the best hope for ridding ourselves of vexing conceptual perplexity is thus to sort through and conceptually model, organize, or "picture" the store of relations and associations so as to resolve conceptual confusions and antinomies, thereby relieving ourselves of the need to summarize or join them in a general explanation of their fundamental nature. So conceived, activities of restructuring, picturing, and formulating the architecture of concepts (*Begriffsbildung*) provide new ways of conceiving fundamental problems and notions, and are to be contrasted with the activity of discovery of truths by reasoned deduction from first principles. As Hertz wrote in his introduction to his *Principles,* "When these painful contradictions are removed, the question as to the ultimate nature of [e.g.] force will not have been answered; but our minds, no longer vexed, will cease to ask illegitimate questions." The allusion to Kant's critique of human claims to know the *Ding an sich,* the ultimate essence of things apart from human conditions of representation, is clear in this quotation.

Yet Wittgenstein pressed to a logical conclusion Kant's talk about "limits to knowledge," his distinction between "things in themselves" and things conceived as appearances to us, subject to human forms of sensibility. Wittgenstein saw the Kantian attempt to fashion theoretically justified limits to thought and/or knowledge by articulating general principles of knowledge and/or experience as nonsensical, for there is no making sense of a "transcendental" standpoint on our knowledge that lies at or beyond its limits. For Wittgenstein the logical paradoxes indicated that no such maximally general standpoint on knowledge can be made sense of. The best that can be done is to engage in an immanent exploration of the very notion of a *limit* to thought or to language.

Thus Wittgenstein radicalized Kant's idea that the laws of logic and/or mathematics require philosophical analysis because of our tendency to misconstrue them as unconditionally true of an ahumanly conceived domain of necessary truth. For Wittgenstein, mathematics and logic are to be seen not only as sciences of truth or bodies of knowledge derived from basic principles, but also as evolving activities and techniques of thinking and expressing ourselves; they are complex, applied, human artifacts of language. The logicism of Frege and Russell was right to see the application and content of logic and arithmetic as internal to their nature; mathematics and logic cannot be accounted for wholly in terms of merely formal rules. Yet what it is to be part of arithmetic, grammar, or logic cannot, Wittgenstein believed, be understood in terms of an appeal to a set of primitive truths or axioms, or knowledge of logical objects. The best way to understand the apparently unique epistemic roles of logic and mathematics is to forgo the quest for a unified epistemology in favor of a detailed investigation of how these activities and artifacts shape and are shaped by our evolving language.

In the end, Wittgenstein's work must be measured not in terms of its direct contributions to knowledge—there were no theorems, definitions, or results of any special importance in it—but in terms of the critical power of the widely various arguments, analogies, terms, questions, suggestions, models, and modes of conceptual investigation he invented in order to understand the post-Fregean place of mathematics and logic within philosophy as a whole. Wittgenstein offers us problems rather than solutions, new ways of thinking rather than an especially persuasive defense or application of any particular philosophical thesis about the nature of language, logic, or mathematics. Despite the "negativity" and "reactionary" flavor that such a statement may suggest to some,[5] these ways of thinking mark a decisive step in the history of philosophy, in particular in relation to the legacies of Kant, Frege, Russell, and the Vienna Circle, and inaugurated new and fruitful ways of investigating phenomena of *obviousness* and *self-evidence* in connection with the notion of *logic*.

After Wittgenstein, traditional logical and philosophical vocabulary involving fundamental, apparently a priori categorial notions such as *proposition, sentence, meaning, truth, fact, object, concept, number* and *logical entailment* has more often been relativized by philosophers to particular languages, rather than assumed to have an absolute or universal interpretation a priori. Fundamental logical notions have been conceived as requiring immanent explication, a localized examination of their "logical syntax" in the context of a working, applied language or language game, rather than assumed to reflect independently clear, universally applicable categories capable of supporting or being derived from universally or necessarily true propositions or principles. It is largely to Wittgenstein, especially as inherited by Russell, Carnap and Quine, that analytic philosophy owes its willingness to question the Fregean idea that objectivity requires us to construe collections of various declarative sentences and that-clauses of declarative sentences, whether in the same language or in different ones, as reflecting a determinate content, thought, proposition, sense, or meaning.

In what follows, some major themes governing Wittgenstein's discussions of mathematics and logic are treated against the background of the evolution of his thought. A strong emphasis is placed on the early philosophy, both because this set in place the major problems with which Wittgenstein was to grapple throughout his life, and because it was the early work that placed Wittgenstein in direct confrontation with earlier philosophies of mathematics, those of Frege and Russell in particular. It was also the early philosophy that exerted the most direct and measurable influence on the Vienna Circle. Readers primarily interested in the

[5] See, e.g., remarks in works of Kreisel and in Hintikka (1993).

later philosophy, which has received widespread attention since the late 1970s, are referred to the accompanying bibliography, which provides an overview of current literature.

Evolution of Wittgenstein's Thought

1. The Early Period

It seems that an early effort to solve Russell's paradox was partly responsible for turning Wittgenstein toward philosophy as a vocation. Trained as an engineer at the Technische Hochschule in Berlin (1906–1908), he studied aeronautics in Manchester (1908–1911), there encountering the works of Frege and Russell.[6] By April 1909 he had sent to Jourdain a proposed "solution" to the Russell paradox; Jourdain's response sufficiently exercised him that by 1911 he had drawn up plans for a book and had met with both Frege and Russell to discuss his ideas (Grattan-Guinness 2000, p. 581). At Cambridge (1911–1913) conversations with Moore and especially with Russell stimulated Wittgenstein profoundly, and his final version of the *Tractatus Logico-Philosophicus* was finished by 1918. It may be read, at least in part, as a commentary on the philosophies of Frege, Moore, and Russell, and the place of logicism in relation to the legacy of Kantian idealism. The *Tractatus* thus forms a bridge between the earliest phases of analytic philosophy and the logical positivism of the Vienna Circle (especially as expressed in the work of Carnap), which in many ways it helped to create.

Wittgenstein made his primary impact by pressing to the fore the broadly philosophical question "What is the nature of logic?," a question that naturally emerges if we attempt to gauge the general philosophical significance of Frege's and Whitehead and Russell's "logicist" analyses of arithmetic. Given that it had been shown how formally to derive basic arithmetical truths and principles from basic logical principles, on what grounds may these principles themselves be held to be "purely logical?" Our grasp of Frege's basic laws seem, it is true, to involve no obvious appeal to intuition or empirical knowledge, and his formal proofs (*Aufbauen*) appear to be fully explicit, gap-free logical deductions. On the surface, his basic principles express truths about fundamental notions (such as *concept*,

[6] He also evinced skills as an applied mathematician, patenting a propeller engine that was later used for helicopter propellers by the Austrian army in World War II.

proposition, extension) that had long been acknowledged to be logical in nature. Yet neither Frege, nor after him Whitehead and Russell, provided a satisfactory account of why their systematized applications of the mathematical notion of *function* to the logico-grammatical structure of sentences should compel us to regard their analyses as purely logical in anything more than a verbal sense.

Frege's and Russell's own mark of the logical had been the explicit formulability and universal applicability of its truths: they conceived of logic as a science of the most general features of reality, framing the content of all other special sciences.[7] Their quantificational analysis of generality was what they had on offer to make this conception explicit. But there were internal tensions within this universalist view. Since the content and applicability of logic are assumed by the universalist to come built-in with the maximally general force of its laws, it is difficult to see how to make sense of its application as *application*, for from what standpoint will the application of logic be understood, given that the application of logic is what frames the possibility of having a standpoint? Frege's and Russell's views led them to resist the idea of reinterpreting their quantifiers according to varying universes of discourse, for they conceived their formalized languages as languages whose general truths concern laws governing all objects, concepts, and propositions whatsoever. They had no clear conception of ascending to a metalanguage in the model-theorist's way. And yet, as a result, much of the extrasystematic talk about the application of logic in which Frege and Russell engaged could not, by their own lights, be formalized *within* their respective systems of logic.

Thus, for example, Frege was explicit that he had to fall back on extraformal "elucidations"—hints, winks, metaphors—to explain his basic, purely logical distinctions. He took inferences applying universal instantiation—the move from "$(x)fx$" to "fa"—to reflect a universally valid rule of inference, but the universal validity of that rule *itself* could not be explicitly formulated, on pain of a vicious regress. He denied that "truth" is a genuine property word. The applicability of many of his most fundamental notions and analyses (e.g., the procedures of cut and assertion involved in modus ponens, such presuppositions as that "there are functions," "no function is an object," and "Julius Caesar is not the number one") could not be enunciated in terms of purely logical truths and/or definitions within the language of his logical system. At best, they could be indicated or *shown* by regarding the use of the formal system as the use of a contentful, universally applicable language.

Even more worrisome, the unrestricted application of Frege's primitive notions mired him in contradiction, as Russell (and earlier Zermelo) showed. If the unrestricted application of basic logical principles engendered contradiction, how

[7] See the annotated bibliography below for some key articles explaining the universalist conception and its significance for the history of logic.

were the restrictions placed upon their applications to be defended as purely *logical*? It was difficult to see how Russell's theory of types could be conceived to be "purely logical" on the universalist view, for it fractured the interpretation of quantifier ranges into an infinitely ascending series of stratified levels, forcing readers to understand each statement of a logical law, and each variable, as a "typically ambiguous" expression. Moreover, Whitehead's and Russell's unlimited, ascending series of types cannot be spoken about as a whole, in terms of a meaningfully demarcated range of generality. The universal application of Russell's general requirement of predicativity (associated with the "vicious circle principle") suffers from the same problem. And his axiom of infinity raised the question how *any* claim about the cardinal number of objects in the universe (whether finite or infinite) can possibly be seen to rest on considerations of logic alone. Either this question is one for physics, in asking what mathematics is needed to account for the cosmological structure of the universe, or it begs the question for the logicist in taking for granted that we already have mathematical knowledge of precisely those mathematical structures that the "logical" theorems will have to account for.

The universalist conception of logic's limitless application seemed in the end to leave no room for any model-theoretical approach to logic. This left insufficient room for the kind of rigorization of the notion of *logical consequence* with which we are now familiar. Neither Frege nor Russell had any means of formally establishing that one truth *fails* to follow from another, because they had no rigorous systematization of what in general it *is* for one truth to follow by logic from another. All they had conceptual space for were explicit formulations of the logical laws and rules of inference that they regarded as universally applicable and the display of (positive) proofs in their systems. The *completeness* of these systems with respect to logically valid inference could not be assessed except inductively and/or in general philosophical terms, by pronouncing on the maximal generality of logic in its role of framing the content of all thought.[8] It would take nearly fifty years after Frege's 1879 *Begriffsschrift* for the question of completeness with respect to the notion of *logical validity* to be properly formulated, in part because of the universalist conception that informed his formalization of logic.[9]

In his early work, while responding to the internal conceptual tensions within the universalist view, Wittgenstein began to zero in on the project of isolating a notion of *logical consequence*. This is somewhat ironic, in light of the fact that Wittgenstein's own later remarks about the notion of *following a rule* appear, at least at first blush, to sit uncomfortably with the idea that we *have* a clear intuitive idea of *one sentence's*

[8] Frege (1903, § 17). Compare Whitehead and Russell (1910, p. 95n) and Russell (1919, p. 191).

[9] On this, see Dreben and van Heijenoort (1986), and compare the reply of Dummett (1985).

following with necessity from another.[10] Still, though his philosophy always remained indebted to its universalist origins, he took several decisive steps away from the universalist model, steps that were to influence many who followed him.

Frege, Moore, and Russell had each rejected as unclear the notion of *necessity*, replacing it with that of *universality*; the early Moore and Russell explicitly held that the notion of a *"necessary" proposition* was a contradiction in terms, arguing that all "necessary" or "analytic" truths were in fact tautologies; and since no tautology could express a proposition, all genuine propositions involving predication are synthetic (for a discussion of this prehistory, see Dreben and Floyd 1991). In his early philosophy, Wittgenstein brought necessity and apriority back into view, denying that the notion of universality could replace them in accounting for the nature of logic. At the same time he preserved the idea that *necessity* is not a clear property word or predicate; indeed, he took the Russell–Moore view one step further in calling *all* so-called propositions of logic "tautologies." This had the radical effect of denying that logic consists of genuine propositions (i.e., of sentences capable of being either true or false, depending upon the relevant state of affairs in reality), a sharp break with the universalist view of logic as a science of (maximally general) truths.

Wittgenstein's attempt to distinguish the concepts of *necessity* and *generality* while retaining the universalist idea that logic frames all thought was fraught with difficulty, and his struggles to make sense of this project dictated the course of his future discussions of logic and mathematics. The *Tractatus* countered Russell's oft-repeated insistence on the importance to mathematical logic of the reality of *external* relations (i.e., relations independent both of the mind and of the relata related by the relations) by revisiting the idealist notion of an "internal" relation or property, a notion connected in the Kantian tradition with necessity understood as a reflection of human conditions and forms of knowledge.

[10] It is not certain that Wittgenstein ever read Gödel's completeness theorem (of 1930; in Gödel 1986), and quite certain that he did not appreciate its mathematical significance, given that he died in 1951, before the theorem's significance and fertility were well understood. No explicit discussions of the theorem are so far known in Wittgenstein's works, though it is not impossible they might be found in the future. It would be nice if they were, for it is the completeness theorem, much more than Gödel's incompleteness theorems, that would seem to be most difficult for Wittgenstein to interpret philosophically, given his later rule-following discussions.

Wittgenstein may have learned of the completeness theorem from Turing and Watson, with whom he had discussions in 1937 that we have good reason to believe touched on Church's and Turing's 1936 results concerning the undecidability of first-order validity (cf. Watson [1938]). Goodstein (1957, 1972) mentions that by the early 1930s Wittgenstein knew that the notion of *finite* could not be expressed in an axiomatic system; this is a consequence of compactness in Gödel's 1930 paper on the completeness theorem (cf. Gödel [1986]).

All logical concepts and relations were for Wittgenstein internal, and vice versa. Internal relations and properties are necessary features of the objects, facts, or properties they relate in the sense that we cannot conceive these to be what they are, were their internal relations and properties to differ. An example is the coordinate (2, 2) in a Cartesian coordinate system: it is part of what it is to *be* this particular coordinated point that its first coordinate be 2 rather than 1, that it have two coordinating numbers, that it differ from the point (1, 1), and so on.[11] To ask, for example, whether its first coordinate *must* be the number 1 is an ill-posed question, to evince a misunderstanding, both of this particular coordinate and of the framework within which it figures. In the *Tractatus* all such necessities are taken to reflect structures of possible thought or modes of representation of facts via propositions, not independent metaphysical facts or substances; they shape *how* we see facts holding or not holding in the world, but they are not articulable in terms of propositions telling us truly *what* objects really, ultimately are. The holding of an internal property or relation thus does not turn on any facts being or not being the case, but on logical features of a proposition internal to its modes of possible expression. These are not describable with a proposition, true or false, for they reflect conditions of possible description that must be reflected in any description. For this reason Wittgenstein calls "internal" or "formal" properties and relations *pseudo concepts* (i.e., not notions serviceable in the description of reality).

Wittgenstein used this idea of the internality of logical characteristics to reject the Hegelian tradition's insistence that the internal relatedness of subject and predicate in a proposition could be used to establish monism as a metaphysical truth. This was, in effect, to revitalize Kant's view that logic, in being a constitutive framework of thought as such, is factually empty, giving us no absolute or substantial knowledge of reality or thought as they are in and of themselves (the Hegelians had granted this for formal or "general" logic, but not for substantial, true logic). It was also to revitalize the Kantian idea that logic is in its nature a fertile source of dialectical illusion, constantly tempting us to confuse our forms of representation with necessary features of reality, and therefore demanding careful critique for its appropriate application.

Each of these ideas directly conflicts with the universalist standpoint of Frege and Russell, according to which logical laws are truths, full stop.

Wittgenstein's emphasis on internal relations was, however, also symptomatic of his departure from Kant's way of critiquing general logic. Kant set out a "transcendental" logic of appearances that displayed the conditions of human knowledge in a

[11] Another example is the expression of happiness in a particular facial expression: the facial expression could not be the expression of happiness it is apart from expressing what it expresses. Yet another example is the numberhood of the number 2: 2 could not be what it is and fail to be a natural number, nor could the concept *natural number* be what it is without applying to the number 2.

set of synthetic a priori principles. Wittgenstein's conception of logical relations as internal to their relata was intended to allow him to undercut any such independently given principled or mentalistic role for the transcendental. He aimed to display the limits of sense from within language itself, by spelling out that which is internal to the basic notion that we express ourselves meaningfully and communicatively, sometimes truly and sometimes falsely. This was the ultimate point at which Wittgenstein could hope to show that neither *necessity* nor *universality* can be seen aright without surrendering the idea that they are substantial concepts or properties.

Universal instantiation is as good an example as any. In the *Tractatus* the logical relation between "$(x)fx$" and "fa," in being "necessary" and "universally applicable," expresses an *internal* logical relation between these two propositions: neither proposition could be what it is apart from the second sentence's "following by logic" from the first. This indicates the idea that it is nonsensical to ask, given the assumption that "$(x)fx$," whether or not "fa" is true—as nonsensical as it would be (on Wittgenstein's view) to ask whether the coordinate (2, 2) has as its first coordinate the number 2, or whether the number 2 is a natural number. The *logical* or *formal* character of the relation (or property) comes in, ready-made, with our representation of its elements. Differently put, the logical content and force of generality is for Wittgenstein something shown in our language, but not said; it is applied, but not described or established by a truth.

Wittgenstein's chief philosophical difficulty lay in working out a conception of logic that would display all logical features of propositions as internal to them, and display all necessities as logical. His *Notebooks 1914–1916* had begun with the remark that "logic must take care of itself," an insistence intended to preempt as nonsensical any appeal to the extralogical in understanding the scope and nature of logic. This insistence incorporated strands of the universalist point of view, but pointed in a new direction. Wittgenstein was simultaneously rejecting an empty formalism about logic that leaves its application to a theory of interpretations (in the universalist spirit) and at the same time rejecting all theories—including certain aspects of Frege's and Russell's philosophies of logic—that rely on an appeal to purely mental activity—or to any extralogical facts—to account for the application of logic in judgment.[12] If logic could be *shown* to "take care of itself,"

[12] The theories he had in mind to undercut included Kant's transcendental doctrine of judgment as synthesis, Frege's notion of assertion as an act working on a content, Moore's and Russell's analyses of judgment as constituted by a relation between a mind and a proposition, and Russell's later multiple relation theory of judgment. All these theories fall back on an appeal to an extralogical mental activity to account for the nature, structure, and objectivity of truth in judgment. The effects of Wittgenstein's reactions (1913 and earlier) to Russell's theories of judgment on the *Tractatus* notion of *picture* are greatly illuminated by Pears (1977), Ricketts (1996b), and Proops (2002), and analogous effects of his reactions to Frege by Ricketts (2002) and Sullivan (2001b).

then the pseudo character of all attempts to *justify* its application and our obligations to it could be unmasked, precisely by showing them to purport to have achieved a perspective *outside* of logic from which to explain it. Of course, strictly thought through, this conception would impugn itself, in that its own remarks *about* logical relations would positively invite themselves to be read as substantial truths about a reality underlying logic. In the *Tractatus* Wittgenstein bites the bullet, declaring his own sentences as, by his own lights, nonsensical, to be surrendered in the end as themselves misleading, as neither true nor false to any kind of fact. Their use was to portray much traditional epistemic and metaphysical talk of necessity and reality as equally so by freeing us from misleading construals of logical grammar.

Wittgenstein's most basic conceptual move, expressed in his *Tractatus* conception of sentences as "pictures" or "models" of states of affairs, was to take the unit of the perceptibly expressed judgment—the *Satz* or "proposition" as an applied sentence, true or false—to be logically fundamental. The challenge was to contrive a presentation of logic that would allow this notion to remain basic to the presentation of what logic is, and to show how to unfold from it, without further ontological or epistemological appeal, all fundamental logical notions and distinctions. Logical form, on this view, was not to be construed as anything entity- or property-like, but instead as expressed in the ways in which we (take ourselves to) apply and logically operate with propositions. If this could be accomplished, Wittgenstein could surrender as otiose the whole idea of a logical fact or law, and still view logic as in some way limitlessly applicable.

The analogy between pictures and propositions intuitively suggests this line of thinking. A picture (a proposition) is conceived not as an object, but as a complex structure, a fact laid up against reality as a standard without intervention of an intermediate entity—something like a yardstick. The variety of different ways in which one might model (or measure) the same particular fact, configuration of objects, situation, or state of affairs—for example, by means of colored dots on a grid, or by means of words, or by means of pencils and toy people (by means of miles, feet, inches, or nanometers)—tempts us to conceive of a "something that these models have in common with one another" (a "real length" of an object independent of how it is measured on one or more occasions). Yet there is no such notion to be had apart from the applicability of some particular system of measurement (there is no notion of *length in itself*, unindexed to such a system). This leaves us with the idea that the "it" that is common, for example, to all possible depictions of a particular person's stance at a particular moment in time, is their common depiction of that particular person's stance, and no other, independent kind of fact. But then the "it" of a picture's content, the essence of its particular comparison with the way the world is—analogously, a proposition's logical form—is on this view not object-like, or to be understood apart from our understanding and application of it (i.e., of the propositional signs that, in application, are true or false in expressing it). Thoughts and senses, identified by Wittgenstein with applied propositional expressions, are on this analogy equally unobject- or unentity-like: they are *shown* or *expressed in* our activities of arguing, agreeing, and disagreeing on the truth of particular propositions. On Wittgenstein's conception, there is thinking without (Fregean) thoughts: what it is to take ourselves objectively to communicate is given through our acceptance of different complexes of signs (in different mouths of different speakers) to express the same propositions or symbols. But this is simply to redescribe or reiterate as basic the notion of a proposition as an applied sentence, true or false.

The weight of Wittgenstein's overarching conception of logic was thus not taken to rest so much on its ability to uncover a true logical structure of the world (e.g., an ontology of objects and facts suitable to justify the priority of truth-functional and/or quantificational logic) or a model-theoretic semantics of possible states of affairs to ground an intuitive notion of *logical consequence* or *entailment* (though, as Wittgenstein was aware, one may easily interpret some of his remarks in this way). Instead, he took his conception to rest wholly on its ability to display all such (very natural, indeed perhaps inevitable) talk of necessity in connection with the relation of logical consequence as in the end nothing but a repeating back or reflection of the acknowledgment—to be recognized as fundamental rather than defended as a further independent proposition—that we express propositions, some of which are true and some of which are false. This unmasked as nonsensical (*unsinnig*) the idea that the kind of categorical talk most natural to general philosophical discussion of logic could be viewed as contributing to an independent body of metaphysical truth about the most general features of reality.

Wittgenstein used the truth-table notation—implicit in Frege's and Russell's logic—to set forth his conception of the logical. He construed a truth-table as a fully complete expression of a proposition or thought, and aimed to unfold from this diagram the most fundamental logical distinctions. First, he identified the "sense" (*Sinn*) of a proposition with the truth-functional dependence of the proposition's truth-value upon the different possible assignments of truth and falsity to its elementary propositional parts. This extensionalized the notion, allowing him to dispense with Frege's (intensional) way of defining a *thought* as the *Sinn* of a proposition, the "mode of presentation" of a truth-value. It also vividly displayed Wittgenstein's idea of *bipolarity*, the notion that what it is to *be* a proposition—to express a definite sense—is to be an expression that is true or false, depending upon how the facts are. It depicted as internal to the proposition its combinatorial capacity to contribute to the expression of further truth-functionally compounded propositional structures. And it gave Wittgenstein a way to tie the notion of *sense* to that of *logical equivalence* in terms of purely truth-functional structure: "$\neg p \vee q$," "$p \supset q$," and "$(p \,\&\, p) \supset q$" all express the same sense in being logically equivalent, and vice versa, for the truth-table form of writing each of these shows that they each express the very same dependence of truth-value upon assignments to "*p*" and "*q*," regardless of the fact that different truth-functions appear in their expression. It is, in other words, the way in which the final column of the truth table relates assignments of truth and falsity to the elementary parts that ultimately (logically) is what matters, and not the classification of a proposition's logical form as that of a conditional, a conjunction, or a disjunction. Meaning and sense, in their logical aspects, are thus extensionalized.

This conception yielded a way to distinguish, on the basis of the truth-table notation alone, the "propositions" of logic: "tautologies" are *defined* as having "T" in every row of the final column, "contradictions" as having "F" in every such row, and the "propositions" of logic are just these sentences. This portrays logic as empty of factual content, for the truth-values of such sentential forms may be seen not to *depend* upon any particular truth-assignment to their elementary parts; they hold no matter what assignment is chosen. (As Wittgenstein remarked, to say that it is either raining or not raining right now tells me nothing about the weather.) These purely logical "propositions" lack bipolarity, and hence sense: Wittgenstein declares them *sinnlos*, regarding them as limiting cases of propositions that are not really either true or false, that carry "zero" information.

This characterization of logic was purely "formal" or "extensional" that is, it worked without any independent appeal to extralogical meanings or truths, and solely on the basis of reflection on the propositional sign, the truth-table notation itself. On Wittgenstein's view the logical status of logical sentences emerges solely through their internal logical structure, a structure that cancels out the depicting features of the sentence's elementary parts, but not through any general truths or through their gap-free derivability from basic logical laws. The truth-functional connectives are themselves reduced, in this notation, to what is displayed in a truth-table, particular dependencies of truth-values upon truth-assignments. The interdefinability of the truth-operations indicates that they do not contribute to any factual aspect of a proposition. Thus Wittgenstein remarks that his fundamental thought (*Grundgedanke*) is that the logical constants do not refer to or serve as proxies for logical objects. There is, then, no room for a genuine disagreement over the truth of individual logical laws. Any purported justification (e.g., of bivalence, excluded middle, or noncontradiction) would be at best another way of spelling out one's prior acceptance of the truth table as a suitable expression of the proposition itself, and hence no justification at all. In inviting us to suppose that one can step outside of logic to assess the truth of its laws, any such justification is nonsense. There is no question of generality here at all, only a question of one's understanding of the notion of *proposition* properly.

This understanding of the notion of *proposition*, built upon Wittgenstein's "extensionalism" about logic, was thus tied to a rejection of Frege's and Russell's respective applications of the distinction between function and argument (Frege's distinction between concept and object), a distinction intrinsic to the quantificational conception of generality that appeared to give the universalist conception its plausibility. As we have seen, Wittgenstein took the ability of a truth table to stand in as an expressive replacement for (or a definition of) a truth-functional sign (for example, in showing the expressive adequacy of "¬" and "&" relative to the other truth-functions) to indicate that the signs for the truth-functions are not *names* of logical objects or functions, as Frege and Russell had held. Instead, he conceived the connections as part of a framework for expressing logical dependencies, "internal" (nongenuine) relations between sentences. This demanded, for a proper treatment of the logical signs, the introduction of a new grammatical category distinct from that of *function* and *name*, namely, that of an *operation*. The depiction of objects and functions by configurations of names and concept words is part and parcel of the content of propositions for Frege, Russell, and Wittgenstein. But for Wittgenstein operations, being *merely* formal, are simply a way of operating with propositional signs; they do not stand proxy either for concepts (functions) or for objects. They are instead like punctuation marks: an operation sign forms part of the way in which a particular sense happens to be expressed, but in no way can a particular logical connective be taken to reflect what a sentence means or is about. Wittgenstein takes part of his task to be to show how we may regard all of the usual logical signs as operation signs (i.e., as truth-operational features internal to the expression of propositions) rather than features referring to that about which the propositions in which they explicitly figure speak.[13]

Wittgenstein's construal of logical features as emerging from what we do with propositions—reflecting internal features of our ways of seeing the world rather than properties or functions—remained a central leitmotif in all of his subsequent work. Over and over again he warns against the tendency to reify our means of expression, showing how the oversimplification of logical grammar naturally leads to the posing of pseudo-questions. He saw it as all too easy to assume that objectivity demands there must be an object or meaning corresponding to every working sign in the language. Frege's so-called context principle ("never ask for the meaning of a word in isolation, but only within the context of a proposition") and Russell's notion of an incomplete symbol (his use of the theory of descriptions to analyze away apparently denoting phrases in favor of quantificational structures) were moving in the right critical direction, but did not go far enough in undercutting our tendency to confuse elements of our expressive powers with entities corresponding to elements of our language. For Frege and Russell failed to apply their own principles to the logical elements of language. To Wittgenstein, their uniform extensions of the distinction between function and argument across logical notions were conceptually procrustean, mired in an overly optimistic extension of (an updated form of) the ancient duality between subject and predicate.

[13] Hylton (1997) and Floyd (2001a) examine the distinction between functions and operations in the *Tractatus* in relation to Frege's and Russell's philosophies of logic.

Difficulties about the analysis of generality were, however, central to Wittgenstein's problem, and these could not be faced by means of the truth-table and the picturing ideas alone. It was alien to Wittgenstein, as much as it had been to Frege and Russell, to conceive of the quantifiers as reinterpretable according to arbitrarily chosen universes of discourse; the whole point was to see logic growing seamlessly out of the given application of sentences in a language. But he conceived of this growth in radically new terms. The decision procedure (*algorithm*) for truth-functional validity, satisfiability, and implication that Wittgenstein exhibited with the truth-tables in the *Tractatus* allowed him to make this view plausible for part of logic: as he pointed out, in cases where the generality symbol is not present, we can see that so-called "proof" *within* pure logic is a mere calculation, a "mechanical" expedient allowing us to display for ourselves the tautologousness (or contradictoriness) of a logical proposition simply by rewriting it. The idea that the truth-functional connectives refer to separable entities, or that we can *prove* the logicality of a sentence (or the validity of a proof through its conditionalization) via rational inferences from truth to truth, justified step-by-step by the application of logical laws to general features of objects and concepts, is thus—for Wittgenstein—otiose.

But how was this conception to be extended to quantification theory, where (as Church was to establish in 1936) we *have* no decision procedure for validity, and where (as Gödel was to show in 1930) we *do* have a systematic search procedure for it?

The purely philosophical motivation for extending Wittgenstein's view across "all of logic" was clear. His emphasis on the mechanical aspect of proof in pure logic was motivated by his critical reactions to the wedding of Frege's and Russell's philosophical arguments for logicism with their success in having formalized logic. They had each argued, against formalists and traditional algebraists of logic, that logic and mathematics are more than merely mechanical, formal games of calculation—they are applicable and contentful sciences of truth. Frege explicitly insisted that arithmetic was more than a *mere* game like chess *because* its propositions express senses, thoughts [Frege 1903, §91]. For Wittgenstein, this kind of argument rang hollow. Arguably what Frege and Russell had *done*, in spite of (or perhaps because of) their universalist conceptions, was to show how the purely logical character of an argument *can* be verified mechanically, by tracing through its formalized structure. This appeared to reinforce the algebraical view that logic is essentially empty, a calculus with uninterpreted signs. Wittgenstein did not wish to fall back on an uncritical formalism. But he did wish to portray the debate between Frege and Russell and their formalist forebears as mired in confusions over what the application of logic and mathematics involves.

Wittgenstein understood quantification as something operational, that is, as belonging to the way a proposition is expressed and applied, rather than to a distinctive descriptive or functional element of a proposition. The *Tractatus* claims that all quantificational forms of sentence may be generated purely operationally, and uses for this Wittgenstein's operator *N*, a generalized form of the Sheffer stroke of joint denial that negates elementary propositions either singly or jointly. Since he was writing at a time when the distinction between first- and second-order logic had not been formulated, we may assume that Wittgenstein intended to use operator *N* to express *all* quantificational structures. But his notational suggestions have never been made wholly precise. For the first-order case, however, at least this much is clear: "fx," a propositional variable indicating a function, stipulates a range of elementary propositions (i.e., those in which the function figures with an object); "$\neg(\exists x)fx$" is expressed through operator N by jointly (generally) denying the elementary propositions presented by "fx"; and "$(x)fx$" is generated by jointly (generally) denying the elementary propositions presented by "$\neg fx$." The logical step from "$(x)fx$" to "fa" is thereby seen to be an immediate, purely formal operation rather than an instantiation in accordance with a general logical rule or law of inference; generality contributes nothing in and of itself to the material content of a proposition. The sign for generality is thus not to be read as a function symbol, but as a way in which a collection of elementary propositions is operated upon, or displayed. In a sense, generality is already expressed in "fx," and therefore, in a sense, the "(x)" part of the generality notation is unnecessary or misleading.[14] In virtue of the fixed domain of (elementary) sentences that are presupposed from the outset, this sign can at best be seen to contribute as an index to the applications of operator N at work in cases of multiple generality. But it does not express a genuine kind word, second-order function, or concept in its own right.

This is very far from the universalist's conception of generality. Wittgenstein is conceiving generality to be expressed in something like the way a genre painting expresses an archetypal feature or scene. Such a painting is applicable to an aspect of each concrete situation exemplifying the relevant genre features. That is what its being a genre painting is, and that is what makes any exemplar an exemplar of *it*. Its own way of representing is, of course, not reducible to the depiction of any *one* such example, but each such example exemplifies on its own, without any intermediary principle. This is reflected in the fact that the very *same* picture could be used to depict (could be inspired initially as a depiction of) a *particular* scene, and also a genre scene; there is nothing in its internal structure that *says* how it should be interpreted. That comes out in our uses of it.

Wittgenstein's treatment is not intended to give a unified account of the notion of *generality*. His distinction between function and operation bifurcates it into two different notions, the *material* generality of concepts involved as functions at the atomic level in elementary propositions, and the *operational* generality of formal concepts. As we have seen, he conceives of formal concepts as pseudo

[14] Compare Kremer (1992).

concepts because of our tendency to assimilate them to genuine concepts or kind words when we ought, as he thinks, to focus on the operational, procedural aspects of our hold on them: formal notions do not sort independently given objects into kinds (they are neither true classifiers nor sortals), but instead are *shown* or *expressed* through the kinds of applications we make of them. Their applications are internal to them, and the applications are, in turn, what they are insofar as they successfully display or express the formal notions.

This gives Wittgenstein a way to dissolve Russell's paradox. According to the "picture" idea, the elementary proposition is a fact in which names serve as proxies for objects via their structural, configuration within the sentence as it is used to depict the holding of a state of affairs. Material ("propositional") functions (concepts and relations) are expressed via fixed modes of positioning names within the projected propositional sign. This exhausts the articulation of functional and objectual material content (and generality) at work in language. Wittgenstein holds that no *material* function can apply to itself: no sentence can say of itself that it is true, for the sentence is a proposition only insofar as it is compared, through its functional articulation via a structure of names, with something *else* that it models truly or falsely. He is thinking here in terms of an analogy with the positionality notation of our decimal system: *that* our numerals are written in the decimal system, and not some other one, is presupposed in our understanding of how to read the positional notation. This is not something that the numerals *say*; it is something that our operations with them *show*. Given a fixed system of positionality to express material functions, writing down "$f(f(x))$" in effect gives rise to a claim about two *different* material functions, much as the digit "2" reflects something different in each of its occurrences within the (decimal) notation "222."[15] This reflects Wittgenstein's idiosyncratic reading of the generality of the variable: "fx," on his view, presents a fixed range of elementary propositions ("fa," "fb," and so on) that cannot be changed by trying to reapply "f." In "$f(fx)$" we thus have two occurrences of the same sign functioning as different

[15] Wittgenstein's idea that functionality might be expressed through positional structure within a sentence resembles the positionality conventions at work in our decimal notation in another way: a decimal numeral is not, from this perspective, just a name, but a structure or an aspect of pictorial form. This is to rework Frege's analogy between the functional articulateness of the complex arithmetical term (such as "$2^2 + 3^5$") and the sentence, using the analogy to reject Frege's view that numerals and sentences are object expressions (names).

On Wittgenstein's analogy, the formally characterizable difference between an elementary proposition and a molecular proposition emerges through applications of truth-operations in just the way that the difference between decimal numerals and complex arithmetical terms emerges through the application of arithmetical operations (addition, multiplication, subtraction, and exponentiation).

symbols; in principle we could simply use different signs (say, with indices) to distinguish them. This is Wittgenstein's "solution" to Russell's paradox.[16]

Yet Wittgenstein does not rule out a purely formal, "self-referential," or recursive series of nested applications, as in $O'x$, $O'(O'x)$, $O'(O'(O'x))$.... What he does insist is that if we imagine such a series being logically connected according to repeated application of a single formal rule, then we are treating "O" as a sign for an *operation*, not as a genuine (propositional, material) function symbol. Operations, unlike (material) functions, are in their very nature recursively iterable procedures generating collections of instances that are internally ordered in what Wittgenstein calls formal series. Wittgenstein's notation for this kind of (recursive) generality uses square brackets: in, for example, "$[a, x, O'x]$," "a" stands for the basis step of the series, x for an arbitrary member of the series, and "$O'x$" for the result of applying an operation O to x. This bracket notation is thus equivalent to writing down "a, $O'x$, $O'O'x$, $O'O'O'x$...," an expression with an ellipsis. It expresses a rule whose applications are internal to its expression, but the generality of the rule is *shown* or *indicated* through the manner of presenting its instances, not said or described. Furthermore, Wittgenstein allows that the basis might *itself* be given through a formal series, so these bracketed expressions may themselves be iterated. On this view, the procedure of moving from type to type in Russell's hierarchy is *operational*, something done formally, without any appeal to the facts. It cannot be summed up in the quantificational manner of Frege and Russell without confusion, for the generality of the formal series, ordered by an "internal" rule of the series, is fundamentally different from the generality of, for example, the (material) concept *horse*.

We have seen that the notion of *proposition* is held by Wittgenstein to be a *formal* (pseudo)concept. This is evinced in his treating it as fully displayed through applications of the (formal) operator N. Wittgenstein holds that *all* propositions are results of applying iterations of operator N to a basis of elementary propositions. No proposition talks about the general form of *proposition*, for this is a formal notion, given through a recursive template or rule, not via the functionally articulate name of a class or totality. Beginning with some basis of elementary propositions, p, the idea is that by a finite number of applications of operator N, we *see* that we shall be able to generate all propositions. Wittgenstein writes what he calls "the general form of proposition" in terms of his recursive notation: "$[\bar{p}, \bar{\xi}, N(\bar{\xi})]$," where "$\bar{p}$" is a schema for a basis propositions, "$\bar{\xi}$" a variable standing for all results of some finite number of applications of operator N to elements in that basis, and "$N(\bar{\xi})$" for the totality of results of applying operator N.

[16] Sullivan (2000) shows how understanding the cohesion of the *Tractatus*'s technical and philosophical treatment of this distinction is crucial for understanding his responses to Russell's type-stratification of the universe. For another useful treatment, see Potter (2000).

The *Grundgedanke* of the *Tractatus* is that the so-called logical constants (the logical expressions) do not refer: all logical expressions indicate operations, according to Wittgenstein. Now we see that all operations are logical in being purely formal, recursively iterable rules. And operations are always tied, directly expressed through or indirectly, to our application of propositions in language, via truth-operations on the elementary propositions. Nothing that is a proposition fails to be subject to logical operations.

There has been a recent debate over the question whether and in what sense this *Tractatus* notation for the general form of *proposition* is expressively adequate, even for first-order logic alone, and, therefore, whether and on what grounds Wittgenstein might have been committed, wittingly or unwittingly, to the existence of a decision procedure for all of logic.[17] Wittgenstein's situation with respect to decidability was vexed and complex. In 1913 he was apparently one of the first (wrongly) to conjecture the existence of a decision procedure for all of logic, writing to Russell that the mark of a *logical* proposition was one's being able to determine its truth "from the symbol alone" (cf. CL, pp. 59, 63). Yet it remains unclear on precisely what basis Wittgenstein might still have been committed to this as late as 1918.[18] It is usually held—and it was reported to have been said by Wittgenstein himself—that in the *Tractatus* he conceived of the quantifiers in terms of potentially infinite truth-functional conjunctions and disjunctions, not realizing the intrinsic barrier to extending the truth-table analysis into the unrestricted quantificational domain.

Yet Wittgenstein himself had no interest in setting out a smooth-running codification of quantification theory, and in fact he is not explicit in the *Tractatus* about what semantics or notational details he would require. His central philosophical problem was, after all, to indicate how one might be able to dispense, conceptually, with the idea that general logical laws or rules of inference such as universal instantiation are true in virtue of meanings of the words "all" and "some" that go beyond what is expressed at the elementary level by their instances; he wanted to show that none of the logical constants must be conceived as going proxy for an object or function. Only in this way could he undercut the idea of a stance from which one might *dispute* the correctness or incorrectness of an interpretation of the quantifier's range, of the inference from "$(x)fx$" to "fa," and show that logic is not a science of true propositions or laws at all.

Arithmetic remained, however, as a potential stumbling block. For in rejecting Frege's and Russell's quantificational conception of generality, Wittgenstein was in effect rejecting the means by which they carried through their account of

[17] See Fogelin (1982), Geach (1981), Soames (1983), Sundholm (1990), and Floyd (2001a).

[18] Certainly by the early 1930s he was insisting, in response to conversations with Ramsey, that no such "leading problem" of mathematical logic would help us understand logic's basic nature. But this was in part a criticism of his earlier conception's having (apparently) *made* it a central concern.

arithmetic as logical in nature. The first difficulty he had to face was an account of the universal applicability of the cardinal numbers within extramathematical propositions. Here he proposed making cardinality and numerical identity internal features or forms of the elementary propositions. This was to construe cardinal numbers as undefinable, as internal aspects of our presentation of facts in propositions, rather than as objects or properties of objects. Falling back once again on his rejection of the adequacy of the function/argument distinction, Wittgenstein sharply distinguished (as algebraists of logic had traditionally done) between logical identities and mathematical equations. He then argued that the sign for identity is, in connection with propositional functions, expressively, and therefore logically, unnecessary. It would always be possible, he argued, to devise a language in which there were no redundant names: logic cannot bar this as an expressive option. This would allow us to read numerical identity between material concepts directly off from the expression of an elementary proposition; ascriptions of number were then not (as they had been for Frege and for Russell) logical identities.

On this view, "fa & fb" *shows* without *saying* that there are at least two F's; the number is a depicting feature of the symbolism, and not a separate object referred to by the proposition.[19] Wittgenstein adopts a convention for the variable that allows him to read "$\neg(\exists x, y)(fx \,\&\, fy)$" as saying that there are not two f's.[20] Thus the fact that we use equations in mathematics cannot, for Wittgenstein, be used to argue that the numbers are objects, for though he argues that mathematics consists essentially of equations, he holds that equations are not objectual identities in which genuine (propositional) functions work. Instead, they are part of an operational calculus of signs, working via (what mathematicians had long called) the method of substitution.[21] Probability is also subjected to analysis in terms of the nation of operation in the *Tractatus*, via the truth tables.

[19] The nonlogical character of Russell's Leibnizian second-order analysis of identity in terms of coincidence of all properties is intended to be exposed by this thought experiment: Wittgenstein is arguing that it is perfectly possible that we could conceive that two *different* objects fall under precisely the same concepts. Furthermore, what Russell tries to assert in his axiom of infinity would not be asserted, but *shown* in the use of infinitely many different proper names. And if—as one would expect Wittgenstein to have assumed, given the obvious demands of physics—real numbers were added to the forms of the elementary propositions, this approach would yield elementary propositions with potentially infinite formal complexity. Moreover, such pseudo statements as "$(\exists x)(x = x)$" and "$(\exists x)(x \neq x)$" could not be written in this language as propositions, true or false; that which they try to express as general truths would be shown in the application of names in genuine propositions.

[20] The precise interpretation of this proposal is complex, and still not worked out with full formal precision; for two discussions, see Hintikka (1956) and Floyd (2001a).

[21] The terms in which Wittgenstein treats arithmetic in the *Tractatus* are precisely those used by Whitehead (1898). See Floyd (2001a) for a discussion.

In this way Wittgenstein revitalized in the *Tractatus* an older, algebraical way of conceiving mathematics and logic that Frege and Russell had hoped to wipe out as "formalist," but precisely by tying this language itself back to the logicist idea that our cardinal number words reflect aspects of the logical form of propositions that we apply to the world. A mathematical equation, according to the *Tractatus*, is neither true nor false, it expresses no thought or proposition, but instead sets forth rules for the substitution of one numerical (operational) sign for another, either in other mathematical equations or in genuine propositions (extramathematical ascriptions of number). Ascriptions of number *within* mathematics (e.g., "there is only one solution to $x+1=1$") enunciate symbolic rules for the application of operation signs; in this way the "x" in such an equation is not read as a universally quantified variable, but as a formal operator.[22] Regarded from the perspective of mixed contexts, equations are taken to show us ways we may interchange numerical operations in genuine propositions without affecting the sense expressed. They reflect what we can do with these operations, not underlying general truths.

The pictorial character of number words is part and parcel of Wittgenstein's idea here. Operating on an equation via substitutions of one term (operation sign) for another yields different "pictures" or standpoints from which to consider the operation signs figuring in it: in mathematics we are *shown* different aspects of the logical syntax of number words (cf. TLP 6.2323, 2.173). Arithmetical "proofs" of equalities between particular number words are calculations, manipulations of a series of pictures using operation signs in accordance with "the method of substitution" (6.241). On this view, the numbers are construed neither objectually nor adjectivally, but *practically*, in terms of what we *do* with them. And Wittgenstein has no need to define a *general* notion of (mathematical) *function*.

On this view, second-order logical principles such as Hume's principle[23]—which essentially requires the notion of *identity* for its formulation—are otiose. The equinumerosity of two concepts (whether material or formal) is already

[22] Wittgenstein (1993–), vol. IV, p. 239:

> $x^2=1$ has two roots versus on the table are 2 apples. The former is a grammatical rule of the variable.... Can I determine a variable by saying that its values should be all objects which satisfy a certain function? Not if I don't know this some other way—if I don't, the grammar of the variable is simply not determined (expressed).

[23] "Hume's principle" is Frege's second-order contextual definition of sameness of number, so dubbed in work by Wright (1983) and Boolos and Heck (cf. Boolos [1998, part II] for a discussion). Intuitively speaking, for concepts U and V the principle is that the number belonging to the concept U is the same as the number belonging to the concept V if and only if there exists a one-to-one correlation between the objects falling under each concept.

expressed (applied) in the similarity of form shared by different first-order propositions (pictures) involving these concepts. For example, given Wittgenstein's eliminative proposal about redundant names, "fa & fb" and "gc & gd" show without saying that there are at least as many f's as g's through their shared forms. Given Wittgenstein's exclusive interpretation of the variables "x" and "y," "$(\exists x)(\exists y)(fx \,\&\, fy) \,\&\, \neg(\exists x)(\exists y)(\exists z)(fx \,\&\, fy \,\&\, fz)$" and "$(\exists x)(\exists y)(gx \,\&\, gy) \,\&\, \neg(\exists x)(\exists y)(\exists z)(gx \,\&\, gy \,\&\, gz)$" *show* equinumerosity through their shared forms. Cardinality, for material and for formal concepts, is thus taken by Wittgenstein to be given with a concept's expressive possibilities, possibilities whose applicability must, he believes, already be in place in the use of (nonpurely logical) propositions *before* talk of one-to-one correlation is cogent. The existence of a one-to-one mapping between numerals and objects may be understood either materially or formally, externally or internally. Externally and materially, this is a way in which individual objects may be associated, either with one another or with numerals. But the heart of Wittgenstein's picture idea is that a *mere* association between names and objects is, in and of itself, not something qualifying as propositional at all; only the internal features of a proposition give rise to a logical relation of equinumerosity. The proposition, conceived as an articulate, applied structure or fact, true or false, is fundamental. The upshot is that Wittgenstein treats the cardinal numbers and the notions of *object* and *numerical identity* as indefinable forms, expressive features of the symbolism, rather than as objects or names.[24]

Wittgenstein did not mention the *principle* of mathematical induction in the *Tractatus*—a remarkable thing, given that one of the primary achievements of Frege's and Russell's second-order analysis of arithmetic was to show how to derive this (apparently "synthetic") truth from pure logic alone. He did take the cogency of proof by induction to be a reflection of the logical structure of number words; it is just that his conception of the logical differed from the logicists'. For Wittgenstein an induction "principle" is one whose generality and applicability are shown in the general form *natural number*, which is a formal notion. On this view, mathematical induction's application to the natural numbers is part and parcel of what makes a number a number. Induction is thus not depicted as a separable general truth that could be deduced via a second-order definition of number in terms of equivalence classes. Wittgenstein's own "definition" construes the

[24] Such considerations against the priority of Hume's principle were forwarded by Wittgenstein (cf. WVC) and later, under his influence, by Goodstein (1951) and Waismann (1951, 1982, 1986). Dummett (1978, 1991a) responded critically to Waismann's version of these considerations, which lacked Wittgenstein's more or less explicit reliance on the idea of the proposition as a picture. Marion (1998) contains a useful survey of the issues raised here, which are obviously connected with the notion of a function-in-extension (i.e., with the notion of a function as an arbitrary mapping between individuals or a set of ordered pairs).

numbers recursively, as "exponents" of operations (i.e., as forms common to the development of any formal series).[25] Every formal series has a first, a second, a third member, *and so on*; the notion of "operation" is equivalent, Wittgenstein remarks, to the notion of *and so on*. By writing down "$[0, \xi, \xi + 1]$," a variable ranging over exponents of all operations, Wittgenstein shows the general form of natural number (TLP 6.03). In effect, he simply writes down "$0, 1, 1 + 1, 1 + 1 + 1, \ldots$," without analyzing the ellipsis away explicitly, as is done with the ancestral construction of Frege and Russell.

To the objection that he was opening arithmetic up to an account in terms of synthetic a priori intuition of succession, Wittgenstein replied that language itself would provide the necessary "intuition"—which was just to dismiss any serious independent role for the notion of *intuition*. On his view, the individual arithmetical terms within this series are not freestanding, but emerge through their connection with the notation for the general form of proposition and, hence, through the expressive structure of the operations we perform with elementary propositions. Wittgenstein went so far as to deny (in the early 1930s) that there could be a notation for arithmetic in which numerals functioned as proper names (WVC, p. 226). And he always insisted that the theory of classes is "superfluous" for mathematics (TLP 6.031). Indeed, he considered the ancestral definition of *successor* to suffer from an expressive "vicious circle" (TLP 4.1273). This was not for him an epistemic argument against our claiming to have knowledge of abstract objects or a freestanding preference for predicativity. It formed part of an expressive argument about the philosophical advantages of one notation for the numbers over another, *presupposing* his treatment of generality.

Wittgenstein was thus always sympathetic to the kind of considerations Poincaré brought forward against logicism—namely, that in setting out the formalization of logic within which the logistic reduction would be carried out, Frege and Russell already depended upon their reader's ability to apply arithmetic and inductive inference. But this was not (as it was for Poincaré) intended to refute the notion that arithmetic is in some sense "analytic." Instead, it was to show that the logistic reduction is not of any fundamental epistemic relevance.[26] Wittgenstein feared that Frege and Russell would reinforce the philosophical tendency to look at the logical characteristics of mathematics through the distorting lens of a

[25] Detailed reconstructions of this idea may be found in Frascolla (1994, 1997), Marion (1998), and Potter (2000).

[26] An assessment of the legitimacy of this kind of objection, and the differences between Wittgenstein and Poincaré, lies outside the scope of my discussion here; the adjudication of these issues is complex. For two especially relevant treatments of Poincaré's arguments, see Goldfarb (1988) and Mclarty (1997); on Wittgenstein's criticisms of Frege and Russell's logicism, see Steiner (1975) and, for a contrasting interpretation, Marion (1998).

unified account of generality. From his perspective, the logicistic reduction is likely to reinforce the tendency to look behind mathematical practice for an underlying account of a reality of necessities. The advantage of his recursive notation lay, he believed, in its showing on its face that which is logical in arithmetic, namely, the internality of the relations among arithmetical principles, arithmetical operations, and arithmetical terms, regarded as a reflection of our fundamental ability to ascribe number in empirical contexts. The general notion of *natural number* is not a material predicate or kind word, as he sees it, but, like the notion of *proposition*, is given through our application of recursive operations to expressive features of our language.

The universalist's conception is depicted here as wrongheaded, a needless wrapping up of logic and arithmetic in the guise of talk derived from general principles about objects, functions, and relations. Logic and arithmetic could certainly be developed axiomatically on Wittgenstein's view, but such a development would be, at best, just one mode of exposition among many other possible ones, and at worst, a philosophically misleading exposition. It could never demonstrate to us, as an analysis, the true nature of our concepts of *logic*, *number*, and *arithmetic*, or ground our knowledge of mathematics in absolutely certain or absolutely general principles. It would not show any *deep* facts of deductive determination, but would recapitulate in one style the very same expressive grammatical structure that we could look at differently; that, indeed, is what would make its derivations "purely logical." As Wittgenstein sees it, logic "takes care of itself": a proposition of logic (or, in mathematics, an equation) belongs to logic (or to arithmetic) just as it is, in virtue of its figuring in a calculus of operations that we apply to it. A universalist interpretation in terms of an extensional conception of classes or concepts thus distracts from what is to Wittgenstein's mind truly fundamental to arithmetic and logic—namely, our ability iteratively to operate with (to "follow") a recursive rule in connection with internal necessities of language; and it is on this ability that the applications of logic and mathematics fundamentally rest. Frege and Russell gloss over this in eliding the crucial difference between accidental (material) and nonaccidental (operational, formal) generality, the difference between that which the sentences of our language depict, and that which comes built-in with the logical syntax of any depiction.

2. "Middle" Life and Philosophy (1929–1933)

After the First World War, Wittgenstein largely withdrew from academic philosophy. He did speak with Ramsey in the mid-1920s about the foundations of mathematics (about the notions of *identity* and *cardinality* in particular), and by the late 1920s he had decided to return to Cambridge to attempt a further articulation of his views. He continued to write and discuss philosophy until his death. His "middle" or transitional period (1929–1933) was a period of exploration of his earlier views in light of his own and others' reactions.

After 1935, Wittgenstein came explicitly to advocate philosophical investigation of grammar and concepts (i.e., of logic) in which concepts generally, and mathematical concepts in particular, are treated as a "motley," a many-colored, evolving family of notions, notions that lack sharply definable ranges of application (cf. RFM III, §§46ff.). In a general way, his focus shifted from an emphasis on the notion of a *calculus* or *system* to the broader notion of a *language game*, but this shift was made against the backdrop of an underlying continuity in his thinking about the contributions and errors made by Frege and by Russell. During the initial phase of this latter part of his philosophical life, his ideas were constantly evolving under the pressure of his attempts to clarify his *Tractarian* ideas and shift them in response to recent developments in the foundations of mathematics and logic. What he had eventually to surrender was the idea that the notion of an "internal" property or relation (i.e., his unmasking of the idea that arithmetical and logical participles refer to independently given objects and functions) could be accomplished through the display of mechanically effective algorithms or the notion of a purely "internal" or "formal" relation or technique. In the end, however, his aim of thinking through the nature of logic, and the limitations of the universalist conception, always remained central to his work.

Feigl recounts that Wittgenstein returned to philosophy after hearing a lecture by Brouwer in Vienna in 1927, and some have inferred from this that Wittgenstein should be read as a Brouwerian, but this reading seems very doubtful, both in light of Wittgenstein's insistence on the importance of linguistic expression to our understanding of mathematics (his skepticism about a transcendental, solipsistic self), and in light of his denial (in a sense more radical than Brouwer's) that logic consists of *any* laws or principles.[27] In any case no *one* influence may be said to have governed his thinking: he experienced no radical conversion in his overarching conception of philosophy, but thought through his ideas in a new, far more complicated intellectual setting. By the late 1920s and through the early 1930s he was reacting critically to the *Tractatus*'s positivistic, empiricistic appropriation by the Vienna Circle and attempting to come to terms with recent developments in the foundations of mathematics and epistemology: not only Brouwer's intuitionism but also Russell's *Analysis of Mind*, Ramsey's work on the foundations of mathematics, Hilbert's metamathematics and finitism, discussions by Weyl, and Skolem's analysis of quantifier-free arithmetic (cf. here especially WA, PR, PG, WVC). By 1935 he began to try to come to terms with the diagonalization arguments of Cantor, Gödel, and Turing (Wittgenstein and Turing discussed

[27] Three interesting (and contrasting) discussions of Wittgenstein and intuitionism may be found in Fogelin (1968), A.W. Moore (1989), and Marion (1998).

philosophy and logic between 1937 and 1939). Beginning in 1929, when he first returned to Cambridge, conversations with Ramsey and Sraffa stimulated Wittgenstein, as did his reading of Nicod's *Geometry and Induction*; he was also influenced by his reading of Spengler, Frazer's *Golden Bough*, and Freud. In conversations with the Vienna Circle he discussed, among others, Frege, Russell, Husserl, Hilbert, Weyl, Brouwer, and Heidegger. The grammar of the concept of infinity, both within and outside of mathematics, was a central preoccupation from 1929 to 1934.

Given this complexity and ferment, there are many different ways of understanding the development of Wittgenstein's thought. Here we shall content ourselves with a brief description of the ways he developed just a few *Tractarian* ideas about philosophy, logic, and mathematics; in particular, we shall focus on his treatment of *generality* and *necessity*.

Not until he returned to England in 1929, and to sustained conversations with Ramsey, did Wittgenstein begin seriously to investigate whether and how far his *Tractarian* treatment of arithmetic might be adequate. The difficulty came in assessing how this treatment could be interwoven with his *Tractarian* conception of logic without glossing over his sharp distinction between material and operational generality, and without adopting a substantial metaphysical or epistemic position in discussing the nature of logic and its applicability. Wittgenstein had not been explicit in the *Tractatus* about *which* cardinalities he took to be built into the forms of the elementary propositions and, thus, what *sort* of elementary propositions he had in mind. He discussed the natural numbers and left all further discussion to the side. This left his attitude toward real and imaginary numbers—and the application of mathematics in physics—unclear, and thus, especially in connection with the topic of the infinite, Wittgenstein's elliptical uses of the notion of *operation* came to seem to him too coarse and nebulous. The fundamental problem was a clash of analogies at the heart of the *Tractatus*: he had linked his conception of finite cardinality much too tightly to his view of the truth-table as an adequate display of the logical. This, as Russell and Ramsey helped him to see, left crucial conceptual questions open.

In the *Tractatus* Wittgenstein had suggested that the logical impossibility of an object's being both red and green all over at the same time, or both exactly three inches long and exactly four inches long at the same time, would be shown at the atomic level of analysis. Now he saw that he could not hold on to his views about the application of number without linking his conception of the logical to *systems* of propositions. If, in Wittgenstein's manner, we ascribe a cardinal number x to a material concept fx (e.g., through a proposition like "fa & fb"), then it is immediate (shown within the expression of this proposition) that the application of this number (viz., two) rules out as false the ascription of exactly one to the concept. This treatment of number thus pulled in the opposite direction from the image of an ultimately self-sufficient elementary proposition; what was given already in the *application* of the truth-table were *systems* of propositions, according to Wittgenstein's own construal of the number words as operation signs. Imagining real numbers used in the formulation of elementary propositions makes this point vivid: here we have infinite complexity of a certain kind within the elementary proposition, complexity we understand through our understanding of the system of real numbers itself. The yardstick analogy with the proposition pulled in this direction anyway. For if an object is presented, via a material concept, to be exactly one meter long (or green all over), then it is immediate (internal to the proposition's expression) that the object so described is not three meters long, not four meters long, *and so on*. A yardstick, in other words, figures in a *system* of measurement.

Wittgenstein's treatment of cardinal number left it unclear whether he wished to (or could) make room for the transfinite, as Russell noted in his introduction to the *Tractatus*. It was equally unclear how Wittgenstein's replacement for Russell's axiom of infinity was to work. *How* were we to be shown the cardinal number of all objects as *all* of them? If there were a finite number of objects, n, then on Wittgenstein's view it would be nonsensical to try to say that there were $n+1$ objects; but on what ground could we possibly regard this as something purely logical? And if there were an infinite number of objects, then how would *that* be shown in language? No formal rule could set out the names in advance without destroying Wittgenstein's sharp distinction between names and operations signs, a distinction on which his conception of the universal applicability of the truth table depended. But without such a rule, the totality of elementary propositions remained expressively opaque, in principle impossible to list completely. Russell suggested in his introduction to the *Tractatus* that an infinite hierarchy of languages would defeat Wittgenstein's use of the show/say distinction—perhaps the first place in print where the contemporary idea of a metalanguage was broached. Wittgenstein had no way to rule this out formally, given his own discussion of the movement from type to type in Russell and Whitehead's hierarchy. But the picture of a series of languages raised with a vengeance the question of whether and on what basis one could make sense of the unity of language (i.e., the very notion of a speaker's grasp of *a*—his or her *own*—language.

Nearly immediately upon his arrival in Cambridge in 1929, Wittgenstein's exclusive reliance on the model of the independence of elementary propositions fell by the way, precisely in order that he could retain that which was fundamental to his philosophical conception (cf. WA, PR). For some time he attempted to think through what a logical analysis of the "phenomenological" language of first-person experience would look like, focusing in his discussions with the Vienna Circle on the logical status of hypotheses and the holistic systematicity to be found in descriptions of experience (WVC). He accused himself of having erroneously construed the quantifiers in terms of conjunctions and disjunctions. (see G. E. Moore [1954–1955]). He admitted that he had failed clearly to show how one could view the application of the quantifier in pure number theory (cf. PR p. 130).

Wittgenstein did not surrender the show/say distinction; he elaborated and adapted it to his evolving standpoint. Confronted with recent work on the *Entscheidungsproblem*, he retained, but softened, his earlier tendency to connect this distinction with the display of algorithms and the notion of a *calculus*. Conversations with Ramsey in 1929–30, along with his reading of Hilbert, Weyl, and others, made him critical of the vagueness of his earlier appeal to the notion of *operation*. He now explicitly retreated from embracing algorithmicism as a general account of proof in logic and mathematics. Ramsey had obtained an important result (1928) for a partial case of the *Entscheidungsproblem*, and was engaged in trying to develop an "extensional" account of the foundations of mathematics in terms of the notion of *arbitrary function*, conceiving of his own work as a development of the *Tractatus*'s extensionalized standpoint. In sharply differentiating his philosophical outlook from Ramsey's, Wittgenstein came to radicalize his previous tendency to resist a unified account of generality. By 1934 he was explicitly rejecting two ideas that, as he now saw it, the *Tractatus* had, in its vagueness, invited: (1) the idea that mathematics and/or logic have a unified core or nature; (2) the idea that a philosophical understanding of logic or mathematics could rest upon the solution to a leading or single fundamental problem (such as the *Entscheidungsproblem*). The articulation of these rejections became the hallmark of his later discussions of mathematics and logic.

The evolution of Wittgenstein's thought was, however, piecemeal and uneven during this "middle" period. In 1929 he began to speak (as he had not in the *Tractatus*) of the "sense" of a mathematical proposition or equation as something that is shown in its proof (cf. WA, PR). He still sharply distinguished between equations and propositions, but now began to try to do better justice to the distinctive generality at work in mathematical proof. This comes through in his examination of Skolem's quantifier-free treatment of primitive-recursive arithmetic (PG, pp. 397ff.). Wittgenstein still wished to deny that an induction could be conceived on the model of the demonstration of a true general proposition based on general principles. Now he insisted that such a proof shows on its own the "internality" of the connection between universal quantification and instances by constructing a kind of algorithm or template. In other words, the meaning of "all" in an inductive proof is fully expressed in the giving of the argument itself—it is a reflection not of the existence of a further function or rule, but of the grammar of the formal series of natural numbers. On this view it would be nonsense to ask for a general principle of mathematical induction to justify the generalization: once a property has been shown to hold for zero and to hold for the successor of an arbitrarily chosen *k*, in some sense the application of "all" is fully expressed or exhausted.[28] For Skolem to claim that in an inductive argument a general propo-

[28] For a range of slightly different readings of these remarks, see Waismann (1951), Shanker (1987), and Marion (1998).

sition is proved to be true is, as Wittgenstein sees it, just dispensable and potentially misleading "prose," gas surrounding the hard and genuine core of the proof, which consists in nothing but the construction of a diagram for an algorithm. Thus Wittgenstein conceives the recursive proof as a (schematic) picture, not as a proposition: it shows or directs us in the way an algorithm, table, or rule does. In an inductive proof, something is expressed that cannot be satisfactorily accounted for in terms of the notion of an arbitrary function or an explicit second-order principle of logic. Instead, the generality comes out in our applications of the template or schema. We *see* from the proof (the picture) how to go on.[29]

Wittgenstein's resistance to Ramsey's extensionalist conception of (propositional) functions—a major theme of his middle period—also turns on his adaptation of the *Tractarian* conception of showing. The heart of his unwillingness to follow Ramsey's approach to the foundations of mathematics was that he could not see what made the notion of function-in-extension a *logical* notion. Logic unfolded what could be conceived of as given in the use of propositions, true or false. But then a function-in-extension à la Ramsey could not be taken to be a *propositional* function: it lacked the kind of pictorial complexity Wittgenstein associated with propositions. For Ramsey to allow any arbitrary output whatsoever for an input to such a function left us at sea about the applications of this notion within language, within the expression of propositions. From 1929 through 1934 Wittgenstein set his face against Ramsey's extensionalism about the infinite by speaking of what he called the "intensional" character of the concept of *infinity*, by which he meant a conception of the potential infinite as given through our possible applications of a rule, in contrast to a conception of it as a list or abstract object such as a set. From the *Tractatus* Wittgenstein retained the idea of conceiving the indefinitely extendable as a grammatical rule or elliptical expression we apply in language, as opposed to a property, object, or totality, as well as the idea that the notion of *infinite* is not a specifically logical notion; infinity, he argued, is not expressed in a picture or isolated fact. During his early middle period he tried to investigate in far greater detail than he had before different contexts in which the notion of the *infinite* shows its grammatical face. His discussions, though often read as endorsing a straightforward form of verificationism or intensionalism, are perhaps better seen as a link between his earlier and his later philosophy: a further, more detailed effort to reject an overarching or homogenous categorial structure for all of logic and mathematics. By 1935 Wittgenstein came to see his talk of an "intensional" conception of infinity as too vague and coarse, as liable (just as his earlier talk of operations had been) to misconstrual in

[29] This approach to mathematical induction, not unlike Weyl's, held heuristic value for one of Wittgenstein's students, Goodstein, who went on to develop a "logic-free" system of quantifier-free arithmetic in the 1940s. For relevant discussions, see Goodstein (1951) and Marion (1998).

the hands of intuitionists, finitists, and verificationists. He became explicit that revision of mathematical practice in the light of epistemic or empirical constraints on human modes of knowledge was not his aim—thereby more or less admitting that his earlier discussions had invited such misuses. And he also began to zero in on the ways in which his earlier discussions of operations and intensions had rested on far too general and uncritical an appeal to the notion of *following a rule.* While the latter topic was to become an explicit focus of concern by the mid-1930s, Wittgenstein was never to surrender the idea that the theory of classes (the extensional conception of [propositional] function) was "superfluous" for (i.e., parasitic upon) the working applications of mathematics in language, in its role as part of our framework for giving empirical descriptions. For him, set theory was a strange hybrid of traditional logical notions (*concept, extension, proposition*) and purely mathematical notions about whose mathematical applications he always remained confused and about whose purely philosophical applications and metaphysical motivations he always remained doubtful. This was not to reject set theory's results as genuine parts of mathematics, to but insist that these results require detailed scrutiny (cf. PG, pp. 460ff.; RFM II, §§19ff.; V, §§7ff.; VII, §§33ff.).

To rid himself of spurious philosophical appeals to meanings and objects, Wittgenstein had always emphasized the central role of algorithms and calculation in mathematics, the image of mathematical activity epitomized by the solving of equations according to calculation techniques. He saw no reason to give this strategy up, or to try to justify it. But as a picture of mathematics generally, it was misleading. Mathematics consists in more than the construction of algorithms, procedures, and axiomatic systems; it seems to give insight into structure. In particular, the metamathematics of Hilbert seemed to be able to show us how to mathematize certain notions and philosophical questions that had not been mathematized before. In response, Wittgenstein insisted that metamathematics is not a *theory* establishing true, independently grounding principles for otherwise incomprehensible practices. Instead, it gains its point and purpose from its application to our practices of arithmetical calculation. Metamathematics was for Wittgenstein just another branch of mathematics, an extension of it in a new direction, not a more fundamental supertheory (cf. WVC).

In one way, metamathematics appeared to Wittgenstein to be just a sophisticated way of formally picturing or modeling the grammar already in place in arithmetic. In another, it struck him as misleading prose inessential to the working core of everyday mathematical practices such as devising techniques for solving equations, engineering solutions for the construction of machines, reckoning calculations, and applying geometry. The trouble with metamathematics, for Wittgenstein, is that it tends to mislead philosophers into thinking that the metamathematical language gives us a single way of surveying the core, or interpreting the meaning, of apparently fundamental mathematical and logical notions. But ascent to the metalanguage is just another perspective on practices that gain their character within language from their working applications in human life. Such ascent may change our perspective on our own language, but it grows from our current practices, and is parasitic upon them: it cannot make them more epistemologically certain.

3. The Later Philosophy

Throughout his life Wittgenstein mounted many arguments designed to question how far we may legitimately hold that logical or mathematical necessity (or truth) is lodged in the purely deductive implications of prior intellectual commitments (e.g., commitment to the truth of certain general arithmetical principles, to definitions and laws of inference, or to general rules of grammar previously accepted). His point was not to question arithmetic or the formal derivations to be found in Frege's *Grundgesetze* or *Principia Mathematica*, nor was it to insist, in an irrationalist or mystical vein, that its bases are not in any way cognitive—though he did often highlight the *arationality* of the sometimes rote, unreflective aspects of training on which teaching of logic and mathematics depends. His primary focus was on the misleading pictures of language, understanding, and rationality that emerge from naïve interpretations of such accounts of mathematics. Like the logicists, Wittgenstein took the kind of apparently unrevisable, impartial, universally agreed-upon truth of mathematics to arise from the very grammar of our understanding of the terms involved. At the same time he insisted more and more stridently over time on the philosophical relevance of the fact that this grammar arises in tremendously complicated, partly contingent ways that evolve over time within the natural world, depending upon how a particular sentence is learned, taught, applied, and contextualized within larger mathematical and extra-mathematical linguistic contexts.

A central analogy remaining from Wittgenstein's earliest work was between mathematical theories and systems of measurement. An equation such as "$2+2=4$" may be conceived to serve speakers of the language as an operational recipe, a standard or "paradigm" to which certain activities, if they are to count as arithmetical, must conform (RFM, I). As a purely arithmetical tool licensing the substitution of one numerical term for another, the equation serves a role much like that of a conversion ruler showing how to switch between inches and meters; in its applied role, it is used like a ruler for measuring particular episodes of calculating and counting, giving us something like a rule or model of grammar rather than a particular belief. If one counts two apples and another two apples, without overlap, then there *must be* four apples counted, and if there are not four on a subsequent count, we say either that there has been an error in counting or that a surprising physical event has occurred. The equation's "truth," if we wish to speak this way, holds as much in virtue of our own, contingently evolved commitments to certain methods of representation and ongoing communal practices and needs as it does in virtue of the nature of things. Like a system of measurement, mathematics (like logic) is for Wittgenstein a complex human artifact, situated and created in and for an evolving natural world, and its claims to objectivity and applicability ultimately turn upon our human ability to find one another in sufficiently constant agreement about results of its application to make the practice prove its worth (RFM VI, §§39ff.).

If we insist on speaking of *discovery* in mathematics, Wittgenstein suggests that we ought to allow ourselves to think in terms of an analogy with technological discoveries, which are perhaps better conceived of as *inventions*, like the steam engine or the wheel or the decimal notation or the computer (RFM I, §§168ff.; II). As we have seen, algorithms and fixed procedures of calculation were for Wittgenstein—as they are in any system of measurement—of central importance here, and he always conceived tables, algebraic representations of constructive procedures, algorithms, and calculations to play a central role in mathematics.[30] As time went on, he relied more explicitly on his looser analogy between applied mathematical structures (systems of numerals, geometrical diagrams, equations, episodes of counting, and proofs) and pictures or models (*Bilder*). He explicitly regarded proofs and episodes of establishing equinumerosity as conceptual paradigms, models giving us ways of synoptically expressing reproducible routines for describing and/or constructing empirical events. Insofar as these routines allow us to take in, understand, and make sense of empirical situations, they are to be counted as part of the logical syntax or "grammar" of our language; they fix concepts and their applications precisely by making them synoptically surveyable. Such artifacts are, like systems of measurement, as much invented as discovered, because to understand them, as to understand any human artifact, we must be able to understand their uses in concrete situations we encounter and wish to describe.

The use of such pictures in communicating and establishing mathematical conviction is tied to empirically conditioned perceptual and intuitive aspects of our nature as human beings, aspects that shape their qualities of design. A theorem that may be communicated with an easily surveyable model gives us a kind of understanding and insight (an ability to "take it in," to appreciate and communicate

[30] Marion (1998) argues that Wittgenstein was an *algorithmicist* about mathematics, the kind of finitist or constructivist who sees the construction of algorithms as the core of mathematics. In light of Wittgenstein's later discussions of proof and rule-following, this is not, I believe, all there is to his philosophy of mathematics (see Floyd [2002]), but there is no doubt that he insisted on construing calculation as something extremely important and central to mathematics—something not every mathematician or philosopher does.

the result) that a long, intricate, formalized version of the proof does not. That is not to say that there are no uses for formalized proofs (e.g., in running computer programs). It is to say that a central part of the challenge of presenting proofs in mathematics involves synoptic designs and models, the kind of manner of organizing concepts and phenomena that is evinced in elementary arguments by diagram. This Wittgenstein treated as undercutting the force of logicism as an epistemically reductive philosophy of mathematics: the fact that the derivation of even an elementary arithmetical equation (like $7 + 5 = 12$) would be unperspicuous and unwieldy in *Principia Mathematica* shows that arithmetic has not been reduced in its essence to the system of *Principia* (cf. RFM III, §§25,45–46).

Of course, our ability to use a picture or diagram to express and/or communicate a mathematical principle or concept depends upon our understanding of how to apply it. A central difficulty Wittgenstein raises here is, the question of what is to count as an application (correct or incorrect) of a logical or mathematical rule, concept, or procedure, construed as such a model or picture. Wittgenstein is to be credited with having tied this question to more wide-ranging ones about the nature of concept-possession. In his later work he often drew an analogy between the ability to project a noun or adjective into new contexts and the computation of instances of an elementary arithmetical function, or the drawing of a purely logical inference (cf. RFM I; PI §§186ff.). This was intended to soften his earlier *Tractatus* distinction between material and formal generality. Now he emphasized that the difference between the mathematical and the empirical (or the psychological) was not given by the grammatical structure of sentences: the very same sentence might in one way be viewed as empirical, and in another way as merely conceptual or mathematical (cf. RFM VII, §§20ff.). A fortiori mathematics and logic were to be seen to have no monolithic core, just as—and indeed because—our notion of *that which belongs to language* does not. All we have to rely on in expressing the general applicability of a model, an episode of empirical counting, or a diagram is our language, broadly construed. And yet no image, picture, sentential form, or model applies *itself.* Not any arbitrary application or use one might make of a sentence or diagram can be held to be fitting or appropriate to it if there is to be any coherent notion of *applying* it. Yet how deviant from the communal norm can an application be before we say that it is no longer an *application*? The question is just as difficult (and perhaps just as unanswerable) as the question of when the use of an artifact—such as a knife—ceases to be a use of the artifact as the artifact that it is. One of Wittgenstein's main points, throughout his discussions of rule-following, is to show how an absolute notion of logical *necessity*, treated without qualification, is as much a will-o'-the-wisp as an absolute or general notion of *rule.*[31]

[31] And hence, that no notion of *intention* is likely to be able to determine the applications of these notions. Compare Floyd [1991].

Indeed, in drawing analogies between equations, proofs, numerals, and concepts and rules of grammar, Wittgenstein was suggesting that our cognitive verbs such as "know" and "believe" ultimately work against the grain of our usual traffic in logical and mathematical statements (cf. RFM I, §§106ff.). For instance, to say that a person *believes* that $2+2=4$ appears to leave open the possibility that the sentence "$2+2=4$" might not be true, suggesting that one could understand the equation quite apart from agreeing with it. Wittgenstein's view, in depicting the equation as a rule of language, is that understanding and assent work so closely together in the case of this kind of accepted logical and mathematical sentence that one cannot step outside one's own understanding of the notions and terms at work in them to prescind from one's commitments within our language when considering the equation's truth. To understand "$2+2=4$" in the way we expect an adult speaker of our language to understand it is to be unwilling to grant the sense of circumstances in which this sentence might turn out to be false. To suppose that its truth is a matter of very well confirmed or perhaps even indubitable professional opinion, or that only a proof from more general principles can, ideally, bring about belief or certainty in its truth, is to fail to give this conceptual aspect its proper place.

Proof is thus not sufficiently understood, according to Wittgenstein, by maintaining that it yields firm conviction or certain truth dependent upon knowledge of derivations from prior truths, for it equally involves acceptance of a sentence (or proof) as a standard in one's own language of what does and does not make sense. Proof, on this view, is not a series of sentences meeting purely formal deductive requirements, but an activity of achievement and acceptance constitutively shaping one's conceptions of language and the world, sustained in complicated ways by psychology, natural regularities, and communal practice. To appreciate the necessity and universal applicability of an equation like "$2+2=4$," then, Wittgenstein asked his readers to investigate the role it plays in shaping what does and does not make sense to us, our very notion of understanding.

Wittgenstein tries to display the importance of this idea in his later writings by sketching simplified language games or imagined forms of life—often usefully pictured as those that might be played with children—in order to illustrate how much varied, active cultural training, both rote and reflective, must be in place for human objectivity in mathematics to take the forms that it does. This brings out the relativity to particular practice of such apparently fundamental notions as *object, name, reference, number, proof,* and so on. There is in fact a great variety of empirical, contingent human factors—historical, aesthetic, anthropological, pedagogical, psychological, and physiological—on which the evolution and interest of mathematical objectivity depend. Wittgenstein's point is an anti-reductive one, though it is easy to slide from his remarks to the notion that he was offering a psychologistic, radically empiricistic, purely conventionalist, ethnomethodological or relativistic account of mathematics: indeed, Wittgenstein is often hailed as a hero by social constructivists.[32] His primary aim was, however, to diffuse (without refuting) assumptions about knowledge shared by various forms of skepticism and those anti-skeptics (such as Frege, Russell, and Gödel) who hold that a universally applicable, eternal framework of contents, thoughts, or senses, possibly transcending the reach of humanity's current concepts, must be postulated to account for mathematical objectivity and truth.

Wittgenstein came to treat mathematics and logic as a "motley," then, partly because he came to emphasize the importance to philosophy of the idea that there is no single property, criterion, mental state, fact, or characteristic feature common to all cases of the expression of understanding. Indeed, it is part and parcel of his later investigations of logic to make this image of our concept of *understanding* plausible. Understanding is manifested in many different kinds of ways, even in the case of a particular linguistic form such as "$2+2=4$": sometimes in a characteristic experience of the moment ("Aha!! Now I get it!!!"); sometimes in the ability to pronounce it unhesitatingly (as in the memorization of multiplication tables by children); sometimes in the ability to apply it to solve minimally difficult word problems; sometimes in the ability to set it into a more general, systematic context of definitions and proofs; sometimes in the ability to successfully teach, communicate, and/or defend it to another. Wittgenstein suggested, famously, that the notion of *understanding*—along with such notions as *proof, truth, meaning, mathematics, number, language*, and *game*—has a "family resemblance" character, in that, though no one characteristic might belong to every instance of the general notion, overlapping similarities from case to case would suffice to tie the instances together. His discussion was intended to explore the extent to which the notion of *full* or *complete understanding* of a sentence is itself a potentially misleading idealization, especially for the logician.[33] In this way he retained his early willingness to question the relevance to philosophy of the idea that to every predicate or concept of the language we may associate in the same way a function or an extension (i.e., a sharply defined property or concept-word). He also retained his idea that knowledge and/or understanding ought never to be conceived of as a relation between a person and a proposition without regard to the particular context of utterance or the particular sentence affirmed (cf. RFM I, appendix III).

A major difficulty here is, of course, how to do justice to the intuitive notion of mathematical and logical *truth*, for mathematics is not *just* a game. Wittgenstein's resistance to any doctrine of contents or knowledge that pretends to universal validity, coupled with his insistence that the applications of mathematics are internal to its character, led him to deny that there is a substantial (i.e., more than merely formal or family resemblance) characterization of the notion of *mathematical* (or even just *arithmetical*) *truth*. His conception aimed to deny that

[32] See, for example, works by Bloor.

[33] For an especially lucid article on this idea, see Goldfarb (1992).

the objectivity, sense, and applicability of the most basic notions of logic and mathematics (e.g., *concept, proposition, elementary arithmetical truth, proof, number*, and so on) may be explained by setting forth an axiomatized theory in which truths involving the said notions are explicitly derived from fundamental principles. From this perspective it is not surprising that by the mid-1930s Wittgenstein became fascinated with understanding the fundamental basis of Cantor's and Gödel's limitative arguments, as well as Turing's analysis of computability (cf. RFM I, appendix III; RFM II; RFM VII). Of course, his resistance to making a definition of *truth* central to philosophy was, as we have seen, based on purely philosophical considerations that predated by over a decade his encounter with Gödel's work; in general historical terms it was a hallmark of the Kantian tradition—shared by Frege, among others—that philosophers could learn nothing important from analyzing or defining the concept of *truth*.[34] But Gödel's incompleteness theorems forced Wittgenstein to rethink the role of the notion of *truth* in mathematics.

Wittgenstein's emphasis on the image of the mathematician as *inventor* or fashioner of models, pictures, and concepts was, in the main, directed at the philosophical talk of those, like Hardy and Russell, who insisted on speaking of mathematical reality in a freestanding way, picturing the logician or mathematician as a zoologist embarked on an expedition to new, hitherto unseen lands, analogous to an empirical scientist. Wittgenstein himself explicitly said he did not wish to *deny* that there is a "mathematical reality."[35] But on his view the Hardy–Russell picture of truth tended to preempt as irrelevant to mathematics its evolution as a language, the importance to it of problems of expression, intention, formulation, and construction. For Wittgenstein the mathematician is an inventor, not in the sense of making up truth willy-nilly as he or she goes along, as a pure conventionalist would suppose,[36] but in the sense of engaging in the activities of fashioning proofs, diagrams, notations, routines, or algorithms that allow us to *see* and *accept* (understand, apply) results as answering to what does and does not make sense to us. We "make" mathematics in the sense that we

[34] Cf. Kant on truth at (1781/1787, A58/B82). Frege's argument that truth is a primitive undefinable notion, and not a genuine property word, may be found in Frege (1918). This was an essay that Wittgenstein always disliked, as we now know from the Frege correspondence with him (1989), but Wittgenstein seems to have objected to Frege's handling of idealism, not his treatment of truth (cf. Floyd [1998]). For two useful papers on the role of the concept of *truth* in this tradition, see Ricketts (1996a) and Diamond (2002). So far as we know, Wittgenstein never discussed Tarski's theorem on the undefinability of truth. Ramsey and Turing appear to have agreed with him that the concept of truth was not of central importance.

[35] See, e.g., LFM, pp. 136–141.

[36] For a conventionalist reading of Wittgenstein, see Dummett (1959); compare the discussions of Dummett in Stroud (1965) and in Wright (1980).

make history: as actors within it (cf. WVC, p. 34, n. 1). The "what" in "what makes sense to us" is then evolving, to be understood locally for present purposes, and not in terms of any theory of content or a priori known domain of fact. Wittgenstein still took philosophical considerations ultimately to rest on terms and structures of the language we take ourselves to be speaking right now, but in the end he emphasized the variety of perspectives we may bring to bear on our understanding of language's internal necessities and requirements. That which "belongs to language" is something open for current investigation, not something to be taken as determined, stipulated, or fully circumscribable in advance.

BIBLIOGRAPHY

Bibliographic Essay

The bibliography that follows contains a representative, though by no means exhaustive, sampling of literature relevant to Wittgenstein's philosophy of logic and mathematics. The following remarks are intended to provide an overview of themes treated in recent literature for the interested reader.

Those who wish to learn the biographical details of Wittgenstein's life are advised to consult McGuinness (1988) and Monk (1990), as well as the brief Malcolm (2001).

New Logics

Wittgenstein's philosophy has helped to inspire the construction of new logics that serve to question the philosophical hegemony of first-order logic. Hintikka's development of systems of modal logic, the logic of knowledge, and game-theoretic semantics was inspired in part by the *Tractatus*, and in part by Wittgenstein's notion of a language-game (see Hintikka (1956, 1962, 1973, 1996a, 1996b, 1997). Parikh (1985, 2001) has also stressed connections between logic and games in his formal treatments of logics of knowledge and information. For the development of an inconsistency theory of truth and paraconsistent logic that is relevant to Wittgensteinian questions about whether an inconsistency in arithmetic can be logically contemplated, see Priest (1987, 1994). Tennant (1987) develops a constructive logicist foundation for number theory based on intuitionistic relevance logic, describing his system as a generalization from Wittgenstein. Sundholm

(1994) subsumes an intuitionistic view of truth as existence of proof under a truthmaker analysis, situating his analysis explicitly in relation to logical applications of Wittgenstein's *Tractarian* ideas. Shapiro (1991) makes an interesting application of the rule-following considerations to argue for the expressive superiority of second-order logic over first-order logic.

Rule-Following

The centrality of what are now called "rule-following" considerations to the interpretation of Wittgenstein was initially emphasized by Fogelin—see Fogelin (1987), Wright (1980), and Kripke (1982)—building on Wittgenstein's later writings exploring the analogy between the application of an arithmetical principle and concept-projection; each formed a response to the view, voiced in Dummett (1959), that Wittgenstein was a "full-blooded" conventionalist about logical necessity. Different responses to Dummett may be found in Stroud (1965) and also in Diamond (1989, 1991), who is simultaneously reacting to Wright. A range of differing reactions to Kripke's exegesis of Wittgenstein may be found in Baker and Hacker (1984), Goldfarb (1985), Cavell (1990), Floyd (1991), Minar (1991, 1994), Steiner (1996), and Wright (2001, ch. 7); Holtzman and Leich (1981) is a useful anthology exploring general philosophical implications of this theme. More focused discussions of the implications of these considerations for the notion of *content* and the ascription of meaning via linguistic behavior may be found in Boghossian (1986) and Chomsky (1986). Proudfoot and Copeland (1994) and Shanker (1998) have connected their readings of Wittgenstein on rules with questions about Church's thesis, Turing's mechanism about the mind, and the foundations of cognitive science. "Rule-following" considerations have also been brought to bear on the presumption (e.g., by Quine and Putnam) that the Skolem–Löwenheim theorem can establish inscrutability of reference and/or "internal" realism: on this, see Hale and Wright (1997), Benacerraf (1998), and Wright (2001, part IV), as well as Parikh (2001), which connects these questions with Searle's Chinese Room thought experiment and the theoretical foundations of computer science.

Wittgenstein's Place in the History of Analytic Philosophy

A more detailed historical approach to Wittgenstein's philosophy has characterized much recent work on his philosophy of mathematics, sparked by a growing interest in developing an enriched account of the origins and development of early analytic philosophy in relation to the history of logic. Baker (1988), Coffa (1991), Simons (1992), Friedman (1997), and Hacker (1986, 1996) offer readings of early Wittgenstein through the broad lens of a larger philosophical history of neo-Kantianism, realism, and logical positivism; Potter (2000) offers an especially concise technical reconstruction of the *Tractatus* treatment of logic and number in light of the history of philosophies of arithmetic from Kant to Carnap. Dummett (1991b) examines Wittgenstein's relation to Frege in connection with the theory of meaning. Marion (1998) places Wittgenstein into context against the background of the history of finitism as a mathematical tradition, contrasting Wittgenstein with Brouwer and drawing connections between Wittgenstein's middle-period views on generality and remarks by Weyl and Hilbert. Aside from this work, the only other recent book-length treatments of Wittgenstein's philosophy of mathematics in its historical development, from early through later philosophy, are Shanker (1987) and Frascolla (1994); Shanker (ed., 1986) is a useful anthology of articles covering all phases of Wittgenstein's philosophy of mathematics. Shanker rejects both Kripke's rule-following skepticism and Dummett's use of the realist/anti-realist distinction to interpret Wittgenstein on mathematics, elaborating an alternative reconstruction inspired by the grammatical conceptualist reading of Baker and Hacker; he explores in detail Wittgenstein's conception of proof and his reactions to Skolem and Hilbert, and also explores the significance of computer proofs for Wittgenstein's point of view. Frascolla offers the most detailed formal reconstruction yet of Wittgenstein's *Tractatus* account of arithmetic—cf. Frascolla (1997, 1998) and Wrigley (1998) for a rejoinder; and, like Marion and Shanker, questions the attribution to Wittgenstein of an epistemic perspective that could support strict finitism.

Van Heijenoort (1967) is the locus classicus laying out the "universalist" conception of logic, attributing this view to Frege, Russell, and the early Wittgenstein and contrasting it with the algebraic tradition of "logic as calculus"; this article has profoundly affected how the *Tractatus* show/say distinction has been read, along with the history of the quantifier and the notion of *completeness* (cf. Goldfarb [1979, 2000]; Dreben and van Heijenoort [1986]). The extent of this conception's accuracy to Frege has been disputed by, among others, Dummett (1984); for a discussion of the debate, see Floyd (1996). Van Heijenoort's reading remains especially influential among those interested in the history of early analytic philosophy, and in Wittgenstein in particular. Among others, Hintikka and Hintikka (1986), Weiner (1990, 2001), Ricketts (1985, 1986a, 1986b), and Diamond (2002) have deepened and generalized Van Heijenoort's analysis in interpreting Wittgenstein's relation to Frege. Reck (2002) and Crary and Read (2000) contain further essays interpreting Wittgenstein's philosophy of logic and mathematics in the historical context of early analytic philosophy.

Boghossian and Peacocke (2000) contains several essays touching on Wittgenstein's philosophy of mathematics in relation to the evolution of twentieth-century conceptions of a priori knowledge as they shape contemporary discussion.

Wittgenstein and Gödel

An abiding interest in locating Wittgenstein's philosophy in relation to Gödel began with nearly unanimous condemnation of Wittgenstein: reviews of the earliest published excerpts from his writings on Gödel were roundly dismissed, and skepticism about his remarks on Gödel still remains (see Hintikka [1993]). Klenk (1976) offered an early attempt to defend the cogency of some of Wittgenstein's remarks. In recent years there has been a revisiting of the topic, both philosophically and historically, in light of newly detailed historical questions about the evolution of Wittgenstein's appreciation of the incompleteness theorems and the significance of diagonalization methods. Assessment of the nature and implications of Wittgenstein's attitude toward Gödel's incompleteness theorems has been taken up by Shanker (1988), Floyd (1995, 2001b), Rodych (1999, 2002), Steiner (2000), and Floyd and Putnam (2000), following earlier discussions by Goodstein (1957, 1972) and Wang (1987b, 1991, 1993); Kreisel (1983, 1989, 1998) has also elaborated thoughts on the topic. Webb (1980) and Gefwert (1998) situate remarks of the middle Wittgenstein within the larger history of formalism, mechanism, the *Entscheidungsproblem*, and the effect of Gödel's and Turing's work on Hilbert's program. Maddy (1997) contains a useful discussion of Wittgenstein and Gödel's attitudes toward the philosophy of mathematics. Friedman (1997, 2001) probes Carnap's attitude toward the relevance of Gödel's arithmetization of syntax to Wittgenstein's *Tractatus*. Tymoczko (1984) invokes Gödel's epistemological remarks on the incompleteness theorems to argue for a socially based "quasi-empiricism" he associates with Wittgenstein.

Recent Trends in Interpreting Wittgenstein: New Approaches to the Philosophy of Mathematics

General Themes. Large-scale shifts in the interpretation of Wittgenstein's general philosophy since the mid-1980s have affected and been affected by accounts of his philosophy of logic and mathematics. The presentations of Wittgenstein in Wright (1980) and Kripke (1982) radicalized Dummett's (1959) view of Wittgenstein as a conventionalist, reading him as offering a form of skepticism about the rational necessity of applying a rule. Among those who questioned Dummett's reading while rejecting the attribution to Wittgenstein of skepticism were Stroud (1965) and, beginning in the 1970s, Diamond (see [1989, 1991]); she fashioned a rival account of Wittgenstein's philosophy of logic and mathematics in the context of a wider recasting of discussions of realism in ethics and a stress on the continuity of

Wittgenstein's development (compare Gerrard [1991]). Diamond's approach to Wittgenstein has been especially influential in the United States, along with work of Cavell (e.g., [1976, 1979]), which independently challenged both the then-popular broadly empiricistic reading of Wittgenstein (see, e.g., Pears [1986, 1987, 1988]) and the reading of Wittgenstein's conception of grammar due to Baker and Hacker (e.g., their [1980], [1984], [1985]). Diamond (1999) and Gerrard (1999) reject traditional readings of Wittgenstein as a verificationist and explore Wittgenstein's treatment of the notion of *proof* in connection with his later discussions of the notion of *meaning*. Readers influenced by Cavell and Diamond have inspired recent heated interpretive debate about a "New Wittgensteinian" approach (see Crary and Read [2000]); this kind of reading has been applied to Wittgenstein's later discussions of mathematics by Floyd (1991), Conant (1997), and Putnam (1996). Questions surrounding this "new" interpretation in relation to the *Tractatus* conceptions of analysis and realism have been pursued by Goldfarb (1997), Diamond (1997), McGinn (1999), Floyd (1998), Proops (2001), Ostrow (2002), and Sullivan (2002a); Floyd (2001) and Kremer (2002) examine its implications for our understanding of the *Tractatus* treatment of arithmetic. Wright (1993) sets the verifiability principle into conceptual context against the backdrop of different strands of Wittgenstein's philosophy of mathematics, and outlines in (2001) challenges he takes to be faced by the "noncognitivist" approach to mathematical proof he associates with the later Wittgenstein.

Constructivism. The grounds on which Wittgenstein might be taken to have held a substantial constructivist attitude toward mathematics—despite his claim not to be forwarding "theses" in philosophy—have received substantial investigation in recent years. Marion (1995), Sahlin (1994), Wrigley (1995), and especially Sullivan (1994) analyze in useful detail Wittgenstein's debates with Ramsey in the late 1920s with an eye toward understanding the character of Wittgenstein's interest in the infinite; Noë (1994), Hintikka (1996a), and Marion (1998) offer general philosophical interpretations of Wittgenstein's transitional period writings on the philosophy of mathematics. Mancosu and Marion (2002) examine Wittgenstein's effort to "constructivize" Euler's proof of the infinity of primes in light of the immediate historical context of constructivist discussions in Vienna in the late 1920s and early 1930s.

Finitism. Kreisel's (1958) and Wang's (1958) association of Wittgenstein with "strict finitism" or "anthropologism" about mathematics spawned in the hands of Dummett (1970), a "finitist" reading of Wittgenstein that has received scrutiny and detailed development since the mid-1980s. Kielkopf (1970) associates the term "strict finitism" with a view of mathematical necessities as a posteriori, but more frequently the term is associated, as in Dummett (1970), with the form of constructivism that rejects constructions that outstrip what is too complex or too lengthy for an individual or community to actually carry out in practice—a restriction on methods of proof more severe than any suggested by the intuitionists. It is unclear what precise form such a strict finitism should ideally take; Wright (1982) defends the program's internal coherence. Wang (1958) associated the notion of human "feasibility" with Wittgenstein's remarks on the surveyability of proof, suggesting that an anti-reductionist investigation of complexity would be a fruitful foundational approach. He also emphasized connections between Wittgenstein's conception of the surveyability of proof and his critical comments about *Principia*'s "reduction" of arithmetic to logic. Wright (1980), Steiner (1975), and Marion (1998) discuss the character and validity of these criticisms of logicism from three rather different points of view.

Despite the continued association of Wittgenstein with finitism, as well as his critical attitude toward set theory, there are some (e.g., A.W. Moore [1990] and Kanamori [2003]) who have suggested that Wittgenstein's grammatical investigations of infinity and the limits of sense may be the best kind of philosophical approach to a proper conception of set theory and the higher infinite.

Social Constructivism. Wittgenstein's emphasis on the social aspects of language use have inspired some to develop his ideas into a social constructivist philosophy of mathematics: Bloor (1983, 1991) and Ernest (1998) offer recent accounts with a strongly sociological orientation that take up the relation of this approach to philosophy of mathematics education. Tymoczko ([1993]; ed., [1986]) are also relevant, rejecting foundationalist programs and calling for a view of mathematical practice as a community-based art rather than a science.

Abbreviations of Wittgenstein Works, Lecture Notes, and Transcriptions

CL	*Cambridge Letters* (1995)
LFM	*Wittgenstein's Lectures on the Foundations of Mathematics, Cambridge 1939* (1989)
MN	"Notes Dicated to G.E. Moore in Norway, April 1914," appendix II in NB (in 1979a)
NB	*Notebooks 1914–1916* (1979a)
PG	*Philosophical Grammar* (1974)
PI	*Philosophical Investigations*, 2nd ed. (1958a)
PR	*Philosophical Remarks* (1975)
RFM	*Remarks on the Foundations of Mathematics*, rev. ed. (1978)
SRLF	"Some Remarks on Logical Form" (1929)
TLP	*Tractatus Logico-Philosophicus* (1921)

WA *Wiener Ausgabe* (1993–)
WVC *Ludwig Wittgenstein and the Vienna Circle* (1973)

Ambrose, A., and Lazerowitz, M., eds. 1972. *Ludwig Wittgenstein, Philosophy and Language*, New York, Humanities Press.

Baker, G.P. 1988. *Wittgenstein, Frege and the Vienna Circle*, Oxford, Blackwell.

Baker, G.P., and Hacker, P.M.S. 1980. *Wittgenstein: Understanding and Meaning. An Analytical Commentary on the Philosophical Investigations*, vol. 1, Oxford, Blackwell; Chicago, University of Chicago Press.

———. 1984. *Scepticism, Rules and Language*, Oxford, Blackwell.

———. 1985. *Wittgenstein: Rules, Grammar and Necessity. An Analytical Commentary on the Philosophical Investigations*, vol. 2, Oxford, Blackwell.

Benacerraf, P. 1998. "What Mathematical Truth Could Not Be—I," in M. Schirn, ed., 1998, pp. 33–75.

Bernays, P. 1959. "Comments on Ludwig Wittgenstein's *Remarks on the Foundations of Mathematics*," *Ratio* 2: 1–22.

Bloor, D. 1983. *Wittgenstein: A Social Theory of Knowledge*, London, Macmillan.

———. 1991. *Knowledge and Social Imagery*, 2nd rev. ed., Chicago, University of Chicago Press.

Boghossian, P. 1986. "The Rule-Following Considerations," *Mind* 93: 507–549.

Boghossian, P., and Peacocke, C., eds. 2000. *Essays on the A Priori*, Oxford, Oxford University Press.

Boolos, G. 1998. *Logic, Logic and Logic*, Cambridge, MA, Harvard University Press.

Cavell, S. 1976. *Must We Mean What We Say? A Book of Essays*, Cambridge, Cambridge University Press.

———. 1979. *The Claim of Reason: Wittgenstein, Skepticism, Morality and Tragedy*, New York, Oxford University Press.

———. 1990. *Conditions Handsome and Unhandsome: The Constitution of Emersonian Perfectionism*, Chicago, University of Chicago Press.

Chomsky, N. 1986. *Knowledge of Language*, New York, Praeger.

Coffa, A. 1991. *The Semantic Tradition from Kant to Carnap: To the Vienna Station*, Cambridge, Cambridge University Press.

Conant, J. 1997. "On Wittgenstein's Philosophy of Mathematics," *Proceedings of the Aristotelian Society* 97 (2): 195–222.

Crary, A., and Read, R., eds. 2000. *The New Wittgenstein*, New York, Routledge.

Diamond, C. 1989. "Rules: Looking in the Right Place," in *Wittgenstein, Attention to Particulars: Essays in Honour of Rush Rhees*, ed. D.Z. Phillips and P. Winch, London, Macmillan, pp. 12–34.

———. 1991. *The Realistic Spirit: Wittgenstein, Philosophy and the Mind*, Cambridge, MA: MIT Press, pp. 226–260.

———. 1997. "Realism and Resolution," *Journal of Philosophical Research* 22: 75–86.

———. 1999. "How Old Are These Bones? Putnam, Wittgenstein and Verification," *Proceedings of the Aristotelian Society* supp. 73: 99–134.

———. 2000. "Does Bismarck Have a Beetle in His Box? The Private Language Argument in the *Tractatus*," in Crary and Read, eds., pp. 262–292.

———. 2002. "Truth Before Tarski: After Sluga, After Ricketts, After Geach, After Goldfarb, Hylton, Floyd, and Van Heijenoort," in Reck, ed., pp. 252–283.

Dreben, B., and Floyd, J. 1991. "Tautology: How Not to Use a Word," *Synthèse* 87(1): 23–50.

Dreben, B., and van Heijenoort, J. 1986. "Introductory note to 1929, 1930 and 1930a," in *Gödel 1986*, pp. 44–59.

Dummett, M. 1959. "Wittgenstein's Philosophy of Mathematics," *Philosophical Review* 68: 324–348; reprinted in Dummett, 1978, pp. 166–185.

———. 1970. "Wang's Paradox," *Synthèse* 30: 301–324, reprinted in Dummett, 1978, pp. 248–268.

———. 1978. *Truth and Other Enigmas*, Cambridge, MA, Harvard University Press.

———. 1981. "Frege and Wittgenstein," in Dummett, 1991b, pp. 237–248.

———. 1984. "An Unsuccessful Dig: Critical Notice of G. Baker and P. Hacker, *Logical Excavations*," in Wright, ed., *Frege: Tradition and Influence*, pp. 194–226.

———. 1987. "Review of *Kurt Gödel Collected Works*," *Mind* 96: 570–575.

———. 1991a. *Frege: Philosophy of Mathematics*, Cambridge, MA, Harvard University Press.

———. 1991b. *Frege and Other Philosophers*, Oxford, Clarendon Press.

Ernest, P. 1998. *Social Constructivism as a Philosophy of Mathematics*, Albany, State University of New York Press.

Floyd, J. 1991. "Wittgenstein on 2,2,2...: On the Opening of *Remarks on the Foundations of Mathematics*," *Synthèse* 87(1): 143–180.

———. 1995. "On Saying What You Really Want to Say: Wittgenstein, Gödel and the Trisection of the Angle," in Hintikka, ed., pp. 373–426.

———. 1996. "Frege, Semantics and the Double-Definition Stroke," in *The Story of Analytic Philosophy: Plot and Heroes*, ed. Anat Biletzki and Anat Matar, London, Routledge, pp. 141–166.

———. 1998. "The Uncaptive Eye: Solipsism in Wittgenstein's *Tractatus*," in *Loneliness*, ed. Leroy S. Rouner, Notre Dame, IN: University of Notre Dame Press, pp. 79–108.

———. 2001a. "Number and Ascriptions of Number in Wittgenstein's *Tractatus*," in Floyd and Shieh, eds., pp. 145–191; reprinted in Reck, ed., 2002, pp. 308–352.

———. 2001b. "Prose Versus Proof: Wittgenstein on Gödel, Tarski and Truth," *Philosophia Mathematica* 3rd ser., 9: 901–928.

———. 2002. "Critical Study of Mathieu Marion, *Wittgenstein, Finitism, and the Philosophy of Mathematics*," *Philosophia Mathematica* 10(1): 67–88.

Floyd, J., and Putnam, H. 2000. "A Note on Wittgenstein's 'Notorious Paragraph' About the Gödel Theorem," *Journal of Philosophy* 45(11): 624–632.

Floyd, J., and Shieh, S., eds. 2001. *Future Pasts: The Analytic Tradition in Twentieth-Century Philosophy*, Oxford, Oxford University Press.

Fogelin, R. 1968. "Wittgenstein and Intuitionism," *American Philosophical Quarterly* 5: 267–274, reprinted in Shanker, ed., 1986, pp. 228–241.

———. 1982. "Wittgenstein's Operator *N*," *Analysis* 42(3): 124–128.

———. 1983. "Wittgenstein on Identity," *Synthèse* 56: 141–154.

———. 1987. *Wittgenstein*, 2nd ed., New York, Routledge & Kegan Paul. (1st ed. 1976).

Frascolla, P. 1994. *Wittgenstein's Philosophy of Mathematics*, New York, Routledge.

———. 1997. "The *Tractatus* System of Arithmetic," *Synthèse* 112: 353–378.

———. 1998. "The Early Wittgenstein's Logicism. Rejoinder to M. Wrigley," *Acta Analytica* 21: 133–137.

Frege, G. 1893. *Grundgesetze der Arithmetik*, vol. 1, Jena, H. Pohle; reprinted with vol. 2, Hildesheim: Georg Olms, 1966; partially translated into English (through §52) by M. Furth as *The Basic Laws of Arithmetic*, Berkeley, University of California Press, 1964.

———. 1903. *Grundgesetze der Arithmetik*, vol. 2, Jena, H. Pohle.

———. 1918. "Der Gedanke," *Beiträge zur Philosophie des deutschen Idealismus* 1: 58–77, translated in Frege, 1984.

———. 1984. *Collected Papers on Mathematics, Logic, and Philosophy*, ed. B. McGuinness, trans. M. Black et al., Oxford, Blackwell.

———. 1989. "Briefe an Ludwig Wittgenstein aus den Jahren 1914–1920," ed. A. Janik, in McGuinness and Haller eds., pp. 5–83; English trans. by J. Floyd and B. Dreben forthcoming in J. Hintikka and E. de Pelligrin, eds., *In Memory of Professor Georg Henrik von Wright*, Dordrecht, Kluwer.

Friedman, M. 1997. "Carnap and Wittgenstein's *Tractatus*," in Tait, ed., pp. 213–248; reprinted in Friedman, 1999, pp. 177–197.

———. 1999. *Reconsidering Logical Positivism*, Cambridge, Cambridge University Press.

———. 2001. "Tolerance and Analyticity in Carnap's Philosophy of Mathematics," in Floyd and Shieh, eds., pp. 223–256; reprinted in Friedman, 1999, pp. 198–234.

Geach, P.T. 1981. "Wittgenstein's Operator *N*," *Analysis* 41(4): 168–171.

Gefwert, C. 1998. *Wittgenstein on Mathematics, Minds and Mental Machines*, Brookfield, VT, Ashgate.

Gerrard, S. 1991. "Wittgenstein's Philosophies of Mathematics," *Synthèse* 87(1): 125–142.

———. 1996. "A Philosophy of Mathematics Between Two Camps," in Sluga and Stern, eds., pp. 171–197.

———. 1999. "Reply to Cora Diamond's 'How Old Are These Bones? Putnam, Wittgenstein and Verification,'" *Proceedings of the Aristotelian Society* supp. 73: 135–150.

Gödel, K. 1986. *Kurt Gödel, Collected Papers, vol. 1.*, eds. Solomon Feferman, John W. Dawson, Jr., Stephen C. Kleene, Gregory H. Moore, Robert M. Solovay, Jean van Heijenoort, New York, the Clarendon Press.

———. 1987. "Letter to Menger," quoted in Wang, 1987a, p. 49.

Goldfarb, W. 1979. "Logic in the Twenties: The Nature of the Quantifier," *Journal of Symbolic Logic* 44: 351–368.

———. 1983. "I Want You to Bring Me a Slab," *Synthèse* 56: 265–282.

———. 1985. "Review of Kripke's *Wittgenstein on Rules and Private Language*," *Journal of Philosophy* 82: 471–488.

———. 1988. "Poincaré Against the Logicists," in *History and Philosophy of Modern Mathematics*, ed. W. Aspray and P. Kitcher, Minnesota Studies in the Philosopy of Science, 11, Minneapolis, University of Minnesota Press, pp. 61–81.

———. 1992. "Wittgenstein on Understanding," in *The Wittgenstein Legacy*, ed. P.A. French, T.E. Uehling, Jr., and H.K. Wettstein, Midwest Studies in Philosophy, 17, pp. 109–122. Notre Dame, IN, University of Notre Dame Press.

———. 1997. "Metaphysics and Nonsense: On Cora Diamond's 'The Realistic Spirit,'" *Journal of Philosophical Research* 22: 57–73.

———. 2001. "Frege's Conception of Logic," in Floyd and Shieh, eds., pp. 25–42.

———. 2002. "Wittgenstein's Understanding of Frege: The Pre-Tarctarian Evidence," in Reck, ed., pp. 185–200.

Goodstein, R.L. 1951. *Constructive Formalism*, Leicester, UK, Leicester University Press.

———. 1957. "Critical Notice of *Remarks on the Foundations of Mathematics*," *Mind*(66): 549–553.

———. 1965. *Essays in the Philosophy of Mathematics*, Leicester, UK, Leicester University Press.

———. 1972. "Wittgenstein's Philosophy of Mathematics," in Ambrose and Lazerowitz, eds., pp. 271–286.

Grattan-Guinness, I. 2000. *The Search for Mathematical Roots 1870–1940*: Logics, Set Theories and the Foundations of Mathematics from Cantor Through Russell to Gödel, Princeton, NJ, Princeton University Press.

Hacker, P.M.S. 1986. *Insight and Illusion: Themes in the Philosophy of Wittgenstein*, rev. ed., New York, Oxford University Press.

———. 1997. *Wittgenstein's Place in Twentieth-Century Analytic Philosophy*, Cambridge, MA, Blackwell.

Hale, R.V. 1997. "Rule-Following, Objectivity and Meaning," in Hale and Wright, eds., pp. 369–396.

Hale, R.V., and Wright, C. 1997. "Putnam's Model-Theoretic Argument Against Metaphysical Realism," in Hale and Wright, eds., pp. 427–457.

Hale, R.V., and Wright, C., eds. 1997. *A Companion to the Philosophy of Language*, Oxford, Blackwell.

Hintikka, J. 1956. "Identity, Variables and Impredicative Definitions," *Journal of Symbolic Logic* 21: 225–245.

———. 1962. *Knowledge and Belief*, Ithaca, NY, Cornell University Press.

———. 1973. *Logic, Language-Games, and Information: Kantian Themes in the Philosophy of Logic*, Oxford, Clarendon Press.

———. 1993. "The Original *Sinn* of Wittgenstein's Philosophy of Mathematics," in *Wittgenstein's Philosophy of Mathematics*, ed. Klaus Puhl, Vienna, Hölder-Pichler-Tempsky, pp. 24–51; reprinted in Hintikka, 1996a.

———. 1996a. *Ludwig Wittgenstein: Half-Truths and One-and-a-Half Truths*, Vol. 1, Selected Papers, Dordrecht, Kluwer.

———. 1996b. *The Principles of Mathematics Revisited*, Cambridge, Cambridge University Press.

———. 1997. "Game-Theoretical Semantics," in *Handbook of Logic and Language*, ed. J. van Benthem and A. ter Meulen, pp. 361–410. Amsterdam, Elsevier; Cambridge, MA, MIT Press.

Hintikka, J., ed. 1995. *From Dedekind to Gödel: Essays on the Development of the Foundations of Mathematics in the Twentieth Century*, Dordrecht, Kluwer.

Hintikka, J., and Hintikka, M.B. 1986. *Investigating Wittgenstein*, New York, Blackwell.

Hintikka, J., and Puhl, K., eds. 1995. *The British Tradition in Twentieth Century Philosophy: Proceedings of the 17th International Wittgenstein Symposium*, Vienna, Hölder-Pichler-Tempsky.

Holtzman, S., and Leich, C., eds. 1981. *Wittgenstein: To Follow a Rule*, London, Routledge & Kegal Paul.

Hylton, P. 1990. *Russell, Idealism, and the Emergence of Analytic Philosophy*, New York, Oxford University Press.

———. 1997. "Functions, Operations and Sense in Wittgenstein's *Tractatus*," in Tait, ed., pp. 91–106.

Jacquette, D., ed. 2002. *Philosophy of Mathematics: An Anthology*, Blackwell.

Kanamori, A. 2003. *The Higher Infinite: Large Cardinals in Set Theory from Their Beginnings*, 2nd ed., New York, Springer-Verlag.

Kant, I. 1781/1787. *Kritik der reinen Vernunft*. English translation by Norman Kemp, Smith, *Critique of Pure Reason*, New York, St. Martin's, 1965.

Kielkopf, C.F. 1970. *Strict Finitism: An Examination of Ludwig Wittgenstein's Remarks on the Foundations of Mathematics*, The Hague, Mouton.

Klenk, V. 1976. *Wittgenstein's Philosophy of Mathematics*, The Hague, Martinus Nijhoff.

Kreisel, G. 1958. "Review of Wittgenstein's 'Remarks on the Foundations of Mathematics,'" *British Journal for the Philosophy of Science* 9: 135–158.

———. 1959. "Wittgenstein's Theory and Practice of Philosophy," *British Journal for the Philosophy of Science* 11: 238–252.

———. 1978. "The Motto of 'Philosophical Investigations' and the Philosophy of Proofs and Rules," *Grazer philosophische Studien* 6: 13–38.

———. 1983. "Einige Erläuterungen zu Wittgensteins Kummer mit Hilbert und Gödel," in *Epistemology and Philosophy of Science. Proceedings of the 7th International Wittgenstein Symposium*, ed. P. Weingartner and J. Czermak, Vienna, Hölder-Pichler-Tempsky, pp. 295–303.

———. 1989. "Zu einigen Gesprächen mit Wittgenstein," in *Ludwig Wittgenstein Biographie-Philosophie-Praxis*, Vienna, Catalog for the exhibition at the Wiener Secession, September 13–October 29, Vienna, Wiener Secession, pp. 131–143.

———. 1998. "Second Thoughts Around Some of Gödel's Writings: A Non-Academic Opinion," *Synthèse* 114(1): 99–160.

Kremer, M. 1992. "The Multiplicity of General Propositions," *Noûs* 26(4): 409–426.

———. 2002. "Mathematics and Meaning in the *Tractatus*," *Philosophical Investigations* 25: 272–303.

Kripke, S. 1982. *Wittgenstein on Rules and Private Language*, Cambridge, MA, Harvard University Press.

Maddy, P. 1997. *Naturalism in Mathematics*, Oxford, Clarendon Press.

Malcolm, N. 2001. *Ludwig Wittgenstein: A Memoir*, Oxford University Press, 2nd edition.

Mancosu, P., and Marion, M. 2002. "Wittgenstein's Constructivization of Euler's Proof of the Infinity of Primes," in *Vienna Circle Institute Yearbook*, ed. F. Stadler, Dordrecht, Kluwer.

Marion, M. 1995. "Wittgenstein and Ramsey on Identity," in Hintikka, ed., pp. 343–371.

———. 1998. *Wittgenstein, Finitism, and the Foundations of Mathematics*, Oxford, Clarendon Press.

McGinn, M. 1999. "Between Metaphysics and Nonsense: Elucidation in Wittgenstein's *Tractatus*," *The Philosophical Quarterly* 49(197): 491–513.

McGuiness, B. 1988. *Wittgenstein, A Life: Young Ludwig 1889–1921*, Berkeley: University of California Press.

McGuiness, B., and Haller, R., eds. 1989. *Wittgenstein in Focus-im Brennpunkt: Wittgenstein*, Amsterdam, Rodopi.

Mclarty, C. 1997. "Poincaré: Mathematics and Logic and Intuition," *Philosophia Mathematica* 3(5): 97–115.

Minar, E. 1991. "Wittgenstein and the Contingency of Community," *Pacific Philosophical Quarterly* 72(3): 203–234.
———. 1994. "Paradox and Privacy," *Philosophy and Phenomenological Research* 54(1): 43–75.
Monk, R. 1990. *Ludwig Wittgenstein: The Duty of Genius*, New York, Free Press.
Moore, A.W. 1989. "A Problem for Intuitionism: The Apparent Possibility of Performing Infinitely Many Tasks in a Finite Time," *Proceedings of the Aristotelian Society* 90: 17–34; reprinted in Jacquette, ed., pp. 312–321.
———. 1990. *The Infinite*, New York, Routledge.
Moore, G.E. 1954–1955. "Wittgenstein's Lectures in 1930–33," *Mind* 63(1954): 1–15 (part I); 289–315 (part II); *Mind* 64(1955): 1–27 (part III); and 264(remark); reprinted in Wittgenstein, 1993a, pp. 46–114.
Noë, R.A. 1994. "Wittgenstein, Phenomenology and What It Makes Sense to Say," *Philosophy and Phenomenological Research* 1(4): 1–42.
Ostrow, M.B. 2002. *Wittgenstein's Tractatus: A Dialectical Interpretation*, Cambridge, Cambridge University Press.
Parikh, R. 1985. "The Logic of Games," *Annals of Discrete Mathematics* 24: 111–140.
———. 2001. "Language as Social Software," in Floyd and Shieh, eds., pp. 339–350.
Pears, D. 1979. "The Relation Between Wittgenstein's Picture Theory of Propositions and Russell's Theories of Judgment," in Luckhardt, ed., 1979, pp. 190–212; reprinted in Shanker, ed., 1986, vol. 1, pp. 92–107.
———. 1986. *Wittgenstein*, Cambridge, MA, Harvard University Press. 1st ed., New York, Viking, 1970.
———. 1987. *The False Prison: A Study of the Development of Wittgenstein's Philosophy*, vol. 1, Oxford, Clarendon Press.
———. 1988. *The False Prison: A Study of the Development of Wittgenstein's Philosophy*, vol. 2, Oxford, Clarendon Press.
Potter, M. 2000. *Reason's Nearest Kin: Philosophies of Arithmetic from Kant to Carnap*, New York, Oxford University Press.
Priest, G. 1987. *In Contradiction: A Study of the Transconsistent*, Dordrecht, Martinus Nijhoff.
———. 1994. "Is Arithmetic Consistent?," *Mind* 103: 337–349.
Proops, I. 2001. "The New Wittgenstein: A Critique," *European Journal of Philosophy* 9(3): 375–404.
———. 2002. "The *Tractatus* on Inference and Entailment," in Reck, ed., pp. 283–307.
Proudfoot, D., and Copeland, B.J. 1994. "Turing, Wittgenstein and the Science of the Mind," *Australasian Journal of Philosophy* 72(4): 497–519.
Putnam, Hilary. 1996. "On Wittgenstein's Philosophy of Mathematics," *Proceedings of the Aristotelian Society supp.* 70: 243–264.
Ramsey, F.P. 1928. "On a Problem of Formal Logic," *Proc. of London Mathematical Society (Ser. 2)* 30 (Part 4): 338–384.
Reck, E.H., ed. 2002. *From Frege to Wittgenstein: Perspectives on Early Analytic Philosophy*, Oxford, Oxford University Press.
Ricketts, T. 1985. "Frege, the *Tractatus*, and the Logocentric Predicament," *Noûs* 19: 3–14.
———. 1986a. "Objectivity and Objecthood: Frege's Metaphysics of Judgment," in *Frege Synthesized*, ed. L. Haaparanta and J. Hintikka, Dordrecht, D. Reidel, pp. 65–96.

———. 1986b. "Generality, Meaning, and Sense in Frege," *Pacific Philosophical Quarterly* 67: 172–195.

———. 1996a. "Logic and Truth in Frege," *Proceedings of the Aristotelian Society supp.* 70: 121–140.

———. 1996b. "Pictures, Logic, and the Limits of Sense in Wittgenstein's *Tractatus*," in Sluga and Stern, eds., pp. 55–99.

———. 2002. "Wittgenstein Against Frege and Russell," in Reck, ed., pp. 227–252.

Rodych, V. 1999. "Wittgenstein's Inversion of Gödel's Theorem," *Erkenntnis* 51: 173–206.

———. 2002. "Wittgenstein on Gödel: The Newly Published Remarks," *Erkenntnis* 56: 379–397.

Russell, B. 1903. *Principles of Mathematics*, Cambridge, Cambridge University Press; 2nd. ed. New York, Norton, 1938.

———. 1919. *Introduction to Mathematical Philosophy*, London, Allen and Unwin; 2nd ed. 1920.

———. 1984. *The Theory of Knowledge: The 1913 Manuscript*, ed. E.R. Eames and K. Blackwell, London, Routledge.

Sahlin, N.E. 1995. "On the Philosophical Relations Between Ramsey and Wittgenstein," in Hintikka and Puhl, eds., pp. 150–162.

Schirn, M., ed. 1998. *The Philosophy of Mathematics Today*, Oxford, Clarendon Press.

Shanker, S.G. 1987. *Wittgenstein and the Turning Point in the Philosophy of Mathematics*, Albany, State University of New York Press.

———. 1988. "Wittgenstein's Remarks on the Significance of Gödel's Theorem," in Shanker, ed., pp. 155–256.

———. 1998. *Wittgenstein's Remarks on the Foundations of AI*, New York, Routledge.

Shanker, S.G., ed. 1986. *Ludwig Wittgenstein: Critical Assessments, vol. 3, From the Tractatus to Remarks on the Foundations of Mathematics: Wittgenstein on the Philosophy of Mathematics*, London, Croom Helm.

———. 1988. *Gödel's Theorem in Focus*, London, Croom Helm.

Shapiro, S. 1991. *Foundations Without Foundationalism: A Case for Second-Order Logic*, Oxford, Clarendon Press.

Simons, P. 1992. *Philosophy and Logic in Central Europe from Bolzano to Tarski*, Dordrecht, Kluwer.

Sluga, H., and Stern, D.G., eds. 1996. *The Cambridge Companion to Wittgenstein*, Cambridge, Cambridge University Press.

Soames, S. 1983. "Generality, Truth Functions, and Expressive Capacity in the *Tractatus*," *The Philosophical Review* 92(4): 573–587.

Steiner, M. 1975. *Mathematical Knowledge*, Ithaca, NY, Cornell University Press.

———. 1996. "Wittgenstein: Mathematics, Regularities and Rules," in *Benacerraf and His Critics*, ed. Adam Morton and Stephen Stich, Cambridge, MA, Blackwell, pp. 190–212.

———. 2000. "Wittgenstein as His Own Worst Enemy: The Case of Gödel's Theorem," *Philosophia Mathematica* 3(9): 257–279.

Stroud, B. 1965. "Wittgenstein and Logical Necessity," *Philosophical Review* 74: 504–518; reprinted in Shanker, ed., 1986, pp. 289–301 and in Stroud, 2000, pp. 1–16.

———. 2000. *Meaning, Understanding and Practice: Philosophical Essays*, Oxford, Oxford University Press.

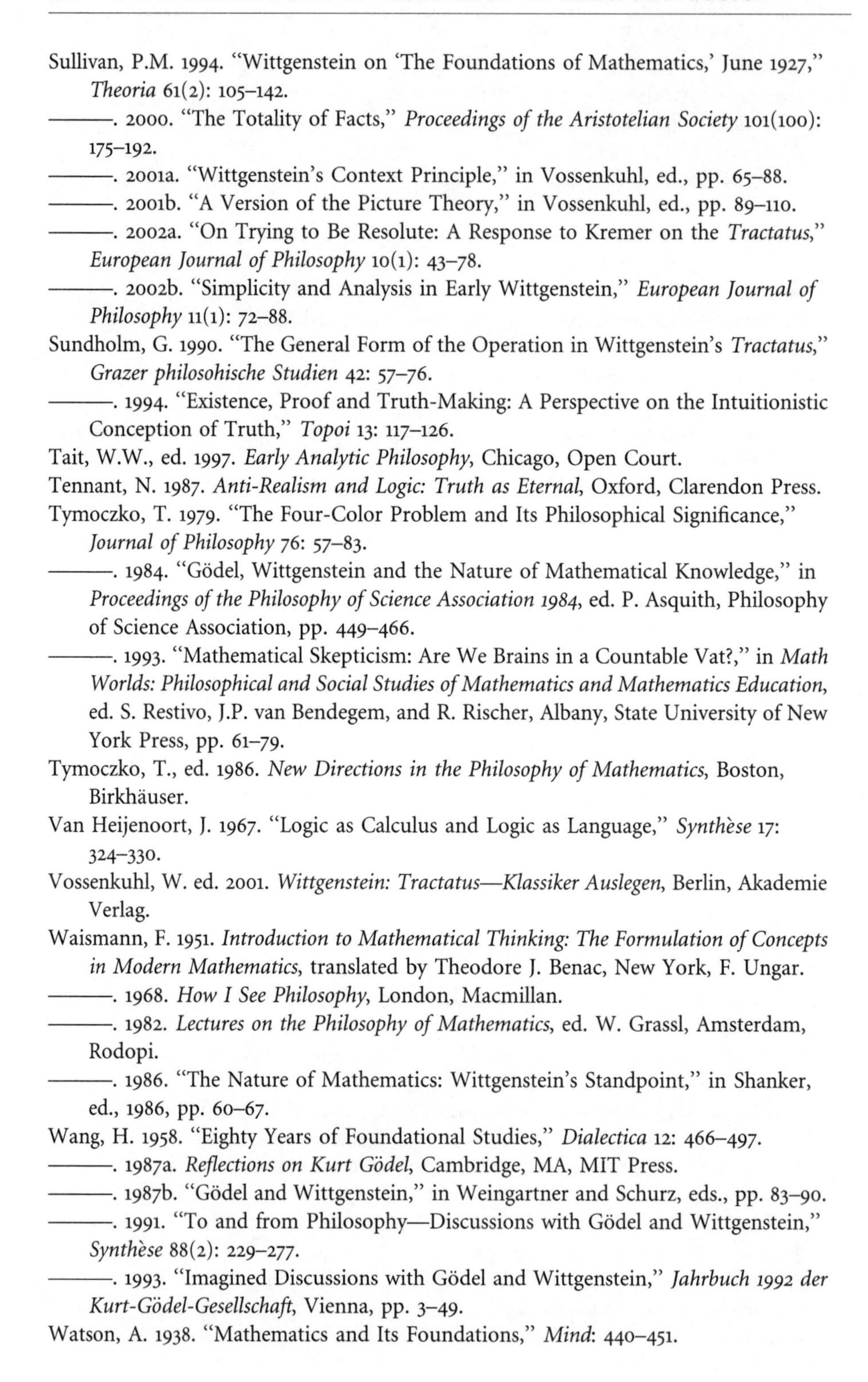

Sullivan, P.M. 1994. "Wittgenstein on 'The Foundations of Mathematics,' June 1927," *Theoria* 61(2): 105–142.

———. 2000. "The Totality of Facts," *Proceedings of the Aristotelian Society* 101(100): 175–192.

———. 2001a. "Wittgenstein's Context Principle," in Vossenkuhl, ed., pp. 65–88.

———. 2001b. "A Version of the Picture Theory," in Vossenkuhl, ed., pp. 89–110.

———. 2002a. "On Trying to Be Resolute: A Response to Kremer on the *Tractatus*," *European Journal of Philosophy* 10(1): 43–78.

———. 2002b. "Simplicity and Analysis in Early Wittgenstein," *European Journal of Philosophy* 11(1): 72–88.

Sundholm, G. 1990. "The General Form of the Operation in Wittgenstein's *Tractatus*," *Grazer philosohische Studien* 42: 57–76.

———. 1994. "Existence, Proof and Truth-Making: A Perspective on the Intuitionistic Conception of Truth," *Topoi* 13: 117–126.

Tait, W.W., ed. 1997. *Early Analytic Philosophy*, Chicago, Open Court.

Tennant, N. 1987. *Anti-Realism and Logic: Truth as Eternal*, Oxford, Clarendon Press.

Tymoczko, T. 1979. "The Four-Color Problem and Its Philosophical Significance," *Journal of Philosophy* 76: 57–83.

———. 1984. "Gödel, Wittgenstein and the Nature of Mathematical Knowledge," in *Proceedings of the Philosophy of Science Association 1984*, ed. P. Asquith, Philosophy of Science Association, pp. 449–466.

———. 1993. "Mathematical Skepticism: Are We Brains in a Countable Vat?," in *Math Worlds: Philosophical and Social Studies of Mathematics and Mathematics Education*, ed. S. Restivo, J.P. van Bendegem, and R. Rischer, Albany, State University of New York Press, pp. 61–79.

Tymoczko, T., ed. 1986. *New Directions in the Philosophy of Mathematics*, Boston, Birkhäuser.

Van Heijenoort, J. 1967. "Logic as Calculus and Logic as Language," *Synthèse* 17: 324–330.

Vossenkuhl, W. ed. 2001. *Wittgenstein: Tractatus—Klassiker Auslegen*, Berlin, Akademie Verlag.

Waismann, F. 1951. *Introduction to Mathematical Thinking: The Formulation of Concepts in Modern Mathematics*, translated by Theodore J. Benac, New York, F. Ungar.

———. 1968. *How I See Philosophy*, London, Macmillan.

———. 1982. *Lectures on the Philosophy of Mathematics*, ed. W. Grassl, Amsterdam, Rodopi.

———. 1986. "The Nature of Mathematics: Wittgenstein's Standpoint," in Shanker, ed., 1986, pp. 60–67.

Wang, H. 1958. "Eighty Years of Foundational Studies," *Dialectica* 12: 466–497.

———. 1987a. *Reflections on Kurt Gödel*, Cambridge, MA, MIT Press.

———. 1987b. "Gödel and Wittgenstein," in Weingartner and Schurz, eds., pp. 83–90.

———. 1991. "To and from Philosophy—Discussions with Gödel and Wittgenstein," *Synthèse* 88(2): 229–277.

———. 1993. "Imagined Discussions with Gödel and Wittgenstein," *Jahrbuch 1992 der Kurt-Gödel-Gesellschaft*, Vienna, pp. 3–49.

Watson, A. 1938. "Mathematics and Its Foundations," *Mind*: 440–451.

Webb, J. 1980. *Mechanism, Mentalism, and Metamathematics: An Essay on Finitism*, Dordrecht, Reidel.

Weiner, J. 1990. *Frege in Perspective*, Ithaca, NY, Cornell University Press.

———. 2001. "Theory and Elucidation: The End of the Age of Innocence," in Floyd and Shieh, eds., 2001, pp. 43–66.

Weingartner, P. and Schurz, G., eds. 1987. *Logik, Wissenschaftstheorie und Erkenntnistheorie: Acts of the 11th International Wittgenstein Symposium*, Vienna, Hölder-Pichler-Tempsky Verlag.

Whitehead, A.N. 1898. *A Treatise on Universal Algebra with Applications*, Cambridge, Cambridge University Press.

Whitehead, A.N., and Russell, B. 1910. *Principia Mathematica*, vol. 1, Cambridge, Cambridge University Press.

Wittgenstein, L. 1913. "On Logic and How Not to Do It," in Wittgenstein, 1993, pp. 1–2.

———. 1921. "Logische-philosophische Abhandlung," final chapter in Ostwald's *Annalen der Naturphilosophie*; English translation as *Tractatus Logico-Philosophicus* by C.K. Ogden, New York, Routledge and Kegan Paul, 1922; a critical edition in German containing the *Prototractatus* is Wittgenstein 1989a.

———. 1926. *Wörterbuch für Volksschulen*, Vienna, Hölder-Pichler-Tempsky, reprinted 1977; preface reprinted in Wittgenstein, 1993, pp. 12–27.

———. 1929. "Some Remarks on Logical Form," *Proceedings of the Aristotelian Society supp.* 9: 162–171; reprinted in Wittgenstein, 1993a, pp. 28–35.

———. 1933. Letter to the editor, *Mind* 42(167): 415–416.

———. 1958a. *Philosophical Investigations*, 2nd ed., ed. G.E.M. Anscombe and R. Rhees, trans. G.E.M. Anscombe, Oxford, Blackwell.

———. 1958b. *The Blue and Brown Books*, Oxford, Blackwell.

———. 1969. *On Certainty*, ed. G.E.M. Anscombe and G.H. von Wright, New York, Harper Torchbooks.

———. 1973. *Ludwig Wittgenstein and the Vienna Circle*, ed. B. McGuinness and J. Schulte, trans. B. McGuiness, Oxford, Blackwell. Shorthand notes recorded by F. Waismann.

———. 1974. *Philosophical Grammar*, ed. R. Rhees, trans. A.J.P. Kenny, Berkeley, University of California Press.

———. 1975. *Philosophical Remarks*, ed. R. Rhees and R. Hargreaves, trans. R. White, Chicago, University of Chicago Press.

———. 1978. *Remarks on the Foundations of Mathematics*, 3rd ed., rev., G.H. von Wright, R. Rhees, and G.E.M. Anscombe, trans. G.E.M. Anscombe, Oxford, Blackwell.

———. 1979a. *Notebooks 1914–1916*, ed. G.H. von Wright, G.E.M. Anscombe, trans. G.E.M. Anscombe, 2nd edition, Oxford, Blackwell.

———. 1979b. *Wittgenstein's Lectures, Cambridge, 1932–1935: From the Notes of Alice Ambrose and Margaret Macdonald*, ed. Alice Ambrose, Chicago, University of Chicago Press.

———. 1980. *Wittgenstein's Lectures, Cambridge 1930–32: From the notes of John King and Desmond Lee*, ed. Desmond Lee, Oxford, Blackwell.

———. 1989a. *Logisch-philosophische Abhandlung-Tractatus Logico-Philosophicus, Kritische Edition*, ed. F. McGuinness and J. Schulte, Frankfurt, Suhrkamp.

———. 1989b. *Wittgenstein's Lectures on the Foundations of Mathematics, Cambridge 1939*, ed. C. Diamond, Chicago, University of Chicago Press. (1st published by Harvester, 1976.).

———. 1993. *Ludwig Wittgenstein: Philosophical Occasions 1912–1951*, ed. J. Klagge and A. Nordmann, Indianapolis, Hackett.

———. 1993– . *Wiener Ausgabe*, ed. Michael Nedo, vols. 1–5, Vienna, Springer-Verlag.

———. 1995. *Ludwig Wittgenstein, Cambridge Letters: Correspondence with Russell, Keynes, Moore, Ramsey and Sraffa*, ed. B. McGuinness and G.H. von Wright, Oxford, Blackwell.

———. 2000. *The Collected Manuscripts of Ludwig Wittgenstein on Facsimile CD-ROM*, Oxford, Oxford University Press.

———. 2001. *Philosophische Untersuchungen, Kritisch-genetische Edition*, ed. J. Schulte et al., Frankfurt am Main, Suhrkamp.

Wright, C. 1980. *Wittgenstein on the Foundations of Mathematics*, Cambridge, MA, Harvard University Press.

———. 1982. "Strict Finitism," *Synthèse* 51: 203–282; reprinted in Wright, 1993.

———. 1983. *Frege's Conception of Numbers as Objects*, Scots Philosophical Monographs no. 2, Aberdeen, Aberdeen University Press.

———. 1993. *Realism, Meaning and Truth*, 2nd ed., Oxford, Blackwell.

———. 2001. *Rails to Infinity: Essays on Themes from Wittgenstein's Philosophical Investigations*, Cambridge, MA, Harvard University Press.

Wright, C., ed. 1984. *Frege: Tradition and Influence*, Oxford, Blackwell.

Wrigley, M. 1980. "Wittgenstein on Inconsistency," *Philosophy* 55: 471–484; reprinted in Shanker, ed., 1986, pp. 347–359.

———. 1995. "Wittgenstein, Ramsey and the Infinite," in Hintikka and Puhl, eds., pp. 164–169.

———. 1998. "A Note on Arithmetic and Logic in the *Tractatus*," *Acta Analytica* 21: 129–131.

CHAPTER 5

THE LOGICISM OF FREGE, DEDEKIND, AND RUSSELL

WILLIAM DEMOPOULOS
PETER CLARK

Philosophy confines itself to universal concepts; mathematics can achieve nothing by concepts alone but hastens at once to intuition, in which it considers the concept in concreto, though not empirically, but only in an intuition which it presents a priori, that is, which it has constructed, and in which whatever follows from the universal conditions of the construction must be universally valid of the object thus constructed.

—(Kant, *Critique of Pure Reason* A716/B744)

[*W*]*e see how pure thought, irrespective of any content given by the senses or even by an intuition a priori, can, solely from the content that results from its own constitution, bring forth judgements that at first sight appear to be possible only on the basis of some intuition. This can be compared with condensation, through which it is possible to transform the air that to a child's consciousness appears as nothing into a visible fluid that forms drops. The propositions . . . developed in what follows far surpass in generality all those that can be derived from any intuition.*

—(Frege, *Begriffsschrift* §23)

Introduction and Scope of the Study

Our discussion is organized around logicism's answers to the following questions: (1) What is the basis for our knowledge of the infinity of the numbers? (2) How is arithmetic applicable to reality? (3) Why is reasoning by induction justified? Although there are, as we will see, important differences, the common thread that runs through all three of our authors is their opposition to the Kantian thesis that reflection on our reasoning with mere concepts (i.e., without attention to intuitions formed a priori) can never succeed in providing us with satisfactory answers to these three questions. This description of the core of the view differs from more usual formulations which represent the opposition to Kant as an opposition to the contention that mathematics in general, and arithmetic in particular, are synthetic a priori rather than analytic. From our perspective, such a formulation is not sufficiently general: it fails to capture Dedekind, who does not use the terminology of analytic vs. synthetic judgments at all; it overlooks the fact that Frege relies on this terminology only in *Gl*, eschewing it in his other major writings, *Bg* and *Gg*.[1] And it is not easily squared with the fact that during his philosophically most productive period, Russell used the terminology only to express his *agreement* with Kant that mathematics is *synthetic*, while holding that, contrary to Kant, logic is *as synthetic as* mathematics.[2] What *is* essential to the logicism of history (if not the logicism of legend) is its opposition to the incursion of Kantian intuition into the content of arithmetical theorems or their justification: the thesis that all our authors share is the contention that the basic truths of arithmetic are susceptible of a justification that shows them to be more general than any truth secured on the basis of an intuition given a priori.

One final note on our selection of topics. The more purely philosophical dimension to Frege's program centered on the problem of explaining how the numbers are "given to us," if we have neither experience nor intuitions of them (*Gl* §62). The solution to this problem, from a contemporary viewpoint, would consist in an explanation of our reference to abstract objects—objects which subsist "outside" of space and time. In our view, it is in this context that the "Julius Caesar problem" becomes especially pressing. Since our concern is to expound those aspects of logicism that are more closely connected with its status as a

[1] A point observed by Dummett ([1991a], p. 28). Our references to Frege's *Begriffsschrift*, *Grundlagen*, and *Grundgesetze* are abbreviated *Bg*, *Gl*, and *Gg*, respectively.

[2] *Principles*, p. 457. For a discussion of Russell and Kant, see Coffa [1981].

contribution to the philosophy of the exact sciences, we have largely exempted these issues from the purview of our discussion.[3]

Frege's Logic and Theory of Classes

From *Bg* we can extract a presentation of second-order logic based on a language containing a single binary relation symbol. The central innovation of the work is the analysis of generality by means of the introduction of function/variable notation; it is this that is alluded to in the subtitle of *Bg* (*A Formula Language Modeled upon That of Arithmetic for Pure Thought*) as the characteristic feature of the language of arithmetic on which the work is modeled. When combined with expressions for relations, the result is a language of great expressive power, even when restricted to its "first-order fragment." Part III of *Bg* is devoted to the "theory of sequences." It is here that Frege presents his celebrated definition of "following in a sequence"—the ancestral of a relation, in the terminology of Whitehead and Russell or, in more modern terms, the transitive closure of a relation—and proves generalized analogues of mathematical induction, as well as various formal properties of the ancestral.

With the exception of his reliance on an implicit rule of substitution, Frege's formulation of his logic achieved a degree of precision that was not to be equaled for several decades. *Bg*'s implicit principle of substitution is equivalent to the second-order Comprehension Scheme:

$$\exists P\, \forall x(Px \equiv \Phi(x)),$$

where $\Phi(x)$ is a second-order formula in one free variable in which P does not occur free. The scheme tells us that for every condition $\Phi(x)$ which can be formulated in the language of *Bg* there is a corresponding property P. *Bg* makes no assumptions concerning the connection between properties or, as Frege says, "concepts," and the "extensions" or classes associated with them. Such a connection is introduced in *Gg* with Basic Law V, which we formulate as the universal closure of

$$\{x\colon Fx\} = \{x\colon Gx\} \equiv \forall x\, (Fx \equiv Gx),$$

[3] For a discussion of the "Julius Caesar problem" from the perspective of the present article, see Demopoulos [in press].

where the expression "$\{x: Fx\}$" is a class abstraction operator forming, for a concept F, the value range or extension of F. Frege took the class abstraction operator as primitive; given it, we can define the notion of membership by

$$x \in y \equiv \exists H(Hx \wedge y = \{x: Hx\}).$$

In the context of second-order logic, this definition of membership, together with Basic Law V, implies the Naive Comprehension Axiom:

$$\forall P \exists z \forall x(x \in z \equiv Px).$$[4]

[4] The proof depends on the following lemma (cp. *Gg* I §54):

$$\forall x(x \in \{y: Py\} \equiv Px).$$

Naive Comprehension immediately follows since, by generalizing on the term $\{y: Py\}$, we have

$$\exists z \forall x(x \in z \equiv Px);$$

and since P is arbitrary,

$$\forall P \exists z \forall x(x \in z \equiv Px).$$

Proof of the Lemma: Suppose $x \in \{y: Py\}$. By the definition of $\in$,

$$x \in \{y: Py\} \equiv \exists H(\{y: Py\} = \{y: Hy\} \wedge Hx).$$

Thus, Hx and $\{y: Py\} = \{y: Hy\}$; whence, by the left-to-right (problematic) direction of Basic Law V, $\forall y(Py \equiv Hy)$, and thus $Px \equiv Hx$, and therefore, Px.

Next, suppose Px. By the right-to-left (benign) direction of the following instance of Basic Law V,

$$\forall y(Py \equiv Py) \equiv \{y: Py\} = \{y: Py\},$$

$\{y: Py\} = \{y: Py\}$; and hence,

$$\exists z(z = \{y: Py\}).$$

Let $z = u$. Then $(u = \{y: Py\} \wedge Px)$ and, therefore,

$$\exists P(u = \{y: Py\} \wedge Px),$$

whence, by the definition of $\in$,

$$x \in \{y: Py\}. \ \square$$

On its intended interpretation, Naive Comprehension asserts that for every property there is a class which consists of precisely the individuals having the property. It will be observed that the axiom is a *sentence* of second-order logic, and that it implies the first-order instances of the following Naive Comprehension Scheme:

$$\exists z \forall x(x \in z \equiv \varphi(x)),$$

where $\varphi(x)$ is a first-order formula in one free variable, not containing the variable z. The Russell–Zermelo paradox follows immediately from the instance

$$\exists z \forall x(x \in z \equiv x \notin x),$$

which asserts the existence of a class consisting of all and only those sets which are not members of themselves.

It is a relatively recent discovery (see Parsons [1987]) that in the context of first-order logic, any collection of instances of Basic Law V is consistent; that is, the first-order fragment of *Gg* is consistent; moreover, any first-order theory can be conservatively extended by the addition of such instances (Bell [1995]). Striking as these technical results are, their philosophical significance is limited, since without the (second-order definable) relation of membership, it is wholly unclear in what sense the notion of the extension of a concept is interchangeable with the familiar notion of a class.

For Frege, concepts are thoroughly extensional, so that if F and G apply to precisely the same objects, they are the same concept. Thus, in the direction

$$\forall x (Fx \equiv Gx) \rightarrow \{x\colon Fx\} = \{x\colon Gx\},$$

Basic Law V may be seen to assert the functionality of the relation which associates every concept with an extension; the direction

$$\{x\colon Fx\} = \{x\colon Gx\} \rightarrow \forall x\ (Fx \equiv Gx)$$

asserts that this function is one-one. But the Russell–Zermelo paradox shows that there is no such function. Indeed, if we further assume that the totality of extensions of concepts corresponds to the power set of the set of objects falling under them, then Basic Law V clearly contravenes Cantor's Theorem that the power set of a set cannot be injected into that set.

In the Appendix that was added to *Gg* in response to Russell's letter informing him of the contradiction, Frege himself showed that no map from concepts to objects can be one-one. Since Basic Law V asserts the existence of at least one such map, the system of *Gg* does not have a model—no classes or sets could possibly satisfy it. Thus, while Frege's development of second-order logic is perfectly consistent, its elaboration, in *Gg*, to include a theory of concepts and their extensions, foundered on the Russell–Zermelo paradox. However, both the mathematical development of Frege's theory of the natural numbers, and a significant component of his philosophy of mathematics, may be rendered completely independently of his theory of extensions.

Frege's Analysis of the Natural Numbers

Frege tells us (*Gg*, p. ix) that the "fundamental thought" on which his analysis of the natural or "counting" numbers is based, is the observation that a statement of number involves the predication of a concept of another concept; numerical concepts are concepts of "second level," which is to say, concepts under which concepts (of first level) are said to fall. This yields an analysis of the notion of a numerical property, as when we predicate of the concept *horse which draws the King's carriage* the property of having four objects falling under it. Such properties are first-order definable in terms of the numerically definite quantifiers. In order to pass from the analysis of numerical properties to the numbers, Frege introduced a "cardinality operator," which he "defined" contextually by the principle

$$\mathrm{N}xFx = \mathrm{N}xGx \equiv F \approx G,$$

that is, the number of *F*s is the same as the number of *G*s if and only if the *F*s and the *G*s are in one-to-one correspondence. (It is well known that $\approx$ is definable in second-order logic.) Again, as with the class abstraction operator, here also it is simply assumed that $\mathrm{N}x \ldots x$ is a term-forming operator. This contextual definition and the fundamental thought yield Frege's account of the applicability of mathematics. In this, the simplest case for which the question arises—the application of the cardinal numbers—the solution is that arithmetic is applicable to reality because the concepts, under which things fall, fall under numerical concepts. Thus it is possible to prove in second-order logic that $\exists^n xFx \equiv n = \mathrm{N}xFx$ (i.e., F falls under the numerical property expressed by the numerically definite quantifier $\exists^n x$ if and only if the Frege number of F is n, where n is defined in terms of the cardinality operator). It follows that the fact that the number of horses drawing the king's carriage is four is interderivable with the fact that there are four horses drawing the king's carriage. We will return to the Fregean solution to the applicability of arithmetic.

The contextual definition is the basic principle upon which Frege's development of his theory of the natural numbers was based; it is not a definition in the proper sense, not least in the sense that it is not conservative over the language $L = \{zero, successor, natural\ number\}$ of second-order arithmetic. Indeed, the contextual definition allows for the proof *both* of the existence of the (Dedekind infinite) sequence of natural numbers *and* of an infinite cardinal (called *endlos* in *Gl*).

In the recent secondary literature,[5] the contextual definition has come to be called "Hume's principle," since in introducing it (*Gl* §63), Frege quotes from Hume's *Treatise* (I, iii, 1). Notice that Hume's principle bears an obvious formal resemblance to Basic Law V: both relate an identity involving singular terms to an equivalence relation on concepts. And both postulate a type *reducing* correspondence between Fregean concepts and objects, in the case of Basic Law V, one which associates distinct objects with noncoextensive concepts, with its attendant difficulties, but in the case of Hume's principle, merely one which associates distinct objects with nonequinumerous concepts. We will return to the significance of type reduction in a moment.

For Frege, Hume's principle provides the necessary "criterion of identity" for numbers; that is, it gives us a statement, in other words, of the condition under which we should pronounce as true certain "recognition statements" (as Frege calls them in *Gl* §62) involving numerical singular terms, namely, statements of the form

$$\mathrm{N}xFx = \mathrm{N}xGx,$$

for arbitrary sortal concepts F and G. (Hume's principle is clearly at best a *partial* account of the identity of numbers because it fails to settle the truth-value of statements of the form $a = \mathrm{N}xFx$ where a is not given in the form $\mathrm{N}xHx$ for some H.) By contrast with Basic Law V, the second-order theory, whose sole axiom is Hume's principle, is consistent, with the sequence $\langle 0, 1, \ldots, \omega \rangle$ forming the domain for a model of the theory, each concept F interpreted as a subset of ω, and each term of the form $\mathrm{N}xFx$ interpreted as the cardinal of the subset corresponding to F.

Frege derives Hume's principle from his inconsistent theory of concepts and their extensions by appealing to his well-known explicit definition of the cardinality operator:

$$\mathrm{N}xFx = \{G\colon G \approx F\}$$

(*Gl* §73). This is the "Frege–Russell definition of the cardinal numbers" as classes of equinumerous concepts (or sets). The characterization "Frege–Russell" slurs

[5] See especially Boolos [1990].

over the fact that for Russell the number associated with a set (concept of first level) is an entity of higher type than the set itself. Beginning with individuals—entities of lowest type—we proceed first to concepts or sets of individuals and then to classes of such sets (corresponding to Frege's concepts of second level). For Russell, numbers, being classes, are of higher type than sets. But for Frege, extensions, and therefore numbers, belong to the totality of objects *whatever the level of concept with which they are associated.* Thus, while Russell and Frege both subscribe to *some* version of Hume's principle, their conceptions of the logical form of the cardinality operator—and, therefore, of the principle itself—are quite different: the operator is type-*raising* for Russell, and type-*lowering* for Frege. This difference is fundamental, since it enables Frege to establish—on the basis of Hume's principle—those of the Peano–Dedekind axioms of arithmetic which assert that the system of natural numbers is Dedekind infinite. By contrast, when the cardinality operator is type-raising, Hume's principle is rather weak, allowing for models of it of every finite power.

The intuitive idea[6] underlying Frege's proof of the infinity of the numbers is the definition of an appropriate sequence of representative concepts (for concepts we use square brackets, and we use curly brackets for classes):

N_0 $[x\colon x \neq x]$
N_1 $[x\colon x =$ the number of the concept $N_0]$
N_2 $[x\colon x =$ the number of $N_0 \vee x =$ the number of $N_1]$
N_3 $[x\colon x =$ the number of $N_0 \vee x =$ the number of $N_1 \vee\ x =$ the number of $N_2]$
and so on.

The existence of numbers for each of the concepts $N_0, N_1, \ldots$ is established using only logically definable concepts and those objects whose existence and distinctness from one another can be proved on the basis of Hume's principle. Nowhere in this construction is it necessary to appeal to extensions of concepts. It is, however, possible to make use of a consistent fragment of the theory of extensions, and when we do so, an interesting comparison with later developments emerges. To see this, suppose we limit the introduction of extensions to those belonging to the concepts of the sequence $\langle N_0, N_1, \ldots \rangle$. This yields a new sequence of concepts:

N_0 $[x\colon x \neq x]$
E_1 $[x\colon x =$ the extension of the concept $N_0]$
E_2 $[x\colon x =$ the extension of $N_0 \vee x =$ the extension of $E_1]$
E_3 $[x\colon x =$ the extension of $N_0 \vee x =$ the extension of $E_1 \vee x =$ the extension of $E_2]$
and so on;

[6] Here we follow the presentation in Demopoulos [1998].

the *extensions* of these concepts are the finite von Neumann ordinals:

0 $\emptyset$
1 $\{\emptyset\}$
2 $\{\emptyset, \{\emptyset\}\}$
3 $\{\emptyset, \{\emptyset\}, \{\emptyset, \{\emptyset\}\}\}$
and so on.

In its mathematical development Frege's theory thus provides an elegant reconstruction of our understanding of the cardinal numbers, one which can be carried out quite independently of the portion of his system which led to inconsistency.

The Problem of Applicability

A remark should be made on the application problem and its relation to the "Julius Caesar" problem. It is certainly true that Frege put the application problem at the heart of his philosophy of mathematics. To the question "Why does arithmetic apply to reality?" the logicist provides the clear answer "Because it applies to everything that can be thought; it is the most general science possible." The partial contextual definition, provided by Hume's principle, and the fundamental thought that numerical concepts are second-level concepts yields Frege's account of *how* arithmetic is applicable to reality. The central result, the theorem that $\exists^n xFx \equiv n = \mathrm{N}xFx$ (i.e., F) falls under the property expressed by the numerically definite quantifier $\exists^n x$ if and only if the Frege number of F is n, answers this question by connecting a fact involving a mathematical object—the fact that the number of *F*s is n—with one that does not, namely, the fact that there are n *F*s. Thus, if F is the property of being a book on the desk, the "mathematical fact" is that the number of books on the desk is three (say), and the "physical fact" is that there are three books on the desk. The proof of the above theorem for the case $n=3$ shows how the two are connected, and thus answers the question of how arithmetic applies to the physical world.

But the very possibility of framing the question of the applicability of arithmetic raises problems of its own. Hume's principle provides at best a partial contextual definition. The principle cannot settle the truth conditions of sentences of the form $a = \mathrm{N}x\,Fx$ where a is not given in the form $\mathrm{N}xHx$ for some H. This of course is the Julius Caesar problem. In the case of pure arithmetic the Julius Caesar problem cannot be raised, since there are no singular terms like a in the language of pure arithmetic. But once the language is expanded to include empirical singular terms, as it would in the language of applied arithmetic, Hume's principle will no longer settle the sense of all numerical identities and the Julius

Caesar problem can no longer be ignored. Of course this issue does not arise in the formal language of *Gg*, since in that language all the objects it is possible to refer to are already given as extensions (value ranges), the identity conditions for which are purportedly given by Basic Law V. Michael Potter comes close to making the right observation when he writes:

> [A] formal language in which Julius Caesar cannot be spoken of is one in which he cannot be counted, and in such a language the applicability of arithmetic remains unexplained. At some stage in the development we shall have to extend the formal language by adding some empirical vocabulary, and we shall then have to address the Julius Caesar problem just as before. (2000, p. 108)

But the difficulty is *not* that the applicability of arithmetic remains unexplained: it is explained perfectly by Frege's account. The difficulty is, rather, that any language for which the problem of the application of arithmetic can be nonvacuously posed is one in which the Julius Caesar problem must *also* be addressed. And this latter problem appears to have eluded Frege.

Frege's Account of Reasoning by Mathematical Induction

Frege's earliest contribution to the articulation of logicism consisted in showing that the validity of reasoning by induction can be accounted for on the basis of our general knowledge of principles of reasoning discoverable in *every* domain of inquiry. This directly engages the Kantian tradition in the philosophy of the exact sciences, according to which principles of general reasoning peculiar to our understanding must be supplemented by a faculty of intuition if we are to account for arithmetical knowledge. We are inclined today to view the answer to Kant as requiring the demonstration that mathematical reasoning—in this case, reasoning about the natural numbers—is recoverable as part of logical reasoning.

In *Bg*, part III, Frege showed how the ancestral of a relation—intuitively, the property of, for example, following the number 0 in the sequence of natural numbers after "some number of relative products of successor"—might be defined without reference to number. The (necessarily second-order) definition is easily presented in three simple steps.[7] Frege's account is quite general; he proceeds "schematically" by

[7] See Boolos [1985], whose presentation we follow.

taking a fixed, but arbitrary, binary relation f, and then shows how to form its ancestral. Suppressing explicit reference to the relation f, we abbreviate "F is hereditary in the f-sequence" by Her(F), and define it by the condition

$$\forall u\, \forall v(ufv \wedge Fu \rightarrow Fv),$$

given by proposition 69 of *Bg*. For "F is inherited from a in the f-sequence" we write In(a, F), and define this as

$$\forall z(afz \rightarrow Fz).$$

For "y follows x in the f-sequence"—the ancestral of f—we write

$$xf^{*}y.$$

With these abbreviations, Frege's definition of the ancestral becomes

$$xf^{*}y \equiv \forall F(\mathrm{Her}(F) \wedge \mathrm{In}(x,F) \rightarrow Fy).$$

Given the relation *(immediate) successor* (P) and the number *zero* (0), this definition is easily adapted to a definition of the property *natural number* by the condition

$$y = 0 \vee 0P^{*}y. \qquad (*)$$

This definition implies another, more tractable, condition:

$$\forall F(\mathrm{Her}(F) \wedge F0 \rightarrow Fy). \qquad (\S)$$

That is, y is a natural number—NNy—if and only if y has every hereditary property of 0. (The proof that (*) implies (§) is very simple: If $F0$ and Her(F), then $F1$. Since 1 is the *only* successor of 0, In(0, F), and hence, by our hypothesis (viz., $0P^{*}y$) and the definition of the ancestral, Fy.)

For the philosophical purpose of showing that from the definition of *natural number* we can prove the Principle of Mathematical Induction (MI)—thus showing induction to be a species of general reasoning—our simpler condition (§) suffices. First, we formulate MI in the relational form:

$$\frac{F0 \;\&\; \text{For any } m \text{ and } n, \text{ if } mPn \text{ and } Fm, \text{ then } Fn}{\text{For all } n,\ Fn.}$$

Here, "relational form" simply means that we take *successor* to be a binary relation and assume the presence of those axioms which ensure that it is a one-one function. To prove MI, we assume the hypothesis of MI:

$F0$ & For any m and n, if mPn and Fm, then Fn,

and derive the conclusion

For all n, Fn,

using only our definition of *natural number* given by the condition (§). That is, we have to show, for an arbitrary individual y, that y has F, when y has all the hereditary properties of 0. The proof is completely trivial: since the hypothesis of MI is simply the hypothesis of the generalized conditional of our definition—namely, $F0$ and Her(F)—we have Fy. Since y was arbitrary, we have shown that for all n, Fn. And since the property $0P^*x$ is hereditary with respect to successor (cf. *Bg* 96), MI implies our *original* definition of NNy in terms of the condition (∗), thus completing the circle and showing the equivalence of our conditions (∗) and (§) to one another and to MI.

Comparison with Dedekind's Chains

It has long been known that what Frege calls the property *following a in the f-sequence*—af^*x—is closely related to Dedekind's notion *the chain of an element a*, introduced in *Was sind und was sollen die Zahlen?* To elucidate the connection, let f be a one-one relation in $Y \times Y$. We will say that X *covers* f if $\text{Dom}(f) \subseteq X$ and $\text{Rg}(f) \subset X$. For $A \subseteq \text{Dom}(f)$, A nonempty, the *chain* or *closure* $C_f(A)$ *of* A *with respect to* f is given by

$\cap\{X\text{: } X \textit{ covers } f \wedge A \subseteq X\}$;

the *chain of* a is just $C_f\{a\}$. To see the connection between Dedekind's chains and Frege's ancestral, notice that on the assumption that f is one-one, the chain of a is just the extension of the concept $af^*x \vee x = a$. If A covers f, then $A = C_f(A)$. Thus, in particular, what Dedekind calls the "system" N of natural numbers covers *successor*, which is one-one, so N can be defined—as Dedekind defines it—as the chain of its initial element with respect to *successor*. The connection with Frege's definition of the property NNx of being a natural number is this: the success of Frege's definition of NNx in terms of the condition

$$\forall F(\text{Her}(F) \wedge F0 \rightarrow Fx)$$

depends on the fact that among the hereditary properties of 0 (hereditary with respect to *successor*) are those whose extensions are, in Dedekind's sense, chains of 0.

A Remark on Frege and Kant

For Kant, the concept of the number 7 is available to the Understanding without the aid of Sensibility and, hence, without intuitions. But its schematization—the provision of a reference for any concept-expression associated with it—*does* require intuition. This is not only immediate from the definition of "schematization"; it is also required by the fact that many statements of number involve equations and inequations, and their reference must also be provided for. In Kant's terminology, this means that the categories of unity and plurality must be schematized, and this requires addressing how unity and plurality relate to intuitions. We will return to this latter problem of the schematization of the categories. In the case of the concept of 7, the schematization is given by counting the objects to which it is applied. Counting may be likened to the construction of a geometric figure: both relate to empirical intuition, to the data of possible experience. There are, however, also objects of pure intuition; although such objects are not given in empirical intuition, they are the product of the *form* of empirical intuition. In the case of geometrical objects, it seems clear that this form is spatial, and that it constitutes a framework within which constructions involving ruler and compass, together with their possible combination and iteration, are possible. It would be tendentious to say that in the arithmetical case the relevant form of empirical intuition is time, and that it is a *temporal* framework that makes counting possible, just as it is a *spatial* one that makes ruler and compass constructions possible. But perhaps we can say that in the arithmetical case, the constructive procedure which is made possible is just the process of iteration itself. The possibility of indefinitely iterating any procedure, whether in counting or in the construction of figures, is guaranteed by the spatial and temporal forms of empirical intuition. To return to the topic of schematization, insofar as counting is iterative, it and the schematization of the various number concepts depend on the form of our intuition, and therefore on space and time.

From this brief sketch we can extract several desiderata for an account of number that could be held to be independent of Kantian intuition. It is clear, first of all, that such an account must explain the *applicability* of numerical concepts without invoking any intuitions beyond those that are demanded by the concepts to which the numerical concepts are applied. Second, it must explain the *reference* of numerical equations and inequations without recourse to intuition, perhaps in terms of relations among concepts, or in terms of "logical" objects—objects that are transparent to Reason itself. Finally, it must recover any piece of arithmetical reasoning which rests on the possibility of indefinitely iterating an operation as a species of general reasoning. Frege's articulation of his fundamental thought in *Gl*, according to which a statement of number involves the predication of something of a concept, clearly addresses the first desideratum: the application of *there are exactly four moons of Jupiter* must not involve any intuitions beyond those demanded by *x is a moon of Jupiter*. *Gl*'s criterion of identity for number—the contextual definition—addresses the second: provision of a criterion of identity for number is the Fregean analogue of the schematization of the categories of unity and plurality. The theory of classes of *Gg* is an extension of this attempt, mandated by the fact that the contextual definition of numerical identity and distinctness (the unity and plurality of number) in terms of the relation of one-one correspondence between concepts is at best *partial*. Of course the definition of the ancestral, given in *Bg*, is the key to Frege's solution to reasoning by induction, the paradigm of iterative reasoning.

Summary

The dependence relations among Frege's definitions as they appear in *Gl* are summarized in the following chart:[8]

$$F \approx G$$

$$\downarrow$$

$$CN\kappa \text{ (iff } \exists F(\kappa = \mathrm{N}x\,Fx), \text{ §72)} \Leftarrow \mathrm{N}x\,Fx \Rightarrow 0 = \mathrm{N}x[x \neq x](\text{§74}) \Rightarrow \exists^0 xFx$$

$$\Downarrow$$

$$uPv$$

$$\text{(iff } \exists F\,\exists y(Fy \wedge \mathrm{N}x[Fx \wedge x \neq y] = u \wedge \mathrm{N}x\,Fx = v), \text{ §76)}$$

$$\Downarrow$$

$$xP^{*}y(\text{from } Bg)$$

$$\Downarrow$$

$$n{+}1 = Nn[0P^{*}n \vee n = 0] \text{ (§79)} \Rightarrow \exists^{n+1}xFx$$

$$\Downarrow$$

$$NNn \text{ (§§81,83)}$$

[8] Adapted from Dummett ([1991a], p. 31).

$A \Rightarrow B$ is read: B is defined in terms of A. $\exists^0 x$, $\exists^1 x$, ... are the "numerically definite quantifiers" (also first-order definable in terms of identity and the usual logical operators and quantifiers), and are read: there are exactly zero *F*s, there is exactly one *F*, and so on. Only the first link in this chain (indicated by the single arrow, ↓) is a contextual "definition"; all the rest are proper explicit definitions. Note that the explicit definition of the cardinality operator does not appear here, since it is not required by the mathematical development of *Gl.* Notice also that there are two definitions of number, one of cardinal number ($CN\kappa$) given directly in terms of the cardinality operator, and the other, a definition of natural number (NNn), given in terms of the ancestral of the successor relation. The first defines cardinal numbers—answers to the question "How many?"—irrespective of whether they are finite or infinite. The second definition (that of NNn) splits into two: the "pure" definition, given above—which is independent of the cardinality operator and defines *natural number* purely ordinally—and the definition (of §§81, 83) which adapts this general strategy to the case where *successor* is defined in terms of the cardinality operator and, thus, defines *natural number* in terms of cardinality. Although the successor relation is defined in terms of the cardinality operator, the definition of the ancestral is, of course, completely general and does not in any way require the cardinality operator.

In conclusion, let us note that the second-order *predicative fragment* of *Gg* results when the Comprehension Scheme $\exists P \forall x(Px \equiv \Phi(x))$ is restricted to formulas $\Phi(x)$ in the language of *Gg* which contain no bound occurrences of second-order variables. By contrast with the first-order fragment of *Gg*, the second-order predicative fragment has a notion of membership and, therefore, a genuine notion of class. Extending methods of Parsons [1987], Heck [1996] has shown that the predicative fragment of *Gg* is consistent. This is intuitively plausible, since by so weakening Comprehension, there are vastly fewer properties for which it is necessary to find a representative extension. Nevertheless, the significance of the consistency of such an introduction of classes is unclear. For Frege, NNy is a *defined* property—indeed, its definition marks a high point of his logicism. But since Frege's definition of NNy is *im*predicative, this definition cannot occur as a condition in the second-order predicative fragment of *Gg*; and, therefore, the class which NNy determines cannot occur with its intended defining condition. Hence, whatever foundational interest the consistency of the predicative fragment of *Gg* might have for others, its interest for Frege is, at best, limited.

Assessment

Notice that the proof of MI, like the proof of the infinity of the number sequence, rests on Frege's use of second-order quantification. So also do two further key theorems that Frege establishes in part II of *Gg*: the proof of the validity of definition by induction over the natural numbers (theorem 263), and the proof that the relations together with their fields which satisfy Frege's axioms for the natural numbers are isomorphic.[9] As is well known, both theorems were first established by Dedekind in *Was sind*, where they occur as theorems 126 and 132, respectively; that analogues of them appear in *Gg* is less well known. The proof of the infinity of the number series requires treating numbers as objects and needs the additional presence of Hume's principle, and therefore, unlike the proofs of MI, categoricity, and the justification of definition by induction, it is not purely a result of second-order logic.

A difficulty which is often urged against Frege's account of general principles is that it relies on second-order features of his logic. The assessment of Frege's success in addressing the basis for our knowledge of the infinity of the numbers and the justification of induction, by contrast with the assessment of his success in connection with the applicability of arithmetic, has stalled on the question of whether second-order logic is really logic. We believe this to be unfortunate, since it has deflected attention from Frege's claim to have shown that reasoning by induction and our knowledge of a Dedekind infinite system depend on principles which, whether they are counted as logical or not, are appropriately *general* in their application and, therefore, do not rest on a notion of Kantian intuition of the sort Frege sought to refute. It may be necessary to concede that these demonstrations rest on intuition in a familiar, psychologistic, sense of the term; what has been missed is how little is conceded by such an admission. When this contention is, as we believe, rightly emphasized, a question emerges that leads naturally to Russell's contributions to logicism.

Although a consistent theory of arithmetic can be extracted from *Gl*, one with Hume's principle as its only nonlogical axiom, it remains unclear to what extent the full conceptual apparatus of *Gl* (or its elaboration in *Gg*) is entirely coherent—even when the notion of class is excised from it. As Frege tells us in his introduction to *Gl*, one of the methodological pillars of the work is the distinction between concepts and objects. But Frege's notion of object is of necessity one of great generality; if it were not, it could not be maintained with confidence that the development of arithmetic that *Gl* advances is independent of intuition. Russell's discovery that Frege's notion of class is paradoxical unless somehow restricted was problematic for Frege for just this reason: to preserve the idea of founding arithmetic on classes, it would be necessary to qualify the generality of the notion of an object, and it would then be unclear whether logicism represented the advance on Kant that it had promised to provide.

The mathematically central feature of a Fregean object is that it is a possible argument to a concept of first level. That numbers are objects in this minimal sense is all that is needed for the proof of Frege's theorem. The suggestion we have

[9] See (Heck [1995], pp. 315–316) for a discussion and references.

been pursuing in our exposition is that the Fregean framework of concepts and objects—without the notion of class—is a sufficient basis for arithmetic since, among other things, it allows us to recover the basic laws of arithmetic. Numbers are unproblematic precisely because their criterion of identity—namely, Hume's principle—requires only that the relation between concepts and objects should be functional; by contrast, the criterion of identity for classes (Basic Law V) demands that it should be injective. The rejection of classes is relatively unproblematic once it is seen that arithmetic can be developed without their use. But as we will show in the next section, there are, in addition to classes, other Fregean "objects" that will have to go if the framework of concepts and objects is to be coherent. The cumulative effect of successive prunings is that at some point it must become questionable whether arithmetic is being developed within an appropriately general conception of concepts and objects, or whether the notion of a framework of concepts and objects is being tailored for the development of arithmetic. Neither project is without philosophical significance and interest, but only the successful execution of the first can count as an unequivocally compelling answer to Kant.

Russell's Propositional Paradox and Fregean Thoughts

At the conclusion of *The Principles of Mathematics* (indeed, on the last two pages of the final appendix of the book), Russell derives a paradox that has come to be known as the propositional paradox. *Principles* contains the only mention of the propositional paradox in Russell's published writings. However, it occupies a significant position in his correspondence with Frege, where it led Frege to extend the theory of sense and reference which he had articulated in *Über Sinn und Bedeutung* to include a more elaborate development of the theory of indirect reference.[10] The issues raised by the paradox are so central to the interpretation of Frege and Russell that it will be worthwhile to go into the paradox in some detail.[11]

Like Russell's paradox of the class of all classes that are not members of themselves, the propositional paradox illustrates how "[f]rom Cantor's proposition that any class contains more subclasses than objects we can elicit constantly

[10] See especially Frege's letter to Russell of December 28, 1902, reprinted in McGuinness ([1980], pp. 152–154).

[11] Our presentation follows Demopoulos [2001], which in turn draws from Potter [2000]. See also Linsky [1999].

new contradictions."[12] While one can extract from the derivation of the paradox a proof that there are more properties of propositions than there are propositions, its main interest, like the interest of the Russell–Zermelo paradox, consists in the fact that it challenges a number of basic assumptions, in this case, assumptions about the nature of propositions and propositional identity—and by an extension we will explore presently, Fregean thoughts.

The formulation of the paradox we will follow[13] focuses on propositions of the form

$$\forall p(\varphi(p) \rightarrow p).$$

In a propositional theory like Russell's, such a proposition is understood as saying that every proposition with property φ is true. Notice that in order for "$\varphi(p)$" to express a propositional function, and for "$\varphi(p) \rightarrow p$" to express a proper truth function of $\varphi(p)$ and p, "p" must have what we will call a *designative occurrence* in "$\varphi(p)$", that is, substituends for "p" must designate propositions. By contrast, the occurrence of "p" in the consequent of "$\varphi(p) \rightarrow p$" must not be designative, but *expressive*; that is, a substituend for "p" must express a proposition. There is a systematic ambiguity between designative and expressive occurrences of "p" in the argument to the paradox, but there is no confusion of designative and expressive occurrences. As remarked by Church ([1984], p. 520), the paradox is derivable for a wide choice of propositions. In the extension to Frege's theory of thoughts given below, we derive the paradox using a proposition whose universal quantifier binds only designative occurrences of "p."

Although, as we have noted, "$\forall p(\varphi(p) \rightarrow p)$" is naturally taken to express the proposition that every proposition with property φ is true, the notion of a truth *predicate* does not enter into the derivation of the paradox. Nor does the paradox require for its formulation the availability of a semantic relation of designation, a point first made explicit in Church [1984]. Church derives the paradox within a theory consisting of the simple theory of types augmented by the axioms of propositional identity that are implicit in *Principles.* In the context of the simple theory, it can be shown (see Church [1984], p. 520) that the analogue of our (Injectivity), below, is derivable from these axioms of propositional identity. Thus, in terms of a well-known division of the paradoxes into those belonging to logic and mathematics "in their role as symbolic systems" versus those which depend on the bearing of logic on "the analysis of thought" (Ramsey [1931], p. 21), the propositional paradox falls squarely into the second category; yet, for the reasons just given, it is not straightforwardly *semantic.*

[12] Russell's letter to Frege of November 11, 1902, in McGuinness ([1980], p. 147).

[13] See Russell's letter to Frege of May 24, 1903, in McGuinness ([1980], pp. 148–150).

The paradox arises as follows. Let φ be a property of propositions; in Russell's terminology, φ is a propositional function which takes propositions as arguments and yields propositions as values. Consider the function f from properties of propositions to propositions defined by

$$f\varphi =_{\mathrm{Df}} \forall p(\varphi(p) \rightarrow p).$$

The notation is justified by Russell's theory of propositions: since φ is a constituent of $f\varphi$, f is a many–one relation, and on the assumption that for every φ there is such a proposition $f\varphi$, f is a function. Since φ is a constituent of the proposition $f\varphi$, it follows from Russell's conception of propositions and propositional identity (see Church [1984], p. 520) that

$$(f\varphi = f\psi) \rightarrow (\varphi = \psi),$$

a consequence we will call "(Injectivity)." Now let ψ be the property of propositions given by

$$\psi(p) =_{\mathrm{Df}} \exists\varphi(p = f\varphi \wedge \neg\varphi(p)),$$

and suppose $\neg\psi(f\psi)$:

$$\neg\exists\varphi(f\psi = f\varphi \wedge \neg\varphi(f\psi));$$

that is,

$$\forall\varphi(f\psi = f\varphi \rightarrow \varphi(f\psi)),$$

and therefore, $\psi(f\psi)$.

Next suppose $\psi(f\psi)$:

$$\exists\varphi(f\psi = f\varphi \wedge \neg\varphi(f\psi)).$$

Then, by (Injectivity),

$$\exists\varphi(\varphi = \psi \wedge \neg\varphi(f\psi));$$

whence $\neg\varphi(f\varphi)$.

Our exposition brings out how little the paradox requires from the theory of Russellian propositions developed in *Principles.* Thus, in the context of Russell's conception of propositions, (Injectivity) is required by the assumption that the propositional function φ is a constituent of $f\varphi$; and the assumption that for every φ there is a proposition $f\varphi$—the proposition $\forall p(\varphi(p) \rightarrow p)$—is so intuitively obvious that it would seem to be a necessary component of any theory of propositions. Assuming that for every φ there is a proposition of the form $\forall p(\varphi(p) \rightarrow p)$, all we require is that the relation which associates $\forall p(\varphi(p) \rightarrow p)$ with φ (the inverse of the relation determined by f), should be functional. We will return to this latter point in the context of Frege's theory of thoughts.

The solution mandated by the ramified theory of types depends on observing the restrictions it imposes on the "orders" of the propositions and propositional functions to which the derivation of the paradox appeals.[14] The technical idea behind the ramified theory's stratification of propositions into orders is that the order of a proposition, p, must exceed the order of any proposition that falls within the range of one of its quantified propositional variables, and therefore, by the restrictions the theory imposes, p cannot itself fall within the range of one of its quantified propositional variables. By an extension of terminology, the order of a propositional variable is the order of the propositions which belong to its range of values. According to the ramified theory, the proposition $q =_{\mathrm{Df}} \forall p(\psi(p) \rightarrow p)$ and the propositional function ψ, defined in the course of the derivation of the paradox, are such that if the order of the bound propositional variable in q is n, the orders of q and ψ must be $n+1$, so that q is not itself a possible argument for the function ψ. Since the contradiction was derived from the supposition that $\psi(q)$ or $\neg\psi(q)$, the argument collapses once the restrictions imposed by ramification are taken into account. Notice that this solution preserves the idea that φ is a constituent of $f\varphi$, as well as the idea that there always is a proposition $f\varphi$ corresponding to φ.[15] However, the generality of these assumptions is the restricted generality of the ramified theory: for every ψ of order $n+1$, there is a proposition of order $n+1$ which asserts the truth of every proposition of order n for which ψ holds.

[14] The remarks which follow are merely illustrative. The subject of Russell's ramified theories of types is subtle and complex, and deserves a separate exposition. (We return to it below.) See in this connection the admirable review of type theory in Urquhart [2003]; see also Chihara [1972, 1973].

[15] Russell's remarks in the correspondence with Frege leave it unclear whether the solution to the paradox just outlined is the one he came to favor. Fuhrmann [2002] argues that Russell saw the paradox's resolution to lie in the simple rejection of (Injectivity). But merely rejecting this assumption fails to address the problem of providing a positive theory of propositions; it also overlooks the fact that a solution in terms of ramification preserves the central features of the theory of propositions—subject, of course, to the restrictions noted in the text. A more serious difficulty with Fuhrmann's analysis is that the rejection of (Injectivity), while plausible on one formulation of the paradox, is not plausible on another, as Fuhrmann himself concedes (p. 209).

As is well known, there are two hierarchies that emerge from Frege's philosophical reflections on logic and language: the hierarchy of functions (or concepts) and objects, and the hierarchy of senses. The first hierarchy has a simple type-theoretical structure and, like the simple theory of types, is naturally motivated by considerations of predicability: first-level functions are predicated of objects, second-level functions of first-level functions, third-level functions of second-level functions, and so on. There is therefore no analogue in Frege's hierarchy of functions of Russell's paradox of the propositional function that is predicable of all and only those propositional functions which are not predicable of themselves. But as we have seen, the propositional paradox can be derived from the theory of Russellian propositions of *Principles* even when this theory is formulated within a simple type-theoretical formalism. It is therefore natural to ask whether Frege's theory of sense, and in particular his theory of thoughts, fares just as badly—to ask, in other words, whether Fregean thoughts, which are the proper correlates of Russellian propositions, admit the construction of an analogue of the propositional paradox.

Potter [2000, p. 135] claims that the propositional paradox can be extended to Fregean thoughts only if we ignore the "degree of indirectness" of a thought and its constituent senses. When degrees are taken into account, the argument can be seen to fail because it rests on the identification of thoughts of different degree. Potter does not explain the principle controlling the assignment of degree indices; for that, one must look to Frege's letter to Russell of December 28, 1902, which tells us that the degree of a thought is assigned on the basis of the number of embeddings into opaque contexts a canonical linguistic expression of the thought exhibits. The details of this need not concern us. The important point is that however high its degree, a thought remains a possible argument to a concept of first level. As we will now show, this is all the extension of the paradox to Frege's theory of thoughts requires.

Passing to the hierarchy of senses, notice that insofar as senses are objects, they, too, fall under concepts. Thoughts, being the senses of complete expressions, are objects. Among thoughts, some are *universal* (i.e., are composed of the senses for the universal quantifier and a concept-expression; let us call the concept to which such a concept-expression refers, the *principal concept of the universal thought.* Although Frege's hierarchy of *concepts* escapes Russell's paradox of the predicate which holds of all predicates that are not predicable of themselves, there appears to be nothing to prohibit the concept χ that holds of all universal *thoughts* that do not fall under their principal concept. Does the universal thought expressed by, for instance, "Everything is χ" fall under χ? If it does, it violates the defining characteristic of χ, and so does not fall under χ. And if it does not, it violates the universality of χ, and therefore, *does* fall under χ.

The informal argument we have just given can be found in Myhill [1958]. More explicitly, define a one–many relation h on the domain of Fregean concepts which relates every concept φ to a universal thought $\forall p(\varphi(p))$ of which it is the principal concept: φhq if, and only if, $q = \forall p(\varphi(p))$; h is one–many since there may be distinct universal thoughts whose constituent senses determine φ as their principal concept. At the level of the *senses* φ_s of concepts φ, the situation for Fregean thoughts exactly parallels Russellian propositions and propositional functions: the theory implies that the relation between universal thoughts and the *senses* of their principal concepts maps one–one the domain of such senses onto the class of universal thoughts. But the derivation of a paradox in Frege's theory requires only that h satisfy the following two conditions: (1) every universal thought is in the range of the relation h, and (2) whenever φhp, ψhq, and $p = q$, $\varphi = \psi$. It is clear that (1) is satisfied. As for (2), if the universal thoughts p and q are the same, so also are their constituent senses φ_s and ψ_s, and since sense *determines* reference, their principal concepts must also be the same.

Now let $q = \forall p(\chi(p))$, where

$$\chi(p) =_{\mathrm{Df}} \exists\varphi(\varphi hp \wedge \neg\varphi(p)).$$

Then χhq. Now suppose that $\chi(q)$, that is,

$$\exists\varphi(\varphi hq \wedge \neg\varphi(q)).$$

It then follows that

$$\exists\varphi(\chi hq \wedge \varphi hq \wedge \neg\varphi(q)),$$

so that, using (2),

$$\exists\varphi(\chi = \varphi \wedge \neg\varphi(q)),$$

whence $\neg\chi(q)$.

Next suppose $\neg\chi(q)$:

$$\forall\varphi(\varphi hq \rightarrow \varphi(q)).$$

Thus, in particular, $\chi hq \Rightarrow \chi(q)$. Since $q = \forall p(\chi(p))$, χhq; and therefore, $\chi(q)$.

Klement ([2001], [2002], ch. 5) formalizes a fragment of the theory of sense and reference sufficient to secure a rigorous derivation of the "the concept/*Gedanke* antinomy" within this formalization of the theory, and argues that since the fragment does not employ Basic Law V, the paradox reveals an inconsistency in *Gg*'s philosophy of *language*, one that arises independently of its theory of *classes*. Klement is certainly correct to observe that the difficulty goes beyond Frege's theory of classes. But the source of the difficulty is a general and pervasive feature of the theory of concepts and objects, a feature that lies at the basis of both paradoxes, yielding one inconsistency in the context of classes and another in the context of thoughts. Because classes and thoughts are themselves objects, we can always "diagonalize out" of any attempt to specify a representative object for each concept—the difficulty is the same whether that object is taken to be a thought or a class. The essential similarity between the concept/class and concept/thought paradoxes can be made intuitively compelling as follows: Recall that an object belongs to a class if it falls under the concept whose extension the class is. Similarly, let us say that an object *belongs to a universal thought* if it falls under the principal concept of the thought. Then the concepts expressed by "x is the class of all classes that do not belong to themselves" and "x is the universal thought of all universal thoughts that do not belong to themselves" both lack objectual representatives.

As we remarked earlier, Frege shows that, given any map f from concepts to objects, we can always find concepts φ and ψ such that $f\varphi = f\psi$ but $\varphi(f\varphi) \wedge \neg\psi(f\varphi)$;[16] not only are distinct concepts necessarily mapped onto the same object, but we can actually specify an object which distinguishes φ and ψ in terms of the mapping f. Thus, the fact that there cannot be an injection of concepts into objects is entailed by a consistent fragment of *Gg*, and this shows that the difficulty posed by Basic Law V is an instance of a more general incoherence in the Fregean theory of concepts and objects.

In the case of thoughts, Frege might argue that they, at least, are not *proper* objects—not arguments to concepts of first level. Thoughts are *like* objects in being "saturated"—in corresponding to "complete" expressions; but this is only one component of what characterizes something as an object. The most important component of being an object—that of being an argument to a concept of first level—may not be a feature that thoughts enjoy; rather, they might form a different hierarchy from concepts and objects proper. But an examination of the argument leading to the paradox shows it to require only that thoughts fall under the concepts appropriate to them and is unaffected if thoughts and the concepts they fall under form a separate hierarchy from ordinary objects and concepts.[17]

Frege's proof of the infinity of the numbers does not require any assumptions about the objectual character of thoughts, but rests, as we saw, on the thesis that *numbers* are objects. Frege's proof is therefore in no way compromised if the hierarchy of thoughts and the concepts they fall under is distinguished from the realm of objects and the hierarchy of concepts under which they fall. But the foundational significance of Frege's account of our knowledge of the infinity of the numbers does rest on the *generality* of the framework of concepts and objects within which his account is presented. The Russell–Myhill paradox shows that

[16] See the appendix Frege added to *Gg* in response to Russell's letter of June 16, 1902, informing him of the contradiction.

[17] We are indebted to Kevin Klement for this observation.

this generality may need to be constrained, and this inevitably raises a question regarding the basis for settling on one or another set of restrictions.

Let us briefly review the situation we have arrived at. The consistency of *Bg* shows that the generality of the Comprehension Scheme for properties is not by itself problematic. The consistency of the first-order fragment of *Gg* shows that the mere combination of concepts and objects does not necessarily run into difficulty. The consistency of the predicative fragment of *Gg* shows that when the notion of a Fregean concept is suitably restricted, every allowable concept can have a *class* as its (bi-unique) representative object. Given the Comprehension Scheme of *Bg*, a difficulty arises if the objects which represent concepts are in turn capable of falling under the concepts of which they are the objectual representative. Whether these objects are taken to be *classes* is quite irrelevant to the derivation of an inconsistency. That thoughts and classes should exhibit this feature appears unavoidable in view of the logicist's conception of their generality. As Frege says in the motto we have taken from *Bg*, thoughts have a generality that sets them apart from intuitions, even intuitions given a priori. And a comparable generality is arguably also unavoidable in any notion of class capable of supporting the idea that classes are *logical* objects. Number escapes paradox without sacrificing its claim to generality because the numerical correlates of concepts are not representative of them—they are given by a relation which is many–one, but not one–one.

Dedekind's Account of Our Knowledge of the Infinity of the Numbers

To a great extent, Dedekind's and Frege's treatments of the natural numbers are on a par. Frege's theorem shows that if the finite cardinals are introduced by way of Hume's principle and the successor relation is defined in terms of the cardinality operator, then the finite cardinals together with *successor* satisfy the Dedekind–Peano axioms. Conversely, Dedekind shows (theorem 120) that the version of Hume's principle which results when the concept variables range over only *finite* concepts, is obtainable in the system of *Was sind.* But despite these formal similarities, there are considerable differences in their accounts of our knowledge of the existence and infinity of the natural numbers. We have described Frege's methodology at some length. Let us now consider Dedekind's.

Dedekind defines a system to be (*Dedekind*) *infinite* if and only if it can be put into one-to-one correspondence with a proper subsystem of itself. He defines a *simply infinite* system to be any system which forms the domain of a relation satisfying the Dedekind–Peano axioms. He then proves that every infinite system contains a simply infinite subsystem (theorem 72). Now suppose that we have already proved that there is an *infinite* system. Then we know there is a simply infinite system which, together with its relation, models the Dedekind–Peano axioms. Once we are in possession of a simply infinite system, we

> entirely neglect the special character of [its] elements, simply retaining their distinguishability and taking into account only the relations to one another in which they are placed.... [T]hese elements [are] called natural numbers or ordinal numbers or simply numbers.... With reference to this freeing [of] the elements from every other content (abstraction) we are justified in calling numbers a free creation of the human mind. (*Was sind*, p. 68)

We should record an important difference between Frege's account of number and the account of Dedekind. For Dedekind, the exhibition of a simply infinite system is designed to yield the *concept* of number; it does not purport to fix the identity of the individual numbers—indeed, Dedekind is opposed to any such extension of his ideas. By contrast, Frege does attempt this in order to secure their status as "self-subsistent objects." On Dedekind's analysis, the existence of numbers is a consequence of a mental power to abstract from particular characteristics of the elements of a simply infinite system to its ordinal structure.

The success of Dedekind's methodology hinges on the existence of an *infinite* system. As Dummett ([1991 a], pp. 49ff.) has observed, Dedekind sought to secure the existence of such a system by the provision of a concrete example, one from which the concept of number could be viewed as having been abstracted in the manner just reviewed. It would then follow by his categoricity theorem (theorem 132) that the peculiarities of the starting point of the construction can be discounted, and the *generality* of the construction of the infinite system secured. This is the goal of theorem 66, whose proof runs as follows:

> The world of my thoughts, i.e., the totality S of all things that can be objects of my thought, is infinite. For if s denotes an element of S, then the thought s', that s can be an object of my thought, is itself an element of S. If s' is regarded as the image $\varphi(s)$ of the element s, then the mapping φ on S determined thereby has the property that its image S' is a part of S and indeed S' is a proper part of S, because there are elements in S (e.g., my own ego), which are different from every such thought s' and are therefore not contained in S'. Finally it is clear that if a, b are different elements of S, then their images a', b' are also different, so that the mapping φ is distinct (similar). Consequently, S is infinite, q.e.d.

Following Frege,[18] we may say that for Dedekind the realm of thoughts is an objective one, one that exists independently of us. Our access to this realm is taken for granted, since thoughts are transparent to our reason. The realm of thoughts

[18] As, indeed, Frege [1897] suggests; see Hermes et al. ([1980], p. 136).

yields an exemplar of an infinite system because the relation, *x is an object of my thought y*, is a one-one function which maps a subsystem of objects of my thought onto a proper part of itself. From this exemplar we abstract the concept of an infinite system, and thus of number. For Dedekind's proof to succeed, it is essential that thoughts be *proper* objects—arguments to concepts of first level, in fact to the concept *x is an object of my thought*, one that holds of thoughts and of objects which are not thoughts. Dedekind says very little about the nature of thoughts, so it can hardly be urged against him that his account rests on an inconsistent theory of thoughts. But were we to regard Dedekind's tacit theory of thoughts as essentially Frege's—together with the further condition that thoughts are proper objects—the concept/thought paradox would be an obstacle to the success of his account of our knowledge of the infinity of the numbers: if the proof of theorem 66 fails, we lack an exemplar from which the concept of number can be abstracted and cannot be said to have an account of how we come to know the numbers' most salient property.

Russell's Logicism and the Rejection of Denoting Concepts

For Russell, no satisfactory solution of the problem of our knowledge of arithmetic is possible so long as the theory of concepts and objects is not on a secure footing. The character of this footing will determine to what extent logicism can be judged a success. The first steps toward a solution came with the discovery of the theory of descriptions. Its contributions to the philosophy of arithmetic were at least twofold: (1) the clarification of the theory of generality and the consequent simplification of the theory of concepts and objects by the elimination of denoting concepts; (2) the method of contextual analysis.

The *Principles* theory of generality was based on the notion of a denoting concept. Such concepts were important because without them, Russell believed, our knowledge of the infinite combinations of things that correspond to words of quantification—hence also our ability to express, by finite means, thoughts involving infinite totalities—would be severely limited:

> Indeed it may be said that the logical purpose which is served by the theory of denoting is, to enable propositions of finite complexity to deal with infinite classes of terms: this object is effected by *all*, *any*, and *every*, and if it were not effected, every general proposition about an infinite class would have to be

> infinitely complex. Now, for my part, I see no possible way of deciding whether propositions of infinite complexity are possible or not; but this at least is clear, that all the propositions known to us (and, it would seem, all propositions that we *can* know) are of finite complexity. It is only by obtaining such propositions about infinite classes that we are enabled to deal with infinity; and it is a remarkable and fortunate fact that this method is successful. (*Principles* §141)

But Russell came to view denoting concepts as bringing with them many of the difficulties he associated with Fregean senses. A denoting concept is like the predicable notion of a concept in applying to many things, but in some cases, at least, it is like an object or "term" insofar as it is expressed or "indicated" by a singular term. The central case is that of the denoting concepts which we express with definite descriptions. If such concepts could be eliminated, this would mark a significant simplification and would secure the theory of concepts and objects by removing one potential source of paradox. The means by which this was achieved by the theory of descriptions—the method of contextual analysis—constitutes the second contribution of the theory to the articulation of Russell's logicism.

It has for a long time been supposed that the central achievement of the theory of descriptions was the elimination of the *denotations* of denoting phrases that express vacuous or nondenoting denoting concepts. But the theory of *Principles*, and almost all of those theories, such as Frege's, to which it bore any resemblance, were perfectly capable of representing the possibility that a denoting phrase might express a sense while lacking a reference. The really significant and novel achievement of Russell's theory was its elimination of denoting *concepts*.[19] The way this was achieved by the method of contextual analysis is as follows: Given a sentence S containing, for example, a definite description, the theory specifies how to transform S into a logically equivalent sentence S^T in which the descriptive phrase does not occur. S^T gives the correct logical form of the proposition expressed by S, showing it to be quantificational (in fact of the form $\exists\forall$) rather than subject–predicate, and showing, as a consequence, that it was illusory to have sought the denotation of a denoting concept at all, since S^T shows that there is not, in the proposition expressed by S, anything that even purports to be such a symbolic entity as a denoting concept. A contextual analysis differs from an ordinary analysis, such as one involving an explicit definition, by showing how to transform any "context"—any sentence—in which the expression to be analyzed occurs into an equivalent sentence, without thereby assuming that the expression contributes an entity to the proposition expressed by the sentences in which it occurs.

On Russell's theory, our *understanding* of the proposition expressed by S proceeds by our being acquainted with its constituents; what these constituents

[19] As has been made particularly clear by Noonan [1996].

are, we can know only relative to an analysis. The theory of descriptions shows that when a sentence contains a description, the proposition it expresses contains only ordinary, predicable concepts and individuals—not denoting concepts; in particular, therefore, our understanding of such propositions does not require acquaintance with denoting concepts. An important and insufficiently appreciated aspect of Russell's theory is the manner by which it allows us to distinguish propositions that we understand from those that, because they contain among their constituents things which fall outside our acquaintance, can only be asserted.[20] We succeed in knowing *about* such things because the propositions we *do* understand contain propositional functions that hold uniquely of them. To take a simple, nonmathematical illustration of Russell's theory of propositional understanding, since I am not acquainted with Bismarck, any proposition I can understand which purports to say something about him must contain a propositional function which holds uniquely of Bismarck and is composed wholly of constituents with which I am acquainted. An equivalent proposition which contains Bismarck is not one I can *understand*; but in view of the fact that there is a proposition that I do understand, one that contains a suitable propositional function which holds only of Bismarck, the limitations on the scope of my acquaintance need not impose a limitation on what I can have knowledge *about*.

The logical theory consisting of some sort of hierarchy of types of propositional functions seeks to preserve certain features of the theory of propositional understanding that, together with the theory of descriptions, forms the backbone of Russell's theory of meaning. As Ramsey [1931] was perhaps the first to make fully explicit, what is distinctive of this logical theory is its division of propositional functions into *two* hierarchies: one satisfying an intricate system of restrictions whose natural justification is that they are forced by reflection on how propositional functions are known by us, and another, unconstrained by considerations of this sort and reflecting the purely logical restrictions of the simple theory of types.[21]

The immediate inspiration for ramified-type theory is the intuitive, informal principle that Russell seems to have drawn from Poincaré's diagnosis of the source of the logical paradoxes, the Vicious Circle Principle: "Whatever involves all of a collection must not be one of the collection" (Russell [1908], p. 63). Russell gives a number of glosses on this idea. In a footnote he adds, "When I say that a collection has no total, I mean that statements about all its members are nonsense"

[20] The distinction between asserted and expressed propositions and its importance for Russell's theory of propositional understanding is further elaborated in Demopoulos [1999].

[21] Here we are indebted to a suggestion of Goldfarb [1989].

(Russell [1908], p. 63).[22] And again, "No totality can contain members defined in terms of itself[; . . . w]hatever contains an apparent variable must not be a possible value of that variable" (Russell [1908], p. 75). In *Principia Mathematica* (*PM*) Russell gives the following formulation:

> Given any set of objects such that, if we suppose the set to have a total, it will contain members which presuppose this total, then such a set cannot have a total. By saying that a set has "no total" we mean, primarily, that no significant statement can be made about "all its members." (p. 37)

The theory of order effects the main constraint of the Vicious Circle Principle since it rules out the impredicative propositional functions which Russell thought were directly implicated in the paradoxes, where an *impredicative propositional function* is one that contains a quantifier that binds a variable whose range contains the propositional function itself as a member. Individuals have order 0, as do the variables ranging over them. Propositional functions of individuals of order 1 then take individuals as arguments. The arguments of propositional functions of individuals of order 2 include propositional functions of order 1; those of order 3 include propositional functions of order 2, and so on. If a propositional function φ of individuals is presented in terms of quantification over other propositional functions of individuals, φ will fail to respect the principle underlying ramification if it lies within the range of one of its constituent variables. More generally, the presentation of a function φ—whether of functions or of individuals—of order $n > 1$ depends on the presentation of functions of order less than n, since φ quantifies over variables within whose range such functions lie. φ is precluded from falling within the range of one of its quantified variables because the presentation of φ is understood as following on the presentation of its constituents. The principle underlying ramification is entirely plausible and in no way ad hoc if the ramified hierarchy is held to reflect the fact that our epistemic access to functions higher in the hierarchy depends on our access to those below it. A violation of such a constraint on epistemic access would indeed be viciously circular.

In this connection, it is important to appreciate that the propositional functions of *PM* are not fixed in advance: its "symbolic resources" are not bounded in the manner that is familiar to us from our experience with formalized languages,

[22] The expression "has no total" is an allusion to the idea of "indefinitely extensible" concepts or domains. Building on Russell, Dummett has made much of the notion of indefinitely extensible domains, seeing in their existence the origin of the contradiction which afflicted Frege's theory of classes and an independent argument for intuitionistic logic as the natural logic of classes (Dummett [1991c], ch 17; [1994]). We haven't space to go into this subject here. For an extended commentary see Boolos [1993], Clark [1993], [1994], [1998], and Oliver [1998].

but are susceptible of indefinite extension by the addition of new functions as applications of the framework may demand. Although our *access* to a function φ is mediated by quantification over other functions, this in no way precludes the *existence*, within the hierarchy, of extensionally equivalent *predicative functions* or functions of order 1—that is, functions that do not contain quantification over other functions. The restrictions respected by predicative functions are merely the type restrictions of the *simple* hierarchy. The orders that ramification imposes constrain our epistemic access to propositional functions of order greater than 1. The hierarchy of orders of the ramified theory is compatible with the existence, independently of our knowledge of them, of predicative (i.e., order 1) equivalents of functions of all orders. But a predicative function may be such that what knowledge we have *about* it is mediated by access to a logically equivalent function of order greater than 1. The assumption that for every propositional function of order n there is an equivalent propositional function of order 1, is an *axiom of reducibility*. The infinite collection of such axioms—one for every order of propositional function greater than 1—constitutes "the" axiom of reducibility of *PM*.

There is a suggestive comparison between the ramified theory with reducibility and Russell's theory of propositional understanding: Not being acquainted with an individual *i*, reference to it requires a description (i.e., a propositional function which holds uniquely of *i*—a "descriptive function" for *i*. An "axiom of acquaintance" that parallels reducibility would assert that every descriptive function for an individual *i* has an equivalent that is composed wholly of constituents with which I am acquainted. The axiom thus enforces a concordance between the domain of individuals picked out by descriptive functions—presumably all individuals—and those picked out by such "good" descriptive functions as the axiom postulates. The functions corresponding to good descriptive functions are the analogues in the theory of propositional understanding of the predicative functions of the ramified hierarchy. Ordinary descriptive functions correspond to propositional functions of order greater than 1; individuals correspond to classes. The axioms of acquaintance and reducibility postulate (respectively) the possibility of knowing individuals and classes in terms of functions that possess a certain epistemic transparency, a transparency embodied by acquaintance in the one case and the absence of complex forms of quantification in the other. Classes occur in the hierarchy only under the guise of predicative functions, which are the means by which they are known. Reducibility thus postulates a concordance between mathematical reality and our knowledge of it that the ramified theory is otherwise unable to demonstrate. It is precisely for this reason—its possession of the virtues of theft over honest toil—that the axiom's necessity was regarded as a serious flaw. An important foundational question arises in this connection, one which brings into sharp relief a basic and pervasive difference between Frege's and Dedekind's developments of logicism and Russell's. This is the analysis of mathematical induction and the definition of the set of natural numbers, to which we now turn.

Russell's Account of Reasoning by Induction

The tradition of predicative analysis that originates with Weyl [1987][23] proceeds on the assumption that the system N of natural numbers is not in need of definition or reconstruction but is available to us. The considerable achievements of this tradition would not have been possible without this assumption. By contrast, logicism sought to account for N, impredicatively in the case of Frege and Dedekind, predicatively in the case of Russell.

The Frege–Dedekind definition of x *is natural number* amounts to the condition, called (§) above, that x is a natural number if and only if x has every hereditary property of 0. But this is excluded by ramification since "every hereditary property of zero" quantifies over properties of 0 of *all* orders. The proper formulation in *Principia Mathematica* (*PM*) would be "x has every predicative property possessed by 0." But if we know that for each propositional function there exists a predicative propositional function coextensive with it, then the two conditions "every hereditary property of zero" and "every predicative hereditary property of zero" are, in fact, logically equivalent and condition (§) is simply the principle of mathematical induction in full generality. That for each propositional function there exists a coextensive predicative function is guaranteed by an axiom of reducibility.

In his Appendix B to the 1925 edition of *PM*, Russell attempted to dispense with this axiom in the justification of the definition of *natural number.* Gödel ([1944], p. 145) noted a mistake in Russell's proof of the fundamental lemma, *89.16. In the terminology of Hellman and Feferman ([1995], p. 4), Russell's error occurs in the context of his attempt to prove that every subclass of a finite set is itself finite. (This lemma is equivalent to an axiom of Hellman and Feferman's Elementary Theory of Finite Sets and Classes, which they use to give a "predicative" characterization of N.) Myhill [1974] later showed that without reducibility, the set of natural numbers is not definable in the system of *PM* (1910).[24]

[23] See chapter 19 in this volume for a discussion and references.

[24] Landini [1996] does not contradict Myhill [1974]. Indeed, according to Landini (personal communication—W.D.), the system Landini [1996] attributes to the 1925 *PM* contains axioms of extensionality equivalent to reducibility.

Russell on the Dedekind Infinity of the Numbers

Call a set x *inductive* if it belongs to every class A of sets such that A contains the empty set Λ of individuals, and for any set y and individual a, if $y \in A$, then $y \cup \{a\} \in A$. (Compare, an inductive *number* is one which belongs to every set x such that $0 \in x$ and for any n, if $n \in x$, so is $n+1$.) x is *noninductive* if it is not inductive, and x is *reflexive (Dedekind infinite)* if it can be injected into a proper subset of itself. *PM*'s axiom of infinity tells us that the set V of individuals is noninductive. The axiom says nothing about the infinity or otherwise of the reconstruction of the natural numbers as "Russell-numbers" in *PM* (a reconstruction to which we will soon come). By postulating the axiom of infinity, Russell did *not* take the infinity of the Russell-numbers for granted; instead, he derived their *Dedekind* infinity from the existence of a noninductive set of individuals by a beautiful classical argument based on the following:

Lemma. *If x is noninductive, then the natural numbers can be injected into PPx, the power set of the power set of x.*

Proof. Observe first that Λ is a subset of x, so x has a subset of cardinality 0. Suppose there is a subset y of x of cardinality n. Since x is noninductive, there is an $a \in x-y$. Hence $y \cup \{a\}$ is a subset of x of cardinality $n+1$, and we may conclude that for every n, there is a subset of x of cardinality n. Hence we may define, for each natural number n, a *Russell-number* S_n, given by

$$S_n = \{y: y \subseteq x \text{ and } y \text{ has cardinality } n\}.$$

For each n, S_n is nonempty, $S_n \subseteq Px$, and for $m \neq n$, $S_m \neq S_n$, so that the map, sending n to S_n, injects the natural numbers into PPx, as required. □

It follows from the lemma that the Russell-numbers form a Dedekind infinite subclass of PPx since the "successor" map, that sends S_n to S_{n+1}, is evidently one-one from the S_n into a proper subclass of Russell-numbers. More generally, we have

Russell's theorem: If x is infinite, PPx is Dedekind infinite,

since if a class contains a Dedekind infinite subclass, it is itself Dedekind infinite (compare *PM* II *124.57).

In Russell's development of logicism, the proper analogue of Frege's theorem is this proof of the Dedekind infinity of the Russell-numbers.[25] Russell's analysis

[25] This point is emphasized by Boolos [1994], whose discussion we have followed.

shares with Dedekind's a reliance on an assumption regarding the infinity of a set of objects not internal to arithmetic. Dedekind, as we saw, sought to discharge this assumption by a proof. Russell and Whitehead explicitly left their assumption unproven, conceding that it could not be secured by completely general considerations regarding the nature of concepts and the objects which fall under them. But in the rush to object to the incursion of an evidently empirical assumption into the foundations of arithmetic, critics of *PM* have overlooked the subtlety of its analysis of the infinity of the numbers.

Final Assessment

We have based our discussion of the logicism of Frege, Dedekind, and Russell around its answers to three questions: (1) What is the basis for our knowledge of the infinity of the numbers? (2) How is arithmetic applicable to reality? (3) Why is reasoning by induction justified? It is an extraordinary achievement of these authors that they were able to throw such light upon these questions and to produce such mathematically fruitful answers. It is possible to portray logicism as a failure because of the Zermelo–Russell contradiction and its need to rely on nonlogical principles. And of course it is now entirely clear that even if one thinks of logic as second-order logic, the basic laws of arithmetic are not deducible by logical means alone from the truths of logic; at present it is at best unclear as to whether, in any sense demanded by Frege, numbers can be seen as logical objects, or that our knowledge of the infinity of the numbers is purely logical knowledge. While all that is so, it does absolutely nothing to detract from the depth of the logicist analysis of number. Frege's basic idea gave us quantification theory and the notion of the ancestral; Dedekind's methods provided a general mathematical technique, based upon an informal notion of set (or "system") which in subsequent refinements became extended to the notions of ideal and lattice, both central mathematical tools; and with its development in *PM*, Russell showed just what a powerful method type theory really was. It is possible to see *PM* as a work not in the logicist tradition of Frege and Dedekind but in what is now called "reverse mathematics," that is, as asking and answering the questions "What basic principles of a general kind are needed to develop the whole of classical mathematics?" and "Can these principles be incorporated into a general theory of concepts and objects which is free of the logical paradoxes?" (as Whitehead and Russell themselves expressed their achievement; see *PM* I, p. 59).

The monumental achievement of the work was to exhibit such principles and to actually show how mathematics—not merely elementary arithmetic—might be reconstructed from them. Not only did logicism show how few were the nonlogical assumptions necessary for the development of arithmetic, but particularly in the work of Dedekind and Russell it began to emerge how surprisingly few principles were required to develop all of the working mathematician's cannon, culminating ultimately in the system Z of Zermelo [1908]. The penetration to the mathematical heart of the problem achieved by the logicists is in stark contrast to the sterile, mathematically empty views which preceded them. We owe to logicism first the completion of the revolution in rigor which was so important a part of nineteenth-century mathematics; we owe to it second the discovery that many arithmetical concepts are purely logical concepts; and we owe to it third the most sustained analysis of the relation between thought in general and mathematics in particular that has ever been provided.

Acknowledgments We are particularly indebted to John L. Bell for his comments on earlier drafts; many thanks also to Kevin Klement and Peter Koellner for numerous suggestions that greatly improved our presentation. Stewart Shapiro has been both a patient editor and a valuable source of advice. W.D. thanks the Social Sciences and Humanities Research Council of Canada for support of his research, and P.C. thanks the Leverhulme Foundation for support of his.

BIBLIOGRAPHY

Bell, J.L. [1995]: "Fregean extensions of first-order theories," in Demopoulos [1995], pp. 433–437.

Boolos, G. [1985]: "Reading the *Begriffsschrift*," in Demopoulos [1995], pp. 163–181, and in Boolos [1998], pp. 155–170.

Boolos, G [1990]: "The standard of equality of numbers," in Demopoulos [1995], pp. 234–254, and in Boolos [1998], pp. 202–219.

Boolos, G. [1993]: "Whence the contradiction?," in Boolos [1998], pp. 220–236.

Boolos, G. [1994]: "The advantages of honest toil over theft," in Boolos [1998], pp. 255–274.

Boolos, G. [1998]: *Logic, logic, and logic*, Cambridge, MA: Harvard University Press.

Brandl, J., and P. Sullivan (eds.) [1998]: *New essays on the philosophy of Michael Dummett, Grazer Philosophische Studien*, vol. 55, Amsterdam: Rodopi.

Chihara, C.S. [1972]: "Russell's theory of types," in Pears [1972], pp. 245–289.

Chihara, C.S. [1973]: *Ontology and the vicious circle principle*, Ithaca, NY: Cornell University Press.

Clark, P. [1993]: "Sets and indefinitely extensible concepts and classes," *Proceedings of the Aristotelian Society*, 97, pp. 235–249.

Clark, P. [1994]: "Poincaré, Richard's paradox and indefinite extensibility," *Proceedings of the 1994 Biennial Meeting of the Philosophy of Science Association*, vol. 2, pp. 227–235.

Clark, P. [1998]: "Dummett's argument for the indefinite extensibility of *set* and *real number*," in Brandl and Sullivan [1998], pp. 51–63.
Church, A. [1984]: "Russell's theory of identity of propositions," *Philosophia Naturalis*, 21, pp. 513–522.
Coffa, A. [1981]: "Russell and Kant," *Synthése*, 46, pp. 246–264.
Crossley, J. [1973]: "A note on Cantor's theorem and Russell's paradox," *Australian Journal of Philosophy*, 73, pp. 70–71.
Dedekind, R. [1963]: *Essays on the theory of numbers*, W.W. Beman (ed. and trans.), New York: Dover.
Demopoulos, W. [1994]: "Frege and the rigorization of analysis," in Demopoulos [1995], pp. 68–88.
Demopoulos, W. (ed.) [1995]: *Frege's philosophy of mathematics*, Cambridge, MA.: Harvard University Press.
Demopoulos, W. [1998]: "The philosophical basis of our knowledge of number," *Noûs*, 33, pp. 481–503.
Demopoulos, W. [1999]: "On the theory of meaning of 'On denoting,'" *Noûs*, 33, pp. 439–458.
Demopoulos, W. [2001]: "Michael Potter: *Reason's nearest kin: Philosophies of arithmetic from Kant to Carnap*," *British Journal for the Philosophy of Science*, 52, pp. 599–612.
Demopoulos, W. [in press]: "Our knowledge of numbers as self-subsistent objects," *Dialectica*.
Dummett, M. [1991a]: "Frege and the paradox of analysis," in Dummett [1991b], pp. 17–52.
Dummett, M. [1991b]: *Frege and other philosophers*, Oxford: Clarendon Press.
Dummett, M [1991c]: *Frege: Philosophy of mathematics*, Cambridge, MA: Harvard University Press.
Dummett, M [1993a]: "What is mathematics about?," in Dummett [1993b], pp. 429–445.
Dummett, M. [1993b]: *The seas of language*, New York: Oxford University Press.
Dummett, M. [1994]: "Chairman's address: Basic Law V," *Proceedings of the Aristotelian Society, 94*, pp. 243–251.
Feferman, S., and Hellman, G. [1995]: "Predicative foundations of arithmetic," *Journal of philosophical logic*, 24, pp. 1–17.
Frege, G. [1879]: *Begriffsschrift, eine der arithmetischen nachgebildete Formelsprache des reinen Denkens*, English translation: *Begriffsschrift, a formula language, modeled upon that of arithmetic, for pure reason*, S. Bauer-Mengelberg (trans.), in van Heijenoort [1967], pp. 1–82, Cambridge, MA: Harvard University Press.
Frege, G. [1884]: *Die Grundlagen der Arithmetik: Eine logisch-mathematische Untersuchung über den Begriff der Zahl*, Breslau: W. Koebner. English translation: *The foundations of arithmetic: A logico-mathematical enquiry into the concept of number*, 2nd rev. ed., J.L. Austin (trans.), Evanston, IL: Northwestern University Press, 1968.
Frege, G. [1892]: "Über Sinn und Bedeutung." English translation: "On sense and reference," M. Black (trans.), in Geach and Black [1952], pp. 56–78.
Frege, G. [1893–1903]: *Grundgesetze der Arithmetik*, 2 vols., Jena: H. Pohle. Partial English translation: *The basic laws of arithmetic: Exposition of the system*, M. Furth (trans.), Berkeley: University of California Press, 1964.

Frege, G. [1897]: "Logic," in Hermes et al. [1980], pp. 126–151.

Fuhrmann, A. [2002]: "Russell's way out of the paradox of propositions," *History and Philosophy of Logic*, 23, pp. 197–213.

Geach, P., and M. Black (eds.) [1952]: *Translations from the philosophical writings of Gottlob Frege*, Oxford: Basil Blackwell.

Gödel, K. [1944]: "Russell's mathematical logic," in Schilpp [1944], pp. 123–154.

Goldfarb, W. [1989]: "Russell's reasons for ramification," in Savage and Anderson [1989], pp. 24–40.

Heck, R.G. [1995]: "Definition by induction in Frege's *Grundgesetze der Arithmetik*," in Demopoulos [1995], pp. 295–333.

Heck, R.G. [1996]: "The consistency of predicative fragments of Frege's *Grundgesetze der Arithmetik*," *History and Philosophy of Logic*, 17, pp. 209–220.

Hermes, H., et al. (eds.) [1980]: *Posthumous writings of Gottlob Frege*, Chicago: University of Chicago Press.

Kitcher, P., and W. Aspray (eds.) [1988]: *History and philosophy of modern mathematics*, *Minnesota Studies in the Philosophy of Science*, vol. 11, Minneapolis: University of Minnesota Press.

Klement, K.C. [2001]: "Russell's paradox in Appendix B of *The principles of mathematics: Was Frege's response adequate?*," *History and Philosophy of Logic*, 22, pp. 13–28.

Klement, K.C. [2002]: *Frege and the logic of sense and reference*, London: Routledge.

Landini, G. [1996]: "The *definability* of the set of natural numbers in the 1925 *Principia*," *Journal of Philosophical Logic*, 25, pp. 597–614.

Lewis, H.D. (ed.) [1979]: *Bertrand Russell: Memorial volume*, London: George Allen and Unwin.

Linsky, B. [1999]: *Russell's metaphysical logic*, Stanford, CA: CSLI Publications.

Marsh, R. (ed.) [1956]: *Logic and knowledge*, London: Allen and Unwin.

McGuinness, B. (ed.) [1980]: *Philosophical and mathematical correspondence of Gottlob Frege*, H. Kaal (trans.), Oxford: Blackwell.

McGuinness, B. (ed.) [1984]: *Gottlob Frege: Collected papers on mathematics, logic and philosophy*, Oxford: Blackwell.

Monk, R., and A. Palmer (eds.) [1996]: *Bertrand Russell and the origins of analytical philosophy*, Bristol, UK: Thoemmes Press.

Myhill, J. [1958]: "Problems arising in the formalization of intensional logic," *Logique et analyse*, 1, pp. 78–83.

Myhill, J. [1974]: "The undefinability of the set of natural numbers in the ramified *Principia*," in Nakhnikian [1974], pp. 19–27.

Myhill, J. [1979]: "A refutation of an unjustified attack on the Axiom of Reducibility," in Lewis [1979], pp. 81–90.

Nakhnikian, G. (ed.) [1974]: *Bertrand Russell's philosophy*, London: Duckworth.

Noonan, H. [1996]: "The 'Gray's Elegy' argument and others," in Monk and Palmer [1996], pp. 65–102.

Oliver, A. [1998]: "Hazy totalities and indefinitely extensible concepts," in Brandl and Sullivan [1998], pp. 25–50.

Parsons, C. [1965]: "Frege's theory of number," reprinted in Demopoulos [1995], pp. 182–210.

Parsons, T. [1987]: "On the consistency of the first-order portion of Frege's logical system," in Demopoulos [1995], p. 422–431.

Pears, D.F. (ed.) [1972]: *Bertrand Russell. A collection of critical essays*, New York: Anchor Books.

Potter, M. [2000]: *Reason's nearest kin: Philosophies of arithmetic from Kant to Carnap*, Oxford: Oxford University Press.

Ramsey, F.P. [1931]: "The foundations of mathematics," in *The foundations of mathematics and other logical essays*, R.B. Braithwaite (ed.), London: Routledge and Kegan Paul.

Russell, B. [1903]: *The principles of mathematics*, London: Allen and Unwin.

Russell, B. [1905]: "On denoting," in Marsh [1956], pp. 41–56.

Russell, B. [1908]: "Mathematical logic as based on the theory of types," reprinted in Marsh [1956], pp. 59–102.

Russell, B. [1919]: *Introduction to mathematical philosophy*, London: Allen and Unwin.

Russell, B. [1994]: *The collected papers of Bertrand Russell, Vol. IV, Foundations of logic 1903–05*, McMaster University ed., A. Urquhart and A.C. Lewis (eds.), London and New York: Routledge.

Savage, C.W., and C.A. Anderson (eds.) [1989], *Rereading Russell: Essays in Bertrand Russell's metaphysics and epistemology, Minnesota Studies in the Philosophy of Science*, vol. 12, Minneapolis: University of Minnesota Press.

Schilpp, P.A. (ed.) [1944]: *The philosophy of Bertrand Russell*, New York: Harper Torchbooks.

Shapiro, S. [2000]: *Thinking about mathematics*, Oxford: Oxford University Press.

Stein, H. [1988]: "Logos, logic and logistiké: Some philosophical remarks on the nineteenth-century transformation of mathematics," in Kitcher and Aspray [1988], pp. 238–259.

Stein, H. [1998]: "Logicism," in *Routledge encyclopedia of philosophy*, E. Craig (ed.), pp. 811–817. London and New York: Routledge.

Urquhart, A. [2003]: "The theory of types," in *The Cambridge companion to Russell*, N. Griffin (ed.), pp. 286–309. Cambridge: Cambridge University Press.

Van Heijenoort, J. (ed.) [1967]: *From Frege to Gödel: A source book in mathematical logic, 1879–1931*, Cambridge, MA: Harvard University Press.

Weyl, H. [1987]: *The Continuum. A Critical Examination of the Foundations of Analysis.* English translation of 1918 edition by S. Pollard and T. Bole, Kirksville, MA: Thomas Jefferson Press.

Whitehead, A.N., and B. Russell [1910–1913]: *Principia mathematica*, Vols 1–3, Cambridge: Cambridge University Press.

Zermelo, E. [1908]: "Investigations in the foundations of set theory I," in Van Heijenoort [1967], pp. 199–215.

CHAPTER 6

LOGICISM IN THE TWENTY-FIRST CENTURY

BOB HALE

CRISPIN WRIGHT

1. Frege's Philosophy of Mathematics and the Neo-Fregean Program

Frege believed, at least for most of his career, that the fundamental laws of elementary arithmetic—the theory of the natural numbers (finite cardinals)—and real analysis are *analytic*, in the sense he explained in *Grundlagen* §3—that is, provable on the basis of general logical laws together with suitable definitions. Since his defense of this thesis depended upon taking those theories to concern a realm of independently existing objects (*selbständige Gegenstände*)[1]—the finite cardinal numbers and the real numbers—his view amounted to a *Platonist* version

[1] "Independently existing" means, roughly, "not dependent for their existence upon objects of any other kind"—for a somewhat fuller expression of the idea, see Hale and Wright (2001), essay 14, sec. 5. It does not, as some structuralists have supposed, mean that the natural numbers, for example, exist independently of *one another*!

of *logicism*. Neo-Fregeanism[2] holds that he was substantially right in both these components of his philosophy, but takes a more optimistic view than Frege himself did of the prospects for contextual explanations of fundamental mathematical concepts—such as those of cardinal number and real number—by means of what are now widely called *abstraction principles*. Generally, these are principles of the form

$$\forall\alpha\,\forall\beta(\Sigma(\alpha) = \Sigma(\beta) \leftrightarrow \alpha \approx \beta),$$

where Σ is a term-forming operator applicable to expressions of the type of α,β and $\approx$ is an equivalence relation on entities denoted by expressions of that type. The type of α,β may be that of singular terms, in which case $\approx$ is a first-level relation and the resulting abstraction principle represents a *first-order* abstraction on *objects*. A well-known example, in terms of which Frege conducted most of his own discussion of this type of explanation, is the Direction Equivalence.[3]

> The direction of line a = the direction of line b if and only if lines a and b are parallel.

A potentially more important class of *higher-order* abstractions results from taking the type of α and β to be that of first (or higher)-level predicates, with $\approx$ a correspondingly higher-level relation. One much discussed—and, for our purposes, crucial—principle of this kind is what has come to be known as *Hume's Principle*: the abstraction by means of which, on the proposal reviewed in the central sections (§§60–68) of *Grundlagen*, the concept of (cardinal) number may be explained:

> The number of *F*s = the number of *G*s if and only if there is a one-one correlation between the *F*s and the *G*s

As Frege went on to point out (*Grundlagen* §§71–72), what is required for there to be a one-one correlation between the *F*s and the *G*s may itself be explained in purely logical terms.[4]

As is well known, Frege very swiftly came to the conclusion that the explanatory purport of abstraction principles is severely qualified by an objection which had earlier led him, at *Grundlagen* §56, to reject as inadequate an attempt to

[2] Often called "neologicism," and sometimes—better but more cumbersome—"neo-Fregean Logicism." The approach to be discussed in what follows differs in fundamental respects from other recently defended forms of logicism, such as the quantificational approach of Bostock (1974, 1979) and the abstraction-free approach of Tennant (1987). These are well worth separate study but the requisite comparisons would take us too far afield here.

[3] More accurately, its universal closure; likewise with Hume's Principle as stated below.

[4] In standard notation, thus: $\exists R\,\forall x(Fx \rightarrow \exists y(Gy \wedge \forall z(Rxz \leftrightarrow z = y)) \wedge (Gx \rightarrow \exists y(Fy \wedge \forall z(Rzx \leftrightarrow z = y)))$.

define number by recursive definition of the series of numerically definite quantifiers of the form "there are exactly n Fs" As he there puts it:

> ... we can never—to take a crude example—decide by means of our definitions whether any concept has the number JULIUS CAESAR belonging to it, or whether that same familiar conqueror of Gaul is a number or not.[5]

In his view, the very same difficulty fatally infects the attempt to define *number* by means of Hume's Principle: the proposed definition provides us with a means to decide whether the number of *F*s = q, when q is given in the form "the number of *G*s"—but it apparently entirely fails to do so, when q is not a term of that form (nor a definitional abbreviation of such a term). Perceiving no way to solve the problem, Frege concluded that an altogether different form of definition is needed and opted for his famous explicit definition in terms of extensions:

> The number of *F*s = the extension of the concept "equal to the concept *F*"

in effect, that the number of *F*s is the class containing exactly those concepts *G* such that the *G*s are one-one correlated with the *F*s. To derive the fundamental laws of arithmetic with the aid of this definition, Frege obviously required an underlying theory of extensions or classes. This he sought to provide, in *Grundgesetze*, by means of his Basic Law V, which governs what Frege called value ranges of functions (concepts being a particular kind of function, on Frege's account). As regards extensions of concepts, what Basic Law V asserts is that the extensions of two concepts are identical just in case those concepts have the same objects falling under them, that is (in class notation):

$$\forall F \forall G(\{x: Fx\} = \{x: Gx\} \leftrightarrow \forall x(Fx \leftrightarrow Gx)).$$

As very soon became apparent, this is disastrous, since Basic Law V leads to Russell's Paradox. Frege sought at first to modify his law V so as to avoid the inconsistency, but to no avail, and finally abandoned his attempt to provide a logical foundation for arithmetic and analysis as a complete failure.

Neo-Fregeanism holds that Frege need not have taken the step which led to this unhappy conclusion. At least as far as the theory of natural numbers goes, Frege's central mathematical and philosophical aims may be accomplished by basing the theory on Hume's Principle, adjoined as a supplementary axiom to a suitable formulation of second-order logic. Hume's Principle cannot, to be sure, be taken as a definition in any strict sense—any sense requiring that it provide for the eliminative paraphrase of its definiendum (the numerical operator, "the number of...") in every admissible type of occurrence. But this does not preclude its being viewed as an *implicit* definition, introducing a sortal concept of cardinal number

[5] Frege (1884), p. 68.

and, accordingly, as being analytic of that concept—and this, the neo-Fregean contends, coupled with the fact that Hume's Principle so conceived requires a prior understanding only of (second-order) logical vocabulary, is enough to sustain an account of the foundations of arithmetic that deserves to be viewed as a form of logicism which, while not quite logicism in the sense of a reduction of arithmetic to logic, or as a demonstration of its analyticity in Frege's own strict sense,[6] preserves the essential core and content of Frege's two fundamental theses.

Restricting attention for the time being to elementary arithmetic, there are two main claims—one logical, the other more purely philosophical—which must be seen to hold good, if the neo-Fregean's leading thesis is to be sustained. The logical claim is that the result of adjoining Hume's Principle to second-order logic is a consistent system which suffices as a foundation for arithmetic, in the sense that all the fundamental laws of arithmetic are derivable within it as theorems.[7] The philosophical claim is that if that is so, that constitutes a vindication of logicism, on a reasonable understanding of that thesis.

In our view, investigations have now reached a point at which at least the technical part of the logical claim may be taken to have been established.[8] Hume's Principle, added to a suitable system of second-order logic, does indeed suffice for a proof of the Dedekind–Peano axioms.[9] Moreover, as far as consistency is concerned, we now have as much assurance as it seems reasonable to demand. In "The Consistency of Frege's *Foundations of Arithmetic*," Boolos (1987) presents a formal theory FA (Fregean Arithmetic), incorporating an equivalent of Hume's Principle, which captures the mathematical content of the central sections of Frege's *Grundlagen* (§§63–83) and proves not just that we can give a model for Hume's Principle, but also that the informal model-theoretic proof can be replicated both in standard set theory and in the weaker theory known as "second-order

[6] For some reasons in favor of construing analyticity more generously than Frege does, see Hale and Wright (2001), pp. 12–14.

[7] Formulating the logical claim in this way—rather than as that every truth of arithmetic be derivable as a theorem of the system—avoids an obvious clash with Gödel's first incompleteness theorem. For some discussion of the claim that Gödel's theorem shows that the logicist project was doomed to failure right from the outset, see Wright (1983), p. 131.

[8] It would be a mistake, of course, to think that the issues here are entirely technical. The formal result is strictly only that Hume's Principle plus higher-order logic permit the derivation of a theory which *allows of interpretation* as arithmetic. It is a philosophical issue to defend that interpretation, pursued to some extent below and much more fully in several of the essays collected in Hale and Wright (2001)—see especially essay 13.

[9] This result, following a suggestion of George Boolos, is now often referred to as *Frege's Theorem*. The derivability of Frege's Theorem seems to have been first explicitly noted in Charles Parsons (1964); see remark at p. 194. For detailed accounts of the proof, see Wright (1983), §xix; Boolos (1987 and 1990, appendix); and Boolos and Heck (1998).

arithmetic" or "analysis," the consistency of which seems well beyond serious question.

The *philosophical* significance of the situation is another matter entirely. If the neo-Fregean is to justify his contention that Frege's Theorem can underwrite a viable Platonistic version of logicism, then, even if he restricts that claim to elementary number theory, he has much philosophical work to do on several fronts; and if he additionally aspires, as we do, to sustain a more inclusive version of logicism encompassing at least the theory of real numbers and perhaps some set theory, then he must take on an additional range of tasks—some technical, some philosophical.

In the four immediately following sections, we shall confine ourselves to issues which neo-Fregeanism must address, even if the scope of its leading claims is restricted to elementary arithmetic. Many of these concern the capacity of *abstraction principles*—centrally, but not only, Hume's Principle itself—to discharge the implicitly definitional role in which the neo-Fregean casts them, and thereby to subserve a satisfactory apriorist epistemology for (at least part of) mathematics. Others, to be briefly reviewed in section 8, concern the other main assumption that undergirds the specifically logicist aspect of the neo-Fregean project (and equally, of course, Frege's original project): that the logic to which abstraction principles are to be adjoined may legitimately be taken to include *higher-order*—at the very least, second-order—logic without compromise of the epistemological purposes of the project. In between, in sections 6 and 7 respectively, we shall canvass some of the issues that attend a neo-Fregean construction of analysis and of set theory.

2. Abstraction Principles—Ontology and Epistemology

Frege's Platonism is the thesis that number words have reference, and that their reference is to objects—objects which, on any reasonable account of the abstract-concrete distinction,[10] must be reckoned to lie on the abstract side of it. Why it was so crucial for Frege that numbers be recognized as objects becomes clear when we consider how he proposes to prove that every finite number is immediately followed by another (and hence, in the presence of others of the Dedekind–Peano axioms, which Frege can straightforwardly establish, that the sequence of finite numbers is infinite). The idea of Frege's proof is to show that for each finite *n*, the

[10] Exactly how this distinction is best drawn is a delicate issue. One of the best modern discussions is Dummett (1973), ch. 14. See also Noonan (1976) and Hale (1987), ch. 3.

number belonging to the concept *finite number ancestrally preceding or equal to n* itself directly follows after *n*. To prove this, in the manner sketched in *Grundlagen* §§82–83, requires applying Frege's criterion of identity for cardinal numbers, as encoded in Hume's Principle, to first-level concepts of the type just indicated—that is, concepts under which numbers, and indeed only numbers, fall—and thus requires numbers to be objects.

It may be rejoined that this gives Frege a *motive* to treat numbers as objects, but not a *justification* for doing so. However, while *Grundlagen* provides no systematic argument for this apparently crucial thesis, Frege did at least provide hints, and some of the materials, on the basis of which a case for it may be constructed. We start from two ideas. First: *objects*, as distinct from entities of other types (properties, relations, or, more generally, functions of different types and levels), just are what (actual or possible) singular terms refer to. Second: no more is to be required in order for there to be an at least prima facie case that a class of apparent singular terms have reference, than that they occur in certain true statements free of all epistemic, modal, quotational, and other forms of vocabulary standardly taken to compromise straightforward referential function. In particular, if certain expressions function as singular terms in various true first-order atomic contexts, there can be no further question that they have reference and, since they are singular terms, refer to objects. The underlying thought is that—from a semantic point of view—a singular term just *is* an expression whose function is to effect reference to an object, and that a wide class of statements containing such terms cannot be true unless those terms successfully discharge their referential function. Provided, then (as certainly appears to be the case), there are such true, suitable statements so featuring numerical singular terms, there are objects—numbers—to which they make reference.[11]

This simple argument, of course, is not conclusive. At most it creates a presumption in favor of numbers' existence as a species of objects. And there are various ways in which that presumption might be defeated. An opponent might agree that the statements to which appeal is made have, superficially, the right logical form, but deny that appearances are trustworthy. Or, more radically, she

[11] This argument requires, of course, that we should be able to identify expressions functioning as singular terms without appeal to the assumption that they refer—or purport reference—to objects. We hold that this can be done, employing criteria of the kind originally proposed by Dummett (1973, ch. 4) to the effect that singular terms may be discriminated by their distinctive inferential role—in the sense that certain simple patterns of inference, such as the natural language analogue of existential generalization, are valid when identified positions in their premises and/or conclusions are occupied by singular terms, but otherwise invalid. The exact formulation of suitable broadly inferential tests is a matter of some delicacy, involving complications we cannot review here. For discussion of them, and defense of the proposal against some criticisms (e.g., Wetzel (1990)), see Hale and Wright [2001], essays 1 and 2.

may accept those appearances as a fair guide to truth-conditions (at least as far as purely arithmetical statements go), but deny that any of the relevant statements are actually—taken literally and at face value—*true.*[12] It might be conceded that the argument makes a prima facie case, but argued that the relevant statements cannot really be seen as involving reference to numbers, on the ground that there are (allegedly) insuperable obstacles in the way of making sense of the very idea that we are able to engage in identifying reference to, or thought about, any (abstract) objects to which we stand in no spatial, causal or other natural relations, however remote or indirect. A closely related objection—originating in Paul Benacerraf's much discussed dilemma for accounts of mathematical truth, and subsequently pressed in a revised form by Hartry Field—has it that a Platonist account of the truth-conditions of mathematical statements puts them beyond the reach of humanly possible knowledge or reliable belief, and must therefore be rejected.

The simple argument—it is worth noting—makes no appeal to the specific possibility of explaining singular terms for natural numbers, along with the corresponding sortal concept, via Hume's Principle. Of course, the case for the existence of numbers *can* be made on the basis of Hume's Principle, and it is important to the neo-Fregean that this should be so, precisely because it provides for a head-on response to the epistemological challenge posed by Benacerraf's dilemma. Hume's Principle, taken as implicitly defining the numerical operator, fixes the truth-conditions of identity-statements featuring canonical terms for numbers as those of corresponding statements asserting the existence of one-one correlations between appropriate concepts. Given statements of the latter sort as premises, the truth of such identities (and hence the existence of numbers) may accordingly be inferred. We thus have the makings of an epistemologically unproblematic route to the existence of numbers and a fundamental species of facts about them. The key idea is clearly present in Frege's metaphorical explanation of how the Direction Equivalence may serve to introduce the concept of direction: "we carve up the content [expressed in the statement that lines *a* and *b* are parallel] in a new way, and this yields us a new concept" [i.e., of direction] (Frege 1884, §64).

> More generally, it is open to us, by laying down an abstraction principle $\forall\alpha\forall\beta(\Sigma(\alpha)=\Sigma(\beta)\leftrightarrow\alpha\approx\beta)$, to re-describe or re-conceptualise the type of state of affairs apt to be depicted by sentences of the form $\alpha\approx\beta$—we may so reconceive such states of affairs that they come to constitute the identity of a new kind of object of which, by that very stipulation, we introduce the concept.[13]

[12] This is the central negative thesis of Hartry Field's atraditional version of Nominalism, first advocated in his (1980) and subsequently defended in various papers, mostly collected together in his (1989). Our own objections to Field's position are presented in Hale (1987), ch. 5; Hale (1990); Hale and Wright (1992) (to which Field (1993) replies), and Hale and Wright (1994). For discussion and other recent references, see Colyvan (2000).

[13] For fuller explanation of this idea, see Hale and Wright (2001), essays 4, 8, and 12.

This is all quite consistent with Platonism, modestly and soberly conceived. For it is no part of the proposal that objects of the new kind—abstract objects such as directions, numbers, or whatever—are creations of the human mind, somehow brought into being by our stipulation. The sense in which they are new is merely that the (sortal) concept under which they fall is newly introduced by our abstractive explanation. In and of itself, the abstraction does no more than introduce that concept and establish a use for a corresponding range of singular terms by means of which its instances, if any, may be designated, and involves no attempt to guarantee that the concept does indeed have instances. The existence of objects of the new kind (e.g., directions) depends—and, if the explanation is accepted, depends exclusively—upon whether or not the relevant equivalence relation (e.g., parallelism) holds among the entities of the presupposed kind (e.g., lines) on which that relation is defined.[14] And, to return to the case that most concerns us here, provided that facts about one-one correlation of concepts—in the basic case, sortal concepts under which only concrete objects fall—are, as we may reasonably presume, unproblematically accessible, we gain access, via Hume's Principle and without any need to postulate any mysterious extrasensory faculties or so-called mathematical intuition, to corresponding truths whose formulation involves reference to numbers.[15]

Both this form of the case for the existence of numbers and its claim to provide trouble-free access to a basic class of facts about them may, like the simple argument, be challenged. For example, it may be claimed that if statements having the form of the left-hand side of Hume's Principle involve reference to numbers, then they *cannot* legitimately be accepted as equivalent to corresponding statements having the form of the right-hand side, since the latter involve no such reference and may accordingly be true under circumstances where their left-hand counterparts are not (viz. if there are no numbers). On this basis, it may be held that Hume's Principle—along with abstraction principles quite generally—is acceptable only *either* if it is so construed that the apparent reference to numbers presented by the left-hand sides of its instances is treated as merely apparent, *or* it is amended so as to explicitly allow for the presupposition that numbers exist.

[14] Whether or not that condition is fulfilled may be a contingent matter, or it may be a matter of necessity—or even, as in the case of abstractive explanation of *number* via Hume's Principle, a matter of logical necessity. Either way, stipulative introduction of a concept by abstractive reconceptualization is one thing and ensuring that it has a nonempty extension is another. There is thus plenty of clear water between Fregean abstractions, on the one hand, and, on the other, arguments such as Anselm's first Ontological Argument, with which the former have been invidiously compared (by Hartry Field in, e.g., (1989), essay 5) in one form of "Bad Company" objection. For other forms of this objection, see section 4 below.

[15] For a fuller account, see Hale & Wright (2002), esp. secs. 5 and 6.

An advocate of the first alternative—likely to find favor with orthodox nominalists—will insist that we may accept abstraction principles as explanatory of the truth conditions of their left-hand sides only if we treat the latter as devoid of semantically significant syntax, beyond the occurrences of the term or predicate variables α and β. The only legitimate reading of Hume's Principle, on this view, is an *austere* one according to which it merely serves to introduce a new, semantically unstructured two-place predicate:

thenumberof . . . isidenticalwiththenumberof__

as alternative notation for the equivalence relation that figures on the right-hand side: "... corresponds one-one with __." But why insist upon austerity? The seemingly straightforward answer—that since the statement "F corresponds one-one with G" plainly involves no reference to numbers, it can be taken as explaining the truth conditions of "the number of Fs = the number of Gs" only if the latter likewise involves no such reference—implicitly relies upon a question-begging assumption. It is certainly true that statements of one-one correspondence between concepts involve no terms purporting reference to numbers. And it is equally clearly true that two statements cannot have the same truth-conditions if they differ in point of existential commitment. But the premise the nominalist requires—if he is to be justified in inferring that statements of one-one correspondence can be equivalent to statements of numerical identity only if the latter are austerely construed—is that statements of one-one correspondence do not demand the existence of numbers. And this premise does not follow from the acknowledged absence, in such statements, of any explicit reference to numbers. The statement that a man is an uncle involves no explicit reference to his siblings or their offspring, but it cannot be true unless he has a nonchildless brother or sister.

It may be objected that while no one would count as understanding the statement that Edward is an uncle if she could not be brought to agree that its truth requires the existence of someone who is brother or sister to Edward and father or mother of someone else, it is quite otherwise with statements of one-one correspondence. Someone may perfectly well understand the statement that the Fs correspond one-one with the Gs without being ready to acknowledge—what the nominalist denies—that its truth requires the existence of numbers. The neo-Fregean reply is that so much is perfectly correct, but insufficient for the nominalist's argument. It is *correct* precisely because—just as the neo-Fregean requires—understanding talk of one-one correspondence between concepts does not demand possession of the concept of *number.* But it is *insufficient* because the question that matters is, rather, whether one who is fully cognizant of *all* the relevant concepts—that is, the concepts of *one-one correspondence between concepts* and *number*—could count as fully understanding a statement of one-to-one correspondence between concepts without being ready to agree that its truth called for the existence of numbers. If the concept of *number* is implicitly defined, as the neo-Fregean proposes, by Hume's Principle, she could not. Insisting at this point that no such explanation can be admitted, because only an austere reading of Hume's Principle is permissible, amounts to no more than an unargued—and now explicitly question-begging—refusal to entertain the kind of explanation on offer.

Nominalists of the less traditional kind represented, most prominently, by Hartry Field[16] are more likely to favor the second alternative: to allow that abstraction principles are all very fine as stipulative explanations, provided (what the nominalist views as) their existential presuppositions are made properly and fully explicit. Hume's Principle, in particular, should be replaced by a properly *conditionalized* version along the lines of "If there exist such things as the number of *F*s and the number of *G*s, then they are identical if and only if the *F*s and *G*s are one-one correlated." Such conditionalized principles would, of course, be inadequate to the neo-Fregean's purposes.

There is, however, an immediate difficulty confronting the proposal to replace Hume's Principle by such a conditionalized version.[17] How are we to understand the antecedent condition? In rejecting Hume's Principle as known a priori, Field holds that the obtaining of a one-one correlation between a pair of concepts cannot be regarded as *tout court* sufficient for the truth of the corresponding numerical identity. So, since that was an integral part of the proposed implicit definition of the concept of number, his position begs some other account of that concept—an intelligible doubt about the existence of numbers requires a concept in terms of which the doubt might be framed. More specifically, if we take it that the hypothesis that numbers exist may be rendered as "$\exists F \exists x\, x = \mathrm{N}yFy$," then in order to understand the condition under which Field is prepared to allow that Hume's Principle holds, we must *already* understand the numerical operator. But it was the stipulation—unconditionally—of Hume's Principle which was supposed to explain it. That explanation has lapsed; but Field has put nothing else in its place.

Is there perhaps a better way to implement Field's conditionalization strategy, one which avoids this difficulty? A familiar way of thinking about the manner in which theoretical scientific terms acquire meaning[18] holds that we should view a scientific theory, embedding one or more novel theoretical terms, as comprising two components: one encapsulating the distinctive empirical content of the theory without deployment of the novel theoretical vocabulary, the other serving to fix the

[16] Cf. Field (1984).

[17] This difficulty was stressed, originally, in Wright (1983), pp. 148–152, and Wright (1988), sec. V.

[18] The tradition includes Bertrand Russell (1927); F.P. Ramsey, "Theories," in Ramsey (1931); Rudolf Carnap (1928); and David Lewis (1970).

meaning(s) of the theoretical term(s). The theory's total empirically falsifiable content is, roughly, that there exist entities of a certain kind, namely, entities satisfying (a schematic formulation of) the (basic) claims of the theory. This can be expressed by the theory's *Ramsey sentence* (i.e., roughly, an existential generalization obtained from the original formulation of the theory employing the new theoretical terms by replacing each occurrence of each new term with a distinct free variable of appropriate type, and closing the resulting open sentence by prefixing the requisite number of existential quantifiers). Thus if, focusing for simplicity on the case where a single new theoretical term, "*f*," is introduced, the undifferentiated formulation of the theory is "$\Theta(f)$", then its empirical content is exhaustively captured by its Ramsey sentence, "$\exists x \Theta(x)$," where the new variable "*x*" replaces "*f*" throughout "$\Theta(f)$." The new term "*f*" can then be introduced, by means of what is sometimes called the *Carnap conditional*[19]: "$\exists x \Theta(x) \rightarrow \Theta(f)$," as denoting whatever (if anything) satisfies "Θ" (on the intended interpretation of the old vocabulary from which it is constructed). This conditional expresses, in effect, a convention for the use of the new term "*f*." Being wholly void of empirical content, it *can* be stipulated, or held true a priori, without prejudice to the empirical disconfirmability of the theory proper. Alternatively, the definitional import of the theory might be seen as carried, rather, by a kind of *inverse* of its Carnap conditional—in the simple instance under consideration, something of the shape "$\forall x(x=f \rightarrow \Theta(x))$." We shall follow this proposal in the brief remarks to follow.[20]

This may seem to offer Field a way around the difficulty: treat the system consisting of Hume's Principle and second-order logic as a "theory," in a sense inviting comparison with empirical scientific theories, whose capacity to introduce theoretical concepts by implicit definition is uncompromised by the fact that they may turn out to be false. We may indeed think of this theory as *indirectly* implicitly defining the concept of cardinal number; however the real vehicle of this definition is not Hume's Principle but the corresponding inverse-Carnap conditional:

(HP*) $$\forall F \forall G \forall u \forall v((u = \mathrm{N}xFx \wedge v = \mathrm{N}xGx) \rightarrow (u = v \leftrightarrow F1\text{-}1G)).$$

And now the complaint that Field has put nothing in place of Hume's Principle to enable us to construe the condition on which he regards it as a priori legitimate to affirm Hume's Principle is met head-on. HP*, Field may say, tells us what

[19] It is called the Carnap conditional by Horwich. In Lewis (1970) it is called the Carnap sentence.

[20] For more on this suggestion, see Hale and Wright (2001), pp. 141–142. The principal advantage of the inverse formulations over the Carnap conditionals is in the provision of space for the theoretical terms they define to express concepts of natural kinds, whose essential nature may not be disclosed by the particular theory in question—which may indeed be partially incorrect— rather than concepts of functional (explanatory) role.

numbers are in just the way that the inverse-Carnap conditional for any (other) scientific theory tells us what the theoretical entities it distinctively postulates are—by saying what (fundamental) law(s) they characteristically satisfy, *if* they exist. That there are numbers is itself no conceptual or definitional truth; it is, rather, the content of a theory (in essence, a theory given by the Ramsified version of Hume's Principle: $\exists\eta\,\forall F\,\forall G(\eta F = \eta G \equiv F1\text{-}1G)$), which may perfectly well be—and in Field's view is—false.[21]

Should this proposal be accepted? Let's allow that *if* there is good reason to insist that an implicit definition of the numerical operator should proceed, not through an outright stipulation of Hume's Principle but through something more tentative, then one plausible shape for the stipulation is an inverse Carnap formulation of the kind suggested. But is there any such reason?

The comparison with the empirical scientific case cannot provide one. Conditionalization is called for in the scientific case in order to keep open the possibility of empirical disconfirmation. Fixing the meaning of "f" by stipulating the truth of the whole conditional "$\forall x(x = f \rightarrow \Theta(x))$" leaves room for acknowledgment that the antecedent (more precisely, its existential generalization "$\exists x\, x = f$") might turn out false—grounds to think it false being provided by empirical disconfirmation of the consequent (more precisely, of the theory's Ramsey-sentence "$\exists x\,\Theta(x)$"). But there is no possibility of empirical disconfirmation of the Ramsified Hume's Principle if, as seems reasonable to suppose, Hume's Principle is *conservative* over empirical theory—as Field allows, indeed argues—since in that case it will have no proper empirical consequences.

It does not, of course, follow that there cannot be other reasons to insist upon a more cautious, conditionalized form of stipulation. Let us say that a stipulation of a sentence as true is *arrogant* if its truth requires that some of its ingredient terms have reference in a way that cannot be guaranteed just by providing them with a sense. An example would be the stipulation "Jack the Ripper = the perpetrator of the Whitechapel murders" offered by way of introduction of the term "Jack the Ripper." Were the outright stipulation of Hume's Principle rightly regarded as arrogant, that would—plausibly—provide an independent reason for insisting that it cannot be known a priori just on the basis of its meaning-conferring credentials and for adopting something like the distinction between

[21] This idea—or something closely akin to it—seems to be what Kit Fine has in mind when he suggests, in *The Limits of Abstraction*, that Field can explain number using Hume's principle while denying the existence of numbers "by treating Hume's Law as an explanation of a variable number operator. The existence of numbers may then intelligibly be denied"—Fine claims—"since the denial simply amounts to the claim that there is no operator that conforms to the law. If we regard Hume's Law as part of a 'scientific' theory, then this response is equivalent to a Ramsey-style treatment of theoretical terms" (cf. Fine (2002), p.524, fn. 10).

factual content, carried by Ramsified Hume, and meaning-fixing—accomplished by something appropriately conditionalized—proposed in the scientific-theoretical case. The crucial question is, then, whether an outright stipulation of Hume's Principle is indeed guilty of this failing. We think not. In general, it would seem, it suffices for an implicitly definitional stipulation to avoid arrogance that it does no more than lay down introduction and elimination conditions for the definiendum in terms themselves free of any problematic existential presupposition. And this will surely be so, if the import of the stipulation may be parsed into introductory and/or eliminative components, each conditional in form, prescribing which true statements free of occurrences of the definiendum are to be respectively sufficient and/or necessary for true statements variously embedding it.

But that is exactly what Hume's Principle, proposed as a stipulation, does. The principle does not just assert the existence of numbers as "Jack the Ripper is the perpetrator of this series of killings" asserts the existence of the Ripper. What it does is to fix the truth conditions of identities involving canonical numerical terms as those of corresponding statements of one-one correlation among concepts (compare the schematic stipulation "*a* is the single perpetrator of these killings if and only if *a* is Jack the Ripper"). Its effect is that one kind of context free of the definiendum—a statement of one-one correlation between suitable concepts—is stipulated as sufficient for the truth of one kind of context embedding the definiendum: that identifying the numbers belonging to those respective concepts. That is its introductory component. And, conversely, the latter type of context is stipulated as sufficient for the former. That is the principle's eliminative component. All thus seems squarely in keeping with the constraint that in order to avoid arrogance, legitimate implicit definitions must have an essentially conditional character.[22] If the additional conditionalization in HP* is proposed in the interests of avoiding arrogance, it thus merely involves a condition too many.[23]

Obviously the general notion of implicit definition invoked here calls for further clarification and defense.[24] But even if it is accepted that this is one way in which a statement might qualify as analytic in something akin to the general spirit of Frege's notion, if not its detail, there are several more specific grounds on which

[22] It is, of course, true—and essential to the case for number-theoretic logicism—that the truth of certain instances of the right-hand side of Hume's Principle is a matter of logic. In particular, it is vital that the existence of a one-one correlation of the non-self-identicals with themselves should be—as it is—a theorem of second-order logic. More generally, if F is any sortal concept, then, as a matter of logic alone, the identity relation correlates F's instances one-one with themselves, so that, applying Hume's Principle, we have it that $\mathrm{N}xFx = \mathrm{N}xFx$, whence $\exists y\, y = \mathrm{N}xFx$.

[23] The point applies to abstraction principles quite generally: their biconditional character ensures that they are never arrogant per se.

[24] See Hale and Wright (2001), essay 5.

it might be denied that Hume's Principle in particular can function as such an implicit definition. In the next three sections we summarize the most important of them, with just the briefest indication how the neo-Fregean may respond.

3. Abstraction Principles and Julius Caesar

One paramount such ground is, of course, the Julius Caesar problem. The concept of *number* to be explained, in both Frege's and the neo-Fregean view, is a *sortal* concept. Mastery of any general concept applicable to objects involves knowing what distinguishes objects to which it applies from those to which it does not—involves, that is, a grasp of what Dummett has called a *criterion of application* for the concept. What sets sortal concepts apart from others is that full competence with a sortal concept involves, in addition, knowing what settles questions of identity and distinctness among its instances—what determines, given that *x* and *y* are both *F*s, whether they are one and the same, or distinct, *F*s. In other words, sortal concepts are distinguished from others by their association with what Frege called a *criterion of identity*.[25] Hume's Principle appears, at least, to take care of a necessary condition for *number* to be a sortal concept in this sense—by supplying a criterion of identity—and so to constitute at least a partial explanation of a sortal concept of number. The neo-Fregean, however, makes a stronger claim—that by stipulating that the number of *F*s is the same as the number of *G*s just in case the *F*s are one-one correlated with the *G*s, we can set up *number* as a sortal concept (i.e., that Hume's Principle *suffices* to explain the concept of *number* as a sortal concept). This stronger claim appears open to an obvious objection: simply, that Hume's principle cannot be sufficient, since there is also required, as with any other concept, a criterion of application. Since such a criterion would determine which sorts of things are *not* numbers, it would also settle the truth-conditions of any identity statement of the form "the number of $Fs = q$," where "q" marks a place for a term explicitly purporting to denote a thing of some previously understood sort. In short, the Caesar Problem, in the form in which Frege expresses it at *Grundlagen* §66, is nothing other than the problem of supplying a criterion of application for *number* and thereby of setting it up as the concept of a genuine sort of object. But that it is such a concept is the cardinal thesis of (neo-) Fregean Platonism.

[25] This is only an initial characterization; for some refinements, see Hale and Wright (2001), essay 14, sec. 6.

For this reason,[26] we believe it is no option for the neo-Fregean to declare "mixed" identity statements of the troublesome sort out of order or somehow ill-formed. Rather, he must grant that they raise perfectly genuine, legitimate questions and seek to explain how, within the resources at his disposal, their truth-values may in principle be resolved. In short, he needs a positive solution to the Caesar Problem. The best hope for providing one remains, in our view, the consideration that there are at least some restrictions on the extension of any given sortal concept which are mandated by the type of canonical ground—criterion of identity—associated with identity statements concerning its instances; and, more specifically, that because they are canonically decided by reference to independent considerations, statements of numerical and of, say, personal identity must be reckoned to concern different categories (or ultimate sorts) of objects.[27]

4. Abstraction Principles and Bad Company

Frege's Basic Law V, like Hume's Principle, is a second-order abstraction, and thus provides a sharp reminder that not every imaginable abstraction principle constitutes an acceptable means of introducing a concept. Some constraints are needed—with consistency, obviously, a minimum such constraint. But some critics have thought the inconsistency reveals far more—not merely that Law V cannot function as an explanation of a coherent concept of extension or set, but also that abstractive explanations are defective quite generally. If it were possible satisfactorily to explain the concept of *number* by laying down Hume's Principle, the objection goes, then it ought to be possible to do the same for the concept of *extension* by laying down Basic Law V. Since the concept of *extension* cannot be satisfactorily so explained, neither can the concept of *number*, and nor can any other concept.

This is the Bad Company objection in its simplest, most sweeping, and—in our estimation—least challenging form. The objection as stated seems to assume that if a pattern of concept explanation—in this case, explanation by means of an abstraction principle—can be exploited to set up the resources for derivation of

[26] There are other reasons, one of which we touch upon below in section 5. For fuller discussion, see Hale and Wright (2001), essay 14.

[27] We lack space to develop this thought here. For more detailed discussion, defense of it, and references to further work, see Hale and Wright (2001), essay 14.

an antinomy, that somehow shows not just that that particular attempt at explaining a concept aborts, but that the pattern itself is defective. This extreme claim should be rejected. Allowed a sufficiently free hand in explaining predicates by laying down their satisfaction conditions, we can explain a predicate applicable to predicates by stipulating that a predicate satisfies "x is heterological" if and only if it does not apply to itself, and speedily arrive at the well-known contradiction that "heterological" is heterological just in case it is not. No one seriously supposes that the moral in that case is that we can *never* satisfactorily explain a predicate by saying what it takes for an object to satisfy it. Why suppose it any less unreasonable to draw the parallel moral in the case in hand—that one can never explain a sortal concept by means of a Fregean abstraction? If it should be demanded that the neo-Fregean point to some relevant difference between good and bad cases of would-be abstractive explanations, it would seem, so far, a perfectly adequate reply that Basic Law V, for example, is known to be inconsistent, whereas Hume's Principle is not. Indeed, he can in this case make a stronger reply—although it is not clear that he needs to make it, in order to dismiss the objection as stated—that Hume's Principle is very reasonably believed *not* to be inconsistent.

A potentially more interesting form of Bad Company objection, due to George Boolos,[28] allows that the possibility of inconsistent abstractions does not discredit abstractive explanation as such, and grants that an explanation based on Hume's Principle may justifiably be presumed at least consistent, but points to the possibility of other equally consistent abstraction principles which are, however, inconsistent with Hume's Principle. Hume's Principle has models, but none which have less than countably infinite domains. By taking a different equivalence relation on (first-level) concepts, such as one which holds between a pair of concepts F and G if and only if just finitely many objects are either F-but-not-G or G-but-not-F, we can frame an abstraction—the *Nuisance Principle*[29]—which has it that certain objects $\nu(F)$ and $\nu(G)$—the "nuisances" of F and G, respectively—are one and the same just in case F and G are so related. The Nuisance Principle is, like Hume's Principle, consistent. The trouble is that while it has models, it can be shown to have only *finite* models. What this threatens is not so much the capacity of Hume's Principle to function as an explanation, as its title to be regarded as known in virtue of being analytic of the concept of *number*. For does not the Nuisance Principle enjoy an equally good title to be regarded as analytic of the concept of *nuisance*? Yet inconsistent principles cannot both be

[28] Boolos (1990), p. 273.

[29] As it is dubbed in Hale and Wright (2001), essay 12. The Nuisance Principle is a close relative of Boolos's Parity Principle, in terms of which his objection was originally formulated.

true, much less known. Pending exposure of some relevant disparity, should we not conclude that *neither* principle can, after all, be known on the basis of its stipulation?

If the neo-Fregean view of the philosophical significance of Frege's Theorem is to be sustainable, there must *be* a relevant difference: some further constraint(s), in addition to any which may be needed to ensure satisfaction of the minimal (and of course fundamental) condition of consistency, with which good abstractions must comply, but which the Nuisance Principle breaches. One natural proposal is to the effect that no acceptable explanation—whether it proceeds through an abstraction principle or takes some other shape—can carry implications for the size of extensions of concepts other than, and quite unconnected to, the concept to be explained. The requirement this suggests is that an acceptable abstraction principle should be *conservative* in a sense closely akin to that deployed by Hartry Field in his defense of Nominalism, according to which a mathematical theory M is conservative if its adjunction to a nominalistically acceptable theory N has no consequences for the ontology of N which are not already consequences of N alone.[30] Since the Nuisance Principle constrains the extensions of all other concepts to be at most finite, it fails this requirement and cannot, therefore, constitute an acceptable explanation. Evidently the same will go for any other "limitative" abstraction principle (i.e., any abstraction which, while consistent, can have no models exceeding some assignable cardinality.[31]

There is a wider question here which merits investigation in its own right, quite apart from its obvious bearing on the prospects for extending the neo-Fregean approach beyond elementary arithmetic. What conditions, in general, are necessary and sufficient for an abstraction principle to be acceptable? The apparent need for conservativeness constraints, over and above any restrictions which may be needed to avoid inconsistency, is one important aspect of this larger question. Another issue concerns the requirement of consistency itself. If an appreciable body of mathematics can be generated from acceptable abstraction principles, we should not expect to be able to formulate a general theory of abstraction that is *provably* consistent—save, perhaps, relative to the assumption

[30] This is an approximation. The correct formulation of the notion of conservativeness relevant to our purposes is actually a little tricky. For details, see Hale and Wright (2001), essay 13, n. 21.

[31] There is no conflict here with the Upward Löwenheim–Skolem Theorem—remember that we are concerned here, in effect, only with abstractions which are (at least) second-order.

The introduction of a constraint of conservativeness does not dispose of all forms of "nuisance," for, as Alan Weir has observed, there are pairwise incompatible but individually conservative abstractions. For a review of the issues, see Weir (2003).

of the consistency of some yet more powerful theory. It may, nevertheless, be possible to articulate some general restrictions which must be observed if inconsistency is to be avoided. One obvious danger here arises from the fact that an equivalence relation defined on the concepts on a specified underlying domain of objects may partition those concepts into more equivalence classes than there are objects in the underlying domain, so that a second-order abstraction may "generate" a domain of abstracts strictly larger than the initial domain of objects. This, in itself, need be no bad thing—indeed, it is essential, if there is to be a neo-Fregean abstractionist route to (classical) analysis. But it raises the specter of something like Cantor's Paradox. If that is to be avoided, it is natural to think, then we must bar all "inflationary" abstractions—abstractions which make impossible demands upon the cardinality of any determinate objectual domain.[32] But how precisely such a restriction should be formulated is a delicate and difficult question which we cannot further pursue here.[33]

There are several other grounds on which the analyticity of Hume's Principle may be, and indeed has been, disputed, to which we can here merely draw the reader's attention. Most obviously, perhaps, it may be claimed that no statement encumbered with significant existential implications can possibly be analytic, in any reasonable sense. Hume's Principle, however, certainly does carry such implications, since it entails the existence of infinitely many objects—each and every one of the natural numbers, at least. Again, while we know that Fregean arithmetic is consistent if (and only if) second-order arithmetic is consistent, we have no absolute guarantee that either theory is consistent; but if we don't know that Fregean arithmetic is consistent, how can we justifiably claim that Hume's Principle is analytic? Or—focusing again upon the principle's existential implications—it may be contended that, quite apart from the fact that no analytic principle ought to carry existential commitments at all, Hume's Principle carries particular such commitments which are, at best, highly questionable. For given that the principle, in conjunction with the second-order logical truth that there is a one-one map from any empty concept to itself, ensures the existence of zero defined as $Nx{:}x \neq x$, it would seem that the neo-Fregean can have no ground for refusing to coinstantiate its second-order variables F and G by the complementary predicate of *self-identity*, so that the principle entails (together with another obvious second-order logical truth) the existence of the universal number $Nx{:}x = x$—the number, "anti-zero," of all the things that there are. But the existence of such a number is doubtful at best—perhaps worse, since under standard definitions, it is provable

[32] This is the driving thought behind the treatment developed in Fine (1998). Cf. the diagnosis offered of Russell's paradox in Boolos (1993).

[33] For further discussion see, in addition to the items cited in note 32, Hale and Wright (2001), essay 15, sec. IV; Cook (2002); and Hale (2000a).

in Zermelo–Fränkel set theory (ZFC) that there can be no cardinal number of all the sets there are.[34]

5. Abstraction Principles and Impredicativity

Formulated in the language of second-order logic, and with quantifiers made explicit, Hume's Principle is

$$\forall F\, \forall G[\mathrm{N}x{:}Fx = \mathrm{N}x{:}Gx] \leftrightarrow \exists R\, \forall x((Fx \rightarrow \exists!y(Gy \wedge Rxy)) \wedge (Gx \rightarrow \exists!y(Fy \wedge Ryx))).$$

If we are to be able to prove, by means of this principle, the distinctness of the individual natural numbers as Frege proposed to define them ($0 =_{\mathrm{df}} \mathrm{N}x{:}x \neq x$, $1 =_{\mathrm{df}} \mathrm{N}y{:}(y = \mathrm{N}x{:}x \neq x)$, etc.), and to prove there to be infinitely many of them by proving, as Frege envisaged, that each natural number n is immediately followed by the number of numbers up to and including n, then it is essential that the principle's initial second-order quantifiers be taken as ranging over concepts under which numbers themselves fall, and hence that its first-order quantifiers be taken as ranging over numbers as well as objects of other kinds. In short, if Hume's Principle is effectively to discharge the role in which the neo-Fregean casts it, its first-order quantifiers must be construed as ranging over, inter alia, objects of that very kind whose concept it is intended to introduce. Is that a cause for concern?

That it is—that the unavoidable impredicativity of Hume's Principle amounts to some sort of vicious *circularity*—has long been the general thrust of much of Michael Dummett's objection both to Frege's philosophy of arithmetic and, more recently, to neo-Fregean efforts to resuscitate it. Indeed, while Dummett has

[34] And if so, presumably, provable in ZFC with ur-elements that there is no cardinal number of all the things (sets and ur-elements) there are. However, we here record a misgiving not developed in the sequel: the "standard definitions" referred to identify each cardinal number in the standard series with the smallest ordinal having that cardinal number of predecessors. Thus everything treated as a cardinal in ZFC is a certain kind of ordinal, and anti-zero could be a cardinal of this kind only if some ordinal were preceded by universe-many others—on pain of the Burali–Forti paradox. But where does ZFC state or imply that *all* cardinals are ordinals of this kind?

The objections noted in this paragraph, along with some others, have all been pressed by George Boolos in various of his writings. Hale and Wright (2001), essay 13, responds to them.

developed a number of quite distinct lines of criticism of that enterprise,[35] it seems clear that it is the impredicativity of Hume's Principle which forms the focus of his principal complaint. It is, however, neither obvious that impredicativity is always harmful, nor easy to be clear why, exactly, Dummett takes it to be a fatal defect in this case. There is, of course, the comparison with Basic Law V—the form of Bad Company objection noted earlier. It is true enough that both principles are impredicative, and true also that the derivation of Russell's contradiction exploits precisely this feature of Law V—predicativity restrictions would certainly block the derivation, and in that sense, impredicativity may be held to be at least partly to blame for the contradiction. But impredicativity does not always lead to contradiction, and Hume's Principle is, as we have emphasized, at least very plausibly taken to be consistent. Dummett's thought seems to be that impredicativity is objectionable, regardless of whether it actually results in inconsistency. But why?

There seem to be a number of distinguishable lines of objection suggested by Dummett's discussion. One is that, for Frege at least, it was necessary to ensure that every statement about numbers possesses a determinate truth-value, and that the impredicativity of Hume's Principle obstructs this.[36] Another is that quantification over a domain of objects is in good order only if it is possible to supply an independent, prior characterization of the domain, and that this cannot be done for the domain over which Hume's Principle's impredicative first-order quantifiers are required to range.[37] Finally, there is the thought that the impredicative character of Hume's Principle disqualifies it from functioning as an explanation of the numerical operator because, considered as such, it would be viciously circular.[38] None of these objections, in our view, carries conviction. The first appears to rely either upon the questionable assumption that Frege's unswerving commitment to bivalence is indispensable to anything worth describing as a version of logicism or upon an unjustifiably exacting interpretation of the requirements for contextual explanation of a sortal concept. The second invokes an exorbitant—and indeed incoherent—condition on the intelligibility of quantification. And granted—crucially—the availability of a positive solution to the Caesar Problem, it does seem possible, contra the third, to explain how intelligent reception of Hume's Principle as an explanation could enable one, impredicativity notwithstanding, to understand numerical statements of arbitrary complexity.[39]

[35] What we take to be Dummett's other main objections are discussed in Hale and Wright (2001), essays 8 and 9.

[36] Cf Dummett (1991), p. 226.

[37] Cf. Dummett (1991), p. 236.

[38] Cf. Dummett (1991), p. 236.

[39] Assuming, of course, an understanding of any other concepts involved. For fuller discussion of Dummett's objections, see Hale and Wright (2001), essays 8, 10, and 11.

6. Neo-Fregean Real Analysis

The minimal formal prerequisite for a successful neo-Fregean foundation for a mathematical theory is to devise presumptively consistent abstraction principles strong enough to ensure the existence of a range of objects having the structure of the objects of the intended theory—in the case of elementary arithmetic, for instance, the existence of a series of objects having the structure of the natural numbers (i.e., constituting an ω sequence). Second-order logic, augmented by the single abstraction, Hume's Principle, accomplishes this formal prerequisite. The outstanding question is then whether Hume's Principle, beyond being presumptively consistent, may be regarded as acceptable in a fuller, *philosophically interesting* sense. That raises the intriguing complex of metaphysical and epistemological issues just reviewed.

In parallel, the minimal formal prerequisite for a successful neo-Fregean foundation of real analysis must be to find presumptively consistent abstraction principles which, again in conjunction with a suitable—presumably second-order—logic, suffice for the existence of an array of objects that collectively comport themselves like the classical real numbers; that is, compose a complete, ordered field. Recently a number of ways have emerged for achieving this result.

One attractive approach is known as the *Dedekindian Way.*[40] We start with Hume's Principle plus second-order logic. Then we use the *Pairs* abstraction,

$$(\forall x)(\forall y)(\forall z)(\forall w)(\langle x,y\rangle = \langle z,w\rangle \leftrightarrow x = z \wedge y = w),$$

to arrive at the ordered pairs of the finite cardinals so provided.[41] Next we abstract over the *Differences* between such pairs,

$$\mathrm{Diff}(\langle x,y\rangle) = \mathrm{Diff}(\langle z,w\rangle) \leftrightarrow x + w = y + z,$$

and proceed to identify the *integers* with these differences. We then define addition and multiplication on the integers so identified and, where m, n, p, and q are any integers, form *Quotients* of pairs of integers in accordance with this abstraction:

$$\mathrm{Q}\langle m,n\rangle = \mathrm{Q}\langle p,q\rangle \leftrightarrow (n = 0 \wedge q = 0) \vee (n \neq 0 \wedge q \neq 0 \wedge m \times q = n \times p).$$

[40] This is explored in work of Stewart Shapiro and Crispin Wright. See Shapiro (2000) and Wright (2000).

[41] Shapiro's description in his (2000) does not make direct use of the *Pairs* abstraction, but moves directly to abstraction principles which "operate on objects taken two at a time." However, since the order in which the objects are taken matters for these principles, it seems better to signal the assumptions involved in an explicit principle and to treat their abstractive domains as composed by the appropriate ordered pairs delivered by it.

We may now identify a *rational* with any quotient $Q\langle m,n\rangle$ whose second term n is nonzero. Then, defining addition and multiplication and the natural linear order on the rationals so generated, we can move on to the objects which are to compose the sought-for completely ordered field via the Dedekind-inspired *Cut Abstraction*:

$$(\forall P)(\forall Q)(\text{Cut}(P) = \text{Cut}(Q) \leftrightarrow (\forall r)(P \leq r \leftrightarrow Q \leq r),$$

where "r" ranges over rationals and the relation $\leq$ holds between a property, P, of rationals and a specific rational number, r, just in case any instance of P is less than or equal to r under the constructed linear order on the rationals. Cuts are the same, accordingly, just in case their associated properties have exactly the same rational upper bounds. Finally, we identify the *real numbers* with the cuts of those properties P which are both bounded above and instantiated in the rationals.

On the Dedekindian Way, then, successive abstractions take us from one-to-one correspondence on concepts to cardinals, from cardinals to pairs of cardinals, from pairs of finite cardinals to integers, from pairs of integers to rationals, and finally from concepts of rationals to (what are then identified as) reals. Although the path is quite complex in detail and the proof that it indeed succeeds in the construction of a completely ordered field is at least as untrivial as Frege's Theorem, it does make for a near-perfect neo-Fregean capture of the Dedekindian conception of a real number as the cut of an upper-bounded, nonempty set of rationals. True, the abstractions involved do not provide for the transformability of any statement about the reals, so introduced, back into the vocabulary of pure second-order logic with which we started out. But that—*pure* logicist—desideratum was already compromised in the construction of elementary arithmetic on the basis of Hume's Principle. As in that case, a weaker but still interesting version of logicism remains in prospect. If each of the successive abstractions invoked on the Dedekindian Way succeeds as an implicit definition of the truth-conditions of contexts of the type schematized on its left-hand side, then there is a route of successive concept formations that starts within second-order logic and winds up with an understanding of the Cuts and a demonstration of the fundamental laws of a canonical mathematical theory of them: in effect, a foundation for analysis in second-order logic and (implicit) definitions. Dedekind did not have the notion of an abstraction principle. But it seems likely that his logicist sympathies would have applauded this construction and its philosophical potential.[42]

[42] An illuminating brief discussion of Dedekind's "logicism" may be found in Shapiro (1997), pp. 170–176.

The Dedekindian Way contrasts significantly with another route explored by one of the present authors in recent work.[43] In claiming to supply a foundation for analysis—in particular, in claiming that the series of abstractions involved effectively leads to the real numbers—the Dedekindian Way may be viewed as resting on an essentially *structural* conception of what a real number is: in effect, the idea of a real number merely as a location in a certain kind of (completely) ordered series. For one following the Dedekindian Way, success just consists in the construction of a field of objects—the Cuts, as defined—having the structure of the classical continuum. Against that, we may contrast what is accomplished by Hume's Principle in providing neo-Fregean foundations for elementary arithmetic. The corresponding formal result is that Hume's Principle plus second-order logic suffices for the construction of an ω sequence. That is certainly of mathematical interest. But what gives Frege's Theorem its distinctive and additional *philosophical* interest is that Hume's Principle also purports to give an account of *what cardinal numbers are*. The philosophical payload turns not on the mathematical reduction as such but on the specific content of the abstraction by which the reduction is effected.

To enlarge: Hume's Principle effectively incorporates a variety of philosophical claims about the nature of number for which Frege prepares the ground philosophically in the sections of *Grundlagen* preceding its first appearance—for example, the claims (1) that number is a second-level property, a property of concepts, and it is concepts that are the things that *have* numbers, which is incorporated by the feature that the cardinality operator is introduced as taking concepts for its arguments; and (2) that the numbers themselves are objects, which is incorporated by the feature that terms formed using the cardinality operator are singular terms. And in addition, of course, Hume's Principle purports to explain (3) what sort of things numbers are. It does so by framing an account of their criterion of identity in terms of when the things that have them have the same one: numbers, according to Hume's Principle, are the sort of things that concepts share when they are one-to-one correspondent.

Now one could, if one wanted, read a corresponding set of claims about real number off the Cut Abstraction principle featured in the Dedekindian Way. One would then conclude, correspondingly, that real numbers are objects, that the things which have real numbers are properties of rationals, and that real numbers are the sorts of things that properties of rationals share just when their instances have the same rational upper bounds. One could draw these conclusions. But, apart from the first, they are unmotivated-seeming conclusions to draw. There is no philosophical case that real number is a property of properties of rationals which stands comparison with Frege's case that cardinal number is a property of

[43] Hale (2000). This was the first neo-Fregean treatment of the real numbers.

(sortal) concepts. On the contrary, the intuitive case is that real number belongs to things like *lengths, masses, temperatures, angles,* and *periods of time.* We could conclude that the Dedekindian Way incorporates poor answers to questions whose analogues about the natural numbers Hume's Principle answers relatively well. But a better conclusion is that the Dedekindian Way was not designed to take *those* kinds of questions on.

In order to understand what is at stake here, we need to appreciate that Hume's Principle accomplishes two quite separate foundational tasks. There is, a priori, no particular reason why a principle intended to incorporate an account of the nature of a particular kind of mathematical entity should also provide a sufficient axiomatic basis for the standard mathematical theory of that kind of entity. It's one thing to characterize what kind of entity we are concerned with, and another thing to show that and why there are all the entities of that kind that we standardly take there to be, and that they compose a structure of the kind we intuitively understand them to do. The two projects may be expected to interact, of course. But they are distinct. It is a peculiar feature of the standard neo-Fregean foundations for elementary arithmetic that the one core principle, Hume's Principle, discharges *both* roles. This is not a feature which we should expect to be replicated in general when it comes to providing neo-Fregean foundations for other classical mathematical theories. What the reflections of a moment ago suggest is that the Dedekindian Way, for its part, is best conceived as addressing only the second project.

It is the distinction between these two projects—the metaphysical project of explaining the nature of the objects in a given field of mathematical inquiry and the epistemological project of providing a foundation for the standard mathematical theory of those objects—that, so far as one can judge from the incomplete discussion in *Grundgesetze,* seems to have been a principal determinant of the approach taken by Frege himself. Real numbers, as remarked, are things possessed by lengths, masses, weights, velocities, and such—things which allow of some kind of magnitude or *quantity.* Quantities are not themselves the reals, but the things which the reals *measure.* As Frege says:

> . . . the same relation that holds between lines also holds between periods of time, masses, intensities of light, etc. The real number thereby comes off these specific kinds of quantities and somehow floats above them. (*Grundgesetze* §185)

So if we want to formulate an abstraction principle incorporating an answer to the metaphysical question "What kind of things are the reals?" after the fashion in which Hume's Principle incorporates an answer to the metaphysical question "What kind of things are the cardinal numbers?" then quantities will feature, not as the domain of reference of the new singular terms which that abstraction will introduce, but rather as the abstractive domain;—that is, as the terms of the abstractive relation on the right-hand side.

However, it's clear that individual quantities don't have their real numbers after the fashion in which a particular concept, say *Julio-Claudian Emperor of Rome*, has its cardinal number. We are familiar with different systems of measurement, like the imperial and metric systems for lengths, volumes, and weights, or the Fahrenheit and Celsius systems for temperature, but there is no conceptual space for correspondingly different systems of counting. Of course, there can be different systems of counting *notation*: we can count in a decimal or binary system, for instance, or in Roman or Arabic numerals. But if they are used correctly, they won't differ in the cardinal number they deliver to any specified concept, but only in the way they name that number. By contrast, the imperial and metric systems do precisely differ in the real numbers they assign to the length of a specified object. One inch is 2.54 centimeters. The real number properly assigned to a length depends on a previously fixed unit of comparison. Thus real numbers are *relations* of quantities, just as Frege says.

These reflections seem to enforce a view about what a principle would have broadly to be like whose metaphysical accomplishment for the real numbers matches that of Hume's Principle for the cardinals. Where Hume's Principle introduces a monadic operator on concepts, our abstraction for real numbers will feature a dyadic operator taking as its arguments a pair of terms standing for quantities of the same type; more specifically, it will be a *first-order* abstraction:

Real Abstraction $\quad \mathrm{R}\langle a,b\rangle = \mathrm{R}\langle c,d\rangle \leftrightarrow \mathrm{E}(\langle a,b\rangle,\langle c,d\rangle)$,

where a and b are quantities of the same type, c and d are quantities of the same type (but not necessarily of the type of a and b), and E is an equivalence relation on pairs of quantities whose holding ensures that a is proportionately to b as c is to d. In effect, the analogy is between the abstraction of cardinal numbers from one-one correspondence on concepts, and abstraction of real numbers from equiproportionality on pairs of suitable quantities.

A neo-Fregean who wishes to pursue this antistructuralist, more purely Fregean, conception of a foundation for analysis must therefore engage with each of the following three tasks.

First, a philosophical account is owing of what in the first place a *quantity* is—what the ingredient terms of the abstractive relation on the right-hand side of the Real Abstraction principle are.

Second—if the aspiration is to give a logicist treatment in the extended sense in which Hume's Principle provides a logicist treatment of number theory—it must be shown that, parallel to the definability of one-one correspondence using just the resources of second-order logic, both the notion of *quantity* and the relevant equivalence relation, E, allow of (ancestral)[44] characterization in (second-order) logical terms.[45]

Third, a result needs to be established analogous to Frege's Theorem: specifically, it needs to be shown that there are sufficiently many appropriately independent truths of the type depicted by the right-hand side of Real Abstraction to ground the existence of a full continuum of real numbers. And while, as stressed, Hume's Principle itself suffices for the corresponding derivation for the natural numbers, here it is clear that additional input—the construction of a complete domain of quantities—is going to be required to augment the Real Abstraction principle.[46]

We have no space here to go further into the philosophical issues at stake between the Dedekindian Way and the more purely Fregean approach. A key question is the cogency, in the context of foundations for analysis, of what we term *Frege's Constraint*: that a philosophically satisfactory foundation for a mathematical theory must somehow intimately build in its possibilities of application. This may be taken to require that an abstraction to the special objects of a mathematical theory is satisfactory only if the abstractive domain on its right-hand side comprises the kinds of things that the mathematical objects in question are *of*—the kinds of things of which they provide a mathematical measure. A structuralist may be expected to reject Frege's constraint, or anyway this understanding of it. The matter is open.[47]

[44] "Ancestral" characterization in the sense that a chain of effective implicit definitions, eventually grounding in concepts of second-order logic, may be reckoned good enough, even if it does not provide the resources for eliminative paraphrase of the definienda. This, of course (as noted), is the most that is achieved by the Dedekindian Way. But there is still a disanalogy: no issue arose, on that approach, concerning the logical character of the items in the abstractive domain for the abstraction that yields the reals. On the Dedekindian Way, those items were concepts (of ancestrally logical objects). On the more purely Fregean route they are (pairs of) *quantities.*

[45] Should it prove impossible to do this, that would not necessarily deprive the abstractionist project of interest. But the point would have to be faced that an abstractionist treatment of analysis would apparently have to originate in a special *nonlogical* subject matter, with significant possible impact on the epistemological payoff of the project.

[46] For further discussion and steps toward the accomplishment of each of these three tasks, see Hale (2000).

[47] For further discussion, see Wright (2000).

7. Neo-Fregean Set Theory

Much further work is needed here. We give but the briefest sketch of some salient approaches and their problems.

An abstraction principle which can plausibly be seen as implicitly defining the concept of *set* will do so by fixing the identity conditions for its instances, and will—on pain of changing the subject—take the identity of sets to consist in their having the same members. So if we assume—what does not seem seriously disputable—that any plausible candidate will be a higher-order abstraction, then the equivalence relation involved had better be coextensiveness of concepts. What we are looking for is, accordingly, a consistency-preserving restriction of Basic Law V, where the restriction is given by imposing a constraint on which concepts are to have properly extensionally behaved sets corresponding to them.[48]

Let's schematically represent the sought-after constraint using a second-level predicate "Good." Then two natural ways to restrict Basic Law V are

$$\text{(A)} \quad \forall F\,\forall G[(\text{Good}(F) \vee \text{Good}(G)) \rightarrow (\{x|Fx\} = \{x|Gx\} \leftrightarrow \forall x(Fx \leftrightarrow Gx))]$$

and

$$\text{(B)} \quad \forall F\,\forall G[\{x|Fx\} = \{x|Gx\} \leftrightarrow (\text{Good}(F) \vee \text{Good}(G) \rightarrow \forall x(Fx \leftrightarrow Gx))].$$

The main difference between these is that (A) is a conditionalized abstraction principle, whereas (B) is unconditional, with the restriction built into the relation required to hold between F and G for them to yield the same set—fairly obviously, the resulting relation is an equivalence relation.[49] Consequently, (B) yields a "set" for every F, regardless of whether it is Good or not, whereas (A) yields a set $\{x|Fx\}$ only if we have the independent input that F is Good. If neither F nor G is Good,

[48] Frege's own hasty reaction to the paradox was of course different: not to restrict which concepts determine sets, but to relax the requirement of strict coextensiveness, allowing that sets could be the same merely when the corresponding concepts were coextensive up to, but possibly excluding, the sets themselves. For a useful review of the shortcomings of the approach, see Resnik (1980), pp. 211–220.

[49] Reflexivity and Symmetry *are* obvious. For Transitivity, suppose (a) $\text{Good}(F) \vee \text{Good}(G) \rightarrow \forall x(Fx \leftrightarrow Gx)$ and (b) $\text{Good}(G) \vee \text{Good}(H) \rightarrow \forall x(Gx \leftrightarrow Hx)$. If the antecedents of both (a) and (b) are both false, then $(\text{Good}(F) \vee \text{Good}(H))$, whence (c) $\text{Good}(F) \vee \text{Good}(H) \rightarrow \forall x(Fx \leftrightarrow Hx)$. Likewise, if the consequents of both (a) and (b) are true, then $\forall x(Fx \leftrightarrow Hx)$, whence (c). If (a)'s antecedent is false but (b)'s consequent is true, then $\text{Good}(H)$, whence $(\text{Good}(F) \vee \text{Good}(H))$, and so (c) again. Similarly if (a)'s consequent is true but (b)'s antecedent is false. Essentially this is the proof given in Wright (1997), fn.32.

the right-hand side of (B) holds vacuously, so we get that $\{x|Fx\} = \{x|Gx\}$—regardless of whether F and G are coextensive. That is, we get the same "set" from all Bad concepts. We get real sets via (B)—that is, objects whose identity is determined by their membership—only from Good concepts.

So what might it be for a concept to be Good? Various suggestions have been canvassed. One general approach, first proposed by Boolos,[50] picks up on the well-entrenched "limitation of size" idea, that the set-theoretic paradoxes stem from treating as sets "collections" which are in some sense "too big"—the collection of all sets, of all sets that are not members of themselves, of all ordinals, and so on. Following Boolos, define a concept F to be *Small* = Good if there is a one-one function taking the Fs into the concept *self-identical* but no one function taking the self-identicals into the Fs. The resultant restricted set abstraction in the style (B) is what Boolos called New V (and Wright called VE[51]):

New V $\forall F \forall G[\{x|Fx\} = \{x|Gx\} \leftrightarrow (\text{Small}(F) \vee \text{Small}(G) \rightarrow \forall x(Fx \leftrightarrow Gx))]$.

Boolos showed that a significant amount of set theory can be obtained in the system consisting of New V and second-order logic.[52] But there are two serious problems with it. The first is that we don't get *enough* set theory. As Boolos showed, neither an axiom of infinity (that there is an infinite, properly behaved *set*) nor the power set axiom can be obtained as theorems on this basis—so the theory affords not even a glimpse of Cantor's Paradise.

A further, quite different, problem relates to the constraints needed to differentiate between acceptable abstraction principles and unacceptable ones. Earlier we noted the need for a constraint of a certain kind of *conservativeness*: acceptable abstraction principles should do no more than fix the truth-*conditions* of statements involving the entities whose concept they introduce. They should have nothing to say about the truth-*values* of statements whose ontology is restricted to entities of other kinds. The Nuisances abstraction, for instance, was faulted on precisely these grounds. Unfortunately, New V appears to be in difficulty for the same reason. The basic point is simple enough.[53] If the concept *ordinal* is Small, New V yields a properly behaved set of all the ordinals and will therefore generate the Burali–Forti contradiction. Hence *ordinal* must be Big. But in that case it is exactly as big as the universe (i.e., there is a one-one correspondence between *ordinal* and a(ny) universal concept, say self-identity. And

[50] Boolos (1987).

[51] Wright (1997), p. 300.

[52] Boolos (1987). For a general discussion of the potentialities of this style of abstraction for varying interpretations of 'good,' see Shapiro (2003).

[53] The point is first made in Shapiro and Weir (1999). Boolos noted New V's implication of global well-ordering in his (1989).

this, together with the fact that the ordinals are well-ordered by membership, entails global well-ordering—the existence of a well-ordering of the universe, and hence of any subuniverse consisting just of the ontology of a prior theory: a result which may well be independent of a suitably chosen such theory. So New V must be reckoned nonconservative.

This problem may be remediable by a suitable redefinition of Smallness. Reinterpret Goodness as *Double Smallness*, where a concept is doubly small if and only if it is (strictly) smaller than some concept which is itself (strictly) smaller than some concept; that is:

$$\text{Small}^2(F) \leftrightarrow \exists G \exists H(F < G < H).$$

This blocks the reasoning which shows that New V, as originally understood, implies global well-ordering. We can still show, of course, that *ordinal* cannot be Good—that is, now, Small^2—since if it were, we would have the Burali–Forti paradox, just as before. So we have to agree that *ordinal* is Bad. But that just means that it is not Small^2, and from this we cannot infer that it is bijectable onto any universal concept (or, indeed, onto any concept).[54]

The other, and more serious, of the two problems New V faces, as we noted, was its weakness: its inability to deliver either an axiom of infinity or a power set axiom as theorems. The same will go, of course, for New V reinterpreted with Good as Small^2 and Small^2 V. However, this need not be a crippling drawback from the neo-Fregean's point of view, if he can justify supplementing New V, or Small^2 V, with other principles—most especially, other abstraction principles—which compensate. On this more catholic approach, we separate two distinct roles one might ask a set-abstraction principle to discharge: *fixing the concept* of set, on the one hand, and, on the other, serving as a *comprehension principle*. The claim would then be that New V's, or Small^2 V's, shortcomings as a comprehension principle need not debar it from successfully discharging a concept-fixing role—of

[54] It may be observed (and was, by Roy Cook) that while the global well-ordering result is blocked, one still gets a significant result—that, since *ordinal* is not Small^2, there cannot be any concept that is bigger than *ordinal* but smaller than the universe. Though a weaker result, this is independent of second-order ZFC. Perhaps so, but it is not a nonconservativeness result for a set theory based on New V with Small interpreted as Small^2 (or Small^2 V, that is, Small V with Small so reinterpreted) in the sense in which global well-ordering is a nonconservativeness result for original New V. Global well-ordering implies well-ordering for the "old" ontology (i.e., the universe of objects as a whole, including those not included among the abstracts provided by New V). By contrast, if our set theory is to be based on Small^2 V, then the theory of ordinals will be naturally construed as a part of it, and the ordinals themselves will be a species of the *new* abstracts so introduced. That this species is at most singly small would seem to have no bearing on anything—cardinal or ordinal—essentially to do with the old ontology.

serving as a means of introducing the concept, while leaving its extension to be filled out, largely or even entirely,[55] by other principles. An obvious first port of call for this approach would be to consider a system consisting of second-order logic augmented by, say, Small^2 V, Hume's Principle, and an appropriate range of instances of a schematic form of the Cut Abstraction principle. Notice that it would be perfectly legitimate, for the purposes of this project, to restrict the domain of the $<$ relation, featured in the definition of Small^2, to concepts instantiated exclusively by the objects given by these privileged prior abstractions.[56]

A quite different direction, but still in keeping with the overall approach, would be to interpret Goodness not in terms of size but in terms of *definiteness*. Here definite concepts contrast with those which Michael Dummett has termed "indefinitely extensible." An indefinitely extensible concept is one, roughly, for which any attempt at an exact circumscription of the extension immediately gives rise to further intuitively acceptable instances falling outside that circumscription. One classic line of diagnosis of the set-theoretic paradoxes, originating in effect with Russell, has it that they arise through reasoning about indefinitely extensible concepts such as *ordinal number*, *cardinal number*, and *set* itself as if they were definite, and in particular by unguardedly assuming that such concepts give rise to collections of entities apt to constitute *sets*. If it were possible to give sufficient of an exact characterization of indefinite extensibility to make that thought good, there would be both a sense of having accomplished a solution to the paradoxes and a clear motive for an appropriate version of either schema (A) or schema (B) which restricted the concepts determining sets to those which are definite. And if the sought-for characterization could somehow be given using only resources available in higher-order logic, we might have the basis for a strong and well-motivated neo-Fregean set theory.

It must be confessed that the hope at present seems somewhat utopian. No sufficiently clear account of the notion of indefinite extensibility, still less one deploying only logical resources, has yet been achieved. And even if one can be, it is by no means clear in advance that it will not be beset by problems of weakness analogous to those of the various Smallness abstractions. The crucial question, as far as overcoming the problem with infinity is concerned, is whether there is any

[55] Largely, but not entirely, if one works with a (B)-type abstraction, such as New V with Good understood as Small^2, but entirely if one works with a conditionalized, (A)-type abstraction, such as Small^2 V. We make no attempt to adjudicate here whether there are compelling reasons to favor one approach over the other. Very roughly, the fragment of standard set theory which one can recover without appeal to non-set-theoretic abstraction principles—though existentially very weak—is larger if one works with a (B)-type principle such as New V rather than an (A)-type principle such as Small^2 V. This might be thought a reason for preferring New V over Small^2 V.

[56] Some of the relevant issues are reviewed in Hale (2000a).

acceptable account of indefinite extensibility which, while embracing the usual suspects—*ordinal*, *cardinal*, and *set*—ranks *natural number* as definite. Work so far has not found one. And even if that trick can be turned, it will be necessary that the characterization of natural number as definite on that basis can proceed on the basis of considerations formalizable in higher-order logic. Otherwise there will be no payoff in terms of the proof-theoretic strength of the abstraction principle in question.

8. Neo-Fregean Logic

The classic logicist thesis about a particular mathematical theory is that its fundamental laws are obtainable on the basis just of definitions and logic. It is still, at the time of writing, a justifiable complaint that while much attention has been paid by neo-Fregeans, and their critics, to the first component in the recipe—issues to do with abstractions in general and Hume's Principle in particular—comparatively little has been given to the second component: the demands, technical and philosophical, to be made on the logical system which is to provide the medium for the proofs the neo-Fregean needs. That system will be either, to all intents and purposes, the higher-order logic pioneered in Frege's *Begriffsschrift* or some substantial fragment of it. Even if it is granted that all the outstanding questions concerning abstractions may be settled in the neo-Fregean's favor, the interest of the neo-Fregean reconstructions of classical mathematical theories will very substantially depend upon their being possible to return logicist-friendly answers to a number of fundamental questions about logical systems of this kind.

The significance which a successful logicist treatment of a particular mathematical theory was traditionally believed to carry was based on the assumption that logic is interestingly set apart from, and somehow more fundamental than, other formal a priori disciplines—that it is marked off by, for instance, its generality, its implication in anything recognizable as rational thought, and by the special character of its distinctive vocabulary—the connectives and other logical constants. It is only in the setting of an acceptance that logic is, in some such way, metaphysically and epistemologically privileged that a reduction of mathematical theories to logical ones—more accurately, to logical theories and legitimate (implicit) definitions—can be philosophically any more noteworthy than a reduction of any mathematical theory to any other. The great issue raised is therefore whether the philosophy of logic can indeed provide a clear account answering to our sense of the fundamentality of logic and its separation from other a priori disciplines and, more specifically, whether—in the context of such an account—higher-order logic will indeed be classified as properly so conceived.

A key issue here concerns the *ontology* of higher-order logic. When Quine famously quipped that he regarded higher-order logic as "set theory in sheep's clothing," his point was not that, in his view, set theory might have a case to count as logic; it was, rather, that higher-order logic had better be regarded as mathematics. However, someone who was persuaded that Quine was wrong about that—that, qua logic, higher-order logic loses nothing to first-order logic—might still be concerned about the ontology apparently demanded by higher-order quantification. Such a thinker might thus be persuaded that there was a *logicist* insight to be recovered by neo-Fregeanism but still thoroughly skeptical whether anything to render the ontology of traditional mathematical theories more palatable can be achieved by the neo-Fregean's abstractionist story, the invocation of the Context Principle, and the rest, when the logical component of the reducing theory—higher-order logic plus abstraction principles—incorporated such indigestible ontological presuppositions of its own.

A frequent type of response would be to accept the problem at face value—to accept that any ontology which was allowed to be distinctively involved in higher-order quantification would indeed be problematical—and try to maintain that in contrast to first-order quantification, higher-order quantification actually incorporates no special ontological commitments: that it is best construed substitutionally, for instance. Or else second-order quantification might be understood in the manner, explored by George Boolos, whereby the only entities involved are those already quantified over at first order (viz., objects). The meaning of higher-order quantifiers is determined by postulating an analogy between their relationship to plural denoting expressions and the relationship between first-order quantification and ordinary singular terms.[57] Each of those proposals, however, would have important limitations from the neo-Fregean perspective. A substitutional interpretation of higher-order quantification looks obviously insufficient to sustain the demands placed on higher-order logic by the kind of neo-Fregean treatment of analysis canvassed above, for instance, while Boolos's construal extends no further than existential quantification into the places marked by monadic predicates.[58]

Ingenuity may come up with further possibilities. For instance, it might be possible to reduce relational quantification, of arbitrary degree, to monadic quantification in the context of a first-order ontology enriched by appropriate *n*-tuples of objects, yielded in turn by appropriate first-order abstraction principles.[59]

[57] Cf. Boolos (1975, 1984).

[58] It thus provides no resources for the construal of universal higher-order quantification for one not inclined to accept the classical interdefinability of the quantifiers; and it says nothing about the proper interpretation of higher-order *relational* quantification—of which, of course, Hume's Principle itself provides a signal example.

[59] Cf. Shapiro and Weir (2000).

But without prejudging the prospects for such reductionist approaches, we ourselves are inclined to favor a different direction. The thought is intuitively compelling that quantificational statements cannot import a *type* of ontological commitment that is not already present in their instances. Thus, if the quantifiers of higher-order logic were indeed best interpreted as calling for an ontology of sets, or setlike entities of some sort, the same should be said about the simple predications which instantiate them. Alternatively, it might be argued that since simple predication, unlike singular reference, incorporates no special ontology, higher-order quantification doesn't either.[60] That thought might be taken to demand a substitutional interpretation of higher-order quantification. But there are at least two other possibilities to explore. First, suppose we can lucidly conceive of thoughts—propositions—as literally internally structured entities.[61] Then we can grasp the idea of one thought resulting from another merely by variation on some internal structural theme they both exhibit, and perhaps also the idea of *all* thoughts which so result (or at least, all which satisfy some other specific restriction). And in that case there has to be a way of understanding quantification which liberates it from the idea of *range*—the association with a domain of entities of some kind or other which is the (Quinean) root of the ontological problems with higher-order logic. Simply, let

$$(\forall F)\mathrm{Q}(F)$$

be the weakest thought whose truth suffices for that of any thought accessible by playing the prescribed kind of variation on the place marked by the dots in

$$\mathrm{Q}\ldots$$

Likewise, $(\exists F)\mathrm{Q}(F)$ will be the strongest thought for whose truth the truth of any such thought suffices.

This construal has a kind of substitutional flavor, but it doesn't have the usual limitation of substitutional quantification of being confined to a particular linguistic repertoire. It can literally be of unrestricted generality, or as unrestricted a generality as makes thinkable sense. Whether higher-order quantification implicates an ontology now depends on whether or not the kind of "thought-component" for which the dots mark a place does. If that component involves direction upon an object—the thought-analogue of singular reference—then of course it will. So first-order quantification does. But it is only if one has argued independently that the thought-components that correspond to predication are entity-involving that higher-order quantification will be entity-involving.

[60] This is a point emphasized in Rayo and Yablo (2001).

[61] Along the kind of lines that Evans supposes in the way he formulates the Generality Constraint. See Evans (1982), pp. 100–102.

The other possibility is more Fregean and more squarely in keeping with the neo-Fregean approach to abstract ontology in general. It is to apply the Context Principle, conceived as a principle concerning *Bedeutung*, to incomplete expressions as well as to complete ones. This approach would hold that, just as the occurrence of a singular term in a true (extensional) statement demands an object as its referent, so the occurrence of an *n*-place predicate in such a statement demands a concept under which the object(s) referred to therein are thereby brought. And just as, in the case of reference to abstract objects, any sense of epistemic impasse is to be offset by explaining what is to refer to and know about such objects in terms of competence in the discourse in which their names occur, so the epistemology of the entities answering to predication would be accounted for in terms of competence in the use of the predicates associated with them. In brief, the proposal would be that higher-order logic should be accepted at face value, as quantifying over entities which constitute the semantic values of incomplete expressions, and that worries about the character of such entities, and about their epistemic acceptability, are to be addressed in exactly the same kind of way, mutatis mutandis, as the way in which neo-Fregeanism approaches the issues to do with abstract singular reference. Of course, a number of issues immediately loom large, not least the paradox of "the Concept *Horse*," the nature of the semantic relationship between incomplete expressions and the entities over which higher-order logic quantifies, and how (finely) these entities are individuated. But perhaps enough has been said to indicate why someone who is content that the neo-Fregean approach to abstract singular reference is broadly along the right lines might be hopeful that no specially opaque or intractable issue might be posed by higher-order quantification as such.

Finally, one more technical matter arising in this context concerns the varying demands placed on the underlying logic by different phases of the neo-Fregean reconstruction of mathematics. J. L. Bell (1999) has shown that the higher-order logic necessary for the demonstration of Frege's Theorem on the basis of Hume's Principle is in fact quite weak.[62] By contrast, it's obvious enough that any abstraction which is to generate a classically uncountable infinity of objects—like that canvassed for the real numbers—must draw on a classically uncountable higher-order domain. So the intriguing and difficult issue therefore arises whether it is possible to justify a conception of the higher-order ontology which can sustain this demand; and whether, in particular, it is possible to do so on the basis

[62] Bell shows that we can restrict the range of the higher-order variables to the finite concepts and relations on the domain in question, so that no more than a countable population of higher-order entities is demanded. (It would follow that if one was otherwise attracted to such a thing, substitutional interpretation of the higher-order quantifiers in neo-Fregean arithmetic would be viable. This consideration very effectively addresses the line of concern developed in Clark (1993)).

of the approaches to higher-order quantification canvassed a moment ago, which inevitably effect a very close tie between (n-adic) *concept* and *(item of a) thinkable predication.* The possibility must be taken seriously, it seems to us, that the most tractable philosophies of higher-order logic will have the side effect that a neo-Fregean treatment of analysis will be constrained to be *constructivist*, and that the classical continuum will be out of reach. (Whether that would be cause for regret is, of course, a further question.)

REFERENCES

Bell, John L. (1999) "Frege's Theorem in a Constructive Setting" *Journal of Symbolic Logic* 64(2), pp. 486–488.

Boolos, George (1975) "On Second-Order Logic" *Journal of Philosophy* 72, pp. 509–527.

——— (1984) "To Be Is to Be the Value of a Variable (or To Be Some Values of Some Variables)" *Journal of Philosophy* 81, pp. 430–450.

——— (1987) "The Consistency of Frege's Foundations of Arithmetic" in J. Thomson (ed.) *On Being and Saying: Essays for Richard Cartwright.* Cambridge, MA: MIT Press, pp. 3–20.

——— (1989) "Iteration Again" *Philosophical Topics* 17, pp. 5–21.

——— (1990) "The Standard of Equality of Numbers" in Boolos (ed.) *Meaning and Method: Essays in Honor of Hilary Putnam.* Cambridge: Cambridge University Press, pp. 261–277.

——— (1993) "Whence the Contradiction?" *Proceedings of the Aristotelian Society* supp. 67, pp. 213–233.

Boolos, George, and Heck, Richard, Jr. (1998) "Die Grundlagen der Arithmetik §§82–83" in Schirn, pp. 407–428.

Bostock, David (1974) *Logic and Arithmetic: Natural Numbers.* Oxford: Clarendon Press.

——— (1979) *Logic and Arithmetic: Rational and Irrational Numbers.* Oxford: Clarendon Press.

Carnap, Rudolf (1928) *Der logische Aufbau der Welt.* Berlin: Weltkreis.

Clark, Peter (1993) "Logicism, the Continuum and Anti-realism" *Analysis* 53(3), pp. 129–141.

Colyvan, Mark (2000) "Conceptual Contingency and Abstract Existence" *Philosophical Quarterly* 50, pp. 87–91.

Cook, Roy (2002) "The State of the Economy: Neo-Logicism and Inflation" *Philosophia Mathematica* 3rd ser. 10, pp. 43–66.

Dummett, Michael (1973) *Frege: Philosophy of Language.* London: Duckworth; 2nd ed. 1981.

——— (1991) *Frege: Philosophy of Mathematics.* London: Duckworth.

Evans, Gareth (1982) *The Varieties of Reference.* Oxford: Oxford University Press.

Field, Hartry (1980) *Science Without Numbers* Oxford: Blackwell.

——— (1984) "Platonism for cheap? Crispin Wright on Frege's context principle" *Canadian Journal of Philosophy* 14, pp. 637–662; reprinted in Field (1989).

——— (1989) *Realism, Mathematics and Modality.* New York: Blackwell.
——— (1993) "The Conceptual Contingency of Mathematical Objects," *Mind* 102, pp. 285–299.
Fine, Kit (2002) *The Limits of Abstraction.* Oxford: Clarendon Press.
Frege, Gottlob (1884) *Die Grundlagen der Arithmetik* Breslau: W. Koebner; reprinted with English translation by J. L. Austin as *The Foundations of Arithmetic.* Oxford: Blackwell, 1950.
Hale, Bob (1987) *Abstract Objects.* Oxford: Blackwell.
——— (1990) "Nominalism," in A. D. Irvine (ed.) *Physicalism in Mathematics* Dordrecht: Kluwer.
——— (2000) "Reals by Abstraction" *Philosophia Mathematica* 3rd ser., 8, pp. 100–123, reprinted in Hale and Wright (2001).
——— (2000a) "Abstraction and Set Theory" *Notre Dame Journal of Formal Logic* 41(4), pp. 379–398.
Hale, Bob, and Wright, Crispin (1992) "Nominalism and the Contingency of Abstract Objects" *Journal of Philosophy* 89, pp. 111–135.
——— (1994) "A Reductio ad Adsurdum?: Field on the Contingency of Mathematical Objects" *Mind* 103, pp. 169–184.
——— (2001) *The Reason's Proper Study: Essays Towards a Neo-Fregean Philosophy of Mathematics.* Oxford: Clarendon Press.
——— (2002) "Benacerraf's Dilemma Revisited" *European Journal of Philosophy* 10, pp. 101–129.
Heck, Richard, Jr., ed. (1997) *Language, Thought and Logic.* Oxford: Oxford University Press.
Lewis, David (1970) "How to Define Theoretical Terms" *Journal of Philosophy* 67, pp. 427–446.
Noonan, Harold (1976) "Dummett on Abstract Objects" *Analysis* 36, pp. 49–54.
Parsons, Charles (1965) "Frege's Theory of Number" in Max Black (ed.) *Philosophy in America.* London: Allen and Unwin.
Ramsey, F.P. (1931) *The Foundations of Mathematics and Other Logical Essays.* London: Routledge and Kegan Paul.
Rayo, Agustín, and Yablo, Stephen (2001) "Nominalism Through De-nominalisation" *Noûs* 35, pp. 74–92.
Resnik, Michael (1980) *Frege and the Philosophy of Mathematics.* Ithaca: N.Y.: Cornell University Press.
Russell, Bertrand (1927) *The Analysis of Matter.* London: Kegan Paul, Trench and Trubner.
Schirn, Matthias (ed.) (1998) *The Philosophy of Mathematics Today.* Oxford: Clarendon Press.
Shapiro, Stewart (1997) *Philosophy of Mathematics: Structure and Ontology.* New York: Oxford University Press.
——— (2000) "Frege Meets Dedekind; a Neo-logicist Treatment of Real Analysis" *Notre Dame Journal of Formal Logic* 41(4), pp. 335–364.
——— (2003) "Prolegomenon to Any Future Neo-logicist Set-Theory: Abstraction and Indefinite Extensibility" *British Journal for the Philosophy of Science* 54, pp. 59–91.
Shapiro, Stewart, and Weir, Alan (1999) "New V, ZF and Abstraction" *Philosophia Mathematica* 3rd ser., 7, pp. 293–321.

——— (2000) "Neo-logicist' Logic Is Not Epistemically Innocent" *Philosophia Mathematica* 3rd ser., 8, pp. 160–189.

Tennant, Neil (1987) *Anti-realism and Logic.* Oxford: Clarendon Press.

Weir, Alan (2003) "Neo-Fregeanism: An Embarrassment of Riches?" *Notre Dame Journal of Formal Logic,* 44(1), pp. 13–48.

Wetzel, Linda (1990) "Dummett's Criteria for Singular Terms," *Mind* 99, pp. 239–254.

Wright, Crispin (1983) *Frege's Conception of Numbers as Objects.* Aberdeen: Aberdeen University Press.

——— (1988) "Why Numbers Can Believably Be: A reply to Hartry Field" *Revue Internationale de Philosophie* 42(167), pp. 425–473.

——— (1997) "The Philosophical Significance of Frege's Theorem" in Heck (1997), pp. 201–244, reprinted in Hale and Wright (2001), pp. 272–306.

——— (2000) "Neo-Fregean Foundations for Real Analysis: Some Reflections on Frege's Constraint" *Notre Dame Journal of Logic* 41(4), pp. 317–334.

CHAPTER 7

LOGICISM RECONSIDERED

AGUSTÍN RAYO

This chapter is divided into four sections.* The first two identify different logicist theses, and show that their truth-values can be established given minimal assumptions. Section 3 sets forth a notion of "content-recarving" as a possible constraint on logicist theses. Section 4—which is largely independent from the rest of the paper—is a discussion of "neologicism."

1. Logicism

1.1. What Is Logicism?

Briefly, logicism is the view that mathematics is a part of logic. But this formulation is imprecise because it fails to distinguish among the following three claims:

1. *Language-Logicism*
   The language of mathematics consists of purely logical expressions.
2. *Consequence-Logicism*
   There is a consistent, recursive set of axioms of which every mathematical truth is a purely logical consequence.

* I am grateful to Matti Eklund, Stewart Shapiro, and Gabriel Uzquiano for many helpful comments.

3. *Truth-Logicism*
   Mathematical truths are true as a matter of pure logic.

In fact, this is still not what we want. Consider *Language-Logicism* as an example. When taken at face value, standard mathematical languages do not consist of purely logical expressions. (For instance, when taken at face value, the standard language of arithmetic contains the individual constant "0," which is not a purely logical expression.) So if *Language-Logicism* is to have any plausibility, it should be read as the claim that mathematical sentences can be *paraphrased* in such a way that they contain no nonlogical vocabulary. Similarly for *Consequence-Logicism* and *Truth-Logicism*: if they are to have any plausibility, they should be read, respectively, as the claim that mathematical truths can be paraphrased in such a way that they are a purely logical consequence of the relevant set of axioms, and as the claim that mathematical truths can be paraphrased in such a way that their truth is a matter of pure logic. Let us therefore distinguish the following logicist theses (relative to a mathematical language $\mathcal{L}$ with an intended model $\mathcal{M}$):

> *Notation*: A *paraphrase-function* $\star$ is a function such that, for any sentence ϕ in the domain of $\star$, ϕ^* is a paraphrase of ϕ.

1. *Language-Logicism*
   There is a paraphrase-function $\star$ such that, for any sentence ϕ of $\mathcal{L}$, ϕ^* contains no nonlogical vocabulary.
2. *Consequence-Logicism (Semantic Version)*
   There is a paraphrase-function $\star$ and a consistent, recursive set of sentences $\mathcal{A}$ such that, for any sentence ϕ of $\mathcal{L}$ which is true (false) according to $\mathcal{M}$, ϕ^* (the negation of ϕ^*) is a semantic consequence of $\mathcal{A}$.
3. *Consequence-Logicism (Syntactic Version)*
   Like the semantic version, except that "semantic consequence of $\mathcal{A}$" is replaced by "derivable from $\mathcal{A}$ on the basis of purely logical axioms and rules of inference."
4. *Truth-Logicism (Semantic Version)*
   There is a paraphrase-function $\star$ such that, for any sentence ϕ of $\mathcal{L}$ which is true (false) according to $\mathcal{M}$, ϕ^* (the negation of ϕ^*) is a logical truth.
5. *Truth-Logicism (Syntactic Version)*
   Like the semantic version, except that "logical truth" is replaced by "derivable from the empty set on the basis of purely logical axioms and rules of inference."

In stating these five logicist theses, we have left open two important issues. First, there is the question of what it takes for one sentence to be a *paraphrase* of another. Is synonymy required, or could we settle for less? Second, there is the question of what should be counted as *logic*. Which expressions count as logical? Which axioms and rules of inference are logical? Which sentences are logical truths? These are difficult questions, which we cannot hope to address here. In section 2 we will evaluate the logicist theses on the basis of different assumptions about logic and paraphrase, but refrain from assessing the assumptions themselves.

1.2. Why Bother?

The five logicist theses are connected to important philosophical problems. Here are some examples:

1. On the assumption that statements of pure logic carry no commitment to abstract objects, *Language-Logicism* might be used as part of a nominalist account of mathematics, that is, an account of mathematics according to which statements of pure mathematics carry no commitment to abstract objects. (See chapters 15 and 16 in this volume.[1])
2. (a) If the semantic version of *Consequence-Logicism* is true, then every mathematical sentence has a determinate truth-value, provided only that the relevant axioms are determinately true and that semantic consequence preserves determinate truth. Such a result is important for certain brands of structuralism, which face the challenge of explaining how it is that finite beings like ourselves can succeed in picking out the standard mathematical structures. (See chapters 17 and 18 in this volume.)
   (b) If, in addition, the syntactic version of *Consequence-Logicism* is true, then one can come to know (at least in principle) any mathematical truth by deriving it from an appropriate set of axioms, provided only that the axioms are known to be true and that derivations preserve knowledge.
3. (a) If the semantic version of *Truth-Logicism* holds, then every mathematical truth is a logical truth.
   (b) If, in addition, the syntactic version of *Truth-Logicism* holds, then we get a very impressive result: one can come to know (at least in principle) any mathematical truth by carrying out a purely logical derivation, provided only that logical derivations preserve knowledge.

1.3. Some Brief Historical Remarks

An important focus of traditional logicist projects was *Truth-Logicism*. Famously, Frege's *Foundations of Arithmetic* and *Basic Laws of Arithmetic* attempted to

[1] For more on nominalistic accounts of mathematics, see Burgess and Rosen (1997).

establish a version of *Truth-Logicism* according to which mathematical truths can be proved merely on the basis of "general logical laws and definitions." The precise motivation behind Frege's defense of *Truth-Logicism* is a matter of scholarly debate.[2]

Frege's project collapsed with the discovery that one of its Basic Laws leads to contradiction. In the latter part of the twentieth century, however, the project was revitalized in a somewhat different form by Crispin Wright and Bob Hale, largely as an attempt to secure a version of the epistemological result in 3(b) of section 1.2. (For more on neo-Fregeanism, see section 4 of the present text and chapter 6 in this volume.)

The logical empiricists of the first half of the twentieth century were also interested in *Truth-Logicism*. Their program was based on the doctrine that any meaningful statement concerns either a "matter of fact" (and, if true, is knowable only a posteriori), or a "relation of ideas" (and, if true, is "true in virtue of meaning" and consequently knowable a priori).[3] In order to resist the idea that mathematical statements concern "matters of fact," it was therefore crucial to logical empiricism that every mathematical truth be shown to be "true in virtue of meaning" and consequently knowable a priori. But if every mathematical truth can be known on the basis of a purely logical derivation, as in 3(b), then mathematical truths are indeed "true in virtue of meaning" and knowable a priori—assuming that logical truths are "true in virtue of meaning" and that purely logical derivations preserve a priori knowledge, as logical empiricists believed.

It is not clear, on the other hand, whether *Truth-Logicism* was Russell's aim in *Principia Mathematica*. It might well have consisted of *Language-Logicism* together with *Consequence-Logicism*.[4] (For more on the logicism of Frege and Russell, see chapter 5 in this volume.)

2. An Assessment of Logicism

2.1. Logic and Paraphrase

In stating our five logicist theses in section 1.1, we left open the question of what should be counted as logic. Two possible answers are that logic is first-order logic, and that logic is higher-order logic. More precisely,

[2] See, for instance, Benacerraf (1981).

[3] See, for instance, the preface to the first edition of Ayer (1946).

[4] See, for instance, Boolos (1994).

The First-Order View

A sentence is a logical truth just in case it can be paraphrased as a first-order sentence which is true in every model of the standard first-order semantics.[5] An axiom is logical just in case it is a logical truth. A rule of inference is logical just in case it preserves logical truth. Logical expressions are those that can be paraphrased as an expression of pure first-order logic.

The Higher-Order View

Like the first-order view, except that "first-order" is everywhere replaced by "higher-order."[6] (For more on higher-order languages, see chapters 25 and 26 in this volume.)

In order to read the logicist theses in accordance with the first-order view, all we need to do is insist that the relevant paraphrase-functions deliver first-order sentences as outputs, and understand logical notions in accordance with the standard semantics for first-order languages. Similarly for the higher-order view.

These are certainly not the only ways of answering the question of what should be counted as logic—one might think, for instance, that an infinitely long sentence can be logically true.[7] But they are the only ones we will consider here.

A second issue which we left open is the question of what it takes, in the context of logicism, for one sentence to count as a paraphrase of another. This is a difficult problem. But, whatever the constraints on paraphrases turn out to be, it is reasonable to expect that any legitimate paraphrase-function will be recursive and will preserve truth-values. And we will see there is a lot that can be learned about the logicist theses on the basis of these rather minimal constraints.

Formally, let us say that a paraphrase-function $\star$ is *minimally adequate* for a mathematical language $\mathcal{L}$ with an intended model $\mathcal{M}$ just in case the following conditions are satisfied:

1. Every sentence of $\mathcal{L}$ is in the domain of $\star$.
2. The restriction of $\star$ to sentences of $\mathcal{L}$ is recursive.

[5] Similarly, a sentence ϕ is a semantic consequence of a set of sentences Γ just in case there is a first-order sentence ϕ' and a set of first-order sentences Γ' such that (*a*) ϕ' is a paraphrase of ϕ, (*b*) every sentence in Γ' is a paraphrase of some sentence in Γ, and (*c*) ϕ' is true in every model of the standard first-order semantics which makes every sentence in Γ' true. In order not to count "$\exists x\ (x = x)$" as a logical truth, we must avoid the simplifying assumption that the domain of a model is always nonempty.

[6] Unless a suitable Reflection Principle holds, the standard semantics for higher-order languages is inadequate for languages whose intended domain contains too many objects to form a set (such as the language of set-theory). To avoid the problem, one can use a higher-order semantics, such as the one developed in Rayo and Uzquiano (1999).

[7] Yablo (2002) provides an interesting paraphrase function from arithmetic and set theory to an infinitary version of first-order logic.

3. There is a model $\mathcal{S}$ (intuitively, the intended model of the paraphrases) such that, for any sentence ϕ of $\mathcal{L}$, $\models_{\mathcal{M}} \phi$ if and only if $\models_{\mathcal{S}} \phi^*$.

2.2. The First-order View

When the logicist theses are read in accordance with the first-order view, the following relations obtain: (see figure 7.1)

[FOV-1] The syntactic and semantic versions of *Truth-Logicism* and *Consequence-Logicism* are equivalent, since first-order logic is sound and complete according to the first-order view.

[FOV-2] *Truth-Logicism* implies *Consequence-Logicism*, since *Truth-Logicism* is the special case of *Consequence-Logicism* when $\mathcal{A}$ is empty.

[FOV-3] When paraphrase functions are assumed to be minimally adequate, *Language-Logicism* implies *Consequence-Logicism*.[8]

These implications suffice to decide the question of whether the five logicist theses are true, in the context of the first-order view. For, assuming paraphrase-functions are minimally adequate, it is a consequence of Gödel's Incompleteness Theorem that the syntactic version of *Consequence-Logicism* is false for any interesting fragment of mathematics.[9] And since each of the other logicist theses implies the syntactic version of *Consequence-Logicism*, it follows that each of the other logicist theses must also be false.

[8] This follows from the fact that any consistent first-order theory T containing only sentences with no nonlogical vocabulary has a recursive axiomatization: if T has a finite model, then every sentence in T follows from a sentence stating that there are precisely n objects, for some n; if T has an infinite model, then every sentence in T follows from a set of sentences to the effect that the universe is infinite.

[9] *Proof Sketch*: Let T be a sufficiently interesting mathematical theory (i.e., a theory in which Robinson Arithmetic can be interpreted), and suppose there is a minimally adequate paraphrase-function $*$ which delivers the syntactic version of *Consequence-Logicism* with respect to a recursive axiomatization $\mathcal{A}$ of T. Then there is a decision procedure for Robinson Arithmetic. (For ϕ a sentence of Robinson Arithmetic, let ψ be the interpretation of ϕ in T. Since $*$ is minimally adequate, it is recursive. And since there is a recursive function listing the theorems derivable from $\mathcal{A}$, the set S of syntactic consequences of $\mathcal{A}$ is recursive. But since *Consequence-Logicism* holds, ψ^* is in S if ψ is true, and the negation of ψ^* is in S if ψ is false.) This contradicts Gödel's Incompleteness Theorem, which implies that that there is no decision procedure for Robinson Arithmetic.

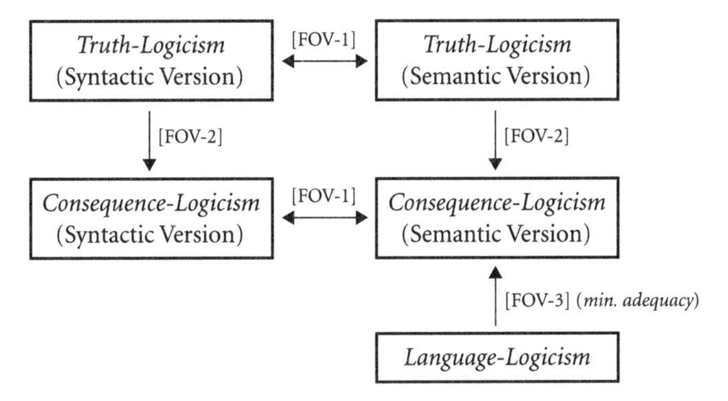

Fig. 7.1. Implications among the five logicist theses, when read in accordance with the first-order view. (Arrows go from *implicans* to *implicatum*.)

2.3. The Higher-order View

When the logicist theses are read in accordance with the higher-order view, the following relations obtain (see figure 7.2):

[HOV-1] The syntactic versions of *Truth-Logicism* and *Consequence-Logicism* imply the corresponding semantic versions, since higher-order logic is sound according to the higher-order view. (The converse implications do not hold because higher-order logic is incomplete.)

[HOV-2] Each version of *Truth-Logicism* implies the corresponding version of *Consequence-Logicism*, since *Truth-Logicism* is the special case of *Consequence-Logicism* when $\mathcal{A}$ is empty.

[HOV-3] If the semantic version of *Consequence-Logicism* obtains on the basis of a finite axiomatization, then there is a minimally adequate paraphrase-function which makes *Language-Logicism* true.[10]

The final implication holds because, when the semantic version of *Consequence-Logicism* obtains on the basis of a finite axiomatization, $\mathcal{A}$, the paraphrase-function

[10] It is worth emphasizing that whereas [FOV-3] describes an implication from *Language-Logicism* to *Consequence-Logicism*, [HOV-3] describes an implication from *Consequence-Logicism* to *Language-Logicism*. Our proof of [FOV-3] cannot be reproduced in the context of the higher-order view because it makes use of the Downward Löwenheim-Skolem Theorem, which holds for first-order languages but not their higher-order counterparts. Our proof of [HOV-3] cannot be reproduced in the context of the first-order view because, although the result of Ramsifying a higher-order sentence is always a higher-order sentence, the result of Ramsifying a first-order sentence is not always a first-order sentence.

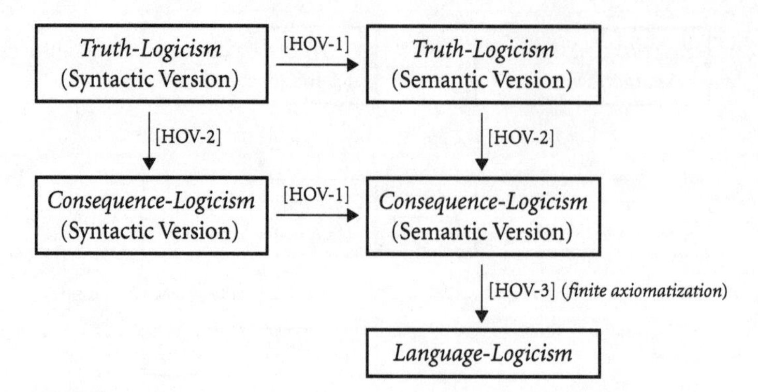

Fig. 7.2. Implications among the five logicist theses, when read in accordance with the higher-order view. (Arrows go from *implicans* to *implicatum*.)

taking a higher-order sentence $\ulcorner \phi \urcorner$ to the result of universally Ramsifying $\ulcorner \mathcal{A} \rightarrow \phi \urcorner$ is minimally adequate.[11] And the result of universally Ramsifying $\ulcorner \mathcal{A} \rightarrow \phi \urcorner$ is a higher-order sentence containing no nonlogical vocabulary.

Let us now turn to an assessment of the five logicist theses, from the perspective of the higher-order view.

2.3.1. *The Syntactic Versions of Consequence- and Truth-Logicism*

When it comes to the syntactic versions of *Consequence-* and *Truth-Logicism*, the higher-order view constitutes no improvement over the first-order view. Assuming paraphrase-functions are minimally adequate, it is a consequence of Gödel's Incompleteness Theorem that the syntactic version of *Consequence-Logicism* is false for any interesting fragment of mathematics.[12] And, since the syntactic version of *Truth-Logicism* implies the syntactic version of *Consequence-Logicism*, it follows that the syntactic version of *Truth-Logicism* must also be false.

2.3.2. *The Semantic Version of Consequence-Logicism*

Because there are semantic consequences of higher-order theories which are not derivable on the basis of any sound deductive system, nothing we have said so far entails that the *semantic* version of *Consequence-Logicism* must fail in the context of the higher-order view. But does it?

For the special case of arithmetic, we have a positive result, thanks to semantic completeness: the second-order Dedekind–Peano Axioms semantically imply every true sentence of the language of pure second-order arithmetic.[13] From this the

[11] The result of *universally Ramsifying* $\ulcorner \mathcal{A} \rightarrow \phi \urcorner$ is the universal closure of the result of uniformly substituting variables of the appropriate type for all nonlogical primitives in $\ulcorner \mathcal{A} \rightarrow \phi \urcorner$.

[12] The proof is analogous to that in footnote 9.

[13] The original result is due to Dedekind. For a proof see Shapiro (1991), theorem 4.8.

semantic version of *Consequence-Logicism* follows immediately, in the context of the higher-order view.

Similar results hold for other branches of pure mathematics.[14] In the case of set theory, however, there is a problem. Unlike the second-order Dedekind–Peano Axioms, the axioms of second-order ZFC are not guaranteed to be semantically complete: there may be a sentence of the language of pure second-order set theory such that neither it nor its negation is a semantic consequence of the axioms. (This will be the case if there are models of second-order set theory whose domain is larger than the first strongly inaccessible cardinal.) All we have is a partial result: if we add to the axioms of second-order ZFC a sentence specifying how many sets there are, then we are guaranteed semantic completeness.[15]

Thus, one possibility is to enrich the axioms of second-order ZFC with a sentence S_μ of pure second-order logic to the effect that there are no inaccessible cardinals. That would certainly yield semantic completeness. Unfortunately, S_μ is not universally regarded as true. Many set theorists endorse a Global Reflection Principle, to the effect that there is a set α such that every true sentence in the language of higher-order set theory is true when the quantifiers are restricted to the elements of α. And the Global Reflection Principle implies that S_μ is false because it implies that there is an inaccessible cardinal (since the standard set-theoretic axioms are true, the Reflection Principle implies that they are also true when the quantifiers are restricted to the members of some set, and this is possible only if there is an inaccessible cardinal.)

As it turns out, matters are worse still: if the Global Reflection Principle is true, then no recursive set of true sentences in the language of higher-order set theory semantically implies every true sentence in the language of higher-order set theory.[16] Thus the Global Reflection Principle implies that *there is no hope* of producing a semantic completeness result in the case of set theory.

[14] For a proof of the categoricity of analysis, see Shapiro (1991), theorem 4.10.

[15] The result is due to Zermelo. See Zermelo (1930).

[16] *Proof Sketch*: Suppose, for *reductio*, that $\mathcal{A}$ is such a set. Define, within the language of higher-order set theory, the notion of a (set-sized) model for the language of higher-order set theory, and a corresponding satisfaction predicate. Let $\ulcorner \bar{\mathcal{S}} \models^+ \bar{\phi} \urcorner$ be a sentence of the language of higher-order set theory stating that every (set-sized) model in which every sentence whose Gödel number is in $\bar{\mathcal{S}}$ is satisfiable is also a model in which the sentence with Gödel-number $\bar{\phi}$ is satisfiable. Since $\mathcal{A}$ is recursive, "$x \in \mathcal{A}$" is expressible in second-order ZFC. Thus $\ulcorner \bar{\mathcal{A}} \models^+ \bar{\phi} \urcorner$ is a truth predicate for the language of higher-order set theory. (By our assumption, every true sentence of the language of higher-order set theory is a semantic consequence of $\mathcal{A}$, and hence true on every (set-sized) model; and, by the Global Reflection Principle, there is a (set-sized) model according to which every untrue sentence of the language of higher-order arithmetic is false.) But it follows from Tarski's Theorem that no language can express its own truth-predicate.

So much for pure mathematics. In the case of *applied* mathematics it is unreasonable to ask for a semantic completeness result, since no set of mathematical axioms should imply, for instance, that Mars has two moons (or that it doesn't). But one might hope for a *relative* semantic completeness result, to the effect that there is a set of axioms which, together with the set of true sentences containing no mathematical vocabulary, semantically implies every true sentence in some particular fragment of applied mathematics, and the negation of every false one. As it turns out, it is possible to prove a relative completeness result for higher-order applied arithmetic.[17] Thus *Consequence-Logicism* holds for higher-order applied arithmetic, modulo the set of nonarithmetical truths.

2.3.3. *Language-Logicism*

When all we require of paraphrase-functions is minimal adequacy, it follows from [HOV-3] that any higher-order theory with a finite and complete axiomatization can be paraphrased using no nonlogical vocabulary. Thus, thanks to the semantic completeness results discussed in the preceding section, *Language-Logicism* can be shown to be true for the case of pure higher-order arithmetic.

But minimal adequacy is a very weak constraint on paraphrase functions, and some advocates of logicism might hope for more. One way to strengthen the constraint is by making use of semantic notions. For instance, one might require paraphrase-functions to preserve *content.* We will say something about this in section 3, but for the moment it is best to focus on a syntactic constraint instead. Informally, say that a paraphrase-function $\star$ is *compositional* just in case the compositional structure of $\phi^\star$ mirrors the compositional structure of ϕ, for every ϕ in the domain of $\star$. (For a formal characterization of this constraint, see the appendix.) Paraphrase-functions based on Ramsey-conditionals are not compositional. Thus, when compositionality is required, Ramsey-conditionals do not deliver *Language-Logicism.*

For the case of pure higher-order arithmetic, however, there are compositional paraphrase functions which deliver formulas of pure higher-order logic as outputs. One is implicit in Whitehead and Russell's *Principia* (see chapter 5 in this volume). More recently, Harold Hodes has set forth a paraphrase-function that works by treating first-order quantifiers ranging over natural numbers as *third-order* quantifiers ranging over finite cardinality object-quantifiers (that is, second-order concepts true of all and only first-order concepts under which precisely n objects fall, for some n).[18] Hodes's paraphrase-function can be shown to be

[17] See Rayo (2002a). It is worth noting that whereas completeness results are usually based on categoricity theorems, the proof in Rayo (2002a) relies on an essentially different method.

[18] See Hodes (1984, 1990). See also Wright (1983, pp. 36–40); Bostock (1979, vol. II, ch. 1); and Rayo (2002a).

minimally adequate on the assumption that the universe is infinite, and its restriction to pure higher-order arithmetic always delivers formulas of pure higher-order logic as outputs. Thus, when no more is required of paraphrase-functions than compositionality and minimal adequacy, *Language-Logicism* can be shown to be true for the case of pure higher-order arithmetic in the context of the higher-order view.

2.3.4. *The Semantic Version of* Truth-Logicism

We noted in section 2.3.1 that the syntactic version of *Truth-Logicism* fails in the context of the higher-order view. But, for all that has been said so far, it might still be the case that the *semantic* version of *Truth-Logicism* holds.

None of the paraphrase-functions we have discussed so far deliver this result. Consider, for instance, the paraphrase-function $\mathfrak{R}$, which takes each sentence $\ulcorner \phi \urcorner$ of pure second-order arithmetic to the result of universally Ramsifying $\ulcorner \mathcal{A} \rightarrow \phi \urcorner$ (where $\ulcorner \mathcal{A} \urcorner$ is the conjunction of the Dedekind–Peano Axioms). In order for $\mathfrak{R}$ to deliver *Truth-Logicism*, $\phi^{\mathfrak{R}}$ would have to be a logical truth whenever ϕ is true on the intended interpretation, and a logical falsehood whenever ϕ is false on the intended interpretation. This is certainly the case when we restrict our attention to models with infinite domains. But it is not the case when models with finite domains are allowed, since $\phi^{\mathfrak{R}}$ is true on any model with a finite domain, independently of whether ϕ is true on the intended interpretation.

A similar problem afflicts Hodes's paraphrase-function. When we restrict our attention to models with infinite domains, the Hodes-paraphrase of ϕ is a logical truth just in case ϕ is true on the intended interpretation, for any sentence ϕ in the language of higher-order arithmetic. But this is not the case when models with finite domains are allowed.

Might other paraphrase-functions do better? In an important sense, the answer is "no." On the assumption that paraphrase-functions are required to be compositional, the semantic version of *Truth-Logicism* must fail for any interesting fragment of mathematics, in the context of the higher-order view. (See appendix for proof.)

A final point is worth mentioning. So far we have made the (standard) assumption that first- and higher-order quantifiers are *domain-relative*. In other words, we have presupposed that which individuals the truth-value of a quantified sentence depends on is not a logical matter and, accordingly, that the range of the quantifiers may vary from model to model. But Timothy Williamson has recently made a case for *logically unrestricted* quantifiers: quantifiers whose range consists, as a matter of logic, of absolutely everything.[19] If Williamson's logically unrestricted quantifiers are allowed, then the fact that there are at least two objects in

[19] See Williamson (1999) and Rayo and Williamson (2003).

existence suffices to guarantee that when the quantifiers are regarded as logically unrestricted, "$\exists x \exists y (x \neq y)$" is logically true. Similarly, the fact that there are infinitely many objects in existence suffices to guarantee that when the quantifiers are regarded as logically unrestricted, a second-order sentence stating that the universe is infinite is logically true.

When logically unrestricted quantifiers are allowed, the status of *Truth-Logicism* changes dramatically. The fact that there are infinitely many objects ensures that when quantifiers are regarded as logically unrestricted, the Hodes-paraphrase of ϕ is a logical truth (falsehood) just in case ϕ is true (false) on the intended interpretation, for any ϕ in the language of pure higher-order arithmetic. It also ensures that, when quantifiers are regarded as logically unrestricted, $\phi^{\Re}$ is a logical truth (falsehood) whenever ϕ is true (false) on the intended interpretation, for any ϕ in the language of pure second-order arithmetic. Finally, when the quantifiers are regarded as logically unrestricted, one can show that there is a minimally adequate (but non-compositional) paraphrase-function verifying the semantic version of *Truth-Logicism* for the language of pure higher-order set theory, given the assumption that there are strongly inaccessibly many objects.[20]

This concludes our assessment of the five logicist theses. Figure 7.3 summarizes some of the main results.

2.4. What Have We Learned?

We have seen that *Truth-Logicism* fails on natural assumptions about logic and paraphrase. Although this casts doubt on the view that every mathematical truth is a truth of logic, it does not constitute a decisive refutation. One way of maintaining *Truth-Logicism* is by adopting a conception of logic distinct from the first- and higher-order views. Frege, for example, took Basic Law V—the principle that the extension of the Fs equals the extension of the Gs just in case the Fs are the Gs—to be "purely logical."[21] (Though of course he later discovered that Basic Law V leads to contradiction.) Another way of resisting the first- and higher-order views is by adopting Williamson's logically unrestricted quantifiers. But it is

[20] Consider the paraphrase-function $*$, which takes each sentence $\ulcorner\phi\urcorner$ of the language of higher-order set theory with ur-elements to the result of universally Ramsifying $\ulcorner \text{ZFCU} \rightarrow \phi \urcorner$ (where $\ulcorner\text{ZFCU}\urcorner$ is the conjunction of the axioms of second-order ZFC with ur-elements plus an axiom to the effect that the ur-elements form a set). If there are strongly inaccessibly many objects, it follows from the result in McGee (1997) that when quantifiers are regarded as logically unrestricted, ϕ^* is a logical truth (falsehood) whenever ϕ is true (false) on the intended interpretation for every ϕ in the language of *pure* higher-order arithmetic.

[21] See Frege (1893, 1903), vol 1, pp. vii.

	First-order View (allowing only minimally adequate paraphrase-functions)	Higher-order View (allowing all and only compositional, minimally adequate paraphrase-functions)
Language-Logicism	False	True[a]
Consequence-Logicism (Semantic Version)	False	True[b]
Consequence-Logicism (Syntactic Version)	False	False
Truth-Logicism (Semantic Version)	False	False[c]
Truth-Logicism (Syntactic Version)	False	False

Fig. 7.3. Summary. [a] for the case of pure higher-order arithmetic. [b] for the case of pure and applied higher-order arithmetic (*modulo* the set of non-arithmetical truths). [c] true for the case of pure higher-order arithmetic when logically unrestricted quantifiers are allowed.

important to be clear that logically unrestricted quantifiers have an epistemological cost, since their legitimacy would destroy any hope of establishing a straightforward connection between logical truth and a priori knowledge.[22] The failure of *Truth-Logicism* on the first- and higher-order views makes it seem implausible that one could acquire a priori knowledge of the standard mathematical axioms merely by carrying out logical derivations. (We will see in section 4 that neo-Fregeans have argued for the weaker claim that one can acquire a priori knowledge of the standard arithmetical axioms merely by setting forth an appropriate stipulation and going on to carry out appropriate logical derivations.)

We have also seen that, on the higher-order view, the semantic version of *Consequence-Logicism* holds for the case of (pure and applied) higher-order arithmetic. This delivers the result that every arithmetical sentence is either determinately true or determinately false, provided only that appropriate arithmetical axioms are determinately true and that higher-order consequence preserves determinate truth.[23] As noted in section 2.3.2, the Global Reflection Principle implies that there is no hope of producing a semantic completeness result for the case of set theory, but it is worth noting that it does not preclude a determinacy result. By working within the language of set theory with ur-elements, McGee (1997) and Uzquiano (2002) have found axiomatizations of second-order set theory with the feature that the pure sets of any two models are isomorphic, provided only

[22] See, however, Williamson (1999). It is also worth noting that because of Gödel's incompleteness results, the legitimacy of higher-order quantification should be enough to make us suspicious of a straightforward connection between logical truth and a priori knowledge.

[23] All of this, modulo indeterminacy in nonmathematical expressions.

that the models in question have domains consisting of everything there is.[24] Accordingly, by insisting that one's quantifiers are to range over absolutely everything,[25] one could carry out a recursively specifiable stipulation capable of delivering the result that every sentence in the language of pure set theory is either determinately true or determinately false. As before, this relies on the assumption that the appropriate set-theoretic axioms are determinately true and that higher-order semantic consequence preserves determinate truth.

Finally, we have seen that on the higher-order view, *Language-Logicism* holds for the case of pure higher-order arithmetic. This result might be used as part of a nominalist account of mathematics. (And, of course, Nominalism is of interest even if one is not a Nominalist; for instance, a Nominalist account of higher-order applied arithmetic would conclusively undermine the thought that we are justified in believing standard mathematical theories because they are indispensable for the natural sciences.)

3. Recarving Contents

The only *semantic* constraint on paraphrase-functions we have considered so far is preservation of truth-value. Are there other semantic constraints one might reasonably impose?

One option, of course, is to require strict synonymy. But perhaps a more liberal constraint can be extracted from the following passage in Frege's *Foundations of Arithmetic*:

> The judgement "line *a* is parallel to line *b*", or, using symbols, "*a*//*b*", can be taken as an identity. If we do this, we obtain the concept of direction, and say: "the direction of line *a* is identical with the direction of line *b*". Thus we replace the symbol // by the more generic symbol =, through removing what is specific in the content of the former and dividing it between *a* and *b*. We carve up the content in a way different from the original way, and this yields a new concept. (§64)

Thus, the sentences "line *a* is parallel to line *b*" and "the direction of line *a* is identical to the direction of line *b*" are taken to express the same content, only "carved up" in a different way. And, as Frege observed, something similar might be said for

[24] For discussion on quantifying over everything see Parsons (1974); Dummett (1981), chs. 14–16; Cartwright (1994); Boolos (1998b); Williamson (1999); McGee (2000); the postscript to Field (1998) in Field (2001); Rayo (2002b), Rayo (2003b); Rayo and Williamson (2003); Glanzberg (forthcoming); and Williamson (2003).

[25] The possibility of doing so determinately is defended in McGee (2000) and Rayo (2003b).

the case of *arithmetic*: the sentences "the *F*s can be put in one-one correspondence with the *G*s" and "the number of the *F*s is identical to the number of the *G*s" might be taken express the same content, only "carved up" in a different way.

In this section we will consider the following three questions:

1. Is there an interesting way of making the notion of content recarving precise?
2. Is there a paraphrase-function $\mathcal{I}$ such that, for any sentence ϕ in some interesting fragment of mathematics, ϕ and $\phi^{\mathcal{I}}$ express the same content, only "carved up" in a different way?
3. If $\mathcal{I}$ exists, does it verify any of the logicist theses?

It is best for expositional purposes to begin with question 2, while working with the informal characterization of content-recarving provided by Frege's remark. Later we will return to question 1, and say something about how the notion of content-recarving might be made precise. We end with question 3.

3.1. The Second Question

First, some notation. We let "$F \approx G$" be a second-order formula expressing one-one correspondence between the *F*s and the *G*s,[26] and interpret "$\mathbf{N}(n,F)$" as "n numbers the *F*s."[27]

Next, we set forth a paraphrase-function $\mathcal{F}$. Intuitively, $\mathcal{F}$ works by paraphrasing talk of the *number* of the *F*s by talk of the *F*s themselves.[28] Formally, $\mathcal{F}$ may be characterized as follows, where "m_1," "m_2,"... are first-order variables ranging over numbers, and "Z_1," "Z_2,"... are second-order variables ranging over an unrestricted domain:

- $\ulcorner \forall m_i(\phi)\urcorner^{\mathcal{F}} = \ulcorner \forall Z_i \urcorner \frown \ulcorner(\phi)\urcorner^{\mathcal{F}}$;
- $\ulcorner m_i = m_j \urcorner^{\mathcal{F}} = \ulcorner Z_i \approx Z_j \urcorner$;
- $\ulcorner \mathbf{N}(m_i, X)\urcorner^{\mathcal{F}} = \ulcorner Z_i \approx X \urcorner$;
- $\ulcorner \neg\phi \urcorner^{\mathcal{F}} = \text{``}\neg\text{''} \frown \ulcorner \phi \urcorner^{\mathcal{F}}$;
- $\ulcorner \phi \wedge \psi \urcorner^{\mathcal{F}} = \ulcorner \phi \urcorner^{\mathcal{F}} \frown \text{``}\wedge\text{''} \frown \ulcorner \psi \urcorner^{\mathcal{F}}$;
- $\ulcorner \phi \rightarrow \psi \urcorner^{\mathcal{F}} = \ulcorner \phi \urcorner^{\mathcal{F}} \frown \text{``}\rightarrow\text{''} \frown \ulcorner \psi \urcorner^{\mathcal{F}}$.

[26] That is,
$F \approx G \equiv_{df} \exists R\ [\forall w(Fw \rightarrow \exists! v(Gv \wedge Rwv)) \wedge \forall w(Gw \rightarrow \exists! v(Fv \wedge Rvw))]$.

[27] "**N**" is a second-level predicate because it takes a second-order variable in one of its argument-places. For a discussion of second-level predicates see Rayo (2002).

[28] I use plurals for expository purposes only. I do not presuppose that second-order quantifiers should be understood plurally, as in Boolos (1984).

As an example, consider the sentence "the number of the *F*s = the number of the *G*s." In our notation, it can be formulated as

$$\forall m_1 \, \forall m_2(\mathbf{N}(m_1, F) \wedge \mathbf{N}(m_2, G) \rightarrow m_1 = m_2).$$

The result of applying $\mathcal{F}$ is

$$\forall Z_1 \, \forall Z_2(Z_1 \approx F \wedge Z_2 \approx G \rightarrow Z_1 \approx Z_2),$$

which is equivalent to

$$F \approx G.$$

Thus, the result of applying $\mathcal{F}$ to "the number of the *F*s = the number of the *G*s" is equivalent to "$F \approx G$." In a Fregean spirit, one might be tempted to say that the application of $\mathcal{F}$ has resulted in a recarving of content.

What happens when we apply $\mathcal{F}$ to other arithmetical formulas? For instance, what happens when we apply it to a sentence such as "the number of the planets = 9"? When numerical predicates are defined in the obvious way,[29]

$$\begin{aligned}
0(m) &\equiv_{df} \forall X(\mathbf{N}(m, X) \leftrightarrow \neg \exists x(X(x)));\\
1(m) &\equiv_{df} \forall X(\mathbf{N}(m, X) \leftrightarrow \exists!_1 x(X(x)));\\
2(m) &\equiv_{df} \forall X(\mathbf{N}(m, X) \leftrightarrow \exists!_2 x(X(x)));\\
&\vdots
\end{aligned}$$

"the number of the planets = 9" can be formulated as

$$\forall m_1(\mathbf{N}(m_1, \hat{x}[\textsc{Planet}(x)]) \rightarrow 9(m_1)).^{30}$$

[29] As usual,

$$\begin{aligned}
\exists!_1 x(\phi(x)) &\equiv_{df} \exists x(\phi(x) \wedge \neg \exists y(y \neq x \wedge \phi(y))),\\
\exists!_{n+1} x(\phi(x)) &\equiv_{df} \exists x(\phi(x) \wedge \exists!_n y(y \neq x \wedge \phi(y))).
\end{aligned}$$

[30] Syntactically, an expression of the form "$\hat{x}[\phi(x)]$" takes the place of a monadic second-order variable. But the result of substituting "$\hat{x}[\phi(x)]$" for "Y" in a formula "$\Psi(Y)$" is to be understood as shorthand for

$$\forall W(\forall x(W x \leftrightarrow \phi(x)) \rightarrow \Psi(W)).$$

Additional clauses must be added to the definition of $\mathcal{F}$ before it can be applied to sentences containing "$\hat{x}\ [\phi(x)]$," but they are all trivial. See Rayo (2002a) for details.

And the result of applying $\mathcal{F}$ to this sentence is equivalent to

$$\forall Z_1(Z_1 \approx \hat{x}[\text{PLANET}(x)] \rightarrow \exists!_9 x(Z_1(x))),$$

which is in turn equivalent to

$$\exists!_9 x(\text{PLANET}(x)).$$

Thus, the result of applying $\mathcal{F}$ to "the number of the planets = 9" is equivalent to "$\exists!_9 x$(Planet(x))." One might again be tempted to say, in a Fregean spirit, that the application of $\mathcal{F}$ has resulted in a recarving of content.

It is easy to define "PLUS" and "TIMES" in terms of $\mathbf{N}(n,F)$. Consider "PLUS" as an example:

$$\text{PLUS}(m_1,m_2,m_3) \equiv_{df}$$
$$\forall X \forall Y \forall Z\,(\mathbf{N}(m_1,X) \wedge \mathbf{N}(m_2,Y) \wedge \mathbf{N}(m_3,Z) \rightarrow \text{JOIN}(X,\ Y,\ Z)),$$

where

$$\text{JOIN}(X,Y,Z) \equiv_{df}$$
$$\exists X'(X' \approx X \wedge \forall x(X'(x) \rightarrow \neg Y(x)) \wedge \hat{x}[X'(x) \vee Y(x)] \approx Z).$$

A sentence such as "$\forall n \forall m\ (n+m=m+n)$" can then be formulated as

$$\forall m_1 \forall m_2 \forall m_3(\text{PLUS}(m_1,m_2,m_3) \rightarrow \text{PLUS}(m_2,m_1,m_3)).$$

The result of applying $\mathcal{F}$ is equivalent to[31]

$$\forall Z_1 \forall Z_2 \forall Z_3(\text{JOIN}(Z_1,Z_2,Z_3) \rightarrow \text{JOIN}(Z_2,Z_1,Z_3)).$$

And, again, in a Fregean spirit, one might be tempted to say that the application of $\mathcal{F}$ has resulted in a recarving of content.

As it turns out, $\mathcal{F}$ can be made to encompass every sentence of pure and applied nth-order arithmetic, and it can be done in such a way that the result of applying $\mathcal{F}$ to a sentence of pure n-th order arithmetic is always a sentence of pure second-order logic. As one might have hoped, $\mathcal{F}$ is a compositional and minimally adequate paraphrase-function.[32]

[31] Additional clauses must be added to the definition of $\mathcal{F}$ before it can be applied to this sentence, but they are all trivial. See Rayo (2002a) for details.

[32] See Rayo (2002a).

3.2. The First Question

Is there any interesting way of making the notion of content-recarving precise? Hale (1997) sets forth one proposal, which is intended as a defense of the Neo-Fregean Program. Here we will consider another, which is not.

I would like to suggest that one must ask not whether ϕ expresses a recarving of the content expressed by ψ' when ϕ and ψ' are considered in isolation, but whether ϕ expresses a recarving of the content expressed by ψ when ϕ is considered in the context of a set of sentences A and ψ' is considered in the context of a set of sentences B. More specifically, the proposal is this:

> The content expressed by ϕ is a recarving of the content expressed by ψ (when ϕ is considered in the context of a set of sentences A and ψ' is considered in the context of a set of sentences B) just in case the content expressed by ϕ *relative to* A is the same the content expressed by ψ *relative to* B.

What is it to express a content relative to a set of sentences? Let me begin with an informal explanation. Suppose it is right to say that ϕ and ψ express the same content *simpliciter* just in case (1) ϕ and ψ have the same logical structure; and (2) nonlogical primitives occupying corresponding places in the logical structure of ϕ and ψ have the same semantic-value. One can then say that the content expressed by ϕ relative to A is the same as the content expressed by ψ relative to B just in case (1) the logical structure, LS^{ϕ}_{A}, that is "displayed" by ϕ in the context of A is isomorphic to the logical structure, LS^{ψ}_{B}, that that is "displayed" by ψ in the context of B; and (2) if E is an expression in ϕ that "functions" in the context of A as a nonlogical primitive and occupies a certain position in LS^{ϕ}_{A}, and E' is an expression that "functions" in the context of B as a nonlogical primitive and occupies a corresponding position in LS^{ψ}_{B}, then the semantic-value of E is "analogous" to the semantic-value of E'.

The notion of an $AB_{\mathcal{I}}$-recarving embodies a rigorous version of this informal explanation:

> Let A and B be sets of formulas, let $\mathcal{I}$ be a one-one function from A onto B, and suppose that the following conditions are met:
>
> 1. $\mathcal{I}$ preserves the *logical networking* of A.
> 2. For any $\phi \in A$, ϕ and $\phi^{\mathcal{I}}$ are *equisatisfiable.*
>
> Then we shall say that for any $\phi \in A$, ϕ is an $AB_{\mathcal{I}}$-recarving of $\phi^{\mathcal{I}}$.

The requirement of preservation of logical networking is a way of capturing the thought that the logical structure "displayed" by a sentence ϕ in A is isomorphic to the logical structure "displayed" by $\phi^{\mathcal{I}}$ in *B*. And the requirement of equisatisfiability is a way of capturing the thought that any expression in ϕ that "functions" in the context of *A* as a nonlogical primitive is mapped onto an expression of $\phi^{\mathcal{I}}$ that "functions" in the context of *B* as nonlogical primitive with an "analogous" semantic-value.

Although the appendix provides a precise characterization of the notions of preservation of logical networking and equisatisfiability, an example should suffice to convey the basic idea.

Let *A* be the set of arithmetical formulas,[33] and let *B* be the result of applying $\mathcal{F}$ (from section 3.1) to formulas in *A*. $\mathcal{F}$ preserves the logical networking of *A* because it satisfies three conditions. First, $\mathcal{F}$ is compositional. Second, $\mathcal{F}$ respects the logical connectives of sentences in *A* (e.g., $\ulcorner\neg\phi\urcorner^{\mathcal{F}} =$ "$\neg$" $\frown \ulcorner\phi\urcorner^{\mathcal{F}}$). Third, $\mathcal{F}$ maps identity-statements in *A* to formulas that "function" as identity statements in *B*. For instance, $\mathcal{F}$ maps "$m_1 = m_2$" to "$Z_1 \approx Z_2$," which "functions" as an identity statement in *B* because (1) "$\approx$" is reflexive, symmetric, and transitive, and (2) for any sentence $\ulcorner\phi\urcorner$ of *B* containing only $\ulcorner Z_i\urcorner$ free, the universal closure of

$$Z_i \approx Z_j \rightarrow (\phi \leftrightarrow \phi[Z_j/Z_i])$$

is true.[34] (It is important to note that although "$Z_1 \approx Z_2$" "functions" as an identity statement within the context of *B*, it doesn't do so within the context of the entire language, since condition (2) does not generally hold for sentences outside *B*. For instance, the universal closure of "$Z_1 \approx Z_2 \rightarrow (Z_1(a) \leftrightarrow Z_2(a))$" is always false.)

[33] More specifically, *A* is the set of formulas of a second-order language *L*, containing the following kinds of variables: first-order arithmetical variables, "m_1," "m_2," ...; first-order general variables, "x_1," "x_2," ...; and, for *n* a positive integer, *n*-place second-order general variables $\ulcorner X_1^n\urcorner$,$\ulcorner X_2^n\urcorner$, ... To avoid variable clashes, monadic second-order general variables are divided in two groups: the $\ulcorner X_{2i}^1\urcorner$—which are abbreviated $\ulcorner Z_i\urcorner$—are associated with first-order arithmetical variables by $\mathcal{F}$; the $\ulcorner X_{2i+1}^1\urcorner$—which are abbreviated $\ulcorner X_i\urcorner$—are used for more general purposes. We assume that *L* has been enriched with a single higher-level predicate "N" taking a first-order arithmetical variable in its first argument-place, and a monadic second-order general variable of the second group in its second argument-place. The well-formed formulas of *L* are defined in the usual way, with the proviso that an atomic formula can contain arithmetical variables only if it is of the form $\ulcorner m_i = m_j\urcorner$ or $\ulcorner \mathrm{N}(m_i, X_j)\urcorner$.

[34] "$\phi[Z_j/Z_i]$" is the result of substituting $\ulcorner Z_j\urcorner$ for $\ulcorner Z_i\urcorner$ in $\ulcorner\phi\urcorner$, and possibly relabeling bound variables to avoid clashes.

Let us now turn to equisatisfiability. Consider the following sentence, which is sometimes called "HP":

$$\forall F \forall G(\hat{m}[\mathbf{N}(m, F)] = \hat{m}[\mathbf{N}(m, G)] \leftrightarrow F \approx G).^{35}$$

Assuming HP, ϕ and $\phi^{\mathcal{F}}$ can be shown to be equisatisfiable for any $\phi \in A$. In essence, this is because two conditions are met. First, $\mathcal{F}$ is compositional. Second, we have the following:

> Let $\phi(m_i)$ be a formula in A containing only $\ulcorner m_i \urcorner$ free, and let $\psi(Z_i)$ be the result of applying $\mathcal{F}$ to $\phi(m_i)$. Then $\phi(m_i)$ is true of the *number* of the Gs just in case $\psi(Z_i)$ is true of the Gs.

Putting all of the above together, we get the result that on the assumption that HP is true, ϕ is an $AB_{\mathcal{F}}$-recarving of $\phi^{\mathcal{F}}$ for any ϕ in A. A similar result can be proved for Hodes's paraphrase-function, described in section 2.3.3.

3.3. The Third Question

What can be concluded about the logicist theses? Since the result of applying $\mathcal{F}$ to a sentence of pure n-th order arithmetic is always a sentence of pure second-order logic, $\mathcal{F}$ can be shown to verify *Language-Logicism*, in the context of the higher-order view. In addition, $\mathcal{F}$ can be shown to verify the semantic version of *Consequence-Logicism* for the case of pure and applied nth-order arithmetic.[36, 37] When the intuitive notion of content-recarving is cashed out in terms of $AB_{\mathcal{F}}$-recarvings and

[35] When HP is formulated within a one-sorted higher-order language, it entails that the universe is infinite. But when it is formulated in a two-sorted higher-order language (such as the language described in note 33), it is compatible with a finite universe (it entails only that *either* the range of the arithmetical variables and the range of the general variables are both infinite, *or* there is one more object in the range of the arithmetical variables than in the range of the general variables). It is worth noting that the equisatisfiability result for $\mathcal{F}$ requires no more than the two-sorted reading of HP.

In the context of our discussion of the Neo-Fregean Program in section 4, however, it is crucial that HP be read as formulated within a one-sorted higher-order language. This is because, unlike its two-sorted counterpart, the one-sorted version of HP entails a version of the Dedekind–Peano Axioms. For more on formulations of HP within multisorted languages, see Heck (1997b).

[36] This follows from the completeness result in Rayo (2002a).

[37] Since $\mathcal{F}$ is a compositional paraphrase-function, the results in section 2.3 imply that it cannot verify any of the three remaining logicist theses. However, if logically unrestricted quantifiers are allowed, then $\mathcal{F}$ verifies the semantic version of *Truth-Logicism*.

HP is assumed, this yields the result that *Language-* and *Consequence-Logicism* are both verified by a paraphrase-function which preserves *content*, modulo recarvings.

4. The Neo-Fregean Program

This final section will be devoted to the Neo-Fregean Program (or "neologicism," as it is sometimes called). Despite its roots as an attempt to rescue Frege's logicist project from inconsistency, neo-Fregeanism has increasingly developed a life of its own, and must be assessed on its own terms.[38] This makes the present section largely independent from the rest of the paper.

The core of the Neo-Fregean Program is the contention that it is possible to acquire an a priori justification for arithmetical statements in a special kind of way. Let ⟨N, HP⟩ be the following linguistic stipulation:

> The second-level predicate "N" is to be used in such a way that HP (introduced in section 3.2) turns out to be true.

Neo-Fregeans believe that merely as a result of setting forth ⟨N, HP⟩, one can acquire an a priori justification for the belief that HP is true. But HP deductively implies (definitional equivalents of) the second-order Dedekind–Peano Axioms.[39] So, if the Neo-Fregean story is right, then ⟨N, HP⟩ yields an a priori justification for the second-order Dedekind–Peano axioms.[40]

The Neo-Fregean program has given rise to many objections and replies. We will make no attempt to survey the literature here.[41] Instead, we will focus on a problem which I believe is especially important, and surprisingly underdeveloped. (For a broader perspective on the Neo-Fregean program, see chapter 6 in this volume.)

The problem arises as follows. Say that a linguistic stipulation is *weakly successful* if it has the effect of rendering its definienda meaningful. Say that a linguistic stipulation is *strongly successful* if it has the effect of rendering its definienda meaningful in such a way that the sentence targeted by the stipulation turns out to be true. ⟨N, HP⟩ is of little interest to the Neo-Fregean unless it is strongly

[38] The original incarnation of the Neo-Fregean Program is Wright (1983). For a collection of more recent essays, see Hale and Wright (2001a).

[39] The result, which relies on a one-sorted formulation of HP (see note 35), is known as *Frege's Theorem.* It was originally proved by Frege (by making what was later shown to be a nonessential use of Basic Law V), and more recently rediscovered and cleaned up from contradiction by Crispin Wright. See Frege (1893, 1903) and Wright (1983).

[40] A different question is whether HP is what underlies our actual knowledge of arithmetic. See Heck (1997a) and Demopoulos (2000).

[41] For a comprehensive survey, see MacBride (2003).

successful. (If it is only weakly successful, then it doesn't have the effect of rendering HP true, and a fortiori the Neo-Fregean's proposal cannot yield the result that HP is *known* to be true.) But HP can be true only if there are infinitely many objects.[42] So ⟨N, HP⟩ can only be strongly successful if there are infinitely many objects.

So far, all we have is the conclusion that either there are infinitely many objects or ⟨N, HP⟩ is not strongly successful. But this gives way to a natural worry: Isn't an *antecedent* justification for the infinity of the universe required for ⟨N, HP⟩ to deliver a justification for the belief that HP is true? After all, consider the following stipulation:

> The singular term "Zack" is to be used in such a way that the sentence "Zack is the man currently in my kitchen" turns out to be true.

Nobody would think that this stipulation can deliver a justification for the belief that Zack is the man currently in my kitchen unless one has an *antecedent* justification for the belief that there is a (unique) man currently in my kitchen. Why should the situation be any different when it comes to ⟨N, HP⟩? More generally, the following proposal suggests itself:

> *The Straight Proposal*
> If a linguistic stipulation is to be strongly successful, the world must cooperate. In particular, the result of existentially Ramsifying the sentence targeted by the stipulation must be true.[43] This is because the target sentence can be true only if its existential Ramsification is true,[44] and strong success can obtain only if the target sentence is rendered true.
>
> In order for the setting forth a stipulation to deliver a justification for believing the stipulation's target sentence, one must have an *antecedent* justification that the world cooperates. In particular, one must have an *antecedent* justification for believing the existential Ramsification of the stipulation's target sentence.
>
> What is true in general is true for ⟨N, HP⟩. Since the existential Ramsification of HP is equivalent (modulo choice principles) to the infinity of the universe, the setting forth of ⟨N, HP⟩ can deliver a justification for believing HP only if one has an *antecedent* justification for the infinity of the universe.

[42] This is true only on a one-sorted reading of HP. But on a two-sorted version of HP (see note 35), HP does not imply the Dedekind–Peano Axioms, so the Neo-Fregean Proposal looses interest.

[43] The result of existentially Ramsifying the sentence targeted by the stipulation is the existential closure of result of uniformly replacing the definienda in the target sentence with variables of the appropriate type.

[44] This makes the simplifying assumption that the target sentence contains no intensional vocabulary, but a similar proposal could be made to accommodate intensional cases.

The first paragraph of the Straight Proposal should be uncontroversial, even by the lights of Neo-Fregeans. But the final two paragraphs are not, and Neo-Fregeans must find an alternative proposal if their views are to have any plausibility. The problem of finding a plausible Neo-Fregean alternative to the Straight Proposal is the main focus of the present section.[45]

4.1. Conditional Stipulations

Neo-Fregeans are keen to observe that ⟨N, HP⟩ is, in a certain sense, *conditional*.[46] More specifically, they argue that ⟨N, HP⟩ succeeds in assigning a *concept* to "N" independently of how matters stand in the world. Facts about the world—in particular, facts about one-one correspondence—enter the picture by determining what *extension* the concept will take.

Whatever the value of this observation in other contexts, it is not helpful as a rebuttal of the Straight Proposal. For if there are only finitely many objects, then the existential Ramsification of HP is false, and there can be no concept (whatever its extension) satisfying the constraints imposed by ⟨N, HP⟩. Thus, insofar as proponents of the Straight Proposal are worried about whether one can be justified in thinking that ⟨N, HP⟩ is strongly successful in the absence of an antecedent justification for the belief that the universe is infinite, they will also be worried about whether one can be justified in believing that ⟨N, HP⟩ succeeds in creating a concept of the requisite kind in the absence of an antecedent justification for the belief that the universe is infinite.

4.2. The Abstraction Thesis

Some presentations of the Neo-Fregean program make heavy use of content recarving.[47] Specifically, they rely on the view that as a result of setting forth ⟨N, HP⟩, $\ulcorner \hat{m}[\mathbf{N}(m, F)] = \hat{m}[\mathbf{N}(m, G)] \urcorner$ comes to have the same content as $\ulcorner F \approx G \urcorner$, carved up in a different way. Call this the *Abstraction Thesis.*

It is important to be clear that the Abstraction Thesis is a *special case* of the view the ⟨N, HP⟩ is strongly successful, since it is a special case of the view that as a result of setting forth ⟨N, HP⟩, HP is meaningful and true. Thus, insofar as proponents of the Straight Proposal are worried about whether one can be justified in

[45] Much of the material in this section is based on the more detailed discussion in Rayo (2003a).

[46] See, for instance, Hale and Wright (2000), pp. 315–316.

[47] See, for instance, Wright (1997).

thinking that ⟨N, HP⟩ is strongly successful in the absence of an antecedent justification for the belief that the universe is infinite, they will also also be worried about whether one can be justified in accepting the Abstraction Thesis in the absence of an antecedent justification for the belief that the universe is infinite.

The Abstraction Thesis is therefore not a good way of countering the Straight Proposal, in the absence of further argumentation. In fact, commitment to the Abstraction Thesis can only complicate the task of offering a Neo-Fregean argument for the view that ⟨N, HP⟩ is strongly successful. For however difficult the task might have been when all ⟨N, HP⟩ is expected to deliver is a material equivalence, matters can only be worse if it is also expected to deliver sameness of content.

It is worth noting that in section 3.2 we relied on the assumption that HP is true to show that ϕ is an $AB_{\mathcal{F}}$-recarving of $\phi^{\mathcal{F}}$. In particular, we relied on the assumption that HP is true to show that $\ulcorner \hat{m}[\mathbf{N}(m, F)] = \hat{m}[\mathbf{N}(m, G)] \urcorner$ is an $AB_{\mathcal{F}}$-recarving of $\ulcorner F \approx G \urcorner$. Thus, in the absence of further argumentation, the proposal about content-recarving we set forth in section 3 does *not* deliver a justification for believing that ⟨N, HP⟩ is strongly successful. It does, however, supply a way of understanding the Abstraction Thesis according to which the Abstraction Thesis is true on the assumption that ⟨N, HP⟩ is strongly successful.

Why, then, is the Abstraction Thesis of any interest? In connection with logicism, there is at least this: if the Abstraction Thesis is true, then a content-preserving paraphrase-function verifies *Language-Logicism* for the special case of numerical identities. Neo-Fregeans sometimes sound as if they believe something further: that if the Abstraction Thesis is true, then there is a sense in which every theorem of arithmetic is "logical." Perhaps the paraphrase-function $\mathcal{F}$ (from section 3.1) supplies a way of spelling out this stronger claim.

4.3. Success by Default

One way of offering an alternative to the Straight Proposal is by defending a *default-epistemology*, according to which it may be sufficient for a belief to be justified that there be no reasons to the contrary.[48] In particular, Neo-Fregeans might attempt to defend something like the following:[49]

> *Success by Default*
> In the absence of reasons for thinking that a stipulation fails to meet the adequacy conditions in a certain set *S*, we are justified in thinking that it is strongly successful.

[48] I believe this point is due to Crispin Wright. Unfortunately, it does not seem to have been explicitly defended in print.

[49] In a talk titled "Implicit Definition and Abstraction" (University of St Andrews, October 30, 1999), Stewart Shapiro discussed a principle similar to *Success by Default.*

If *Success by Default* is true, then we needn't have an antecedent justification for the belief that the universe is infinite in order for ⟨N, HP⟩ to deliver a justification for HP, as proponents of the Straight Proposal would have it. All we need is the absence of reasons for thinking that ⟨N, HP⟩ fails to meet the conditions in *S*.

Unfortunately, *Success by Default* is far from being an uncontroversial principle. In order to defend it, Neo-Fregeans must carry out the following two tasks:

1. A specification of the conditions in *S* must be given.
2. It must be argued that a default epistemology is appropriate for linguistic stipulations.

Much discussion in the Neo-Fregean literature has centered upon task 1. The main adequacy condition in *S* is the requirement that the stipulation in question involve an *abstraction principle*, that is, a principle of the form

$$\forall\alpha\,\forall\beta(\Sigma(\alpha) = \Sigma(\beta) \leftrightarrow R(\alpha, \beta)),$$

for R an equivalence relation.[50] Unfortunately, this requirement will not do on its own. There are a number of abstraction principles giving rise to stipulations that Neo-Fregeans cannot regard as successful. This is what has come to be known as the *Bad Company Objection*. The simplest example of a problematic abstraction principle is Frege's Basic Law V, which is inconsistent and, hence, unsatisfiable. Neo-Fregeans must therefore add a requirement of consistency to *S*. But consistency cannot be the end of the matter. Heck (1992), Boolos (1997), and Weir (2003), among others, have set forth an array of consistent but pairwise incompatible abstraction principles. Since *Success by Default* can be taken to provide support for only one or the other of a pair of consistent but incompatible abstraction principles, Neo-Fregeans are forced to postulate additional adequacy conditions. Recent efforts include conservativeness,[51] modesty,[52] stability,[53]

[50] Alternatively, the stipulation might involve a pair of *rules* leading from each side of an abstraction principle to the other.

[51] Roughly, an abstraction principle is conservative if it entails nothing new about objects other than the referents of terms in the principle's left-hand side. See Wright (1997) and Shapiro and Weir (1999).

[52] According to appendix 1 of Wright (1999), "an abstraction [principle] is Modest if its addition to any theory with which it is consistent results in no consequences—whether proof- or model-theoretically established—for the ontology of the combined theory which cannot be justified by reference to its consequences for its own abstracts." Here an abstract is the referent of a term in the left-hand side of the relevant abstraction principle.

[53] An abstraction principle is stable if, for some cardinal κ, it is true at all and only those cardinalities $\geq \kappa$. See Fine (1998), Shapiro and Weir (1999), and Weir (2003).

and irenicity.[54] But Neo-Fregeans have yet to find a fully satisfactory set of adequacy conditions.

What about task 2? The literature on skepticism offers some general discussion on default epistemologies, mostly in connection with perceptual beliefs.[55] It is argued, for example, that, in the absence of reasons to the contrary, I am now justified in thinking that here is a hand and here is another. Unfortunately, there is no obvious way of extending these conclusions to the realm of linguistic stipulation. And, as far as I know, the adequacy of a default-epistemology for the case of linguistic stipulation has never been defended in print. Unless this omission can be remedied, default-epistemologies are unlikely to pose a serious threat to the Straight Proposal.

4.4. Further Challenges

The Neo-Fregean program faces at least two further challenges. One concerns second-order logic. Since Neo-Fregeans wish to show that ⟨N, HP⟩ yields an a priori justification for the Dedekind–Peano Axioms, they must argue not only that HP can be known a priori, but also that the second-order derivation of the Dedekind–Peano Axioms from HP preserves a priori knowledge. Assessing the status of second-order quantification gives rise to difficult questions which cannot be reviewed here. (See chapters 25 and 26 in this volume.)

Neo-Fregeans also face the *Indeterminacy Challenge.* It arises from the observation that many different assignments of semantic-value to the second-level predicate "N" are compatible with the truth of HP. In particular, it is compatible with the truth of HP that the members of any ω-sequence whatsoever serve as referents for the finite numerals (which can easily be defined on the basis of "N"). But it would seem to follow, at least at first sight, that the success of ⟨N, HP⟩ does not by itself suffice to determine whether the referent of "0" is, for instance, Julius Caesar. This is bad news for the Neo-Fregean on the assumption that the following principle obtains:

> (∗) An expression cannot be meaningful if there are several different assignments of semantic value, all of which are equally acceptable by the lights of our linguistic practice.[56]

[54] An abstraction principle is irenic if it is conservative and compatible with any conservative abstraction principle. See Weir (2003).

[55] See, for instance, Wright (2000a, 2000b, 2004) and, for a related proposal, Pryor (2000).

[56] Here "linguistic practice" must be understood broadly enough to include both mental states and environmental factors.

In addressing a related challenge,[57] Hale and Wright have offered an argument to the effect that—contrary to what one might have thought—the success of ⟨N, HP⟩ does, in fact, suffice to determine that the referent of "0" is not Julius Caesar. It is based on the idea that ⟨N, HP⟩ establishes one-one correspondence as the "criterion of identity" for the sortal concept *number*. If it is conceded that something other than one-one correspondence is the "criterion of identity" belonging to the sortal concept *person*, and that no object can fall under sortal concepts with different "criteria of identity," then the referent of "0" (which falls under the sortal *number*) must be distinct from Julius Caesar (which falls under the sortal *person*)—or so the argument goes.

Whether or not this sort of argument can be made to work (and whether or not it succeeds in answering the the challenge it was intended to address), it does not suffice to answer the Indeterminacy Challenge. For, as Hale and Wright are aware, their story is compatible with the possibility that the success of ⟨N, HP⟩ does not determine whether the referent of "0" is the number 7, or any other object falling under the sortal *number*. And, with (*) on board, this is enough to give rise to trouble. It is important to note, however, that the Indeterminacy Challenge is an instance of a general problem in the philosophy of mathematics,[58] and Neo-Fregeans are free to emulate responses offered by some of their rivals. For example, they can offer resistance to (*) by arguing that it threatens the meaningfulness of ordinary English.[59]

Appendix

Compositionality

We say that a function $*$ is *compositional* with respect to a set S of sentences in a first-order language $\mathcal{L}$ just in case there are

- a function $\top$;
- a higher-order formula $\xi_\wedge$ containing θ_1 and θ_2 as subformulas and no free occurrences of variables outside θ_1 and θ_2;
- a higher-order formula $\xi_\neg$ containing θ_1 as a subformula and no free occurrences of variables outside θ_1;

[57] See Hale and Wright (2001b). The related challenge concerns a cluster of interconnected issues which is usually referred to as "the Caesar-Problem."

[58] The standard source is, of course, Benacerraf (1965).

[59] See McGee (1993) and Burgess and Rosen (1997), I.A.2.d.

- a higher-order formula $\xi_\exists$ containing θ_1 as a subformula and no free occurrences of variables outside θ_1;
- for each n-place atomic predicate P_i^n of $\mathcal{L}$, a formula $\xi_{P_i^n}$ containing free occurrences of all and only the variables $\alpha_1, \ldots, \alpha_n$;

such that the following conditions obtain:[60]

- $\ulcorner \phi \wedge \psi \urcorner^\top = \xi_\wedge[\theta_1/\ulcorner \phi \urcorner^\top][\theta_2/\ulcorner \psi \urcorner^\top]$
- $\ulcorner \neg\phi \urcorner^\top = \xi_\neg[\theta_1/\ulcorner \phi \urcorner^\top]$
- $\ulcorner (\exists x_i)(\phi) \urcorner^\top = \xi_\exists[\alpha/\ulcorner x_i \urcorner^\top][\theta_1/\ulcorner \phi \urcorner^\top]$
- $\ulcorner P_i^n(x_{j_1}, \ldots, x_{j_n}) \urcorner^\top = \xi_{P_i^n}[\alpha_1/\ulcorner x_{j_1} \urcorner^\top] \ldots [\alpha_n/\ulcorner x_{j_n} \urcorner^\top]$
- For any variable $\ulcorner x_i \urcorner$ in $\mathcal{L}$, $\ulcorner x_i \urcorner^\top$ is a variable.
- If ϕ is a sentence of S, then $\phi^\top$ is a sentence.
- For any sentence of S in the range of $*$, $\phi^* = \phi^\top$.

where $\xi[\theta/\phi]$ is the result of replacing every occurrence of θ in ξ by ϕ, and $\xi[\alpha/v]$ is the result of replacing every occurrence of α in ξ by v.

Proposition

Let $\mathcal{L}$ be a first-order language with an intended model $\mathcal{M}$ and suppose that all numerical identities can be interpreted in the theory of $\mathcal{M}$. Let $*$ be a function such that (1) every sentence in $\mathcal{L}$ is in the domain of $*$, (2) there is a higher-order language $\mathcal{L}'$ such that every sentence in the range of $*$ is a sentence of $\mathcal{L}'$, and (3) $*$ is compositional with respect to $\mathcal{L}$. Then there is no model $\mathcal{S}$ such that $\mathcal{S}$ has a finite domain and, for any sentence ϕ of $\mathcal{L}$, $\models_\mathcal{M} \phi$ if and only if $\models_\mathcal{S} \phi^*$.

Proof: If ϕ is a formula of $\mathcal{L}'$ with one free first- or higher-order variable α, say that the *extension* of ϕ relative to $\mathcal{S}$ is the set of semantic values $s(\alpha)$ for s a variable-assignment satisfying ϕ in $\mathcal{S}$. Since $\mathcal{S}$ has a finite domain, there cannot be an infinite set C such that (a) for some first- or higher-order variable α every member of C is a formula containing only α free, and (b) no two elements of C have the same extension relative to $\mathcal{S}$.

Assume, for *reductio*, that there is a model $\mathcal{S}$ such that $\mathcal{S}$ has a finite domain and such that, for any sentence ϕ of $\mathcal{L}$, $\models_\mathcal{M} \phi$ if and only if $\models_\mathcal{S} \phi^*$. Since all numerical identities can be interpreted in the theory of $\mathcal{M}$, a formula

$$(\exists x_1)(x_1 = \bar{n} \wedge x_1 = \bar{m})$$

[60] For the sake of simplicity, we assume that $\mathcal{L}$ contains no individual constants or function letters and that all connectives and quantifiers are defined in terms of "$\wedge$," "$\neg$," and "$\exists$," in the usual way.

is true in $\mathcal{M}$ whenever $n = m$ and false in $\mathcal{M}$ whenever $n \neq m$. Call this formula $\chi_{n,m}$. By our assumption, $\chi^*_{n,m}$ is true in $\mathcal{S}$ whenever $n = m$, and false in $\mathcal{S}$ whenever $n \neq m$.

Let $n \neq m$. We know that $*$ is compositional with respect to $\mathcal{L}$. Let $\top$ be as in the definition of compositionality. By the definition of compositionality, $\ulcorner x_1 = \bar{n} \urcorner^{\top}$ and $\ulcorner x_1 = \bar{m} \urcorner^{\top}$ must contain free occurrences of the variable $\ulcorner x_1 \urcorner^{\top}$, and no free occurrences of other variables. Moreover, compositionality ensures that $\chi^*_{n,m}$ is the result of substituting $\ulcorner x_1 = \bar{n} \urcorner^{\top}$ for occurrences of $\ulcorner x_1 = \bar{m} \urcorner^{\top}$ in $\chi^*_{n,m}$. Since $\chi^*_{m,m}$ is true in $\mathcal{S}$ and $\chi^*_{n,m}$ is false in $\mathcal{S}$, it follows that the extension of $\ulcorner x = \bar{n} \urcorner^{\top}$ relative to $\mathcal{S}$ is different from the extension of $\ulcorner x_1 = \bar{m} \urcorner^{\top}$ relative to $\mathcal{S}$.

Consider the set C of formulas $\ulcorner x_1 = \bar{n} \urcorner^{\top}$, for n a natural number. We know that no two formulas in C have the same extension relative to $\mathcal{S}$. But we had noted above that this is impossible on account of $\mathcal{S}$'s finite domain. □

Recarving Contents

Let L be an interpreted higher-order language, and allow the variables of L of any given type to be classified according to different "groups." Let $\ulcorner \alpha_i^{n,m,k} \urcorner$ be the i-th m-place nth-order variable of the k-th group in L.

Say that a formula "$\alpha_1^{n,m,k} \equiv \alpha_2^{n,m,k}$" is an *identity-predicate* for variables of kind $\ulcorner \alpha_1^{n,m,k} \urcorner$ with respect to a set S just in case (a) "$\alpha_1^{n,m,k} \equiv \alpha_2^{n,m,k}$" is reflexive, transitive, and symmetric, and (b) the following condition obtains for any formulas $\ulcorner \phi \urcorner$ and $\ulcorner \psi \urcorner$ in S:

> If $\ulcorner \psi \urcorner$ is the result of replacing every free occurrence of $\ulcorner \alpha_i^{n,m,k} \urcorner$ in $\ulcorner \phi \urcorner$ by $\ulcorner \alpha_j^{n,m,k} \urcorner$ (and perhaps relabeling variables to avoid clashes), then the universal closure of $\ulcorner \alpha_i^{n,m,k} \equiv \alpha_j^{n,m,k} \rightarrow (\phi \leftrightarrow \psi) \urcorner$ is a true sentence of L for any i and j.

Let A and B be sets of formulas of L, and let $\mathcal{I}$ be a compositional,[61] one-one function from A onto B. We shall say that $\mathcal{I}$ *is an identity-isomorphism* with respect to A and B just in case there are formulas "$F(\alpha_1^{n,m,k}, \alpha_1^{n',m',k'})$," "$\alpha_1^{n,m,k} \equiv \alpha_2^{n,m,k}$," and "$\alpha_1^{n',m',k'} \equiv' \alpha_2^{n',m',k'}$" of L such that the following conditions obtain:

- "$\alpha_1^{n,m,k} \equiv \alpha_2^{n,m,k}$" is an identity-predicate for variables of kind $\ulcorner \alpha_i^{n,m,k} \urcorner$ with respect to A.

[61] Although the definition of compositionality above is framed in terms of first-order languages, it can easily be generalized so as to encompass higher-order languages.

- "$\alpha_1^{n',m',k'} \equiv' \alpha_2^{n',m',k'}$" is an identity-predicate for variables of kind $\ulcorner\alpha_i^{n',m',k'}\urcorner$ with respect to B.
- When $\top$ is as in the definition of compositionality, $\ulcorner\alpha_i^{n,m,k}\urcorner^{\top} = \ulcorner\alpha_i^{n',m',k'}\urcorner$.
- The universal closures of the following are true sentences of L:

$$\forall\alpha_1^{n,m,k}\exists\alpha_1^{n',m',k'}(F(\alpha_1^{n,m,k},\alpha_1^{n',m',k'}))$$
$$\forall\alpha_1^{n',m',k'}\exists\alpha_1^{n,m,k}(F(\alpha_1^{n,m,k},\alpha_1^{n',m',k'}))$$
$$F(\alpha_1^{n,m,k},\alpha_1^{n',m',k'}) \wedge F(\alpha_2^{n,m,k},\alpha_2^{n',m',k'}) \rightarrow (\alpha_1^{n,m,k} \equiv \alpha_2^{n,m,k} \leftrightarrow \alpha_1^{n',m',k'} \equiv' \alpha_2^{n',m',k'}).$$

For $\ulcorner\phi\urcorner \in A$, we say that $\ulcorner\phi\urcorner$ and $\ulcorner\phi\urcorner^{\mathcal{I}}$ are *$AB_{\mathcal{I}}$-equisatisfiable* with respect to variables of kind $\ulcorner\alpha_i^{n,m,k}\urcorner$ if $\mathcal{I}$ is an identity-isomorphism with respect to A and B, and the following additional condition obtains:

- Let $\ulcorner\alpha_1^{n,m,k}\urcorner \ldots \ulcorner\alpha_l^{n,m,k}\urcorner$ be the free variables in $\ulcorner\phi\urcorner$. Then the universal closure of

$$F(\alpha_1^{n,m,k}, \alpha_1^{n',m',k'}) \wedge \cdots \wedge F(\alpha_l^{n,m,k}, \alpha_l^{n',m',k'}) \rightarrow (\phi \leftrightarrow \phi^{\mathcal{I}})$$

is a true sentence of L, where $\ulcorner\phi^{\mathcal{I}}\urcorner$ is $\ulcorner\phi\urcorner^{\mathcal{I}}$.

When there is no risk of confusion, we use "$\ulcorner\phi\urcorner$ and $\ulcorner\phi\urcorner^{\mathcal{I}}$ are *equisatisfiable*" as an abbreviation for "$\ulcorner\phi\urcorner$ and $\ulcorner\phi\urcorner^{\mathcal{I}}$ are *$AB_{\mathcal{I}}$-equisatisfiable* with respect to every kind of variables occurring in A."

Finally, we say that $\mathcal{I}$ preserves the *logical networking* of A if $\mathcal{I}$ is an identity-isomorphism with respect to A and B, and $\mathcal{I}$ preserves logical connectives.

REFERENCES

Ayer, A. J. (1946) *Language, Truth and Logic*, Gollancz, London, 2nd ed.

Beall, J., ed. (2003) *Liars and Heaps*, Oxford University Press, Oxford.

Benacerraf, P. (1965) "What Numbers Could Not Be," Reprinted in Benacerraf and Putnam (1983).

Benacerraf, P. (1981) *Frege: The Last Logicist.* In French et al. (1981). Reprinted in Demopoulos (1995).

Benacerraf, P., and H. Putnam, eds. (1983) *Philosophy of Mathematics*, Cambridge University Press, Cambridge, 2nd ed.

Boghossian, P., and C. Peacocke, eds. (2000) *New Essays on the A Priori*, Clarendon Press, Oxford.

Boolos, G. (1984) "To Be Is to Be a Value of a Variable (or to Be Some Values of Some Variables)," *The Journal of Philosophy* 81, 430–449. Reprinted in Boolos (1998a).
Boolos, G. (1994) "The Advantages of Honest Toil over Theft." In George (1994). Reprinted in Boolos (1998a).
Boolos, G. (1997) "Is Hume's Principle Analytic?" In Heck (1997c), 245–261. Reprinted in Boolos (1998a), 301–314.
Boolos, G. (1998a) *Logic, Logic and Logic*, Cambridge, Mass., Harvard University Press.
Boolos, G. (1998b) "Reply to Parsons' "Sets and Classes"". In Boolos (1998a), 30–36.
Bostock, D. (1979) *Logic and Arithmetic*, 2 vols., Clarendon Press, Oxford.
Bottani, A. et al., eds. (2002) *Individuals, Essence, and Identity: Themes of Analytic Metaphysics*, Kluwer Academic, Dordrecht.
Burgess, J., and G. Rosen (1997) *A Subject with No Object*, Oxford University Press, New York.
Cartwright, R. (1994) "Speaking of Everything," *Noûs* 28, 1–20.
Dales, H. G., and G. Oliveri, eds. (1998) *Truth in Mathematics*, Oxford University Press, Oxford.
Demopoulos, W., ed. (1995) *Frege's Philosophy of Mathematics*, Harvard University Press, Cambridge, Mass.
Demopoulos, W. (2000) "On the Origin and Status of Our Conception of Number," *Notre Dame Journal of Formal Logic* 41, 210–226.
Dummett, M. (1981) *Frege: Philosophy of Language*, 2nd ed., Harvard University Press, Cambridge, Mass.
Field, H. (1998) "Which Undecidable Mathematical Sentences Have Determinate Truth Values?" In Dales and Oliveri (1998). Reprinted in Field (2001).
Field, H. (2001) *Truth and the Absence of Fact*, Oxford University Press, Oxford.
Fine, K. (1998) "The Limits of Abstraction." In Schirn (1998), 503–629.
Frege, G. (1884) *Die Grundlagen der Arithmetik*. English Translation by J.L. Austin, *The Foundations of Arithmetic*, 2nd rev. ed., Northwestern University Press, Evanston, Ill., 1980.
Frege, G. (1893, 1903) *Grundgesetze der Arithmetik*. 2 vols. English Translation by Montgomery Furth, *The Basic Laws of Arithmetic*, University of California Press, Berkeley and Los Angeles, 1964.
French, Peter A., Theodore E. Uchling Jr., and Howard K. Wettstern. (1981) *The Foundations of Analytic Philosophy*, University of Minnesota Press, Minneapolis.
George, A., ed. (1994) *Mathematics and Mind*, Oxford University Press, Oxford.
Glanzberg, M. "Quantification and Realism," Forthcoming.
Hale, B. (1997) "Grundlagen §64," *Proceedings of the Aristotelian Society* 97, 243–261. Reprinted in Hale and Wright (2001a), 91–116.
Hale, B., and C. Wright (2000) "Implicit Definition and the A Priori." In Boghossian and Peacocke (2000), 286–319. Reprinted in Hale and Wright (2001a), 117–150.
Hale, B., and C. Wright (2001a) *The Reason's Proper Study: Essays Towards a Neo-Fregean Philosophy of Mathematics*, Clarendon Press, Oxford.
Hale, B. and C. Wright (2001b) "To Bury Caesar. . . ." In Hale and Wright (2001a), 335–396.
Heck, R. (1992) "On the Consistency of Second-Order Contextual Definitions," *Noûs* 26, 491–494.

Heck, R. (1997a) "Finitude and Hume's Principle," *Journal of Philosophical Logic* 26, 589–617.
Heck, R. (1997b) "The Julius Caesar Objection." In Heck (1997c), 273–308.
Heck, R., ed. (1997c) *Language, Thought and Logic*, Oxford University Press, Oxford.
Hodes, H.T. (1984) "Logicism and the Ontological Commitments of Arithmetic," *Journal of Philosophy* 81:3, 123–149.
Hodes, H.T. (1990) "Where Do Natural Numbers Come From?" *Synthese* 84, 347–407.
MacBride, F. (2003) "Speaking with Shadows: A Study of Neo-Fregeanism," *British Journal for the Philosophy of Science* 54, 103–163.
McGee, V. (1993) "A Semantic Conception of Truth?" *Philosophical Topics* 21, 83–111.
McGee, V. (1997) "How We Learn Mathematical Language," *Philosophical Review* 106, 35–68.
McGee, V. (2000) " "Everything." " In Sher and Tieszen (2000).
McManus, D., ed. (2004) *Wittgenstein and Skepticism*, Routledge, London.
Parsons, C. (1974) "Sets and Classes," *Noûs* 8, 1–12. Reprinted in Parsons (1983).
Parsons, C. (1983) *Mathematics in Philosophy*, Cornell University Press, Ithaca, NY.
Pryor, J. (2000) "The Skeptic and the Dogmatist," *Noûs* 34, 517–549.
Rayo, A. (2002a) "Frege's Unofficial Arithmetic," *Journal of Symbolic Logic* 67, 1623–1638.
Rayo, A. (2002b). "Word and Objects," *Noûs* 36, 436–464.
Rayo, A. (2003a) "Success by Default?" *Philosophia Mathematica* 11, 305–322.
Rayo, A. (2003b) "When Does 'Everything' Mean Everything?" *Analysis* 63, 100–106.
Rayo, A., and G. Uzquiano (1999) "Toward a Theory of Second-Order Consequence," *Notre Dame Journal of Formal Logic* 40, 315–325.
Rayo, A., and T. Williamson (2003) "A Completeness Theorem for Unrestricted First-Order Languages." In Beall (2003).
Schrin, M., ed. (1998) *Philosophy of Mathematics Today*, Clarendon Press, Oxford.
Shapiro, S. (1991) *Foundations Without Foundationalism: A Case for Second-Order Logic*, Clarendon Press, Oxford.
Shapiro, S., and A. Weir (1999) "New V, ZF and Abstraction," *Philosophia Mathematica* 7(3), 293–321.
Sher, G., and R. Tieszen, eds. (2000) *Between Logic and Intuition*, Cambridge University Press, New York and Cambridge.
Uzquiano, G. (2002) "Categoricity Theorems and Conceptions of Set," *Journal of Philosophical Logic* 31:2, 181–196.
Weir, A. (2003) "Neo-Fregeanism: An Embarrassment of Riches?" *Notre Dame Journal of Formal Logic* 44, 13–48.
Whitehead, A., and B. Russell (1910–1913) *Principia Mathematica*, Cambridge University Press, Cambridge, 2nd ed. 3 vols.
Williamson, T. (1999) "Existence and Contingency," *Proceedings of the Aristotelian Society*, supp. 73, 181–203. Reprinted with printer's errors corrected in *Proceedings of the Aristotelian Society* 100 (2000), 321–343 (117–139 in unbound edition).
Williamson, T. (2003) "Everything," *Philosophical Perspectives*, Volume 17, *Language and Philosophical Linguistics*. Edited by Dean Zimmerman and John Hawthorne, 415–465.

Wright, C. (1983) *Frege's Conception of Numbers as Objects*, Aberdeen University Press, Aberdeen.
Wright, C. (1997) "The Philosophical Significance of Frege's Theorem." In Heck (1997c), 201–244.
Wright, C. (1999) "Is Hume's Principle Analytic," *Notre Dame Journal of Formal Logic* 40, 6–30. Reprinted in Hale and Wright (2001a), 307–332.
Wright, C. (2000a) "Cogency and Question-Begging: Some Reflections on McKinsey's Paradox and Putnam's Proof," *Philosophical Issues* 10, 140–163.
Wright, C. (2000b) "Replies," *Philosophical Issues* 10, 201–219.
Wright, C. (2004) "Wittgensteinian Certainties." In McManus (2004).
Yablo, S. (2002) "Abstract Objects: A Case Study," *Noûs* 36 supp. 1, 220–240. Originally appeared in Bottani et al. (2002).
Zermelo, E. (1930) "Über Grenzzahlen und Mengenbereiche. Neue Untersuchungen über die Grundlagen der Mengenlehre," *Fundamenta Mathematicae* 16, 29–47.

CHAPTER 8

FORMALISM

MICHAEL DETLEFSEN

1. Introduction: The Formalist Framework

VIEWED properly, formalism is not a single viewpoint concerning the nature of mathematics. Rather, it is a family of related viewpoints sharing a common framework—a framework that has five key elements.

Among these is its revision of the traditional classification of the mathematical sciences. From ancient times onward, the dominant view of mathematics was that it was divided into different sciences. Principal among these were a science of magnitude (geometry) and a science of multitude (arithmetic).

Traditionally, this division of mathematics was augmented by an ordering of the two parts in terms of their relative basicness and which was to be taken as the more paradigmatically mathematical. Here it was geometry that was given the priority.[1] The formalist outlook typically rejected this traditional ordering of the mathematical sciences. Indeed, from the latter half of the nineteenth century onward, it typically reversed it. This reversal is the first component of the formalist framework.

A second component is its rejection of the classical conception of mathematical proof and knowledge. From Aristotle on down, proof and knowledge were

[1] Priority in what sense? Here there were different views. Some adopted a more metaphysical understanding of the ordering, others a more epistemological one. More on this later.

conceived on a *genetic* model. According to this model, we know a thing best when we know it through its "cause." In mathematics, this cause was taken to reside in basic definitions and principles of construction (e.g., Euclid's first three postulates). Generally speaking, a theorem concerning a given type of object was taken to be properly proved or demonstrated by showing how to *construct* objects of its type *as* objects having the property asserted of them in the theorem. This allowed one to witness the formation of the objects from their rudiments, and this exhibition of the formation of objects became an important part of the traditional standard for rigor.

These were the essential elements of the classical conception of proof. Formalists rejected both ideas. They rejected the genetic conception of proof by denying that the only proper knowledge of a thing comes through knowledge of its causes.

They rejected as well the traditional "presentist" conception of rigor which saw it as consisting primarily in the keeping of an object continuously before the visual imagination or intuition of the prover during the course of a proof. Indeed, they moved towards a conception of rigor that emphasized abstraction from rather than immersion in intuition and meaning. This change in the conception of rigor is the third component of the general formalist framework.

The fourth and perhaps most distinctive component of the formalist framework was its advocacy of a nonrepresentational role for language in mathematical reasoning. This idea reached full consciousness in Berkeley. Being particularly impressed with the algebraists' use of imaginary elements, he came to the general view that there are uses of expressions in reasoning whose utility and justification is independent of the (semantic) contents of those expressions.

The fifth and final component of the formalist framework is what I call its *creativist* component. This is the idea that the mathematician, qua mathematician, has a freedom to create instruments of reasoning that promise to further her epistemic goals (e.g., the improvement of cognitive efficiency). This was used by formalists in the nineteenth century to counter charges that they would make mathematics an essentially mechanical matter, the exercise of sheer technique (*technē*). Formalists agreed with nonformalists that the end of mathematical reasoning is the acquisition of genuine knowledge. They did not agree, however, that the only proper, or even the best, way to pursue this end was through the exclusive use of contentual reasoning.

The above are the chief characteristics of the historically and philosophically important variants of formalism. They are all, in one way or another, related to two important developments in the history of mathematics. One of these was the emergence and rapid development of algebraic methods from the seventeenth century onward. The second was the erosion of the authority of classical geometry and its methods throughout much of the same period.

The former gained impetus from various sources. Wallis, for example, argued for a reversal of the priority traditionally granted to geometry over arithmetic. In particular, he argued that many of the traditional geometrical laws were based on arithmetical rather than geometrico-constructional properties of geometrical figures.

The latter derived in part, of course, from the discovery of non-euclidean geometries in the early part of the nineteenth century. This discovery suggested that (classical) geometrical intuition might not be the final authority even in geometry. This was further supported by the discoveries of Bolzano and Weierstrass (and others) in the mid-nineteenth century that suggested that geometrical intuition was not a *reliable* guide to the development of analysis. Mathematicians were therefore forced to look to arithmetic in order to find a foundation for analysis.

The remainder of this essay is largely an attempt to further our understanding of the five elements of the formalist framework just summarized. We begin, in the next section, with an overview of the classical genetic conception of proof. Following that, we describe the main challenges to the classical view that emerged in the late Middle Ages and the Renaissance. In the fourth section we consider the emergence of algebra and its two motivating ideals—what we call the *invariantist* and *symbolical* ideals. We argue there that formalism amounts to more than just an emphasis on algebraic methods or a general retreat from intuition. In the fifth section we identify the main factors comprising modern formalism, the formalism which found expression in the writings of Peacock and the Cambridge algebraists and was later developed more fully by Hilbert. In the final section, we consider a variety of challenges to formalism and suggest some general conclusions.

2. The Traditional Viewpoint

2.1. The Aristotelian Division of Mathematics

Aristotle (cf. book Delta of the *Metaphysics*, 1020a7–14) defined mathematics as the study of quantity (*poson*[2]), with quantity being generally conceived as that which is divisible into constituent parts. The division of a quantity was taken to admit of two basic forms: (1) division into discrete parts and (2) division into parts continuous with one another. Quantity admitting of division of the first type was termed *plēthos* by Aristotle (loc. cit.). The usual English translation is

[2] Others, for example, the Pythagoreans, treated *poson* as the "how many" and *pēlikon* as the "how much."

multitude or *plurality*, which is intended to suggest that it is a type of quantity whose amount is to be determined by *counting* its parts.

Quantity of the second sort was termed *megethos* (loc. cit.), generally translated as *magnitude*. It was understood as being divisible into parts (each itself a magnitude) joined together by common boundaries. The location of each part with respect to others was taken to be discernible, including, of course, parts joined to a given part by a common border. The extent of a magnitude was to be determined by *measurement*, a comparison of a magnitude with a unit of the same type.

Aristotle's division between basic types of quantities gave rise to a corresponding division between two branches of mathematics—*arithmetic* (*arithmetikē*), or the science[3] of multitude, and *geometry* (*geōmetria*), or the science of magnitude.[4]

Beginning with the discovery of incommensurables,[5] there was a long tradition of regarding magnitude as the more basic of the two types of quantity. In ancient times, numbers were mainly regarded as whole numbers, or numbers used in counting. They were also, however, related to geometrical objects, as in the Pythagoreans' theory of *figured* (*figurate*) numbers, and their ratios or proportions were studied in the ancient theory of *means* and *proportions*.

[3] Our use of the term "science" is not casual here. Rather, it is intended to comport with its classical use, a use that distinguished between the *science* of number and the *art* of reckoning or calculation. The science (*epistēmē* in the Greek writers, *scientia* in the Latin commentators) of number was *arithmetikē*. It inquired into the *causes* of arithmetic facts. The art (*technē* in the Greek writers, *ars* in the Latin commentators) of calculation, on the other hand, was *logistikē*, which broadly concerned practical means of reckoning number.

[4] Barrow [*Barrow 1664*, cf. lecture II, pp. 14ff.) attributed the original distinction to the Pythagoreans and viewed Aristotle's distinction as a modification of their scheme.

This is somewhat misleading however. The Pythagoreans apparently divided mathematics into a quadrivium of subjects (cf. [*Proclus*], 29–31): arithmetic (*arithmētikē*), music (*mousikē*), geometry, and spheric, in that order. Arithmetic and music dealt with *poson*; arithmetic with *poson* in itself, music with *poson* taken in relation to sound. Geometry and spheric dealt with magnitude (*pēlikon*); geometry with magnitude considered in isolation from its motion, spheric with magnitude in motion.

Plato had a somewhat different scheme still (cf. [*Plato.*], bk. VII, 525a–530d), adding a subject called *stereometry*, which had clear overlaps with the geometry of the Pythagoreans.

In addition to these, there were yet other divisions of mathematics proposed by Anatolius and Geminus (cf. [*Proclus*], 31–35).

[5] The question of who discovered incommensurables is famously a matter of controversy. One view is that it was Hippasus of Metapontum, in the latter half of the fifth century B.C. A case for this was presented in von Fritz [*von Fritz 1945*]. Zeuthen [*Zeuthen 1896–1897*], Frank [*Frank 1923*], Burkert [*Burkert 1972*], Reidemeister [*Reidemeister 1949*], and Vogt [*Vogt 1910*], however, argue for other alternatives (e.g., that Pythagoras himself or Theodorus of Cyrene deserves precedence). Acceptance of irrational quantities as numbers on which arithmetical operations might be performed seems to have originated with the Hindus of medieval times (cf. [*Cantor 1894*], vol. I).

Irrational quantities were different. They did not represent the results of counting, they did not suggest the shapes of geometrical figures, and they did not represent proportions expressible as ratios of (whole) numbers.[6]

Ratios of incommensurables were thus quantities that admitted of representation as geometrical quantities but not as numbers or arithmetical quantities. Since all arithmetic quantities (i.e., all numbers) were geometrically representable, this suggested that magnitude was the more fundamental of the two basic types of quantity: all quantities could be represented as magnitudes, but not all quantities could be represented as multitudes (or "figures" or ratios thereof).

Of multitude and magnitude, then, magnitude appeared to be the more basic. This led to a corresponding priority of the *science* of magnitude (geometry), and its methods, to the *science* of number (arithmetic), and its methods.[7]

2.2. Traditional Ideals of Proof

The discovery of incommensurables thus supported the establishment of classical geometrical method as the paradigm for mathematical method. It also played a role in determining a leading methodological ideal of classical geometry—what I will call the *constructive ideal.*

The constructive ideal had both a *genetic* component and a *constructional* component. Both represented epistemic ideals. According to the genetic component, knowledge of a geometrical object was at its best when it came through knowledge of that object's *cause*—knowledge, that is, of its *genesis,* or how it came to be the object that it is. According to the constructional component, this knowledge of cause or genesis was at its highest when it consisted in knowledge of how to construct the object in question.

Aristotle gave the classical statement of the genetic ideal.

> We suppose ourselves to possess unqualified scientific knowledge of a thing, as opposed to knowing it in the accidental way in which the sophist knows, when we think that we know the cause (*aitia*) on which the fact depends as the cause

[6] There were no general rules for calculating with irrational quantities, and that, too, distinguished them from numbers. It is true, of course, that Eudoxus developed a scheme for "calculating" with irrational quantities. His scheme was, however, composed of geometrical rather than symbolic or algebraic forms of manipulation, and so, at least in the minds of many, it supported the view that geometrical quantities and operations were more fundamental than arithmetical quantities and operations.

[7] This was not the universal view of the Greeks. Aristotle, for example, famously maintained the priority of arithmetic over geometry. It was, however, the predominant view of postclassical times—in particular, of Renaissance and modern times. It's this dominance that is of principal interest to us here.

> of the fact and of no other, and, further, that the fact could not be other than it is. [*An. post.*], I, 2 (71b9–11)[8]

This was, moreover, taken to apply as much to mathematical as to nonmathematical objects. The question was what cause might come to in the case of mathematical objects. To help answer this question, Aristotle famously distinguished four notions of cause—formal, material, efficient, and final.

> Evidently we have to acquire knowledge of the original causes (for we say we know each thing only when we think we recognize its first cause), and causes are spoken of in four senses. In one of these we mean the substance, i.e. the essence (for the "why" is referred finally to the formula [*logos*], and the ultimate "why" is a cause and principle [*archē*]; in another, the matter or substratum; in a third the source of the change; and in the fourth the cause opposed to this, that for the sake of which and the good (for this is the end of all generation and change). [*Metaph.*, Rev.], I, 3 (983a25–31)

Aristotle believed that it was the first or *formal* notion of cause that applied to mathematical knowledge.

The salient feature of a formal cause was that it did not depend upon or consist in any type of motion. Rather, it was a form of causation that reflected the inherent *nature* of the object itself. The idea was that the nature or essence of an object itself exerts a type of determinative force on its becoming and its development.

One could thus acquire *epistēmē* of a mathematical object by knowing its formal cause—that is, its nature or essence. The nature or essence of an object, however, was supposed to be given by a proper definition of that object. Grasp of the nature of a mathematical object should therefore consist in grasp of its proper definition. From this it follows that, at bottom, knowledge of cause in mathematics consists in knowledge of a definition. As Aristotle so succinctly put it:

> The "why" is referred ultimately . . . in mathematics . . . to the "what" (to the definition of straight line or commensurable or the like) . . . . [*Physics*, Rev.], II, 7 (198a16–18)

The genetic component of the constructive ideal in geometry thus called for knowledge of the definitions of mathematical objects. In itself, however, this does not indicate a constructive approach to proof. The constructive element entered, rather, through the second component of the constructive ideal—what I am calling the *constructional* component. This component amplifies the genetic component by expressing an ideal for its realization. It maintains that the best kind of definition of a geometrical object is one that indicates how it is to be constructed.

This constructional component of the constructive ideal was not particularly Aristotelian in character. Still less, of course, was it Platonist. It is therefore perhaps surprising that the Neoplatonist Proclus should have attributed a systematic role

[8] Cf. [*Physics*], II, 3 (194b16–19); [*Metaph.*], I, 3 (983a24–25).

to construction in his account of the ancient distinction between *problems* and *theorems.*[9]

> ...the propositions that follow from the first principles he [Euclid] divides into problems and theorems, the former including the construction of figures, the division of them into sections, subtractions from and additions to them, and in general the characters that result from such procedures...the latter concerned with demonstrating inherent properties belonging to each figure. Just as the productive sciences have some theory in them, so the theoretical ones take on problems in a way analogous to production. [*Proclus*], 63

The overall purpose of construction, according to Proclus, was to take the reasoner from the *given* of a problem/theorem to that which is *sought* in it ([*Proclus*], 159).

> [Geometry] calls "problems" those propositions whose aim is to produce, bring into view, or construct what in a sense does not exist, and "theorems" those whose purpose is to see, identify, and demonstrate the existence or nonexistence of an attribute. Problems require us to construct a figure, or set it at a place, or apply it to another, or inscribe it in or circumscribe it about another, or fit it upon or bring it into contact with another, and the like; theorems endeavor to grasp firmly and bind fast by demonstration the attributes and inherent properties belonging to the objects that are the subject-matter of geometry. [*Proclus*], 157–158

Constructions were taken to be built up from (applications of) *postulates* in something like the same way in which proofs were taken to be built up from axioms ([*Proclus*], 163).

They were also, however, taken to play a role in proof. One particular such role was that of establishing existence. The idea was that before a proof can rightly appeal to a geometrical entity (e.g., a point), the existence of that entity must be established, and the way to establish that existence was to construct the entity or, perhaps, to give directions for its construction.

In this vein, speaking of Euclid's proof of proposition 4, book I, Proclus wrote:

> The propositions before it have all been problems...our geometer follows up these problems with this first theorem.... For unless he had previously shown the existence of triangles and their mode of construction, how could he discourse about their essential properties and the equality of their angles and sides? And how could he have assumed sides equal to sides and straight lines equal to other straight lines unless he had worked these out in the preceding problems and devised a method by which equal lines can be discovered?...It is to forestall such

[9] Proclus's view of the nature of the distinction and which sciences fell under which sides of it was in rough (though only rough) agreement with Aristotle's conception of the distinction between *productive* and *theoretical* knowledge. Cf. [*Topics*], bk. VI, ch. 6 (145a15–16); [*Topics*], bk. VIII, ch. 1 (157a8–10); [*Metaph.*], bk. VI, ch. 1 (1025b18–27, 1026a 6–17); [*Metaph.*], bk. XI, ch. 7 (1064a10–32).

> objections that the author of the *Elements* has given us the construction of triangles.... These propositions are rightly preliminary to the theorem....
> [*Proclus*], 182–183

Proclus thus seems to have seen constructions as, among other things, the means of establishing the existence of items dealt with in proofs.[10]

As noted above, however, this was not the settled view of the Greeks. Indeed, it was a matter of considerable controversy.[11]

Plato famously held that to be knowable, an object must be eternal and immutable. Knowledge, for him, was a contemplation of essence, not of genesis (cf. [*Plato.*], 526e). Geometrical knowledge, in particular, was "the knowledge of that which always is, and not of a something which at some time comes into being and passes away... geometry is the knowledge of the eternally existent" ([*Plato.*, 527b).

Plato criticized the common view and practice of geometry as being confused on this point. The "language" of the usual geometer, he wrote, "is most ludicrous... for they speak as if they were doing something and as if all their words were directed toward action. For all their talk is of squaring and applying and adding and the like, whereas in fact the real object of the entire study is pure knowledge." ([*Plato.*, 527a).

Plato thus regarded the bulk of geometers as confused about their subject. They conceived of their subject as centrally concerned with the generation or production of objects, when in truth it was a science of the eternally existing and ungenerated. For Plato, then, constructed objects were *produced* and not eternal, and they could not, for that reason, be the objects of geometrical knowledge. They could at best serve as *representations* of real objects and provide some sort of practical guide to their knowledge. As Proclus, describing the views of the Platonists Speusippus and Amphinomus, put it:

> There is no coming to be among eternals... hence a problem [as opposed to a theorem] has no place here [in the *theoretical* science of geometry], proposing as it does to bring into being or to make something not previously existing—such as to construct an equilateral triangle, or to describe a square when a straight line is given, or to place a straight line through a given point. Thus it is better, according to them [Speusippus and Amphinomus], to say that all these objects exist and that we look on our construction of them not as making, but as understanding them, taking eternal things as if they were in the process of coming to be. [*Proclus*], 64

Menaechmus, a student of Eudoxus and an important figure in the early Academy, was among those geometers holding the view thus criticized. He took

[10] It is perhaps worth noting that this was also the view famously taken at the end of the nineteenth century by H.G. Zeuthen in his widely cited and discussed [*Zeuthen 1896–1897*] (cf. p. 223).

[11] For a useful discussion of this controversy see Bowen [*Bowen 1983*].

construction to be central to geometry, and conceived of it generally as a process by which objects were produced or generated.[12]

Proclus tried to harmonize these views. The school of Speusippus was right, he said (cf. [*Proclus*], 64), because the problems of geometry are different from those of mechanics in that the latter but not the former "are concerned with perceptible objects that come to be and undergo all sorts of change" (ibid.).

The school of Menaechmus, on the other hand, was also right because "the discovery of theorems does not occur without recourse to ... intelligible matter" (ibid.).[13] Proclus's suggestion was thus to offer a distinction between the *understanding* (*gnōstikōs, dianoia*) and the *imagination* (*phantasia*) and to say that the understanding keeps the definition (*logos*) of the eternal, changeless geometric objects fixedly in mind while the imagination, "projecting its own ideas" (ibid.) is a *production* (*poiētikōs*) of figures, sectionings, superpositions, comparisons, additions, subtractions, and the other acts of construction (*poiētikōs, sustasis*). The change that takes place is therefore in the actions of the imagination of the constructor, which actions form the efficient cause of the figure constructed. It is not in the understanding, whose "action" remains that of a Platonic contemplation of an eternal form which serves as the formal cause of the figure constructed.[14] The figures produced by the efficient causation of the imagination are thus constructions *of* or *from* the figures contemplated in the understanding.[15]

[12] Cf. [*Proclus*], 64. Menaechmus is widely credited with the discovery of conic sections (cf. Allman [*Allman 1889*], where Menaechmus is listed as a member of the group within the Academy known as the "school of Cyzicus"). Allman, citing Boeckh [*Boeckh 1863*], maintains that the reputation of the Academy as a center for mathematical and astronomical learning was chiefly due to the members of this school (*loc. cit.*, 177–178). Proclus ([*Proclus*], 166) also mentions a genetic tradition extending from Oenopides through Zenodotus to Poseidonius that saw a problem as a query concerning the condition under which a thing exists.

[13] Intelligible matter is distinct from sensible matter. Aristotle describes it as follows:

> ... some matter is perceptible and some intelligible, perceptible matter being for instance bronze and wood and all matter that is changeable, and intelligible matter being that which is present in perceptible things not *qua* perceptible, i.e. the objects of mathematics. [*Metaph.*], 1036a8–13

[14] The formal cause of a thing is "... the form or the archetype, i.e. the definition of the essence, and its genera ... and the parts in the definition" ([*Physics*, Rev.], 194b27–29). The efficient cause is "... the primary source of the change or rest; e.g. the man who deliberated is a cause, the father is cause of the child, and generally what makes of what is made and what changes of what is changed ([*Physics*, Rev.], 194b30–32).

[15] Seen this way, the constructive activity of the geometer is based not on a genetic or generational definition of a figure (i.e., a definition specifying a means of constructing that figure), but on what some (e.g., [Molland, 1976]) have called *definition by property*—that is, a specification of the essence of the figure. Since, however, the definition by property, in the account sketched above, serves as the formal cause of the figure (as the figure having the properties it has), the definition by property is itself a kind of genetic definition.

Indeed, the constructor (or his constructive actions) becomes the efficient cause of a triangle in his imagination by virtue of his having the form (*logos*) of the triangle in his understanding. Thus, the triangle the diagrammer constructs is in one sense a product of his diagramming actions. In another it is a product of the form he holds in his mind:

> ... from art proceed the things of which the form is in the soul of the artist. (By form I mean the essence of each thing and its primary substance.)
> [*Metaph.*], 1032a32–b1

If Proclus's account of the controversy within the early Academy concerning the place of constructions in geometry is correct, the constructive ideal had a significant following among ancient geometers.[16] Another indication of this is the above-mentioned distinction in the *Elements* between postulates and axioms.[17] Other ancient geometers, too, gave construction a prominent place in their work. Among these were Apollonius (e.g. in his constructive definition of the cone (cf. [*Apollonius*], 1)) and Archimedes (e.g. in his operations on circles, spheres and cylinders (cf. [*Archimedes*], e.g. Prop. 5)). This tradition continued throughout the Middle Ages and Renaissance and into the modern era.

In addition to the criticisms of the Platonists noted above, there were other criticisms of the constructive ideal. These included criticisms of both its constructional and its genetic components. On the constructional side, this led to a discussion of the forms of motion that ought to be admissible in construction. Discussion focused on the use of "mixed motion" or "mixed line" (cf. [*Proclus*], 84–86, 211ff.) constructions. This was construction via lines or points moving at equal or fixed proportional rates of speed (e.g. Hippias's quadratrices). This type of construction (commonly termed "mechanical" or "instrumental" (*organikos* or *mēchanikos*), as opposed to properly "geometrical" (*geometrikos*)) was believed by some to require the marking of motion in both the imagination and the understanding. That is, it required essential use of or reference to motion not only to visualize but also to understand. As such, it violated the terms of Proclus's Platonistic compromise by not confining motion to what properly belonged only to the imagination.

[16] Proclus, too, seems to have given it an important place. See, for example, his definition of the cylindrical helix ([*Proclus*], 85).

[17] Euclid's first three postulates all seem to be constructional in character. They are (1) to draw a straight line from any point to any point; (2) to produce a finite straight line continuously in a straight line; (3) to describe a circle with any center and distance. Euclid's fourth and fifth postulates do not have this constructive character. Euclid's fourth postulate, as Proclus (citing Geminus) remarked, has the character of an axiom (cf. [*Proclus*], 147). The fifth has neither the constructive character of a postulate nor the self-evident character of an axiom. Rather, it has the character of a theorem (cf. [*Proclus*], 150).

The controversy concerning the propriety of "mechanical" construction and the proper separation of it from truly geometrical construction continued for a long time. During part of this time, there also arose a controversy concerning the propriety of the genetic element of the constructive ideal—that is, the idea that there is, properly speaking, *causation* in geometry (and in mathematics generally). This was generally regarded as posing a deeper challenge to the classical constructive ideal than the controversy over the proper office of construction in geometry. It is to this discussion that we now turn.

3. Medieval and Renaissance Challenges to the Ancient Ideals

A remnant of the constructional component of the constructive ideal was thus preserved by the Neoplatonists of late antiquity. Various of the Jesuit commentators of the late Middle Ages and Latin Renaissance, however, launched sweeping criticisms of the constructive ideal, including its genetic component. Of these, the critique of Ludovicus Carbone in his 1599 treatise *Introductio in universam philosophiam* seems to have been particularly influential.[18] This treatise offered a classification of the sciences based on Aristotle's conception of *scientia* (*epistēmē*) as knowledge based on demonstration or knowledge of cause. According to this classification, there were five speculative sciences and two practical sciences.[19] The speculative sciences were a science of God, a science of intelligences, a science of being in common, a science of natural bodies, and a science of quantity. The two practical sciences were logic and ethics.

The sciences in this classification were ranked according to their nobility and their certitude. In the ranking according to certainty, mathematics was at the top. In the ranking according to nobility, it was at the bottom. This agreed with the relatively widespread belief, shared by Carbone, that there are no genuine definitions, no genuinely causal connections, and, hence, no truly genuine demonstration or *scientia* in mathematics (cf. Wallace [*Wallace 1984*], 131–132).

A concise statement of this general viewpoint was given by the Jesuit Benedictus Pererius in his 1576 treatise *De communibus omnium rerum naturalium principiis et affectionibus.* He there argued for a more Platonistic viewpoint according to which mathematical objects are or should be treated as having no essential connection to motion (efficient cause) or any other genuine type of cause.

[18] This treatise is widely believed to have been plagiarized from notes of lectures delivered in 1588 by Paulus Valla. Cf. Wallace [*Wallace 1984*], ch. 1, for more on this.

[19] Cf. Wallace [*Wallace 1984*], 130.

> My opinion is that mathematical disciplines are not proper sciences ... To have science (*scire*) is to acquire knowledge of a thing through the cause on account of which the thing is; and science (*scientia*) is the effect of demonstration: but demonstration (I speak of the most perfect kind of demonstration) must be established from those things that are *per se* and proper to that which is demonstrated, but the mathematician neither considers the essence of quantity, nor treats of its affections as they flow from such essence, nor declares them by the proper causes on account of which they are in quantity, nor makes his demonstrations from proper and *per se* but from common and accidental predicates. [*Pererius 1576*], i.12, p. 24[20]

A central example used to illustrate this claim was Euclid's proof that the interior angles of a triangle sum to two right angles (cf. prop. 32 in book I of the *Elements*). In his proof, Euclid both extended the base of the triangle and constructed a line parallel to one of its sides. For Pererius, this made the proof noncausal (and hence nonscientific) because neither the extension of the base of a triangle nor the construction of a line parallel to one of its sides is an essential property of the triangle in itself or of the construction of the triangle itself. This being so, Euclid's proof could not rightly be seen as providing a causal demonstration (i.e., a demonstration from the essence of the triangle).[21]

The charge that mathematics did not preserve the genetic ideal was familiar to Christopher Clavius and the other mathematicians at the Collegio Romano[22] in the latter part of the sixteenth century. Clavius and his student Josephus Blancanus systematically opposed the charge, insisting that there is genuine causation and, hence, genuine demonstration and genuine *scientia* in mathematics. To accept anything less, they believed, was to jeopardize the place of mathematics in the curriculum of the Jesuit universities. Their reasoning was essentially this: if there is not genuine causal demonstration in mathematics, then mathematical knowledge is not knowledge of the highest type; and if mathematical knowledge is not knowledge of the highest type, then it does not deserve a prominent place in a university curriculum. Clavius and Blancanus therefore viewed the Platonistic arguments of Carbone, Piccolomini, and Pereira as a serious threat to the place of

[20] Quoted from the English translation of this passage in Crombie [*Crombie 1977*], 67. Alessandro Piccolomini in his *Commentarium de certitudine mathematicarum disciplinarum* (1547) gave an earlier statement of this view. Various of the Coimbran Jesuits (e.g., Toletus) also expressed the view at roughly the same time as Pererius. Gassendi (cf. [*Gassendi 1972*], 104–108) gave a later statement. See Crombie [*Crombie 1977*]; Wallace [*Wallace 1984*], ch. 3; Dear [*Dear 1987*]; and Mancosu [*Mancosu 1996*], ch. 1 for useful discussions of some of the controversy in the Collegio surrounding the question of mathematics' status as *scientia*.

[21] Essentially the same argument was given by Proclus (cf. [*Proclus*], 161–162).

[22] The Collegio Romano was founded by Ignatius in 1550 and was the principal Jesuit university of the late sixteenth century, and the one where mathematics was best represented. It was also well-represented at La Flèche, Ingolstadt, and Würzburg, however.

mathematics in the universities, and they called upon the "teachers of philosophy" to stop characterizing mathematics as falling short of the genetic ideal.

> It will ... contribute much ... if the teachers of philosophy abstain from those questions which do not help in the understanding of natural things and very much detract from the authority of the mathematical disciplines in the eyes of the students, such as those in which they teach that the mathematical sciences are not sciences.... [*Clavius 1586*], quoted in [*Crombie 1977*], 66[23]

More substantively, Blancanus argued that the quantities studied in geometry and arithmetic are formed from *intelligible matter*, which is abstracted from sensible matter. Because of this, he maintained, it is possible to define the essences of objects composed of such matter, and these definitions will in turn provide the middle terms necessary for true formal-causal (*logos* or *archē*) demonstration. He also argued that intelligible matter, like sensible matter, is composed of parts, and that such parts are capable of providing the *material* causes of the properties of intelligible objects as well as their formal causes. He thus maintained that both material and formal causation operate in mathematics.[24]

In addition to Clavius and Blancanus, Galileo too shared this view, at least during the earlier part of his life. So, too, it seems, did many other mathematicians and philosophers. So much so, in fact, that it became the prevailing view of mathematical knowledge in the seventeenth century and remained influential for a considerable time thereafter. The following two remarks, one by a mathematician and one by a philosopher, are thus representative of the common view of seventeenth-century thinkers regarding the nature of mathematical knowledge.

> ... Mathematical Demonstrations are eminently Causal ... because they only fetch their Conclusions from Axioms which exhibit the principal and most universal Affections of all Quantities, and from Definitions which declare the constitutive Generations and essential Passions of particular Magnitudes. From whence the Propositions that arise from such Principles supposed, must needs flow from the intimate Essences and Causes of the Things. [*Barrow 1664*], 83[25]

> ... demonstrations are flawed unless they are scientific, and unless they proceed from causes, they are not scientific. Further, they are flawed unless their conclusions are demonstrated by construction, that is, by the description of figures,

[23] Clavius and Blancanus were largely successful in their efforts to defend the scientific status of mathematics and to give it a central place in the Jesuit curriculum of the seventeenth century. The ideal curriculum outlined in the 1599 *Ratio studiorum* gave it a prominent place. See Wallace [*Wallace 1984*], 139–141, for more on this.

[24] See [*Wallace 1984*], 142–143 for a brief but useful discussion of Blancanus's views on these points.

[25] The occasion was Barrow's installation as the first Lucasian chair of mathematics at Cambridge. The lectures were published in Latin (under the title *Lectiones mathematicae*) in 1683 and in English translation in 1734.

that is, by the drawing of lines. For every drawing of a line is motion, and so every demonstration is flawed, whose first principles are not contained in the definitions of the motions by which figures are described. [*Hobbes 1666*], vol. 4, 421[26]

4. The Emergence of Formalism

Modern formalism emerged against the backdrop of two noteworthy developments. One was a general decline in the importance of intuition as a guide to proof.[27] The other was the general increase in popularity of algebraic methods that began in the seventeenth century.

Modern formalism historically accompanied these developments. This notwithstanding, there is no necessary conceptual connection between them. The decline of intuition was a broad and complex phenomenon that left ample room for nonformalist responses. Similarly for the ascent of algebraic method. It arose from a variety of motives, only some of which were formalist.

Such, at any rate, is the argument I will sketch in the remainder of this section. The remaining element in the emergence of formalism—the *symbolic* element—will be discussed in section 5.

4.1. The Decline of Intuition

It has been suggested that the fundamental influence shaping nineteenth-century formalism was the general loss of confidence in geometric intuition that reached a critical point in the late nineteenth century. The following remark by the early twentieth-century mathematician–philosopher Hans Hahn gives a fairly typical assessment of this crisis.

> Because intuition turned out to be deceptive in so many instances and because propositions that had been accepted as true by intuition were repeatedly proved false by logic, mathematicians became more and more sceptical of the value of intuition. They learned that it is unsafe to accept any mathematical proposition, much less to base any mathematical discipline, on intuitive convictions. Thus a demand arose for the expulsion of intuition from mathematical reasoning and for the complete formalization of mathematics.... every new mathematical

[26] Quoted in [*Jesseph 1993*], p. 312, n. 8.

[27] This was brought on primarily by the early nineteenth-century discovery of non-Euclidean geometries and the later discovery of curves with counterintuitive properties (e.g., having no gaps but also having no tangents).

concept was to be introduced through a purely logical definition; every mathematical proof was to be carried through by strictly logical means.

...

It is not true, as Kant urged, that intuition is a pure *a priori* means of knowledge... rather... it is force of habit rooted in psychological inertia. [*Hahn 1980*], 93, 101

Hahn makes two charges here that it is important to distinguish. One is the sweeping claim that belief in a mathematical proposition ought never to be based on intuition. The other is the more cautious claim that reliance on intuition ought rightly to be banished from mathematical *reasoning* or inference. Commitment to the latter does not entail commitment to the former.

The latter is primarily a concern for rigor; a concern which, moreover, was clearly in evidence prior to the nineteenth-century developments mentioned above. It was widely known among seventeenth-century geometers that the proofs in the *Elements* were not so rigorous as had commonly been supposed. Specifically, they were known to rely upon assumptions that were not stated in the axioms and postulates. Lamy and Wallis, to mention but two seventeenth-century figures, each noted a number of such lapses.

Lambert, a century later, offered both a diagnosis and a cure. His diagnosis was that lapses in rigor were principally due to the tacit smuggling of undeclared (even unrecognized) elements of our intuitive grasp of geometrical figures into the inferences of geometric proofs. His prescribed cure was direct. He called for proof to proceed solely on the basis of its "algebraic" character (i.e., the syntactic character of the expressions occurring in it) and in *abstraction from* its subject matter. In determining whether a given proposition follows from others, he thus maintained, we should abstract away from everything having to do with the representation (*Vorstellung*) of the object or *thing(s)* (*Sache*) with which the proof is concerned. In this respect, a proof should "never appeal to the thing itself... but be conducted entirely symbolically [*durchaus symbolisch Vortrage*]" ([*Lambert 1786*], 162). Thus, a proof should treat its premises "like so many algebraic equations [*algebraische Gleichungen*] that one has ready before him and from which one extracts *x*, *y*, *z*, etc. without looking back to the object itself [*ohne daß man auf die Sache selbst zurücke sehe*]" (ibid.).

This was in radical opposition to the traditional conception of rigor. It also anticipated the formal conception of rigor and reasoning that was to be put into effect in the geometrical work of Pasch and Hilbert a century later (cf. [*Pasch 1882*] and [*Hilbert 1899*]). Its main contention was that rigorous verification of the validity of an inference should not require appeal to its subject matter. As Pasch later put it:

...if geometry is to be genuinely deductive, the process of inferring must always be independent of the sense of geometrical concepts just as it must be

> independent of diagrams. It is only relations between geometrical concepts that should be taken into account in the propositions and definitions dealt with. In the course of the deduction, it is certainly legitimate and useful, though by no means necessary, to think of the reference of the concepts involved. In fact, if it is necessary so to think, the deficiency of the deduction and the inadequacy of the method of proof is thereby revealed.... [*Pasch 1882*], 98

The modern conception of rigor thus required that both intuition and meaning be eliminable from the verification of the validity of the *inferences* that are made in a proof. It did not, however, require and did not justify a similar eliminability of intuition from the verification of the *propositions* that are used as the *premises* of such proofs. It did not, therefore, justify anything as radical as Hahn's claim that "it is unsafe to accept any mathematical proposition, much less to base any mathematical discipline, on intuitive convictions" ([*Hahn 1980*], 93). Nor was it generally claimed to do so.

The justification of this more radical proposal to eliminate intuition was generally based not on lapses in traditional rigor but on the more challenging discoveries of Bolzano and Weierstrass and others of curves that "cannot possibly be grasped by intuition" ([*Hahn 1980*], 87) and "can only be understood by logical analysis" (ibid.). The existence of such curves suggested, if it did not strictly imply, that logical analysis was superior to intuition as a means of determining the truth of mathematical propositions. This being so, logical analysis should replace intuition everywhere possible in order to minimize the chances of error.

This, of course, raises the question of whether use of logical analysis instead of intuition is always or even generally a means of minimizing the likelihood of error. My interest in formalism does not require that I offer an answer to this question here. The formalist can treat it as an open question. This is because he does not require the exclusion of intuition from mathematics but, rather, the *inclusion* of the nonintuitive (even the *noncontentual*) in it.[28] The elimination of intuition is therefore not an essential element of *formalism*. Hilbert made this clear in his criticism of the logicists Dedekind and Frege when he said that their attempts were "bound to fail" ([*Hilbert 1926*], 376) because the very use of logical inference and the performance of logical operations require that something be given to intuitive representation prior to all thinking.

The elimination of intuition is therefore not a necessary ingredient of formalism. Neither, I will argue, is it a sufficient condition. It is both a historical and a philosophical mistake to think otherwise. The pivotal commitment of formalism is not the negative one of eliminating intuition. Rather, it is a view concerning the nature of language—namely, that it can serve as a guide to thought even when it does not function descriptively. This general view was put forward by Berkeley in

[28] In separating the nonintuitive from the noncontentual, I am allowing what Pasch allowed—namely, that not all content in mathematics need be intuitive.

the eighteenth century, partly as an attempt to account for the usefulness of imaginary expressions in algebra. We will discuss it further in section 5. For the moment, however, we will consider more closely the questions of what might constitute a retreat from intuition and whether formalism is essentially a case of such.

4.2. Formalism and the Retreat from Intuition

In the remark quoted from Lambert, we already see a tendency to associate formalism with what might be called the *algebraic* way of thinking, where this is taken to consist in some type of abstraction from the intuitive contents of geometrical objects. In this subsection, I want to consider what such abstraction might, or perhaps *should*, be seen as coming to. It is not, I think, so clear as it may first appear to be. Locating some of the unclarities will help us to see better why formalism is not to be identified in any simple way with a retreat from intuition.

To begin with, let's try to get clearer on what is to count as *intuition*. The classical conception of intuition emphasized its immediacy. Plato first, but later Descartes (1596–1650), too, identified it with immediate knowledge of the (purely) *intelligible*, or what the Greeks referred to as *noēsis* (where this was distinguished from knowledge arising from the senses, or *aesthesis*). Descartes described this noetic conception of intuition as follows:

> By *intuition* I understand, not the fluctuating testimony of the senses, nor the misleading judgment that proceeds from the blundering constructions of imagination, but the conception which an unclouded and attentive mind gives us so readily and distinctly that we are wholly freed from doubt about that which is understood. . . . *intuition* is the undoubting conception of an unclouded (purae) and attentive mind. . . . [*Descartes 1970a*], 7 (rule III)

Locke (1632–1704) in his *Essay Concerning Human Understanding* also emphasized the immediacy of intuition, as well its clarity and certainty.

> . . . if we . . . reflect on our own ways of thinking, we will find, that sometimes the mind perceives the agreement or disagreement of two ideas *immediately by themselves*, without the intervention of any other: and this I think we may call *intuitive knowledge*. . . . this kind of knowledge is the clearest and most certain that human frailty is capable of. . . . [it] is irresistible, and, like bright sunshine. [*Locke 1689*], vol. II, pp. 176–177

Unlike Descartes, however, Locke found the central qualities of clarity and certainty not in the *noemata* of pure intellection (i.e., the ideas of reason or reflection) but in the ideas of sense. He thus maintained a basically empiricist conception of intuition while Descartes maintained a rationalist conception.

Kant (1724–1804) fashioned a view lying somewhere between these extremes. Specifically, he divided intuition into two types (or, perhaps better, two components): the *pure*, which supplied the *form(s)* of experience, and the *empirical*, which supplied its matter or content. He employed this division to obtain a conception of intuition that allowed it to possess robust content while at the same time retaining an element of *apriority*.

> In whatever manner and by whatever means a mode of knowledge may relate to its objects (*Gegenstände*), intuition (*Anschauung*) is that through which it is in immediate relation to them, and to which all thought (*Denken*) as a means is directed. But intuition takes place only in so far as the object is given to us. This... is only possible... in so far as the mind (*Gemüt*) is affected in a certain way. The capacity for receiving representations (*Vorstellungen*) through the mode in which we are affected by objects, is entitled *sensibility*. Objects are *given* to us by means of sensibility, and it alone yields *intuitions*....
>
> The effect (*Wirkung*) of an object upon the faculty of representation... is *sensation* (*Empfindung*). That intuition which is in relation to the object through sensation, is entitled *empirical* (*empirisch*). The undetermined object of an empirical intuition is entitled *appearance* (*Erscheinung*).
>
> That in the appearance which corresponds to sensation I term its *matter* (*Materie*); but that which so determines the manifold of appearance (*Mannigfaltige der Erscheinung*) that it allows of being ordered... I term the *form* (*Form*) of appearance. That in which alone the sensations can be posited and ordered... cannot itself be sensation; and therefore, while the matter of... appearance is given to us *a posteriori* only, its form (*Form*) must lie ready (*bereitliegen*) for the sensations *a priori* in the mind.... The pure form of sensible intuitions... must be found in the mind *a priori*. This pure form of sensibility may... be called *pure intuition*. [*Kant 1781*], A19–20, B33–35

It is important for our purposes to consider the significance of Kant's distinction between *pure* and *empirical* intuition. What, we should ask, is the essence of this distinction? Our answer is that it is an attempt to distinguish the variable (*Materie*) from the invariable (*Form*) elements of intuition and to allign mathematics with the invariable.

We thus already see in Kant's distinction between the form and matter of intuition a certain type of retreat from intuition—namely, a retreat from its variable elements. We see something of the same in Locke's distinction between primary (invariant) and secondary (variant) qualities of objects and in his identification of objective knowledge with knowledge of the former. In each case, there is a variant type of intuition from which the enterprise of knowledge is separated. At the same time, there is an invariant intuition to which knowledge remains bound. Knowledge generally and, in the case of Kant, at least, mathematical knowledge in particular, is identified with the grasp of certain invariants of experience.

4.2.1. *The Invariantist Ideal*

Commitment to invariantist ideals was already clear in the methodological writings of various mathematicians and mathematically engaged philosophers in Locke's day. It took a variety of forms, however, one of which was a shift of focus, generally attributed to Viète (1540–1603) and Descartes, from particular quantities to quantities in general or, to use the traditional terminology, from particular quantities to *species* of quantities.[29]

The leading idea of this generalized arithmetic was that while some properties of mathematical quantities are due to those quantities being the particular quantities they are, others of their properties are due to more general features that those quantities share with other quantities. Viète was interested in this latter type of property. In this spirit, he inquired into the relations that exist between the coefficients and the roots of equations generally, and discovered that there are invariances in such relations. One example is the general relation that exists between the coefficients and roots of a second-degree equation when the coefficient of the second term of the equation is the negative of two numbers whose product is the third term. Under this condition, the two numbers in question are roots of the equation, regardless of the particular values of the quantities to which the coefficients may be attached. This suggests that certain relationships between coefficients and roots of quadratic equations are due not to the particular quantities with which an equation is concerned, but to certain standing properties that endure regardless of the particular quantities dealt with.

Descartes intensified this search for the most general laws of quantity. It was, indeed, the *leitmotif* of his *mathesis universalis.* He believed that these laws would illuminate not only arithmetic but geometry. Indeed, he believed that it was these laws that guided the work of ancient geometers. Unfortunately, Descartes maintained, the ancients coveted the admiration of others more than they loved truth and, so, hid their methods to keep others from seeing how easy and unremarkable many of their discoveries really were. This was a familiar theme among the early developers of algebra (cf. the quote from Pedro Nuñez' algebra in [*Wallis 1685*], ch. II, 3, the remark by Viète) in ch. V [*Viète 1983*], p. 27 and the claim by Wallis in ch. II of [*Wallis 1685*], p. 3).

The secret of the Greeks was *Algebra*, and it was this realization, said Descartes, that recalled him from

> ...the particular studies of Arithmetic and Geometry to a general investigation of Mathematics, and thereupon... to determine what precisely was universally meant by that term... as I considered the matter carefully it gradually came to light that all those matters only were referred to Mathematics in which order and measurement are investigated, and that it makes no difference whether it be in numbers, figures, stars, sounds or any other object that the question of measurement arises. I saw consequently that there must be some general science to explain that element as a whole which gives rise to problems about order and

[29] Viète commonly referred to algebra as the "logistic of species," and represented species (or abstract, general quantities) by writing them as letters. Others (e.g., Wallis) referred to algebra as the "arithmetic of species" or "specious arithmetic."

> measurement, restricted as these are to no special subject matter. [*Descartes 1970a*], 13

This general or universal mathematics (*mathesis universalis*) "contained everything on account of which the others [that is, the particular mathematical sciences] are called parts of Mathematics" (ibid.). A genetically penetrating proof of a truth of this general mathematics would therefore not depend upon the special properties of any particular quantity or kind of quantity.

Seen this way, algebra was intended to serve the genetic ideal. In addition to this, however, it also promised to make mathematical thinking more economical by keeping it focused. Descartes expressed this idea as follows:

> By this device [symbolic algebra] not only shall we economize our words, but, which is the chief thing, display the terms of our problem in such a detached and unencumbered way that, even though it is so full as to omit nothing, there will nevertheless be nothing superfluous to be discovered in our symbols, or anything to exercise our mental powers to no purpose, by requiring the mind to grasp a number of things at the same time. [*Descartes 1970a*], 67

The Oxford mathematician John Wallis (1616–1703) made similar points in his 1685 *Treatise of Algebra.* As regards the simplification of thinking, he wrote:

> This Specious Arithmetick, which gives Notes or Symbols (which he [Viète] calls Species) to Quantities both known and unknown, doth (without altering the manner of the demonstration, as to the substance,) furnish us with a short and convenient way of Notation; whereby the whole process of many Operations is at once exposed to the Eye in a short Synopsis. [*Wallis 1685*], preface, a5

Wallis did, then, consider (some kind(s) of) brevity or efficiency to be a prime virtue of algebraic reasoning—particularly as a means of what he called *investigation* (i.e., "finding out of things yet unknown," [*Wallis 1685*], 305–306) as distinct from *demonstration.* Wallis's view of algebra was, however, also deeply motivated by what he took to be its capacity to *genetically* improve upon the demonstrations of classical geometry. He argued, in particular, that some of the deepest properties of geometrical figures are due not to their manner of construction but to more basic algebraic or arithmetic characteristics which hold of them not as *geometrical* quantities but as quantities per se:

> ... beside the supposed construction of a Line or Figure, there is somewhat in the nature of it so constructed, which may be abstractly considered from such construction; and which doth accompany it though otherwise constructed than is supposed. [*Wallis 1685*], 291

Again we see here an invariantist ideal—the idea that geometry ought to be concerned with the most invariant features of geometrical figures. In Wallis's view, these were (at least in some instances) *arithmetical* features of those figures and not features dictated by the manner of their construction. He thus proceeded

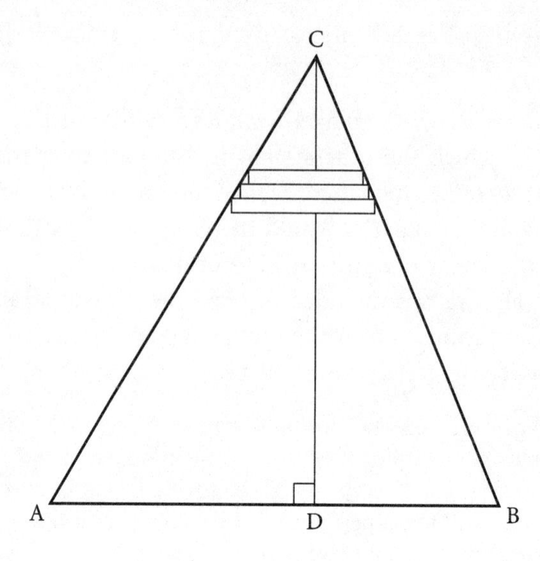

Fig. 8.1. **Area of Triangle.**

to offer arithmetic or algebraic proofs—proofs appealing only to a figure's general characteristics as a *quantity*—of various classical properties of geometrical figures.[30] He applied this to the classical problem of quadrature. A good example was his "demonstration" of the classical law of area for the triangle($\triangle = \frac{1}{2}bh$).[31]

The basic idea of the proof is that the interior of a triangle *ABC* can be approximated by a stack of thin rectangles (call them *composing rectangles*) that approximates ever more closely the interior of *ABC* as the common height of the rectangles is allowed to go to 0.[32] (See figure 8.1.) Thus, let the length of the base *AB* of *ABC* be b, and its height (i.e., the length of *CD*) be h. As h_c (the common height of the composing rectangles) goes to 0, the sum of the heights of the composing rectangles approaches h, and the sum of their areas approaches the area of *ABC*.

How is the sum of the areas of the composing rectangles to be reckoned? Wallis reasoned that since the sides of a triangle are straight lines (and hence of constant slope) and the composing rectangles are of uniform height, every move from one rectangle to the next one down in the stack will result in the same change in the length of the base of that rectangle. For each value of h_c, then, the

[30] This was a main preoccupation of his 1656 treatise *Arithmetica infinitorum*.

[31] The argument here can be found both in [*Wallis 1656*], 366, and, in its essentials, in [*Wallis 1685*], 285f. It was closely patterned after a general technique of exhaustion designed fifty years earlier by Cavalieri (cf. [*Geometria*]). Wallis detailed his debt to Cavalieri in chapter 74 of [*Wallis 1685*], whose title is "Of Cavallerius His Method of Indivisibles." He also described his method as a "Method of Exhaustion" (cf. [*Wallis 1685*], 286–287).

[32] A similar argument approximating *ABC* from the "outside" can also be given.

lengths of the bases of the composing rectangles will form an arithmetic progression.[33] This, in turn, implies that for each value of h_c the areas of the composing rectangles will form an arithmetic progression.[34]

This being so, we can use the law concerning sums of arithmetic progressions to find the joint area of the composing rectangles. The sum of an arithmetic progression is the average size of its terms times their number. The average size of the terms of a progression is the average of its largest and smallest terms. The sum of an n-element arithmetic progression whose smallest element is k_1 and whose largest element is k_n is thus $(k_1 + k_n)/2 \times n$.

Letting n_c be the number of composing rectangles, the sum of their areas will then be $\frac{1}{2}(n_c h_c s + n_c h_c b)$, where s is the length of the base of the smallest rectangle. Since s approaches 0 as h_c approaches 0, and since the product $n_c h_c$ approaches h as h_c approaches 0, we can say that the quotient $\frac{1}{2}(n_c h_c s + n_c + h_c + b)$ approaches $\frac{1}{2}(h0 + hb) = \frac{1}{2}bh$ as h_c approaches 0.

Using Wallis's arithmetical computation for the area of a triangle, then, we can always come closer than any assigned value to the value determined by the classical law for the area of a triangle (i.e., $\triangle = \frac{1}{2}bh$). By the principle of exhaustion, it then follows that the value determined by Wallis's procedure and the value determined by the classical law are the same. Q.E.D.[35]

Wallis took this and similar proofs to show that the genetic ideal is not always well served by joining it to a constructive ideal. The real "causes" of truths concerning geometrical figures do not always reside in the manner of their construction, but in certain of their construction-invariant *arithmetical* features. Because these arithmetical invariants are more persistent than their geometrical counterparts, the features they signify are deeper and more revealing of the essence of geometrical objects than are the associated constuctional properties.

[33] An arithmetic progression is a sequence in which each term after the first differs from the previous one by a constant amount, called the *common difference.*

[34] This is so because multiplying the terms of any arithmetic progression by a constant (in this case, h_c) produces a sequence whose terms also form an arithmetic progression.

[35] Wallis himself (cf. [*Wallis 1685*], 286–287) characterized his method as "the Method of Exhaustion." The Method of Exhaustion is, of course, a well-known ancient method of proof based on the idea that if two quantities differ by less than any assignable quantity, they are then equal. Heath ([*Heath 1921*], vol. I, 327f.) notes that Archimedes attributed the first proof of it to Eudoxus, though Democritus is said to have been the first to have discovered it. Hankel ([*History*], 122) cites evidence that it was formulated and used by Hippocrates before Eudoxus. It is sometimes referred to as *Archimedes' Axiom* (op. cit., 217), one variant of which is stated as def. IV in bk. V of *The Elements*, and used, in effect, to prove another variant—namely, that given in prop. 1 in bk. X.

What is supposedly new in Wallis's application of exhaustion is its use to force equality between an arithmetically determined value and a geometrically determined value.

This posed a challenge to the traditional Aristotelian injunction against *metabasis*, and it did so in the name of genetically deeper proof. Wallis thus maintained that his algebraic methods were not only satisfactory but, in many cases, genetically superior to the methods of classical geometry:

> ... this *Abstractio Mathematica* (as the Schools call it,) is of great use in all kind of Mathematical considerations, whereby we separate what is the proper Subject of Inquiry, and upon which the Process proceeds, from the impertinences of the matter (accidental to it,) appertaining to the present case or particular construction.
>
> For which reason, whereas I find some others (to make it look, I suppose, the more Geometrical) to affect Lines and Figures; I choose rather (where such things are accidental) to demonstrate universally from the nature of Proportions, and regular Progressions; because such Arithmetical Demonstrations are more Abstract, and therefore more universally applicable to particular occasions. Which is one main design that I aimed at in this Arithmetick of Infinites. [*Wallis 1685*], 292[36]

For Wallis, then, as also for Descartes, fidelity to the genetic ideal did not entail fidelity to the constructive ideal. In particular, it devalued that intuition which consists in the use of vision or the visual imagination to construct geometric figures. It did so on the grounds that preoccupation with such intuition can be, and oftentimes is, preoccupation with the less rather than the more invariant and, as such, is preoccupation with the genetically less deep. Thus it was that adherence to the classical genetic ideal itself provided a motive for the retreat from intuition that began in earnest in the seventeenth century.

Something of the same spirit animated that nineteenth-century "revival" of synthetic method in geometry known as *projective geometry*. We now briefly consider this.

4.2.2. *Projective Geometry and the Principle of Continuity*

There was, of course, a need to be clearer about invariance and its connection with objectivity, and this became a central concern of eighteenth- and nineteenth-century

[36] Wallis's views are complicated by the unclarity of his position on the classical distinction between "art" and "science." At one place (cf. [*Wallis 1685*], 290), for example, he describes algebra as being not a "science" capable of producing demonstrations (explanations), but only an "*art*" of "invention" capable of identifying truths without demonstrating or explaining them. He was also less than perfectly consistent in the statement of his purpose(s). Replying to objections of Fermat's, for example, he wrote:

> ... he doth wholly mistake the design of that Treatise; which was not so much to shew a Method of Demonstrating things already known; (which the Method that he commends, doth chiefly aim at,) as to shew a way of *Investigation* or finding out of things yet unknown: (Which the Ancients did studiously conceal.) For which he doth admit this (*Epist. 12*), if warily used, to be a good Method; and therefore should not have found fault with it, when applyed to such a purpose. [*Wallis 1685*], 305–306

geometry. Jean-Victor Poncelet (1788–1867), one of the founders of projective geometry, proposed a solution in the form of a principle—the so-called *Principle of Continuity* (PC).[37] He presented it in the preface to the 1822 edition of his famous *Traité des propriétés projectives des figures*:

> Let us consider some geometrical diagram, its actual position being arbitrary and in a way indeterminate with respect to all the possible positions it could assume without violating the conditions which are supposed to hold between its different parts. Suppose now that we discover a property of this figure.... Is it not clear that if, observing the given conditions, we gradually alter the original diagram by imposing a continuous but arbitrary motion on some of its parts, the discovered properties of the original diagram will still hold throughout the successive stages of the system always provided that we note certain changes, such as that various quantities disappear, etc.—changes that can easily be recognized *a priori* and by means of sound rules? [*Poncelet 1822*], vol. 1, p. xii

There are, I think, few clearer examples of philosophical ambition in the history of mathematics. Poncelet asserts nothing less than that properties of figures that change across what he calls "gradual" transformations cannot properly be regarded as being among their *real* properties.

But what, really, does this say? Two readings, at least, seem possible. One is *metaphysical*, the other *methodological*. On the metaphysical reading, the PC says that any figure $\mathcal{F}_D$ (=*d*erived figure) obtained from another figure $\mathcal{F}_O$ (=*o*riginal figure) by means of a "gradual" transformation is co-real with, and has the same lawlike features as, $\mathcal{F}_O$.

Clearly, such properties as the size of an angle or the length of a line segment are not invariant under gradual transformation. We can increase or decrease the sizes of both angles and line segments by degrees. They are therefore not among the properties that the PC would count as *real* or *objective* properties of figures.

The PC thus demotes to nonreal or nonobjective various properties counted as real or objective by traditional geometry. Curiously, it also gives rise to entities that could only be regarded as imaginary from an intuitive standpoint. A classical example is the so-called *point at infinity* of plane projective geometry. The following reasoning illustrates its interaction with the PC.

Consider a pair of lines L_1, L_2 arranged in nonparallel fashion (figure 8.2). Call this the *original* figure. This figure can be changed by "gradual" transformation into the figure shown in figure 8.3, in which L_1 and L_2 are parallel. This can be

[37] The name calls to mind Leibniz's *Law of Continuity*, presented more than a century before Poncelet's principle. Leibniz (1646–1716) wrote: "Everything goes by degrees in nature, and nothing by leaps, and this rule of changes is part of my *law of continuity*" ([*Leibniz 1704*], bk. IV, ch. xvi, §12; emphasis added). Leibniz's Law of Continuity seems to have been intended as a law of nature, however, and not as a principle concerning proper mathematical method.

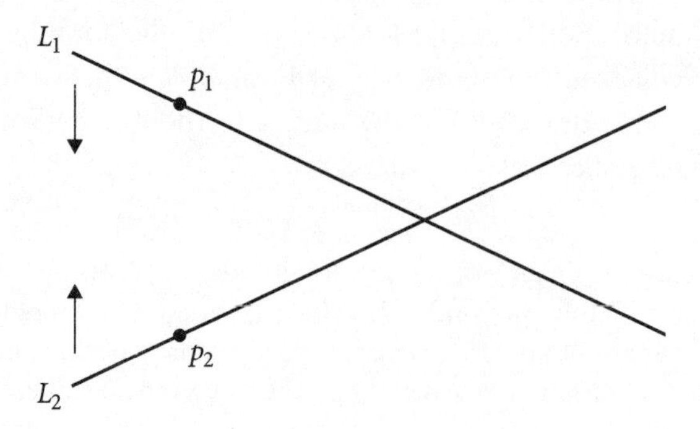

Fig. 8.2

done, for example, by pivoting L_1 and L_2 as arranged in figure 8.2 about the points p_1 and p_2 in the direction indicated by the arrows.

The metaphysical interpretation of the PC imples that the lawlike properties that hold of the original figure should also hold of the parallel figure shown in figure 8.3. One of the lawlike properties of the original figure, however, is that L_1, L_2 have exactly one point of intersection. At every point in the transformation of figure 8.2 into figure 8.3, L_1 and L_2 have a unique point of intersection. Moreover, continuing the rotation of L_1 and L_2 about p_1 and p_2 "past" their parallel orientation in figure 8.3, L_1 and L_2 once again continuously have a unique point of intersection. If, therefore, the parallel orientation of L_1 and L_2 in figure 8.3, is not to violate the PC by representing a "jump" or discontinuity in the continuous transformation of the original figure, there must be a unique point at which L_1, L_2 intersect in it, too. The PC thus implies the existence of a point of intersection for parallel lines.

On its metaphysical reading, then, the PC can dictate the introduction of what, from an intuitive vantage, could only be regarded as imaginary or ideal (=unreal) elements.

On the *methodological* reading, the PC has a different function—namely, that of a directive that enforces a type of simplicity. On this reading, the PC says (roughly) that if one figure can be obtained from another by means of gradual

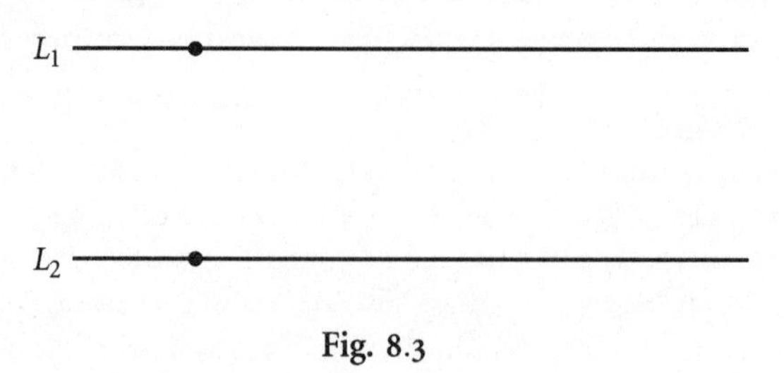

Fig. 8.3

transformation, then, even if there is no supporting intuition (and perhaps even if there is contrasting intuition), it is nonetheless *good method* to ascribe the same general features to the one figure as to the other.

The "dualities" of projective geometry bear witness to the fruitfulness of this understanding of the PC. They arrange theorems into pairs, each element of which can be obtained from the other by a uniform scheme of substitutions. The well-known *point/line* duality of plane projective geometry, for example, allows one to obtain the theorem

(a) For every two lines, there is exactly one point that is their meet

from the theorem

(b) For every two points, there is exactly one line that is their join,

and vice versa. One need only replace every occurrence of the term "point" (resp. "line") with an occurrence of the term "line" (resp. "point"), and every occurrence of the term "meet" (resp. "join") with an occurrence of the term "join" (resp. "meet") in order to obtain the one from the other.

The dualities suggest that certain terms (e.g., "point," "line," "meet," "join," etc.) that function as *constants* in ordinary (i.e., metric) geometry function as *variables* in projective geometry. This, in turn, indicates a greater level of generality or abstraction—hence a greater genetic depth—for projective geometry as over against ordinary metric geometry.

The status and/or plausibility of the PC is not, however, our main concern here. Rather, we are concerned with the invariantist ideal—the idea that invariance is the mark of the *objective*—and what the relationship between the PC and this ideal may be. We believe that the PC embodies an invariantist ideal—one which contends that only those properties of geometrical figures are real that are invariant under continuous transformation. It thus also embodies a retreat from intuition—specifically, a retreat from that type of intuition that is particular to a particular intuitive viewpoint or vantage. The PC not only retreats from intuition, however. It also violates it by giving rise to such counter-intuitive results as the existence of points at infinity.

The invariantist ideal of knowledge became a mainstay of mathematical method of the later nineteenth and early twentieth centuries. This is illustrated by Cayley's work in geometry in the mid-nineteenth century in which (cf. [*Cayley 1859*]) he identified a function which possessed the same algebraic features as a metric or distance function, but which was nonetheless invariant under the transformations of projection and section. Cayley took this to show that metric geometries could be derived from purely nonmetric properties of descriptive geometry. He concluded that "The metrical properties of a figure are not the properties of the figure considered *per se* apart from everything else, but its properties when considered in connexion with another figure, viz. the conic termed the Absolute" ([*Cayley 1889*], vol. 2, 92).

The invariantist ideal was developed further by Felix Klein in his well-known *Erlanger Program.* Klein went beyond Cayley to show how to systematically generate non-Euclidean geometries from projective geometry using different choices of conics for the Absolute (cf. [*Klein 1871*]).[38] Projective geometry thus came to be seen as providing a general framework for metric geometry.[39]

The general epistemology of this approach was summed up well by the twentieth-century philosopher-mathematician Friedrich Waismann. He wrote:

> The properties which projective geometry brings to light are more intimately tied to the space structures and harder to destroy than the properties considered in the usual geometry.... on ascending from the usual or "metric" geometry to affine and then to projective geometry, the differences of the configurations vanish more and more, and the deeper lying, more general, characteristic features come to the fore instead. [*Waismann 1951*], 177

4.3. Conclusion

In closing, let me summarize the argument of this section. Our question has been: To what extent should formalism be seen as consisting in a retreat from intuition? Our answer, generally, has been "to no very great extent." In saying this, however, we must be careful to distinguish the intuition of classical geometry, with its emphasis on visualizability, from the very different type of intuition that Kant labeled a priori. A priori intuition emphasized certain invariances in our visual experience that are due to the existence of various standing conditions which order it. Kant may not have correctly identified which conditions these are, but in emphasizing their invariant nature, he anticipated the leitmotif of the geometry that was to come after him. This is what I will call the *invariantist ideal*, an ideal which, in one form or another, has been a leading ideal of mathematics since the late seventeenth century.

What, at bottom, is this ideal? One way of describing it is as a commitment to the classical ideal of genetic depth in proof. But what does this commitment represent? At bottom, it represents a commitment to objectivity—that is, a commitment to finding the properties of geometric figures that are *in* the

[38] The Absolute for three-dimensional Euclidean space is the plane. But the Absolute can also take hyperbolic or elliptical forms, and when it does, non-Euclidean geometries are the result.

[39] The classic statement of the Erlanger Program is [*Klein 1872*]. See [*Richards 1988*], ch. 3, for a brief but useful discussion of Klein's program and related developments.

figures themselves, as opposed to those properties that are "in" the mind of the geometer. This was at least among the principal motives driving Descartes's development of algebra, and it seems to have been a principal motive guiding Wallis as well.

Kant changed the goal from objectivity to intersubjectivity, but this preserved the emphasis on invariance since the intersubjective is that which is invariant from subject to subject.

Formalism is not, I believe, primarily the development of an invariantist motif. It is centered instead on an instrumentalist conception of language and its use in reasoning. So, at any rate, I will argue.

5. Symbolic Formalism

The instrumentalist conception of language allows for purely *symbolic* uses of signs in our reasoning—uses that do not depend in any essential way on the semantic content of the signs involved or on their even having such content. This is the central doctrine of the position I am calling *symbolic formalism* or, more simply, *formalism*. This central doctrine is itself composed of two key elements. The first is what I will call the *creative* or *creativist element*—the idea that the mathematician is free to introduce or "create" methods out of considerations of convenience or efficiency as distinct from evaluation of content. The second is the distinctively *symbolic* element—namely, that nonsemantical uses of signs may, at least on occasion, constitute such conveniences. In the next few subsections, I'll reprise the basic history of these ideas leading up to the mature formalism of Hilbert. I'll then close the section with a discussion of Hilbert's views.

5.1. The Berkeleyan Conception of Language and Reasoning

Symbolic formalism received its chief historical impetus from the rapid development of algebra in the sixteenth and seventeenth centuries. Impressed by this development, Berkeley (1685–1753), in the early eighteenth century, attempted to sketch a general philosophical framework for it. The cornerstone of this framework was a conception of language and thought that contrasted strongly with the "received opinion" of the day—namely, that language gains its cognitive significance by semantic means, specifically, by its expression of ideas and their combinations (cf. [*Berkeley 1710*], 36–37).

Against this view, Berkeley urged a broadly instrumentalist function of language. He took the practice of algebraists to illustrate this, though he regarded the general phenomenon as much broader (cf. [*Berkeley 1732*], vol. 3, 307).[40]

> ...it is a received opinion, that language has no other end but the communicating [of] our ideas, and that every significant name stands for an idea....a little attention will discover, that it is not necessary (even in the strictest reasonings) [that] significant names which stand for ideas should, every time they are used, excite in the understanding the ideas they are made to stand for: in reading and discoursing, names being for the most part used as letters are in *algebra*, in which though a particular quantity be marked by each letter, yet to proceed right it is not requisite that in every step each letter suggest to your thoughts, that particular quantity it was appointed to stand for....the communicating of ideas marked by words is not the chief and only end of language, as is commonly supposed. [*Berkeley 1710*], 37[41]

Berkeley gave various examples intended to clarify the instrumental use of language. One of these was the use of counters or chips in a card game. He noted, first, that the chips were not used "for their own sake" (cf. [*Berkeley* 1732], 291), as it were, but for the sake of determining how money was to be exchanged between the players at game's end. Chips or counters therefore ultimately had semantic significance. This notwithstanding, the players were not obliged to keep the semantic values of the counters before them as they played the game. Nor was there any advantage in doing so. It was enough that, at the end of the game, the counters could be converted to the money they were agreed to signify. As Berkeley put it:

> ...it is sufficient the players at first agree on their [the counters'] respective values, and at last substitute those values in their stead....words may not be insignificant, although they should not, every time they are used, excite the ideas they signify in our minds; it being sufficient that we have it in our power to substitute things or ideas for their signs when there is occasion. It seems to follow that there may be another use of words besides that of marking and suggesting distinct ideas, to wit, the influencing [of] our conduct and actions, which may be done...by forming rules for us to act by....A discourse...that directs how to act or excites to the doing or forbearance of an action may...be useful and significant, although the words whereof it is composed should not bring each a distinct idea into our minds. [*Berkeley* 1732], 291–292, (brackets added)

There is an important formalist message in this passage—namely, that reasoning can, at some level, proceed on the basis of syntactically marked regularities of expressions, without adversion to supposed semantic contents. This is,

[40] Page references here and throughout are to the reprinting of [*Berkeley 1732*] in [*Berkeley 1948–1957*].

[41] Emphasis Berkeley's, brackets mine.

moreover, true even of *rigorous* reasoning, though this ran strongly counter to the prevailing conception of rigor in Berkeley's day.

That conception of rigor was what I will call the *presentist* conception. It saw rigor as consisting in the keeping of the subject matter of a piece of reasoning continually before the mind—or, more commonly, before the *intuition*—of the reasoner. Poncelet expressed the idea as follows.

> In ordinary geometry, which one often calls synthetic . . . the figure is described, one never loses sight of it, one always reasons with quantities and forms that are real and existing, and one never draws consequences which cannot be depicted in the imagination or before one's eyes by sensible objects. [*Poncelet 1822*], xix[42]

The paradigm of presentist procedure was the constructive proof of classical synthetic geometry. The diagrams and diagrammatic operations of classical synthetic proof were believed to offer the best means of keeping the subject matter of a proof before the prover's mind. They did so by "mirroring" the objects and constructions of classical geometry. This relationship of mirroring or resemblance was, however, utterly absent from the symbolic methods of the algebraists. The rigor of those methods was therefore open to serious question. The semantic mysteriousness of various of the algebraists' symbols (e.g., $\sqrt{-1}$) only added to the dubiousness of their methods. The Scottish cleric and mathematician John Playfair summed up the general situation this way:

> The propositions of geometry have never given rise to controversy, nor needed the support of metaphysical discussions. In algebra, on the other hand, the doctrine of negative quantity and its consequences have often perplexed the analyst, and involved him in the most intricate disputations. The cause of this diversity in sciences which have the same object must no doubt be sought for in the different modes which they employ to express our ideas. In geometry every magnitude is represented by one of the same kind; lines are represented by lines, and angles by an angle, *the genus is always signified by the individual, and a general idea by one of the particulars which fall under it.* By this means all contradiction is avoided, and the geometry is never permitted to reason about the relations of things which do not exist, or cannot be exhibited. In algebra again *every magnitude being denoted by an artificial symbol, to which it has no resemblance*, is liable, on some occasions, to be neglected, while the symbol may become the sole object of attention. It is not perhaps observed where the connection between them ceases to exist, and the analyst continues to reason about the characters after nothing is left which they can possibly express; if then, in the end, the

[42] Poncelet did not ultimately endorse this conception of proof. Rather, he emphasized its inconvenience, saying that its genetic imperative made it laborious. This laboriousness he saw as principally due to the perceived need for the prover to take things all the way back to the rudimentary constructions—to, in his words, "reproduce the entire series of primitive arguments from the moment where a line and a point have passed from the right to the left of one another, etc." ([*Poncelet 1822*], xix).

> conclusions which hold only of the characters be transferred to the quantities themselves, obscurity and paradox must of necessity ensue. [*Playfair 1778*], 318–319 (emphasis added)[43]

On the presentist conception, then, rigor was compromised at any point in an algebraic argument where there was an expression whose object was not directly present to the reasoner's mind or intuition, and this is something that occurs frequently in algebraic reasoning. Indeed, one might say that it is typical, even paradigmatic, of algebraic reasoning. The constructional proofs of classical synthetic geometry were taken to avoid this by dint of a relationship of *resemblance* between the diagrams used in them and the geometrical objects with which the proofs were ultimately concerned. Such, at any rate, was the prevailing view.

Mathematicians of Berkeley's day often emphasized the reliability of judgments based on the above-mentioned resemblance. Baron Francis Maseres (1731–1824), for example, posited a distinction between two cognitive faculties, a faculty of *understanding* and a faculty of *sense* or intuition, claiming that the latter was more reliable than the former. The latter he took to be the faculty employed in classical geometric proof. The former he took to be the faculty used in algebraic reasoning. He thus emphasized

> . . . the greater facility with which a writer may impose upon himself as well as his readers, and fancy he has a meaning where, in reality, he has none, in treating of an abstract science, such as Algebra, that addresses itself only to the understanding, than in treating of such a science as Geometry, that addresses itself to the senses as well as to the understanding; for the impossibility, or difficulty, of representing a false, or obscure, conception in lines to the eye, would immediately strike either the writer or the reader, and make them perceive its falsehood or its ambiguity; whereas when things are expressed only in words, or in any abstract notation, wherein the senses are not concerned, men are much more easily deceived: and for the same reasons, a defect of proof, or a hasty extension of a conclusion justly drawn in one case to several other cases that bear some resemblance, but not a complete one, to it, may be much more easily perceived in Geometry than in Algebra. [*Maseres 1758*], ii–iii[44]

[43] This paper, the first mathematical paper published by Playfair, was submitted while he was a parish minister. In 1785 he was appointed Joint Professor of Mathematics at the University of Edinburgh.

[44] Maseres was not fatalistic about this difference. Indeed, his motive for writing his treatise was "to raise [algebra] to a level with Geometry, in respect both of perspicuity of conception, and accuracy of reasoning" ([*Maseres 1758*], iii). His general strategy for effecting this elevation centered on the provision of intuitive interpretations for algebraic expressions (especially for those that involved the "negative sign"). Other thinkers, equally committed to the algebraic way of doing things, questioned the propriety, or at least the superiority, of intuitional grasp as a basis for mathematical knowledge. Lacroix, for example, cautioned against "borrowing from appearances and sensations those things that can be drawn from judgment alone" in mathematics (cf. [*Lacroix 1805*], 174).

Berkeley's claim that reasoning in which the reasoner's mind was not continually fixed on the meanings of the expressions involved could nonetheless be rigorous was thus counter to the traditional view of rigor.[45]

This, however, only committed him to the view that signs need not, in *all* their uses, function to express ideas. It is a further step, though perhaps not a great one, to say that there are signs that do not in *any* of their uses express ideas.

Berkeley seems to have taken this further step in arriving at his general view of the significance of algebraic language:

> ... the true end of speech, reason, science, faith, assent, in all its different degrees, is not merely, or principally, or always, the imparting or acquiring of ideas, but rather something of an active operative nature, tending to a conceived good: which may sometimes be obtained, not only although the ideas marked are not offered to the mind, *but even although there should be no possibility of offering or exhibiting any such idea to the mind: for instance, the algebraic mark, which denotes the root of a negative square, hath its use in logistic operations, although it be impossible to form an idea of any such quantity.* [*Berkeley 1732*], 307 (emphasis added)[46]

[45] It was not unprecedented, however. Leibniz, for instance, in a well-known letter to Fr. Des Bosses, praised algebraic methods for their usefulness while, at the same time, labeling them fictions:

> To speak philosophically, I would advocate neither infinitely small nor infinitely large magnitudes; nor infinitely numerous nor infinitely scarce. To put it compendiously, I hold these to be mental fictions [*pro mentis fictionabus*], suited for use in calculations, like the imaginary roots in algebra. And yet I have demonstrated that these expressions [*expressiones*] are a great aid in shortening thought [*ad compendium cogitandi*] and also in discovery [*ad inventionem*], and it is not possible that they should lead us into error [*in errorem ducere non posse*]. Letter to Fr. Des Bosses, March 17, 1707, in [*Leibniz 1959*], 436

Leibniz generally attributed the economizing power of algebra to the fact that it "unburdens the imagination" (cf. [*Leibniz 1704*], bk. IV, ch. 17), where the term "imagination" signified a capacity to visualize.

Before Leibniz, too, the usefulness of algebraic methods had been emphasized. Clear expressions of this are found, for example, in the writings of Viète, Descartes, and Wallis. Berkeley's contribution was to link this to a broader instrumental conception of language and reasoning.

[46] Berkeley's use of the term "logistic" agreed with the then current usage. Barrow, shortly before him, characterized logistic as "... a kind of Artifice for designing Magnitudes and Numbers by certain Notes or Symbols, and collecting and comparing their Sums and Differences ..." ([*Barrow 1664*]), 28).

Hutton, shortly after, identified it with "... the rules of computations in Algebra, and in other species of Arithmetic: witness the Logistics of Viète and other writers ([*Hutton 1795–1796*], vol. II, 51–52).

The cognitive significance of language was therefore not, in Berkeley's view, exclusively a result of its *semantic* usage. It could, in addition, have a kind of *logistic* use whose aim was to assist the mind in reasoning, judging, extending, recording, and communicating knowledge even when it was not used to express ideas (cf. [*Berkeley* 1732], 304–305). As Berkeley put it:

> A discourse . . . that directs how to act or excites to the doing or forbearance of an action may . . . be useful and significant, although the words whereof it is composed should not bring each a distinct idea into our minds. [*Berkeley* 1732], 292

This, as we will now see, became a popular view among mathematicians of the eighteenth and nineteenth centuries.

5.2. The Usefulness of Algebra

Despite their ultimate preference for classical synthetic proof, eighteenth-century mathematicians widely acknowledged the usefulness of algebraic methods. A typical expression of this is the following remark of MacLaurin (1698–1746) in reference to the methods employed by Wallis in [*Wallis 1656*].

> His demonstrations, and some of his expressions . . . have been excepted against. But it was not very difficult to demonstrate the greatest part of his propositions in a stricter method. . . . He chose to describe plainly a method which he had found very commodious for discovering new theorems; and . . . this valuable treatise contributed to produce the great improvements which soon followed after. . . . In general, it must be owned, that if the late discoveries were deduced at length, in the very same method in which the ancients demonstrated their theorems, the life of man could hardly be sufficient for considering them all: so that a general and concise method, equivalent to theirs in accuracy and evidence, that comprehends innumerable theorems in a few general views, may well be esteemed a valuable invention. [*MacLaurin 1742*], 114

Here, as in the remark by Leibniz quoted earlier, we see admiration for both the efficiency and the reliability of algebraic methods. This did not, however, imply an unqualified acceptance of them. MacLaurin, indeed, said that they generally lack the clarity and plainness that geometric methods ought ideally to possess. His overall assessment was thus that algebraic methods

---

Logistic was, of course, a Greek idea. Plato, for example, spoke of it in *The Republic* (cf. Bk. VII), the *Gorgias* (cf. 451b, c) and the *Theaetetus* (e.g. 145a, 198a). There is also a substantial description of it in a scholium to the *Charmides*, 165e (the relevant part of which is reprinted and translated in [*Thomas 1980*], 16–19). Heath notes that the pre-Socratic thinker Archytas used it in the same sense (cf. [*Heath 1921*], I, 14). See [*Heath 1921*], I, 14–16 for further sources.

> ... may be of use, when employed with caution, for abridging computations in the investigation of theorems, or even of proving them where a scrupulous exactness is not required; and we would not be understood to affirm, that the methods of indivisibles and infinitesimals, by which so many uncontested truths have been discovered, are without a foundation. But geometry is best established on clear and plain principles; and these speculations [the infinitistic speculations of Wallis] are ever obnoxious to some difficulties.... Geometricians cannot be too scrupulous in admitting of infinites, of which our ideas are so imperfect. [*MacLaurin 1742*], 113

This was a common view among mathematicians of the eighteenth and early nineteenth centuries. There were, however, exceptions to this common view and exceptions of three types. On the one hand, there were those who thought it did not go far enough in its approval of algebraic methods. Wallis was a prime example of this. As noted earlier, he regarded his methods as more than mere instruments. They were, in his view, methods of genuine proof. Indeed, they were superior methods of proof because they provided a truer and more probing genetic account of many of the theorems of classical geometry than did the traditional constructional proofs.

Others—for example, Maseres (cf. [*Maseres 1758*]) and William Frend (1757–1841) (cf. [*Frend 1796*])—regarded the prevailing view as too tolerant of algebraic methods. They believed this to be particularly true of its attitude toward the use of negative and imaginary quantities.

More common and influential than either of the above, though, were attempts to justify imaginary elements by comparing them to more commonly accepted elements of the mathematical universe. These attempts were not as ambitious as Wallis's. They did not, in particular, argue that use of negative, imaginary, and infinite numbers are necessary for genetically penetrating proofs. They did, however, try to put them on an equal footing with more commonly accepted numbers.

Gauss's celebrated defense of the imaginary and complex numbers is a case in point (cf. [*Gauss 1831*]). This defense centered on the provision of a geometrical interpretation for these numbers—an interpretation which Gauss believed provided a basis in classical geometrical intuition for their acceptance.

The basics of this interpretation are as follows. Each complex number c is first represented as an arithmetic sum, $a + bi$ (where a, b are real numbers and i is the imaginary number $\sqrt{-1}$). $a + bi$ is thus uniquely determined by the pair of reals (a, b), and can be interpreted as the point in the (Cartesian) *plane* whose real coordinate x is a and whose complex coordinate iy (measuring the number of units of i) is b. The value of $a + bi$ is then taken to be the Euclidean distance of $a + bi$ from the origin O of this plane, which, by the Pythagorean metric, is $\sqrt{a^2 + b^2}$ $(= |a + bi|)$.

Given this interpretation of distance (or absolute value), addition and multiplication can be defined as geometrical operations. Addition is planar translation (vector addition) and operates according to the so-called *parallelogram law.* The sum of $c_1(= a_1 + b_1 i)$ and $c_2(= a_2 + b_2 i)$ is the point whose x coordinate is $a_1 + a_2$ and whose iy coordinate is $b_2 + b_2$. Multiplication is a more complex operation, but it, too, can be represented geometrically as a composition of rotations (about O) and dilations.[47]

Gauss initially maintained that his interpretation justified the use of (expressions for) imaginary and complex numbers as means of proof:

> ... the arithmetic of the complex numbers is ... capable of the most intuitive sensible representation [*der anschaulichsten Versinnlichung*], and although the author in the present account has pursued a purely arithmetical treatment, he nevertheless also gave the necessary hints for this sensible representation—a sensible representation which makes the insight come to life, and is therefore to be recommended.... By this device the effect of the arithmetical operations on the complex quantities becomes capable of sensible representation that leaves nothing to be desired.... in this way the true metaphysics of the imaginary quantities is placed in a bright new light. [*Gauss 1831*], 174–175[48]

He did not remain confident in this view, however. A few years following the publication of the geometrical interpretation, he voiced doubts concerning the legitimacy of imaginary numbers as means of *arithmetic* proof. His misgivings seem to have been motivated by an ideal of genetic purity reminiscent of Aristotle, who believed that one ought not to cross genetic lines in demonstrations. Gauss seems to have held a similar scruple, though he applied it in the opposite direction. That is, instead of disallowing arithmetic reasoning in geometry, as Aristotle had done, he questioned the use of spatial (i.e., geometrical) intuition in arithmetic. He expressed these misgivings in a letter to Moritz Drobisch (August 14, 1834):

[47] As is well known, there were others who discovered such interpretations prior to the publication of Gauss's *Anzeige* in *1831*. These include C. Wessel, J. R. Argand, J. Warren, and C. V. Mourey. Accounts of these and related developments are widely available. Three brief but useful ones are [*Beman 1897*], [*Cajori 1912*], and [*Windred 1929*]. Somewhat more wide-ranging are the discussions in [*Lewis 1994*] and [*Novy 1973*]. Wessel, a surveyor, presented his ideas to the Royal Academy of Denmark in 1797. They were published in 1799 (English trans. in [*Smith 1929*]). Argand published his interpretation in 1806, and Warren and Mourey each published theirs in 1828. Attempts at planar represenation of the complexes seem to have had their start with Wallis ([*Wallis 1685*], ch. LXVI), who suggested an inadequate interpretation. Euler, too, presented indications of such a view (cf. [*Euler 1749*]). A letter of Gauss to Bessel in 1811 (cf. [*Gauss 1870–1927*], vol. VIII, p. 90) suggests that he (Gauss) may also have had the idea of planar representation in mind by that time.

[48] Gauss also expressed his acceptance of imaginary elements as means of demonstration in an entry of October 1797 in his *Tagebuch* (cf. [*Gauss 1903*], #80, p. 18). In his dissertation ([*Gauss 1799*]), however, he conspicuously avoided use of imaginary numbers.

> The representation (*Darstellung*) of the imaginary quantities as relations of points in the plane is not so much their essence (*Wesen*) itself, which must be grasped in a higher and more general way, as it is for us humans the purest or perhaps a uniquely and completely pure example of their application. [*Gauss 1870–1927*], vol. X, 106

Gauss expressed his misgivings more publicly in the 1849 *Jubiläumschrift*, where, commenting on proofs of the existence of roots of equations, he wrote:

> At bottom the real content (*eigentliche Inhalt*) of the entire argumentation belongs to a higher realm of the general, abstract theory of quantity (*abstracten Grössenlehre*), independent of the spatial. The object of this domain, which tracks the continuity of related combinations of quantities, is a domain about which little, to date, has been established and in which one cannot maneuver much without leaning on a language of spatial pictures (*räumlichen Bildern*). [*Gauss 1870–1927*], vol. III, 79[49]

This suggests that Gauss lost confidence in his geometrical interpretation as a means of revealing the essential nature of the complex numbers. His remark to Drobisch that his interpretation provided a "uniquely pure" example of the application of the complexes might nonetheless imply that he took his interpretation as establishing their consistency.

Whatever Gauss's own views of his interpretation may have been, however, it was widely viewed by others as justifying the use of complex expressions in proofs.

5.3. Peacock's Formalism

At roughly the same time, there was a revival of formalist ideas. This was perhaps most prominent in England, where the leading figure was George Peacock, a key figure in the Cambridge School of the 1830s and 1840s.[50] He believed that interpretation was necessary for the "final result" of reasoning ([*Peacock 1830*], xiv), but he did not believe that such interpretation need be applied to the "intermediate part[s] of the process of reasoning which lead[s] to" (ibid.) that result.

[49] For more on Gauss's changing views of imaginary numbers, see [*Schlesinger 1912*].

[50] "Cambridge algebraists" is the title Hankel used to refer to the school of algebraic thinkers to which Peacock belonged (cf. [*Hankel 1867*], 15). Others have used the term "English School" or "British School." The central figures of this movement were Peacock, D. F. Gregory, and A. De Morgan. Their work derived inspiration, however, from the earlier ideas and methods of Babbage, Woodhouse, and Lacroix (the latter through the publication of [*Lacroix 1802*]). Various of the ideas of this group were also emphasized in the work of Condillac and Condorcet. In Germany, Martin Ohm (1792–1872) advanced similar ideas.

The similarity to Berkeley's views stands out. According to Boole, something like this view had become common among early and mid-nineteenth-century thinkers.

> That language is an instrument of human reason, and not merely a medium for the expression of thought, is a truth generally admitted.... whether we regard signs as the representatives of things and of their relations, or as the representatives of the conceptions and operations of the human intellect, in studying the laws of signs, we are in effect studying the manifested laws of reasoning. [*Boole 1854*], 24[51]

This was the conception of mathematical reasoning adopted by Peacock and the Cambridge algebraists. We will call it the symbolic conception. According to it, mathematical reasoning need not involve the constructional manipulation of intuitions or the logical manipulation of propositions. It can instead consist in the syntactic manipulation of uninterpreted (even uninterpretable) signs.

Symbolic reasoning is thus very different from *contentual* reasoning—that is, reasoning with the semantic contents of expressions, be they intuitive or conceptual in character. That notwithstanding, it is subject to constraints that are based on concern for its use in producing contentually correct results. For Peacock, this generally meant that a system of algebraic laws must preserve the laws of contentual arithmetic:

> Algebra may be considered, in its most general form, as *the science which treats of the combinations of arbitrary signs and symbols by means of defined though arbitrary laws:* for we may *assume* any laws... so long as our assumptions are... not inconsistent with each other: in order, however, that such a science may not be one of useless and barren speculations, we choose some subordinate science as the guide merely, and not as the foundation of our assumptions, and frame them in such a manner that Algebra may become the most general form of that science... and as Arithmetic is the science of calculation, to the dominion of which all other sciences, in their application at least, are in a greater or less degree subject, it is the one which is usually, because most usefully, selected for this purpose. [*Peacock 1830*], §78[52]

There are a number of points in this passage that call for comment. The first concerns Peacock's use of the terms "Algebra" and "Arithmetic." The former is here used synonymously with "Symbolical Algebra," a term that Peacock employed widely throughout his writings. "Algebra" is thus to be contrasted with "Arithmetic" or "Arithmetical Algebra." Peacock took the latter to be composed of evident necessary *truths.* He regarded the former as composed of evidentially arbitrary but epistemically useful *assumptions* (cf. [*Peacock 1830*], §§ 71–77) or *conventions* (cf. [*Peacock 1830*], viii).

[51] This is a striking early statement of a view that was later advanced by Hilbert (cf. [*Hilbert 1928*], 475) and that formed a fundament of his proof-theory.

[52] See [*Peacock 1834*], 200–201 for a similar statement of the consistency requirement.

A second point concerns what Peacock meant by "defined though arbitrary" laws. Of the notion of *definition* involved here, he says that we must take the signs involved in such laws "not merely as the general representatives of numbers, but of every species of quantity" ([*Peacock 1830*], ix) and that we must "give a form to the definitions of the operations of Algebra, which must render them independent of any subordinate science" (ibid.). Whatever the "definitions" of the signs of a symbolical algebra are, then, Peacock regarded them as independent of the subject matter of any "subordinate" science which may suggest it. Specifically, they are not deducible from the definitions of the suggesting science. Later, we will consider another crucial difference separating the definitions of a symbolical algebra from the definitions of its suggesting science(s) as well. This is that, in a suggesting science, the definitions of terms are taken to determine the rules concerning their use; in a symbolical algebra, on the other hand, "the rules determine the meanings of the operations, or more properly speaking, they furnish the means of interpreting them" ([*Peacock 1834*], 200).

A third and final point concerns the constraints Peacock placed on the choice of laws for a symbolical algebra. My comment here will of necessity be longer and more involved than those just given. It descends to the very elements of Peacock's views of algebra.

Peacock explicitly required that the laws of algebra be consistent with each other. A little reflection reveals, however, an additional consistency constraint, one implicit in Peacock's requirement that symbolical algebra be *useful*, or, in his words, that it not be a body of "useless and barren speculations" ([*Peacock 1830*], §78).

Peacock saw the usefulness of symbolical algebra as depending upon its preservation of certain arithmetic laws—namely, those vital to its capacity as a "science of calculation, to the dominion of which all other sciences... are in a greater or lesser degree subject" ([*Peacock 1830*], §78).[53] In contemporary terminology, the usefulness of symbolical algebra depends upon its being a *conservative extension* of its "science of suggestion," the suggesting science of arithmetic.[54]

The idea that a symbolical extension of a "suggesting" science[55] should conserve its laws (or the laws that underwrite its computational utility, at any rate)

[53] Arithmetical algebra was not the only science of suggestion that a symbolical algebra might have. Indeed, Peacock expressly mentioned geometry, mechanics, and dynamics in this connection (cf. [*Peacock 1830*], xxi). Generally speaking, he broadened the class of possible suggesting sciences to include "every... branch of natural philosophy, which can be made to depend, by approximation, at least, upon fixed and invariable principles" ([*Peacock 1830*], xxi).

[54] Conservative extension of the suggesting science of arithmetic need not be taken to imply conservative extension of the whole of arithmetic. There might be theorems of the latter that are not important to the utility of arithmetic as a computing instrument and, so, not theorems of the former.

[55] Also termed a "subordinate" science by Peacock (cf. [*Peacock 1830*], §132).

was enshrined in a general methodological principle that Peacock formulated in 1830 and which, in one form or another, became influential in the later nineteenth and early twentieth centuries. Peacock called it the *Principle of the Permanence of Equivalent Forms* (*PPEF*, for short) and stated it as follows:[56]

> ...if we discover an equivalent form in Arithmetical Algebra or any other subordinate science, when the symbols are general in form though specific in their nature, the same must be an equivalent form, when the symbols are general in their nature as well as in their form.[57] [*Peacock 1830*], §132

Two clarificatory remarks are in order. The first concerns Peacock's distinction between generality of *form* and generality of *nature* or *value* (cf. [*Peacock 1842–1845*], vol. II, §631). The second concerns his notion of equivalence between algebraic expressions, specifically expressions containing variables.

In referring to the *form* of a symbol as "general," Peacock meant that it was a variable rather than a constant. In calling its *nature* or *value* general he seems to have meant that it could serve as a variable ranging not only over a given particular type of quantity, but over any quantity of any type. A symbol general as to both form and nature, then, would be a variable ranging over all quantities of all types.

As regards equivalence, Peacock regarded two variable-containing expressions, "$\mathcal{F}$" and "$\mathcal{G}$," as equivalent when "$\mathcal{F}=\mathcal{G}$." There are questions, however, as to what such an equation could or should mean.

When the variables contained in "$\mathcal{F}$" and "$\mathcal{G}$" vary over arithmetical entities (e.g., numbers, or some more particular kind of number), it may seem that there is a readily available answer: namely, that "$\mathcal{F}=\mathcal{G}$" is true when each instance of "$\mathcal{F}=\mathcal{G}$" is a truth of arithmetic.[58]

[56] He also called it the *Law of the Permanence of Algebraic Forms* (cf. [*Peacock 1830*], §133). Peacock was the first to call attention to the principle and to name it. He was not, however, the first to use it. Nor is it clear that he was the first to have recognized its main idea. Engel and Stäckel (cf. [*Stäckel and Bolyai 1913*], 35) point out an earlier claim by Wolfgang Bolyai that seems to have had the same general tendency.

[57] This is one of two clauses in the original statement of the principle in [*Peacock 1830*]—the clause which Peacock referred to as the "Converse Proposition" or "Converse Proportion." The other clause, the "Direct Proposition" or "Direct Proportion," was this: "Whatever form is Algebraically equivalent to another, when expressed in general symbols, must be true, whatever those symbols denote" ([*Peacock 1830*], §132). Peacock later dropped the Direct Proposition from the statement of the principle and identified it simply with the Converse Proposition ([*Peacock 1842–1845*], §§630–631): "Whatever algebraical forms are equivalent, when the symbols are general in form but specific in value, will be equivalent likewise when the symbols are general in value as well as in form" ([*Peacock 1842–1845*], §§630–631).

[58] Depending upon the relationship that arithmetic is taken to bear to logic, one might also include here the instances of "$\mathcal{F}$" and "$\mathcal{G}$" that are logical truths.

Such a view is not, however, without problems. There might, for example, be arithmetic equalities that do not become algebraic equalities (i.e., laws of symbolical algebra) when their variables, which are specific as to nature (i.e., are restricted to numbers), are replaced by variables that are general as to nature (i.e., range over quantities of all types). There is thus a need for a more precise identification of those arithmetical laws the conservative extension of which is supposed to be the basis of the fruitfulness of symbolical algebra.

This problem notwithstanding, the central thrust of the *PPEF* seems clear. It is intended to mark a distinction between two types or levels of algebra, the *arithmetical* and the *symbolical*, and to make clear that to be fruitful, the latter must substantially conserve the laws of the former. The *PPEF* thus requires that the laws of symbolical algebra be consistent not only with themselves but also with the laws of arithmetical algebra.

Peacock emphasized, however, that satisfaction of this constraint does not by itself provide a *justification* or *foundation* for symbolical algebra.

> [The] ... derivation of Algebra from Arithmetic, and the close connection which it has been attempted to preserve between those sciences ... has led to ... the opinion, that one is really founded upon the other: there is one sense, which we shall afterwards examine, in which this opinion is true: but in the strict and proper sense in which we speak of the principles of a demonstrative science, which constitute the foundation of its propositions, it would appear from what we have already stated, that such an opinion would cease to be maintainable ... [the relationship of] the principles and operations of Arithmetic to those of Algebra ... is not necessary but conventional.... Arithmetic can only be considered as a Science of Suggestion, to which the principles and operations of Algebra are adapted, but by which they are neither limited nor determined. [*Peacock 1830*], vii–viii (brackets added)

> The rules of symbolic combination which are thus assumed have been *suggested only* by the corresponding rules in arithmetical algebra. They cannot be said to be *founded* upon them, for they are not *deducible* from them; for though the operations of addition and subtraction, in their arithmetical sense, are applicable to all quantities of the same kind, yet they necessarily require a different meaning when applied to quantities which are different in their nature, whether that difference consists in the kind of quantity expressed by the unaffected[59] symbols, or in the different signs of affection of symbols denoting the same quantity; neither does it necessarily follow that in such cases there *exists* any interpretation which

[59] Here's the characterization of the term "affected" from [*Hutton 1795–1796*]:

> The term, affected, is ... used sometimes in algebra, when speaking of quantities that have coefficients. Thus in the quantity $2a$, a is said to be affected with the coefficient 2. It is also said, that an algebraic quantity is affected with the sign + or −, or with a radical sign; meaning no more than that it has the sign + or −, or that it includes a radical sign.

This seems to be Peacock's meaning as well.

> can be given of the operations, which is competent to satisfy the required symbolical conditions. [*Peacock 1834*], 197–198[60]

Symbolic algebra was thus, in Peacock's view, *suggested* by arithmetical algebra, not *founded* (in Peacock's strict deductive sense) by it. Its laws were *conventional* or *assumptive* in character—even, in a sense, *arbitrary*.

> In arithmetical algebra, the definitions of the operations determine the rules; in symbolical algebra, the rules determine the meanings of the operations, or more properly speaking, they furnish the means of interpreting them.... We call those rules... *assumptions*, in as much as they are not deducible as conclusions from any previous knowledge of those operations which have corresponding names: and we might call them *arbitrary* assumptions, in as much as they are *arbitrarily* imposed upon a science of symbols and their combinations, which might be adapted to any other assumed system of consistent rules. [*Peacock 1834*], 200–201

For Peacock, then, the laws of symbolical algebra were not *read off* a subject matter they were intended to describe. Rather, they preceded any such interpretation and, indeed, *determined* it. This idea—that the laws of symbolical algebra are essentially *assumptions* or *postulates* that precede any interpretation or definition of the quantities and operations with which they deal—is what I will call the *creativist* element of Peacock's viewpoint.

It is important to note, however, that Peacock took this view of the status of symbolical algebra only as regards its *strict* (deductive) foundation. He allowed that the laws of symbolical algebra are founded upon the laws of arithmetical algebra in a "looser" sense. Indeed, this is required by his basic distinction between a symbolic algebra and its so-called *suggesting* science. Symbolical algebra is not deducible from arithmetical algebra, but it is suggested by it.

> But though the science of arithmetic, or of arithmetical algebra, does not furnish an adequate foundation for the science of symbolical algebra, it necessarily *suggests* its principles, or rather its laws of combination; for in as much as symbolical algebra, though arbitrary in the authority of its principles, is not arbitrary in their application, being required to include arithmetical algebra as well as other sciences, it is evident that their rules must be identical with each other, as far as those sciences proceed together in common; the real distinction between them will arise from the *supposition or assumption that the symbols in symbolical algebra are perfectly general and unlimited both in value and representation, and that the operations to which they are subject are equally general likewise.* [*Peacock 1834*], 195 (emphasis Peacock's)

[60] Peacock went on to say, however, that "... the investigation of... interpretations, when they are discoverable, becomes one of the most important and most essential of the deductive processes which are required in algebra and its applications" ([*Peacock 1834*], 198).

Peacock thus attached importance to the matter of whether a system of laws was justified by appeal to a prior scheme of definitions or meanings. He articulated his ideas in the form of a distinction between *definition* and *interpretation.*

Arithmetical algebra, he maintained, begins with definitions—definitions intended to capture the established meanings of the terms involved. These definitions *determine* the rules or laws of arithmetical algebra (cf. [*Peacock 1834*], 200).

> To *define*, is to assign beforehand the meaning or conditions of a term or operation.... It is for this reason, that we *define* operations in arithmetical algebra conformably to their popular meaning.... [*Peacock 1834*], 197, note *

Symbolical algebra, on the other hand, begins with conditions or laws. These conditions determine the meanings or interpretations assignable to them:

> ... to *interpret*, is to determine the meaning of a term or operation conformably to definitions or to conditions previously given or assigned.... we *interpret* [terms or operations] in symbolical algebra conformably to the symbolical conditions to which they are subject. [*Peacock 1834*], 197, note *

In sum:

> In arithmetical algebra, the definitions of the operations determine the rules; in symbolical algebra, the rules determine the meaning of the operations ... we might call them *arbitrary* assumptions, in as much as they are *arbitrarily* imposed upon a science of symbols and their combinations, which might be adapted to any other assumed system of consistent rules. [*Peacock 1834*], 200–201

We see in Peacock's conception of symbolical algebra the seeds of the more mature variants of formalism that were to follow. They adopted, albeit with modification, his essentially creativist conception of symbolic instruments. They also retained his basic conception of the constraints that should govern the use of such instruments—specifically, that they should be consistent and should preserve the laws of basic arithmetic. This notwithstanding, there were also differences. It is to these differences that we now turn.

5.4. The Axiom of Solvability: Basic Character

By the latter half of the nineteenth century, there was widespread acceptance of Peacock's Principle of Permanence.[61] Different variants of the principle emerged

[61] It enjoyed widespread acceptance in England (cf. the writings of Boole, Gregory, De Morgan, et al.), on the Continent (cf. the writings of Hankel, Peano, Durège, Hilbert et al.) and in America (cf. the basic algebra text [*Fine 1890*] by H. B. Fine). See, in particular, Hankel's endorsement ([*Hankel 1867*], 11); Peano's discussion ([*Peano 1910*], 224); Durège's statement of the principle as that to which mathematics owes its consistency

over time, but they shared a common theme; namely, that Algebra should preserve (to the greatest extent possible) the arithmetic laws of the simplest quantities, the positive whole numbers. We'll refer to this general principle as the *Principle of Permanence.*

Preservation of the laws of the whole numbers was not, however, the only factor influencing the development of algebra. Were it otherwise, there would not have been a reason to extend the basic domain (i.e., the domain of the positive whole numbers) in the first place. To explain why arithmetic was extended to successively broader domains requires something other than the Principle of Permanence.

This is where the *Axiom of Solvability* seems to have entered the scene. It was, according to Hilbert, a central precept of correct mathematical method:

> However unapproachable . . . problems may seem to us and however helpless we stand before them, we have, nevertheless, the firm conviction that their solution must follow by a finite number of purely logical processes.[62] Is this *axiom of the solvability* (emphasis added) of every problem a peculiarity characteristic of mathematical thought alone, or is it possibly a general law inherent in the nature of the mind, that all questions which it asks must be answerable?[63] . . . The conviction of the solvability of every mathematical problem is a powerful incentive to the worker. We hear within us the perpetual call: There is the problem. Seek its solution. You can find it by pure reason, for in mathematics there is no *ignorabimus.* [*Hilbert 1901*], 444–445

The drive to solve problems was thus, in Hilbert's view, *the* (or at least a) central drive of the mathematician. It impelled him to create. Nowhere is this fundamental urge more in evidence than in the historical development of the number-concept, the following "summary of which expresses the basic ideas well:

> One . . . begins by premissing certain propositions whose validity is in no way doubtful.

---

and much of whatever simplicity it has ([*Durège 1882*], 8, 10); and H. B. Fine's defense ([*Fine 1890*], §§12, 15, 31, 36). Its influence continued into the twentieth century (cf. the applications by E. B. Wilson [*Wilson 1912*], 2); and J. W. Young ([*Young 1911*], 102, 111–113); E. W. Hobson's statement and employment ([*Hobson 1926*], 20–21); and Hilbert's reference ([*Hilbert 1930a*], 231); and, indeed, continues to the present day (cf. the endorsement in [*Courant and Robbins 1947*], ch. II, §5.1).

The principle has also had its detractors, however. See, for example, [*Russell 1903*], 376–377.

[62] To avoid misunderstanding, it is important to note that Hilbert allowed as a "solution" to a problem, a proof of the impossibility of a direct answer to it under a specified set of conditions or "hypotheses." Cf. [*Hilbert 1901*], 444. More on this below.

[63] This raises the question of whether, contrary to Kant's critical epistemology, the Axiom of Solvability *is* a *general* principle of human reason.

> These are:
> There does not exist a number (from the sequence 0, 1, . . .), which when added to 1 gives 0.
> There does not exist a number (integral), which multiplied by 2 gives 1.
> There does not exist a number (rational), whose square is 2.
> There does not exist a number (real), whose square is -1.
> Then one says: in order to overcome such an inconvenience, we extend the concept of number, that is, we introduce, manufacture, create (as Dedekind says) a new entity, a new number, a sign, a sign-complex, etc., which we denote by -1, or 1/2, or $\sqrt{2}$, or $\sqrt{-1}$, which satisfies the condition imposed. [*Peano 1910*], 224[64]

Each of Peano's "premises" amounts to an observation concerning the nonexistence of a number to serve as a solution to a given "problem." The extension of the number-concept consists in the successive addition of these "solutions."

This extension reaches a natural end with the addition of the complexes. Unlike the naturals, integers, rationals, and reals, the complexes form an *algebraically closed* set—that is, a set that contains at least one solution (i.e., root) for every polynomial whose coefficients are elements of that set. This is essentially what the *Fundamental Theorem of Algebra* (in one of its forms) tells us: every polynomial with complex coefficients *has* a complex solution. There is therefore no "unresolved problem"—no rootless polynomial—to serve as a basis for extending the number concept beyond the complexes.

Together, the Axiom of Solvability and the Principle of Permanence guided the progressive extension of the number-concept. The Axiom of Solvability expressed the mathematician's goal to solve problems. The Principle of Permanence acted as a constraint upon the application of this axiom. It required that newly introduced numbers preserve the basic laws of arithmetic. More precisely, it required that the laws governing new numbers be *consistent with* the laws governing the old ones.

A tidy picture. Tidier, indeed, than the truth. We'll consider some of the complications shortly, after a brief survey of the historical background against which this conception of method arose.

5.5. The Axiom of Solvability: Historical Background

In Hilbert's own thinking, the Axiom of Solvability was a challenge to the epistemic pessimism voiced by certain leading scientists of his day. I refer, of course, to such

[64] Peano's ordering is not the historical one. The following description by Gauss is historically more accurate.

> Starting originally from the notion of absolute integers, it has gradually enlarged its domain. To integers have been added fractions, to rational quantities the irrational, to positive the negative, and to the real the imaginary. [*Gauss 1831*], 175

figures as Emil and Paul du Bois-Reymond, Rudolf Virchow ([*Virchow 1878*]), and even Brouwer. All of these, in one way or another, argued for limitations on human knowledge. In the case of Virchow, especially, these "limitations" became "restrictions." In particular, they became proposals for reforming the curricula in German gymnasia and universities. Roughly, the proposal was that only *fact*, not *theory*, was to be taught. Darwin's *theories*, in particular, were not to be taught.

Thus arose the well-known *Ignorabimusstreit* of late nineteenth-century Germany. The more pessimistic (some would say the more modest) parties to this controversy advocated a doctrine of ignorance. Specifically, they advocated the *Ignorabimus*[65] doctrine—the view that there are permanent, irremediable limitations on our knowledge, including our scientific and mathematical knowledge. The leading exponent of this doctrine was the eminent German physiologist Emil du Bois-Reymond (1818–1896). In an influential set of essays and scientific addresses in the 1870s, he argued, in particular, that a complete formulation of the truths of mechanics is impossible. His mathematician brother, Paul (1831–1889), extended the idea to mathematics, arguing that there are inherent limitations on our knowledge of the continuum.[66]

Such large-scale, principled pessimism was, of course, a theme familiar from Kant's critical epistemology. Kant began the *First Critique* with the observation that there are questions we cannot avoid asking, but also can not, without antinomy, answer:

> Human reason has this peculiar fate that in one species of its knowledge it is burdened by questions which, as prescribed by the very nature of reason itself, it is not able to ignore, but which, as transcending all its powers, it is also not able to answer. [*Kant 1781*], A vii

Hilbert's *Axiom of Solvability*—the claim that every exactly formulated problem can be solved—announced his opposition to this pessimism and became a central theme of his early foundational work. It reflected the general optimism of the algebraic movement in mathematics. Viète (1540–1603) expressed this optimism well in one of the founding statements of the algebraic movement:

> ... the analytic art ... claims for itself the greatest problem of all, which is:
> TO LEAVE NO PROBLEM UNSOLVED [*Viète 1983*], 32[67]

[65] Translated literally, *Ignorabimus* means "we shall never know."

[66] For a useful discussion of the *Ignorabimusstreit* in mathematics, see [*McCarty 2004*].

[67] Viète traced the root idea of his Art to Plato, writing:

> There is a certain way of searching for the truth in mathematics that Plato is said first to have discovered. Theon called it analysis, which he defined as assuming that which is sought as if it were admitted and working through the consequences of that assumption to what is admittedly true.... The ancients propounded only two types of analysis ... I have added a third.... [T]he whole analytic art ... may be called the science of correct discovery in mathematics. [*Viète 1983*], 12–13

The optimism expressed in Hilbert's Axiom of Solvability thus reflected the general optimism of the algebraic method. As we will now see, there was not only optimism in Hilbert's Program but also determination—a determination which also ran parallel to the thinking of the early algebraists.

To see this, recall the antipathy expressed by Descartes toward the ancient geometers. In their writings, he said, they presented themselves as proceeding by *synthesis* alone. They did so, however, "not because they were wholly ignorant of the analytic method" ([*Descartes 1970b*], 49), but "because they set so high a value on it that they wished it to themselves as an important secret" (ibid.).[68] The Greek geometers thus used algebra to discover their theorems, but they attempted to conceal this fact from posterity in order to make their work appear more ingenious than it truly was. Such, at any rate, was Descartes's view.

Wallis expressed similar sentiments [*Wallis 1685*]:

> It [the present treatise] contains an Account of the Original, Progress, and Advancement of (what we now calle) *Algebra*, from time to time shewing its true Antiquity (as far as I have been able to trace it;) and by what Steps it hath attained to the Height at which now it is.
> That it was in the use of old among the *Grecians*, we need not doubt; but studiously concealed (by them) as a great Secret.
> Examples we have of it in *Euclid*, at least in *Theo*, upon him; who ascribes the invention of it (amongst them) to *Plato*.
> Other Examples we have of it in *Pappus*, and the effects of it in *Archimedes*, *Apollonius*, and others, though obscurely covered and disguised. [*Wallis 1685*], preface, a3[69]

> It is to me a thing unquestionable, That the Ancients had somewhat of like nature with our *Algebra*; from whence many of their prolix and intricate Demonstrations were derived. And I find other modern Writers of the same opinion with me therein. . . .
> But this their *Art of Invention*, they seem very studiously to have concealed: contenting themselves to demonstrate by *Apogogical* Demonstrations, (or reducing to Absurdity, if denied,) without shewing us the method, by which they first found out those Propositions, which they thus demonstrate by other ways.
> Of which, Nuñes or Nonius in his *Algebra* (in Spanish) *fol. 114. b.* speaks thus: *O how well had it been if those Authors, who have written in Mathematics, had delivered to us their Inventions, in the same way, and with the same Discourse, as they were found! And not as* Aristotle *says of Artificers in Mechanics, who shew us the Engines they have made, but conceal the Artifice, to make them the more*

[68] See also rule IV of the *Regulae*, where Pappus and Diophantus are mentioned by name.

[69] Elsewhere (cf. [*Wallis 1685*], 290), Wallis sarcastically remarked that Apollonius would indeed have deserved his title "Magnus Geometra" had he indeed "discover[ed] all those Propostions, and perplex demonstrations, in the same order they are delivered, without some such *Art of Invention*, as what we now call *Algebra*."

> *admired! The method of Invention, in divers Arts, is very different from that of Tradition, wherein they are delivered. Nor are we to think, that all these Propositions in* Euclid *and* Archimedes *were in the same way found out, as they are now delivered to us.* [*Wallis 1685*], 3

Hilbert held a similar, though not identical, view with regard to the reactionary proposals of Kronecker and Brouwer. He did not accuse them of guile (i.e., trying to hide a method of discovery) or vanity (i.e., claiming a degree of respect they knew they didn't deserve). He did, however, see them as advocating unnecessarily restrictive methods of proof—methods which made the discovery of proof more difficult. As he put it:

> What Weyl and Brouwer do amounts in essence to following the path formerly trod by Kronecker. They seek to found mathematics by throwing everything overboard that makes them uncomfortable and erecting a dictatorial prohibitionism (Verbotsdiktatur). This means, however, to mutilate (zerstückeln) and mangle (verstümmeln) our science, and if we follow such reformers (Reformatoren), we run the risk of losing a great part of our most valuable treasures. [*Hilbert 1922*], 159

The Axiom of Solvability expressed his determination to overcome such arbitrariness and to claim the full measure of his freedom as a mathematician. The freedom of the mathematician permits him to solve by creation. This freedom is limited only by the need to be consistent and fruitful. For Peacock and other formalists of the early nineteenth century, fruitfulness was a matter of preserving computational utility, and preserving computational utility was essentially a matter of preserving the basic algebraic laws of number. In other words, it was adherence to the Principle of Permanence. Fruitfulness, in other words, required consistency with the basic laws of arithmetic. Later formalists, as we will soon see, broadened the conception of fruitfulness.

5.6. Complications

I alluded above to possible difficulties in the view that formalism (and, in particular, Hilbert's formalism) consists essentially in joint application of the Axiom of Solvability and the Principle of Permanence—that the principal freedom claimed by the formalist (in arithmetic) is the freedom to solve arithmetic problems in any way compatible with the basic laws of the whole numbers. I will now consider some of these difficulties.

One of them centers on the question of what is to count as a *solution* to a problem. Hilbert began his answer to this question by saying:

> ... every definite mathematical problem must necessarily be susceptible of an exact settlement, either in the form of an actual answer to the question asked, or

> by the proof of the impossibility of its solution and therewith the necessary failure of all attempts. [*Hilbert 1901*], 444

He thus allowed that a proof that no "actual answer to the question asked" ([*Hilbert 1901*], 444) is possible be a type of "solution" to that question. However, if a proof that a question is unanswerable is allowed to count as an answer to that question, one can only wonder how much difference there is between *Ignorabimus* and *kein Ignorabimus* (cf. [*Hilbert 1901*], 445). The du Bois-Reymonds offered *arguments* for their limitative theses. Is the difference between them and Hilbert that Hilbert called for *proof* where the du Bois-Reymonds offered only *argument*? Perhaps. But, if so, there is room to wonder whether this isn't making too much of the possible differences between proof and (mere) argument.

There are also questions concerning how to rate the severity of departures from the standard laws of arithmetic. The Principle of Permanence says that the laws of basic arithmetic are to be retained *to the fullest extent possible.* But which departures from standard arithmetic law are to be rated as *possible* (i.e., admissible) and which as *impossible* (i.e., inadmissible)?

It is well known that extension of the number-concept to quantities of dimension greater than 2 (the complexes being viewed as quantities of dimension 2 in their representation as pairs of reals) require substantial departures from the standard laws of arithmetic. Hamilton's *quaternions*, numbers or quantities of dimension 4, and Graves's and Cayley's *octonions*, numbers or quantities of dimension 8, both involve violations of basic arithmetic laws. Specifically, quaternions violate the law of commutativity of multiplication, and octonions violate, in addition, the law of associativity of multiplication.

But even the complexes involve violation of conditions that have some right to be regarded as standard arithmetic laws. In particular, they violate the law that any number multiplied by itself is either 0 or positive. More accurately, they lack the conjunction of the following two properties:

(I). For every x, either $x<0$ or $x=0$ or $x>0$,

and

(II). For every x, y, if $x, y>0$, then $x+y>0$ and $xy>0$.

All the "preceding" number systems satisfy this conjunction.[70] What, then, should we say? Do the complexes violate the Principle of Permanence or do they not? One is tempted, perhaps, to say that they violate the Principle of Permanence

[70] (I) doesn't strictly hold for the postive integers because "$x<0$" isn't defined. There is, however, an evident disjunctive property that one could put in place of (I) to accommodate this fact.

but only minimally, or not in a way that matters. But the difficulty then arises of saying which exceptions to the standard laws of arithmetic matter, and why.[71]

That departures from the standard laws of arithmetic arise with the introduction of the complexes is significant. It is significant because it suggests a possible conflict between the Principle of Permanence and the Axiom of Solvability. Because the complexes are algebraically closed, the Axiom of Solvability does not enjoin extension of the number-concept beyond them. It does, however, urge extension *to* the complexes because, with such extension, certain previously unsolvable polynomials become solvable. The Principle of Permanence, on the other hand, is resistant to extension to the complexes because it leads to failure of a standard arithmetic law, namely, the conjunction of (I) and (II) above.

Certain facts concerning the extension of the number-concept to the complexes thus suggest a possible conflict between the Axiom of Solvability and the Principle of Permanence. The situation, however, is even more complicated. For if we do not extend the number-concept to include complexes, then the Fundamental Law of Algebra does not hold. Does this imply that failure to extend the number-concept to the complexes would violate the Principle of Permanence?

Seemingly not. The Fundamental Law of Algebra does not appear to be a law of the same sort or in the same sense as other basic laws of arithmetic (including, I would say, the conjunction of (I) and (II) above). Indeed, it isn't a "law" for the systems "preceding" the complexes. In particular, it isn't a law for the reals. It does not, therefore, appear to be the type of law the preservation of which the Principle of Permanence was intended to cover.

It does, however, indicate a third type of methodological principle in mathematics—one distinct from the Axiom of Solvability and the Principle of Permanence. This is a preference for simplicity. This preference directs us to extend the number-concept not only in such ways as maximize the solution of problems or maximize preservation of the basic laws of arithmetic, but in ways that maximize certain types of simplicity—in this case, the simplicity borne of algebraic closure.

Let me close this subsection by mentioning one final puzzle—this time, a puzzle concerning the Principle of Permanence. I mentioned above that the Axiom of Solvability does not compel extension of the number-concept beyond the level of the algebraically closed system of the complexes. For various thinkers (e.g., Euler and Hamilton, to mention but two), however, the imaginary number $i = \sqrt{-1}$, and the complex numbers constructed from it, did not constitute genuine solutions to problems. The reason was that they made use of the supposed number or quantity $\sqrt{-1}$, and this, for them, was an absurdity. It followed that any number making use of it was likewise an absurdity and, thus, not a

[71] To say that the complexes violate the standard laws *as little as possible* raises other questions—for example, the question of what this might or should mean.

genuine solution to a problem. Viewed correctly, then, the Axiom of Solvability did not sanction the introduction of the complex numbers.[72]

In 1837, Hamilton (cf. [*Hamilton 1837*], sec. 3)[73] produced an analysis of the complex numbers that he believed rescued the complexes from the charge of absurdity. On this analysis, complex numbers were *two-dimensional* quantities, each of whose dimensions was real-valued. As such, he maintained, they were fully intelligible:

> In the THEORY OF SINGLE NUMBERS, the symbol $\sqrt{-1}$ is *absurd*, and denotes an IMPOSSIBLE EXTRACTION, or a merely IMAGINARY NUMBER; but in the THEORY OF COUPLES, the same symbol $\sqrt{-1}$ is *significant*, and denotes a POSSIBLE EXTRACTION, or a REAL COUPLE.... In the latter theory, therefore, though not in the former, this sign $\sqrt{-1}$ may properly be employed....
> [*Hamilton* 1837], 417–418

In Hamilton's view, then, the meaningfulness of $\sqrt{-1}$ and signs for complexes generally depended upon how they were conceived. More particularly, it depended upon what we might call the *dimensionality* of their conception. Conceived as one-dimensional quantities, they're meaningless and "absurd." Conceived as two-dimensional quantities, they're fully intelligible. The same holds for the more extensive class of imaginary units used in defining higher-dimension hypercomplexes (e.g., quaternions and octonions). Viewed as one-dimensional quantities, they're absurd; viewed as multidimensional quantities, they're not. Such, at any rate, was Hamilton's view.

We argued above that the failure of the complexes to preserve (I) and (II) can be seen as a violation of the Principle of Permanence. Hamilton suggests that this is valid at most for the one-dimensional conception of the complexes. This raises the question of what should be counted as the *permanent* "laws" of multidimensional quantities—the laws to be preserved under extension from one dimensionality to another. Alternatively, it raises a question as to whether the Principle of Permanence should be taken as a valid methodological principle governing the extension of multidimensional concepts of number or quantity.

The basic point can be illustrated by considering the law of multiplicative commutativity. Hamilton's treatment of two-dimensional hypercomplexes (i.e., the complexes) preserved it as a law. His treatment of higher-dimensional hypercomplexes (viz., the quaternions) did not. One would like to know what he took

[72] Similar arguments were made with respect to the negative numbers. See, for example, [*Frend 1796*] and [*Maseres 1758*].

[73] Hamilton had reported the chief ideas of his treatment three years earlier (cf. [*Hamilton 1834*], published in 1835. H. G. Grassmann formulated a more extensive and more general multidimensional approach to quantities ([*Grassmann 1844*]). It wasn't until Hankel's discussion of Grassmann's work ([*Hankel 1867*]), however, that Grassmann's study gained significant recognition.

the significance of this to be. Hamilton gave us little to go on. He argued for a reconception of number based on an intuition of time. Given the inherent directedness of time, it would not be surprising for this to lead to an element(s) of *order-sensitivity* in arithmetic and, hence, a limited role for such order-insensitive laws as commutativity. One would not, however, expect preservation of such order-*insensitive* laws as multiplicative commutativity more in the case of two-dimensional quantities than in the case of higher-dimensional quantities.

The multidimensional approach to quantity thus raises questions concerning the application (even the applicability) of the Principle of Permanence. There is a clear link between the move to the multidimensional analysis of quantity and the violation of the basic laws of algebra for one-dimensional quantities. It was Hankel who, in 1867, first succeeded in proving a link of this type. Hankel presented his theorem as an answer to a question raised by Gauss—namely: "...why the relations between things which represent a multiplicity of more than two dimensions cannot provide other sorts of permissible quantities in general arithmetic" (cf. [*Gauss 1870–1927*], vol. II, p. 178). He stated it as follows:

> A higher complex number system, whose formal operations of calculating are fixed by the conditions given in §28, and whose products of unity are particular linear functions of the original unities, and in which no product can vanish without one of its factors being zero, contains within itself a contradiction and cannot exist. [*Hankel 1867*], 107[74]

In more modern terms, *Hankel's Theorem*, as we'll call it,[75] says that the field $\mathbb{C}$ is, up to isomorphism, the only commutative field obtainable by adding roots of polynomials with complex coefficients to the field $\mathbb{C}$. In still other words, any attempt to develop the (or a) number-concept in a way that includes $\mathbb{C}$ but goes beyond it will inevitably lead to (further) violations of the Principle of Permanence. Understood in this way, Hankel's Theorem is a kind of *completeness* result for $\mathbb{C}$. It says that $\mathbb{C}$ is the most complete (or most nearly complete) of all the multidimensional elaborations of the number concept in the sense that it preserves the greatest portion of the standard laws of number. We noted earlier that since the complex numbers form an algebraically closed system, the Axiom of Solvability would seem not to compel extension of the number concept beyond the complexes. Hankel's Theorem, it seems, promises problems for the Principle of Permanence if the number concept *is* so extended. The complex numbers thus seem to form *maxima* for both the Axiom of Solvability and the Principle of

[74] In addition to the conditions named by Hankel in the above statement, the conditions in his §28 are essentially that addition and multiplication are commutative and associative (cf. [*Hankel 1867*], 99) and that multiplication distributes over addition (102).

[75] Remmert (cf. [*Ebbinghaus et al. 1991*], 119) observes that the theorem was used by Weierstrass in lectures dating from 1863. Hankel is therefore only the first to have published a proof of it, not the first to have thought of it.

Permanence. The maxima are of different types, however. The complexes are the maximum required by the Axiom of Solvability. They are the maximum permitted by the Principle of Permanence.

5.7. Hilbert's Formalism

One might think that Hankel's Theorem would have been the end of the Principle of Permanence. Such, however, was not the case. It continued to have many proponents, not the least of whom was Hilbert.[76]

What seems to have changed is the emphasis placed on the qualifying phrase "to the fullest extent possible." This is illustrated by Peano's criticism of Hermann Schubert's formulation of the principle. Quoting the (corrected) French edition of Schubert's "Grundlagen der Arithmetik" (from the *Encyclopädie der mathematischen Wissenschaften*, vol. I), he wrote: "...one must be 'guided by a concern for keeping the formal laws *as much as possible*.'" [*Peano 1910*], 225 (emphasis added).

Peano argued that this was the strongest plausible form of the principle. His reasoning, presented in the form of a criticism of Schubert's misstatement of it, was that a stronger form requiring complete preservation of laws would lead to nonsense (e.g., the identity of the integers with the rational numbers):

> This principle of permanence reached its apogee with Schubert, who, in the *Encyclopädie der mathematischen Wissenschaften*, affirmed that one must "prove that for numbers in the broad sense, the same theorems hold as for numbers in the narrow sense." Now, if all the propositions which are valid for the entities of one category are valid also for those of a second, then the two categories are identical. Hence—if this could be proved—the fractional numbers are integers! [*Peano 1910*], 225[77]

These days, of course, we would insist on a more careful drawing of certain distinctions (e.g., between propositions that are *true of* a set of items versus the laws or theorems that can be *proved of* them). There is also a disconnect between Peano's criticism of Schubert and his insistence on the qualified version of the Principle of Permanence. The right idea here, it seems, is not to point to the importance of the qualificatory phrase "as much as possible," but to highlight the

[76] Other late nineteenth- and early to mid twentieth-century proponents included H. Durège ([*Durège 1882*], 8ff.), J. W. Young ([*Young 1911*], 102, 111–113), H. B. Fine ([*Fine 1890*], §§12, 19), E. B. Wilson ([*Wilson 1912*], 2, 478), E. W. Hobson ([*Hobson 1926*], 20–21), and R. Courant and H. Robbins ([*Courant and Robbins 1947*], 88–89).

[77] Schubert's statement in the *Encyclopädie* is on p. 11 of IA1. It reads: "...zu beweisen, dass für die Zahlen im erweiterten Sinne dieselben Sätze gelten, wie für die Zahlen im noch nicht erweiterten Sinne...."

importance of distinguishing between an extension's preserving the laws of its predecessors and its having the same laws. The Principle of Permanence does not prohibit an extension from adding to the laws of its predecessors. Nor does it make minimization of such additions an ideal. The minimization it seeks is minimization of *deletions from* (or contradictions with) the laws of the predecessor theories.

This said, Peano's basic point—that the Principle of Permanence can be seen and appreciated as a methodological principle urging *maximal possible* preservation of the basic or original laws of number—is sensible enough. It is also significant because it expresses a preference for freedom (i.e., for the Axiom of Solvability, or freedom to create a solution to a problem) over preservation of the familiar. A little more accurately, it implies, modulo consistency, a preference for the Axiom of Solvability over the Principle of Permanence.

Hilbert's position is more complicated. He wrote approvingly of the Principle of Permanence as a possible way of working toward completeness in certain theories—particularly what he referred to as theories of "higher domains," (cf. [*Hilbert 1930a*], 231). He did not say what he meant by "higher domains" but it is reasonable to take him as including both theories of higher-dimension quantities and such abstract basic theories as set theory.[78] He maintained that the Principle of Permanence could be used to provide a motivation for new axioms.

This, however, was the least important of Hilbert's applications of something like a Principle of Permanence. Of much greater significance were the broad applications he directed toward the laws of set theory and classical logic.

Concerning the former, he wrote:

> We shall carefully investigate those ways of forming notions and those modes of inference that are fruitful; we shall nurse them, support them, and make them usable, wherever there is the slightest promise of success. No one shall be able to drive us from the paradise that Cantor has created for us. [*Hilbert 1926*], 375–376[79]

His attempt to preserve the classical laws of logic was even more self-consciously undertaken on analogy with other, more familiar applications of the Principle of Permanence.

[78] Hilbert contrasted the so-called theories of "higher domain" with arithmetic and analysis, writing:

> It is generally maintained that the axiom system for number theory as well as for analysis are complete. . . . The case of the consistency of S as well as of $\neg S$ would be conceivable in higher domains. In that case the assumption of one of the two statements S and $\neg S$ as axiomatic is to be justified by methodological advantages (principle of permanence of laws, further possibilities of construction, etc.). [*Hilbert 1930a*], 231

[79] Here we see the connection of the ideal of permanence with that of fruitfulness, a connection Peacock also stressed.

At all events we observe the following. In the domain of finitary propositions, in which we should, after all, remain, the logical relations that prevail are very imperspicuous, and this lack of perspicuity mounts unbearably if "all" and "there exists" occur combined or appear in nested propositions. In any case, those logical laws that man has always used since he began to think, the very ones that Aristotle taught, do not hold. Now one could attempt to determine the logical laws that are valid for the domain of finitary propositions; but this would not help us, since we just do not want to renounce the use of the simple laws of Aristotelian logic, and no one, though he speak with the tongues of angels, will keep people from negating arbitrary assertions, forming partial judgments, or using the principle of excluded middle. What then shall we do?

Let us remember that *we are mathematicians* and as such have already often found ourselves in a similar predicament, and let us recall how the method of ideal elements, that creation of genius, then allowed us to find an escape. I presented some shining examples of the use of this method at the beginning of my lecture. Just as $i = \sqrt{-1}$ was introduced so that the laws of algebra, those, for example, concerning the existence and number of the roots of an equation, could be preserved in their simplest form, just as ideal factors were introduced so that the simple laws of divisibility could be maintained even for algebraic integers (for example, we introduce an ideal common divisor for the numbers 2 and $1+\sqrt{-5}$, while an actual one does not exist), so we must here *adjoin the ideal propositions to the finitary ones* in order to maintain the formally simple rules of ordinary Aristotelian logic. [*Hilbert 1926*], 379 (emphasis Hilbert's)[80]

Hilbert thus saw the relationship between the Axiom of Solvability and the Principle of Permanence differently than Peano. He seems to have regarded the Axiom of Solvability as a means of securing adherence to the Principle of Permanence (or some principle like it). We introduce $i = \sqrt{-1}$ because it allows us to preserve a law—the law (the Fundamental Theorem of Algebra) concerning the existence and number of roots of an equation—in an attractively simple form. Even more importantly, we create the ideal propositions in order to preserve the

[80] Hilbert makes similar statements elsewhere. One in which he again asserts the parallel between the use of the so-called *ideal propositions* and the use of more familiar ideal elements in mathematics is the following:

> ... if to the real propositions we adjoin the ideal ones, we obtain a system of propositions in which all the simple rules of Aristotelian logic hold and all the usual methods of mathematical inference are valid. Just as, for example, the negative numbers are indispensable in elementary number theory and just as modern number theory and algebra become possible through the Kummer–Dedekind ideals, so scientific mathematics becomes possible only through the introduction of ideal propositions. [*Hilbert 1928*], 471

simple, psychologically efficient laws of classical logic as the laws governing logical inference in mathematics.[81, 82]

Psychological convenience—or avoidance of the type of inconvience represented by the "unbearable" awkwardness of finitary logic—thus became for Hilbert a prime objective of proper mathematical method. When faced with awkwardness or inconvenience in the realm of *real* mathematics, the mathematician is urged to remember that she is free to create a more convenient environment in which to conduct her reasoning.

Hilbert codified this freedom in the form of a methodological principle he termed the *Creative Principle* (das *schöpferische Prinzip*).

> Having arrived at a certain point in the development of a theory, I may designate (*bezeichnen*) a further proposition as correct (*richtig*) as soon as it is recognized (*erkannt*) that its introduction results in no contradiction with propositions previously admitted as correct.... [This is] the creative principle which, in its freest use, justifies us in introducing ever newer concept-formations (*Begriffsbildungen*), the only restriction being that we avoid contradiction. [*Hilbert 1905*], 135–136

This was an affirmation of Hilbert's basic belief in the mathematician's *freedom*—"the freedom of concept-formation and of the methods of inference ought not to be limited beyond what is necessary" ([*Hilbert 1920*]). Restrictions on concept-formation and inference "must be formulated in such a way that the contradictions are eliminated but everything valuable remains" (ibid.). Hilbert asserted this laissez-faire approach to mathematics as a method greatly to be

[81] Hilbert's conception of the *real* vs. *ideal* distinction was essentially this:

> ... elementary mathematics contains, first, formulas to which correspond contentual communications of finitary propositions ... and which we may call the *real propositions* of the theory, and second, formulas that ... in themselves mean nothing but are merely things that are governed by our rules and must be regarded as the *ideal objects* of the theory.... if to the real propositions we adjoin the ideal ones, we obtain a system of propositions in which all the simple rules of Aristotelian logic hold and all the usual methods of mathematical inference are valid.... scientific mathematics becomes possible only through the introduction of ideal propositions. [*Hilbert 1928*], 469–71

[82] One might also think that, in some sense, Hilbert intended to preserve the laws of finitary reasoning. He did not, however, advocate their preservation on grounds of their computational utility or general utility for efficient mathematical reasoning. Rather, he required their preservation because of their evidentness. Their preservation was thus not to be secured by application of a methodological principle such as the Principle of Permanence.

preferred to the more restrictive approaches of the intuitionists (viz., Brouwer, Poincaré, Weyl, and especially Kronecker).[83]

The picture of mathematics and its method that Hilbert constructs is thus reminiscent of Peacock's. Like Peacock, Hilbert made a distinction between *real* and *ideal* elements in mathematics.[84] Also like Peacock, he saw this distinction as being at least partially a distinction between those parts of mathematics that purport to express an independently given reality, and those parts of mathematics that are "creations" whose purpose is to preserve mathematical reasoning in a simple and efficient form.[85]

Further like Peacock, Hilbert subscribed to an essentially Berkeleyan conception of language; a conception according to which the cognitive significance of some parts or uses of language lies in their capacity to guide reasoning or inference *without* essential adversion to an interpretation. Finally, like Peacock, Hilbert accepted the existence of two basic constraints on the use of symbolical methods in mathematics—namely, that they be (1) consistent[86] and (2) fruitful.[87]

There were thus strong similarities between the formalist viewpoints of Hilbert and Peacock. There were also significant differences. Perhaps the most important of these is that concerning Hilbert's distinctive view of *Begriffsbildung*

[83] The remarks quoted in this paragraph are taken from §4 of [*Hilbert 1920*], a section titled "The Method of Extreme Prohibition of Kronecker and Poincaré" ("Die Methode der extremen Verbote von Kronecker und Poincaré"). The German of the larger passage is

> Wir sagen vielmehr, die Verbote müssen so gemacht werden, dass die Widersprüche beseitigt werden, dass aber doch alles Wertvolle bestehen bleibt, und zwar nicht nur alle wertvollen Resultate müssen erhalten bleiben, sondern auch die Freiheit der Begriffsbildung und der Schlussmethoden soll nicht über das Mass des Notwendigen hinaus beschränkt sein. [*Hilbert 1920*], 19–20

[84] Peacock expressed his distinction between the "real" and "unreal" elements of algebra in this way.

> Quantities and their symbols are said to be *real* or *possible*, when they can be shewn to correspond to real or possible existences: in all other cases, they are said to be *unreal*, *impossible* or *imaginary*. [*Peacock 1842–45*], vol. II, §557 (emphasis Peacock's)

[85] Peacock said that, in algebra, the laws or axioms *precede* their interpretation. Hilbert and Bernays offered a similar view in the context of distinguishing formal axiomatization from various other types of axiomatization (cf. [*Hilbert and Bernays 1934*], 7).

[86] Consistent themselves, individually and collectively, and consistent with the *real* (i.e., the semantically interpreted) part(s) of mathematics).

[87] The Principle of Permanence and the consistency constraint are, or at least can be, formulated in such a way as to be, related. Specifically, satisfaction of the Principle of Permanence for a new theory might be seen as essentially requiring consistency with the laws of an older theory.

(concept-formation or -construction) in mathematics. Kant had famously maintained that

> ... mathematical knowledge is the knowledge gained by reason from the construction of concepts. To *construct* a concept means to exhibit a priori the intuition which corresponds to the concept. For the construction of a concept we therefore need a *non-empirical* intuition.... Thus I construct a triangle by representing the object which corresponds to this concept either by imagination alone, in pure intuition, or in accordance therewith also on paper, in empirical intuition—in both cases completely a priori, without having borrowed the pattern from any experience. [*Kant 1781*], A713, B741

For Kant, this meant that the introduction of concepts in mathematics proceeds in an essentially different way from the way in which it proceeds elsewhere. Generally speaking, that is, outside mathematics, concepts may be introduced without being *constructed.* Indeed, they may be introduced at will, subject only to a condition of consistency:

> ... I can think (*denken*) whatever I want, provided only that I do not contradict myself, that is, provided my concept is a possible thought. This suffices for the possibility of the concept, even though I may not be able to answer for there being, in the sum of all possibilities, an object corresponding to it. Indeed, something more is required before I can ascribe to such a concept objective validity (that is, real possibility; the former possibility was merely logical). [*Kant 1781*], p. xxvi, preface to the second edition[88]

Introduction of concepts in mathematics, on the other hand, required "something more" for Kant than mere consistency. It required *construction*, that is, a priori exhibition of an intuition corresponding to the concept.

Hilbert's view of concept-formation (*Begriffsbildung*) in mathematics contrasted in clear ways with Kant's. He (Hilbert) accepted that some concepts (viz., those belonging to what he referred to as *real* mathematics) were constructed in the Kantian manner (or something close to it). He explicitly allowed, however, that other concepts—specifically, the concepts belonging to *ideal* mathematics—may be introduced without assignment of *any* interpretation, much less assignment of an intuitive interpretation reached through a priori construction. He saw the formal axiomatic method as a tool for doing this:

[88] The German is

> Aber denken kann ich, was ich will, wenn ich mir nur nicht selbst widerspreche, d. i. wenn mein Begriff nur ein möglicher Gedanke ist, ob ich zwar dafür nicht stehen kann, ob im Inbegriffe aller Möglichkeiten diesem auch ein Objekt korrespondiere oder nicht. Um einem solchen Begriffe aber objective Gültigkeit (reale Möglichkeit, denn die erstere war bloß die logische) beizulegen, dazu wird etwas mehr erfördert.

> When we are engaged in investigating the foundations of a science, we must set up a system of axioms which contains an exact and complete description of the relations subsisting between the elementary ideas of that science. The axioms so set up are at the same time the definitions of those elementary ideas; and no statement within the realm of the science whose foundation we are testing is held to be correct unless it can be derived from those axioms by means of a finite number of logical steps. . . . if it can be proved that the attributes assigned to the concept can never lead to a contradiction by the application of a finite number of logical processes, I say that the mathematical existence of the concept . . . is thereby proved. [*Hilbert 1901*], 447–448[89]

In Hilbert's view there was thus, in addition to Kant's construction of concepts, a second and radically different way to introduce concepts into mathematics. This was the means provided for by the *axiomatic method.* Hilbert described this method (in the case of the theory of the "number-concept") as follows.

> We think (*denken*) a system of things. We call these things numbers and signify (*bezeichnen*) them by *a*, *b*, *c*, ... We think (*denken*) these numbers in certain mutual relationships whose exact (*genaue*) and complete (*vollständige*) description (*Beschreibung*) is accomplished (*geschiet*) by the . . . axioms. . . . [*Hilbert* 1900], 242

For Hilbert, then, concept-introduction in mathematics behaved more like Kant's "*denken*" than his "*anschauen.*" Because of this, his view of mathematical reasoning ran contrary to the classical view. Like Berkeley, he allowed reasoning to include noncontentual manipulation of signs, and he specifically did not require that all reasoning be with items having intuitional content.

It is also significant that he described the axioms of a theory as constituting an exact and *complete* description of the things thought about. The completeness spoken of here is of particular interest. It does not seem to represent so much a descriptive goal or ideal of axiomatic concept-introduction as a *prerogative* of it. It expresses the idea, that is, that the mathematician has a freedom or authority to stipulate that the concepts she introduces have exactly the properties that are provided for them by the axioms by which they are introduced.

It is difficult to know for sure what Hilbert had in mind. The final aim seems to have been categoricity, a property which (see [*Hilbert 1900*]) Hilbert seems to have thought was achievable by *postulation.*[90] This is how he used the term "completeness" in [*Hilbert 1900*], at any rate. In particular, he *posited* his so-called *Vollständigkeitsaxiom* (i.e., Completeness Axiom) to ensure the categoricity of his axiom system. It was fairly common practice for early twentieth-century

[89] Hilbert made the same point in a variety of places. See, for example, [*Hilbert 1900a*], 242, and Hilbert's letter of Dec 29, 1899 to Frege (cf. [*Frege 1980*], 38–41).

[90] A theory is categorical if all of its models are isomorphic. Veblen introduced the term "categoricalness" in [*Veblen 1904*], though he attributed the term to Dewey (cf. p. 346, note ‡).

writers to use the term "complete" with this meaning (cf. [*Wilson 1907–1908*], [*Wilson 1904*]).[91] The completeness of Hilbert's early writings thus seems to have been (some kind of) completeness *as a definition.*

Hilbert's later writings (e.g., [*Hilbert and Ackermann* 1928], [*Hilbert 1930a*]) modified this somewhat. In [*Hilbert 1930a*], he noted that categoricity was not a finitary notion and he introduced the notion of *negation-completeness* (not Hilbert's term) as a finitary surrogate for it.[92] He claimed that the usual systems of arithmetic and analysis were generally regarded as being complete in this sense. He also suggested, however, that such completeness was not a forgone conclusion for theories of "higher domains" (*Hilbert 1930a*, 231). Additional axioms might be needed in order to attain completeness in these theories. Choice of such axioms was to be guided by methodological principles such as the "principle of the permanence of laws." These remarks raise some difficult questions concerning the late Hilbert's understanding of and attitude toward completeness.

These are heightened by the fact that, in his very latest writings, Hilbert suggested a more Peacockean conception of axiomatization:

> In formal axiomatization ... the basic relations are not taken as having already been determined contentually. Rather, they are determined implicitly by the axioms from the very start. And in all thinking with an axiomatic theory only those basic relations are used that are expressly formulated in the axioms. [*Hilbert and Bernays 1934*], 7

The mathematician, on this view, is free to stipulate of a concept she's introducing that it have exclusively the properties provided for by the axioms she uses to introduce it. There is no content belonging to a concept introduced in this way except that which is provided for by the introducing axioms. The axioms used to introduce a concept thus guaranteedly take on a kind of completeness—what might be called *constitutive completeness* because, together, they *constitute* the "content" (i.e., the role in reasoning) of the concept.

Concepts are thus identified with the *roles* they occupy in mathematical thinking (*denken*). They do not have to have an intuitive content in order to be significant.[93] Hilbert believed that the inferential roles of concepts are determined not by contents given prior to the axioms which introduce them, but by those introducing axioms themselves. Indeed, all reasoning concerning a concept

[91] [*Huntington 1902*] (cf. p. 264) used "complete" (he also said "sufficient") in a more comprehensive way, to mean that the axioms of a system were consistent, mutually independent, and categorical.

[92] A formal system T whose language is $\mathcal{L}$ is said to be *negation-complete* when, for every sentence $\mathcal{S}$ of $\mathcal{L}$, either $\mathcal{S}$ or $\neg\mathcal{S}$ is provable in T.

[93] This is not to say that there are no concepts whose significance rests upon their having intuitive content. Hilbert believed that the cognitive significance of finitary concepts did rest on their intuitive contents.

is *restricted* to that which is provided for by its introducing axioms. It follows that concepts function as signs or sign-complexes in formal axiomatic systems:

> Herein lies the firm philosophical viewpoint that I take to be necessary for the grounding of pure mathematics, and for scientific thinking, understanding and communication generally: *in the beginning*—as we mean it here—*is the sign.* [*Hilbert 1922*], 163 (emphasis Hilbert's)[94,95,96]

[*Remark*: There are, I realize, other possible readings of Hilbert's "In the beginning is the sign." One is that it suggests his finitist outlook—specifically, his belief that the justification of a mathematical theory ultimately rests on judgments concerning the shapes of signs. I do not reject such a reading. Indeed, it is strongly suggested by various things Hilbert said (cf. [*Hilbert 1926*], 376, where he speaks of a "content" in mathematics secured independently of all logic). I see no problem, however, in combining this reading with the reading just sketched. Indeed, the combination of the two readings makes it more reasonable for Hilbert to sum up his foundational viewpoint in this single phrase.]

The Hilbertian formalist's claim is thus, at bottom, a claim about the nature of mathematical reasoning—including, perhaps especially, a view of the nature

[94] The German is

> Hierin liegt die feste philosophische Einstellung, die ich zur Begründung der reinen Mathematik—wie überhaupt zu allem wissenschaftlichen Denken, Verstehen und Mitteilen—für erforderlich halte: *am Anfang*—so heißt es hier—*ist das Zeichen.*

[95] Hilbert's "in the beginning is the sign" is, of course, a reference to St. John's "In the beginning was the Word..." (John 1:1). (In Luther's German translation, the verse reads: "Im Amfang war das Wort, und das Wort war bei Gott, und Gott war das Wort.") It is also reminiscent of Goethe's "in the beginning was the act" and, as such, makes a lot of sense. Taken thus, it would represent Hilbert's siding with God rather than the devil—the devil, in this case, being personified by Brouwer and his intuitionism. Recall that Brouwer defined his intuitionism as consisting essentially in two "acts." The first of these—the so-called *first* act of intuitionism—was intended to completely divorce mathematics from language. As such, it can be seen as the antithesis of Hilbert's views.

[96] Peano (cf. [*Peano 1910*], 224) and Poincaré made similar suggestions, though Poincaré's suggestion was made out of different motivations. He wrote:

> To sum up, the mind has the faculty of creating symbols, and it is thus that it has constructed the mathematical continuum, which is only a particular system of symbols. The only limit to its power is the necessity of avoiding all contradiction.... [*Poincaré 1905*], 27

He went on to add a kind of fruitfulness condition, namely, that "the mind only makes use of it [i.e. the faculty of creating symbols] when experiment gives a reason for it" (ibid.).

of concept-formation and/or concept-introduction. It allows signs and sign-complexes to play significant roles in mathematical reasoning independently of any interpretation that might be given to them. Indeed, when concepts are *given* exclusively by formal axioms and are regarded as having only such contents as those axioms provide, there is no significant difference between the roles of those concepts and the roles of signs. They function essentially as signs in a formal axiomatic system. One can do with them only that which is permitted by the introducing axioms. There is no preaxiomatic grasp or understanding that the axioms are intended to capture, and there is no extra-axiomatic model or structure that might be consulted in a search for new axioms or to correct current axioms.

The history of mathematics contains examples of concepts that appear to have been introduced in this way. The imaginary and complex numbers are, I believe, one such example. They were introduced as items obeying certain laws. These laws were not selected, however, because they captured a preaxiomatic structure given to intuition or to our understanding. Rather, the complexes were assigned laws that maximized their utility as computational instruments and that, at the same time, permitted us to preserve the Fundamental Theorem of Algebra in a pleasingly simple form. They did not simultaneously admit of both a notion of inequality and notions of addition and multiplication that were normal. They did, however, admit of a consistent axiomatization of enough of the usual laws of arithmetic computation to make them useful as computational devices. The fact that they were later given a geometrical interpretation by Gauss and others does not alter these facts, which are the basic facts of interest to the formalist.

They were, indeed, of exceptional utility as computational instruments. As various sixteenth-century mathematicians (e.g., Cardano, Bombelli, and Viète) noticed, this included rules—specifically, Cardano's Rule—useful for finding even *real* roots of polynomials.[97] Cardano's Rule is given by the formula: $x =$

[97] Sixteenth century algebraists also became aware of instances of the general cubic where use of complex expressions posed problems for their solution via Cardano's Rule. These were the instances of the so-called *casus irreducibilis* or *irreducible case*—the case of $x^3 - px + q = 0$ where $\sqrt{\frac{q^2}{4} - \frac{p^3}{27}}$ is negative. The irreducible case was interesting because it represents a case where a cubic has three real roots but where those roots can not be expressed, by Cardano's Rule, without adversion to complex numbers. The problem caused by this is that extraction of the roots of the imaginary quantities involved in Cardano's Rule in this case itself depends upon the solution of a cubic equation. Hence, the solution of the original problem is not in this case "reduced" by application of Cardano's Rule. In 1892, Otto Hölder (cf. [*Hölder 1892*]) showed what algebraists for nearly three hundred years had strongly suspected—namely, that there is no general formula for solving the cubics of the irreducible case that makes use of only the usual operations of algebra (i.e., addition, multiplication, subtraction, division and exponentiation) plus root extraction for reals.

$\sqrt[3]{-\frac{q}{2}+\sqrt{\frac{q^2}{4}-\frac{p^3}{27}}}+\sqrt[3]{-\frac{q}{2}-\sqrt{\frac{q^2}{4}-\frac{p^3}{27}}}$, for the so-called *general* cubic (i.e., a cubic of the form $x^3-px+q=0$, where p, q are real). The general cubic is so-called because, by making appropriate substitutions, any cubic can be put into this (or a related) form.[98]

Of a more "theoretical" nature, perhaps, is the need for complex numbers in preserving the Fundamental Theorem of Algebra in a simple form. We desire the simplicity that the Fundamental Theorem of Algebra (in the forms made possible by the addition of imaginary and complex numbers) brings to the determination of the numbers of roots of equations. In all cases, however, both practical and theoretical, the use of complex numbers is motivated by a desire to preserve arithmetical laws in forms that are convenient to our reasoning. Something similar, I submit, holds of the introduction of ideal elements in other fields of mathematics (e.g., the introduction of points at infinity in projective geometry). When convenience makes their introduction desirable, we introduce ideal elements if we consistently can. In modern mathematics, we introduce them by giving a system of axioms that completely specifies their place in our reasoning.

Being of this mind, it was only natural that Hilbert should urge a *noncontentual* role for signs and/or concepts in mathematical thinking—a role that does not make use of any pre- or extra-axiomatic content that might be attached to them:

> To make it a universal requirement that each individual formula ... be interpretable by itself is by no means reasonable;[99] on the contrary, a theory by its very nature is such that we do not need to fall back upon intuition or meaning in the midst of some argument. What the physicist demands precisely of a theory is that particular propositions be derived from laws of nature or hypotheses solely by inferences (*allein durch Schlüsse*), hence on the basis of a pure formula game (*Formelspiel*), without extraneous considerations being adduced. Only

[98] Frege also took an interest in the *casus irreducibilis*. Consistent with his antipathy towards the formalist or Berkeleyan conception of reasoning, he saw it as a reason to require that there be a magnitude that the sign "$\sqrt{-1}$" refer to.

> ... it is especially important for the derivation of many theorems that there be ... higher numbers. And just as for many proofs in geometry, one needs points or lines which do not occur in the theorems themselves, and just as in each of these cases it is then necessary to show that there are such auxiliary points or lines; so too in arithmetic, many theorems are proved with the aid of $\sqrt{-1}$, where this magnitude does not itself occur in the theorems. Now if there were simply no number whose square is 1, these proofs would collapse. [*Frege 1885*], 117

Though I will not develop the idea here, I think the correct response would be to press the Berkeleyan conception of reasoning and to challenge the aptness of Frege's analogy.

[99] The German is "Die Forderung, wonach dabei jede einzelne Formel für sich allein deutbar sein soll, allgemein aufzustellen, ist keineswegs vernünftig. ..."

> certain combinations and consequences of physical laws can be checked by experiment—just as in my proof theory only the real propositions are directly capable of verification. [*Hilbert 1928*], 475[100]

This, in its elements, was Hilbert's Berkeleyan conception of mathematical reasoning. It saw mathematical reasoning as allowing interludes of pure symbol-manipulation (what we'll generally refer to as "ideal reasoning") within larger, generally contentual reasoning environments. Signs were allowed to "come before" their meanings in the sense that reasoning was, at least sometimes, a matter of *technique.* Indeed, in his later work, Hilbert seemed to favor the view that, at a deep level, a great deal of reasoning of all types *is* technique (what the Greeks might have termed *technē*)—a manipulation of signs or symbols in accordance with psychologically convenient rules. As he put it.

> In our theoretical sciences we are accustomed to the use of formal thought processes (*formaler Denkprozesse*) and abstract methods... [but] already in everyday life one uses methods and concept-formations (*Begriffsbildungen*) which require a high degree of abstraction and which only become intelligible by means of an unconscious (*unbewußte*) application of axiomatic methods. Examples include the general process of negation and, especially, the concept of infinity. [*Hilbert 1930c*], 380

The mathematician is free to exploit the psychological advantage that this type of reasoning offers. Her only restriction is that she not produce a body of reasoning that leads to conclusions that contradict the findings of contentual reasoning. Ideal reasoning does not, therefore, displace contentual reasoning. Rather, it supplements it by adding signs that produce an overall body of reasoning that is psychologically more natural and epistemically more productive than contentual reasoning itself.

The question, of course, as Frege repeatedly emphasized, is how purely symbolical reasoning can be epistemically productive. To this we answer that symbolical reasoning does not, by itself, justify a contentual conclusion to which it leads. Rather, it is supplemented by contentual metamathematical judgments to secure contentual conclusions.

Hilbert's "metamathematics" or "proof theory" was intended to supply the means for such an application. One would show by contentual metamathematical means that symbolical (what Hilbert called "ideal") reasoning to contentual conclusions is reliable. This would be possible because the rules for the usage of the noncontentual signs, which form the technique of our thinking, would be precisely stated in their axiomatic presentation.

> The formula game that Brouwer so deprecates has, besides its mathematical value, an important general philosophical significance. For this formula game is carried out according to certain definite rules, in which the technique of our thinking is expressed. These rules form a closed system that can be discovered

[100] See [*Peacock 1834*], 198 (quoted above), for a similar statement by Peacock.

> and definitively stated. The fundamental idea of my proof theory is none other than to describe the activity of our understanding, to make a protocol of the rules according to which our thinking actually proceeds. Thinking, it so happens, parallels speaking and writing: we form statements and place them one behind another. If any totality of observations and phenomena deserves to be made the object of a serious and thorough investigation, it is this one.... [*Hilbert 1928*], 475

Hilbert's attempt to execute this "serious and thorough investigation" of our mathematical *technē* came to be known as *Hilbert's Program.* It was barely begun when the announcement of Gödel's Incompleteness Theorems called its overall realizability into question. We'll discuss this challenge after considering other challenges to formalism, most notably those developed by Frege in the late nineteenth and early twentieth centuries.

6. Challenges to Formalism

What seems to have been the most influential variety of formalism of the late nineteenth and early twentieth centuries—the neo-Berkeleyan, neo-Peacockean formalism of Hilbert—was thus essentially a view concerning the nature of mathematical reasoning and, in particular, of mathematical concept-formation and/or -introduction. There were, however, other variants of formalism, some of which had a certain degree of notoriety among foundational thinkers of the late nineteenth and early twentieth centuries.

Chief among these is what I will refer to as *empiricist* formalism.[101] This is a formalism that drew its inspiration from empiricist views of existence in mathematics. Its chief idea was to reduce the question of the existence of mathematical entities—in particular, the existence of numbers—to the question of the existence of perceivable signs. It is this variant of formalism that became the focus of Frege's engagement with and well-known criticisms of formalist ideas. Because of Frege's great influence on Anglo-American analytic philosophy, the formalism he criticized has sometimes been identified with formalism per se. This, I believe, is a mistake. There were thinkers who held positions like those criticized by Frege. They did not, however, represent the mainstream of formalist thinking in mathematics—the tradition coming down from Peacock and various others of the Cambridge algebraists to various German foundational thinkers, most particularly Hilbert.

Frege described what he took to be the core element of formalism in his 1891 lecture "Function and Concept." That was "a tendency not to recognize as an object

[101] Others (cf. Resnik in [*Resnik 1980*], ch. 2) have called this type of formalism *game* formalism. I prefer my title because it suggests the basic epistemological motive behind the view.

anything that cannot be perceived by means of the senses" ([*Frege 1984*], 138). He regarded this tendency as "very widespread" (ibid.), and said that it was "often met with in mathematical works, even those of celebrated authors" (ibid.). Among the authors Frege had in mind were Helmholtz and Kronecker (cf. [*Frege 1984*], 139).[102] He also named a number of lesser but still well-known figures as having held this or related views. These included Cantor (cf. [*Frege 1893*], vol. II, §§68–85), Husserl (cf. [*Frege 1984*], 208), Hankel (cf. [*Frege 1884*], §§98–99; [*Frege 1893*], vol. II §§141–42), Heine (cf. [*Frege 1893*], vol. II, §§59–60, 86–87, 104–105), Stolz (cf. [*Frege* 1893], vol. II, §§143–145), Thomae (cf. [*Frege 1893*], vol. II, §§86, 88–103, 106–137), Schubert (cf. "On Mr. Schubert's Numbers," in [*Frege 1984*], 249–272) and Korselt (cf. the 1906 essay "On the Foundations of Geometry: Second Series," in [*Frege* 1984], 293–340).

Heinrich Eduard Heine, a well-known mathematician at the University of Halle,[103] stated the key idea of empiricist formalism as follows:

> Suppose I am not satisfied to have only positive rational numbers. I do not answer the question, What is a number?, by defining number conceptually, say by introducing irrationals as limits whose *existence* is presupposed. I adhere to the definition of the purely formal standpoint (*rein formalen Standpunkt*), in which what I call numbers are certain tangible (*greifbare*) signs (*Zeichen*) so that the existence of these numbers does not, therefore, stand in question. The main emphasis is to be put on the calculating operations, and the number-signs (*Zahlzeichen*) must be selected in such a way or furnished with such an apparatus as provides a clue to the definition of the operations. [*Heine 1872*], 173[104]

[102] Frege's view of Hilbert's ideas is more difficult to determine, and there is not enough space to explore the subtleties here. There is little indication, though, that Hilbert's formalism was motivated by empiricist concerns or that Frege believed it to be so.

[103] He is the Heine whose name is commonly associated with Borel's in conjunction with the well-known Heine–Borel covering theorem (i.e., the theorem that a subset of the reals is compact just in case it is closed and bounded). Arthur Schoenfließ is the first to have mentioned Heine's name in conjunction with this theorem. Heine did not prove this theorem, however. Nor did he formulate the covering property. What he did was to formulate a notion of uniform continuity and prove the uniform continuity, in his sense, of every continuous function on a closed, bounded interval. This theorem, which, unbeknownst to Heine, had been proved by Dirichlet approximately ten years earlier, contained a significant core of the ideas of Borel's proof.

[104] The German is "Die Frage, was eine Zahl sei, beantworte ich, wenn ich nicht bei den rationalen positiven stehen bleiben will, nicht dadurch dass ich die Zahl begrifflich definiere, die irrationalen etwa gar als Grenze einführe, deeren *Existenz* eine Voraussetzung wäre. Ich stelle mich bei der Definition auf den rein formalen Standpunkt, indem ich gewisse greifbare Zeichen Zahlen nenne, so dass die Existenz dieser Zahlen also nicht in Frage steht. *Ein Hauptgewicht ist auf die Rechenoperation zu legen*, und das Zahlzeichen muss so gewählt, oder mit einem solchen Apparate ausgerüstet werden, dass es einen Anhalt zur Definition der Operationen gewährt." Heine also remarked that he had employed this viewpoint in his lectures on algebraic analysis for many years.

Heine seems to have been motivated by epistemological concerns—specifically, a concern that *our knowledge of mathematical existence be as unproblematic as our knowledge of the existence of empirically observable objects.* This is the central idea of Heine's formalism, and the (or at least a) chief focus of Frege's criticism of it.

Johannes Thomae, Heine's colleague at Halle for five years and Frege's colleague at Jena for thirty-five, gave a similar indication of epistemological motivation, though not one so clearly empiricist in character:[105]

> The formal conception of the numbers accepts more modest limits than the logical. It does not ask what the numbers are or must be; rather it asks what one requires of the numbers in arithmetic. For the formalist conception, arithmetic is a game (*Spiel*) with signs (*Zeichen*) that are well referred to as empty, by which one means to say that they (in the calculating game (*Rechenspiel*)) have no content (*Inhalt*) given to them beside that which comes from their role in the rules of combination (rules of the game).
>
> . . .
>
> The formal standpoint allows us to throw off all metaphysical difficulties (*metaphysischen Schwierigkeiten*). That's the gain it offers us. [*Thomae 1898*], 3

Frege went on to elaborate the "difficulties" referred to.

> The difficulties spoken of here may well be those with which we became acquainted in our reflection on Cantor's theory, namely, the grasping of actual numbers (*eigentlichen Zahlen*) and the proof of their availability (*Vorhandensein*). In formal arithmetic we don't need to ground the game-rules (*Spielregeln*)—we simply declare them. We don't need to prove that numbers have certain properties; we just introduce figures (*Figuren*) for whose handling we give rules. We then regard these rules as properties of the figures and thus are able to freely create—seemingly at least—things (*Dinge*) having the desired properties. In this way it is evident that we at least save ourselves intellectual work (*geistige Arbeit*). [*Frege 1893*], §89

Frege admitted that there were differences in the viewpoints of Heine and Thomae. Specifically, he took Thomae to essentially reject the question concerning the nature of numbers. Heine, on the other hand, he believed to accept the question. He answered it, though, by identifying numbers with signs (cf. [*Frege 1893*]; 98). In the end, however, Frege saw little difference between the two viewpoints because "both agree that arithmetic has to occupy itself with signs" (ibid.).

Frege's criticisms of empiricist formalism centered on this alleged occupation with signs. Without attempting to survey all the different criticisms he gave, I'll focus on the one I take to be of the most fundamental and far-reaching significance—one reaching, in particular, to Hilbert's type of formalism.

[105] Frege himself observed this, remarking, "Heine sets out the basic idea more sharply than does Thomae" (cf. [*Frege 1893*], vol. II, 96).

This is a criticism that was briefly stated in the 1885 essay on formal theories of arithmetic (cf. [*Frege 1885*], 114ff.) and in the correspondence with Hilbert (cf. [*Frege 1980*], letter of December 27, 1899). It was developed at greater length in [*Frege 1903*] and in the second volume of the *Grundgesetze* (cf. §91). It received what was perhaps its most mature statement in ([*Frege 1906*]).

The basic argument was essentially this: formalism requires that arithmetic terms be treated as not having reference; but if arithmetic terms have no reference, arithmetic sentences do not express *thoughts* (in the Fregean sense); if arithmetic sentences do not express thoughts, then there can be no genuine demonstration or proof in and no application of arithmetic; if there is no genuine proof in and no application of arithmetic, then it can not rightly be considered a *science* (in the classical sense of the term); to retain arithmetic as a genuine science, therefore, one must abandon formalism.

The core element of this argument was Frege's anti-Berkeleyan conception of reasoning and proof, a conception according to which (1) genuine *inference* or reasoning cannot consist in a manipulation of signs, and (2) a genuine *premise* can not be a mere formula. Frege expressed these ideas in the following remarks:

> ... an inference does not consist of signs. We can only say that in the transition from one group of signs to a new group of signs, it may look now and then as though we are presented with an inference. An inference simply does not belong to the realm of signs; rather, it is the pronouncement of a judgment made in accordance with logical laws on the basis of previously passed judgments. Each of the premises is a determinate thought recognized as true; and in the conclusion, too, a determinate thought is recognized as true....
>
> What is a formal inference? We may say that in a certain sense, every inference is formal in that it proceeds according to a general law of inference; in another sense, every inference is non-formal in that the premises as well as the conclusions have their thought-contents which occur in this particular manner of connection only in that inference. [*Frege 1906*], 387

> From the fact that the pseudo-axioms do not express thoughts, it ... follows that they cannot be premises of an inference-chain.... one ... cannot call ... groups of audible or visible signs ... premises anyway, but only the thoughts expressed by them. Now in the case of the pseudo-axioms, there are no thoughts at all, and consequently no premises. Therefore when it appears that Mr. Hilbert nevertheless does use his axioms as the premises of inferences and apparently bases proofs on them, these can be inferences and proofs in appearance only. [*Frege 1906*], 390

Frege's most fundamental challenge to formalism was thus directed at its Berkeleyan conception of reasoning. He held that mathematical reasoning and proof are inherently and essentially contentual in character. They are processes whose purpose is to produce warranted judgments (i.e., affirmations of propositional contents or thoughts) called "conclusions." They are to achieve this purpose by exhibiting the thought forming the content of a conclusion as logically implied by the contents of a group of antecedently justified judgments called "premises." These premises are ultimately comprised of a group of judgments called "axioms"—the contents of which are self-evidently true. The axioms thus form the contents of the fundamental judgments of the subject being investigated.

Various of Heine's and Thomae's ideas appear to disagree with the idea that the premises of proofs are to be axioms in the sense described and with the idea that the purpose of proof is to exhibit a thought as a logical consequence of such premises. They may even disagree (albeit perhaps unintentionally) with the still more basic idea that the ultimate aim of proof is to produce warranted judgment (i.e., a warrantedly believed propositional content).

Hilbert's formalism was different. It did not dispute the idea that the purpose of proof is ultimately to produce warranted judgment. Nor did it dispute the idea that to achieve this end, a proof must exhibit a logical relationship between contentual premises and a contentual conclusion. It took exception only to the idea that the premises of a proof must be axioms of the subject being investigated. It allowed as well that the premises of genuine proofs should sometimes be metamathematical propositions concerning a formal system (and/or various of its items), and not axioms of that system. Since these metamathematical propositions could be directed at noncontentual as well as at contentual axioms, it (i.e., Hilbert's formalism) allowed that the axioms with which they were concerned should be noncontentual in nature.

This means, of course, that such axioms cannot be premises of proofs where "proof" is understood in Frege's sense. Hilbert permitted noncontentual formulas to be premises of a certain type of "proof" that he called *ideal* proof. But here we have what is essentially a terminological difference concerning the use of the term "proof." Frege insisted that it be reserved for what Hilbert generally referred to as contentual proof. Hilbert allowed it to refer to noncontentual thinking as well.

Unfortunately, this terminological dispute seems to have dominated much of Frege's side of the Frege–Hilbert correspondence and controversy. This is no doubt in part due to the fact that their correspondence occurred at a time when Hilbert had not yet arrived at a clear understanding of his proof-theoretic conception of consistency. This kept him from responding in a clear and satisfying way to Frege's claim that the only way to prove a body of formal reasoning consistent was to *interpret* it—that is, to transform it into a body of contentual reasoning through the assignment of an interpretation to the formulas that made their truth evident.

In a letter of January 6, 1900, Frege thus remarked to Hilbert:

> What means have we of demonstrating that certain properties, requirements (or whatever else one wants to call them) do not contradict one another? The only means I know is this: to point to an object that has all those properties, to give a case where all those requirements are satisfied. It does not seem possible to demonstrate the lack of contradiction in any other way. [*Frege 1980*], 43

He pressed the point again in a letter nine months later:

> I believe I can deduce, from some places in your lectures, that my arguments failed to convince you, which makes me all the more anxious to find out your counter-arguments.
>
> It seems to me that you believe yourself to be in possession of a principle for proving lack of contradiction which is essentially different from the one I formulated in my last letter and which, if I remember right, is the only one you apply in your *Foundations of Geometry.* If you were right in this, it could be of immense importance, though I do not believe in it as yet, but suspect that such a principle can be reduced to the one I formulated and that it cannot therefore have a wider scope than mine. It would help to clear up matters if in your reply to my last letter—and I am still hoping for a reply—you could formulate such a principle precisely and perhaps elucidate its application by an example. [*Frege 1980*], 49

Frege's reference to ([*Hilbert 1899*], §9) is to Hilbert's proof of the consistency of his axiomatization of geometry by provision of an interpretation of it in the real numbers. It was indeed, as Frege suggests, the only type of proof of consistency that Hilbert offered at that time. It would be another twenty years before he would be able to describe even the outlines of a new conception of consistency proofs—namely, that arising from his more mature proof-theoretic program. Frege was right to say that this was a development of potentially "immense importance."

The "principle" to which Frege referred was a principle for inferring the existence of an object having a set of properties from the mere fact that a set of such attributions is not contradictory. Frege dealt with it in the above-mentioned letter of January 6, 1900:

> Suppose we knew that the propositions
>
> (1) A is an intelligent being
> (2) A is omnipresent
> (3) A is omnipotent
>
> together with all their consequences did not contradict one another; could we infer from this that there was an omnipotent, omnipresent, intelligent being? This is not evident to me. The principle would read as follows:
>
> If the propositions
>
> "*A* has property Φ"
> "*A* has property Ψ"
> "*A* has property *X*"
>
> together with all their consequences do not contradict one another in general (whatever *A* may be), then there is an object which has all the properties Φ, Ψ and *X*. This principle is not evident to me, and if it was true it would probably be useless. Is there another means of demonstrating lack of contradiction besides

> pointing out an object that has all the properties? But if we are given such an object, then there is no need to demonstrate in a roundabout way that there is such an object by first demonstrating lack of contradiction. [*Frege 1980*], 47

This passage illustrates the firmness of Frege's belief that the only way to demonstrate consistency was to exhibit a witnessing object or interpretation. He believed that any potential alternative means of demonstrating noncontradiction would eventually reduce to this.[106] One might therefore just as well produce the witnessing object directly, without detouring through the supposed alternative means, which at some point, if it is to be effective, must produce the witnessing object anyway.

It also serves as a powerful illustration of the radical character of Hilbert's later proof-theoretic conception of consistency. Frege can hardly be blamed for shortsightedness or narrow-mindedness in this matter. He failed to see only what almost everyone else failed to see as well.

In closing, let me briefly mention two further challenges to formalism—challenges that seem for the most part to have escaped the attention of those writing on the subject. The first derives from certain remarks of Gödel concerning the basis for certainty in mathematics. The certainty of mathematics, he maintained,

> ... is to be secured not by proving certain properties by a projection onto material systems—namely, the manipulation of physical symbols—but rather by cultivating (deepening) knowledge of the abstract concepts themselves which lead to the setting up of these mechanical systems, and further by seeking, according to the same procedures, to gain insights into the solvability, and the actual methods of solution, of all meaningful mathematical problems. [*Gödel 1986–1995*], vol. III, 383[107]

A challenge for formalism arises from this and Gödel's further description of how cultivation of a knowledge of abstract concepts is to proceed and what its results may be expected to be. He cites as his inspiration Husserl's phenomenological method, which he describes as follows:

> ... there exists ... the beginning of a science which claims to possess a systematic method for ... clarification of meaning, and that is the phenomenology founded by Husserl. Here clarification of meaning focuses more sharply on the concepts concerned by directing our attention in a certain way, namely, onto our own acts in the use of these concepts, onto our powers in carrying out our acts, etc. [This] is a procedure or technique that should produce in us a new state of consciousness in which we describe in detail the basic concepts we use in our thought, or grasp other basic concepts hitherto unknown to us. I believe there is

[106] As he saw it, moreover, the reasoning leading from consistency to existence of such a witness was dubious.

[107] This is from the 1961 essay "The Modern Development of the Foundations of Mathematics in the Light of Philosophy."

> no reason at all to reject such a procedure at the outset as hopeless. . . . on the contrary one can present reasons in its favor. . . . In fact, one has examples where, even without the application of a systematic and conscious procedure . . . a further development takes place . . . one that transcends "common sense." Namely, it turns out that in the systematic establishment of the axioms of mathematics, new axioms, which do not follow by formal logic from those previously established, again and again become evident . . . it is just this becoming evident of more and more new axioms on the basis of the meaning of the primitive notions that a machine cannot imitate. [*Gödel 1986–1995*], vol. III, 384–385

What Gödel suggests is that there is something either in the concepts represented by a set of axioms, or in our grasp of these concepts, that is not captured by the axioms themselves. To the extent that this is true, Hilbert's (and other creativists') idea that the axiomatic method expresses our freedom to *stipulate* the contents of concepts would seem to be either incorrect or of limited interest. An important part of the value of a set of axioms, in Gödel's view, resides in what arises from our turning our attention to our own acts in using them. Such reflection, says Gödel, cultivates or deepens our understanding of the concepts represented by the axioms, and the fruit of this cultivation is the discovery of new axioms which were in some sense latent in the original ones without being logically contained in them. To the extent that this is typical, it would be wrong or of little interest to claim that we have a freedom to create concepts by stipulating exactly what conditions they are to satisfy. Such stipulations would be routinely superseded as a result of proper reflection on our use of the concepts concerned. In other words, latent in the use of a set of axioms that is used to introduce a concept or sign are further axioms the uncovering of which will make the original axioms obsolete and cause us not to view them as the stipulations fixing the concept we originally took them to fix.

The second challenge is what I will call the *Division Problem*. This is a problem for those varieties of formalism which admit a distinction between parts of mathematics (typically referred to as "real") that behave in a contentual, non-algebraico-symbolical way and those parts of mathematics that function essentially as algebraico-symbolical instruments ("ideal" mathematics, in Hilbert's terminology). It concerns how to determine, in a nonarbitrary and descriptively adequate way, where the dividing line between real and ideal mathematics ought properly to lie. More specifically, it concerns the problem of whether a real/ideal distinction can be designed in such a way as to facilitate a proof of the consistency of the ideal side of the division by methods that can be distinguished in a principled way from the methods of the ideal side. A genuine proof of consistency must proceed by contentual methods. If, however, there is no principled way to distinguish between the methods used in the consistency proof and the methods whose consistency it is supposed to establish, then a formalist approach to the field in question would make little sense.

What we are calling the Division Problem raises the question of whether *there is* a plausible, principled way to distinguish between real and ideal methods in mathematics. It has been suggested by a number of nineteenth- and twentieth-century mathematicians that there is not. The American mathematician James Pierpont raised the problem in an interesting way in his widely used 1914 text *Functions of a Complex Variable.*

> The complex numbers are often called *imaginary numbers*, and we have in the present work followed usage as far as to call . . . numbers . . . purely imaginary, the number i the imaginary unit, and the axis of ordinates the imaginary axis. For the beginner the term imaginary is most unfortunate; and if it had not become so ingrained in elementary algebra, much would be gained if it could be dropped and forgotten.
>
> The use of the term imaginary in connection with the number concept is very old. At first only positive integers were regarded as true numbers. To the early Greek mathematician the ratio of two integers as $\frac{4}{5}$ was not a number. After rational numbers had been accepted, what are now called negative numbers forced themselves on the attention of mathematicians. As their usefulness grew apparent they were called fictitious or imaginary numbers. To many an algebraist of the early Renaissance it was a great mystery how the product of two such numbers as $-a$, $-b$ could be the real number ab.
>
> Hardly had the negative numbers become a necessary element to the analyst when the complex numbers pressed for admittance into the number concept. These in turn were called imaginary, and history repeated itself. How many a boy to-day has been bothered to understand how the product of two imaginary numbers ai and bi can be the real number $-ab$. As well ask why in chess the knight can spring over a piece and why the queen cannot. . . .
>
> The symbols (a, a') or $a + a'i$ are mere marks until their laws of combination are defined, they then become as much a number as 2/3 or 5. *The student must realize that all integers, fractions, and negative numbers are imaginary. They exist only in our imagination. Five horses, three quarters of a dollar, may have an objective existence, but the numbers 5 and 3/4 are imaginary. Thus all numbers are equally real and equally imaginary. Historically we can see how the term imaginary still clings to the complex numbers; pedagogically we must deplore using a term which can only create confusion in the mind of the beginner.* [*Pierpont 1914*], 10 (emphasis in the last paragraph added)

In Pierpoint we see an example of a mathematician who questioned the possibility of plausibly making a principled distinction between real and ideal (or imaginary) methods in mathematics and who moved from that question to the view that basically all mathematical objects are ideal.

There were, however, at least as many mathematicians who took the opposite tack—that is, they denied the possibility of making a principled distinction between a real and an ideal part of mathematics, but moved from there to the view that all of mathematics is real. This was essentially Gauss's view.

> Our general arithmetic, so far surpassing in extent the geometry of the ancients, is entirely the creation of modern times. Starting originally from the notion of absolute integers, it has gradually enlarged its domain. To integers have been added fractions, to rational quantities the irrational, to positive the negative, and to the real the imaginary. This advance, however, has always been made at first with timorous and hesitating step. The early algebraists called the negative roots of equations false roots, and these are indeed so when the problem to which they relate has been stated in such a form that the character of the quantity sought allows of no opposite. But just as in general arithmetic no one would hesitate to admit fractions, although there are so many countable things where a fraction has no meaning, so we ought not to deny to negative numbers the rights accorded to positive simply because innumerable things allow no opposite. The reality of negative numbers is sufficiently justified since in innumerable other cases they find an adequate substratum. This has long been admitted, but the imaginary quantities—formerly and occasionally now, though improperly, called impossible—as opposed to real quantities are still rather tolerated than fully naturalized, and appear more like an empty play upon symbols to which a thinkable substratum is unhesitatingly denied by those who would not depreciate the rich contribution which this play upon symbols has made to the treasure of the relations of real quantities.
>
> The author has for many years considered this highly important part of mathematics from a different point of view, where just as objective an existence may be assigned to imaginary as to negative quantities, but hitherto he has lacked opportunity to publish these views, though careful readers may find traces of them in the memoir upon equations which appeared in 1799 and again in the prize memoir upon the transformation of surfaces. In the present paper the outlines are given briefly....
>
> ... in the language of the mathematician ... $+i$ is a geometric mean between $+1$ and -1, or corresponds to the sign $\sqrt{-1}$.... Here, therefore, an intuitive meaning of $\sqrt{-1}$ is completely established, and one needs nothing further to admit this quantity into the domain of the objects of arithmetic. [*Gauss 1831*], 175–177

The Division Problem thus raises a difficult and important challenge to all varieties of formalism which admit of a distinction between real and ideal methods: namely, to provide a plausible, principled way of forging a real vs. ideal distinction.

Supposing this to be manageable, the Division Problem raises another question as well—namely, whether within any plausible, principled division of real mathematics from ideal mathematics, real methods can possess enough strength to be capable of proving the consistency of the ideal methods. This is a pressing problem since if there is a way to make a principled distinction between real and ideal methods at all, that distinction will quite likely take hold at a relatively basic and elementary place in the hierarchy of number systems.

Hilbert, as is well known, identified the border between finitary and non-finitary methods as the border of the real/ideal distinction. He did so in part because he sought a corrective to Kronecker's reactionary response to the loss of geometric intuition in analysis that he (Kronecker) would be obliged to credit. What we now see is that there may be another, and perhaps even more potent, reason for choosing the division between finitary and nonfinitary reasoning as the place at which to divide real and ideal methods—namely, that there may be no place other than this at which a principled division of real and ideal methods can be made. Indeed, if Gauss and Pierpont are right, there may be no place at all for such a distinction. The formalist thus faces a stern challenge.

So, too, however, does the nonformalist. For unless a philosophy of mathematics provides for a division between real and ideal elements, it is doubtful that it can be adequate to those historical "data" by which alone the subject is given.

This brings me to my closing comments, which concern the effects of Gödel's Incompleteness Theorems on formalism. Since I have written elsewhere on this topic, and since my basic views have not changed, I will be brief here.

The formalist creates instruments and, when he does, he is obliged to establish their consistency. Do Gödel's Incompleteness Theorems—in particular, his Second Incompleteness Theorem—show this to be impossible? In general, my answer is "no." There are many aspects to the argument behind this answer. Some of these are discussed at greater length in [*Detlefsen 1986*] and [*Detlefsen 2001*]. Here I will focus on matters connected to the following two general facts: (1) that the formalist's purpose is to create *useful instruments*, and (2) that his consistency proofs ought therefore in principle be restricted to the theorems proved by ideal proofs that are useful (viz., ideal proofs of real theorems that are, in some appropriate sense, "easier" than the "easiest" of the non-instrumental or real proofs of the theorems they prove).

There are, of course, well-known formalisms (e.g., PA, PA^2, and ZF) that include a wide range of formalist instruments. If we assume that for each such theory T, every formula of the language of T that expresses T's consistency is subject to Gödel's Second Theorem, we must accept that there is no proof of T's consistency that is provable by real means codifiable in T. The question remains, however, whether there is a proper subsystem T_U of T such that (1) all the instrumentally useful proofs in T of real theorems are contained in T_U, and (2) the consistency of T_U is provable by real means. The answer to this question is not, I think, generally known.

One reason why is that it requires general knowledge of the theories T for which there exist subsystems of the type of T_U, and this seems generally not to be available. A related reason is that it requires clarification of the exact form(s) that instrumental usefulness is supposed to take. Is it, for example, to reside in proofs that are easier to *verify* as proofs? Or should it rather reside in proofs that are easier to *discover* and/or easier to *understand* or otherwise grasp? A final reason is that it requires partial solution of the above-mentioned Division Problem, and such solution does not currently exist. Gödel's Second Theorem, together with other current knowledge, does not, therefore, provide a refutation of formalism.

Acknowledgments I would like to thank audiences at the University of Helsinki, the University of California-Irvine, Senshu University, Keio University, Tokyo Metropolitan University, the Japanese Association for the Philosophy of Science, the Hilbert Workshop, the First Iranian Seminar in the Philosophy of Mathematics (Shahid Beheshti University), Northwestern University and the University of Notre Dame for helpful discussions on various parts of this material. Among individuals, I am particularly indebted to my students Andrew Arana and Sean Walsh, and to Stewart Shapiro, Timothy McCarthy, Paolo Mancosu, Richard Zach, Juliette Kennedy, Jouko Väänänen, Panu Raatikainen, Jaakko Hintikka, Hiroshi Kaneko, Kazuyuki Nomoto and Mitsuhiro Okada for useful comments.

REFERENCES

Allman, G.J., *Greek Geometry from Thales to Euclid*, Longmans, Green, London, 1889.

Apollonius, *Treatise on Conic Sections*, T. Heath (ed.), Barnes and Noble, New York, 1961.

Archimedes, *The Works of Archimedes*, T. Heath (ed.), Dover, Mineola, NY, 2002.

Aristotle, *Metaphysics*, Oxford Translations, W.D. Ross and J.A. Smith, eds., Oxford University Press, Oxford, 1908–1954.

Aristotle, *Physics*, Oxford Translations, W.D. Ross and J.A. Smith eds., Oxford University Press, Oxford, 1908–1954.

Aristotle, *Metaphysics, The Complete Works of Aristotle*, vol. II. Revised Oxford Translations, J. Barnes ed., Oxford University Press, Oxford, 1984.

Aristotle, *Physics, The Complete Works of Aristotle*, vol. I. Revised Oxford Translations, J. Barnes ed., Oxford University Press, Oxford, 1984.

Aristotle, *Posterior Analytics*, Oxford Translations, W.D. Ross and J.A. Smith eds., Oxford University Press, Oxford, 1908–1954.

Aristotle, *Topics*, Oxford Translations, W.D. Ross and J.A. Smith eds., Oxford University Press, Oxford, 1908–1954.

Barrow, I. *The Usefulness of Mathematical Learning Explained and Demonstrated: Being Mathematical Lectures Read in the Publick Schools at the University of Cambridge.* Lectures given in 1664. Originally published in 1734. New imprint by Frank Cass, London, 1970.

Beman, W.W. "A Chapter in the History of Mathematics," *Proceedings of the American Association for the Advancement of Science* 46 (1897): 33–50.

Berkeley, Bishop G. *A Treatise Concerning the Principles of Human Knowledge* (1710). In [*Berkeley 1948–1957*], vol. II.

Berkeley, Bishop G. *Alciphron: Or the Minute Philosopher* (1732). In [*Berkeley 1948–1957*], vol. III.

Berkeley, Bishop G. *The Works of George Berkeley, Bishop of Cloyne*, A. Luce and T. Jessop eds., 9 vols., Thomas Nelson, Edinburgh, 1948–1957.

Boeckh, A. *Über die vierjährigen Sonnenkreise der Alten, vorzüglich den Eudoxischen: Ein Beitrag zur Geschichte des Zeitrechnung und des Kalenderwesens der Aegypter, Griechen und Römer*, G. Reimer, Berlin, 1863.

Boole, G. *An Investigation of the Laws of Thought*, Walton and Maberly, London, 1854.
Bowen, A. "Menaechmus *vs.* the Platonists: Two Theories of Science in the Early Academy," *Ancient Philosophy* 3 (1983): 12–29.
Burkert, W. *Weisheit und Wissenschaft: Studien zu Pythagoras, Philolaos und Platon*, H. Carl, Nuremberg, 1962. English translation by E.L. Minar, Jr., *Lore and Science in Ancient Pythagoreanism*, Harvard University Press, Cambridge, Mass., 1972.
Cajori, F. "Historical Note on the Graphic Representation of Imaginaries Before the time of Wessel," *American Mathematical Monthly* 19 (1912): 167–171.
Cantor, M. *Vorlesungen über Geschichte der Mathematik*, 2nd ed., 4 vols., Teubner, Leipzig, 1894–1908.
Cavalieri, B. *Geometria indivisibilibus continuorum nova quadam ratione promota*, 1653 ed. with corrections to the original 1635 ed., Ex typographia de Duciis, Bononi (Bologna) Microform.
Cayley, A. "A Sixth Memoir upon Quantics," *Philosophical Transactions of the Royal Society of London* 149 (1859): 61–90. Page citations to the reprint in Cayley 1889–1897, vol. 2.
Cayley, A. *The Collected Mathematical Papers of Arthur Cayley*, Cambridge University Press, Cambridge, 1889–1897. 13 vols.
Clavius, C. *Modus quo disciplinae mathematicae in scholis Societatis possent promoveri*, 1586.
Courant, R., and H. Robbins, *What Is Mathematics? An Elementary Approach to Ideas and Methods*, 4th ed., Oxford University Press, Oxford, 1947.
Crombie, A.C. "Mathematics and Platonism in the 16th Century Italian University and in Jesuit Educational Policy," in *Prismata: Naturwissenschaftliche Studien*, Y. Maeyama and W.G. Saltzer eds., Wiesbaden, Franz Steiner Verlag, 1977.
Dear, P. "Jesuit Mathematical Science and the Reconstitution of Experience in the Early Seventeenth Century," *Studies in History and Philosophy of Science* 18 (1987): 133–175.
De Morgan, A. *On the Study and Difficulties of Mathematics*, Chicago, Open Court, 1898. First published as a paper in London in 1831 and as a book in London in 1832 under the superintendence of the Society for the Diffusion of Useful Knowledge.
Descartes, R. *La Géométrie*, 1637. First published as an appendix to his *Discourse on Method*. Page references are to the English translation by D. Smith and L. Latham, Dover, New York, 1954.
Descartes, R. "Rules for the Direction of the Mind," in *The Philosophical Works of Descartes*, vol. I, trans. by E. Haldane and G. Ross, Cambridge University Press, Cambridge, 1970.
Descartes, R. "Reply to the Second Set of Objections" (by Mersenne), in *The Philosophical Works of Descartes*, vol. II, trans. by E. Haldane and G. Ross, Cambridge University Press, Cambridge, 1970.
Detlefsen, M. *Hilbert's Program: An Essay on Mathematical Instrumentalism*, vol. 182 of the *Synthese Library*, D. Reidel Publishing Co., Dordrecht and Boston, 1986.
Detlefsen, M. "What Does Gödel's Second Theorem Say?", *Philosophia Mathematica* 9 (2001): 37–71.
Du Bois-Reymond, E. *Reden von Emil Du-Bois Reymond*, 2 vols., Veit, Leipzig, 1886–1887.
Du Bois-Reymond, P. *Die allegemeine Funktionentheorie*, H. Laupp'schen, Tübingen, 1882. Repinted in 1968 by Wissenschaftliche Buchgesellschaft, Darmstadt.

Durège, H. *Elemente der Theorie der Funktionen einer komplexen veränderlichen Grösse*, 4th ed., Teubner, Leipzig, 1882. Page references are to the English translation of the 4th ed., *Elements of the Theory of Functions of a Complex Variable*, Fisher and Schwatt, trans., Macmillan, London, 1902. 1st ed., Teubner, Leipzig, 1864.

Ebbinghaus, H.-D., Hermes, H., Hirzebruch, F., Koecher, M., Mainzer, K., Neukirch, J., Prestel, A., and Remmert, R. *Numbers*. H.L.S. Orde, trans., Springer-Verlag, New York, 1991.

Euclid, *The Thirteen Books of the Elements*, vol. I, trans., with introduction and commentary, by T.L. Heath, Dover, New York, 1956.

Euler, L. "De la Controverse entre Mssrs. Leibniz et Bernoulli sur les logarithmes des nombres négatifs et imaginaires," *Opera Omnia*, I, ser. XVII, 195–232, 1749. Teubner, Leipzig, 1911.

Ewald, W. *From Kant to Hilbert: A Source Book in the Foundations of Mathematics*, 2 vols., Oxford University Press, Oxford, 1996.

Fine, H.B. *The Number System of Algebra*, 2nd ed., D.C. Heath, Boston/New York/ Chicago, 1890.

Frank, E. *Platon und die sogenannten Pythagoreer*, Niemeyer, Halle, 1923.

Frege, G. *Die Grundlagen der Arithmetik: Eine logisch-mathematische Untersuchung über den Begriff der Zahl*, W. Koebner, Breslau, 1884. Page references are to the English trans. by J.L. Austin, *The Foundations of Arithmetic: a logico-mathematical Enquiry into the Concept of Number*, 2nd rev. ed., Northwestern University Press, Evanston, Ill., 1968.

Frege, G. "On Formal Theories of Arithmetic," *Sitzungsberichte der Jenaischen Gesellschaft für Medizin und Naturwissenschaft* 19 supp. 2 (1885): 94–104. Page references are to the English trans. in [*Frege 1984*].

Frege, G. *Grundgesetze der Arithmetik: Begriffsschriftlich abgeleitet*, 2 vols., H. Pohle, Jena, 1893 (vol. I), 1903 (vol. II).

Frege, G. "Über die Grundlagen der Geometrie," *Jahresbericht der Deutschen Mathematiker-Vereinigung* 12 (1903): 319–24; 368–75. English translation in [*Frege 1971*].

Frege, G. "Über die Grundlagen der Geometrie," *Jahresbericht der Deutschen Mathematiker-Vereinigung* 15 (1906): 293–309, 377–403, 423–430. English translation in [*Frege 1971*].

Frege, G. *Gottlob Frege: On the Foundations of Geometry and Formal Theories of Arithmetic*, trans. and ed., with an introduction, by Eike-Henner W. Kluge, Yale University Press, New Haven, Conn., 1971.

Frege, G. *Gottlob Frege: Philosophical and Mathematical Correspondence*, G. Gabriel et al. eds., University of Chicago Press, Chicago, 1980.

Frege, G. *Collected Papers on Mathematics, Logic and Philosophy*, B. McGuinness, ed., B. Blackwell, Oxford, 1984.

Frend, W. *The Principles of Algebra*, J. Davis, London, 1796.

Fritz, K. von "The Discovery of Incommensurability by Hippasus of Metapontum," *Annals of Mathematics*, 2nd ser. 46 (1945): 242–264.

Gassendi, P. *Exercises Against the Aristotelians*, in *The Selected Works of Pierre Gassendi*, Craig Brush, trans. and ed., Johnson Reprint, New York, 1972.

Gauss, C. "Demonstratio nova theorematis omnem functionem algebraicam rationalem integram unius variabilis in factores reales primi vel secundi gradus resolvi posse," in [*Gauss 1870–1927*], vol. III.

Gauss, C. *Theoria residuorum biquadraticorum, commentatio secunda (Anzeige)*, Göttingische gelehrte Anzeigen, 1831. Reprinted in [*Gauss 1870*–1927], vol. II. Page references are to this reprinting. English trans. of excerpts can be found in [*Ewald 1996*], vol. I.

Gauss, C. *Werke*, 12 vols., Königlichen Gesellschaft der Wissenschaften, Göttingen, *1870–1927*.

Gauss, C. "Gauß' wissenschaftliches Tagebuch 1796–1814," edited with a commentary by F. Klein, *Mathematische Annalen* 57 (1903–1904): 1–34.

Gödel, K. *Kurt Gödel: Collected Works*, 3 vols., Oxford University Press, Oxford, 1986–1995.

Grassmann, H. *Die lineale Ausdehnungslehre: Ein neuer Zweig der Mathematik dargestellt und durch Anwendungen auf die übrigen Zweige der Mathematik, wie auch auf die Statik, Mechanik, die Lehre vom Magnetismus und die Krystallonomie erläutert*, Otto Wigand, Leipzig, 1844. English trans. in *A New Branch of Mathematics: The Ausdehnungslehre of 1844, and Other Works*. L. Kannenberg, ed. and trans., Open Court, Chicago and LaSalle, Ill., 1995.

Hahn, H. "The Crisis in Intuition," in *Krise und Neuaufbau in den exakten Wissenschaften*, F. Deuticke, Leipzig and Vienna, 1933. Page references are to the English translation in [*Hahn 1980*], 73–102.

Hahn, H. *Empiricism, Logic and Mathematics*, D. Reidel, Dordrecht, London, and Boston, *1980*.

Hamilton, W.R. "On Conjugate Functions, or Algebraic Couples, as Tending to Illustrate Generally the Doctrine of Imaginary Quantities, and as Confirming the Results of Mr Graves Respecting the Existence of Two Independent Integers in the Complete Expression of an Imaginary Logarithm," *British Association Report* 1834, published 1835, pp. 519–523.

Hamilton, W.R. "Theory of Conjugate Functions, or Algebraic Couples; with a Preliminary and Elementary Essay on Algebra as the Science of Pure Time," *Transactions of the Royal Irish Academy* 17 (1837): 293–422.

Hamilton, W.R. "On Quaternions; or on a New System of Imaginaries in Algebra," *The London, Edinburgh and Dublin Philosophical Magazine and Journal of Science*, 3rd ser., published in 18 installments over the years 1844–1850.

Hankel, H. *Theorie der complexen Zahlensysteme: Insbesondere der gemeinen imaginären Zahlen und der Hamilton'schen Quaternionen nebst ihrer geometrischen Darstellung*, Leopold Voss, Leipzig, 1867.

Hankel, H. *Zur Geschichte der mathematik in alterthum und mittelalter*, Teubner, Leipzig, 1874.

Heath, Sir Thomas, *A History of Greek Mathematics*, 2 vols., Clarendon Press, Oxford, 1921. Page references are to the 1981 Dover reprint.

Heine, E. "Die Elemente der Funktionenlehre," *Journal für die reine und angewandte Mathematik* 74 (1872): 172–188.

Henkin, L. "Completeness in the Theory of Types," *Journal of Symbolic Logic* 15 (1950): 81–91.

Hilbert, D. *Die Grundlagen der Geometrie*, in *Festschrift zur Feier der Enthullung des Gauss–Weber Denkmals in Göttingen*, Teubner, Leipzig, 1899.

Hilbert, D. "Über den Zahlbegriff," *Jahresbericht der Deutschen Mathematiker-Vereinigung* 8 (1900): 180–194. Reprinted in [*Hilbert 1930b*]. Page references are to this reprinting.

Hilbert, D. "Mathematische Probleme," *Archiv der Mathematik und Physik*, 3rd ser., 1 (1901): 44–63, 213–237. English trans. in *Bulletin of the American Mathematical Society* 8 (1902): 437–479. Page references are to this translation.

Hilbert, D. "Über die Grundlagen der Logik und der Arithmetik," *Verhandlungen des Dritten Internationalen Mathematiker-Kongresses in Heidelberg vom 8. bis 13. August 1904*, Teubner, Leipzig, 1905. Page references are to the English translation in [*van Heijenoort 1967*].

Hilbert, D. "Probleme der mathematischen Logik," notes by M. Schönfinkel and P. Bernays of a course of lectures by Hilbert in the summer semester of 1920. A brief excerpt from these lectures is translated in [*Ewald 1996*], vol. II, 943–946.

Hilbert, D. "Neubegründung der Mathematik. Erste Mitteilung," *Abhandlungen aus dem mathematischen Seminar der Hamburgischen Universität* 1 (1922): 157–77. Page references are to the reprinted version in [*Hilbert 1935*], vol. III.

Hilbert, D. "Über das Unendliche," *Mathematische Annalen* 95 (1926): 161–190. Quotations are from and page references are to the English translation "On the Infinite" in [*van Heijenoort 1967*].

Hilbert, D. "Die Grundlagen der Mathematik," *Abhandlungen aus dem mathematischen Seminar der Hamburgischen Universität* 6 (1928): 65–85. Quotations are from and page references are to the English translation in [*van Heijenoort 1967*].

Hilbert, D. "Problems of the Grounding of Mathematics," Eng. trans. of "Probleme der Grundlegung der Mathematik," *Mathematische Annalen* 102 (1930): 1–9. Page references are to the printing in P. Mancosu, ed. and trans., *From Brouwer to Hilbert: The Debate on the Foundations of Mathematics in the 1920s*, Oxford University Press, Oxford, 1998.

Hilbert, D. *Die Grundlagen der Geometrie*, 7th ed., Teubner, Leipzig and Berlin, 1930.

Hilbert, D. "Naturerkennen und Logik," *Die Naturwissenschaften* 18 (1930): 959–963. Page references in text are to the reprinted version in [*Hilbert 1935*], vol. III.

Hilbert, D. *Gesammelte Abhandlungen*, 3 vols., Springer-Verlag, Berlin, 1932–1935.

Hilbert, D., and W. Ackermann, *Grundzüge der theoretischen Logik*, Springer Verlag, Berlin, 1928. English trans. of the second edition by L. Hammond, G. Leckie, and F. Steinhardt, edited and with notes by R. Luce, Chelsea, New York, 1958.

Hilbert, D., and P. Bernays, *Grundlagen der Mathematik*, vol. I, Springer-Verlag, Berlin, 1934.

Hobbes, T. "De principiis et ratiocinatione geometrarum. Ubi ostenditur incertitudinem falsitatemque non minorem inesse scriptis eorum, quam scriptis Physicorum & Ethicorum," in *Thomae Hobbes Malmesburiensis Opera Philosophica Quae Latine Scriptsit Omnia in Unum Corpus Nunc Primum Collecta*, W. Molesworth, ed., 5 vols., Andrew Crooke, London, 1666. Reprinted by Scientia Verlag, Aalen, Germany, 1966.

Hobson, E.W. *Theory of Functions of a Real Variable and the Theory of Fourier's Series*, Cambridge University Press, Cambridge, 1926.

Hölder, O. "Über den Casus Irreducibilis bei der Gleichung dritten Grades," *Mathematische Annalen* 38 (1892): 307–312.

Huntington, E.V. "A Complete Set of Postulates for the Theory of Absolute Continuous Magnitudes," *Transactions of the American Mathematical Society* 3 (1902): 264–279.

Hutton, C. *A Mathematical and Philosophical Dictionary*, 2 vols., J. Johnson and G.G. and J. Robinson, London, 1795–1796. Reprinted by G. Olms Verlag, Hildesheim and New York, 1973. Reprinted in 4 vols. by Thoemmes Press, Bristol, 2000.

Jesseph, D. "Hobbes and Mathematical Method," *Perspectives on Science* 1 (1993): 306–341.

Kant, I. *Critique of Pure Reason*, Norman Kemp Smith, trans., and ed., St. Martin's, New York, 1965.

Klein, F. "Über die sogennante Nicht-Euklidische Geometrie," *Mathematische Annalen* 4 (1871): 573–625.

Klein, F. *Vergleichende Betrachtungen über neuere geometrische Forschungen*, Dietrich, Erlangen, 1872. Reprinted in [*Klein 1921–1923*], vol. I, 460–497.

Klein, F. *Gesammelte mathematische Abhandlungen*, J. Springer, Berlin, 1921–23. 3 vols.

Lacroix, S. *Traité élémentaire de calcul différential et du calcul intégral: Précédé de réflexions sur la manière d'enseigner les mathématiques, et d'apprécier dans les examens le savoir de ceux qui les ont étudiées*, Chapelet, Paris, 1802.

Lacroix, S. *Essais sur l'enseignement en général et sur celui des mathématiques en particulier*, V. Courcier, Paris, 1805.

Lacroix, S. *Elémens d'algébre: Á l'Usage de l'École Centrale des Quatre-Nations*, 11th ed., V. Courcier, Paris, 1815. 1st ed. Courcier, Paris, 1808.

Lacroix, S. *An Elementary Treatise on the Differential and Integral Calculus*, trans. of [*Lacroix 1802*] by C. Babbage, J. Herschel, and G. Peacock, J. Smith for J. Deighton and Sons, Cambridge, 1816.

Lacroix, S. *Elements of Algebra*, trans. of [*Lacroix 1808*] by J. Farrar. Hilliard and Metcalf, Cambridge, 1825.

Lambert, J.H. "Theorie der Parallellinien," *Magazin für reine und angewandte Mathematik* 2 (1786): 137–164. Reprinted in P. Stäckel and F. Engel, eds., *Theorie der Parallellinien von Euclid bis auf Gauss*, Teubner, Leipzig, 1895.

Leibniz, G.W.F. *New Essays Concerning Human Understanding*, 1704. English trans. by A. Langley, Open Court Press, LaSalle, Ill., 1949.

Leibniz, G.W.F. *Opera philosophica quae exstant latina, gallica, germanica omnia*, J.E. Erdmann, ed., Scientia, Aalen, 1959.

Lewis, A. "Complex Numbers and Vector Algebra," in *Companion Encyclopedia of the History and Philosophy of the Mathematical Sciences*, vol. II, Routledge, London, 1994.

Locke, J. *An Essay Concerning Human Understanding*, 1689. Reprinted in 2 vols., A. Fraser, ed., Dover, New York, 1959.

MacLaurin, C. *A Treatise of Fluxions*, 2 vols., Ruddimans, Edinburgh, 1742. Page references to the partial reprint in [*Ewald 1996*], vol. I.

Mancosu, P. *Philosophy of Mathematics and Mathematical Practice in the Seventeenth Century*, Oxford University Press, Oxford, 1996.

Maseres, F. *A Dissertation on the Use of the Negative Sign in Algebra*. S. Richardson, London, 1758.

McCarty, D. "David Hilbert and Paul du Bois-Reymond: Limits and Ideals," in *One Hundred Years of Russell's Paradox*, G. Link, ed., DeGruyter, Berlin, 2004.

Molland, A. "Shifting the Foundations: Descartes's Transformation of Ancient Geometry," *Historia Mathematica* 3 (1976): 21–49.

Monge, G. *Géométrie descriptive: Leçons données aux écoles normales, l'an 3 de la République*, Boudouin, Paris, 1798.

Novy, L. *Origins of Modern Algebra*, Noordhoff International, Leiden, and Academia, Prague, 1973.

Pasch, M. *Vorlesungen über neuere Geometrie*, Teubner, Leipzig, 1882.

Peacock, G. *A Treatise on Algebra*, Deighton, Cambridge; Rivington, and Whittaker, Treacher & Arnot, London, 1830.

Peacock, G. "Report on the Recent Progress and Present State of Certain Branches of Analysis," in *Report of the Annual Meeting of the British Association for the Advancement of Science*, J. Murray, London, 1834.

Peacock, G. *Treatise on Algebra*, 2 vols. (vol. I, *Arithmetical Algebra*; vol. II, *On Symbolical Algebra and its Applications to the Geometry of Position*), Deighton, Cambridge; Rivington, and Whittaker, Treacher & Arnot, London, 1842–1845.

Peano, G. "Foundations of Analysis," trans. of "Sui fondamenti dell' analisi," *Mathesis Societá Italiana di Matematica, Bolletino* 2 (1910): 31–37. In H. Kennedy, ed. and trans., *Selected Works of Giuseppe Peano*, University of Toronto Press, Toronto, 1973, 219–226.

Pererius, B. *De communibus omnium rerum naturalium principiis et affectionibus*, 1576.

Pierpont, J. *Functions of a complex variable*, Ginn, Boston and New York, 1914. Reprinted Dover, New York, 1959. Page references are to this reprint.

Plato, *The Republic*, in *Plato: The Collected Dialogues*, P. Shorey, trans., E. Hamilton and H. Cairns, eds., Princeton University Press, Princeton, N.J., 1961.

Playfair, J. "On the Arithmetic of Impossible Quantities," *Philosophical Transactions of the Royal Society of London* 68 (1778): 318–343.

Poincaré, H. *Science and Hypothesis*, W.J. Greenstreet, trans. of the 1905 French original. Dover, New York, 1952.

Poncelet, J-V. *Traité des propriétés projectives des figures; Ouvrage utile à ceux qui s'occupent des applications de la géométrie descriptive et d'opérations géométriques sur le terrain*, Bachelier, Paris, 1822. 2 vols.

Proclus, *A Commentary on the First Book of Euclid's Elements*, G.R. Morrow, trans., Princeton University Press, Princeton, N.J., 1970.

Reidemeister, K. *Das exakte Denken der Griechen: Beitrage zur Deutung von Euklid, Plato, Aristoteles*, Classen and Goverts, Hamburg, 1949.

Resnik, M. *Frege and the Philosophy of Mathematics*, Cornell University Press, Ithaca, NY, 1980.

Richards, J. *Mathematical Visions: The Pursuit of Geometry in Victorian England*, Academic Press, Boston, 1988.

Russell, B. *The Principles of Mathematics*, George Allen and Unwin, London, 1903. Page references are to the seventh impression of the 2nd ed.

Schlesinger, L. "Über Gauss' Arbeiten zur Funktiontheorie," reprinted in [*Gauss 1870–1927*], vol. X.2, 1–122.

Smith, D.E. *A Sourcebook in Mathematics*, McGraw-Hill, New York, 1929. Unabridged reprint by Dover, New York, 1959.

Stäckel, P. *Wolfgang und Johann Bolyai: Geometrische Untersuchungen*, pt. I, Teubner, Leipzig and Berlin, 1913.

Thomae, J. *Elementare Theorie der analytischen Functionen einer complexen Veränderlichen*, 2nd enl. and rev. ed., L. Nebert, Halle, 1898.

Thomas, I., ed. and trans., *Greek Mathematical Works, vol. I, Thales to Euclid*, Loeb Classical Library, vol. 335, Harvard University Press, Cambridge, Mass., 1939. Page references are to the rev. and enl. edition of 1980, reprinted in 1998.

Van Heijenoort, Jean, *From Frege to Gödel: A Sourcebook in Mathematical Logic, 1879–1931*, Harvard University Press, Cambridge, Mass., 1967.

Veblen, O. "A System of Axioms for Geometry," *Transactions of the American Mathematical Society* 5 (1904): 343–384.

Viéte, F. *Introduction to the Analytic Art*, T.R. Witmer, ed. and trans., Kent State University Press, Kent, Ohio, 1983.

Virchow, R. *The Freedom of Science in the Modern State*, English trans. of *Die Freiheit der Wissenschaft in Modernen Staat*, Berlin, Wiegart, Hempel and Parey. London, John Murray, 1878.

Vogt, H. "Die Entdeckungsgeschichte des Irrationalen nach Plato und anderen Quellen des 4. Jahrhunderts," *Bibliotheca Mathematica* 3 (1910): 97–155.

Waismann, F. *Introduction to Mathematical Thinking: The Formation of Concepts in Modern Mathematics*, Ungar, New York, 1951.

Wallace, W. *Galileo and His Sources*, Princeton University Press, Princeton, N.J., 1984.

Wallis, J. *Arithmetica infinitorum*, 1656. In *Opera mathematica*, vol. I of 2 vols. reprinted by Georg Olms, Hildesheim, 1968.

Wallis, J. *A Treatise of Algebra, Both Historical and Practical: Shewing the Original, Progress, and Advancement Thereof, from Time to Time, and by What Steps It Hath Attained to the Height at Which It Now Is.* John Playford, London, 1685.

Wilson, E.B. "The Foundations of Mathematics," *Bulletin of the American Mathematical Society* 11 (1904–1905): 74–93.

Wilson, E.B. "Logic and the Continuum," *Bulletin of the American Mathematical Society* 14 (1907–1908): 432–443.

Wilson, E.B. *Advanced Calculus*, Ginn, Boston and New York, 1912.

Windred, G. "History of the Theory of Imaginary and Complex Quantities," *Mathematical Gazette* 14 (1929): 533–541.

Young, J.W. *Lectures on Fundamental Concepts of Algebra and Geometry*, Macmillan, New York, 1911.

Zeuthen, H.G., "Die geometrische Construction als Existenzbeweis in der antiken Geometrie," *Mathematische Annalen* 47–48 (1896–1897): 222–228.

CHAPTER 9

INTUITIONISM AND PHILOSOPHY

CARL POSY

The Two Phases of Intuitionism

The word "philosophy" in this chapter's title refers to the philosophical views of L. E. J. Brouwer, intuitionism's founder, and also to the views of Brouwer's student Arend Heyting and those of Michael Dummett. Intuitionism has a philosophical program—indeed, a number of apparently diverse programs, as I shall point out—and it has a philosophical history. That is the main point of this chapter. But the title-word "Intuitionism" itself signifies technical intuitionistic mathematics itself and modern intuitionistic logic. For the philosophy or philosophies of intuitionism are inseparable from their technical core: intuitionistic mathematics in Brouwer's case, and intuitionistic logic in the case of Heyting and Dummett.[1]

Israel Science Foundation grant #914/02-1 supported the research for and writing of this chapter. I benefited from helpful comments by David McCarty, and I am especially grateful to Stewart Shapiro whose generous advice improved this chapter and whose encouragement and patience made it possible.

[1] Kino et al. [1970], which reports the proceedings of a conference held in 1968, two years after Brouwer's death, recognized a distinction between the development of Brouwerian themes (even in ways that diverged from Brouwer's own ideas), and quite separate constructivist themes that take inspiration rather than direct influence from Brouwer. The present volume is based on a kindred, though far from identical, division of the field. On the one hand, the present chapter concerns Brouwer's intuitionistic mathematics, its

Intuitionism began in Brouwer's doctoral dissertation (Brouwer [1907]). Brouwer, a Dutch mathematician of extraordinary scope, vision, and imagination, was a major architect of twentieth-century mathematics—his fixed-point theorem, for instance, is a landmark of modern topology—but his dissertation contained a trenchant attack on the foundations of modern mathematics and the seeds of a radical revision of the entire field. His views held then, and hold today, great mathematical and philosophical interest.

Brouwer's foundational work has special *mathematical* interest, because it clashed with the standard mathematics—now misnamed "classical" mathematics—that came to be solidified in the late nineteenth and early twentieth centuries. His theory of the continuum is central here. For in the course of solving a pressing problem about the continuum, he produced a new intuitionistic mathematics that not only refined the classical set-theoretic picture but also sometimes actually contradicted classical mathematics. It denies, for instance, that there are any fully defined discontinuous functions.

Brouwer's work interests *philosophers* because his mathematics rests upon a unique epistemology, a special ontology, and an underlying picture of intuitive mathematical consciousness. Brouwer used this philosophical picture not only to found his positive mathematical work but also to ground a sweeping attack on classical mathematics in general and the classical use of logic and language in particular. The classical mathematician, in Brouwer's view, illicitly appeals to logic and language in order to fill the gaps created by too narrow a view of intuitive construction. Brouwer was particularly harsh toward Hilbert's formalist program in this context.

Had Brouwer won the day, this is the picture we would see of intuitionism: a new mathematics, based on his account of nonlinguistic intuitive construction and largely logic-free. And had he won, this would be our picture of all mathematics.

But Brouwer did not win the larger mathematical battle. Contemporary mathematics is generally nonconstructive,[2] and it is routinely formalized. Moreover, neither did he prevail unilaterally within the intuitionistic camp. Several of Brouwer's mathematical and philosophical doctrines have taken on independent lives and have been developed in ways that actually diverge from his own views.

This is a "post-Brouwerian" phase of traditional intuitionism. Its main architects are Heyting and Dummett. Here, too, there is a technical core and a

---

philosophical underpinnings, and the philosophical issues that arose from post-Brouwerian intuitionism. A number of these issues revolve around Heyting's logic and the subsequent metalogic, so these technical points will briefly be sketched as well. However, chapter 10, by David McCarty, covers the strictly mathematical treatment of topics stemming from the technical core of intuitionism.

[2] The return by E. Bishop [1967] to the idea of a constructive mathematics has much interest but not much following in practice.

philosophical gloss. Technically, phase-II intuitionism centers not on mathematics but rather, as I said, upon Heyting's formal system for "intuitionistic" logic and the associated metalogic for that system. Heyting's logical system—differing though it did from Brouwer's pronouncements—had enormous impact. For it spawned decades of research into intuitionistic logic in its own right, and intensive study of connections between intuitionistic logic and other branches of classical and nonclassical logic. For many logicians, intuitionistic logic and its metalogic are the only part of intuitionism that they know.

Philosophically, the second phase deviates even more radically from Brouwer: Heyting himself jettisoned any metaphysical foundation for mathematics and even adopted a more tolerant attitude to classical mathematics. And, building on Heyting, Dummett justifies formal intuitionistic logic via a general "linguistic anti-realism" that replaces the notion of "truth" in our fundamental conception of language with the notion of "warranted assertability." In so doing, Dummett not only eschews Brouwer's ontological approach to intuitionism and its anti-linguistic bias, but he even abandons the special mathematical focus of Brouwer's philosophy.

Chapter Outline

Part I will show you some main points of Brouwer's mathematics and the philosophical doctrines that anchor it. It will point out that Brouwer's special conception of human consciousness spawns his positive ontological and epistemic doctrines as well as his negative program.[3] Part II focuses on intuitionistic logic: once again a brief picture of the technical field will precede the philosophical analyses—this time those of Heyting and Dummett—of formal intuitionistic logic and its role in intuitionism.

That will describe today's intuitionism. Not a terribly flattering description, I must say: intuitionism will emerge as technically deviant in both mathematics and logic, philosophically divided, and, as we shall see, resting on an irreducible internal tension. Part III, however, aims to show that matters aren't (or needn't be) so bleak. It suggests, in particular, that putting all this in historical perspective—a look at mathematical precedents starting from Aristotle, and logical and philosophical ones from Kant—will show intuitionism as technically less quixotic and philosophically more unified than it had initially seemed. This historical hindsight will also address the internal tension of which I spoke.

[3] Dummett (in [1973] and in [2000], ch. 7) stresses the ontological side of Brouwer's philosophy of mathematics, and Michael Detlefsen in [2002] sees Brouwer as primarily an epistemologist. But it is important to note that phenomenology—rather than either pure ontology or pure epistemology—lies at the base of Brouwer's philosophical worldview.

I. Brouwerian Intuitionism

I.1. Background: The Problem of the Continuum

Let us start with the problem about the continuum that set much of Brouwer's mathematical agenda. The problem emerged from the roil of new mathematics and philosophy that greeted the twentieth century.

This was a period of growing mathematical abstraction: the new functional analysis and its generalizations in algebra and topology were testaments to this trend. They took mathematics irrevocably away from any dependence on perceptual intuition.

It was a period of unification in mathematics: mathematicians finally closed the ancient gap between the notion of number and that of a continuous magnitude like a surface or a smooth motion. Weierstrass and Dedekind were chief architects here. They showed how to build a continuous manifold out of "real numbers" which were themselves defined from sets or sequences of rational (and ultimately, natural) numbers.[4]

And, most important, it was a period in which the notion of infinity finally came of age in mathematics: Weierstrass's convergent sequences and Dedekind's cuts were infinite sets. Indeed, the continuum was thus defined as an infinite collection whose elements are themselves infinite sets each of whose elements in turn is an infinite sequence.[5] Cantor's set theory made these all mathematically rigorous. It made precise, for instance, the heretofore elusive difference between the set of rational numbers and the set of real numbers: the rationals are countable;[6] the reals are not. The reals are closed under convergence; the rationals are not.[7] And, in setting all of this in order, Cantor developed his beautiful theory of transfinite arithmetic and provided tools to study the fine structure and ordering of the continuum.

Brouwer actively embraced the move to abstraction. But the set-theoretic construction of the continuum and Cantorian infinity gave him pause. The problem

[4] The condition that a sequence β converge is that $\forall n \exists m \forall k |\beta(m) - \beta(m+k)| < 2^{-n}$.

[5] The reason for this extra level of infinite set is that a real number has to be an equivalence class of convergent sequences, all of which converge to the same limit. Thus, for instance, the two sequences $\{0.4, 0.49, 0.499, 0.4999, \ldots\}$ and $\{0.5, 0.5, 0.5, \ldots\}$ are distinct as sequences, yet they both converge to the real number ½. See note 13 below.

[6] Cantor proved that the set of rational numbers is countable. That is, it stands in one-to-one-correspondence with the set of natural numbers.

[7] That is, every convergent sequence of real numbers converges to a real number. The sequence $\{1, 1.4, 1.41, 1.414, 1.4141, 1.41412, 1.414121, 1.4141213, 1.41412135, 1.414121356, \ldots\}$ converges to $\sqrt{2}$, and thus shows that not every convergent sequence of rational numbers converges to a rational number.

was not so much that set theory embraced infinite sets and infinite sequences.[8] The problem was, rather, the new set-theoretic penchant toward *arbitrary* sets and sequences: sets that *cannot be described* and sequences whose elements *cannot be calculated.*

Some mathematicians—Brouwer calls them "pre-intuitionists"[9]—attempted to restrict mathematics to finitely describable sets and algorithmically generated sequences. Finite description and algorithmic calculation, they thought, provide the means by which our human intellects can grasp infinite objects. This attempt, however, failed conceptually and mathematically. "Finite description" by itself is a paradoxical notion: think of the smallest set that cannot be finitely described. I have just finitely described it. And "algorithmically generated sequence"—at least in its ordinary sense— is mathematically too weak to provide all the real numbers we need. For the set of all algorithms is a countable set.

And so Brouwer took on the task of providing an epistemologically responsible account of the mathematical continuum. It was a philosophical task ("epistemologically responsible" needs a philosophical gloss) as well as a mathematical one; and, as I said, his philosophical and mathematical solutions went hand-in-hand. On the mathematical side he provided a constructive theory of finite mathematics together with a new set theory encompassing both discrete and continuous infinities. And philosophically he anchored this with special epistemological and ontological doctrines, doctrines that themselves derived from an overall phenomenological outlook. I will briefly describe Brouwer's mathematical solution and the intuitionistic mathematics that he derives from it. Then will I turn to his philosophy of mathematics: his phenomenological theory together with the positive and negative philosophical doctrines it spawned.

I.2. Intuitionistic Mathematics

Brouwer's intuitionistic mathematics naturally divides between discrete—or, as he calls it, "separable"[10]—mathematics on the one hand, and his theory of the continuum on the other.

[8] That was a nineteenth-century problem: Leopold Kronecker notoriously objected to infinity per se in mathematics. But Brouwer did not associate himself with this view.

[9] Brouwer uses the term "pre-intuitionist" in the historical capsules contained in [1952a] and [1954a]. He numbers Poincaré, Lebesgue, and Borel as the main pre-intuitionists. Troelstra [1982] (following Heyting [1934] and [1955] and Bockstaele [1949]) adds Baire, Lusin, and, marginally, Hadamard to the list and uses the term "semi-intuitionists."

[10] Section III.2, below, discusses the reason behind the word "separable" here.

Separable Mathematics

Discrete mathematics begins for Brouwer with the operation of forming ordered pairs of distinguished elements, and continues by considering repeated iterations of that activity. This produces finite mathematics. For such iterations, says Brouwer, generate each natural number; and by abstract manipulations we can derive the standard arithmetic operations, the full complement of integers (i.e., negative whole numbers as well) and the rational numbers as pairs of integers. Brouwer carries forward this picture of construction in the discrete realm in two strong ways.

On the one hand, he entertains increasingly complex and abstract structures. For we can form a set (his word is "species") of objects thus produced or producible, and then take that as an element of a yet more complex constructive continuation. In his dissertation he speaks of constructing sets whose order types—in Cantorian terms—are of the form $m_1\omega^{p_1} + m_2\omega^{p_2} + \ldots$. In subsequent work he specifies even more elaborate ordinal structures.[11]

Yet on the other hand, Brouwer does not countenance the existence of any mathematical entity that cannot be produced in this way (i.e., by a predicative sequential process). This brings him into conflict with Cantor's theory of higher cardinals. To be sure, Brouwer allows the species of all (constructible) infinite sets of natural numbers and accepts Cantor's proof that this set cannot stand in one-one correspondence with the natural numbers. But he will deny that we have thus created a new cardinal number, greater than $\aleph_0$. The species thus constructed are, according to Brouwer, merely "denumerably unfinished."[12]

I should emphasize that despite this anti-Cantorian bent, Brouwer's basic constructions of finite mathematical objects match the ordinary (set-theoretic) picture of the integers and rational numbers fairly well. His theory of the continuum, though, diverges radically from standard mathematics.

The Continuum

To be sure, Brouwer's mature theory, like classical mathematics, builds the continuum from infinite convergent sequences. And like the pre-intuitionists, Brouwer held that a legitimate infinite object must be given by a principle or law. But, unlike these constructivist predecessors, he did not require the generating laws to be fully deterministic. Specifically, a Brouwerian sequence, σ (whose entries are, say, rational numbers), might permit a certain amount of free choice to establish the value of $\sigma(n)$ for some (possibly all) n. The continuum for him was built up from these "choice sequences," and he used this idea to derive its characteristic properties.

Put formally, a choice sequence, σ, is given by a preset finite initial segment $\langle\sigma(1), \ldots, \sigma(n)\rangle$ together with a principle which, given $\langle\sigma(1), \ldots, \sigma(k)\rangle$ determines

[11] See Brouwer [1954a].

[12] See Brouwer [1907], ch. 3, and Brouwer [1908].

the range of possible choices for $\sigma(k+1)$. This principle might allow but a single possible value of $\sigma(k+1)$ for each k; in that case we have an ordinary algorithm. But it might very well allow a broad collection of possible choices. Thus, for instance, we might start a particular sequence α, of rational numbers, with the determination $\alpha(1) = ½$, and then set the rule that for each k, $\alpha(k+1)$ must be a rational number whose distance from $\alpha(k)$ is less than or equal to $(½)^{k+1}$. Here we have an infinite number of possible choices for each $\alpha(k)$.

The definition of a entails that this sequence converges, and thus generates a real number, r_α. The definition, in fact, already shows that r_α is in the interval $\mathbb{R}_{[0,1]}$ (the interval of real numbers between 0 and 1, including the end points).

On the other hand, we cannot tell from this definition whether the real number r_α is less than, equal to, or greater than the real number ½. Indeed, we might never be able to determine that one of these relations holds, for so long as we happen to choose $\alpha(k) = ½$, each one of the possible relations remains an open possibility. This potential for indeterminacy lies at the heart of Brouwer's mathematics.

Actually, Brouwer toyed with other techniques for producing real numbers whose properties may be thus indeterminate. He sometimes used what he called "fleeing properties" to produce such real numbers. Here is an example. Let β be the following sequence of rational numbers:

$\beta(n) = (-½)^n$	if by the nth place in the decimal expansion of π, no sequence of the form 9999999999 has yet occurred.
$\beta(n) = (-½)^k$	if $n \neq k$ and the kth place in the decimal expansion of π is the place at which occurs the last member of the first sequence of the form 9999999999 in that expansion occurs.

So long as no sequence of the form 9999999999 turns up, as we calculate the decimal expansion of π, β oscillates left and right of ½, getting ever closer. It will stop and stay at a particular rational distance from ½ if and when such a sequence occurs. The stopping point will be less than ½ if the k is odd, and greater than ½ if k is even. We can always determine the nth element for any n and can always calculate the corresponding real number to any desired degree of accuracy. So, once again, we know that we are generating a real number that lies in the interval $\mathbb{R}_{[0,1]}$. But, just as in the case of r_α, we cannot tell what relation the number r_β has to the real number ½. Brouwer fully believed that we shall always be able to supply fleeing properties in order to generate such indeterminate real numbers.

Interestingly, examples like these provide a special intuitionistic proof that the continuum, built as it is from such indeterminate sequences, is indeed uncountable.

The proof assumes that a function is defined at a point in its domain only if its value at that point is calculable. So a function, *f*, that is to enumerate, say, $\mathbb{R}_{[0,1]}$, must be calculable at the argument $x = ½$, and at each of its other arguments. And *f* must also be one-to-one. (That is if $y \neq x$, then $f(y) \neq f(x)$.) But no such *f* can meet both of these requirements at the argument $x = r_\alpha$ or $x = r_\beta$, for we cannot determine whether any one of these is equal to ½ or not.

Set Theory and Full Mathematics

Brouwer fashioned an intuitionistic set theory to handle sets that include such unruly sequences, and from this in turn he built a full intuitionistic mathematics.

Brouwer's set theory has, as we saw, species (collections whose elements share a common property),[13] but he also defines constructive sets of sequences that are generated in some mutually common way. He calls these spreads. A spread is set of rules for admissible finite sequences of natural numbers such that each such admissible finite sequence, $\langle n_1, \ldots, n_k \rangle$, has at least one admissible successor $\langle n_1, \ldots, n_k, n_{k+1} \rangle$. An infinite sequence of natural numbers will belong to the spread if each finite subsequence does. We get infinite sequences whose elements are more sophisticated objects by assigning such an object to each admissible finite subsequence. In this way we can get spreads whose elements are convergent sequences of rational numbers,[14] spreads whose elements are sequences of real numbers, and spreads whose elements are even more abstract.

A spread which admits only finitely many successors to each admissible finite sequence is called a *fan*. Brouwer showed that the real numbers falling within a bounded closed interval like $\mathbb{R}_{[0,1]}$ can all be generated by the sequences in a single fan. Thus the following fan generates $\mathbb{R}_{[0,1]}$:

Let η be an enumeration of the rational numbers, such that $\eta(0) = 0$.

- 0 is the only admissible one-member sequence.
- If $\langle n_1, \ldots, n_k \rangle$ is an admissible *k* member sequence, then $\langle n_1, \ldots, n_{k+1} \rangle$ is an admissible $k + 1$ member sequence if and only if there is a natural number *p*, such that $\eta(n_{k+1})$ is of the form $p/2^{k+1}$ and $|\eta(n_k) - \eta(n_{k+1})| \leq 1/2^{k+1}$.
- The fan assigns the rational number $\eta(n_k)$ to the finite sequence $\langle n_1, \ldots, n_k \rangle$.

[13] Thus, for instance, a real number is actually an infinite collection of convergent sequences, each of which stands in the relation of coincidence with each of the others. The relation is defined as follows: $\alpha \sim \beta$ if and only if $\forall n \exists m \forall k |\alpha(m) - \beta(m)| < 2^{-n}$. It amounts to saying that α and β converge to the same limit.

[14] Thus, for instance, we might start a spread by allowing $\langle 0 \rangle$ as the only admissible one-member sequence, and allowing $\langle n_1, \ldots, n_{k+1} \rangle$ to be a successor to $\langle n_1, \ldots, n_k \rangle$ only if the absolute value difference between $n_k/2^{k-1}$ and $n_{k+1}/2^{k-1}$ is less than or equal to 2^{k-1}. If we assign the rational number $n_k/2^{k-1}$ to the admissible sequence $\langle n_1, \ldots, n_k \rangle$ of natural numbers, we will get a spread whose elements are sequences of rational numbers that converge to a real number in $\mathbb{R}_{[0,1]}$.

This fan admits a maximum of three successors to any admissible finite sequence. The real numbers in the interval are given by convergent sequences of rational numbers of the form $p/2^k$.

The mathematics that emerged from these ideas was sometimes weaker than classical mathematics; sometimes it refined classical mathematics; and sometimes actually it contradicted classical theorems.

Here is an example in which intuitionistic mathematics is *weaker*. In the classical set-theoretic continuum the "trichotomy" $[(r_\beta < ½) \vee (r_\beta = ½) \vee (r_\beta > ½)]$ is true, as is its counterpart for r_α. Indeed, these are special cases of the general classical theorem: $(\forall x)(\forall y)[(x < y) \vee (x = y) \vee (x > y)]$. However, since we cannot determine the relation of the real numbers r_α and r_β to ½, Brouwer insists that we pristinely refrain from asserting the corresponding disjunctions. These particular disjunctions, and indeed the full general theorem, are thus classically but not intuitionistically true.

Here is an example of *refinement*. A rational real number, r, is one for which we can give two natural numbers p and q such that $r = p/q$; thus, by intuitionistic lights the real number r_β is not rational. We cannot find natural numbers p and q such that $r_\beta = p/q$. On the other hand, r_β cannot be irrational. (For if there is a sequence of the requisite sort in the decimal expansion p, then r_β will be rational. If there is no such sequence, then $r_\beta = ½$, and is once again rational.) Brouwer calls this a "splitting" of the classical mathematical concept of irrational number: there are numbers that *are not* (so far as we know) rational, and there are numbers that *cannot be* rational. Brouwer points out that this is a case of the failure of the classical principle of double negation: it is impossible for it to be impossible that r_β be rational, but that is not sufficient to claim that r_β is rational.

And here is the most striking—indeed, notorious—*deviation* from classical mathematics, Brouwer's Continuity Theorem: *Fully defined functions of real numbers are always continuous.*

Brouwer actually proves this theorem in two forms. He has a weaker (negative) version saying that *a function cannot be both total* (i.e., have a value at every point in its domain) *and discontinuous at some point in its domain.*[15] And he had a stronger positive version: *every function that is total on* $\mathbb{R}_{[0,1]}$ *is continuous.*

Here, by the way, is another splitting of a classical notion: classically, the two versions of the theorem are equivalent. And, of course, classically they are false. The simple step function defined on $\mathbb{R}_{[0,1]}$ is as follows:

$$\begin{aligned} f(x) &= 1 \quad \text{for } x < ½ \\ &= 2 \quad \text{for } x \geq ½ \end{aligned}$$

[15] Pictorially, to say that a function, f, is discontinuous at a point x is to say that its value "jumps" at x. That is, there is a region of arguments, y, each of which is close to x, such that $f(x)$ is measurably distant from $f(y)$ for each y in that region.

is both total and discontinuous at $x_0 = ½$. But for Brouwer the versions are distinct (in particular, the first does not imply the second), and they require separate proofs.[16]

To prove the negative version, he defines a convergent choice sequence χ so indeterminate that the question of whether its corresponding real number r_χ is less than ½ or equal to ½ remains eternally undecided. This shows that f is not defined at r_χ, and hence is not in fact a total function.

Brouwer's proof of the positive version rests on the assumption that if a function, g (defined on $\mathbb{R}_{[0,1]}$ or alternatively on any set of reals which is generated by a fan), is actually total—that is, if

$$\forall x_{(\in \mathbb{R}[0,1])}\ \exists y[g(x) = y]$$

holds—then there must be a proof of that fact, a proof that makes use only of definitional information about g. Brouwer goes on to analyze the structure of any such proof, and shows that indeed any such proof can be turned into a proof that g is continuous.[17] In particular it will be a proof that if σ is a convergent choice sequence whose corresponding real number r_σ falls in $\mathbb{R}_{[0,1]}$, then $g(r_\sigma)$ must be calculable on the basis of a finite initial segment of σ.[18] And that will guarantee the continuity of g.[19]

This, by the way, provides an even simpler and more straightforward proof of the uncountability of $\mathbb{R}_{[0,1]}$. For, it turns out that any putative function enumerating

[16] Both versions are proved in Brouwer [1927]. Brouwer proved the positive version in earlier papers.

[17] For this theorem to hold, the domain of g must be given by a fan. The proof of the continuity theorem crucially uses this fact in describing the supposed proof that g is a total function.

[18] This in turn shows that for any k there is an n such that an approximation of x up to an accuracy of 2^{-n} will suffice to determine an approximation of y up to 2^{-k}. Such a function, g, must be continuous (i.e., "smooth") because, unlike the step function, f, defined above, one cannot approach any x without the corresponding values approaching $g(x)$.

[19] I need to add three remarks to my quick treatment of the positive continuity theorem:

(1) Brouwer's actual statement of the theorem is stronger. His claim is that every such total function is *uniformly* continuous. In the preceding note, n depends upon k but not upon x. This is what guarantees the *uniform* continuity of g.
(2) Some treatments of this theorem omit Brouwer's argument altogether, and simply assume conditions which entail continuity as part of the claim that g is total function. This, for instance, is the embarkation point in chapter 10, "Intuitionism in Mathematics."
(3) One can find a concise proof of the theorem in Heyting [1966]. Dummett [2000] has an extended discussion of Brouwer's proof and of the underlying "continuity" assumptions on the definition of real-valued and other functions.

$\mathbb{R}_{[0,1]}$ would have to be discontinuous. In particular, the value of such a function at r_χ could not be determined by a finite initial segment of χ.

Brouwer and his students went on to apply these methods to the study of geometry, algebra, calculus, and abstract topology.[20]

The Inconsistency of Classical Mathematics

Finally, a word about Brouwer's technical attack on classical mathematical principles.

In publications starting in [1948] Brouwer used what has come to be called the method of the "creating subject" to form indeterminate sequences.[21] The creating subject is an idealized mathematician who is working on some as yet undecided mathematical proposition, Q (say Goldbach's conjecture[22]). In this case let γ be the following sequence:

$\gamma(n) = (\frac{1}{2})$	if by the nth stage of his research the creating subject has neither proved nor refuted Q;
$\gamma(n) = (\frac{1}{2} - (\frac{1}{2})^k)$	if $n \geq k$ and at the kth stage in his research the creating subject has succeeded in proving Q;
$\gamma(n) = (\frac{1}{2} + (\frac{1}{2})^k)$	if $n \geq k$ and at the kth stage in his research the creating subject has succeeded in refuting Q.

In this case we actually make the strong claim $r_\gamma \neq \frac{1}{2}$ for the real number r_γ, generated by γ. (For $(r_\gamma = \frac{1}{2})$ would entail Q and $\sim Q$.[23]) Just as in the case of fleeing properties, Brouwer fully believed that we shall always be able to supply undecided propositions in order to generate such indeterminate real numbers. Real numbers such as this allowed him to claim that some of the principles of the elementary classical arithmetic of real numbers—principles like $\forall x\, \forall y[x \neq y \rightarrow ((x < y) \vee (x > y))]$—are not merely unprovable, but actually contradictory.[24]

[20] Brouwer used fleeing properties, for instance, in order to invalidate the classical version of his fixed-point theorem, and provided an intuitionistically valid alternative. See Brouwer [1952b].

[21] Brouwer mentioned a version of the method of the creating subject in his Berlin Lectures of 1927 (posthumously published as Brouwer [1992]), but he refrained from putting it in print until 1948.

[22] Goldbach's conjecture is the conjecture that every even number greater than 2 is the sum of two prime numbers. This has been confirmed up to very high even numbers, but has not yet been proven or refuted.

[23] See note 48, below, for Brouwer's reasoning here.

[24] See, e.g., Brouwer [1954b] and Heyting ([1966], sec. 8.1.2). Quite recently Van Dalen has extended Brouwer's method of the creating subject to obtain new results about the structure of the intuitionistic continuum. See van Dalen [1997] and [1999].

I.3. Brouwer's Philosophy: The Phenomenology of Intuition and Constructivity

Each part of Brouwer's intuitionistic mathematics—finite and infinite separable mathematics, the theory of the continuum, and the *reductio* of classical mathematics—has a parallel philosophical foundation. Brouwer proposed epistemic and ontological foundations for the discrete and continuous parts of mathematics, and he authored scathing philosophical critiques of the use of logic and language in classical mathematics. But all these rest upon a pervasive phenomenological base, a picture of a mathematical subject constructing objects from intuitive material abstracted from "primordial consciousness."[25] In this section I'll sketch Brouwer's theory of primordial conscious experience, his specialized phenomenology of mathematics, and the mathematical epistemology and ontology that flow from it. Then I will turn to his philosophical attacks on the classical approach to mathematics.

Brouwer's Phenomenology

General Phenomenology. Primordial consciousness, according to Brouwer, is a "dreamlike" stew oscillating between sensation and rest; and, he says, conscious life would stay that way were it not for "acts of attention" by which the subject focuses on individual changes between different sensory contents. He calls such a change the "falling apart of a life moment." Objective consciousness begins when one focuses on such an event. Through these attentive acts the subject distinguishes individual conscious elements together with their contents and discerns an order between them. The subject iterates this process and thus forms attenuated mental sequences. These in turn are the building blocks of the subject's awareness of ordinary empirical objects.

Brouwer speaks next of what he calls "causal attention." The subject discerns some similarity among distinct conscious moments and then fashions sequences with similar initial parts and similar subsequent parts ("causal chains"). In this way, says Brouwer, the subject produces awareness of the causally ordered world. Reflecting upon this is the root of science; manipulating it is technology.

Three things should be stressed. First, the subject's main productive activity is generating sequences and correlating them with other sequences. Second, the acts

[25] Though Brouwer does not refer to Husserl, he clearly shares Husserl's basic assumption that consciousness itself provides subject matter for disciplined research. Brouwer's phenomenological picture is prominent in [1929], [1933], [1948a], and [1950]. Van Atten [1999], Placek [1999], and Van Atten et al. [2002] focus on the phenomenological side of Brouwer's thought. Van Dalen [2000] describes the origin of this picture in Brouwer's early thought.

forming the world are willful impositions of structure on the original roil of consciousness. To be sure, there may be frustrations, and the formative process includes trial and error; but creative will and the imposition of will are central here. Finally, and most important for us, Brouwer claims that mathematical activity is parallel in both of these respects: It, too, is sequence-centered. It, too, is willful and creative. But, says Brouwer, it is purer, more abstract.

The Phenomenology of Mathematics. Mathematical abstracting takes place at the very beginning, in forming what Brouwer calls the basic intuition of mathematics: The *first act of intuitionism* starts, as in ordinary consciousness, with "the falling apart of a life moment into two distinct things, one of which gives way to the other, but is retained by memory." But it then "divests all quality" from " the two-ity thus born." According to Brouwer, "There remains the empty form of the common substratum of all two-ities. It is this common substratum, this empty form, which is the basic intuition of mathematics."[26]

Empty of sensory content though it may be, this two-ity is the material for all of mathematics. Numbers and finite objects come from manipulating and replicating this empty sequence. This is the source of the individual integers and rational numbers,[27] and it is the phenomenological basis of ordinary arithmetical identities.

It is important to note that as a mathematician I do my abstracting when, after that first "falling apart," I ignore sensory content and form the abstract, empty "two-ity" of which Brouwer speaks. Brouwer does not start his phenomenology of arithmetic with some concrete actual counting or combination, and then abstract the mathematical counterpart. And in particular he does not suppose that mathematical operations on finite objects are schematic skeletons of empirical operations. Mathematics is an independent, empirically empty, process of its own.

This independence is even more evident in what he calls the *second act of intuitionism*, an act which generates infinite mathematical entities in two ways:

> firstly in the form of infinitely proceeding sequences $p_1, p_2, \ldots$, whose terms are chosen more or less freely from mathematical entities previously acquired; in such a way that the freedom of choice existing perhaps for the first element p_1 may be subjected to a lasting restriction at some following p_ν, and again and again to sharper lasting restrictions or even abolition at further subsequent p_ν's, while all these restricting interventions, as well as the choices of the p_ν's themselves, may be made to depend on possible future mathematical experiences of the creating subject...[28]

[26] Brouwer [1952a], 141.
[27] See Brouwer [1907], sec. 1.
[28] Brouwer [1952a], 142.

This is the formation of choice sequences, sometimes yielding determinate sequences (that is, "abolition of freedom" mentioned here), and sometimes, indeterminate ones that depend upon the creating subject.

> secondly, in the form of mathematical species, i.e., properties supposable for mathematical entities previously acquired, and satisfying the conditon that, if they hold for a certain mathematical entity, they also hold for all mathematical entities which have been defined to be equal to it, relations of equality having to be symmetric, reflexive and transitive; mathematical entities previously acquired for which the property holds are called elements of the species.[29]

And here we have the formation of more abstract objects—species—by means of equivalence relations.

The first sort of formation shows how we can grasp infinity: discrete infinity and even the infinity needed to produce a continuum. The second opens the gates to ever increasingly abstract mathematical constructions. Individual real numbers are species of this sort, as are full spaces, and even more abstract products and quotients of spaces. These mathematical constructions parallel empirical sequence formation, but it is in fact empirically empty. There are no infinite sequences in the empirical world, nor are there empirically abstract species.

Thus Brouwer has fashioned a unique phenomenological constructivism. He is not a finitist, nor does he follow the empiricist abhorrence of abstraction. To be sure, his homespun phenomenology and ontogenesis may well grate upon some ears, and I will say something about that a bit later. But if, for now, we grant Brouwer this way of speaking, then we can say that the outcomes of all of these acts of creation are straightforward mathematical objects. Each one, no matter how abstract or complex, is intuitively grasped, and each one exists. From this stylized picture we can derive his specific doctrines about epistemology and ontology and his negative critique of classical mathematics. Let me describe these philosophical "spin-offs" of Brouwer's special constructivism.

Epistemology and Mathematical Intuition

Mathematical knowledge is for Brouwer, as it was for Kant, synthetic a priori knowledge, based on a notion of pure intuition. But Brouwer has his own unique version of mathematical intuition and his own special grounds for *apriority.*

Intuition. Mathematical intuition for Brouwer is temporal, and it is abstract.

Temporality. This is fairly straightforward. The elements distinguished in that first "falling apart" have a temporal order, and that order, together with the order of subsequent pairings, is preserved in the sequences produced by the two acts of intuitionism.

[29] Ibid.

And Brouwer's mathematics tells us that time, on his view, is tensed. It has a sense of past, present and future.[30] Indeed, Brouwerian time has a somewhat open future.[31]

Abstractness. Indeed, not only does Brouwerian mathematics have an intuitive content all its own, empty of sensory, empirical material, but Brouwer's version of mathematical intuition far outstrips any sort of sensory grasp. Brouwer, for instance, will claim that α, β, γ and χ are graspable mathematical objects. And so are the real numbers they generate, as are the spread which gives $\mathbb{R}_{[0,1]}$, the entire continuum, and more and more abstract spaces and points in analysis and topology. There is no limit, Brouwer would say, to the creative ways we might generate the new mathematical objects from the old, so long as they conform to the two laws of intuitionism. And all of these new objects—however abstract and empirically ungraspable they may be—are, in Brouwer's view, given in intuition.

Apriority. Mathematics is, for Brouwer, a priori in the two Kantian meanings of the expression. It is epistemologically independent of sensory empirical experience, and it is a necessary underpinning of empirical science. But Brouwer fleshes out this independence and necessity in ways that differ from Kant's views.

For Brouwer, mathematics is "independent" of empirical experience not because of its abstractness or because of some special notion of mathematical justification. It is independent of experience to simply because it proceeds on a special track of its own. We do not turn to empirical experience to justify, or even to exemplify, the claims of pure mathematics. *Au contraire*, we achieve empirical applications of mathematics by superimposing empirical causal sequences on our independently formed mathematical structures.[32] That, indeed, is also the sense in which mathematics underlies empirical science.

[30] The difference between the initial segment and the tail of a choice sequence shows that Brouwer respects the past/future distinction. And we see the special status of the present from the unique position at which assertions about a sequence are made. (We cannot now say that $\beta > ½$, even though that may be established in the future.) Brouwerian time is thus what MacTaggart calls an A-Series.

[31] Thus, for instance, the sequence β (defined above) has a future that is simply open regarding the determination of whether $r_\beta < ½$ $r_\beta = ½$, or $r_\beta > ½$. But we should be careful here. There are future dependent disjunctions that we can preassert. We haven't yet calculated whether $100^{100} + 1$ is prime or composite. But since we do know how to determine that, and could do so, we can indeed say right now that this number is either prime or composite. The future is *somewhat* open.

[32] See, e.g., Brouwer [1933, 1948a].

Ontology

Unlike Husserl, Brouwer does not separate the question of conscious construction from the issue of objective existence in the world or in mathematics.[33] Each object is given by a particular generating act, and what you build is what there is. This tells us the Brouwerian attitude toward construction and objects, identity and existence.

Construction and Objects. The real numbers r_α, r_β, r_γ and r_χ are perfectly respectable—that is, constructible—real numbers. But their generating constructions do not determine the relation of any one of these to the real number ½, and do not specify their exact location within the interval $\mathbb{R}_{[0,1]}$. Nor do they provide any guarantee that such determinations can ever be made. We have here concrete instances demonstrating that in Brouwer's constructive ontology legitimate objects need not be fully determinate with respect to all of their potential properties.

Brouwer sometimes toyed with the possibility of "second-order" controls on this sort of indeterminacy. Thus, while each of the numbers r_α, r_β, and even r_γ might become determinate at some stage of development; the number r_χ, in his proof of the negative version of the continuity theorem is specifically designed so that its relation to ½ will remain forever undetermined.

Identity. And what goes for general properties and relations clearly goes for identity as well. The numbers r_α, r_β, and r_γ are each formed by distinct generative acts, so they are perhaps intensionally distinguishable. And each one is certainly less determinate than the real number ½. But from a strictly arithmetic point of view, we cannot say that any one of them differs from any one of the others, nor from ½. We have here individual entities without a determination of identity. In Brouwer's constructive ontology, legitimate objects need not be distinguishable one from the other.

Existence. Brouwer, as we have seen, ties existence to constructibility. Putative mathematical entities that aren't constructed in the way that Brouwer describes simply don't exist. His foes here are the well-known indirect existence proofs based on the principle of excluded middle. To be sure, Brouwer has no problem with *reductio ad absurdum* arguments *per se.*[34]And so Brouwer, as I mentioned,

[33] See Husserl [1913] (sec. 18, among several places) for the notion of an ἐποχή (*epokhé*), an agnostic abstention from concern with the existence of consciousness's intentional objects. There is something like an *epokhé* in Brouwer's treatment of other minds in [1948a], but, according to Brouwer there, this does not transfer to a general theory of objects per se.

[34] G. Griss, in a series of papers—[1944], [1946], [1949], [1950] are prime examples—rejected the role of negation in a proper intuitionistic mathematics, and strove to provide a negationless version. Brouwer sharply rejected this position in [1948b].

happily accepts Cantor's diagonal argument as a proof that the real unit interval is uncountable. But, as we saw, he does not conclude from that that there exists an uncountable cardinal number. For, there is no construction here.

Negative Doctrines

From these considerations about intuitive content and intuitive construction Brouwer draws negative conclusions about classical mathematics, about logic and about language.

Intolerance Toward Classical Mathematics. Mathematics for Brouwer has content; it is about something. It is about the objects and structures produced according to his constructive guidelines. Thus, classical mathematics must perforce be talking about that same subject matter. And insofar as it differs from intuitionistic mathematics—for instance, in claims about trichotomy or the existence of discontinuous functions—in these matters it is simply wrong.

Moreover, since mathematics is an a priori science—its truths are necessary truths—to contradict one of these truths is to assert not merely the false but the impossible. That is why Brouwer actually speaks of the "inconsistency" harbored by one or another classical mathematical theory.

To be sure, Brouwer is not claiming that classical mathematics lacks meaning or reference. It refers to mathematical reality, the same mathematical reality to which the intuitionist refers. And it is fully meaningful. It is simply and necessarily false.

Logic. A frequent mistake of classical mathematics is to assert the existence of things that in fact don't exist. Brouwer actually provides an etiology for this classical mistake. He posits an early mathematics concerned only with finite objects and systems, and tells us that at this stage, excluded middle was indeed a valid principle. (Presumably because at this stage, mathematical objects were effectively complete.) Early mathematicians, reflecting on their intuitive constructions, formulated general logical principles—the laws of noncontradiction, syllogism, and excluded middle—and allowed themselves to construct mathematical proofs based on these. However, these rules were simply general properties of constructions, and the derivations were no more than shorthand for possible constructions.

Mathematics evolved. Its objects became infinite, constructably infinite to be sure, but infinite nonetheless; and therefore incomplete. The other logical laws maintained their validity; excluded middle did not. But mathematicians continued to apply the old logic as if nothing had changed. They took logical deduction to be a justification in its own right, the heart of the axiomatic method, but could no longer cash it out in actual constructions. That, says Brouwer, is the heart, and the weakness, of classical mathematics.

In a nutshell: For Brouwer, logic *per se* follows ontology—classical logic is appropriate for finite but not for infinite domains—but it can never lead.

Language. The classical mathematician, says Brouwer, looks to language as a replacement for the now missing intuitive content of mathematics. Language is supposed to be intuitive because it is quasi-perceptual. But for Brouwer language is an illegitimate surrogate, and it has no special intuitive value. It is not especially intuitive because, for him, mathematical intuition is not perceptual. And it is illegitimate because, on Brouwer's phenomenological view, no symbolic notation can ever accurately report the content of a conscious moment.

Brouwer's polemics against formalism in general and Hilbert's program in particular combine all of these criticisms. Hilbert wanted to secure mathematics against paradox and inconsistency by formalizing each part and then using elementary intuitive methods to prove the formal systems consistent. Intuition, for him, was indeed tied to perception: formal languages and formal systems had intuitive content; abstract, infinitary mathematics did not. Given some part of mathematics, Hilbert proposed axiomatizing it informally, transforming the axiomatic theory into a formal system, showing the formal system to be faithful ("sound and complete" in modern terminology), and then proving that formal system to be consistent. Since in his view the formal system is, as I said, intuitively grounded, intuitive ("finitary," ultimately arithmetic) methods would suffice to prove its consistency.[35]

Gödel's theorems challenged the formal viability of Hilbert's program, but well before they appeared, Brouwer objected to this program at every turn. Abstract mathematics, if practiced constructively, is fully intuitive. Translation to a formal language—to any language—is at best questionably accurate. (This, for phenomenological rather than Gödelean reasons.) Formal systems are legitimate mathematical objects, but not necessarily more intuitive than the mathematics they are formalizing. And worst of all, the original axiomatization—and its indiscriminate use of excluded middle—simply perpetuates the original mistakes of classical mathematicians. Indeed, says Brouwer, Hilbert's oft-repeated slogan that every mathematical problem is ultimately solvable is equivalent to the principle of excluded middle and, as such, amounts to an outmoded, ungrounded belief. In the end, for Brouwer, Hilbert's notion of intuition does not exhaust mathematical intuition, and his notion of construction is not constructive.

[35] See Ch. 8 in this volume, by Michael Detlefsen, for a fuller picture of Hilbert's program.

II. Intuitionistic Logic and its Philosophy

II.1. Intuitionistic Logic and Metalogic

This, then, is Brouwer's intuitionism: a revised mathematics resting on a phenomenological philosophy and accompanied by a trenchant critique of logic and language. Thus, one would scarcely expect intuitionists to formalize an intuitionistic logic or to produce formal systems for branches of intuitionistic mathematics built upon that logic. But this is exactly what happened. In the 1930s Arend Heyting published a series of papers presenting formalized intuitionistic logic and formal intuitionistic number theory. These gained immediate attention, for Heyting provided full-dress formal systems for propositional and predicate logic and for intuitionistic number theory.[36] Metamathematical studies of Heyting's systems and extensions of them soon followed. They revealed eye-opening syntactic, algebraic, and topological properties of the system and its refinements. Ultimately they led to the model-theory of intuitionistic logic, initiated by Evert Beth and perfected by Saul Kripke.

All of this acutely raised the question of how such systems and studies fit into the intuitionistic program. After displaying the system and briefly mentioning some of its main metalogical properties, I'll turn to Heyting's own assessment of the philosophical significance of intuitionistic logic, and then to Michael Dummett's treatment of the subject.

The Formal System

Here is an axiomatization of intuitionistic logic (IL) adapted from Kleene [1952] and van Dalen [1986]. The language is the full language of first-order logic, without any abbreviations:[37]

Axiom Schemata

1. $A \to (B \to A)$
2. $(A \to B) \to ((A \to (B \to C)) \to (A \to C))$

[36] Brouwer's 1923 refutation of the principle of double negation inspired a spate of publications about "*la logique de M. Brouwer*" by a number of authors outside the intuitionistic camp. See, for instance, Kolmogorov [1925], Wavre [1926a] and [1926b], Borel [1927], Glivenko [1929]. But Heyting's papers created the field of formal intuitionistic logic.

[37] One cannot interdefine the logical particles by way of DeMorgan equivalences, so the language of IL and any extended intuitionistic formal system must contain a symbol for each of the logical particles explicitly, and the formal system must contain rules and axioms for each logical particle.

3. $A \rightarrow (B \rightarrow (A \& B))$
4a. $(A \& B) \rightarrow A$
4b. $(A \& B) \rightarrow B$
5a. $A \rightarrow (A \vee B)$
5b. $B \rightarrow (A \vee B)$
6. $(A \rightarrow C) \rightarrow ((B \rightarrow C) \rightarrow ((A \vee B) \rightarrow C))$
7. $(A \rightarrow B) \rightarrow ((A \rightarrow \sim B) \rightarrow \sim A)$
8. $A(t) \rightarrow (\exists x)A(x)$
9. $(\forall x)\ A(x) \rightarrow A(t)$ (where t is free for x in A and x is not free in A)
10. $A \rightarrow (\sim A \rightarrow B)$

The inference rules are

1. $\dfrac{A \quad A \rightarrow B}{B}$

2. $\dfrac{A \rightarrow B(x)}{(\forall x)B(x)}$

3. $\dfrac{A(x) \rightarrow B}{(\exists x)A(x) \rightarrow B}.$

This version of the system is formulated so that if Axiom Schema (10) is replaced by

10′. $\sim\sim A \rightarrow A$,

the result will be a system of standard classical logic. (10) is derivable from (10′), so this shows that syntactically intuitionistic logic is a proper subsystem of classical logic. (That is, $\vdash_{\mathrm{IL}}\varphi \Rightarrow \vdash_{\mathrm{CL}}\varphi$ but not, of course, vice versa.)

Add the standard axioms for identity and for number theory, and you get a formal system for intuitionistic number theory, called HA (for "Heyting Arithmetic"). HA is, of course, a proper subsystem of the classical formal system of number theory (called PA, for "Peano Arithmetic"). The 1960s and 1970s saw the further development of intuitionistic formal systems, including formal systems for the theory of choice sequences, and even for the theory of the creating subject.[38]

Modern metamathematical studies of IL followed in quick succession after Heyting's formalizations, and these studies have not abated. Here are some standard syntactic and semantic results:

Syntactic Properties. Intuitionistic logic (and certain extensions) have the "explicit disjunction property" (if $\vdash_{\mathrm{IL}}(\varphi \vee \psi)$, then $\vdash_{\mathrm{IL}}\varphi$ or $\vdash_{\mathrm{IL}}\psi$) and the "explicit definition property" (if $\vdash_{\mathrm{IL}}\exists x\ \varphi(x)$, then for some closed singular term "t," $\vdash_{\mathrm{IL}}\varphi(t)$),

[38] See, e.g., Troelstra [1977] and [1982] on choice sequences, and Kreisel [1967], Van Rootselaar [1970], and Posy [1977] for various axiomatizations of the theory of the creating subject.

results that might well be expected in a system representing constructive mathematical thought.

Gödel showed a connection between IL and the formal system S4 of modal logic,[39] and provided a translation * between formulae such that $\vdash_{CL} \varphi$ holds if and only if $\vdash_{IL}\varphi^{\star}$. A similar property relates HA and formal classical number theory. It can even be shown that for any formula, φ, HA$\vdash\sim\varphi$ if and only if PA$\vdash \sim\varphi$.

That means, in particular, that if a contradiction is derivable in one of these systems, a related contradiction will be derivable in the other. Thus, although, as I said, HA is, strictly speaking, a proper subsystem of PA, it is not "easier" to prove the consistency of HA than it is of PA.[40]

Semantic Properties. There are mappings between the theorems of IL and the elements of various abstract mathematical structures. To give just two examples, the system displays the properties of a particular kind of algebraic "lattice,"[41] and the class of provable formulae has be shown to be parallel to certain properties of the open sets in a topological space.[42]

Philosophically, however, the most interesting semantic treatment of intuitionistic logic is the model-theoretic approach, which interprets quantifiers over domains of objects and connects formulae with truth-values. Indeed, it provided a full, precise semantic notion of truth and of logical validity.[43]

And so—or so the story goes—we get corresponding approaches to "intuitionistic truth" and "intuitionistic validity." Evert Beth published a model-theoretic interpretation of IL in [1947], and Saul Kripke produced one in [1965]. The two approaches differ in details—Kripke's is formally simpler—but their underlying idea is the same: instead of a single model, one has a partially ordered collection of nodes (generally with a base node), each of which is something like a model in its own right. The partial order among nodes is called the "accessibility" relation. Predication at each node can be defined in the standard fashion, but the recursion clauses for the truth of compound statements may well depend upon the truth-values of its component simpler parts at accessible nodes.

Thus, for instance, in Kripke's version

[39] See Gödel, [1933b].

[40] See Gödel, [1933a].

[41] These ideas can be used to show that IL cannot be characterized by any finite set of truth values. See Gödel [1932].

[42] See Rasiowa and Sikorski [1963] for expanded treatments of these algebraic and topological interpretations.

[43] A formal sentence in some mathematical theory will be true *simpliciter* if it is true in the standard model of that theory. A sentence is logically valid if it is true in every possible model, and an inference from *A* to *B* will true if every model in which *A* is true is also a model in which *B* is true.

- $\sim A$ will be true at a node w only if A is not true at any node w' that is accessible to w,
- $(A \vee B)$ will be true at w only if either A is true there or B is,
- $(\exists x)A(x)$ will be true at w only if $A(t)$ is for some t in the domain of w,[44]
- $(\forall x)A(x)$ will be true at w only if for each w' accessible to w, $A(t)$ is true at w' for each t in the domain of w'.

"Truth" for the model structure as a whole can be defined as truth at the base node, and logical validity will be truth in every model. Both the Beth and the Kripke semantics are demonstrably faithful to IL.[45]

Here, using Kripke-models, is a pair of counterexamples to classical logical truths. The first model structure contains three nodes: w_1, w_2, $w_{3.}$ where w_2 and w_3 are accessible to w_1 but not to one another. If A is true at w_2 but not at $w_{3,}$then $(A \vee \sim A)$ will not be true at w_1, thus providing a counterexample to the principle of excluded middle.

The second model structure contains an infinite collection of nodes $\{w_n\}_n$ which is ordered linearly. w_j is accessible to w_k if and only if j is greater than or equal to k. Now suppose that the domain at a node, w_n, is the set of numbers $\{1, \ldots, n\}$ and that $F(t, s)$ holds at any given node, w_n, if and only if both t and s are in the domain and $t < s$. In this case $(\exists y)F(1, y)$ is not true at w_1 (since there is no element of the domain at w_1 that is greater than 1). So clearly $(\forall x)(\exists y)\ F(x, y)$ is not true at w_1[46] However, $(\forall y) \sim F(1, y)$ is clearly not true at w_1, and so neither is $(\exists x)(\forall y)\sim F(x, y)$. This thus shows why $[(\exists x)(\forall y)_{y \neq x} \sim F(x, y) \vee (\forall x)(\exists y)_{y \neq x} F(x, y)]$ is not intuitionistically valid.

II.2. Heyting's Interpretation

Turning to the question of the intuitionistic import of all this, I must say right off that a formal system itself is a perfectly respectable—indeed, a rather tame—intuitionistic object, and that most metamathematical mappings would be legitimate constructions if they were to be carried out within Brouwerian guidelines. The real question before us is the philosophical import of this system and its attendant metalogical studies.

[44] In this version, each node has its own domain of objects. If node w' is accessible to node w, then the domain of w must be a subset of the domain of w'.

[45] The standard proofs of these facts are not intuitionistically acceptable. See De Swart [1976] and [1977], and Veldman [1976], for intuitionistic versions of the model-theoretic semantics together with corresponding completeness theorems.

[46] Indeed, for parallel reasons, the formula $(\forall x)(\exists y)_{y \neq x} F(x, y)$ is not true at any node. Thus $\sim(\forall x)(\exists y)_{y \neq x} F(x, y)$ is true at w_1 , as is $\sim(\exists x)(\forall y)_{y \neq x} \sim F(x, y)$.

We might try to adapt Brouwer's own mode of thought here: logic follows ontology, and so intuitionistic logic might be a formal statement of the special intuitionistic ontology. But in fact, Heyting explicitly abjures this way of thinking. He did this not (merely) on the ground that logic is ontologically neutral, but mainly because he held that intuitionistic mathematics is itself ontologically neutral. Classical mathematics, says Heyting, is metaphysically weighted, and that is its downfall. Intuitionism, he insists, is metaphysically neutral.

Intuitionistic logic, he tells us, reflects something else altogether: the special intuitionistic *meaning* of the logical particles. In particular, for Heyting, intuitionism's central philosophical claim is that mathematics has no unknowable truths. In mathematics, to be true is to be provable; and IL is the result of applying this tenet to the semantics of the connectives and quantifiers. The main idea goes as follows. We set out the conditions under which evidence in an epistemic situation will suffice to count as a proof of a proposition A in that situation. Thus, in particular:

- $A = (B \,\&\, C)$ is proved in a situation if and only if the situation proves B and C;
- $A = (B \vee C)$ is proved in a situation if and only if the situation contains evidence indicating that either B or C will eventually be proved;
- $A = (B \rightarrow C)$ is proved in a situation if and only if the situation contains a method for converting any proof of B into a proof of C;
- $A = \sim B$ is proved in a situation if and only if the situation contains evidence that B can never be proved (i.e., evidence that shows that a proof of B could be turned into a proof of a contradiction);
- $A = (\exists x)\, B(x)$ is proved in a situation if and only if the situation contains evidence indicating that $B(t)$ eventually be proved for some t;
- $A = (\forall x)\, B(x)$ is proved in a situation if and only if the situation contains a method for converting any proof that a given object t is in the domain of discourse into a proof of $B(t)$.

When we replace the standard notion of "truth" in a model with this notion of "proof" in an epistemic situation—or "assertability," as Heyting put it in [1966][47]—we have a philosophical grounding for intuitionistic logic. Certainly, in particular, excluded middle will fail, for there are propositions right now for which we have no evidence that they will ever be decided. And the rest of intuitionistic logic will follow suit, so that logic is the expression of this proof interpretation of the logical particles.

[47] The term "warranted assertability" was actually introduced by John Dewey in [1938], but it has now come to be used generically to cover this sort of epistemic theory of meaning. (Kolmogorov [1932] suggested an interpretation of IL using the notion of mathematical problems, and pointed out that it was equivalent to Heyting's proof interpretation.)

Four things need to be noted. First, Brouwer's work contains ample evidence that he embraced this conception of mathematical truth. His proof of the positive version of the continuity theorem rests precisely on the assumption that truth (in that case, truth of the claim that the function f is a total function) presupposes provability. His use of creating-subject sequences also quite openly involves an assertabilist assumption.[48] And perhaps most telling is his repeated equation of Hilbert's mathematical optimism (Hilbert's belief that every mathematical problem is ultimately solvable) with the principle of excluded middle. This equation presupposes assertabilism. If we assume optimism, then for any given proposition B we have a guarantee that either B or $\sim B$ will eventually be proved. The condition for the assertability of disjunctions will then validate $(B \vee \sim B)$.

Second, if we take this proof interpretation of the logical particles as the central message of intuitionism, then some of the metalogical studies of IL do take on philosophical significance. Thus, in particular, Gödel's mapping between IL and the modal S4 was actually expressed in terms of mathematical provability. Gödel used the symbol "B" (for *Beweissbarkeit*, "provability") rather than the "□" or "◇"of modern modal logic. His idea was that the notion of mathematical provability satisfies the axioms of S4, and that is why S4 is an appropriate mapping for intuitionistic logic. Similarly, Kripke heuristically describes the nodes in an intuitionistic Kripke-model as "epistemic situations," and takes the ordering of these nodes as a model of the monotonic growth of knowledge.

The third point is that Brouwer's insistence on construction is preserved in all this. Or at least it is according to Heyting. For when we speak of "proof" in these clauses, we must in fact keep in mind that, for an intuitionist, construction is the basic proof activity.

And last, though he may keep the emphasis on construction, Heyting jettisons Brouwer's metaphysical ground for intuitionistic logic—indeed, for intuitionistic mathematics altogether. For the proof interpretation is—or should be—ontologically neutral.

I say "should be" because, since proof or evidence is now the criterion of truth, it is easy to slip into the belief that propositions, understood assertabilistically, are *about* proofs. But that is a mistake. The proof interpretation of the logical particles says nothing *per se* about the objects of mathematics. Indeed, the assumption that a theory of truth must be referential—that it must involve a theory of objects—is precisely the assumption that Heyting attributes to the *classical* mathematician. Having made this assumption, the classical mathematician is

[48] In particular Brouwer says that for the real number r_γ, defined above, we know $r_\gamma \neq 1/2$. The basis of this claim is that to assume that $r_\gamma = 1/2$ amounts to assuming that neither Q nor $\sim Q$ will ever be proved. That in turn, he says, is equivalent to assuming the impossible claim that $\sim Q$ and $\sim\sim Q$ are both true. This last move is simply an application of the assertabilist account of negation.

then forced to posit a (Platonistic) world of objects with undecidable properties in order to "correspond" to the demands of his classical logic. This indeed may well be the best way to express Heyting's "anti-metaphysical" philosophy of intuitionism. Though he finds classical mathematics unpalatable, he himself is tolerant of it. For, in Heyting's view, this mathematics has its own subject matter and thus does not conflict with intuitionism.

II.3. Dummett

Since the 1960s, Michael Dummett has been promoting the assertabilist interpretation of the logical particles in a series of very influential papers and books. But Dummett generalizes Heyting's position in a significant way: assertabilist semantics, according to Dummett, applies *to all of language in general*, with the language of mathematics as a single, special case.[49] Shifting to so general a scope for assertabilist semantics means that the game has changed in a fundamental way. There is a new enemy, there are new arguments, and indeed a new playing field altogether.

The Enemy

Because assertabilism now applies in principle to language as a whole, the assertabilist's opponent will be anyone who claims that truth *in general* is somehow not determined by us humans and our actions. Traditional advocates of an overall correspondence theory of truth, for instance, fit into this category. And there are also more focused foes, philosophers who target their "nonhuman" semantic theories in one specific area of discourse or another. The Platonist posited by Heyting is a good example here.

The Arguments

Dummett has provided quite general linguistic arguments designed to support his broadly conceived assertabilism. He has an argument from language acquisition: since, in first acquiring language at all, a child could only be learning the assertability conditions of simple and then complex sentences, these must be the conditions that ultimately define anyone's grasp of the language. And he has a general pragmatic argument: only an assertabilist meaning theory allows a speaker actively to demonstrate that he or she understands the language spoken. According

[49] Dummett's initial formulations of his views on the assertabilist reading of the logical constants can be found in a number of essays collected in his [1978]. His papers [1973] and [1975] speak directly to this issue. The former essay forms the basis of the application of assertabilism to mathematical discourse as it is found in his book [2000]. He has continued to refine and apply his views about the theory of meaning. Two important subsequent books are [1991] and the anthology [1993].

to Dummett, advocates of nonassertabilist conceptions of meaning cannot provide speakers with ways of demonstrating their grasp of simple and complex sentences which are humanly undecidable.

The New Playing Field

This is Heyting's anti-metaphysical bent, writ large. Much of Dummett's philosophical labor aims to show that traditional metaphysical debates about reality and objects can are best translated into modern semantic disputes. Old-fashioned realisms—about empirical objects, about personality, about past times, for instance—are each best replaced by a semantic realism that advocates human-independent truth. And the corresponding idealisms are now to be understood as a semantic "anti-realism," which bases truth upon human knowledge.[50]

In all this one must keep in mind that intuitionistic logic is the natural upshot of a human-based assertabilism. So if one follows Dummett here, intuitionistic and not classical logic will be the logic appropriate not just for mathematics, but for all discourse in which there are undecidable propositions.[51]

And, most important for us, this is how Dummett understands the conflict between classical and intuitionistic approaches to mathematics. Rather than advocating some sort of Platonism or metaphysical realism about mathematical objects, the classical mathematician, on Dummett's picture, is simply an advocate of a nonhuman semantic theory applied to mathematics.[52] This is a basis for mathematical intuitionism that may be more palatable to those who were uncomfortable with Brouwer's elaborate phenomenology and ontogenesis.

III. Looking Backward

III.1. Taking Stock

Here, then, is intuitionism today. Technically, we have Brouwer inventing new analytic and set-theoretic tools in order to solve a particular problem about the continuum. And we have Heyting and those who followed him doing just the

[50] These applications figure prominently in Dummett [1978] and [1993].

[51] To be sure, Dummett questions whether the model-theoretic semantics faithfully conveys the assertabilist meanings of the logical constants. (See, for instance, [2000], sec. 7.4.) However, that does not undermine the link between the assertabilist meaning theory and intuitionistic logic.

[52] Indeed, Dummett labors valiantly to show that even the special results of intuitionistic analysis are derivable ultimately from the special assertabilist meanings of the quantifiers.

opposite: using the full force of traditional existing metamathematical tools to explore and expand intuitionistic logic. Philosophically, we have Brouwer's phenomenology and his attendant metaphysics. And up against this we have Heyting's anti-metaphysical proof interpretation and Dummett's more general assertability semantics.

Credits

Looking at all this positively, we find rich and fine-grained technical studies together with a mature constructivism that paints the human aspect of mathematics in a way that avoids both obsessive finitism and the general empiricist disdain of abstract ideas. Indeed, we have here a school of thought spanning the gamut of modern mathematics (the new analysis, set theory, formal logic, metamathematics) and contemporary philosophy (phenomenology, the latest epistemology, and modern anti-realist semantics).

Debits

I must, however, mention three problems with all this. First, on the technical side, intuitionistic *mathematics*, however bold its deviation from its classical counterparts, and the *logic* of intuitionism, however subtle it may be over against simple classical systems, are just those things: deviations that stand against the modern norm. Practicing mathematicians assume discontinuous functions without a second thought, and intuitionistic logical niceties are viewed as hair-splitting complications in contemporary mathematical logic. Indeed, even the more philosophical sides of logic and semantics often view IL as a messy distraction.[53] The technical side of intuitionism seems a quixotic curiosity.

Second, I must sadly add that the very scope and span of intuitionistic philosophy seems in fact a large liability. For Brouwer's metaphysical grounding—indeed, any metaphysical grounding of logic—is viewed as a very poor bedfellow for the Heyting–Dummett assertabilism. Nowadays we see assertabilist semantics as a replacement—both in content and in style—for an outmoded ontology of objects and their properties, and even for a referential semantics that may base truth on correspondence to objects. When it comes to the justification of its own special logic (or even of its rejection of classical logic), intuitionism is a house divided. And if, like Heyting, we find in Brouwer's work a ground for the assertabilist semantics, then Brouwer himself is internally dissonant.

The third worry stems from a small conundrum in the assertabilist semantics, but in fact betokens an even more profound tension. Suppose, for a minute, that

[53] Thus, for instance, Hilary Putnam ([1978], lecture II) invokes the Gödel translation (mentioned in sec. II.1 above) to argue that assertabilism need not deviate from classical logic. And Timothy Williamson (in [1994]) takes deviation from classical logic to be a prima facie argument against the viability for any natural language semantics.

we were to accept Hilbert's optimistic view and hold that every mathematical problem is ultimately solvable, or that every proposition is ultimately decidable. We needn't suppose that each proposition is already decided by some omniscient knower; it suffices to suppose that each one is in principle decidable. In either case, the assertability clause for disjunction automatically validates the law of excluded middle; there would be no formal difference between the resultant logic and standard classical logic.[54] Brouwer knew this. He repeatedly equated Hilbert's optimism with the validity of the law of excluded middle, and with it, classical logic. But can the intuitionist assert that there are—or even may be—propositions that are fully, eternally undecidable? We can't specify even one of these in mathematics. For were we to do so, then the assertability clause for negation would make us hold this proposition to be false, and thus in effect decide it for us. And so we certainly can't "construct" one.[55] The intuitionist thus finds himself in the unenviable position of depending upon the existence of something—an undecidable proposition—that he cannot in fact construct, and whose possible existence he thus may not assert!

So that is where we are in intuitionism: a house internally divided, bent upon an eccentric technical mission, and based on a fundamental assumption which goes against its own internal standards. In the next section I want to show you, however, that intuitionistic mathematics and logic are far from eccentric curiosities: Brouwer's mathematics, I will suggest, stands in a tradition that goes back at least to Aristotle, and intuitionistic logic makes precise a mode of thought that stems from Kant. I will then go on to show that this same Kantian perspective can serve to reconcile the opposing intuitionistic streams, and can even provide a way to confront that disturbing internal tension.

III.2. Technical Precedents

Aristotle and the Intuitionistic Continuum

Let me briefly address the problem of *mathematical* deviance by first compounding it. Here is a straightforward consequence of Brouwer's continuity theorem: *the*

[54] See Posy [1984] for some refinements of this general point. See also Shapiro ([1997], sec. 6.7).

[55] Posy [1999] shows that a similar difficulty besets Brouwer's proof of the negative uniform continuity theorem. He cannot in fact construct a sequence, χ, so indeterminate that the question of whether its corresponding real number r_χ is less than ½ or equal to ½ remains eternally undecided, as the proof demands. Brouwer never repeated this proof. Indeed, his description in [1952a] of the second act of intuitionism pointedly refrains from mentioning restrictions designed to *enforce* (rather than restrict) freedom of choice. Instead he cites the "creative subject" approach to indeterminacy.

interval $\mathbb{R}_{[0,1]}$ *(indeed, the entire continuum) cannot be split.* That is, there do not exist two disjoint, nonempty sets A and B such that $A \cup B = \mathbb{R}_{[0,1]}$. To see that this is true, consider an apparently natural splitting of $\mathbb{R}_{[0,1]}$, say $A = \{x \mid x < ½\}$ and $B = \{x \mid x \geq ½\}$. If, in fact, $A \cup B = \mathbb{R}_{[0,1]}$ for this A and B, then the "characteristic function" f, defined by

$$\begin{aligned} f(x) &= 0 \quad \text{if } x \in A \\ &= 1 \quad \text{if } x \in B, \end{aligned}$$

will have to be a total function defined on $A \cup B = \mathbb{R}_{[0,1]}$, and thus, by the continuity theorem, will have to be continuous. But this function is clearly discontinuous. It is just the step function of section I.2; and, as we saw there, it is in fact not total. This little argument works for any A and B that purport to split or separate the interval $\mathbb{R}_{[0,1]}$.

Here, by the way, is the origin of Brouwer's term "separable mathematics" for the study of countable collections such as the natural numbers and even the rational numbers: collections such as these are immune from this theorem. They *can* be split or separated into disjoint proper subsets.

Now, this "unsplitability" that characterizes the intuitionistic continuum is a strong topological property.[56] But I want to point out now that in fact this modern Brouwerian concept makes precise a notion that goes back at least to Aristotle.

Aristotle was an architect of the dichotomy between number and magnitude of which I spoke earlier and that characterized mathematics up to the nineteenth century. He thought it the best way to avoid Zeno's paradoxes and the Pythagorean problems of incommensurables. Along with this mathematical split Aristotle offered a conceptual analysis of the notion of continuous magnitude: its parts must share borders.[57] And he initiated an image of fluid motion that remained the hallmark of all continuous magnitudes for millennia. We find the image not only in Aristotle[58] but also in Kant in the eighteenth century,[59] as late as Borel in [1898], and even in Brouwer's dissertation [1907].[60]

As I said, Aristotle's conceptual analysis and his accompanying image of viscosity—you can't cut a continuous medium without some of it clinging to the knife—gain precision in the topological notion of "unsplitability." It is a property

[56] Indeed, it is much stronger than the property of "connectedness," defined as indivisibility into *open* sets A and B. Connectedness is one of the properties that distinguish the continuum from the set of rational numbers. The continuum, even the classical continuum, is connected; the set of rational numbers is not.

[57] See, for instance, *Categories* VI, 4^b25–26 and *Physics* V.3.

[58] *Physics* V4 argues for the continuity or "flowing" of motion, and *Physics* VI. 1 argues that time and (spatial) magnitude must share this same fundamental nature.

[59] See *Critique of Pure Reason* (hereafter abbreviated *CPuR*) A169–170/B211.

[60] See Brouwer [1907] ch. 1, 8–9.

that characterizes the intuitionistic continuum but that is lost in the classical version of the continuum.[61]

So we see, as I claimed above, that far from deviating from the norm, Brouwer's topological view of the continuum, and the mathematics he developed to support that view, actually continue a venerable mathematical tradition.[62]

But here is another twist. From Aristotle onward, mathematicians and philosophers alike have always believed that this viscosity—this topological unsplitability—is inconsistent with the view that a continuum is composed of independent points (atoms). That is to say, they believed that the property of viscosity is inconsistent with the set-theoretic composition of the continuum. This alleged inconsistency is a central theme of Aristotle's physics and metaphysics.[63] Kant endorses the allegation as well,[64] and so does Brouwer, once again, in his early work.[65] To be sure, Cantor, Dedekind, and others succeeded in fashioning a continuum as a collection of independently existing points. But they did so only at the expense of abandoning viscosity. Brouwer has now pushed this all to the side. For, having introduced choice sequences as autonomously existing mathematical entities, and having shown how to construct the continuum as a set derived from such entities in an "epistemologically responsible way," Brouwer has shown us that we can have a viscous continuum and make it out of points as well. This is no windmill-tilting; this is a profound mathematical insight.

Kant and Intuitionistic Logic

As for the forerunners to intuitionistic logic, here I turn to Kant. Though Kant inspired talk of "construction in intuition," it is not his philosophy of mathematics to which I want to turn here. Indeed, on questions of mathematics, Kant's differences from the intuitionists are striking: Brouwer rejects Kant's claims about the *apriority* of space;[66] Kant—quite unlike Brouwer—suggests that mathematical

[61] Classically, for instance, $\mathbb{R}_{[0,1]} = (\{x \mid x < ½\} \cup \{x \mid x \geq ½\})$ does indeed hold.

[62] In making this point, I am suggesting a different topological treatment of Aristotelian continuity than the one proposed by White [1992].

[63] *See Physics VI*.1, 231^a24–26 and *Metaphysics* III.5.

[64] *CpuR*, A169–170/B211.

[65] See [1907], ch. 1, pp. 8–9. Brouwer clung to this opposition to a set-theoretic construction of the continuum out of points at least as late as [1912]. With the series of publications starting from [1918] and [1919], Brouwer came to advocate constituting the continuum out of previously constructed points.

[66] Brouwer points out that non-Euclidean geometries show that there is no uniform set of a priori statements about the nature of space, similar to what can be said about the nature of time. So while accepting Kant's view about time as an a priori intuition, he rejects Kant's parallel claim that space, too, is a form of intuition. Coordinate systems of real numbers will go proxy for geometry, and thus, says Brouwer, we have no need to appeal to an independent spatial intuition. (See Brouwer [1907], ch. 2, and [1912].)

objects have no independent existence;[67] and he will have no truck mathematically or generally with predicatively incomplete objects or undetermined identity.[68] Indeed, I will point out below that for Kant the appropriate logic for mathematics will be classical logic.

No, the place to find origins of intuitionistic logic is rather in Kant's general metaphysics. It is his "transcendental idealism" to which I want to turn, and in particular to the confrontation between transcendental idealism and its rival "transcendental realism," and to the "Antinomy" chapter of the *Critique of Pure Reason,* where Kant brings that confrontation to a head.

In that chapter he claims that the transcendental realist is confronted with the following logical dilemma: on the one hand, the realist accepts—as a logical truth—the claim that the occupied physical world must be either spatially finite or infinite. That is, the realist claims that the following disjunction is a logical truth:

$$(\exists x)(\forall y)[\sim E(x,y) \rightarrow F(x,y)] \vee (\forall x)(\exists y)[\sim E(x,y) \,\&\, F(y,x)]$$

(where x and y range over occupied spatial regions, "$E(x, y)$" means that x and y are equidistant from some fixed central point, and $F(x, y)$ means that x is a further from that point than y).[69] But on the other hand, Kant provides arguments refuting each side of that disjunction.

The transcendental idealist, Kant tells us, escapes this logical dilemma because, for the idealist, this disjunction is not a logical truth in the first place.[70] Here is where intuitionistic logic comes in. For, though the disjunction above is a logical truth under classical logic, it is not intuitionistically valid. Indeed, section II.1 displayed a simple Kripke model that serves as intuitionistic counterexample to this disjunction.

Of course Kant did not express his reasoning here using the formal machinery of intuitionistic logic. But once again we can use intuitionism precisely to express ideas that were less precisely adumbrated in an older philosophical school: Kant's transcendental idealist, we may say, advocates an intuitionistic logic for empirical discourse, while the rival transcendental realist goes for classical logic in the empirical arena.

This is no mere formal analogy. For Kant's logical moves here—and similar moves regarding the world's temporal ordering, causality, and the divisibility of matter—are best explained by an assertabilist view of the meanings of the $\forall\exists$ and $\exists\forall$ quantifier combinations. Indeed, the model in Section II.1 graphically captures Kant's claim that under transcendental idealism, finding an occupied region that is

[67] See *CpuR,* A719/B747.

[68] See *CpuR,* A571–572/B599–600

[69] Actually, to be precise, the realist will accept this as a logical truth after assuming that E is an equivalence relation and that F obeys the axioms of a linear order.

[70] See *CPuR,* 504–505/B532–533.

more distant from a central point than a given region is set as a task, but one cannot assert the existence of that more distant region unless one has actual sensory contact with it.[71] This is paradigmatic assertabilism applied to empirical cosmology. Moreover, Kant's arguments that the world cannot be either spatially infinite or spatially finite are arguments *ad ignorantum,* arguments from the unknowability of an assertion to its actual falsity. Space cannot be infinite because we humans cannot perceive an infinite expanse, and it cannot be finite because we could never verify that we have come to an outer edge.[72] Leaving aside their modern soundness, these are clearly arguments only an assertabilist could accept as valid.

Kant even goes on to espouse the epistemological pessimism that links assertability to intuitionistic logic. In natural science, he says, "there are endless conjectures in regard to which certainty can never be expected."[73]

His reason, interestingly, is that empirical science depends on evidence received passively. (Empirical intuition is "receptive" in Kant's words.) And so if there is an empirical question that is as yet unanswered, then we must simply wait until the appropriate evidence comes in one way or the other; and we have no assurance as to whether or when such evidence will emerge. This is interesting because, mathematics, for Kant, suffers from no such similar passivity. And thus he holds that in mathematics there will be no unanswerable questions. He sides, that is, with Hilbert on this issue of mathematical optimism![74] That is why I said above that, in mathematics, Kant should favor a classical logic.[75]

III.3. Assertability and Ontology

Thus Kant's transcendental idealism is, as I said, an empirical assertabilism; and he derives its intuitionistic logic in that way. Here is a true forerunner to IL— one, to be sure, that appears closer to Dummettian intuitionism than to Brouwer's ontological version. But, in fact, Kant also—indeed, regularly—describes transcendental idealism and its rival transcendental realism as straightforward ontological positions.

The heart of Kant's "Copernican Revolution" is a new theory about the nature of objects: the objects must correspond to our intuitive knowledge and not vice versa.[76] Transcendental idealism, he tells us over and over again, holds that empirical objects are "appearances" and not mind-independent "things in

[71] See *CPuR,* 518–523/B546–551.

[72] See *CpuR,* A426–429/B454–457.

[73] *CPuR,* A481/B509.

[74] See *CpuR,* A480/B508.

[75] And indeed, at *CPuR* A792/B820, Kant comes close to this: he says that only mathematics is entitled to use indirect proofs (i.e., proofs based on the law of excluded middle).

[76] See *CpuR,* Bxvii.

themselves"; and transcendental realism holds that these same objects are in fact things-in-themselves. Indeed, Kant quite explicitly defines truth, the gateway to logic, as "correspondence of cognition with its object."[77] This indeed, is the source of his empirical epistemic pessimism: "for natural appearances are objects that are given to us independently of our concepts," and for that reason there are times in which "no secure information [about them] can be found."[78]

So is it objects, ontology, and reference, or is it evidence and assertability that underlie Kant's idealism and his intuitionistic logic? It is both. This is quite the same mix that we find in intuitionism in general and in Brouwer in particular.

But in fact all of this holds together for Kant. There is no uneasiness to this union, for there is no gap between the ultimate empirical nature of an object and the sum of what we can intuitively (perceptually) know about that object. Once the object is perceived, then all of its properties can be intuitively known. So, similarly, there is no gap between saying, on the one hand, that our judgments are true because of the ontological properties of their objects and claiming, on the other hand, that those judgments are true by virtue of our eventual knowledge.[79] That is the heart of his transcendental idealism.

The same equation characterizes Kant's view of transcendental realism. Transcendental realism, as we saw, applies classical logic in empirical discourse precisely because the realist does not let our receptive and finite epistemic limits influence the conditions of empirical truth. That is the effective content of Kant's claim that "reason" rather than "intuition" underlies the realist's conception of truth. And that, in turn, is the heart of his ontological claim that for the realist, empirical objects are "things-in-themselves."

So for Kant, ontology and assertabilism go hand-in-hand. And we find in Kant a worked-out philosophical position advocating and explaining the very philosophical mix that seemed a problem for the intuitionists.

III.4. The Unknowable

Let's turn then to our final worry, the tension between intuitionism's need to assume the possible existence of unanswerable questions and its inability to do so.

[77] See, for instance, *CpuR*, A58/B82 and A191/B236.

[78] *CpuR*, A481/B509.

[79] This is what underlies Kant's claim at *CpuR*, A571–572/B599–600, mentioned in the preceding section, that empirical objects are predicatively complete. For Kant, once we have evidence to assert the existence of an object, then we have a guarantee that we can uncover its properties. This doesn't, however, tell us in advance of empirical experience which objects will or will not exist. In the end, it is the world that is incomplete, and not the objects existing in it. Kant says as much at A522–533/B550–551. Posy [2000] details Kant's view here, and Posy [2003] traces this view to its Leibnizian roots.

Here again I want to take a cue from Kant to help understand—though not resolve—this tension. For this it is wise to look at Kant's attitude toward that epistemically unruly faculty of "reason."

Transcendental realism's mistake, Kant says, is to allow reason to govern the semantics of empirical discourse. This, we saw, leads the realist to apply classical logic in empirical contexts, and that, says Kant, is untenable. It generates the antinomy sketched above and others like it. But he does not dismiss this realism as incomprehensible or incoherent. Indeed, Kant insists that realism—or, more precisely, the faculty of reason—actually exerts a positive influence upon empirical discourse.

For one thing, reason makes us think about some sort of "intelligible," unperceivable source for our passively received perceptions. For without presuming such a "transcendental object" as source, receptivity itself would make no sense.[80] This is a tension similar to the one we uncovered in intuitionism: the need to assume the existence of an object that cannot be constructed by epistemologically respectable methods.

Kant sets his tension in epistemological terms quite like the ones that pose the intuitionistic tension. Thus he says that in cosmology reason leads us to consider the world-whole and ponder its borders in the first place. And though those limits are unknowable, Kant insists that this sort of thought is empirically essential. For it is this way of thinking that "sets the task" of expanding our empirical knowledge about the extent of the universe.[81] Ultimately it is this "regulative" force of reason that guides scientific research and goads it to extend its range of discoveries. Indeed, it is just this force of reason that aims us at ever more comprehensive scientific theories, even though we know that we will never form an epistemically complete theory.[82]

And, even more closely, the very first line of the *Critique of Pure Reason* expresses our epistemic state: "Human reason has this peculiar fate that in one species of its knowledge it is burdened by questions, which, as prescribed by the very nature of reason itself, it is not able to ignore, but which, as transcending all its powers, it is also not able to answer."[83]

Here, then, is a historical precedent from which we can learn. It doesn't reconcile the intuitionistic tension, but if we follow the Kantian line, it tells us that we shouldn't try to do so. The tension itself defines our human condition.

[80] See *CpuR*, A494/B522ff.

[81] See *CpuR*, A508/B536.

[82] See *CpuR*, A696–697/B725.

[83] *CpuR*, i Avii.

REFERENCES

van Atten, M.S.P.R. [1999] *Phenomenology of Choice Sequences*, Zeno, the Leiden-Utrecht Institute of Philosophy, Utrecht.

van Atten, M., van Dalen, D., and Tieszen, R., [2002] "Brouwer and Weyl: The Phenomenology and Mathematics of the Intuitive Continuum," *Philosophia Mathematica*, 10, 203–226.

Beth, E. [1947] "Semantical Considerations on Intuitionistic Mathematics," *Indagationes Mathematicae*, 9, 572–577.

Bishop, E. [1967] *Foundations of Constructive Analysis*, New York, McGraw-Hill.

Bockstaele, P. [1949] *Het Intuitionisme bij de Franse Wiskundigen, Verhandelingen van de Koninkliijke Vlaamse Academie van Wetenschappen, België*, 11, no. 32.

Borel, E. [1898] *Leçons sur la théorie des fonctions*, Paris, Gauthier-Villars.

Borel, E. [1927] "*A Propos de la récente discussion entre M. R. Wavre et M. P. Levy*," *Revue de métaphysique et de morale*, 34, 271–276.

Brouwer, L.E.J. [1907] "Over de Grondslagen der Wiskunde," Ph.D. Dissertation, University of Amsterdam. Translated as *On the Foundations of Mathematics* in *CW*, pp. 11–101.

Brouwer, L.E.J. [1908] "Die möglichen Mächtichkeiten," *in Atti del IV Congresso Internazionale dei Matematici, Roma, 6–11 Aprile 1908*, Rome, Academia dei Lincei, 569–571. Reprinted in *CW*, pp. 102–104.

Brouwer, L.E.J. [1912] *Intuitionisme en Formalisme*, Clausen, Amsterdam. Translated by A. Dresden as "Intuitionism and Formalism," *Bulletin of the American Mathematical Society*, 20 (1913), 81–96. Reprinted in *CW*, pp. 123–138.

Brouwer, L.E.J. [1918] "Begründung der Mengenlehre unabhängig vom logischen Satz vom ausgeschlossenen Dritten, Erste Teil, Allgemeine Mengenlehre," *Verhandelingen, Koninkliijke Akademie van Wetenschappen te Amsterdam*, 1e sectie, deel XII, no. 5, 1–43. Reprinted in *CW*, pp. 151–190.

Brouwer, L.E.J. [1919] "Begründung der Mengenlehre unabhängig vom logischen Satz vom ausgeschlossenen Dritten, Zweiter Teil, Theorie der Punktmengen," *Verhandelingen, Koninkliijke Akademie van Wetenschappen te Amsterdam*, 1e sectie, deel XII, no. 7, 1–33. Reprinted in *CW*, pp. 191–221.

Brouwer, L.E.J. [1923] "Intuitionistische Zerlegung mathematischer Grundbegriffe," *Jahresbericht der Deutschen Mathematiker Vereiningung*, 33, 251–256. Reprinted in *CW*, pp. 275–280.

Brouwer, L.E.J. [1927] "Über Definitionsbereiche von Funktionen," Mathematische Annalen, 97, 60–75. Reprinted in *CW*, pp. 390–405.

Brouwer, L.E.J. [1929] "Mathematik, Wissenschaft und Sprache," *Monatschefte für Mathematic und Physik*, 36, 153–164. Reprinted in *CW*, pp. 417–428.

Brouwer, L.E.J. [1933]. "*Willen, Weten, Spreken*," *Euclides*, 9, 177–193. Translated as "Will, Knowledge and Speech," in Van Stigt [1990], pp. 418–431.

Brouwer, L.E.J. [1948a] "Consciousness, Philosophy and Mathematics," in *Proceedings of the Tenth International Congress of Philosophy, Amsterdam*, vol. III, North Holland, Amsterdam, 1949, pp. 1235–249. Reprinted in *CW*, pp. 480–494.

Brouwer, L.E.J. [1948b] "Essentieel negatieve Eigenschappen," *Proceedings of the Acad. Amsterdam*, 51, 963–964 (= *Indagationes Mathematicae*, 10, pp. 322–323). Translated as "Essentially Negative Properties" in *CW*, pp. 478–479.

Brouwer, L.E.J. [1950] "Discours final de M. Brouwer," In *Les Méthodes formelles en axiomatique: Colloques internationaux du Centre National de la Recherche Scientifique*, Paris: CNRS, p. 75. Reprinted in *CW*, p. 503.

Brouwer, L.E.J. [1952a] "Historical Background, Principles and Methods of Intuitionism," *South African Journal of Science*, 49: 139–146. Reprinted in *CW*, pp. 508–515.

Brouwer, L.E.J. [1952b] "An Intuitionist Correction of the Fixed-point Theorem on the Sphere," *Proceedings of the Royal Society of London*, A213, 1–2. Reprinted in *CW*, pp. 506–507.

Brouwer, L.E.J. [1954a] "Points and Spaces," *Canadian Journal of Mathematics*, 6, 1–17. Reprinted in *CW*, pp. 522–538.

Brouwer, L.E.J. [1954b] "An example of Contradictority in Classical Theory of Functions," *Proceedings of the Acad. Amsterdam*, ser. A, 57, 104–105 (= *Indagationes mathematicae*, 16, 204–205). Reprinted in *CW*, pp. 549–550.

Brouwer, L.E.J. [1975] *Collected Works*, vol. I, A. Heyting, ed., Amsterdam; North Holland. (Abbreviated as *CW*.)

Brouwer, L.E.J. [1992] *Intuitionismus*, D. van Dalen, ed., Mannheim, Bibliographisches Institut, Wissenschaftsverlag.

van Dalen, D. [1986] "Intuitionistic Logic," in *Handbook of Philosophical Logic*, vol. III, D. Gabbay and F. Guenther, eds., D. Reidel, pp. 225–340.

van Dalen, D. [1997] "How Connected Is the Intuitionistic Continuum," *Journal of Symbolic Logic*, 62, 1147–1150.

van Dalen, D. [1999] "From Brouwerian Counter Examples to the Creating Subject," *Studia Logica*, 62, 305–314.

van Dalen, D. [2000] *Mystic, Geometer and Intuitionist: The Life of L.E.J. Brouwer.* Volume 1: *The Dawning Revolution*, Oxford, Oxford University Press.

Detlefsen, M. [2002] "Brouwerian Intuitionism," in *Philosophy of Mathematics: An Anthology*, D. Jacquette, ed., Oxford, Blackwell, pp. 289–311.

Dewey, J. [1938] *Logic: The Theory of Inquiry*, New York, Holt.

Dummett, M. [1973] "The Philosophical Basis of Intuitionistic Logic," in *Logic Colloquium '73*, H. E. Rose and J. C. Shepherdson, eds., Amsterdam, North Holland, pp. 5–40. Reprinted in Dummett [1978], pp. 215–247.

Dummett, M. [1975] "The Justification of Deduction," *Proceedings of the British Academy*, 59, pp. 201–232. Reprinted in Dummett [1978], pp. 290–318.

Dummett, M. [1978] *Truth and Other Enigmas*, Cambridge, Mass., Harvard University Press.

Dummett, M. [1991] *The Logical Basis of Metaphysics*, Cambridge, Mass., Harvard University Press.

Dummett, M. [1993] *The Seas of Language*, New York, Oxford University Press.

Dummett, M. [2000] *Elements of Intuitionism*, 2nd ed., Oxford, Oxford University Press.

Glivenko, V. [1929] "Sur Quelques Points de la logique de M. Brouwer," *Bulletin, Académie Royale de Belgique*, 15, 183–188.

Gödel, K. [1932] "Zum intuitionistischen Aussagenkalkül," *Anzeiger der Akademie der Wissenschaften in Wien*, 69, 65–66. Reprinted and translated in Gödel [1986], pp. 222–225.

Gödel, K. [1933a] "Zur intuitionistischen Arithmetik und Zahlentheorie," *Ergebnisse eines mathematischen Kolloquiums*, 4, 34–38. Reprinted and translated in Gödel [1986], pp. 286–295.

Gödel, K. [1933b] "Eine Interpretation des intuitionistischen Aussagenkalküls," *Ergebnisse eines mathematischen Kolloquiums*, 4, 39–40. Reprinted and translated in Gödel [1986], pp. 300–303.

Gödel, K. [1986] *Collected Works*, Volume I: *Publications 1929–1936*, S. Feferman, et al., eds., Oxford, Oxford University Press.

Griss, G.F.C. [1944] "Negatieloze intuitionistische Wiskunde," *Verslagen Akad. Amsterdam*, 53, 261–268.

Griss, G.F.C. [1946] "Negationless Intuitionistic Mathematics, I," *Proceedings of the Acad. Amsterdam*, 49, 1127–1133 (= *Indagationes Mathematicae*, 8, 675–681).

Griss, G.F.C. [1949] "Logique des mathématiques intuitionnistes sans négation," *Comptes rendues de Academic des Science, Paris*, 227, 946–947.

Griss, G.F.C. [1950] "Negationless Intuitionistic Mathematics, II," *Proceedings of the Acad. Amsterdam*, 53, 456–463 (= *Indagationes Mathematicae*, 12, 108–115).

Heyting, A. [1930a] "Die formalen Regeln der intuitionistischen Logik," *Sitzungsberichte der preuszischen Akademie der Wissenschaften, phys. math. Kl.*, 42–56.

Heyting, A. [1930b] "Die formalen Regeln der intuitionistischen Mathematik," *Sitzungsberichte der preussischen Akademie der Wissenschaften, phys. math. Kl.*, 158–169.

Heyting, A. [1930c] "Sur la Logique intuitionniste," *AcadémieRoyale de Belguique, Bulletin*, 16, 957–963.

Heyting, A. [1931] "Die inuitionistische Grundlagen der Mathematik," *Erkenntnis*, 2, 106–115.

Heyting, A. [1934] *Mathematische Grundlagenforschung: Intuitionismus, Beweistheorie*, Berlin, Springer.

Heyting, A. [1955] *Les Fondements des mathématiques: Intuitionisme, théorie de la démonstration* (Expanded French translation of [1934]), Paris: Gauthier-Villars.

Heyting, A. [1966] *Intuitionism: An Introduction*, 2nd rev. ed., Amsterdam, North-Holland.

Heyting, A. [1974] "Intuitionistic Views on the Nature of Mathematics," *Synthese*, 27, 99–91.

Husserl, E. [1913] "Ideen zu einer reinen Phänomenologie und phänomenologischen Philosopohie" *Jahrbuch für Philosophie und phänomenologischen Forschung*, 1. Translated by W. Boyce-Gibson as *Ideas: General Introduction to Pure Phenomenology*, New York, Macmillan, 1931.

Kino, A., Myhill, J., and Vesley, R.E. (eds.) [1970] *Intuitionism and Proof Theory*, Amsterdam: North Holland.

Kleene, S.C. [1952] *Introduction to Metamathematics*, Van Nostrand, New York.

Kreisel, G. [1967] "Informal Rigour and Completeness Proofs," in *Problems in the Philosophy of Mathematics*, I. Lakatos, ed., Amsterdam, North-Holland.

Kripke, S. [1965] "Semantical Analysis of Intuitionistic Logic, I," in *Formal Systems and Recursive Functions*, J. N. Crossley and M. Dummett, eds., Amsterdam, North-Holland, pp. 92–130.

Kolmogorov, A.N. [1925] "O principe tertium non-datur," *Matematiceskij sbornik*, 32, 646–667. Translated in *From Frege to Gödel: A Source Book in Mathematical Logic, 1897–1931*, J. van Heijenoort, ed., Cambridge, Mass., Harvard University Press, pp. 416–437.

Kolmogorov, A.N. [1932] "Zur Deutung der intuitionistischen Logik," *Mathematische Zeitschrift*, 35, 58–65.

Placek, T. [1999] *Mathematical Intuitionism and Intersubjectivity*, Amsterdam, Kluwer.

Posy, C.J. [1977] "The theory of Empirical Sequences," *Journal of Philosophical Logic*, 6, 47–81.

Posy, C.J. [1984] "Kant's Mathematical Realism," *The Monist*, 67, 115–134. Reprinted in *Kant's Philosophy of Mathematics: Modern Essays*, C. Posy, ed., Amsterdam, Kluwer Academic Publishers, 1992.

Posy, C.J. [1999] "Epistemology, Ontology and the Continuum," in *Mathematics and the Growth of Knowledge*, E. Grossholz, ed., Amsterdam, Kluwer Academic Publishers, pp. 199–219.

Posy, C.J. [2000] "Immediacy and the Birth of Reference in Kant: The Case for Space," in *Between Logic and Intuition: Essays in Honor of Charles Parsons*, G. Sher and R. Tieszen, eds., Cambridge, Cambridge University Press, pp. 155–185.

Posy, C.J. [2003] "Between Leibniz and Mill: Kant's Logic and the Rhetoric of Psychologism," in *Philosophy, Psychology, and Psychologism*, D. Jacquette, ed., Amsterdam, Kluwer Academic Publishers, pp. 51–79.

Putnam, H. [1978] *Meaning and the Moral Sciences*, London, Routledge and Keegan Paul.

Rasiowa, H., and Sikorski, R. [1963] *The Mathematics of Metamathematics*, Warsaw, Panstwowe Wydawnictwo Naukowe.

van Rootselaar, B. [1970] "On Subjective Mathematical Assertions," in Kino, Myhill, and Vesley, 1970 .

Shapiro, S. [1997] *Philosophy of Mathematics: Structure and Ontology*, New York, Oxford University Press.

van Stigt, W. [1990] *Brouwer's Intuitionism*, Amsterdam, North Holland.

de Swart, H. [1976] "Another Intuitionistic Completeness Proof," *Journal of Symbolic Logic*, 41, 644–662.

de Swart, H. [1977] "An Intuitionistically Plausible Interpretation of Intuitionistic Logic," *Journal of Symbolic Logic*, 42, 564–578.

Troelstra, A.S. [1977] *Choice Sequences: A Chapter of Intuitionistic Mathematics*, Oxford, Clarendon Press.

Troelstra, A.S. [1982] "On the Origin and Development of Brouwer's Concept of Choice Sequence," in *The L.E.J. Brouwer Centenary Symposium*, A.S. Troelstra and D. van Dalen, eds., Amsterdam, North-Holland, pp. 465–486.

Veldman, W. [1976a] "An Intuitionistic Completeness Theorem for Intuitionistic Predicate Logic," *Journal of Symbolic Logic*, 41: 159–166.

Wavre, R. [1926a] "Logique formelle et logique empiriste," *Revue de métaphysique et de morale*, 33, 65–75.

Wavre, R. [1926b] "Sur le Principe du tiers exclu," *Revue de métaphysique et de morale*, 33, 425–430.

White, M. [1992] *The Continuous and the Discrete: Ancient Physical Theories from a Contemporary Perspective*, Oxford, Oxford University Press.

Williamson, T. [1994] *Vagueness*, London, Routledge.

CHAPTER 10

INTUITIONISM IN MATHEMATICS

D. C. McCARTY

Introduction

First, contemporary intuitionism is not a philosophy of mathematics but the mathematics inspired by L.E.J. Brouwer's (1881–1966) ingenious discoveries in topology, analysis, and set theory. In that, contemporary intuitionism is unlike those other "isms"—platonism, nominalism, formalism, structuralism, and anti-realism—with which it is sometimes associated. Intuitionists as such cannot join the ranks of philosophers of mathematics who insist that words in mathematics refer exclusively to sets, categories, structures, or airy nothing. Nor do intuitionists require that mathematical or logical terms be assigned meanings different from those they already possess. They cannot pretend, as do some exponents of the "isms" just mentioned, to take the results of conventional mathematics as true and beyond question. Intuitionists respectfully disagree with their conventional brethern first over mathematical fact, and over meaning or philosophy later, if at all. Often, introductions to intuitionism lead the reader astray in maintaining that intuitionists spurn certain principles of conventional logic and mathematics because of some philosophy or other, an epistemology, ontology, or semantics peculiar to intuitionism. Nothing could be further from the truth. Of course, forays into philosophy may return some profit, but they should not distract us from the central point: ultimately, the intuitionist erects a mathematical edifice, and thereby rejects conventional mathematics, on mathematical foundations and on mathematical foundations alone. A presentation of today's intuitionistic mathematics sets out from meaningful mathematics and, following a clear, bright road of mathematical reasons, arrives at a firm vantage from which conventional mathematics is seen to be generally false and set theory, analysis, arithmetic, algebra—all the fields of modern mathematics—reappear, scoured clean of fallacy and error.

The following is a brief tour of contemporary intuitionism. One convenient point of entry into intuitionistic territory is afforded by logical domains and their attendant mathematics—in particular, the properties of two sets playing prominent roles in intuitionistic reasoning: $P(\mathbb{N})$, the power set of the natural numbers, and Prop, the power set of $\{0\}$. While clearing intuitionistic customs, one sees posted some prohibitions, among them, that the law of the excluded third is illegal. Do not let first impressions mislead you. Not all the souvenirs from our excursion will be equally negative; intuitionism is a set of accomplishments in pure mathematics and not a doctrine of denial. For, from our point of entry, we climb quickly through a more abstract landscape up to the peak that is Brouwer's Theorem in intuitionistic analysis, a positive result from which a large terrain of mathematical domains becomes surveyable. Once on the far side of that peak, we follow a stream that cuts straight through a broad and fertile expanse, tracing the idea that intuitionistic mathematics serves as *internal mathematics* for topological and computability interpretations. This grants intuitionism powers strictly prohibited to conventional mathematics. Finally, in taking leave of the subject, we pause to assess the prospects for anti-realism, a philosophical development on the edge of intuitionism proper, and we attempt to exorcise an old ghost that still haunts the region: the idea that intuitionistic mathematics is weaker than its conventional counterpart.

I. Logical Domains

1. The Uniformity Principle

The extent to which practicing mathematicians of a conventional tendency are already intuitionists is reassuring. Today's mathematicians treat mathematical claims much as Brouwer once did: as independently meaningful efforts to record mathematical facts which are, when true, demonstrable from proofs rooted in basic assumptions or principles. Until they are proven, those assumptions rest upon intuitions: illuminating, at times fallible, insights into the dynamical behaviors of numbers, sets, functions, and operations on them. Much as a conventional set theorist argues for the truth of the foundation axiom by appealing to images of well-founded sets generated from the empty set $\varnothing$, an intuitionist introduces such starting-points as the Uniformity Principle by appealing to intuitions into the natures of numbers and sets.

The Uniformity Principle (UP), which got its first explicit formulation in Troelstra [1973], is the anticonventional claim that every set of natural numbers can be labeled with a number only if some number labels every set. When made available by intuition, UP entails several theorems characteristic of intuitionistic mathematics, among them the invalidity of the *tertium non datur* (TND), or law of the excluded third.

> PRINCIPLE 1 (Uniformity). Whenever an extensional binary relation R links every set of natural numbers X to some natural number n, there is some n that gets related by R to all such sets X. In symbols, $\forall X \exists n\, R(X, n) \rightarrow \exists n\, \forall X\, R(X, n)$.

To see how such a principle might be made plausible, let R be any extensional, binary relation between sets of natural numbers X and natural numbers n such that, for every X, there is an n for which $R(X, n)$. For this liaison between sets and numbers to subsist, there should be a discernible association in virtue of which R links sets to numbers. That association is expressible as a list of instructions α for determining, from sets X, suitable n for which $R(X, n)$. In contrast to the natural numbers and integers, the collection of all sets of natural numbers is not the trace of some recursive generation process. The relation R and the sets themselves are all extensional. Hence, the action of α should not depend upon the fine points of a set's possible specification in language. Further, since α is a rule with which one can *act* on *all* sets X of numbers, the action of α should not depend upon the membership conditions for any particular X. Those conditions might well be so complicated as to elude capture in anything one would rightly call a "rule." The application of α to sets should therefore be uniform: what α does to one set, it does to all. The identity badge of intuitionism as a branch of constructive mathematics is the insistence that every rule underwriting an existential statement about numbers $\exists n\, P(n)$ must provide, if implicitly, an appropriate numerical term t and the knowledge that $P(t)$ holds. Therefore, since α is constructive and labels each set X uniformly with *some* number, α must yield a designation for some particular natural number m uniformly in terms of the Xs. Obviously, for this association to be uniform, m must be the same for every set of numbers X. Hence, there is a number related by R to every set, and UP is seen to hold.

2. Consequences of Uniformity: *Tertium non Datur*, Cantor's Theorem, Indecomposability

UP has a number of easy corollaries, some characteristically intuitionistic and others recognizably conventional. Brouwer proved the first, the invalidity of TND, in his early article "On the Unreliability of the Logical Principles" (Brouwer [1908]). The second was the work of (nonintuitionistic) set theorist Georg Cantor (1845–1918). Cantor's proof differed markedly from the one below, relying upon an impredicative construction that drew critical fire from his contemporaries. No such construction features here. With the letter p a variable ranging over mathematical propositions, the validity of TND can be expressed as

$$\forall p\,(p \vee \neg p).$$

Mathematically, this is no law at all, but a false generality.

Corollary 1 (*TND is invalid*). $\neg\, \forall p\ (p \vee \neg p)$.

Proof. Assume that TND obtains and reason by *reductio*. If TND holds for all mathematical propositions p, it holds a fortiori for any propositions concerning set membership $\in$. Hence, it would be true that for any set X of numbers, either $0 \in X$ or $0 \notin X$. Define a binary relation R between sets X of natural numbers and individual natural numbers as follows. $R(X, n)$ is to obtain just in case either

$$\begin{array}{c} 0 \in X \quad \text{and} \quad n = 1 \\ \text{or} \\ 0 \notin X \quad \text{and} \quad n = 2. \end{array}$$

By assumption, either $0 \in X$ or not. It follows that, given any X, there is an n (either 1 or 2, to be specific) for which $R(X, n)$. UP now guarantees that there is an m such that, for any X, $R(X, m)$. However, this is impossible, because no m can be correlated by R with both the empty set $\varnothing$ and $\{0\}$. Consequently, the original assumption, that TND is valid, must be mistaken. ■

Let $\mathbb{N}$ be the set of natural numbers and $P(\mathbb{N})$ be the power set of $\mathbb{N}$, the set containing all and only subsets of $\mathbb{N}$. A set is *countable* when it can be enumerated by a mathematical function taking only natural numbers as inputs.

Corollary 2 (*Cantor's Theorem*). $P(\mathbb{N})$ is uncountable.

Proof. Recall that legitimate mathematical functions are unambiguous: for any single input, only a single output can be given. Were a function f capable of enumerating $P(\mathbb{N})$ exhaustively, one could find, for each set X, an n such that $f(n) = X$. In other words, X would appear as item n in the enumeration by f. Now, apply UP to the relation R such that $R(X, n)$ if and only if $f(n) = X$, which is the relation *inverse* to function f. Uniformity requires that there be at least one m such that, for *any* X, $f(m) = X$.

Since mathematical functions are by definition unambiguous, there is no number m such that $f(m) = \varnothing$ and, at the same time, $f(m) = \mathbb{N}$. Therefore, no function enumerates $P(\mathbb{N})$ exhaustively. ■

To claim that $P(\mathbb{N})$ is *indecomposable* is to claim that $P(\mathbb{N})$ cannot be carved up mathematically into two nontrivial, disjoint portions. To express the idea succinctly, intuitionists say that a subset Y of a set X is *decidable* on X when the appropriate instance of TND holds, that is, when every element of X is either a member of Y or not a member of Y:

$$\forall x \in X(x \in Y \vee \neg x \in Y).$$

If X is indecomposable, then every decidable subset of X has to be trivial, either $\varnothing$ or X itself.

Corollary 3 (*Indecomposability of* $P(\mathbb{N})$). The powerset of $\mathbb{N}$ is indecomposable.

Proof. Very much the same argument as that employed in Corollary 1. ■

Because $P(\mathbb{N})$ is indecomposable, it has only two decidable subsets: $\varnothing$ and $P(\mathbb{N})$.

3. Quantifiers and Testabililty

Once all agree that no proposition is both true and false, it is readily seen that although the scheme $\phi \vee \neg\phi$ is invalid, its close cousin $\neg\neg(\phi \vee \neg\phi)$ retains validity. Try assuming that $p \vee \neg p$ is false for some p. From this assumption, it follows that p must be false, for were p true, $p \vee \neg p$ would also be true. Similarly, $\neg p$ must be false as well. The reason is the same: were $\neg p$ true, $p \vee \neg p$ would be true. But p and $\neg p$ cannot both be false! Therefore, $\neg\neg(\phi \vee \neg\phi)$ is valid and

$$\forall p \neg\neg(p \vee \neg p)$$

obtains. Consequently, the familiar rule for dropping double negations $\neg\neg\phi \vdash \phi$ is not valid, since it would lead at once from $\neg\neg(\phi \vee \neg\phi)$ to TND. In general, a proposition p and its double negation $\neg\neg p$ are distinct.

The failure of TND and the validity of its double negation have an immediate effect on the rules for quantifiers. Let QNE (*quantifier negation exchange*) be the injunction to interchange universal quantifiers with negations in this way:

$$\neg\forall x\phi \vdash \exists x \neg\phi.$$

Corollary 4 (*QNE is invalid*) $\exists x \neg \phi$ does not follow logically from $\neg \forall x \phi$.

Proof. Because TND is invalid, we know that $\neg \forall p (p \vee \neg p)$. If QNE were valid, then $\exists p \neg (p \vee \neg p)$ would hold. However, as shown above, $\neg \neg (p \vee \neg p)$ is true, since $\neg \neg (\phi \vee \neg \phi)$ is valid. Therefore, QNE is invalid. ■

The intuitionist is naturally prompted to consider a weaker version of TND, *the principle of testability* or TEST:

$$\forall p (\neg p \vee \neg \neg p).$$

UP also suffices to how that TEST is invalid.

Corollary 5 (*TEST is invalid*). $\neg \forall p (\neg p \vee \neg \neg p)$

Proof. We proceed much as in the proof of Corollary 1, but starting from the assumption that TEST holds rather than TND. Construct a relation R between sets X of numbers and individual numbers n so that $R(X, n)$ obtains just in case

$$\neg \neg (0 \in X) \quad \text{and} \quad n = 1$$

$$\text{or}$$

$$0 \notin X \quad \text{and} \quad n = 2.$$

As before, it follows that for every set X, there is an n such that $R(X, n)$. By UP, there is an m such that, for all X, $R(X, m)$. Once again, this is impossible, since R cannot relate both $\varnothing$ and $\mathbb{N}$ to one and the same number. ■

4. Mathematical Propositions and Prop

In our justification for UP, the assumption that sets X were sets of natural numbers specifically played no special role. One suspects that some manner of uniformity should therefore obtain for other nontrivial power sets, other sets of all subsets taken from a given set. To take one example, the set Prop of all truth-values is identified, as in conventional mathematics, with P({0}), the set of all subsets of the singleton set {0}. This identification makes sense, since the propositions or truth-values stand in exhaustive one-to-one correspondence with the subsets of {0}: just pair proposition p with the set

$$\{x : x = 0 \text{ and } p\}.$$

As we shall learn, Prop exerts an intuitionistic fascination much greater than its conventional counterpart.

COROLLARY 6 (*Uniformity for Prop*). If each truth-value in Prop is associated by an extensional relation R to some natural number n, there is at least one number related by R to all elements of Prop. In symbols, $\forall p \, \exists n \, R(p, n) \Rightarrow \exists n \forall p \, R(p, n)$.

Proof. It suffices to note that Prop is the result of intersecting all subsets of $\mathbb{N}$ with the set $\{0\}$. Every proposition p (as a subset of $\{0\}$) can be obtained by intersecting the set $\{n : p\}$ with $\{0\}$. Therefore, if each member p of Prop is related by R to some number n, then every subset X of $\mathbb{N}$ can be related to a number: first intersect X with $\{0\}$, obtain a member p of Prop, and then relate p with some n via R. By UP, there is a number m associated in this fashion with every subset of $\mathbb{N}$. A fortiori every proposition p, as a subset of $\{0\}$, is related by R with m. ∎

COROLLARY 7. Prop is both uncountable and indecomposable.

Proof. Employ Uniformity for Prop and the proof ideas from Corollaries 2 and 3. ∎

But exactly how large is Prop? In conventional mathematics, Prop is finite. It has precisely two elements: $\varnothing$ and $\{0\}$. Intuitionistically, Prop contains at least these two members. But, as we shall now see, Prop is in reality neither finite nor infinite. For intuitionists as for conventional mathematicians, a set X is *finite* when X is enumerable using the numbers less than some particular natural number. X is *infinite* when the set $\mathbb{N}$ can be embedded, via a one-to-one function, into X. Even in conventional mathematics, the notions *finite* and *infinite* so defined are logical contraries, not contradictories: it is possible that sets exist that are neither finite nor infinite. We call such sets *Dedekind sets.*

COROLLARY 8. Prop is a Dedekind set.

Proof. First, since the set of natural numbers less than any given nonzero number is countable, Prop cannot be finite, as the preceding Corollary shows. Second, assume that Prop contains a member p distinct from both $\{0\}$ and $\varnothing$. It follows that $p \neq \{0\}$. Therefore, $0 \notin p$. This means that $p = \varnothing$. But that contradicts the original assumption that p is distinct from both $\{0\}$ and $\varnothing$. Hence, Prop contains at least two truth values but cannot have three. *A fortiori* Prop is not infinite. ∎

The phenomenon represented by Dedekind sets is nowise uncommon in intuitionistic mathematics. There can be uncountably many Dedekind subsets of $\mathbb{N}$. Some of these are strong contenders for the title "potentially infinite": each has no greatest number but none contains all the numbers.

II. Mathematical Domains

1. Natural, Rational, and Real Numbers

Our logical introduction to intuitionism completed, we move into more familiar mathematical territory, the domains of the natural, rational, and real numbers. The facts codified in the axioms of Dedekind and Peano characterize $\mathbb{N}$. Zero is a natural number and not the successor $S(x)$ of any other. The successor function is one-to-one: $m = n$ if $S(m) = S(n)$. Full induction holds; any mathematical property had by 0 and preserved by S is a property of every natural number. For example, it is easy to show by induction that no number is the same as its successor. The primitive recursive functions—including addition, multiplication, exponentiation, and compositions of them—exist with their standard definitions and retain their basic characteristics. A mathematical property or relation R is *decidable* on a set X when, for $x, y \in X$, either $R(x, y)$ or $\neg R(x, y)$; in other words, R satisfies TND relativized to X. The equality and less-than relations on $\mathbb{N}$ are both decidable, as deft use of induction will verify. Incidentally, it is easy to prove, as a consequence of UP, that equality cannot be decidable on either $P(\mathbb{N})$ or Prop.

To say that a set X is *nonempty* is to deny that it is empty: $\neg \forall x.\ x \notin X$. This is not to say that it is *inhabited* (i.e., that X actually has a member): $\exists x.\ x \in X$. Since QNE fails to be valid, the latter is generally stronger than the former. Although mathematical induction holds of $\mathbb{N}$, it is not well-ordered in the downward sense under $\leq$. It is false that every inhabited subset of $\mathbb{N}$ has a least member. This least number principle will hold, however, for decidable, inhabited sets, as a proof by induction can demonstrate.

Theorem 1 ($\mathbb{N}$ *is not well ordered*). It is false that every inhabited subset of $\mathbb{N}$ has a least member.

Proof. Let p be any mathematical proposition and let X be this set of numbers:

$$\{1\} \cup \{n : n = 0 \text{ and } p\}.$$

X is inhabited since 1 is a member of it. However, if X had a least member m, then m must be either 0 or 1. If $m = 0$, then $0 \in X$ and p holds. On the other hand, if $m = 1$, then 1 is the least member of X and $0 \notin X$. In that case, $\neg p$ holds. Therefore, if X had a least member, $p \vee \neg p$ would obtain for every p. Hence, were the least member principle to be true of inhabited sets generally, TND would be valid. ∎

Intuitionistic theories of the integers Z and the rational numbers Q work out much as one would expect. Both $=$ and $\leq$ are decidable relations on Z and Q; they inherit decidability from the corresponding relations on $\mathbb{N}$. As in conventional mathematics, there are a number of ways to define the set $\mathbb{R}$ of real numbers, the numbers of the calculus. For instance, they are definable à la Dedekind as Dedekind cuts or à la Cantor as Cauchy sequences. Roughly speaking, Dedekind cuts are proper, inhabited subsets of Q that have no greatest member and are not bounded below in the usual order on Q. Cauchy sequences are infinite strings of members of Q whose terms draw nearer as one goes further out in the sequence. In this writing, we think of real numbers as given by Cauchy sequences.

2. Principles of Choice

With the *Axiom of Countable Choice* (CC), one can prove that every real number determined by a Dedekind cut is also determined by a Cauchy sequence and conversely. CC is the statement that when an extensional relation R associates, with every natural number n, an $x \in X$ so that $R(n, x)$, there is a function f, a *choice function*, with the property that f chooses for each n a suitable $f(n) \in X$ such that $R(n, f(n))$. In symbols,

$$\forall n \exists x \in X\, R(n, x) \Rightarrow \exists f [\forall n\, R(n, f(n)) \text{ and } f(n) \in X].$$

To give evidence in support of CC, one calls upon an intuition akin to that earlier described. When each n is related by such R to some x in X, there should be a discernible principle α by which members of $\mathbb{N}$ get associated with elements of X via R. Unlike a set of natural numbers, which may well be infinite, individual natural numbers can be completely represented within α. Thus, α is conceivable as a rule tagging each n with a definite member x_n of X and, because R is extensional, it should be true that if $n = m$, $x_n = x_m$. Therefore, α can be construed as yielding a choice function f taking each number n to an allied $x_n \in X$ so that $R(n, x_n)$.

The reader is here cautioned. It is absolutely incorrect to suppose that a choice principle similar to CC will hold *whenever* there are sets X and Y and a relation R such that, for every $x \in X$, there is a $y \in Y$ for which $R(x, y)$. No principle of this kind is licensed by the quantifier combination $\forall\exists$ in all its occurrences. Contrary to an impression that pages 52 and 53 of Dummett [1977] might create, the general *Axiom of Choice* (AC), well known to conventional set theorists, is false. In his [1975], R. Diaconescu showed that AC, even for small sets, implies TND.

Theorem 2 (AC *is false*). It is not the case that $\forall x \in X \exists y \in Y.\ R(x, y) \Rightarrow \exists f \forall x \in X.\ R(x, f(x))$.

Proof. Let p be a proposition and X be a set whose sole members are $a = \{0\} \cup \{n : n = 1 \text{ and } p\}$ and $b = \{1\} \cup \{n : n = 0 \text{ and } p\}$. Each element of

X itself has a member. By AC, there is a function f that chooses a member from each of a and b. Given that the members of a and b are natural numbers and equality is decidable on $\mathbb{N}$, there are only two possibilities for the values of $f(a)$ and $f(b)$: either $f(a) = f(b)$ or $f(a) \neq f(b)$. In the former case, p has to hold. In the latter, a must itself be different from b, so $\neg p$ holds. Therefore, TND follows from AC. ■

3. Real Numbers and Brouwer's Theorem

Familiar basic properties of addition and multiplication on the rational numbers of Q often extend to $\mathbb{R}$. For example, addition is commutative for real numbers and multiplication distributes over addition. But not all the conventional properties of the rationals carry over. For example, we prove, as Corollary 11 below, that equality between real numbers is undecidable. In addition, as was first noted in Brouwer [1921], not every real number is represented by a decimal expansion. Even so, failures in conventional reasoning about $\mathbb{R}$ hardly prevent the development of intuitionistic real analysis. Brouwer showed how to exploit the sorts of intuitions earlier encountered—applied to sequences, rather than to sets—to enrich immeasurably the constructive mathematical theory of $\mathbb{R}$.

As we said, real numbers are given by infinite Cauchy sequences. It would then be natural to think that some principles governing real numbers should be derivable from principles governing infinite sequences. Let Seq be the set of all infinite sequences of natural numbers and let $R(s, n)$ relate members s of Seq to numbers $n \in \mathbb{N}$ in strictly extensional fashion. Assume that for every s there is an n such that $R(s, n)$. As above, there should be a rule α attaching numbers to sequences in accord with R. Sequences from Seq are infinite strings that are represented to α. As with P($\mathbb{N}$), Seq can be obtained neither from recursive generation nor from linguistic specifications of sequences. Furthermore, since R is strictly extensional, α should not depend essentially upon the details of linguistic specifications for sequences. Hence, the members s of Seq can be represented in α only through the numerical values of the terms that comprise their extensions, in the case of s:

$$s_1, s_2, s_3, \ldots, s_n, \ldots.$$

Because rules are finite and not infinite bearers of information, α can contain and operate on only the information presented by some finite number of those terms, perhaps the finite initial segment of s ending with s_m:

$$s_1, s_2, s_3, \ldots, s_m.$$

It should be on the sole basis of such an initial segment that α attaches an n to s so that $R(s, n)$. Moreover, since α can take only finite initial segments of input sequences into consideration, if t is another sequence sharing with s exactly this initial segment—that is,

$$t_1 = s_1,\ t_2 = s_2,\ t_3 = s_3, \ldots,\ t_m = s_m,\text{—}$$

α should attach to t the very same n that it attaches to s. This intuitive reasoning is enshrined in the following Neighborhood Theorem (NT).

> **Theorem 3** (*Neighborhood Theorem*). Assume that each sequence in Seq is related by extensional R to some natural number. Then, for every sequence $s \in$ Seq, there is a finite initial segment u of s and a number n related to s by R such that any sequence sharing u is also related by R to n. ■

Brouwer [1918] contained the intuition behind this justification of NT; Heyting [1930c] gave the theorem a fully explicit formulation. NT is as anti-conventional as UP in that, as one can easily check, ¬TND follows from it. The word "neighborhood" is a term from topology: a topological *neighborhood* is a collection of members of a set, perhaps numbers or sequences, that can be considered approximately the same. Hence, two sequences of Seq are approximately the same or "in the same neighborhood," if they share an initial segment. NT tells us that infinite sequences of natural numbers can be related to individual numbers only by relating them in "clumps" or neighborhoods, so that all sequences which are approximately the same get assigned the same number.

In giving a justification for NT, Brouwer treated members of Seq as *choice sequences.* These are sequences of numbers whose successive terms are conceived as chosen by a mathematician, one by one, over time in a way that may be relatively unregulated by rules or other constraints. In general, all that a mathematician may know of a choice sequence s at a time t is the initial segment comprised of those terms of s chosen up to t. Hence, because a mathematician can apply a rule to an infinite sequence only at some particular time, and in so doing he or she can take cognizance only of the terms of the sequence that have been chosen before that time, mathematical rules act on choice sequences by acting on finite initial segments. This concept of a sequence whose terms are so chosen that there is limited information available about the future behavior of the sequence was first featured in Paul du Bois-Reymond's intriguing book *Die allgemeine Functionentheorie* [1882].

Let X be a set with a *topology*: an appropriate association, with each element $x \in X$, of neighborhoods of x. A function f with inputs from X and outputs in $\mathbb{N}$ is *continuous* when, for each $n \in \mathbb{N}$, $f(x) = n$ implies that there is neighborhood of x such that $f(y) = n$ for any y in that same neighborhood. It is plain that NT entails that every function with inputs from Seq (when initial segments determine neighborhoods) and outputs in $\mathbb{N}$ is continuous.

COROLLARY 9. Every function from Seq to $\mathbb{N}$ is continuous. ■

Our interest in Seq is motivated by the close relation between Seq and $\mathbb{R}$, between sequences and real numbers. In the usual topology on the set of real numbers $\mathbb{R}$, the neighborhoods of a particular real number r are the finite open intervals containing r—that is, sets of the form $\{q \in \mathbb{R} : a < q < b\}$, where a and b are fixed rational numbers.

Theorem 4 (*Continuity Theorem*). Every function from $\mathbb{R}$ into $\mathbb{N}$ is continuous.

Proof. Let f be a function from $\mathbb{R}$ into $\mathbb{N}$. One can turn sequences of Seq into Cauchy sequences in such a uniform fashion β that f will be continuous as a function on $\mathbb{R}$, provided that the function that takes any infinite sequence s of Seq into $f(\beta(s))$ is continuous on Seq. But, as we have seen, every function from Seq into $\mathbb{N}$ is continuous. Hence, the function $f(\beta(s))$ is continuous. It follows that f is continuous. ■

This means that, as with Seq, functions from $\mathbb{R}$ into $\mathbb{N}$ have to be given by neighborhoods: if a real number r maps, under f, into n, there is an entire neighborhood of r that f also maps to the same n.

In a topology, a set is open when it is a union of neighborhoods. Under its usual topology, $\mathbb{R}$ is *connected.* That is, $\mathbb{R}$ cannot be the union of any collection of disjoint, inhabited, open sets. It follows that every function f from $\mathbb{R}$ into $\mathbb{N}$ is constant: for all real numbers r and s, $f(r) = f(s)$.

COROLLARY 10 (*Constancy*). Every function from $\mathbb{R}$ into $\mathbb{N}$ is constant.

Proof. As we have seen, the values of a function f from $\mathbb{R}$ into $\mathbb{N}$ are determined by neighborhoods. Therefore, were f to take on different values in $\mathbb{N}$, $\mathbb{R}$ would have to be the union of the disjoint, inhabited, open sets that determined those distinct values. Because $\mathbb{R}$ is connected, this is impossible. ■

From the statement that every function from $\mathbb{R}$ into $\mathbb{N}$ is constant, it follows that equality on $\mathbb{R}$ is undecidable and $\mathbb{R}$ is indecomposable.

COROLLARY 11 (*Equality is undecidable on* $\mathbb{R}$). It is false that for all r and $s \in \mathbb{R}$, either $r = s$ or $r \neq s$.

Proof. Assume that equality is decidable. Then, for any real number r, either $r=0$ or $r\neq 0$. Therefore, there is a function f from $\mathbb{R}$ into $\mathbb{N}$ such that

$f(r)=0$ if $r=0$
and
$f(r)=1$ otherwise.

However, this function cannot be continuous. Although $f(0)=0$, there is no open interval around 0 such that for every real number s in that interval, $f(s)=0$ as well. ∎

With equal ease, one can show that $\mathbb{R}$ cannot be divided neatly into rational numbers and irrational numbers or into any nontrivial pair of exclusive, exhaustive subsets.

COROLLARY 12 ($\mathbb{R}$ *is indecomposable*). Every decidable subset of $\mathbb{R}$ either is empty or includes all of $\mathbb{R}$.

Proof. Similar to Corollary 1. ∎

In his [1999a], Dirk van Dalen reported on significant and surprising extensions of this indecomposability theorem for $\mathbb{R}$, among them, that the set of irrational numbers is also indecomposable.

Brouwer, in his [1924], claimed that every function mapping the closed unit interval $\{r\in\mathbb{R} : 0\leq r\leq 1\}$ into $\mathbb{R}$ is *uniformly continuous*, a result often called Brouwer's Theorem. A function f from $\mathbb{R}$ to $\mathbb{R}$ is uniformly continuous when its continuity is registered by a rule that calculates uniformly from an open interval i around $f(r)$, an interval j around r that f maps into i. From the truth of Brouwer's Theorem, one can infer that every function from $\mathbb{R}$ to $\mathbb{R}$ is continuous. Intuitionists and other constructivists have established suitable generalizations of Brouwer's Theorem for metric spaces other than $\mathbb{R}$. A set X with a topology is a *metric space* when the neighborhoods of its elements are determined by the values of a binary function or *metric* specifying an abstract distance between elements x and y of X. Naturally, $\mathbb{R}$ with its interval topology is a metric space when the distance between real numbers r and s is the absolute value of the difference $r-s$. For X a metric space, it is true that every function from $\mathbb{R}$ into X is continuous. For details, the reader is advised to consult Bridges and Richman [1987].

Thanks to Brouwer's Theorem, connections between well-structured mathematical domains are forged by functions preserving structure—in this case, continuous functions. Connections between mathematical domains and logical ones like Prop are also of great significance. Clearly, a function f taking X into Prop is a propositional function: f maps each $x\in X$ into a truth-value $f(x)$ in Prop. For example, the function g assigning, to each real number r, the proposition "r is rational" is a propositional function that, one might say, attaches to each real number the "extent" to which it is rational. These propositional functions are the characteristic functions of subsets; g is the characteristic function for the set of rational numbers from $\mathbb{R}$. However, looking in the other direction—from logical domains to mathematical ones—uniformity trims functional connections severely. From the failure of the principle of testability TEST, A.S. Troelstra deduced a general result, described in Troelstra [1980], from which it follows that every function from Prop or $P(\mathbb{N})$ into any metric space has to be constant.

III. Formal Logic and Internal Mathematics

1. Elementary Formal Logic

Here is a cartoon recipe for "creating mathematics" that captures a common misconception: to produce mathematics, turn on the mixer of logic, take the needed assumptions as axioms, feed them into the mixer, press the button marked "derive," and—voilà—the proofs of mathematics will pour from the nether end. The idea is that one starts with logic, a fixed set of rules for deductive reasoning that establish clear norms for thinking rightly about any domain whatsoever. Then, axioms specific to a mathematical domain are added to the rules as premises. (How they get "added" and whether this very process of "addition" might itself prove a legitimate subject for mathematical study are questions not permitted here.) Finally, mathematics is the result of applying the derivational principles of logic, first on those premises and, in turn, on their deductive consequences. According to this parody of creative thought, the vast array of proofs in any particular realm of serious mathematical study arises without muss or fuss as the closure of special axioms under the deductive relations specified by a general logic.

Intuitionism will have none of this, but not merely because intuitionistic logic is incomplete or because intuitionistic mathematics begins with intuitions rather than axioms. In intuitionism there is no logic as caricatured in the cartoon recipe. There is no set of normatively binding principles for right thinking that (1) hold sway over all scientific domains, including mathematical ones, and (2) can be set up once and for all, before we start to pursue mathematics. As for (1), the doctrines of intuitionistic logic are one and all statements of mathematics, largely of an unexciting sort. Ultimately, they owe their correctness to vivid intuition rather than to pale generality. As we have already seen, the claim that TND fails of validity is, as with all other logical results, a theorem of mathematics admitting of mathematical demonstration. As for point (2), one learns of logical principles only after the fact, from truths and proofs in mathematics. Intuitionists intuit, construct, and reason first, and legislate for these activities later.

On the whole, intuitionists do cleave, when they expound or reconstruct proofs, to regular forms of mathematical communication and persuasion, to certain repeated patterns in thinking. The first researcher to do an acceptably comprehensive job of extracting and assembling principles governing that thinking was the intuitionist A. Heyting (1898–1980). His [1930a], [1930b], and [1930c] contain formalisms for intuitionistic propositional and predicate logic, plus arithmetic and set theory. In honor of this achievement, logicians have named the standard formal systems for intuitionistic arithmetic after him: HA for first-order Heyting arithmetic and HAS for its second-order extension. The beginner can most easily discern the main ideas of Heyting's propositional logic in a natural deduction presentation like that given by G. Gentzen in his [1935], on which derivability is defined in terms of introduction and elimination rules for each connective.

Assume that we have an ordinary propositional language, based on a collection of atomic formulas or atoms, and the usual connective signs $\wedge, \vee$, and $\Rightarrow$ in addition to the connective $\bot$ or *absurdum*. Formulas are recursively defined, as is familiar. The inference rules are familiar as well. A conjunction can be introduced by immediate inference from its two conjuncts. It can be eliminated by inverting the process and inferring either conjunct from it. The rules are as follows, with the introduction rule on the left and elimination rules on the right.

$$\frac{\phi \quad \psi}{\phi \wedge \psi} \qquad \frac{\phi \wedge \psi}{\phi} \qquad \frac{\phi \wedge \psi}{\psi}$$

The introduction rule for $\Rightarrow$ is that of conditional proof; it allows one to construct derivations from assumptions made only for the sake of argument. These assumptions are discharged, in the course of the derivation, by employing conditionalization. The elimination rule for $\Rightarrow$ is *modus ponens*. Disjunctions are introduced by inference from either disjunct. The elimination rule for disjunction is argument by cases, which also allows assumptions to be introduced for the sake of argument and to be discharged later. The sign $\bot$ is a formalized stand-in for mathematical absurdity, so there are no introduction rules for it. As one would suspect, its elimination rule is as liberal as its introduction rule is restrictive; it captures the old notion *ex falso quodlibet*:

$$\frac{\bot}{\phi}$$

On this presentation, one thinks of negation $\neg$ as a defined sign: $\neg\phi$ is shorthand for the conditional $\phi \Rightarrow \bot$. Logical rules for negation (the ones that are valid, that is) show up as derived rules. They include the positive form of reductio ad absurdum: when $\bot$ is derived from the assumption ϕ, conclude $\neg\phi$. That this rule is valid is proven via an easy application of $\Rightarrow$ introduction. The injunction "when $\bot$ is derived from assumption $\neg\phi$, conclude ϕ" is reductio ad absurdum in its negative form. It is not licensed here and cannot be derived, as we shall prove in a moment. Intuitionistic predicate logic results from the addition to these rules of standard introduction and elimination rules for the quantifiers: generalization and instantiation for $\forall$, plus generalization and instantiation for $\exists$.

Derivations in the system are finite trees generated by applying introduction and elimination rules recursively to assumed formulas. At the trees' leaves stand active and discharged assumptions, and at their roots, conclusions from those assumptions. As usual, we say that for a set of formulas Γ and a formula ϕ, Γ derives ϕ (in symbols, $\Gamma \vdash \phi$) just in case there is a derivation at whose root is ϕ and whose undischarged assumptions are members of Γ.

Derivation in intuitionistic logic has the pleasant property that any formula ϕ derivable from $\varnothing$ can be derived "directly," without use of lemmas or subproofs requiring derivations of formulas of greater complexity than ϕ itself. Indeed, the deductive system has the *subformula property*: when ϕ is derivable from Γ, there is a derivation for ϕ in which every formula that appears is a subformula either of ϕ or of some member of Γ. Using the subformula property, one can show that the propositional logic is *decidable*: for every formula ϕ, either ϕ is a theorem or it is not.

Among the most significant and far-reaching results on intuitionistic formal logic is the negative translation theorem that K. Gödel reported in his [1933a] and that Gentzen discovered independently (Gentzen [1974]). The theorem reveals the existence of an inferentially accurate picture, under translation, of conventional formal logic within intuitionistic logic. Here, "$g(\phi)$" stands for the translation of a formula ϕ and $g(\Gamma)$ for the set of translations of formulas from Γ.

Translation. If A is an atom but not $\bot$, $g(A)$ is $\neg\neg A$. $g(\bot)$ is $\bot$. And g commutes with $\wedge$, $\Rightarrow$, and $\forall$:

$$g(\phi \wedge \psi) = g(\phi) \wedge g(\psi)$$
$$g(\phi \Rightarrow \psi) = g(\phi) \Rightarrow g(\psi)$$
$$g(\forall x \phi) = \forall x\, g(\phi).$$

The g-translations of $\vee$ and $\exists$ are negative, since they call upon the negative connectives $\neg$ and $\wedge$:

$$g(\phi \vee \psi) = \neg(\neg g(\phi) \wedge \neg g(\psi))$$
$$g(\exists x \phi) = \neg \forall x \neg g(\phi).$$

Let $\vdash_c$ stand for derivability in conventional predicate logic.

Theorem 5 (*Negative translation*). $\Gamma \vdash_c \phi$ if and only if $g(\Gamma) \vdash g(\phi)$.

Proof. Because intuitionistic formal logic is a fragment of its conventional counterpart, it follows from $g(\Gamma) \vdash g(\phi)$ that $\Gamma \vdash_c \phi$. Induction on the structure of derivations for $\Gamma \vdash_c \phi$ proves the converse. ∎

Suitable negative translations, with correlative results linking conventional to intuitionistic derivability, have been devised for higher-order logic, formal arithmetic, type theory, and versions of set theory formalizable within intuitionistic predicate logic.

As far as elementary semantics for intuitionistic propositional logic is concerned, interpretational structures can be defined precisely as in conventional metalogic. A *structure* $\mathfrak{I}$ for a propositional language with Δ as its set of atoms is simple: a function taking inputs from Δ and yielding outputs in the set of propositions Prop such that $\perp$ is assigned the contradictory proposition. $\mathfrak{I}$ is uniquely extendible to a function $\mathfrak{I}^*$ mapping all formulas into Prop by taking $\wedge$ to mean "and," $\vee$ to mean "or," and so on. Then, we say that $\mathfrak{I}$ *satisfies* ϕ ($\mathfrak{I} \models \phi$) when $\mathfrak{I}^*$ takes ϕ into a true proposition. For a collection of formulas Γ, $\mathfrak{I}$ *satisfies* Γ ($\mathfrak{I} \models \Gamma$) just in case $\mathfrak{I}^*$ maps every formula in Γ into a truth. Last, we say that Γ *entails* ϕ ($\Gamma \models \phi$) when every structure that satisfies Γ also satisfies ϕ. Here, as elsewhere, the formal logic is *sound* just in case derivability is certified by entailment: $\Gamma \vdash \phi$ always implies that $\Gamma \models \phi$. It is *complete* when derivability tracks entailment, that is, when the converse of soundness holds for every Γ and ϕ. By induction on derivations, one can easily show that intuitionistic propositional logic is sound. Therefore, since TND is invalid, its formal version $A \vee \neg A$ is underivable. Furthermore, we learn that reductio ad absurdum in its negative form cannot be a derived rule. Were one allowed to infer ϕ from the supposition that $\neg\phi$ leads to $\perp$, one could derive $A \vee \neg A$ from a correct derivation of $\perp$ from $\neg (A \vee \neg A)$. An informal analogue to the latter derivation was our earlier proof that $\neg\neg (\phi \vee \neg\phi)$ is valid.

Only under quite special conditions, conditions independent of the vast bulk of intuitionistic mathematics, is it provable that propositional logic is complete for single formulas or finite sets of them. It is obviously incomplete when it comes to other sets of formulas:

Theorem 6 (*Incompleteness*). Intuitionistic propositional logic is incomplete for arbitrary sets of formulas.

Proof. For a mathematical proposition p, let A_p be a propositional atom so associated with it that distinct propositions are associated with distinct atoms. For each such p, consider the set of formulas $\Gamma_p = \{A_p : p\} \cup \{\neg A_p : \neg p\} \cup \{A_p \vee \neg A_p\}$ and take Γ to be the union of all the individual Γ_p's, so that $\Gamma = \bigcup\{\Gamma_p : p \text{ is a mathematical proposition}\}$. The negative translation theorem shows that Γ is consistent because Γ cannot derive $\bot$. Because Γ contains, for each p, the formula $A_p \vee \neg A_p$ as a member, the assumption that structure $\mathfrak{I}$ satisfies Γ immediately leads to the conclusion that the principle of testability TEST holds. Therefore, $\Gamma \models \bot$, because $\bot$ clearly has no structures that satisfy it. ■

Unless one demands fundamental changes in the interpretations of the connective symbols, as in Veldman [1976], intuitionistic formal logic remains incomplete. For this, intuitionists are generally thankful: the incompleteness of formal logic is further confirmation of their insistence, as in point (2) above, that logic must be ancillary to mathematics.

In his [1932], Gödel proved that intuitionistic propositional logic cannot be finite-valued. No truth-table procedure based on calculations from a finite number of truth-values can provide an adequate semantics for it.

Theorem 7. Intuitionistic propositional logic is not finite-valued.

Proof. Assume that the logic is sound and complete with respect to truth-tables with a finite number k of values. Pick out $k+1$ distinct atoms: A_1, $A_2, \ldots, A_{k+1}$, none of which are $\bot$, and take the formula Φ to be a disjunction of all possible biconditionals constructed from them of the forms

$$A_i \Leftrightarrow A_j,$$

where i and j are different and less than $k+2$. If propositional logic is complete with respect to truth-tables with k truth-values, Φ is a derivable formula, since at least two of the atoms A_i would have to be assigned the same truth-value in any structure. However, by constructing Kripke models (see the next section) and using the relevant soundness theorem, one can show that ϕ is underivable. ■

2. Models and Modality

Conventional mathematical means can prove the completeness of intuitionistic propositional and predicate logic with respect to Kripke models, interpretational structures defined over partially ordered sets. Let P be a set partially ordered by $\leq$, and let α and β be elements or *nodes* of P. A *Kripke model* K (after [Kripke 1965]) over P is given by specifying a relation $\alpha \models \phi$ (read "α forces ϕ") between elements of P and formulas. Attached to each node α of K is an inhabited domain D_α in such a way that when $\alpha \leq \beta$, $D_\alpha \subseteq D_\beta$. For present purposes, we assume that suitable names a for the elements of the domains D_α have been adjoined to the original language; the forcing relation is defined for sentences in this extended language. For atoms A which are not $\perp$, $\alpha \models A$ is defined in any way you like, provided only that when $\alpha \models A$ and $\alpha \leq \beta$, $\beta \models A$. One says that, for atoms, forcing is required to be *persistent*. Understandably enough, $\perp$ is never forced at a node—at least on the standard approach to Kripke models. For more complicated formulas, we say that

$\alpha \models \phi \wedge \psi$ just in case $\alpha \models \phi$ and $\alpha \models \psi$

$\alpha \models \phi \vee \psi$ just in case either $\alpha \models \phi$ or $\alpha \models \psi$

$\alpha \models \phi \Rightarrow \psi$ just in case, for any $\beta \geq \alpha$ in P, if $\beta \models \phi$, then $\beta \models \psi$

$\alpha \models \exists x\, \phi$ just in case there exists $a \in D_\alpha$ such that $\alpha \models \phi(x/a)$ and

$\alpha \models \forall x\, \phi$ just in case, for any $\beta \geq \alpha$ and $a \in D_\beta$, $\beta \models \phi(x/a)$.

Last, $\neg\phi$ is defined as $\phi \Rightarrow \perp$, so we deduce that

$\alpha \models \neg\phi$ just in case, for any $\beta \geq \alpha$, β fails to force ϕ.

We say that $\alpha \models \Gamma$ if α forces every member of Γ. Γ K-entails (in symbols, $\Gamma \models_K \phi$) whenever, for any Kripke model K and node α in K, if $\alpha \models \Gamma$, then $\alpha \models \phi$. A proof by induction on formulas shows that forcing is persistent for all formulas ϕ: if $\alpha \models \phi$ and $\alpha \leq \beta$, then $\beta \models \phi$.

When explaining Kripke models, logicians often employ a metaphor of states of mathematical information. Imagine yourself trying to solve a particular mathematical problem. The mathematically possible states of information you can enter while searching for a solution can be conceived as the nodes of a set P partially ordered by a relation $\leq$ of "at least as much information" determined by the problem under consideration. When nodes α and β stand in the relation $\alpha \leq \beta$, then β is thought of as a possible state of information relative to α: you might, in investigating your problem further, gain information sufficient to take you from state α to state β. The atoms (other than $\perp$) that get forced at α are those simplest propositions that are verified from the information available at α. Forcing is persistent because, once you have gained some information, that information is never lost to you. The clauses in the definition of forcing are reminders that you can, with information α, verify a conjunction $\phi \wedge \psi$ when you can verify both its conjuncts. The only way to verify a disjunction in α is to learn, in α, that one or the other of its disjuncts is verified there. The verification of a conditional $\phi \Rightarrow \psi$ in general requires more information: that no matter when one gains added information β sufficient to verify ϕ, that extra information should also verify ψ, or $\beta \models \psi$. This is required if the information that $\phi \Rightarrow \psi$, once gained, is to persist.

Familiar methods from conventional mathematics yield the result that intuitionistic logic is sound and complete with respect to Kripke models.

Theorem 8 (*Kripke Soundness and Completeness*). $\Gamma \vdash \phi$ if and only if $\Gamma \models_K \phi$.

Proof. Soundness is proved using induction on derivations. Completeness can be proved conventionally by adapting Henkin's proof to Kripke models. This and related completeness results are available from Troelstra and van Dalen [1988]. ■

Strictly intuitionistic means suffice to prove that intuitionistic propositional logic has the *finite model property*: if ϕ is underivable, there is a finite Kripke model and a node in it that fails to force ϕ—in order words, there exists a finite countermodel to ϕ.

Theorem 9 (*Finite Model Property*). When ϕ is not derivable in intuitionistic propositional logic, then ϕ has a finite Kripke countermodel.

Proof. The interested reader can find a proof in Troelstra and van Dalen [1988]. ■

Research into partially ordered models for intuitionistic logic was suggested by another translation proposed by Gödel. This time, propositional logic proves to be the exact replica of a special fragment of Lewis's modal system S4, a system which features the logical sign $\Box$ for alethic modality and requires that $\Box\phi$ always be equivalent to $\Box\Box\phi$. Gödel's translation, here denoted "m," accepts formulas in the language of propositional logic and outputs formulas in the language of the Lewis calculus. The translation m is so defined that, for any atom A, $m(A)$ is $\Box A$. Then, m commutes with $\wedge$ and $\vee$, and $m(\phi \Rightarrow \psi)$ is the modal formula $\Box(m(\phi) \Rightarrow m(\psi))$. In effect, the intercalation of the modal operator $\Box$ into a translated formula enforces the persistence property required by Kripke forcing.

Theorem 10 (*Modal Translation*). $\Gamma \vdash \phi$ if and only if $m(\phi)$ is derivable from $m(\Gamma)$ in S4.

Proof. The "only if" direction yields to proof by induction on derivations. Van Dalen [1994] contains a proof of the "if" direction that uses semantical ideas acceptable to intuitionists. ■

Godel [1933b] presented a modal translation alternative to that defined above and asserted the "only if" direction of the theorem. Gödel there conjectured the other direction of the theorem as well. It was first proved, in McKinsey and Tarski [1948], by algebraic means. The articles in Shapiro [1985] explore extensions to formalized intuitionistic arithmetic and set theory of the modal translation for propositional logic.

3. Internal Mathematics: Realizability and Computability

In justifying UP and NT, we insisted that rules for computation accompany certain mathematical operations. This may suggest that concerns of intuitionistic mathematics stand in proximity to concerns of computability theory. *Realizability* interpretations of formalized intuitionistic logic and mathematics provide partial confirmation for the suggestion. Logician S. Kleene was the first to devise such an interpretation; the original publication was his [1945]. In discovering realizability, Kleene was not inspired so much by the markedly computational air of Brouwer's intuitionism as by D. Hilbert's idea that quantified statements be treated finitistically as expressing "incomplete communications."

As with Kripke models, Kleene's realizability can be glossed using a concept of information: effective information encoded as a natural number and conveyed by a mathematical claim. For an atomic formula ϕ (which, in formal arithmetic, will be decidable) a number n represents the information ϕ conveys or *realizes* ϕ just in case ϕ is true. For conjunctions, n realizes $\phi \wedge \psi$ if n is of the form $2^m \times 3^p$, where m realizes ϕ and p realizes ψ. A conditional $\phi \Rightarrow \psi$ is realized by n just in case n encodes a Turing machine that, if m realizes ϕ, takes m as input, processes for a time, and outputs a number realizing ψ. In other words, a realizing number for a conditional denotes a Turing-computable function converting realizers for its antecedent into realizers for its consequent. Reflecting the constructive character of intuitionistic arithmetic, n is said to realize $\exists x\, \phi(x)$ if and only if n is $2^m \times 3^p$ and m realizes $\phi(x/\underline{p})$, where $\underline{p}$ is the formal numeral denoting p. Therefore, a realizing number for an existential sentence literally contains, as a coded component, a realized instance of that sentence. A number n realizes $\forall x\, \phi(x)$ if and only if n encodes a Turing machine computing a total recursive function such that, for any m, the output of the function on m realizes $\phi(x/\underline{m})$.

In this way, Kleene explained how to associate with each formula ϕ of Heyting arithmetic a set of numbers, perhaps empty, that realize ϕ. He then proved that realizability gives a sound interpretation of that formal system: ϕ is derivable in the formal arithmetic HA only if there is a natural number realizing ϕ. Since 1945, a panoply of realizability interpretations have been devised not only for elementary arithmetic but also for higher-order arithmetic, type theory, set theory, and higher-order logic. Beeson [1985] contains many of the technical details. In Hyland [1982], Kleene's original conception, extended to a higher-order type theory, received an attractive formulation in category-theoretic terms.

Formal versions of many, but not all, results characteristic of Brouwerian intuitionism are validated on Kleene's interpretation. When the original realizability interpretation is applied to higher-order systems, UP, CC, and NT, together with their consequences, are seen to be realized. This yields a consistency proof for them. But there is more: intuitionistic logic provides an *internal logic*, and the mathematics that gets realized provides an *internal mathematics* for the realizability interpretation. This means that the logic and mathematics are sound under the interpretation, and can capture expressively and treat deductively metatheoretic or "external" properties of the interpretational scheme in surprisingly simple and salient ways. In short, facts of pure intuitionistic mathematics become, under realizability, facts of applied computability theory. For example, the falsity of TND under realizability registers the fact that the halting problem is unsolvable, that no Turing machine can decide for every *e* which of these holds:

machine *e* eventually halts when *e* is input

or

machine *e* never halts when *e* is input.

Interpreted using realizability, the Neighborhood Theorem NT becomes a close relative of the theorem on recursive functionals proved in Kreisel et al. [1959].

Yet there are features of Kleene's realizability that are not intuitionistic. These are captured by formal statements that hold under realizability but are not expressions of intuitionistic reasoning. Principal among these are *Church's Thesis* (to be distinguished from the familiar Church-Turing Thesis) and *Markov's Principle.* The former is the claim that every function from $\mathbb{N}$ into $\mathbb{N}$ is total recursive. The latter asserts that if two real numbers are unequal, then they are at least a positive rational distance apart. The former is inconsistent with NT, since a Turing machine for generating a sequence cannot be constructed reliably from a finite initial segment of the sequence alone. Although nonintuitionistic, these elements of realizability mathematics possess some attractive features. For example, Church's Thesis and Markov's Principle together suffice to show that there are no nonstandard models of arithmetic in the realizability interpretation: up to isomorphism, there is at most one set-theoretic structure satisfying the axioms of the first-order theory HA [McCarty 1988].

Once taken, it is not a large step from thinking of numbers encoding abstract programs and realizing formal expressions to thinking of actual computer programs performing a similar function. A realizing number is replaced by a text for a program in a suitable programming language; the formal expression that is realized then becomes a specification; and the program a routine that demonstrably meets the specification. An ordinary metatheoretic proof that every theorem of an intuitionistic formal system is realized now becomes a proof that every theorem of the system is a specification for some program and that, from a formal proof that the specification can be met, a program can be extracted. Such an idea—that computer programs can be extracted automatically by "compiling" intuitionistic proofs—has stimulated a vigorous research program at the interface between mathematical logic and theoretical computer science, a prominent representative of which has been the series of the computer systems designed by R. Constable and coworkers [1986].

4. Internal Mathematics: Topologies and Toposes

A *Boolean algebra* is an inhabited set with operations that obey the laws of conventional propositional logic; its operations, called "meet," "join," and "complement" act in perfect accord with the tautologies governing the logical words "and," "or," and "not," respectively. Boolean algebras, therefore, make up a class of interpretations for theories in conventional propositional logic, as explained in Rasiowa and Sikorski [1963]. In analogous fashion, a *Heyting algebra* is an inhabited set with operations of meet, join, complement, and implication that obey the principles that intuitionistic propositional logic lays down for the logical words "and," "or," "not" and "if...then," respectively. In conventional mathematics, every power set $P(X)$ is a Boolean algebra under the operations of intersection, union, and complement. In intuitionistic mathematics, every power set is a Heyting algebra but not a Boolean algebra. On either approach, the collection of *open sets* of a topology comprise a Heyting algebra. Therefore, as reported in M. Stone [1937] and, independently, in A. Tarski [1938], topological spaces afford natural interpretations for intuitionistic propositional logic. Using conventional mathematics as a metatheory, J. McKinsey and Tarski proved in their [1948] that the logic is sound and complete with respect to interpretations over the usual topology on $\mathbb{R}$.

In these structures, the invalidity of TND registers internally the fact that the real numbers form a connected set. D. Scott, beginning with his [1968], showed how to construct sound topological interpretations for formal systems of intuitionistic analysis and described an interpretational structure validating Brouwer's Theorem. One can view Kripke models as topological interpretations by taking a neighborhood of a node α to be a set of nodes X containing α that is *upward closed*: if $\beta \in X$ and $\beta \leq \gamma$, then $\gamma \in X$. Since every Boolean algebra is a Heyting algebra, not all topological interpretations have intuitionistic mathematics as their internal mathematics; some yield Boolean-valued models of conventional mathematics.

A *topos* is a category, a mathematical structure composed of objects and maps between objects, that satisfies axioms sufficient to guarantee, among other things, that every topos contains objects exhibiting the intuitionistic logical behavior of power sets $P(X)$. Hence, in a topos, the power objects $P(X)$ give rise to topos-internal Heyting algebras, thus affording interpretations of intuitionistic logic and mathematics in strictly categorical terms, as explained in Bell [1988] and MacLane and Moredijk [1992]. W. Lawvere may have been the first to think of categories with extra structure as representations of intuitionism; see his [1976], for example. It turns out that higher-order intuitionistic logic is generally the internal logic of toposes. Researchers have also discovered that topological, Kripke, and realizability interpretations of intuitionistic systems can be incorporated in a natural fashion within the topos-theoretic viewpoint. Nowadays, this subject receives a goodly amount of well-deserved attention from category theorists, logicians, and computer scientists concerned with consistency and independence results, and topos theory often sheds great light on intuitionistic concepts and their formal representations.

IV. Concluding Remarks

1. Intuitionism and Anti-realism

In its bearing upon intuitionism and logic, philosophical anti-realism is supposition or conjecture rather than established fact. It is the supposition that a particular kind of scientifically and philosophically respectable semantical framework can be constructed for proving intuitionistic logic correct and conventional logic incorrect. Anti-realism requires the existence of a semantics S sufficient at least for interpreting standard formal languages and manifesting several properties to which M. Dummett has repeatedly directed the attention of philosophers (see Dummett [1977], [1978] and [1991]). These properties are such that if S possesses them, conventional mathematicians have always been wrong to think that conventional logic, including TND, is cogent and that mathematical statements generally carry the meanings they think them to have. Here are the properties. First, S should support a rigorous proof, acceptable to intuitionists as well as conventional mathematicians, that Heyting's intuitionistic predicate logic is sound and complete relative to S. (As is plain from the foregoing, this implies that S cannot be any simple variant on Kripke models.) Second, to guarantee the desired completeness, it is necessary to show that S invalidates TND. For that purpose, truth according to S is assumed to obey a recognition condition: any semantically competent speaker must be able in principle, when a sentence ϕ obtains in S, to recognize that ϕ thus obtains. When ϕ expresses a statement of pure mathematics, recognition that ϕ is true is to consist in the provision of a proof of ϕ. Third, the interpretations that S attaches to sentences are to express their actual meanings, and sentences are to be valid according to S just in case they are rightly seen as valid. This assumption imposes the further restriction that if one can prove that TND is invalid according to S, then TND is invalid *simpliciter*. It would seem, then, that truth in S should exhibit many of the features one would ordinarily expect of a proper notion of truth. For example, truth in S should commute with disjunction: $\phi \vee \psi$ is true according to S just in case either ϕ is true according to S or ψ is.

It is sometimes imagined that the conjectured anti-realistic argument for the invalidity of TND will proceed along roughly the following lines. Let RH stand for Riemann's Hypothesis, that the real parts of all nonreal zeros of the zeta function have value ½. Today, RH is an open mathematical problem: no one is in a position either to prove RH or to prove its negation. Now, assume that TND is valid. The third property of S listed above implies that TND is valid according to S. RH makes a meaningful mathematical claim; thus RH $\vee \neg$RH is true according to S. Since truth in S commutes with disjunction, either RH is true in S or $\neg$RH is. However, because S satisfies the recognition condition mentioned in the second property, if RH is true in S, then competent speakers should be able to prove that fact. Hence, RH must be proven. On the other hand, if $\neg$RH obtains, then competent speakers will be able to prove that. Hence, RH should have been disproven. Since RH is neither proven nor disproven, TND is invalid.

The steps in this argument trace one part of the philosophical distance that an anti-realist, a proponent for an imagined semantics like S, has yet to travel. He or she must maintain simultaneously the recognition condition and that truth à la S commutes with disjunction. So, if a competent speaker is able to recognize in principle that $\phi \vee \psi$ is true, then $\phi \vee \psi$ will have to be true in S; and since truth in S commutes with disjunction, either the competent speaker can recognize that ϕ is true or she can recognize that ψ is true with respect to S. And this seems false intuitively. Disjunction, when employed, for example, to make indicative statements in the past tense, would likely lose a great deal of its power were speakers obliged to verify, even in principle, one or the other of the disjuncts of every disjunction they assert about the past.

Some investigators have supposed that the clauses of a mathematical definition that first appeared in Heyting [1934] would form a nucleus for anti-realistic semantics S. Here, let e and f stand for pieces of mathematical evidence: proofs, constructions, calculations, or numbers. Heyting's definition can be understood as setting down conditions on which pieces of evidence, or abstract representations of them, *witness* or confirm mathematical sentences. It is presumed that evidence can be analyzed or decomposed into subsidiary pieces of evidence: that any evidence e contains component pieces of evidence e_0 and e_1, obtainable effectively from e. One also takes the natural number 0 to be a piece of evidence comparable with other evidence so that $e = 0$ is a decidable predicate of e. Last, pieces of evidence e are supposed capable of accepting evidence f as input and outputting evidence $e(f)$, the result of applying e to f.

Under those assumptions, the clauses that follow express the relevant portions of Heyting's definition. Evidence e is said to witness a disjunction $\phi \vee \psi$ whenever $e_0 = 0$ and e_1 witnesses ϕ, or $e_0 \neq 0$ and e_1 witnesses ψ. This guarantees that if a disjunction is witnessed, one can tell, via simple operations on the evidence, which of its disjuncts is also witnessed. Next, e witnesses $\neg\phi$ when, if f is evidence witnessing ϕ, $e(f)$ would be evidence that witnesses $\perp$. If one insists, as seems perfectly reasonable, that there be no evidence for $\perp$, then e witnesses $\neg\phi$ just in case there is no witness for ϕ. Finally, e is evidence witnessing $\phi \Rightarrow \psi$ if and only if, when f witnesses ϕ, $e(f)$ witnesses ψ: evidence for a conditional converts evidence for its antecedent into evidence for its consequent.

Difficulties crowd in upon the suggestion that, in the desired S, a sentence ϕ be true if there exists evidence witnessing ϕ in the manner just described. For example, it seems legitimate to ask that the evidence be encoded as natural numbers; that e_0 and e_1 be represented by the usual primitive recursive projection functions applied to e; and that $e(f)$, for e and f natural numbers, be the result, if any, of running the Turing machine with index e on input number f. Barring extra specifications, these requirements can be seen to yield a perfectly reasonable interpretation of Heyting's definition; indeed, it is the Kleene realizability interpretation given above, restricted to propositional logic. However, with conventional logic, one can prove that, for ϕ in the formalism of propositional logic, e witnesses ϕ under this interpretation if and only if ϕ is true conventionally. In particular, one can show that every propositional instance of TND is thus witnessed. Therefore, if S is to feature the clauses of Heyting's definition as the centerpiece of its theory of truth, some further mathematical rider has to be added, a rider that would block the above interpretation. Given that intuitionistic set theory, together with UP and Brouwer's Theorem, are consistent with the truth of every instance of TND in the language of set theory [McCarty 1984], it would seem that this rider would need to be of considerable mathematical strength.

However, one need not suppose that the internal workings of S are built on Heyting's design to come up with a compelling argument to the conclusion that an anti-realist's desire for the semantics S will remain unfulfilled if S has the properties listed at the start of this section. Here is a sketch of such an argument. To begin with, one can insist that for each sentence ϕ, there are mathematical claims Γ and a *stable* assertion γ such that $\phi \vee \neg\phi$ obtains just in case γ is derivable from Γ using predicate logic, that is, $\Gamma \vdash \gamma$. Here, "stable" means that $\Gamma \vdash \gamma$ just in case $\Gamma \vdash \neg\neg\gamma$ This insistence can be made intuitively plausible. Should the formal deductive apparatus of predicate logic not itself be powerful enough to discover the relevant circumstances under which $\phi \vee \neg\phi$ is recognized to be true, there should certainly be information Γ such that, were we to supply that information, the apparatus plus the information Γ would be able to discover those circumstances.

Now, as a matter of logic, $\neg\neg(\phi \vee \neg\phi)$ is valid. It follows that there cannot fail to exist circumstances under which $\phi \vee \neg\phi$ is recognizably true. Therefore, predicate logic plus information Γ cannot fail to detect this fact, that is, $\neg\neg(\Gamma \vdash \gamma)$. The soundness theorem for Heyting's predicate logic with respect to S now implies that the formal argument with premises Γ and conclusion γ cannot fail to be valid in S. Since validity in S is to be coextensive with simple validity, the argument with premises Γ and conclusion $\neg\neg\gamma$ will be valid. Further, given that the S-completeness theorem for Heyting's predicate logic is provable, $\Gamma \vdash \neg\neg\gamma$. Because γ is stable, as described above, γ will itself be derivable from Γ, or $\Gamma \vdash \gamma$. But γ being derivable from Γ marks the obtaining of circumstances under which $\phi \vee \neg\phi$ is recognizably true. Therefore, $\phi \vee \neg\phi$ holds and, since ϕ was arbitrary, TND is valid. Therefore, the anti-realist's suppositions—that semantics S be sound and complete, that truth in S always be recognizable, and that TND fail—seem jointly inconsistent.

2. What Weakness?

Our tour is at an end. It is sad that so much is left unexplored: intuitionistic algebra, measure and integration, projective geometry, large cardinals in set theory. Intuitionistic-type theory and its categorial interpretations offer great benefit to computation theory and theoretical computer science. Higher-order intuitionistic logic reigns supreme as the internal logic of toposes. No area of modern mathematics has proven resistant to intuitionistic incursion; despite initial skepticism, as expressed in G. Hellman [1993], even the foundations of quantum mechanics begin to yield to constructivistic treatment, as F. Richman and D. Bridges have shown in their [1999]. There are also those structures open to intuitionistic investigation but closed to conventional study. For example, in the Kleene realizability structure for set theory, there exist natural models of Church's λ calculus: nontrivial sets X standing in one-to-one correspondence with the collection of all functions taking X into X. This is conventionally impossible, thanks to the conventional form of Cantor's Theorem, but intuitionistically, it affords a lively mathematical prospect.

The efficacy of the negative translation for set theory, even set theory with large cardinals, suggests that any evidence of a first-order sort for conventional set theory would transfer, under negative translation, to intuitionistic set theory. In these respects, intuitionistic mathematics shows itself not as a system of complete renunciation, but as standing on a scientific par with its conventional cousin. These features of intuitionism also draw the teeth of the old objection that intuitionism is somehow scientifically unsatisfactory because its mathematics is disastrously weak. Hilbert lodged the most famous such objection in his [1927], where he wrote, "Taking the principle of excluded middle from the mathematician would be the same, say, as proscribing the telescope to the astronomer or to the boxer the use of his fists" (p. 476). Surely this cannot be a complaint about the fact that TND is underivable in the intuitionistic logic. For TND is provably invalid. One could not reasonably maintain that conventional logic is weak because it fails to derive, for example, the invalid principle $\phi \vee \psi$. Therefore, depriving the intuitionist of TND is not like depriving the astronomer of a telescope, but like depriving the aviator of a diving bell.

Acknowledgments The author wishes to thank Joshua Alexander, Justin Brown, Nathan Carter, Hilmi Demir, Lisa Keele, Stuart Mackenzie, Jeremy McCrary, Carl Posy, Stewart Shapiro, Christopher Tillman, and Ignacio Viglizzo for their insightful comments and suggestions on this and other writings.

BIBLIOGRAPHY

van Atten, M., and D. van Dalen [2002] Arguments for the continuity principle. *Bulletin of Symbolic Logic* 8(September): 329–347.

Awodey, S., and C. Butz [2000] Topological completeness for higher-order logic. *Journal of Symbolic Logic* 65(September): 1168–1182.

Becker, O. [1990] *Grundlagen der Mathematik in geschichtlicher Entwicklung.* Frankfurt am Main, Suhrkamp.

Beeson, M.J. [1985] *Foundations of Constructive Mathematics: Metamathematical Studies.* Berlin, Springer-Verlag.

Bell, J.L. [1988] *Toposes and Local Set Theories.* Oxford, Clarendon Press.

Bell, J.L. [1998] *A Primer of Infinitesimal Analysis.* Cambridge, Cambridge University Press.

Bell, J.L. [1999a] Frege's theorem in a constructive setting. *Journal of Symbolic Logic* 64(June): 486–488.

Bell, J.L. [1999b] Finite sets and Frege structures. *Journal of Symbolic Logic* 64 (December): 1552–1556.

Bridges, D., and F. Richman [1987] *Varieties of Constructive Mathematics.* London Mathematical Society Lecture Notes series 97. Cambridge, Cambridge University Press.

Brouwer, L.E.J. [1908] On the unreliability of the logical principles. [De Onbetrouwbaarheid der logische Principes.] *Tijdschrift voor Wijsbegeerte* 2: 152–158.

Brouwer, L.E.J. [1918] Begründung der Mengenlehre unabhängig vom logischen Satz vom ausgeschlossenen Dritten, I. In A. Heyting (ed.), *L.E.J. Brouwer, Collected Works,* volume I. Amsterdam: North-Holland, 1975, pp. 150–190.

Brouwer, L.E.J. [1921] Besitzt jede reelle Zahl eine Dezimalbruchentwichlung? *Mathematische Annalen* 83: 201–210.

Brouwer, L.E.J. [1924] Beweis dass jede volle Funktion gleichmässigstetig ist. *Koninklijke Nederlandse Akademie van Wetenschappen, Proceedings of the Section of Sciences* 27: 189–193.

Constable, R.L., et al. [1986] *Implementing Mathematics with the Nuprl Proof Development System.* Englewood Cliffs, NJ: Prentice-Hall.

van Dalen, D. [1994] Intuitionistic logic. In D. Gabbay and F. Guenther (eds.), *Handbook of Philosophical Logic,* volume III, *Alternatives to Classical Logic.* Dordrecht, Kluwer Academic Publishers, pp. 225–339.

van Dalen, D. [1999a] From Brouwerian counterexamples to the creating subject. *Studia Logica* 62: 305–314.

van Dalen, D. [1999b] *Mystic, Geometer, Intuitionist: The Life of L.E.J. Brouwer,* volume I, *The Dawning Revolution.* Oxford, Clarendon Press.

van Dalen, D., and M. van Atten [2002] Intuitionism. In D. Jacquette (ed.), *A Companion to Philosophical Logic.* Oxford, Blackwell, pp. 513–530.

Diaconescu, R. [1975] Axiom of choice and complementation. *Proceedings of the American Mathematical Society* 51: 175–178.

Du Bois-Reymond, P. [1882] *Die allgemeine Funktionentheorie.* Tübingen, H. Laupp.

Dummett, M.A.E. [1977] *Elements of Intuitionism.* Oxford, Clarendon Press.

Dummett, M.A.E. [1978] *Truth and Other Enigmas.* Cambridge, MA: Harvard University Press.

Dummett, M.A.E. [1991] *The Logical Basis of Metaphysics.* Cambridge, MA: Harvard University Press.

Dummett, M.A.E. [1993] *The Seas of Language.* Oxford, Clarendon Press.

Fourman, M.P., and D.S. Scott [1979] Sheaves and logic. In M.P. Fourman, C.J. Mulvey, and D.S. Scott (eds.), *Applications of Sheaves. Proceedings, Durham 1977.* Lecture Notes in Mathematics 753, Berlin, Springer-Verlag, pp. 302–401.

Gentzen, G. [1935] Untersuchungen über das logische Schliessen I, II. *Mathematische Zeitschrift* 39: 176–210, 405–431.

Gentzen, G. [1974] Über das Verhältnis zwischen intuitionistischer und klassischer Logik. *Archiv für mathematische Logik und Grundlagenforschung* 16: 119–132.

Gödel, K. [1932] On the intuitionistic propositional calculus. In S. Feferman et al. (eds.), *Kurt Gödel. Collected Works,* volume I, *Publications 1929–1936.* Oxford, Clarendon Press, pp. 222–225.

Gödel, K. [1933a] On intuitionistic arithmetic and number theory. In S. Feferman et al. (eds.), *Kurt Gödel. Collected Works,* volume I, *Publications 1929–1936.* Oxford, Clarendon Press, 1986, pp. 286–295.

Gödel, K. [1933b] An interpretation of the intuitionistic propositional calculus. In S. Feferman et al. (eds.), *Kurt Gödel. Collected Works,* volume I, *Publications 1929–1936.* Oxford, Clarendon Press, 1986, pp. 300–303.

Grayson, R.J. [1979] Heyting-valued models for intuitionistic set theory. In M.P. Fourman, C.J. Mulvey, and D.S. Scott (eds.), *Applications of Sheaves. Proceedings, Durham 1977.* Lecture Notes in Mathematics 753, Berlin, Springer-Verlag, pp. 402–414.

Greenleaf, N. [1981] Liberal constructive set theory and the paradoxes. In F. Richman (ed.), *Constructive Mathematics, Proceedings, New Mexico, 1980.* Springer Lecture Notes in Mathematics 873, Berlin, Springer-Verlag.

Hellman, G. [1993] Gleason's theorem is not constructively provable. *Journal of Philosophical Logic* 22: 193–203.

Heyting, A. [1930a] Die formalen Regeln der intuitionistischen Logik. *Sitzungsberichte der Preussischen Akademie der Wissenschaften, Physikalisch-mathematische Klasse* 42–56.

Heyting, A. [1930b] Die formalen Regeln der intuitionistischen Mathematik I. *Sitzungsberichte der Preussischen Akademie der Wissenschaften, Physikalisch-mathematische Klasse* 57–71.

Heyting, A. [1930c] Die formalen Regeln der intuitionistischen Mathematik II. *Sitzungsberichte der Preussischen Akademie der Wissenschaften, Physikalisch-mathematische Klasse* 158–169.

Heyting, A. [1934] *Mathematische Grundlagenforschung. Intuitionismus. Beweistheorie.* Berlin, Springer-Verlag.

Heyting, A. [1976] *Intuitionism: An Introduction.* 3rd rev. ed. Amsterdam, North-Holland.

Hilbert, D. [1927] The foundations of mathematics. In J. van Heijenoort (ed.), *From Frege to Gödel. A Source Book in Mathematical Logic, 1879–1931.* Cambridge, MA: Harvard University Press, 464–479.

Hyland, M. [1982] The effective topos. In A.S. Troelstra and D. van Dalen (eds.), *The L.E.J. Brouwer Centenary Symposium.* Amsterdam, North-Holland, pp. 165–216.

Joyal, A., and I. Moredijk [1995] *Algebraic Set Theory.* London Mathematical Society Lecture Notes 220. Cambridge, Cambridge University Press.

Kleene, S.C. [1945] On the interpretation of intuitionistic number theory. *Journal of Symbolic Logic* 10: 109–124.

Kreisel, G. [1961] Set theoretic problems suggested by the notion of potential totality. In *Infinitistic Methods. Proceedings of the Symposium on Foundations of Mathematics. Warsaw 1959.* Oxford, Pergamon Press, pp. 103–140.

Kreisel, G. [1962] On weak completeness of intuitionistic predicate logic. *Journal of Symbolic Logic* 27: 139–158.

Kreisel, G., D. Lacombe, and J.R. Schoenfield [1959] Partial recursive functionals and effective operations. In A. Heyting (ed.), *Constructivity in Mathematics: Proceedings of a Colloquium Held at Amsterdam, 1957.* Amsterdam, North-Holland, pp. 195–207.

Kripke, S. [1965] Semantical analysis of intuitionistic logic I. In J. Crossley and M.A.E. Dummett (eds.), *Formal Systems and Recursive Functions.* Amsterdam, North-Holland.

Lawvere, W. [1976] Variable quantities and variable structures in topoi. In A. Heller and M. Tierney (eds.), *Algebra, Topology, and Category Theory.* New York, Academic Press, pp. 101–131.

Leivant, D. [1985] Syntactic translations and provably recursive functions. *Journal of Symbolic Logic* 50: 682–688.

MacLane, S., and I. Moredijk [1992] *Sheaves in Geometry and Logic.* Berlin, Springer-Verlag.

McCarty, D.C. [1984] *Realizability and Recursive Mathematics.* Department of Computer Science, Carnegie-Mellon University, Report CMU-CS-84-131. Pittsburgh, Carnegie-Mellon University.

McCarty, D.C. [1988] Constructive validity is nonarithmetic. *Journal of Symbolic Logic* 53(December): 1036–1041.

McCarty, D.C. [1991] Incompleteness in intuitionistic metamathematics. *Notre Dame Journal of Formal Logic* 32(Summer): 323–358.

McKinsey, J.C., and A. Tarski [1948] Some theorems about the sentential calculi of Lewis and Heyting. *Journal of Symbolic Logic* 13: 1–15.

van Oosten, J. [1996] Intuitionism. CiteSeer, Scientific Literature Digital Library, NEC Research Index. Preprint.

Placek, Tomasz [1999] *Mathematical Intuitionism and Intersubjectivity.* Dordrecht, Kluwer Academic Publishers.

Prawitz, D. [1965] *Natural Deduction. A Proof-Theoretical Study.* Studies in Philosophy 3. Stockholm: Almqvist and Wiksell.

Rasiowa, H., and R. Sikorski [1963] *The Mathematics of Metamathematics.* Warsaw: Panstwowe Wydawnictowo Naukowe.

Richman, F., and D. Bridges [1999] A constructive proof of Gleason's theorem. *Journal of Functional Analysis* 162: 287–312.

Rogers, H., Jr. [1967] *Theory of Recursive Functions and Effective Computability.* New York: McGraw-Hill.

Rose, G.F. [1953] Propositional calculus and realizability. *Transactions of the American Mathematical Society* 75: 1–19.

Scott, D.S. [1968] Extending the topological interpretation to intuitionistic analysis. Part I. *Compositio Mathematica* 20: 14–210.

Shapiro, S. (ed.) [1985] *Intensional Mathematics.* Amsterdam, North-Holland.

Stone, M. [1937] Topological representations of distributive lattices and Brouwerian logics. *Casopis pro pestvovani Matematiky a fysiki cast matematicka* 67: 1–25.

Tarski, A. [1938] Der Aussagenkalkül und die Topologie. *Fundamenta Mathematica* 31: 103–134.

Troelstra, A.S. [1973] Notes on intuitionistic second-order arithmetic. In A. Mathias and H. Rogers (eds.), *Cambridge Summer School in Mathematical Logic.* Berlin, Springer-Verlag, pp. 171–205.

Troelstra, A.S. [1977] *Choice Sequences: A Chapter of Intuitionistic Mathematics.* Oxford, Clarendon Press.

Troelstra, A.S. [1980] Intuitionistic extensions of the reals. *Nieuw Archief voor Wiskunde* 3rd ser., 28: 63–113.

Troelstra, A.S., and D. van Dalen [1988] *Constructivism in Mathematics: An Introduction,* 2 vols. Amsterdam, North-Holland.

Veldman, W. [1976] An intuitionistic completeness theorem for intuitionistic predicate logic. *Journal of Symbolic Logic* 41(March): 159–166.

Veldman, W. [1982] On the constructive contrapositions of two axioms of countable choice. In A.S. Troelstra and D. van Dalen (eds.), *The L.E.J. Brouwer Centenary Symposium.* Amsterdam, North-Holland, pp. 513–523.

Veldman, W., and F. Waaldijk [1996] Some elementary results in intuitionistic model theory. *Journal of Symbolic Logic* 61(September): 745–767.

CHAPTER 11

INTUITIONISM RECONSIDERED

ROY COOK

1. Why Reconsider?

The debate between intuitionists and classical logicians is fought on two fronts. First, there is the battle over subject matter—the disputants disagree regarding which mathematical structures are legitimate domains of inquiry. Second, there is the battle over logic—they disagree over which algebraic structure correctly codifies logical consequence. In this chapter the emphasis is on the latter issue—I shall focus on what the correct (formal) account of correct inference might look like, and, given such an account, how we should understand disagreements regarding the extension of the logical consequence relation.

Intuitionism comes in two or three major forms: L. E. J. Brouwer's Kantian metaphysics (e.g., [1948]), Arend Heyting's semantic elucidation (or perhaps corruption) of Brouwer's view (e.g., [1931]), and Michael Dummett's verificationist stance (e.g., [1975]). Although each rightly deserves the title "intuitionism," it is important to note that there is more than one view on the table. This noted, however, it is useful to isolate what they have in common (i.e., those aspects of the various views that justify the common nomenclature). The reader interested in the nuts and bolts of particular variants of intuitionism is encouraged to read Carl Posy's "Intuitionism and Philosophy" and David McCarty's "Contemporary Intuitionism," chapters 9 and 10 of this volume. The most striking characteristic that unites these views and distinguishes them from classical rivals is their failure to accept certain formulas (or schemata), such as the *law of excluded middle*, as logical truths.

On the face of it, then, the debate seems simple: intuitionists are proposing as correct a logic that is distinct from the more widely accepted classical formalism, and we need only determine which account is correct. Once we scratch the surface a bit, however, we will see that determining where the battle lines of this conflict should be drawn, and even whether there is a conflict at all, is more complicated than it at first appears.

In the next two sections of the chapter, two typical sorts of arguments for intuitionistic logic will be examined. We will then examine exactly what is at stake when one provides a logic as an account of logical consequence. Two ways of understanding the debate, both of which imply that intuitionists and classical logicians are not engaged in the same project, will be rejected. This will provide a set of default assumptions that must be adopted if the intuitionist and the classical logician are giving accounts of the same phenomenon. In the final two sections I will suggest taking a pluralist stance with regard to the debate between intuitionists and classical logicians.[1] Instead of being caught in a dilemma, whereby we are forced to choose between two rival logics, we are instead presented with two equally legitimate codifications of logical consequence. As a result, intuitionistic logic is not in opposition to classical logic, but instead is an alternative (yet, in a sense, still incompatible) formalism that stands alongside it.

2. The Epistemic Argument for Intuitionistic Logic

The first type of argument for intuitionistic logic that we will consider comes in three parts. First, the intuitionist proposes certain epistemic principles that constrain truth. In doing so, the intuitionist of the Brouwer/Heyting school will usually refer to the essentially mental character of the subject matter of mathematics:

> Even if they should be independent of individual acts of thought, mathematical objects are by their very nature dependent on human thought. Their existence is guaranteed only insofar as they can be determined by thought. They have properties only insofar as these can be determined by thought. But this possibility of knowledge is revealed to us only by the act of knowing itself. (Heyting [1931]: 53)

An intuitionist of the more recent Dummettian variety will be concerned with issues of meaning. Dummett argues that the realist position of the classical logician

[1] I use "pluralist" to designate any view that allows the simultaneous acceptance of two distinct logics. This usage should be clearly distinguished from the "logical pluralism" of J. C. Beall and Greg Restall [2000], based on the idea that different logics govern different subject matters, or "cases."

> . . . violates the principle that use exhaustively determines meaning; or, at least, if it does not, a strong case can be put up that it does, and it is this case which constitutes the first type of ground which appears to exist for repudiating classical in favour of intuitionistic logic for mathematics. For, if the knowledge that constitutes a grasp of the meaning of a sentence has to be capable of being manifested in actual linguistic practice, it is quite obscure in what the knowledge of the condition under which a sentence is true can consist, when that condition is not one which is always capable of being recognised as obtaining. (Dummett [1975], p. 106)

Nevertheless, both lines of thought lead to an equation of the truth of a proposition with the existence of a proof.[2]

The intuitionist then argues that certain classical laws, such as *excluded middle*, and certain rules, such as *classical reductio*, are dubious because they are, from an intuitionistic perspective, equivalent to overly optimistic claims about our epistemic powers. Heyting writes:

> Does the *law of excluded middle* hold . . .? When we assert it, this means that for any proposition *A* we can either prove *A* or derive a contradiction from a supposed proof of *A*. Obviously we are not able to do this for every proposition *A*, so the *law of excluded middle* cannot be proved. If we do not know whether *A* is true or not, we better make no assertion about it. ([1974]: p. 87)

Heyting is not asserting the existence of some explicit counterexample to *excluded middle* (nor would any other intuitionist), since in intuitionistic propositional logic the negation of any classical logical truth leads to absurdity.[3] Rather, he is claiming that, from the epistemic stance of the intuitionist, the inference in question makes an unjustified (and possibly unjustifiable) claim about the existence of certain sorts of construction (i.e., the law of *excluded middle*, for the intuitionist, entails global decidability). Along similar lines, Michael Dummett writes:

> Brouwer's reform of mathematics thus led to a revision of the most fundamental part of logic itself, sentential logic. "Or" was to be explained by saying that a proof of "*A* or *B*" must be a proof either of "*A*" or of "*B*," and "if" by saying that a proof of "If *A*, then *B*" is a construction which we can recognise as transforming every proof of "*A*" into a proof of "*B*." These formulations, together with an axiomatic formalisation of intuitionistic logic, were attained by Heyting on the basis of Brouwer's practice. . . . It is evident that many laws of classical logic, such as the *law of excluded middle* and its

[2] There is some question regarding whether various intuitionists regard the relevant notion of proof to entail the actual existence of a (possibly abstract) proof, or merely the possibility of providing one.

[3] Similar, albeit more complicated, negation-related connections hold between the corresponding first-order theories (see Troelstra and Van Dalen [1988] for details).

> generalisation "*A* or (if *A* then *B*)"[4] fail on the intuitionistic interpretation of the logical constants. (Dummett [1980], p. 612)

Finally, once the distinctly classical inferences have been disposed of, these same epistemic principles are used to justify the standard rules of intuitionistic logic.[5]

In the discussion below, our target will be intuitionistic logic motivated by this sort of epistemic worry. There is another much discussed argument for intuitionism, however, so a brief look at it, and at why we will discard it, is in order.

3. The Proof-theoretic Argument for Intuitionistic Logic

The second argument for intuitionistic logic is motivated by an inferentialist view of meaning—the idea that the meaning of various bits of language is to be explained in terms of the rules governing reasoning with those bits of language. Arthur Prior [1960] famously objected to (a naive version of) inferentialism by formulating the *tonk* operator:

$$\text{TONK I} \quad \frac{\Phi}{\Phi \text{ TONK } \Psi} \qquad\qquad \text{TONK E} \quad \frac{\Phi \text{ TONK } \Psi}{\Psi}.$$

If we assume that the consequence relation is transitive, then it is easy to demonstrate that these rules allow us to derive any formula.

In response, inferentialists, and especially intuitionists, have proposed various constraints on the form that such meaning-constitutive rules can take. Michael Dummett calls one such constraint "harmony," and writes:

> The best hope for a more precise characterisation of the notion of harmony lies in the adaptation of the logicians' concept of conservative extension. Given a formal theory, we may strengthen it by expanding the formal language, adding new primitive predicates, terms, or functors, and introducing new axioms or rules of inference to govern expressions formed by means of the new vocabulary.

[4] "$A \vee (A \rightarrow B)$" is a generalization of "$A \vee \neg A$" according to the standard intuitionistic definition of negation in terms of the conditional and an "absurd" formula. For example,

$$\neg A =_{\text{df}} A \rightarrow (0 = 1).$$

See Cook and Cogburn [2000] for a discussion of the drawbacks of this sort of definition.

[5] Although I presume that the reader is familiar with the standard formulations of intuitionistic logic, the rules for intuitionistic versus classical negation are discussed below.

> In the new theory, we can prove much that we could not even express in the old one; but... we can prove in it no statement expressed in the original restricted vocabulary that we could not already prove in the original theory. ([1991], p. 251)

As a result, Dummett requires that the rules for each logical operator be conservative with respect to the other logical operators (and their rules). In the standard natural deduction presentation of classical logic, however, the rules for negation:

$$\neg\text{I}\quad \dfrac{\dfrac{\overline{\Phi}^{(i)}}{\vdots}}{\dfrac{\bot}{\neg_C\Phi}(i)} \qquad \neg\text{E}\quad \dfrac{\Phi \quad \neg_C\Phi}{\bot} \qquad \text{CR}\quad \dfrac{\dfrac{\overline{\neg_C\Phi}^{(i)}}{\vdots}}{\dfrac{\bot}{\Phi}(i)}$$

fail to be conservative. For example, they allow us to prove $(\Phi \rightarrow \Psi) \vee (\Psi \rightarrow \Phi)$, which is not provable from the standard rules for disjunction and implication alone. On the other hand, the intuitionistic rules, obtained by jettisoning *classical reductio* (CR) while retaining negation introduction ($\neg$I) and elimination ($\neg$E), are conservative.

The idea of harmonious introduction and elimination rules providing the meanings of connectives traces back to Gerhard Gentzen's proof-theoretic work, where he notes:

> The introductions *represent*, as it were, "definitions" of the symbols concerned, and the eliminations are no more, in the final analysis, than the consequences of these definitions. This fact may be expressed as follows: In eliminating a symbol, the formula, whose terminal symbol we are dealing with, may be used only "in the sense afforded by the introduction of that symbol." ([1964], p. 295)

Interestingly, Gentzen is referring not to intuitionistic logic here, but to the rules for both intuitionistic and classical logic.

In Gentzen's development of the sequent calculus, the rules for both classical and intuitionistic negation take the form

$$\frac{\Delta, A \Rightarrow \Gamma}{\Delta \Rightarrow \Gamma, \neg A} \Rightarrow\neg \qquad\qquad \frac{\Delta \Rightarrow \Gamma, A}{\Delta, \neg A \Rightarrow \Gamma} \neg\Rightarrow$$

[Δ, Γ sequences of formulas, A a formula].[6]

The difference between intuitionistic logic and classical logic is that in the former (but not the latter) we are restricted in having at most one formula to the right of the inference arrow $\Rightarrow$ in each sequent. Gentzen's rules for negation in either

[6] In a sequent calculus derivation, a sequent $\Delta \Rightarrow \Gamma$ is meant to be read as "If all of the fomulas in Δ are true, then at least one of the formulas in Γ is true."

system are conservative over the (rules for the) other connectives. As a result, the sequent calculus rules, with multiple formulas allowed to the right of the inference arrow, provide a harmonious codification of classical logic.[7] The requirement that logical rules be harmonious and/or conservative does not, therefore, weigh more in favor of intuitionistic logic as opposed to its classical rival.[8] As a result we will constrain our attention in the remainder of this chapter to the epistemic argument. (For a much more in-depth examination of these issues, see Dag Prawitz's "Logical Consequence from a Constructive Point of View," chapter 22 in this volume.)

4. Logics and Logical Consequence

Before examining which of two distinct logics is correct, or whether in some sense both can be legitimate, a bit more needs to be said about what makes a logic "legitimate" and what a logic is meant to provide. Logics are algebraic structures meant to codify the notion of logical consequence. Thus, it is natural to start with a description of logical consequence, and Alfred Tarski provides one:

> Consider any class Δ of sentences and a sentence Φ which follows from the sentences of this class. From an intuitive standpoint it can never happen that both the class Δ consists only of true sentences and the sentence Φ is false. Moreover, since we are concerned here with the concept of logical, i.e. *formal*, consequence, and thus with a relation which is to be uniquely determined by the form of the sentences between which it holds, . . . the consequence relation cannot be affected by replacing the designations of the objects referred to in these sentences by the designations of any other objects. ([1936], pp. 414–415)

There are two ideas at work in this gloss. First, if Φ follows from Δ, then the truth of (all members of) Δ (in some sense) guarantees the truth of Φ. Second, the logical consequence relation is independent of the particular nonformal content of the

[7] Intuitionists are not unaware of such systems. In a paper where he examines natural deduction systems with multiple conclusions, Stephen Read points out that intuitionists

> . . . exclude multiple conclusions from consideration because they allow the assertion of disjunctions neither of whose disjuncts is assertible. But that is to beg the question. The question is whether intuitionistic logic is superior proof-theoretically to classical logic. To exclude forms of proof which are intuitionistically unacceptable is to introduce a circle in the reasoning. ([2000], p. 145)

[8] Of course, one should not conclude that proof-theoretic considerations are irrelevant to determining which logic (if any) is correct.

premises and conclusion—in other words, varying the reference or meaning of nonlogical terms, predicates, and such cannot affect whether the conclusion is a logical consequence of the premise. Thus:

LC: A sentence Φ is a logical consequence of a set of sentences Δ iff the truth of every member of Δ guarantees the truth of Φ, and this guarantee is in virtue of the logical form of Φ and the members of Δ.

Intuitionistic and classical logicians have the same goal in mind: to provide a mathematical structure (a logic) that codifies this notion of "logical consequence." One provides a logic for a natural language discourse D by following something like this recipe:

1. We construct a formal language L out of strings of mathematical symbols.
2. We distinguish some subset LV of L, the *logical vocabulary* of L.
3. We construct a translation function $T: LV \rightarrow D$ mapping the logical vocabulary of L to appropriate bits of D.[9]
4. We define a relation $\Rightarrow$ whose first argument is a subset of L and whose second argument is a member of L.[10]

A logic $\langle L, \Rightarrow \rangle$ (with LV and T) is judged by how well it matches up to logical consequence in the natural language in question. Someone who believes that there is a single logic codifying logical consequence in D will be searching for an $\langle L, \Rightarrow \rangle$ (and T) such that the *correctness principle*

CP: For any function $I: L \rightarrow D$ such that I agrees with T on LV:
For all $\Phi \in L$, $\Delta \subseteq L$:
$I(\Phi)$ is a logical consequence of $I(\Delta)$[11] iff $\Delta \Rightarrow \Phi$.

holds.

Precise definitions of *logical monism* and *logical pluralism* are now straightforward:

LM: There is exactly one $\langle L, \Rightarrow \rangle$ such that *CP* holds.
LP: There is more than one $\langle L, \Rightarrow \rangle$ such that *CP* holds.

[9] On this way of viewing things, there are two distinct logical/nonlogical divides. The first is a purely formal one, given by LV. The second, and more interesting, appears when this division is projected onto natural language through the translation function T.

[10] I will ignore whether the relation is presented as a semantic one or a deductive one, because the extension of the relation is our only concern here.

[11] $I(\Delta)$ abbreviates $\{\Phi : \Phi = I(\Psi) \text{ for some } \Psi \in \Delta\}$.

There are many ways to obtain an uninteresting pluralism, for example, by varying the formal language L.[12] This sort of trivial pluralism is not my interest here, so I assume that the language is fixed. In particular, I shall assume that the formal language in question is the standard propositional one. This constraint provides a ready means for dealing with the idea that logical consequence should hold in virtue of logical form. Since both parties in the intuitionistic/classical dispute agree that the propositional letters are the only nonlogical vocabulary, any logic $\langle L, \Rightarrow \rangle$ must obey the following substitutivity requirement:

SUB: Given any $\Delta \subseteq L$, $\Phi \in L$, $\Psi \in L$, and P_i a propositional letter, if $\Delta \Rightarrow \Phi$, then $\Delta[P_i / \Psi] \Rightarrow \Phi[P_i / \Psi]$.[13]

In what follows, the honorific "logic" is reserved for an algebraic structure $\langle L, \Rightarrow \rangle$ that obeys *SUB*.

5. A Disagreement About Meaning?

It has often been claimed that intuitionists and classical logicians attribute different meanings to the connectives. Proponents of this view will argue that the classical negation of a formula Φ means "Φ is false," while the intuitionistic negation of Φ means something like "Φ is refutable" (where "Φ is false" and "Φ is refutable" are not synonymous). Such a view is strongly suggested by the clauses for the propositional connectives in Heyting semantics for intuitionistic propositional logic (see Dummett [1977], ch. 1):

There is a proof of "$\neg\Phi$" iff there is a procedure for transforming any proof of "Φ" into a proof of "$\bot$."
There is a proof of "$\Phi \wedge \Psi$" iff there is a proof of "Φ" and there is a proof of "Ψ."
There is a proof of "$\Phi \vee \Psi$" iff either there is a proof of "Φ" or there is a proof of "Ψ."
There is a proof of "$\Phi \rightarrow \Psi$" iff there is a procedure for transforming any proof of "Φ" into a proof of "Ψ."

[12] For example, the argument "All men are mortal, Socrates is a man, therefore Socrates is mortal" is invalid in propositional logic yet valid within first-order formalisms.

[13] $\Phi[\Psi/\Sigma]$ is the result of uniformly replacing every occurrence of Ψ in Φ with an occurrence of Σ.

Note that even if we accept the verificationist equation between truth and the existence of a proof, the clauses for negation and the conditional remain distinct from the standard semantic clauses for the corresponding classical connectives.[14]

Viewing the debate as a disagreement about the content ascribed to the various connectives (whether such a view is coherent or not) causes the disagreement no longer to be one about the correct logic. Of course, the disputants will disagree regarding which algebraic structure codifies logical consequence. This surface disagreement stems from a deeper one regarding meaning, however, as Dummett points out:

> Thus the answer to the question how it is possible to call a basic logical law in doubt is that, underlying the disagreement about logic, there is a yet more fundamental disagreement about the correct model of meaning, that is, about what we should regard as constituting an understanding of the statement. (Dummett [1991], p. 17)

If intuitionists and classical logicians map the logical connectives onto different natural language locutions, then it is unsurprising that they go on to conclude that different algebraic structures codify the behavior of these different parts of language. Nevertheless, such a situation is not necessarily one where the disputants are disagreeing about truth-conditions or inference relations within natural language. Instead, the disagreement stems from their failure to find common ground with regard to what parts of natural language are logical—the disputants are disagreeing about what is to be codified.

Nevertheless, there are problems with attributing different meanings to intuitionistic and classical connectives. If the rival logicians mean different things by, for example, negation, and these differences are reflected in the different inference rules given for classical and intuitionistic negation, then one might reasonably suppose that we could construct a formal language containing both connectives, in order to study their interaction. J. H. Harris [1982] has demonstrated the impossibility of such a language, however.[15]

Assume that we have a formal language *L* with two negations, $\neg_I$ and $\neg_C$. $\neg_I$ is governed by the standard natural deduction introduction and elimination rules,

[14] This suggests that glossing the intuitionistic negation in terms of the notion of (classical) refutability is inadequate. Even if truth is equivalent to the existence of proof, refutability does not necessarily need to be equivalent to the existence of the sort of procedure described in the relevant clause of the Heyting semantics.

[15] This interpretation of Harris's result depends on a particular understanding of what it means for a language to "contain" intuitionistic negation. One might claim, based on the translations of intuitionistic logic into *S4*, that *S4* is a language containing both classical and intuitionistic negation, where the latter is understood as the necessitation of classical negation. One problem with this approach is that intuitionistic negation understood in this way does not obey the standard introduction and elimination rules, i.e.:

while $\neg_C$ obeys structurally similar analogues of these rules plus a classical rule such as *classical reductio*. In any such language, the two negation operators will be interderivable,[16] as illustrated by the following proof schemata:

$$\neg_I E \quad \neg_C I \quad \dfrac{\dfrac{\overline{\Phi}^{(1)} \quad \neg_I \Phi}{\bot}}{\neg_C \Phi}(1) \qquad \neg_C E \quad \neg_I I \quad \dfrac{\dfrac{\overline{\Phi}^{(1)} \quad \neg_C \Phi}{\bot}}{\neg_I \Phi}(1)$$

If logical equivalence entails sameness of meaning (at least in the case of logical constants), then we seem forced to the conclusion that intuitionists and classical logicians do not mean different things by negation, or at least that if they do mean different things, then we cannot find neutral territory within which to compare such a difference in meaning.[17] Dummett takes the lesson to be that the intuitionist:

> . . . acknowledges that he attaches meanings to mathematical terms different from those the classical mathematician ascribes to them; but he maintains that *the*

$$\dfrac{\begin{matrix}\overline{\Phi}^{(i)} \\ \vdots \\ \bot\end{matrix}}{K\neg\Phi}(i)$$

is not an admissible rule of *S*4 (unless restrictions are placed on the premises upon which the derivation of absurdity from Φ rests). See Shapiro [1985b] and section 6 below for further discussion of such translations.

[16] The result is actually stronger—in any language that contains two negations that both obey the standard introduction and elimination rules, the two negations will be interderivable, regardless of what other rules are present. It is worth noting that the result can be proved for suitable formulations in the sequent calculus as well.

[17] One possible way of circumventing this argument is to find a set of rules for intuitionistic negation, distinct from the standard rules, that (a) is well-motivated, (b) delivers the same consequences as the standard rules in a language containing only the other intuitionistic connectives, and (c) fails to collapse when combined with classical negation. One set of rules that achieves (b) and (c) is the standard elimination rule, and

$$\neg_I I* \quad \dfrac{\begin{matrix}\overline{\Phi}^{(i)} \\ \vdots \\ \bot\end{matrix}}{\neg_I\Phi}(i)$$

where no undischarged assumption contains any classical connective (such as "$\neg_C$"). Motivating such a set of rules as the "correct" rules of intuitionistic logic (i.e., [a]) is another matter entirely, however.

> *classical meanings are incoherent* and arise out of a misconception on the part of the classical mathematician about how mathematical language functions. ([1991], p. 17; emphasis added)

If the idea of different, but simultaneously legitimate, meanings is incoherent, then we are left with the following possibilities:

[1] Intuitionists and classical logicians attribute the same meanings to the connectives (whatever this meaning might be).
[2] Intuitionists and classical logicians ascribe different meanings to the connectives, but one of the following holds:
  [2a] Intuitionists attribute legitimate meanings to the connectives, while classical logicians are talking nonsense.
  [2b] Classical logicians attribute legitimate meanings to the connectives, while intuitionists are talking nonsense.
  [2c] Both intuitionistic and classical logicians are talking nonsense.

For the purposes of this chapter, we will take an optimistic stance and ignore option [2c].

Regardless of the fact that Dummett (and other intuitionists) explicitly endorse [2a], both [2a] and [2b] are implausible. First, such an interpretation of the situation fails to do justice to the fact that classical logicians can work within intuitionistic systems of mathematics and logic, and seem to understand what they are doing when they do so (and vice versa). While the logician in question might worry about the truth of the theorems thus proved, he does not, as a rule, claim not to understand them, or to be asserting nonsense when proving them. In other words, it is just a datum of logical practice that rival logicians (at least in the intuitionistic/classical case) understand the utterances of their opponents.

Second, the idea that classical logicians ascribe an incoherent meaning to negation does little justice to the (epistemic) argument proposed in favor of intuitionistic logic. The intuitionist, after arguing for the connection between truth and knowability, does not claim that it would be incoherent to accept *excluded middle* (or any classical principle) as a logical truth. Such a claim would be tantamount to arguing that it is incoherent that the correct conception of logic entail global decidability (the intuitionistic "consequence" of *excluded middle*). The idea that global decidability is (a philosophical consequence of) a logical truth is indeed somewhat implausible, but there seems little reason to think that it is a priori incoherent. Similarly, there seems to be no reason (other than Harris's result itself) for the classical logician to ascribe incoherence (instead of just error) to the intuitionistic account of logic.

At any rate, it is highly implausible that an intuitionist can sustain the charge of incoherence that Dummett suggests he level at his classical opponent. Presumably, by the intuitionist's own lights, the claim that the classical interpretation of the connectives is incoherent (that is, *not* coherent) involves (something like) a refutation of the coherence of the (supposed) classical meanings. The possibility of such a proof is hard to square with the mathematical fact that the classical laws of propositional logic are intuitionistically consistent, and in addition are true (even if not logically true) on finite and decidable infinite domains. As a result, we should take it as our default view that intuitionistic and classical logicians agree on the meaning of the connectives—in other words, they agree on the translation function T mapping the logical vocabulary of the logic onto natural language.

6. A Disagreement About What Is Preserved?

Even if the language L and translation function T are fixed, there is another way that intuitionistic and classical logicians might be talking past each other. Intuitionists might be codifying a different notion of logical consequence than that sketched above. Dummett suggests such an interpretation when he writes:

> We must . . . replace the notion of truth, as the central notion of the theory of meaning for mathematical statements, by the notion of proof. . . . As soon as we construe the logical constants in terms of this conception of meaning, we become aware that certain forms of reasoning which are conventionally accepted are devoid of justification. ([1975], p. 107)[18]

Perhaps the intuitionist understands logical consequence as something along the lines of

LC_V: A sentence Φ is a logical consequence of a set of sentences Δ iff the provability[19] of every member of Δ guarantees the provability of Φ, and this guarantee is in virtue of the logical form of Φ and the members of Δ.

On this reading, intuitionist logic would be judged by a different criterion: by whether or not (something like) the following *verification correctness principle*

[18] I am not suggesting that this is the correct interpretation of what Dummett has in mind. One more plausible interpretation is that he is advising the replacement of the classical notion of truth, which he believes has already been shown to be faulty, with a metaphysically more respectable notion of truth *understood* as provability.

[19] "Provability" is intended to be an open-ended notion, not tied to any particular formal system, since provability in, for instance, *Peano Arithmetic*, is, as a result of the Gödel phenomenon, not codified by intuitionistic logic.

CP_V: For any function I: $L \rightarrow D$ such that I agrees with T on LV:
For all $\Phi \in L$, $\Delta \subseteq L$:
The provability of $I(\Phi)$ follows from the provability of $I(\Delta)$ iff $\Delta \Rightarrow \Phi$.

holds.

The idea that classical logicians should understand intuitionists as codifying the preservation of provability, not truth, is suggested by the translations of intuitionistic logic into the classical modal logic *S4*. These mappings associate with each intuitionistic formula, a formula of the classical modal logic *S4*:

$$\mathrm{T} : \{\Phi : \Phi \text{ is a } I\text{-formula}\} \rightarrow \{\Phi : \Phi \text{ is an } S4\text{-formula}\}$$

such that

$$\varnothing \Rightarrow_{\mathrm{I}} \Phi \text{ iff } \varnothing \Rightarrow_{S4} \mathrm{T}(\Phi)$$

The first such mapping was provided by Kurt Gödel [1933], and variations can be found in Tarski and McKinsey [1948]. In addition, some (but not all) of the translations satisfy

$$\Delta \Rightarrow_{\mathrm{I}} \Phi \text{ iff } \{\mathrm{T}(\Phi) : \Phi \in \Delta\} \Rightarrow_{S4} \mathrm{T}(\Phi).$$

Among those satisfying the stronger constraint[20] is

$$\begin{aligned}
\mathrm{T}(A) &= \mathrm{K}A \ (A \text{ a propositional letter})\\
\mathrm{T}(\Phi \wedge \Psi) &= \mathrm{K}(\mathrm{T}(\Phi) \wedge \mathrm{T}(\Psi))\\
\mathrm{T}(\Phi \vee \Psi) &= \mathrm{K}(\mathrm{T}(\Phi) \vee \mathrm{T}(\Psi))\\
\mathrm{T}(\neg\Phi) &= \mathrm{K}\neg(\mathrm{T}(\Phi))
\end{aligned}$$

This suggests a strong connection between the designated value of intuitionistic logic (i.e., intuitionistic "truth") and classical provability.

While few will dispute that there are interesting connections between intuitionistic logic and classical provability, interpreting intuitionistic logic in terms of the preservation of (classical) provability is unsatisfactory. Consider *double negation elimination.* The intuitionist denies that

$$\frac{\neg\neg\Phi}{\Phi}$$

[20] A translation T that preserves logical truth (i.e., one that satisfies the weaker constraint) will preserve consequence if and only if:

For all intuitionistic formulae Φ, Ψ:
[1] $\varnothing \Rightarrow_{S4} \mathrm{T}(\Phi \rightarrow \Psi)$ iff $\varnothing \Rightarrow_{S4} (\mathrm{T}(\Phi) \rightarrow \mathrm{T}(\Psi))$ and
[2] $\varnothing \Rightarrow_{S4} \mathrm{T}(\Phi \wedge \Psi) \leftrightarrow (\mathrm{T}(\Phi) \wedge \mathrm{T}(\Psi))$

is a valid rule of inference. If the interpretation just suggested is correct, the right response would be to claim that the intuitionist is really objecting to

$$\frac{K\neg\neg\Phi}{K\Phi}.$$

On the classical understanding of the connectives, however, the provability of the double negation of *P* does suffice for the provability of *P*. In other words, this is an admissible rule of *S*4.

The problem is that even if the classical logician interprets intuitionistic logic as codifying provability, he must understand it as codifying provability in a language that differs in meaning from the meaning he attributes to his own connectives. Returning to the translations of intuitionistic logic into *S*4, the intuitionist, on this reading, is really objecting to

$$\frac{K\neg K\neg K\Phi}{K\Phi}.$$

This formulation, however, contains not only prefix occurrences of K but also internal ones, suggesting that because of the recursive nature of the translation, we are interpreting the intuitionistic connectives as having a meaning tied up with provability, a meaning different from the one attributed to the connectives by the classical logician. Thus, in order to interpret the intuitionist as codifying, not truth but provability, we must also interpret the intuitionist as attributing different meanings to the connectives. Such a disagreement about meaning, however, was rejected in the previous section.

7. A Disagreement About Truth!

The issues addressed in the last two sections fall under a general topic often called the problem of shared content, or the communication problem. Dummett addressed the problem as follows:

> Intuitionists . . . hold, that certain methods of reasoning actually employed by classical mathematicians in proving theorems are invalid: the premisses do not justify the conclusion. The immediate effect of a challenge to fundamental accustomed modes of reasoning is perplexity: on what basis can we argue the matter, if we are not in agreement about what constitutes a valid argument? ([1991], p. 17)

There are four distinct questions lurking here. The first two can be usefully labeled *the questions of shared content*:

$Q_{SC}1$: Do the disputants in the intuitionism/classical logic debate attach the same content to logical operators (i.e., do they mean the same thing by "and," "or," "not," etc.)?

$Q_{SC}2$: Do the disputants in the intuitionism/classical logic debate attach the same content to the notion of truth?

The second we can call the *questions of communication*:

Q_C1: Do each of the disputants in the debate have the semantic resources to make sense of the content the other attaches to the logical operators (i.e., can each understand what the other means by "and," "or," "not," etc.)?

Q_C2: Does each of the disputants in the debate have the semantic resources to make sense of the content the other attaches to the notion of truth?

An affirmative answer to $Q_{SC}1$ forces an affirmative answer to Q_C1 (and similarly for $Q_{SC}2$ and Q_C2), but not vice versa. Presumably a negative answer to either Q_C1 or Q_C2 renders debate between the disputants pointless. In section 5 we considered, and rejected, an interpretation that can be seen as answering $Q_{SC}1$ negatively and $Q_{SC}2$ positively, and section 6 addressed (and also rejected) the possibility of a positive answer to $Q_{SC}1$ and a negative one to $Q_{SC}2$.[21]

We are left with a default view characterized by two claims. First, intuitionists and classical theorists are attempting to codify logical consequence, and although they disagree regarding how this notion behaves, they are explicating the same notion (a positive answer to $Q_{SC}2$). Second, they attribute the same meanings to the connectives (a positive answer to $Q_{SC}1$), and their disagreement about consequence therefore hinges on whether or not certain unequivocal inferences, such as *excluded middle*, are truth-preserving in virtue of logical form. Neil Tennant, who characterizes the debate in terms of a synthetic principle underlying the strictly classical inferences in question, usefully summarizes what is at issue:

> Indeed, the holding true (as a matter of necessity) of every such instance . . . [of *excluded middle*] . . . expresses an essentially *metaphysical* belief. This belief is to the effect that the world is determinate in every expressible regard. ([1996], p. 213)

I assume, in what follows, that, at least in broad outline, Tennant's description of what is at issue is correct (i.e., the dispute traces directly to a disagreement about the "behavior" of truth). As a result, any pluralistic position must admit the possibility of distinct, incompatible, yet equally legitimate accounts of the very

[21] Presumably we could combine the approaches in the previous two sections to provide an (unsatisfactory) account that answered both $Q_{SC}1$ and $Q_{SC}2$ negatively and Q_C1 and Q_C2 positively.

same phenomenon—preservation of truth in virtue of logical form. In the next section I will sketch (but not defend) such a pluralistic position with regard to logic. In the final section of the chapter I shall argue that the intuitionist, at least, has some reason for adopting such a position.

8. Logic as Modeling

There is an assumption running through most approaches to the philosophy of logic—the *logic-as-description* view: a formal logic is an attempt to describe what is really going on vis-à-vis the truth-conditions, consequence-relation, and so on of the discourse in question. On this view, every aspect of the formalism corresponds to something actually occurring in the phenomenon being formalized, and the logic provides an exact description of correct reasoning. The view incorporates an optimistic epistemological stance, in that adherents of the view implicitly accept that there is such an exact description of logical consequence (i.e., a single, "true" logic), and with sufficient intellectual effort we can arrive at such a description.

There is another way to view the task being undertaken by logical theorists—*logic-as-modeling.*[22] On this view, a logic provides a good model of logical consequence. Logics are not (necessarily) descriptions of what is really occurring, but are instead fruitful ways to represent the phenomenon, that is, formalization is merely one tool among many that can further our understanding of the discourse in question.

The core idea that distinguishes this perspective from accounts falling under the logic-as-description heading is that although there are logics that are better or worse with regard to codifying logical consequence, there might fail to be a single correct, or "best," logic. In fact, there might be situations in which such "failure" is to be expected. The position agrees with other views in that there can be different models of logical consequence depending on our goals (i.e., what discourses or "cases" are of interest, what semantic value we wish to preserve, etc.) In addition, however, the logic-as-modeling framework allows (perhaps even entails) that even when theoretical goals are fixed, there might not be a single, correct codification of inference. In other words, natural language might fail to cooperate fully with attempts to formalize it, and there might be no logic $(L, \Rightarrow)$ such that the *correctness principle* (*CP*) (construed in a precise manner) holds.

[22] The logic-as-modeling view can be traced back to John Corcoran's classic "Gaps Between Logical Theory and Mathematical Practice" [1973], and has found more recent expression in Stewart Shapiro's "Logical Consequence: Models and Modality" [1998]. For an additional case study presented within the logic-as-modeling framework, the reader is encouraged to consult Cook [2002].

If there might be, in principle, no exact fit between natural language and any logic, then we are left with the possibility that some logical disputes, such as that between intuitionists and classical logicians, can be explained away on pluralist grounds. On this picture, the disputants believe that they are disagreeing only because they have accepted the logic-as-description picture, where one (or both) must be endorsing an incorrect codification of the consequence relation. Instead, both are championing equally legitimate accounts (models) of logical consequence. The mismatch between the natural language and any precise logic instead stems from there being no fact of the matter regarding which of two logics is a better model—there need be no more complicated logic that correctly codifies those areas where there is a "mismatch." Instead, there is indeterminacy regarding whether or not certain aspects of the model correctly codify certain aspects of that being modeled (the reader is reminded that in this section I am not yet arguing that such a pluralism is the correct account of the dispute, but only describing a possible understanding of the debate which is permitted by the logic-as-modeling view).[23]

The situation can usefully be described as a case of vagueness. On the one hand, we should accept that intuitionistic logic does not contain any inference that is not a clear, definite case of logical consequence (although it might fail to contain some valid inferences):

CP_I: For any function $I: L \rightarrow D$ such that I agrees with T on LV:
For all $\Phi \in L$, $\Delta \subseteq L$
If $\Delta \Rightarrow_I \Phi$, then $I(\Phi)$ is a logical consequence of $I(\Delta)$.

On the other hand, we can grant that classical logic contains every inference that is a clear, definite case of logical consequence (although it might also contain inferences that are not valid):

CP_C: For any function $I: L \rightarrow D$ such that I agrees with T on LV:
For all $\Phi \in L$, $\Delta \subseteq L$:
If $I(\Phi)$ is a logical consequence of $I(\Delta)$, then $\Delta \Rightarrow_C \Phi$.

There are uncountably many[24] logics between the intuitionistic and classical formalisms. More formally:

[23] There is another way we might countenance multiple models of logical consequence. The difference between two incompatible models might be artefactual, not representative of anything occurring in the phenomenon being modeled. A full description of this idea is beyond the scope of this paper, but the interested reader can consult Shapiro [1998] and Cook [2002] for discussion of the representor/artefact distinction.

[24] There is, of course, only a countable infinity of finitely axiomatizable intermediate logics. For a proof if the uncountability of (a sub-collection) of such logics, see Jankov [1968], pp. 33–34.

Given two logics $\langle L, \Rightarrow_\alpha \rangle$ and $\langle L, \Rightarrow_\beta \rangle$:
$\langle L, \Rightarrow_\alpha \rangle \subseteq \langle L, \Rightarrow_\beta \rangle =_{df}$ For all $\Delta \subseteq L$, $\Phi \in L$, if $\Delta \Rightarrow_\alpha \Phi$, then $\Delta \Rightarrow_\beta \Phi$.[25]
$\langle L, \Rightarrow_\alpha \rangle \subset \langle L, \Rightarrow_\beta \rangle =_{df}$ $\langle L, \Rightarrow_\alpha \rangle \subseteq \langle L, \Rightarrow_\beta \rangle$ and $\Rightarrow_\alpha \neq \Rightarrow_\beta$.
Fact: Given a propositional language L with countably many propositional letters:

$$\{\langle L, \Rightarrow_\alpha \rangle : \langle L, \Rightarrow_I \rangle \subset \langle L, \Rightarrow_\alpha \rangle \subset \langle L, \Rightarrow_C \rangle\}$$

is uncountable.

As a result, the correctness of intermediate logics might be what Crispin Wright [1976] calls "tolerant"—moving to a slightly stronger or weaker logic (where the difference in logical strength is sufficiently small) does not affect the correctness of attributions of logical consequence. [26] We can adopt such a view by accepting one or both of the following *vagueness principles*:

For any $\langle L, \Rightarrow_\alpha \rangle$ such that $\langle L, \Rightarrow_I \rangle \subseteq \langle L, \Rightarrow_\alpha \rangle \subseteq \langle L, \Rightarrow_C \rangle$:
For any function I: $L \Rightarrow D$ such that I agrees with T on LV:
V_1: If $\Delta \Rightarrow_\alpha \Phi$ implies that $I(\Phi)$ is a logical consequence of $I(\Delta)$, then there is some $\langle L, \Rightarrow_\beta \rangle$ such that $\langle L, \Rightarrow_\alpha \rangle \subset \langle L, \Rightarrow_\beta \rangle$ and $\Delta \Rightarrow_\beta \Phi$ implies that $I(\Phi)$ is a logical consequence of $I(\Delta)$.
V_2: If $I(\Phi)$ is a logical consequence of $I(\Delta)$ implies $\Delta \Rightarrow_\alpha \Phi$, then there is some $\langle L, \Rightarrow_\beta \rangle$ such that $\langle L, \Rightarrow_\beta \rangle \subset \langle L, \Rightarrow_\alpha \rangle$ and $I(\Phi)$ is a logical consequence of $I(\Delta)$ implies $\Delta \Rightarrow_\beta \Phi$.[27]

These principles, plus the following (rather plausible) pair of *bounding principles*:

If, for some set of indices X, every $(L, \Rightarrow_\alpha)$ such that $\alpha \in X$ contains only codifications of logical consequences, then, if $\Rightarrow_\beta = \cup_{\alpha \in X} (\Rightarrow_\alpha)$, $(L, \Rightarrow_\beta)$ contains only codifications of logical consequences.

If, for some set of indices X, every $(L, \Rightarrow_\alpha)$ such that $\alpha \in X$ contains all codifications of logical consequences, then, if $\Rightarrow_\beta = \cap_{\alpha \in X} (\Rightarrow_\alpha)$, then $(L, \Rightarrow_\beta)$ contains all codifications of logical consequences.

[25] $\subseteq$ is not the set-theoretic subset-hood relation but is a defined notion that can be glossed as "is a sub-logic of."

[26] Of course, most cases of vagueness involve some sort of metric by which we can measure the "distance" between particular instances and thereby determine when a change is sufficiently small. Such a metric is lacking in the present case. We will see in the final section, however, that a useful notion of sufficiently small change can be resurrected in cases where there are infinite chains of logics that asymptotically approach a given logic.

[27] Because of the non-symmetry of the lattice of logics intermediate between $\langle L, \Rightarrow_I \rangle$ and $\langle L, \Rightarrow_C \rangle$, V_1 and V_2 are not as similar as they might appear. V_2 is rather implausible and is included primarily for the appealing symmetry it provides. V_1 is all that is needed in the final section of this chapter, however.

(classically) imply that, for any $\Phi \in L$, $\Delta \subseteq L$, $\Delta \Rightarrow_C \Phi$ iff $\Delta \Rightarrow_I \Phi$.[28] In other words, this description of classical versus intuitionistic logic provides a novel version of the *Sorites Paradox*.

It does not follow that the situation described above is incoherent, however— any more than the existence of a series of pairwise indistinguishable color patches progressing from a clear case of red to a clear case of orange implies the incoherence of color talk. Instead, whatever account applies to more standard cases of vagueness must also apply to the connection between precise formal logics and the consequence relation of natural language. As a result, on this picture there is no logic between intuitionistic and classical that marks the precise boundary where logical validity stops and invalidity starts.[29]

Pluralism was defined as the claim that at least two distinct logics satisfy the correctness principle *CP*. On the vagueness-inspired picture just discussed, we do not have exactly this, but we can draw the conclusion that for any logic $\langle L, \Rightarrow_\alpha \rangle$, there are distinct logics that are equally good models of logical consequence. Assuming that we judge the quality of a model by the extent to which it agrees with logical consequence in natural language, V_1 implies that there is a logic $\langle L, \Rightarrow_\beta \rangle$ such that $\langle L, \Rightarrow_\alpha \rangle \subset \langle L, \Rightarrow_\beta \rangle$ and $\langle L, \Rightarrow_\beta \rangle$ is at least as good at modeling logical consequence (similar comments apply to the acceptance of V_2). As a result of accepting either V_1 or V_2, either for every logic (no matter how good a model) there is another that is a better model of logical consequence, or there are two distinct logics that are equally good models of logical consequence and are at least as good models as any other logic. Either option seems worthy of the title "pluralism."

At this point we only have a description of pluralism and a defense of its coherence. Nothing yet has been said in favor of such as the right way to view the dispute at hand. In the final section I present some evidence supporting such a pluralist stance.

9. Intuitionism Reconsidered

Assume that we are swayed by the verificationist concerns sketched at the beginning of this chapter (i.e., for the sake of argument, truth entails knowability). Even granting this, the intuitionist has failed to provide an argument that

[28] Restricted to intuitionistic logic, we can derive a contradiction from the claim that there is a $\Phi \in L, \Delta \subseteq L$, such that $\Delta \Rightarrow_I \Phi$ and not $\Delta \Rightarrow_C \Phi$.

[29] For a detailed discussion of vagueness within the logic-as-modeling framework, see Cook [2002]. Much of the treatment there, which is restricted to more pedestrian sorts of vagueness, can be adapted to the present context.

intuitionistic logic is the (only) correct codification of logical consequence. Instead, the intuitionist has good reasons for rejecting (or at least abstaining from) such a monistic position.

It should be emphasized that such a verificationist perspective on truth is consistent with the idea that intuitionists and classical logicians agree on both the semantic value preserved by logic and the meaning of the connectives. Indeed, for the intuitionist to view the situation as a disagreement about logic, he *must* interpret things in this way. Thus, the intuitionist is claiming that (1) the classical logician does (or should) understand the (non-intuitionistic) inferences in question in the same way as the intuitionist, and (2) the principles in question clearly are not (or plausibly could fail to be) instances of logical consequence.[30]

Recall that the (epistemic) argument for intuitionistic logic comes in three parts. The intuitionist (1) proposes certain epistemic principles that constrain truth; (2) argues that these principles imply that certain non-intuitionistic laws such as *excluded middle* should not be accepted; and (3) uses these epistemic principles to justify the standard rules of intuitionistic logic.

This attack is not (by itself) an argument for intuitionistic logic as the unique correct codification of logical consequence. At best, what the intuitionist has argued for is the claim that the correct logic (if there is a unique such thing) is one of the continuum of logics between intuitionistic and classical logic (and is not the latter). The reason is that the intuitionistic objections to strictly classical principles grow weaker as the classical principles in question are weakened. Heyting argued against *excluded middle* as follows:

> When we assert it, this means that for any proposition A we can either prove A or derive a contradiction from a supposed proof of A. Obviously we are not able to do this for every proposition A.... If we do not know whether A is true or not, we better make no assertion about it. ([1974], p. 87)

For the sake of argument, we can agree with Heyting regarding the epistemic implausibility of *excluded middle*, but what about weaker classical principles such as $\neg A \vee \neg\neg A$? According to the intuitionist, accepting this formula (as a logical truth) is equivalent to claiming that we can either refute A or refute the claim that

[30] Tennant [1996] is a defence of this sort of interpretation of the dialectical situation, arguing that classical logicians should accept the implication from truth to knowability, and as a result recognize excluded middle (if in fact true) as a synthetic *a priori* principle implying the determinacy of truth.

Shapiro [2001] contains compelling arguments that such a verificationist interpretation of classical logic is less attractive that it might first appear. Here, however, we agree for the sake of argument with more traditional intuitionists regarding the implausibility of excluded middle on such a reading, and focus instead on the plausibility of weaker, intermediate logics. As a result, we can ignore Shapiro's criticisms.

we can refute A. Would Heyting still be willing to claim that "obviously we are not able to do this for every proposition A"?

More generally, there exists a countably infinite chain of principles, each strictly classical yet intuitionistically weaker than the last (P an arbitrary propositional letter):

$A_0(P) =_{df} \bot$
$A_1(P) =_{df} P$
$A_2(P) =_{df} \neg P$
(for all $n = 1$)
$A_{2n+1} =_{df} A_{2n}(P) \vee A_{2n-1}(P)$
$A_{2n+2} =_{df} A_{2n}(P) \rightarrow A_{2n-1}(P)$

$A_i(P)$ intuitionistically proves $A_j(P)$ iff there is a path from i to j in the digraph

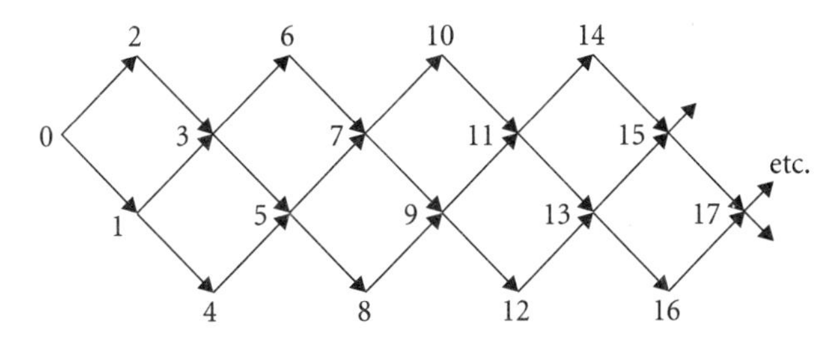

Fig. 11.1.

(For details see Troelstra and van Dalen [1988], p. 49). For all $n > 4$, $A_n(P)$ is a classical tautology, as is $A_3(P)$. Consider $\{A_i(P)\colon i = 2n+1 \text{ for some } n > 0\}$:

$A_3(P) = \neg P \vee P$
$A_5(P) = (\neg P \rightarrow P) \vee (\neg P \vee P)$
$A_7(P) = [(\neg P \rightarrow P) \rightarrow (\neg P \vee P)] \vee [(\neg P \rightarrow P) \vee (\neg P \vee P)]$
$A_9(P) = ([(\neg P \rightarrow P) \rightarrow (\neg P \vee P)] \rightarrow [(\neg P \rightarrow P) \vee (\neg P \vee P)]) \vee$
$\qquad ([(\neg P \rightarrow P) \rightarrow (\neg P \vee P)] \vee [(\neg P \rightarrow P) \vee (\neg P \vee P)])$
etc.

As we travel further down this list, intuitionistic objections to the formula in question become less robust. The relatively strong $A_9(P)$, even given a more intuitive gloss, does not assert the possibility of epistemic achievements that, as Heyting puts it, "obviously we are not able to do." Given this, how is the intuitionist to rule out $A_{999}(P)$, or $A_{99999}(P)$? The intuitionist is faced with seemingly insurmountable difficulties in arguing that his preferred logic has exactly captured logical consequence.

Corresponding to this infinitely descending chain of principles is an infinitely descending chain of intermediate logics asymptotically approaching $\langle L \Rightarrow_I \rangle$:

For all integers i such that $i = 2n+1$ for some $n > 0$, let
$\langle L, \Rightarrow_i \rangle =_{df}$ the logic resulting from adding $A_i(P)$ to $\langle L, \Rightarrow_I \rangle$ and closing under *SUB*.

Theorem: For integers i and j such that $i = 2n + 1$ for some $n > 0$ and $j = 2m + 1$ for some $m > n$:

$$\langle L, \Rightarrow_{\mathrm{I}} \rangle \subset \langle L, \Rightarrow_{\mathrm{j}} \rangle \subset \langle L, \Rightarrow_{\mathrm{i}} \rangle \subset \langle L, \Rightarrow_{\mathrm{C}} \rangle.$$

Even if the intuitionist has legitimate objections to classical logic, it is implausible that he has similarly strong arguments against each (or all) member(s) of this infinite chain of sublogics.[31]

I am not suggesting that there is some $\langle L, \Rightarrow_{\mathrm{i}} \rangle$ that the intuitionist must accept as a correct codification of logical consequence. One the contrary, he might have no reason for accepting any of them. The point is that as i increases, the intuitionist's epistemic reasons for rejecting $\langle L, \Rightarrow_{\mathrm{i}} \rangle$ become correspondingly weaker, and as a result, the intuitionist has no grounds for accepting the following *principle of invalidity recognition*:

IR: For any logic $\langle L, \Rightarrow_{\alpha} \rangle$ such that $\langle L, \Rightarrow_{\mathrm{I}} \rangle \subset \langle L, \Rightarrow_{\alpha} \rangle \subset \langle L, \Rightarrow_{\mathrm{C}} \rangle$, there are reasons for doubting that we can perform the constructions necessary for the (intuitionistic) truth of the axioms of $\langle L, \Rightarrow_{\alpha} \rangle$.

This is a universally quantified statement—its truth, for the intuitionist, is tantamount to the existence of a construction such that, when applied to any intermediate logic. $\langle L, \Rightarrow_{\alpha} \rangle$, it provides reasons for thinking that at least one axiom of $\langle L, \Rightarrow_{\alpha} \rangle$ makes a questionable existence claim about constructions. The points of the last few paragraphs, however, make the existence of such a construction unlikely.

In effect, the intuitionist's own arguments have been turned against him. The intuitionist argues that *excluded middle* (a principle whose logical truth is necessary for the classical logician) expresses an implausible existence claim about

[31] The infinitely descending chain of logics allows us to deal with the worry regarding the lack of a metric noted earlier. Even though there is no obvious metric that gives the actual "distance" between logics, we can note the following:

Given any function δ from pairs of intermediate logics onto the reals such that:

(1) $\delta(\langle L, \Rightarrow_{\alpha} \rangle, \langle L, \Rightarrow_{\alpha} \rangle) = 0$
(2) $\delta(\langle L, \Rightarrow_{\alpha} \rangle, \langle L, \Rightarrow_{\beta} \rangle) + \delta(\langle L, \Rightarrow_{\beta} \rangle, \langle L, \Rightarrow_{\gamma} \rangle) \geq \delta(\langle L, \Rightarrow_{\alpha} \rangle, \langle L, \Rightarrow_{\gamma} \rangle)$
(3) If $\langle L, \Rightarrow_{\alpha} \rangle \subset \langle L, \Rightarrow_{\beta} \rangle$ then $\delta(\langle L, \Rightarrow_{\alpha} \rangle, \langle L, \Rightarrow_{\beta} \rangle) > 0$.
(4) If, for some set of indices X, $\Rightarrow_{\beta} = \bigcap_{\alpha \in X} (\Rightarrow_{\alpha})$, then $\delta(\langle L, \Rightarrow_{\gamma} \rangle, \langle L, \Rightarrow_{\beta} \rangle) = \min \{\delta(\langle L, \Rightarrow_{\gamma} \rangle, \langle L, \Rightarrow_{\alpha} \rangle) : \alpha \in X\}$.

the existence of the infinitely descending chain of logics described above guarantees that:

For every real ε, there is logic $\langle L, \Rightarrow_{\alpha} \rangle$ such that $\langle L, \Rightarrow_{\mathrm{I}} \rangle \subset \langle L, \Rightarrow_{\alpha} \rangle \subset \langle L, \Rightarrow_{\mathrm{C}} \rangle$ and $\delta(\langle L, \Rightarrow_{\mathrm{I}} \rangle \langle L, \Rightarrow_{\alpha} \rangle) < \varepsilon$

In other words, for any reasonable metric on logics intermediate between $\langle L, \Rightarrow_{\mathrm{I}} \rangle$ and $\langle L, \Rightarrow_{\mathrm{C}} \rangle$ (with "reasonable" defined by (1) through (4)), there is are logics arbitrarily close to intuitionistic logic according to that metric (The proof of this fact, however, is strictly classical).

constructions, and thus should not be accepted. Similarly, *IR* (a principle whose truth is necessary for the intuitionist logician's acceptance of logical monism) expresses an implausible existence claim about constructions. This is not to say that the intuitionist should accept that there is some particular logic $\langle L, \Rightarrow_\alpha \rangle$ for which *IR* fails—the evidence marshaled against *IR* falls short of sufficient for the intuitionistic truth of an existential claim. What is more plausible, however, is that the intuitionist has no reason to reject

> There is some $\langle L, \Rightarrow_\alpha \rangle$ such that $\langle L, \Rightarrow_I \rangle \subset \langle L, \Rightarrow_\alpha \rangle$ and $\Delta \Rightarrow_\alpha \Phi$ implies that $I(\Phi)$ is a logical consequence of $I(\Delta)$.

This is a particular instance of V_1 above.

The intuitionist's position at this point is rather subtle. The appropriate intuitionist attitude toward *IR*, and thus logical monism, is analogous to the position he recommends with regard to *excluded middle.* The intuitionist, on pain of contradiction, cannot advise the classical logician to accept the negation of *excluded middle*—he can merely suggest that his rival abstain from asserting the principle in question. Analogously, the intuitionist need not explicitly accept that there is some intermediate logic that is just as legitimate as his own favored codification of logical consequence. He merely ought to refuse to assert that there is no such logic (i.e., refrain from claiming that his logic is the sole correct one). The intuitionist need not explicitly accept *LP*, but he has good reason for not asserting *LM.* This position, while weaker than the more robust description of pluralism in the previous section, is still a far cry from the monism espoused by traditional intuitionists.

Of course, the intuitionist could accept a pluralist view of logic. Whether he does or does not, however, his refusal to assert logical monism should not cause him to stop using intuitionistic logic. This is the whole point of the pluralism sketched above—even if the intuitionist accepts V_1 (and perhaps V_2), this only implies that other logics are equally good codifications of logical consequence, it in no way implies that there is a better codification of this notion. The intuitionist need not abandon intuitionistic logic, but, at worst, he merely needs to abandon some of the metaclaims he might previously have made regarding the connections between his favored logic and logical consequence in natural language. To sum up—intuitionist logicians can retain their use of intuitionistic logic, but they cannot retain logical monism.[32]

[32] An earlier version of this chapter was presented at Queens University Belfast, The University of Stirling, The University of Nottingham, and The University of Aberdeen. Substantial improvements were made based on the resulting discussions. Thanks are also owed to Peter Clark, Neil Cooper, Philip Ebert, Fraser MacBride, Josh Parsons, Nikolaj Pedersen, Agustin Rayo, Marcus Rossberg, Stewart Shapiro, Neil Tennant, Alan Weir, Robert Williams, and Crispin Wright for helpful suggestions and/or guidance. This chapter was written during the tenure of an AHRB research fellowship at Arché: The Centre for the Philosophy of Logic, Language, Mathematics, and Mind.

REFERENCES

Beall, J., and G. Restall [2000], "Logical Pluralism," *Australasion Journal of Philosophy* 78: 475–493.

Benacerraf, P., and H. Putnam, (eds.) [1983], *Philosophy of Mathematics: Selected Readings*, 2nd ed., Cambridge, Cambridge University Press.

Brouwer, L. [1948], "Consciousness, Philosophy, and Mathematics," in Benacerraf and Putnam [1983]: 90–96.

Brouwer, L. [1975], *Collected Works*, ed. A. Heyting, Amsterdam, North-Holland.

Bunge, M. (ed.) [1973], *The Methodological Unity of Science*, Dordrecht, D. Reidel.

Cook, R. [2002], "Vagueness and Mathematical Precision," *Mind* 111: 227–247.

Cook, R., and J. Cogburn [2000], "What Negation Is Not: Intuitionism and '$0 = 1$,'" *Analysis* 60: 5–12.

Corcoran, J. [1973], "Gaps Between Logical Theory and Mathematical Practice," in Bunge [1973]: 23–50.

Dummett, M. [1975], "The Philosophical Basis of Intuitionistic Logic," in Benacerraf and Putnam [1983]: 97–129.

Dummett, M. [1977], *Elements of Intuitionism*, Oxford, Clarendon Press.

Dummett, M. [1980], Review of Brouwer [1975], *Mind* 89: 605–616.

Dummett, M. [1991], *The Logical Basis of Metaphysics*, Cambridge, Mass., Harvard University Press.

Gentzen, G. [1965], "Investigations into Logical Deduction II," *American Philosophical Quarterly* 2: 288–306.

Harris, J.H. [1982], "What's So Logical About the Logical Axioms?" *Studia Logica* 41: 159–171.

Heyting, A. [1931], "The Intuitionistic Foundations of Mathematics," in Benacerraf and Putnam [1983]: 52–61.

Heyting, A. [1974], "Intuitionistic Views on the Nature of Mathematics," *Synthèse* 27: 79–91.

Jankov, V.A. [1968], "The Construction of a Sequence of Strongly Independent Superintuitionistic Propositional Calculus," *Soviet Mathematics Doklady* 9: 806–807.

Keefe, R., and P. Smith (eds.) [1997], *Vagueness: A Reader*, Cambridge, Mass., MIT Press.

Prior, A. [1960], "The Runabout Inference Ticket," *Analysis* 21: 38–39.

Read, S. [2000], "Harmony and Autonomy in Classical Logic," *Journal of Philosophical Logic* 29: 123–154.

Schirn, M. (ed.) [1998], *Philosophy of Mathematics Today: Proceedings of an International Congress in Munich*, Oxford, Clarendon Press.

Shapiro, S. (ed.) [1985a], *Intensional Mathematics: Studies in Logic and the Foundations of Mathematics* 113, Amsterdam, North Holland.

Shapiro, S. [1985b], "Epistemic Arithmetic and Intuitionistic Arithmetic," in Shapiro [1985a]: 11–46.

Shapiro, S. [1998], "Logical Consequence: Models and Modality," in Schirn [1998]: 131–156.

Shapiro, S. [2001], "Why Anti-realists and Classical Mathematicians Cannot Get Along," *Topoi* 20: 53–63.

Tarski, A. [1936], "On the Concept of Logical Consequence," in Tarski [1983]: 409–420.
Tarski, A. [1983], *Logic, Semantics, Metamathematics*, 2nd ed., Indianapolis, Hackett.
Tarski, A., and J. McKinsey [1948], "Some Theorems About the Sentential Calculi of Lewis and Heyting," *Journal of Symbolic Logic* 13: 1–15
Tennant, N. [1996], "The *Law of Excluded Middle* Is Synthetic *a Priori*, if Valid," *Philosophical Topics* 24: 205–229.
Troelstra, A., and D. van Dalen [1988], *Constructivism in Mathematics: An Introduction*, vol. I, Amsterdam, North-Holland.
Wright, C. [1976], "Language Mastery and the Sorites Paradox," in Keefe and Smith [1997]: 151–173.

CHAPTER 12

QUINE AND THE WEB OF BELIEF

MICHAEL D. RESNIK

1. Introduction

When W. V. Quine began his philosophical career, logical positivism and logicism both flourished. The positivists distinguished sharply between truths known empirically through sense experience and truths known a priori or independently of sense experience. But they resolutely rejected a priori intuition, be it Platonic or Kantian, as a source of mathematical knowledge, and they believed that the inadequacy of Mill's empiricist philosophy of mathematics shows that sense perception is no source, too.[1] Fortunately, logicism's new and richer conception of logic and its reduction of mathematics to logic provided the positivists with a ready-made basis for mathematical knowledge: they took it to be a priori knowledge grounded in our conventions for using logical (and mathematical) symbols. And this is just what Quine called into question. Neither their a priori-empirical distinction nor their doctrine of truth by convention survived his criticisms unscathed. Even the thesis that mathematics is logic came to be seen in a different light as a result of Quine's theorizing about logic. In view of this, it is ironic that significant themes from both logical positivism and logicism still ran through Quine's own work.

I am grateful to Matthew Chrisman, Mark Colyvan, and Stewart Shapiro for comments on an earlier draft of this chapter.

[1] For a fuller discussion of Mill's views, see chapter 3 in this volume.

Although his negative work brought Quine into philosophical prominence by the 1950s, he had already developed a positive philosophical vision, which he expanded and refined during the next forty years. Both the power of Quine's criticisms and the depth and scope of his positive views combined to make Quine an influential—perhaps the most influential—philosopher of mathematics today. He set the agenda for many current discussions: the role of convention in logic and mathematics, the nature of the a priori, criteria of ontological commitment, the indispensability of mathematics in science, the reducibility of mathematics to logic, the nature of logic, and the value of ontological parsimony.

Here is a very brief overview of Quine's philosophy of mathematics. Its fundamental feature is a combination of a staunch empiricism with holism. These are the ideas that the ultimate evidence for our beliefs is sensory evidence and that such evidence bears upon our entire system of beliefs rather than its individual elements (whence the phrase "the web of belief"). This means that our evidence for the existence of objects must be indirect, and extracted from the evidence for our system of beliefs. Thus it was essential that Quine develop a criterion for determining which objects our system commits us to (a criterion of ontological commitment). Seeing science as the fullest and best development of an empirically grounded system of beliefs, Quine heralded it as the ultimate arbiter of existence and truth. Mathematics appears to be an indispensable part of science, so Quine concluded that we must accept as true not only science but also those mathematical claims that science requires. According to his criterion of ontological commitment, this also requires us to acknowledge the existence of those mathematical objects presupposed by those claims. Finally, we usually take ourselves to be talking about a definite system of objects. However, there is enough slack in the connection between our talk of objects and the evidence for it that one can uniformly reinterpret us as referring to another system of objects while holding the evidence for the original system fixed. Thus we have ontological relativity: only relative to a fixed interpretation of our beliefs is there a fact as to our ontology.

This chapter will focus on Quine's positive views and their bearing on the philosophy of mathematics. It will begin with his views concerning the relationship between scientific theories and experiential evidence (his holism), and relate these to his views on the evidence for the existence of objects (his criterion of ontological commitment, his naturalism, and his indispensability arguments). This will set the stage for discussing his theories concerning the genesis of our beliefs about objects (his postulationalism) and the nature of reference to objects (his ontological relativity). Quine's writings usually concerned theories and their objects generally, but they contain a powerful and systematic philosophy of mathematics, and the chapter will aim to bring this into focus. Although it will occasionally mention the historical context and evolution of Quine's philosophy, it will not attempt to present a scholarly, complete examination and evaluation of it.

2. Holism and the Web of Belief

2.1. Holism: The Basic Idea

When I speak of holism here, I shall intend epistemic or confirmational holism. This is the doctrine that no claim of theoretical science can be confirmed or refuted in isolation, but only as part of a system of hypotheses. This is different from another view frequently attributed to Quine, namely, meaning holism, which is roughly the thesis that an expression depends upon the entire language containing it for its meaning.

Quine used a number of metaphors to expound his holism. The following passage from "Two Dogmas of Empiricism," which is an early and particularly strong formulation of his view, uses the metaphors of a fabric and a field force:

> The totality of our so-called knowledge or beliefs, from the most casual matters of geography and history to the profoundest laws of atomic physics or even of pure mathematics and logic, is a man-made fabric which impinges on experience only along the edges. Or to change the figure, total science is like a field of force whose boundary conditions are experience. (Quine 1951, 42)

In a later book with Joseph Ullian, the operative metaphor was one of a web, reflected in the book's title, *The Web of Belief* (1970).

Whatever the metaphor, holism is based upon an observation about science and a simple point of logic. The observation is that the statements of any branch of theoretical science rarely imply observational claims when taken by themselves, but do so only in conjunction with certain other statements, the "auxiliary" hypotheses. For example, taken in isolation, the statement that water and oil do not mix does not imply that when I combine samples of each I will soon observe them separate. For the implication to go through, we must assume that the container contains no chemical that allows them to homogenize, that it is sufficiently transparent for me to observe the fluids, that my eyes are working, and so on. Hence—and this is the point of logic that grounds holism—if a hypothesis H implies an observational claim O only when conjoined with auxiliary assumptions A, then we cannot deductively infer the falsity of H from that of O, but only that of the conjunction of H and A, H & A. Furthermore, insofar as observations confirm theories, the truth of O does not confirm H but rather H & A. Strictly speaking, it is systems of hypotheses or beliefs rather than individual claims to which the usual, deductively characterized notions of empirical content, confirmation, and falsification should be applied.

Pierre Duhem expounded these ideas at the beginning of the twentieth century, and defended the law of inertia and similar physical hypotheses against the charges that they have no empirical content and are unfalsifiable. One way of putting the law of inertia, you will recall, is to say that a body remains in a state of uniform motion unless an external force is imposed upon it. Since we can determine whether something is moving uniformly only by positing some observable reference system, this law, taken by itself, implies no observational claims. Furthermore, by appropriately changing reference systems we can guarantee that a body moving uniformly relative to our present system is not relative to the new one, and thereby protect the law against falsifying instances. All this troubled the law's critics, because they believed that it should have an empirical content and be falsifiable. Duhem responded to their worry by observing that the law readily produces empirical consequences when conjoined with auxiliary hypotheses fixing an inertial system; and that in needing auxiliaries to produce empirical consequences, it was no different from many other theoretical principles of science, whose empirical content everyone readily acknowledged. Thus the law's critics could not have it both ways: to the extent that their critique challenged the empirical status of the law of inertia, it also challenged that of most other theoretical hypotheses. (Duhem 1954).

Using logic to extract observational consequences from the law of inertia also depends upon including mathematical principles among the auxiliary hypotheses. Duhem drew no conclusions from this about mathematics. But Quine subsequently did, as the quote above indicates. Using the very strategy Duhem used in defending the law of inertia, he argued that even mathematical principles, which by most accounts are just as unfalsifiable and devoid of empirical content as the law of inertia, share in the empirical content of systems of hypotheses containing them (Quine 1990, 14–16).

In his later writings Quine toned down his holism. In speaking of "the totality of our . . . beliefs," the passage quoted above gives the impression that each of our beliefs and observations is connected logically to every other belief and observation. In *Word and Object*, Quine notes this and qualifies his holism:

> . . . this structure of interconnected sentences is a single connected fabric including all sentences, and indeed everything we ever say about the world; for the logical truths at least, and no doubt many more commonplace sentences too, are germane to all topics and thus provide connections. However, some middle-sized scrap of theory will embody all the connections that are likely to affect our adjudication of a given sentence. (Quine 1960, 12–13)

(Note also the footnote to the first sentence of this passage:

> This point has been lost sight of, I think, by some who have objected to an excessive holism espoused in occasional brief passages of mine. Even so, I think their objections largely warranted. (p.13)

So long as these "middle-sized scraps" of theory contain bits of mathematics, Quine's points about its falsifiability and empirical content will continue to hold.

2.2. Important Consequences of Quine's Holism

Let us note some important consequences of Quine's version of holism. First, and foremost, it entails rejecting the distinction between empirical and a priori truths, where the a priori truths are those that are known independently of experience and immune to revision in the light of it. This is because, for Quine, experience bears upon bodies of beliefs and, insofar as it may be said to bear upon individual beliefs of a system, it bears upon each of them to some extent. Quine rejects other means for distinguishing between the a priori and the empirical, such as the use of a priori intuition, self-evidencing truths, or sentences true by convention or by virtue of the meanings of their component terms. Thus no belief is immune to revision in the face of contrary experience. As Quine famously put it in "Two Dogmas":

> ...it becomes folly to seek a boundary between synthetic statements, which hold contingently upon experience, and analytic statements, which hold come what may. Any statement can be held true come what may, if we make drastic enough adjustments elsewhere in the system. Even a statement very close to the periphery [of experience] can be held true in the face of recalcitrant experience by pleading hallucination or by amending certain statements of the kind called logical laws. Conversely, by the same token, no statement is immune to revision. (Quine 1951, 43)

Second, although Quine acknowledged abstract objects and their perceptual inaccessability, and even spoke of some of our beliefs as arising from observation and others as arising through the exercise of reason, this provided him with no epistemological distinction for privileging statements about abstract mathematical objects. The difference here is simply one of degree rather than of kind.

Mathematics does not yield a priori knowledge, though it seems to proceed largely through the exercise of reason. What, then, of philosophy? Since, for Quine, there is no a priori or conceptual knowledge, any knowledge that philosophy can impart about science must be a piece of science. This is part of what Quine calls naturalism. Here are two passages characterizing it:

> ...naturalism: abandonment of the goal of a first philosophy. It sees natural science as an inquiry into reality, fallible and corrigible but not answerable to any supra-scientific tribunal, and not in need of any justification beyond observation and the hypothetico-deductive method. (Quine 1981b, 72)

> ...naturalism: the recognition that it is within science itself, and not in some prior philosophy, that reality is to be identified and described. (Quine 1981a, 21)

As we will see below, Quine's naturalism is a key component of his argument for mathematical realism (see chapters 13 and 14).

Several interesting questions now arise concerning Quine's own philosophical theory: What is its status? Is it to be a contribution to knowledge? And if it is knowledge, what is its source? Now philosophers commonly distinguish between normative and descriptive epistemology. The former assesses our ways of knowing and systems of beliefs with an eye toward improving upon them; the latter merely describes them. Many of Quine's more recent pronouncements concerning epistemology indicate quite clearly that he takes himself to be pursuing it descriptively. For example, in "Epistemology Naturalized" we read:

> Epistemology, or something like it, simply falls into place as a chapter of psychology and hence of natural science. It studies a natural phenomenon, viz., a physical human subject. The human subject is a accorded a certain experimentally controlled input—certain patterns of irradiation in assorted frequencies, for instance—and in the fullness of time the subject delivers as output a description of the three dimensional world and its history. The relation between the meager input and the torrential output is a relation that we are prompted to study for somewhat the same reasons that have always prompted epistemology; namely, in order to see how evidence relates to theory, in what ways one's theory of nature transcends any available evidence. (Quine 1969b, 82–83)

Can we interpret Quine's doctrine of holism as a piece of descriptive epistemology? As I noted earlier, holism arises in part from observing scientific practice and noting that scientists usually make numerous auxiliary assumptions when designing experiments for testing their hypotheses. That they do so is a straightforward descriptive claim that in turn can be scrutinized scientifically. But this is not enough, since this claim will not yield the conclusion that no statement of science is immune to revision.

Curiously, I don't think that Quine intended his claim that no statement is immune to revision as a description of what scientists have done or as prediction of what they will do. He would be the first to emphasize how radical it would be to revise mathematics in order to save a scientific theory. Perhaps he meant the conclusion as remark concerning the methodological code to which scientists subscribe. This remark could be counted as descriptive epistemology and again be subjected to scientific scrutiny, although, due to its imprecision, the results are likely to be inconclusive. However, I am inclined to read Quine as claiming that not only do scientists use auxiliary assumptions, they must do so to deduce testable conclusions from their hypotheses. If scientists must use auxiliary hypotheses, then it would follow that a negative test result would only call into question the conjunction of the auxiliaries and the main hypothesis rather than the main hypothesis alone. So, absent further specification, revising any component of this conjunction would violate no law of logic. Moreover, given that scientists freely draw our auxiliary assumptions from the entire body of science, we can arrive at the more general conclusion that circumstances could arise in which logic would permit revising any one of our (nonlogical) beliefs.

On the reading of Quine that I am offering, the claim that none of our beliefs is immune to revision amounts to the thesis that from a logical point of view, none of our beliefs is immune to revision. Now, in speaking of what logic permits, I have been using normative terms. Thus one might wonder whether Quine's holism is a piece of normative epistemology after all. I think not. In the case at hand, apparent normative talk of what logic permits is only a metaphorical substitute for descriptive speculation about how various arguments would fare when subjected to standard logical tests. Nor in applying logic do we involve ourselves in the a priori—provided, of course, that with Quine we reject the distinction between a priori and empirical knowledge.

2.3. Holism and Logic

But what of logic itself? Doesn't its role in forging the connections between theory and experience give a kind of a priori status, some immunity to empirical falsification? While we know that in "Two Dogmas" Quine counted the laws of logic as part of the fabric confronting experience, the next passage seems to reflect a change in his view. Here he is discussing the options we have when revising a set of sentences *S* in the face of failed prediction (a "fateful implication"):

> Now some one or more of the sentences in *S* are going to have to be rescinded. We exempt some members of *S* from this threat on determining that the fateful implication still holds without their help. Any purely logical truth is thus exempted, since it adds nothing to what *S* would logically imply anyway.... (Quine 1990, 14)

Quine's point here is that giving up a logical truth to repair *S* is idle. For let *L* be a logical truth and *W* any sentence or conjunction thereof. Then the conjunction of *W* and *L* implies a sentence *O* only if *W* alone implies *O*. To deactivate the fateful implication, we must revamp logic at least to the extent of refusing to recognize that *S* has the implication in question.

Now one might wonder how any revision of logic could even be an option for us. For without logic a failed prediction would be neither connected to a theory nor contrary to it. But this kind of worry can be set aside. Of course, without some logical framework, hypothesis testing could not take place, but that does not mean that the framework and the hypotheses tested cannot both be provisional. Obviously, revisions in the framework must come very gradually, since after changing it, we will need to determine whether previously tested hypotheses still pass muster. Thus, instead of denying all instances of, say, the law for distributing conjunction over alternation, we might reject certain applications of it to quantum phenomena. In this way there would be no danger of lapsing into total incoherence. Nor need we abandon the norms surrounding deduction. While we may change, for example, what counts as an implication or a contrary, we need not abandon norms that commit us to what our theories imply or that prohibit us from simultaneously maintaining two contraries. Logic is revisable, so long as major changes result from the accumulation of minor ones.

2.4. Objections to Holism

Several philosophers have been critical of both the theory of confirmation that seems implicit in holism generally and the account of mathematical evidence that seems implicit in Quine's version of it. Science takes mathematics and logic as fixed points for determining the limits of what we can entertain as serious possibilities (to borrow a phrase from Isaac Levi (1980)). In allowing for experientially motivated revisions of mathematics and logic, Quine appears to be riding roughshod over this feature of scientific practice. Charles Parsons, who has voiced objections of this sort, also points out that Quine seems to provide no place for specific kinds of mathematical evidence, such the intuitive obviousness of elementary arithmetic (Parsons 1979–1980, 1986). In a different vein, Charles Chihara (1990) has observed that in deciding whether to add a new axiom to set theory, no set theorist is going to investigate its benefits for the rest of science. Yet on Quine's approach the ultimate justification of the axiom will rest upon the acceptability of the total system of science to which it belongs. One might add that in developing his own axiomatic set theories, even Quine narrowed his focus to their ability to smoothly reproduce the standard set-theoretic foundations for mathematics while skirting contradictions.

In addition to this, some philosophers continue to hold, contrary to both Quine and Duhem, that observational evidence can be seen to bear upon specific hypotheses instead of whole systems. Elliott Sober, for example, claims that scientific testing consists in deriving incompatible predictions from competing hypotheses. Because these tests share the same auxiliary assumptions, they put specific hypotheses at risk and, consequently, the data they produce reflect upon just these hypotheses and not upon the broader systems to which they belong. Sober also notes that scientific tests never, or hardly ever, put mathematical claims at the risk of being falsified. Because of this, he argues, mathematics cannot share in the confirmation afforded to those hypotheses that do pass such tests. In particular, the mathematical theory of sets, in contrast to, say, the atomic hypothesis, cannot claim empirical support (Sober 1993, 2000).

Now Chihara, Parsons, and Sober are certainly right when it comes to scientific and mathematical practice. Mathematicians tend to keep their focus narrowly mathematical, and their notion of mathematical evidence has much more room for citing obviousness and self-evidence than it does for citing experimental data. But does this refute Quine? Let us take a closer look at these objections. As I see it, they reduce to two points. The first is that Quine is wrong about the revisability of mathematics and logic, and they are a priori, or at least Quine has not shown that they are not. The second is that holism cannot be correct, since it excludes important forms of reasoning used in science and mathematics. Thus Quine's argument, based as it is upon holism, fails to show that mathematics and logic are not a priori.

Sober seems to be urging the first point when he points out that there are very few mathematical statements that we know how to test empirically in any reasonable sense of empirical testing.[2] Now this would tend to favor the apriority of mathematics if we knew how to test empirically almost all statements of theoretical science. But in fact we don't know how to apply the sorts of specific tests that Sober has in mind to the framework principles of the various branches of science. They function as the background principles that we hold fixed when testing lower-level hypotheses. For example, we don't know how to test the hypothesis that space–time is a continuous manifold, and, given quantum mechanics, this may be untestable in principle. (To put my point in Kuhnian terms, the framework principles are part of the paradigms held fixed while we do the testing that is part of ordinary science.) If we take Sober's idea to heart, we will count much more as a priori than fans of the a priori want. Furthermore, we might find ways of testing many more mathematical statements if we tried. At best we have a notion of the a priori that is relative to our current ability to design empirical tests.

On the other hand, one might argue against Quine that as a matter of fact, we have never revised an established branch of mathematics in the face of empirical findings, and thus have little grounds for thinking that it is revisable.[3] Of course, *never revised* does not entail *not revisable*, and even Sober admits that when an observation falsifies a prediction, there is a choice of revising the main hypothesis or the auxiliary assumptions. (Sober 2000, 267). The problem with this response is that no well-formulated methodology recommends taking the choice of revising the mathematics contained in the auxiliary assumptions. Quine's own suggestions (e.g., that we revise so as to obtain the simplest overall theory and try to save as much of our current theory as we can) are too vague, and fail to lead to a unique outcome. Furthermore, we have no reason to believe that revising mathematics or logic will ever lead us to a theory that would even count as optimal in comparison

[2] Many philosophers would argue that no mathematical statement can be tested empirically. However, Sober cites a mathematical conjecture that he takes to have been tested empirically (Sober 2000, 268–269). I have also argued for the empirical testability of certain mathematical statements (Resnik 1997).

[3] Let us set the case of Euclidean geometry to the side, since there is much controversy as to whether in using non-Euclidean geometry in general relativity theory, Einstein falsified a mathematical theory of space.

with its competitors.[4] But, perhaps it is enough that revising logic or mathematics could lead us to a theory that is at least *acceptable*, if not optimal. Ruling this out would appear to beg the question by assuming that any theory arising from revising mathematics or logic is unacceptable. We may have arrived at a standoff here. Quine and his fans see revising mathematics and logic as a live option to be used only when we must take extreme measures. His opponents fail to see how it could ever be appropriate to exercise this option.

We still have to consider the second point: that holism cannot be correct, since it excludes important forms of reasoning used in science and mathematics. The strategy I will use here is to try to show that even within the framework of empiricist holism, one can make sense of the scientific and mathematical practice to which Sober, Chihara, and Parsons have called our attention.

Duhem noted that scientists often leave their auxiliary assumptions unquestioned in the course of testing hypotheses, and thereby take the evidence they obtain as bearing upon the main hypotheses. Duhem said that ordinarily this was just using "good sense," but he added, "These reasons of good sense [for favoring certain hypotheses] do not impose themselves with same implacable rigor that the prescriptions of logic do" (Duhem 1954, 217–218). Holists may readily admit that it is rational for scientists to fix certain hypotheses (as auxiliaries) while testing others, and thus also rational (in the practical sense) for them to act as if the evidence they obtain bears upon the specific hypotheses being tested. Holists can thereby accommodate the type of hypothesis testing that Sober applauds. They will simply deny that, independently of our holding the "auxiliaries" fixed, a logical (or a priori) evidential relationship obtains between the hypotheses tested and the evidence.

Let's develop this point further. In practice the various branches of science take large blocks of theory for granted. Molecular biology, for example, is developed within a framework that draws upon principles of more general theories, such as chemistry, physics, and mathematics. We also find a division of labor in the sciences: mathematicians normally do not meddle in physics nor physicists in mathematics, and biologists and chemists are normally not competent to suggest changes in mathematics or physics even when they might want to see it changed. As a result, when something goes awry in a relatively local science (say, biology), it is not likely that practitioners of more global sciences (say, physics or mathematics) will hear of it, much less be moved to seek a solution through modifying their own more global theories. Nor is it likely that the specialists in a local theory will tinker with global background theories to resolve local anomalies.

This is not just a matter of sociology; it is good sense, too. Practical rationality counsels specialists to attempt to modify more global theories only as a last resort; they probably do not and cannot know enough to tackle the task, and modifying a

[4] Field (1998, 13–14) makes this point.

more global theory is likely to send reverberations into currently quiescent areas of science. Quine has expressed the point by saying that in revising their theories, scientists should minimize mutilation.

Specialization has also fostered local methodologies and standards of evidence. These provisionally override more global and holistic perspectives and declare data, obtained via local methods, to bear on this or that local hypothesis. These will tell us, for example, that we have more reason to be confident of the existence of electrons than of gluons or of the existence of prime numbers than of inaccessible cardinals. Holists will urge, however, that local conceptions of evidence, in particular those that lead us to take data as confirming specific hypotheses, are ultimately justified pragmatically via their ability to promote science as a whole, and not via some a priori basis. Hence the divisions we find in the practice and scope of the various sciences should not be taken as refuting holism or as indicating hard-and-fast epistemic divisions between mathematics and the so-called empirical sciences. Nor do they show that it is invariably irrational to modify some global principle to fix a more local problem.

These reflections apply to our ordinary conception of mathematical evidence as well. Empirical success no more confirms individual mathematical claims than it does individual theoretical hypotheses. However, it does provide a pragmatic justification for positing mathematical objects, truths about them, and principles for applying mathematical laws to experience. It encourages mathematicians to develop their own standards of evidence, so long as the result does not harm science as a whole. Because mathematics is our most global science, we should expect that many mathematical methods and principles would be justified by means of considerations neutral between the special sciences, and thus often pertaining to mathematics alone. In this way we can reconcile holism with the features of mathematical practice that Chihara, Parsons, and Sober have emphasized.

Considering the place of proof in mathematics will illustrate this. Early mathematicians probably took their experience with counting, bookkeeping, carpentry, and surveying as evidence for the rules and principles of arithmetic and geometry that they eventually took as unquestionably true. They began to put more emphasis on deduction after they became aware of the difficulties in deciding certain mathematical questions by appealing to concrete models—which, for example, are notoriously unreliable in deciding geometric questions. By the time of the Greeks, the goal of mathematics was to prove its results. Moreover, proof wins out from the perspective of science as a whole. For requiring mathematicians to give proofs increases the reliability of their theorems, and decreases their susceptibility to experimental refutation.

The development of non-Euclidean geometry and abstract algebra further promoted the purely deductive methodology of the axiomatic method through showing mathematicians how to make sense of structures that might not be realized physically. It also promoted a shift from viewing mathematical sentences as unqualifiedly true or false to regarding them as true or false of structures of various types. These two developments have further insulated mathematics against empirical refutation. To see how, consider the case of Euclidean geometry. General relativity did refute it in its original role as a theory of physical space, but it still has important mathematical models, and survives through being reinterpreted as a theory of Euclidean spaces. A similar move is available when a scientific model incorporating a bit of mathematics proves inadequate to a physical application. It is usually far simpler to save the mathematics from refutation and conclude that the physical situation to which it was being applied failed to exhibit a suitable structure. We can use this technique to rescue any consistent theory—even a so-called empirical theory—by reinterpreting it. However, mathematical theories need no reinterpretation, since they do not assert that the structures they describe are realized in this world.

Of course, more needs to be done to answer fully the objections to the holist account of mathematical evidence. The foregoing paragraphs are offered as an indication of a way of responding to those objections.

3. Ontological Commitment: Recognizing Objects

3.1. Quine's Criterion of Ontological Commitment

One's first thoughts on the recognition of objects might be that we first acknowledge an object and then learn things about it. For example, we first see the tiger as it emerges from the brush, and later realize that it is about to attack us. But this would be contrary to both more sophisticated commonsense and holism. For to see the tiger we must see it as a tiger, and to do that, we must hold many beliefs about it. For example, we probably will believe that it is a large, animate object that looks like other objects we have identified as tigers. According to holism, recognizing the tiger is a matter of modifying our system of beliefs in certain ways; thus our evidence for the existence of the tiger ultimately traces to our evidence for a system of beliefs concerning it. What goes for tigers goes for objects generally: our evidence for them depends upon our evidence for our beliefs countenancing them. For this reason it is essential that we have a means for determining which objects we commit ourselves to in holding a system of beliefs. Quine's criterion of ontological commitment serves this purpose.

Quine's criterion applies directly to sentences and sets thereof (theories), and only indirectly to beliefs. (The transition from beliefs to sentences is based upon assuming we can determine people's beliefs by seeing what sentences they are prepared to affirm.) Then the roughest form of the criterion may be put quite simply: a set of sentences is committed to those entities that must exist in order for the members of the set to be true. But this happens when the sentences in question affirm the existence of things through the use of quantifiers. Thus we may put the criterion more precisely as *Sentences are committed to those entities over which their bound variables must range in order for them to be true.* Here is how Quine puts it in "On What There Is":

> ...We can very easily involve ourselves in ontological commitments by saying, for example, that *there is something* (bound variable) which red houses and sunsets have in common; or that *there is something* which is a prime number larger than a million. But this is essentially the *only* way we can involve ourselves in ontological commitments: by our use of bound variables. (Quine 1948, 12)
>
> ...The variables of quantification, "something," "nothing," "everything," range over our whole ontology, whatever it may be; and we are convicted of a particular ontological presupposition if, and only if, the alleged presuppositum has to be reckoned among the entities over which our variables range in order to render one of our affirmations true. (Quine 1948, 13)

Quine came to emphasize the triviality of what's going on here.

> The artificial notation "$\exists x$" of existential quantification is explained merely as a symbolic rendering of the words "there is something x such that." So, whatever more one may care to say about being or existence, what there are taken to be are assuredly just what are taken to qualify as values of "x" in quantifications. The point is thus trivial and obvious. (Quine 1990, 26–27)

However, often it is far from clear as to what the ontological commitments of a set of sentences are. This is especially true of day-to-day talk in ordinary language.

> The common man's ontology is vague and untidy in two ways. It takes in many purported objects that are vaguely or inadequately defined. But also, what is more significant, it is vague in its scope; we cannot even tell in general which of those vague things to ascribe to a man's ontology at all, which things to count him as assuming.
>
> ...a fenced ontology is just not implicit in ordinary language. The idea of a boundary between being and nonbeing is a philosophical idea, an idea of technical science in a broad sense....
>
> We can draw explicit ontological lines when desired. We can regiment our notation....Then it is that we can say the objects assumed are the values of the variables....Various turns of phrase in ordinary language that seem to invoke novel sorts of objects may disappear under such regimentation. At other points new ontic commitments may emerge. There is room for choice, and one chooses with a view to simplicity in one's overall system of the world. (Quine, 1981a, 9–10)

To illustrate what Quine has in mind, consider this bit of hypothetical dialogue:

> John: I saw a possible car for you.
> Jane: I have many things occupying my time, but for your sake, I will look at it.

John seems to be committed to *possible* cars (he says he saw one). Mary, on the other hand, seems committed to her own time, things that occupy it, and John's sake. But what is a sake? Or a thing occupying a person's time? Or a person's time, for that matter? If we simply paraphrase the dialogue, we can avoid such questions and reduce its apparent ontological commitments.

> John: I saw a car that might do for you.
> Jane: I am very busy now, but I will look at it, since you want me to.

We have done quite a bit to clean up John's and Mary's ontologies! Now we may take John to commit himself to just the car he actually saw instead of a possible one, whatever that might be. Mary, though she is quite busy, need no longer be seen as involved with sakes or things occupying her time. This is just a somewhat humorous illustration of Quine's procedure. For Quine the serious applications are scientific and philosophical theories. By paraphrasing them into the canonical notation of extensional first-order logic, we try to assess and reduce ontological commitments.

> ... But the simplification and clarification of logical theory to which a canonical notation contributes is not only algorithmic; it is also conceptual. Each reduction that we make in the variety of constituent constructions needed in building the sentences of sciences is a simplification. Each elimination of obscure constructions or notions that we manage to achieve, by paraphrase into more lucid elements, is a clarification of the conceptual scheme of science. The same motives that impel scientists to seek ever simpler and clearer theories adequate to the subject matter of their special sciences are motives for simplification and clarification of the broader framework shared by all sciences.... The quest of a simplest, clearest overall pattern of canonical notation is not to be distinguished from a quest of ultimate categories, a limning of the most general traits of reality. (Quine 1960, 161)

For Quine, one of the philosopher's major contributions comes from clarifying the language of science and mathematics in order to assess and reduce the ontological commitments of our theories of the world. Though Quine's criterion may be trivial in itself, the applications one might make of it are far from trivial.

3.2. The Canonical Language and Benefits of Regimentation

As we have seen, before applying Quine's criterion, one must paraphrase the theory to be assessed into "canonical notation." Part of the reason for doing so is to eliminate spurious ontological commitments. For example, even the language of working mathematics contains terms, such as, "1/0" or "(sin x)/x with $x=0$" that appear to be referential, yet may fail to denote anything. Now we might declare sentences containing such terms to be false just as we might declare pieces of fiction false. But then what do we do with truths such as "There is no such number as 1/0"? In the face of this, some philosophers have responded that every name denotes something—even "the largest natural number." But for Quine there is a simpler and more economical solution: we avoid the offending expressions by paraphrasing away names and functional terms altogether. This can be done quite straightforwardly using Russell's theory of descriptions. Quine employs it, or an equivalent device, quite frequently, and in his canonical notation the only singular terms are variables.

Quine sees logic as ending with first-order logic,[5] and his canonical notation also bans other notable adjuncts to first-order logic, such as modal operators and substitutional quantifiers. The modal operator "it is possible that" allows us to formulate the claim "It is possible that there are numbers but there are (in fact) no numbers"—a claim to which some nominalists subscribe. Such talk seems to recognize a kind of existence intermediate between being and nonbeing. Quine has never been able to make sense of this idea, and has made relatively little sense of modal operators themselves. Despite this and Quine's influence, modal operators have made a comeback in recent technical philosophy of mathematics.[6]

Substitutional quantification is also popular with nominalists. Instead of requiring the existence of F, its truth a substitutional "$\mathit{\exists} xFx$" counts as true if and only if "Fx" has a true substitution instance.[7] This will count "$\mathit{\exists} x$(x is a flying horse)" as true so long as we take "Pegasus is a flying horse" as true. Since some philosophers demur at counting the latter as false, they can use substitutional quantification to analyze talk of fictional entities. Fans of modal logic have also used it to deal with problems arising in interpreting quantifications containing modal operators. What is more significant for our purposes is that philosophers of mathematics have proposed using it to gain the formal advantages of having classes without having to pay the price of admitting them into one's ontology.

[5] The acceptability of higher-order logic is a complex and technical issue. Evaluating it and Quine's arguments would take a chapter in itself. Fortunately, this volume contains two chapters devoted to second- and higher-order logic (25 and 26).

[6] See chapters 1, 15, and 16.

[7] I have used an italicized "$\mathit{\exists}$" to distinguish it from the ordinary (or objectual) existential quantifier. A substitutional "$\mathit{\exists} xFx$" can be true without F's existing so long as we count one of its substitution instances as true. (E.g., some philosophers count "$\mathit{\exists} x$(x is a flying horse)" as true by virtue of the supposed truth of "Pegasus is a flying horse.") On the other hand, the objectual "$\exists xFx$" can be true without having a true substitution instance when the Fs are unnamed. Thus "$\exists x$(x is an unnamed real number)" is true while its substitutional counterpart is false.

Although substitutional quantifiers form no part of Quine's canonical notation, he appears to think that, unlike modal operators, they are not intrinsically unacceptable. However, nominalists should draw no comfort from this.

> This does not mean that theories using substitutional quantification and no objectual quantification can get on *without* objects. I hold rather that the question of the ontological commitments of a theory does not properly arise except as that theory is expressed in classical quantificational form, or insofar as one has in mind how to translate it into that form. I hold this for the simple reason that the existential quantifier, in the objectual sense, is given precisely the existential interpretation and no other: there are things which are thus and so. (Quine 1969c, 107)

In Quine's view, rewriting mathematics using substitutional quantifiers would amount to abandoning the quest for ontological economy rather than achieving it.

Today realists in the philosophy of mathematics use Quine's criterion largely to argue for the existence of mathematical entities. "To do science," they claim, "we use variables ranging over mathematical entities and are, consequently, committed to their existence." (More on this in the section 4 below.) Anti-realists, on the other hand, often appeal to Quine's criterion to measure the success of their various attempts at ontological economy. "My system has no variables ranging over mathematical entities," they argue, "so, we need not be committed to their existence." Ironically, it's unclear how successful these anti-realist attempts have been, because they employ languages that exceed Quine's canonical notation.[8]

3.3. Ontological Naturalism: Science Is the Ultimate Arbiter of Existence

Quine often emphasizes that his criterion tells us only what a theory says exists. It does not tell us what does exist. That, you will recall, is the job of science; this is Quine's naturalism.

> ... naturalism: the recognition that it is within science itself, and not in some prior philosophy, that reality is to be identified and described. (Quine 1981a, 21)

But Quine's criterion still has a role to play, since by applying it to the theories that science affirms, we determine what, according to science, exists.

Philosophers are not entirely out of the picture, however. Though they cannot transcend science, they can work within science and propose clarifications and

[8] See chapters 15 and 16 in this volume. Jody Azzouni pursues an atypical anti-realist program through rejecting Quine's criterion and arguing that to "quantify over" mathematical entities is not ipso facto to presuppose their existence (Azzouni, 1998).

ontological reductions. It's in this spirit that Quine regards his own proposals for reducing mathematics to set theory. Here again is part of an earlier quote:

> The same motives that impel scientists to seek ever simpler and clearer theories adequate to the subject matter of their special sciences are motives for simplification and clarification of the broader framework shared by all sciences. . . . The quest of a simplest, clearest overall pattern of canonical notation is not to be distinguished from a quest of ultimate categories, a limning of the most general traits of reality. (Quine 1960, 161)

3.4. Introducing New Objects: Positing

Our focus so far has been on how we justify countenancing various objects. Though one might cite a local conception of evidence as an immediate justification for recognizing an object, ultimately one tacitly appeals to the success of a broader system committed to the type of thing in question. For example, having digested a proof of Cantor's Theorem, we may feel fully justified in countenancing uncountable sets. But this is predicated on our prior acceptance of sets themselves, and of whatever set theory might be needed for carrying out the proof in question and for inferring the existence of uncountable sets. With Quine, we might justify acquiescing in sets by citing their benefits to mathematics. We could in turn justify this by pointing to the importance to science as a whole of having a flourishing mathematics.[9]

Now, what we have observed concerning sets applies to other types of objects—even ordinary physical bodies. You know there's an owl out in the dark, because you heard it screech, and you know an owl's screech. But here you are already presupposing owls and a rich body of beliefs about them, and all these are contained in a framework provided by your beliefs about birds, animals, and physical bodies.

Once we have in place a framework that countenances objects of a given kind, we are in a position to identify and authenticate objects of that kind, whether we do so via observation, instrumentation, or theoretical deduction. But how do we come to add new objects of various types to our ontology in the first place? Even when, according to later theory, we have been observing a type of object with our unaided senses all along, we seem to have a serious problem. For prior to having a framework countenancing the objects in question, we need not even be aware that we are observing any definite thing at all. For example, a layperson looking at the sky on a clear night may be aware of only the stars and the moon, while astronomers will be aware of galaxies, and much more. The difference between the

[9] See, for example, Quine (1981a, 13–16).

layperson and astronomers is that the latter have hypothesized a much richer ontology along with a rich theory of its members' behavior and their observable effects. To introduce an ontology in this way is to *posit* it. But while to posit some things is similar to making up a story about them—at least it is initially—this does not mean that astronomers have created the heavens.

> ... Considered relative to our surface irritations, which exhaust our clues to external physical objects, the molecules and their extraordinary ilk are thus much on a par with the most ordinary physical objects. The positing of these extraordinary things is just a vivid analogue of the positing or acknowledging of ordinary things: vivid in that the physicist audibly posits them for recognized reasons, whereas the hypothesis of ordinary things is shrouded in prehistory. . . .
>
> To call a posit a posit is not to patronize it. . . . Everything to which we concede existence is a posit from the standpoint of a description of the theory-building process, and simultaneously real from the standpoint of the theory that is being built. (Quine 1960, 22)

Quine no more admits existence by fiat than he does truth by fiat.

This has very important consequences for Quine's philosophy of mathematics. Mathematics, at least on the realist reading of it that Quine favors, is about objects that have no place in space or time, and no effects upon our sensory apparatus. How, one wonders, could we have ever come to have any knowledge of such things? Not through intuition or other a priori insight—at least not if Quine is right. But there is nothing mysterious about our building theories that posit mathematical objects, for theory construction itself requires no contact with the things the theory purports to concern. And, if the theory forms a workable part of our overall system, no matter what its subject may be, then the entities to which it is committed (via Quine's criterion) have as much title to existence as "ordinary things" hypothesized "in prehistory."

4. The Indispensability Argument for Mathematical Realism

Everyone grants that mathematics is very useful to the pursuit of science. It gives science the wherewithal for representing empirical findings through statistical and other numerical means and for explaining these findings using such concepts as those of acceleration, state vector, random mating, allelic frequency, expected utility, and welfare function. Moreover, mathematical laws permit scientists to deduce nonmathematical conclusions from assumptions, such as Newton's laws of motion, that are formulated in a mix of scientific and mathematical vocabulary. Eliminating mathematics would thus drastically alter the practice of working science.

But what if the theoretical purposes of mathematics could be accomplished using a more parsimonious ontology without any reduction in the overall simplicity and economy of the resulting scientific theory? Quine would heartily approve, but he would not ask scientists to stop using mathematics. He would merely claim that since mathematics could be excised from the canonical formulation of science, science (and thus we) should no longer acknowledge its truth or ontological commitments.

> ... not that the idioms thus renounced are supposed to be unneeded in the market place or in the laboratory.... The doctrine is that all traits of reality worthy of the name can be set down in idiom of this austere form if in any idiom. (Quine 1960, 228)

Although Quine attempted to eliminate mathematics from science and applauded efforts aimed at showing that the mathematical needs of science can be reduced, he came to believe that most classical mathematics is indispensable to science (Quine 1960, 270).

Since there is, so far as we know, no way of eliminating mathematics from the "austere idiom" of the canonical formulation of science, we are bound to admit the existence of those mathematical objects that science posits. This argument, which is rooted in Quine's writings and was propounded explicitly by Hilary Putnam, has become known as the Indispensability Argument for Mathematical Objects.[10]

We can formulate a more explicit version of an indispensability argument as follows: First, mathematics is an indispensable component of natural science. Second, thus, by holism, whatever evidence we have for science is just as much evidence for the mathematical objects and the mathematical principles it presupposes as it is for the rest of its theoretical apparatus. Third, whence, by naturalism, this mathematics is true, and the existence of mathematical objects is as well grounded as that of the other entities posited by science. I call this the Holism–Naturalism (H–N) Indispensability Argument. It is clearly based upon principles that Quine accepts, although it is not as clear that it accurately paraphrases his or Putnam's arguments.

Now lots of philosophical energy and talent—including some of Quine's—has been spent trying to undermine the first premise of this argument by showing that mathematics is dispensable from science.[11] More recently, however,

[10] Cf. Putnam: "So far I have been developing an argument for realism along roughly the following lines: quantification over mathematical entities is indispensable for science, both formal and physical: therefore we should accept such quantification; but this commits us to accepting the existence of the mathematical entities in question. This type of argument stems, of course, from Quine, who has for years stressed both the indispensability of quantification over mathematical entities and the intellectual dishonesty of denying the existence of what one daily presupposes" (Putnam 1971, 57).

[11] See chapters 15 and 16 in this volume.

philosophers have questioned other aspects of the argument. Neither Penelope Maddy nor Elliott Sober thinks that we can count on science to provide evidence for the truth of mathematics. As we saw earlier, Sober claims that scientific testing fails to confirm the mathematics used in science.

Maddy's criticism is based upon observing that much of the mathematics used in science occurs in theories, such as the ideal theory of gases, that scientists openly acknowledge as false yet still employ. She argues that this raises the possibility that the confirmation coming from membership in scientific theories that are accepted as true covers too little mathematics to be of comfort to mathematical realists. In short, too much of mathematized science may fall outside the scope of the H–N argument's naturalism premise to support mathematical realism (Maddy 1992, 281).

It is not clear how Quine or Putnam would respond to this criticism, for it is not even clear that the H–N argument is theirs. But one can set aside Maddy's worry by moving to another version of the indispensability argument. For whatever attitude scientists take toward their own theories, they cannot consistently regard the mathematics they use as merely of instrumental value. Take Newton's account of the orbits of the planets as an example. He calculated the shape of the orbit of a single planet, subject to no other gravitational forces, traveling about a fixed star. He knew that no such planet exists, but he also believed that there are mathematical facts concerning its orbit. In deducing the shape of such orbits, he presumably took for granted the mathematical principles he used. For the soundness of his deduction depended upon their truth. Furthermore, in using his (mathematical) model to explain the orbits of actual planets, he presumably took its mathematics to be true. For he explained the orbits of planets in our solar system by saying that they approximate the behavior of an isolated system consisting of a single planet orbiting a single star. For this explanation to work, it must be true that the type of isolated system (Newtonian model) has the mathematical properties Newton attributed to it. This illustrates that even when applying mathematics to idealizations or theories they know are wrong, scientists use it in a way that commits them to its truth and ontology.

Reflecting on this leads one to the Pragmatic Indispensability Argument, which runs as follows:

1. In stating its laws and conducting its derivations, science assumes the existence of many mathematical objects and the truth of much mathematics.
2. These assumptions are indispensable to the pursuit of science; moreover, many of the important conclusions drawn from and within science could not be drawn without taking mathematical claims to be true.
3. So we are justified in drawing conclusions from and within science only if we are justified in taking the mathematics used in science to be true.

Notice that, unlike the earlier H–N Indispensability Argument, this one does not presuppose that our best scientific theories are true or even that they are well supported. It applies wherever science presupposes the truth of some mathematics. Thus, as we noted earlier, it applies even to the mathematics contained in those refuted scientific theories that scientists still use and to the mathematics of idealized scientific models. Furthermore, the argument, at least as it stands, contains no claim that the evidence for science is also evidence for mathematics. We can extend this argument to infer that we should acknowledge the truth of mathematics on pragmatic grounds. For given that we are justified in doing science, we are justified in using (and thus assuming the truth of) the mathematics in science, because we know of no other way of obtaining the explanatory, predictive, and technological fruits of science.[12]

Since much standard mathematics is used in science, the indispensability arguments support realism about many parts of mathematics. Yet, as Quine was aware, and Maddy and others have emphasized, indispensability arguments fail to cover the more theoretical and speculative branches of mathematics. Currently science neither needs nor employs this mathematics, and it does not even help in simplifying and systematizing the mathematics that science does apply. Thus it is not part of the Web of Belief, and not connected even indirectly to experience.

5. Ontology and Ontological Relativity

Quine has frequently urged that we take the ontology of mathematics to be one of classes or sets. Numbers, functions, vectors, groups, spaces, and so on are to be reduced to them in the familiar ways.[13] This is because we need to use classes for many purposes, both mathematical and nonmathematical, and we obtain a simpler ontology by having just classes rather than classes plus other mathematical and abstract objects. Here is how Quine argued for countenancing classes in *Word and Object*:

> The versatility of classes in thus serving the purposes of widely varied sorts of abstract objects is best seen in mathematics, but it spills over.... Such is the power of the notion of class to unify our abstract ontology. To surrender this benefit and face the old abstract objects again in their primeval disorder would be a wrench, worth making if it were all. But we must remember that the utility

[12] For further discussion of this argument see Resnik (1997), and for a thorough discussion of indispensability arguments for mathematical realism, see Colyvan (2001).

[13] See Quine (1963).

> of classes is not limited to explication of the various other sorts of abstract objects. The power of the notion on other counts... keeps it in continual demand in mathematics and elsewhere as a working notion in its own right.... Thus it is that one resolves to keep classes and somehow excise the paradoxes. (Quine 1960, 267)

Here Quine was writing both as a philosopher and as a logician, and advocating that we need no abstract objects in all of science but classes. This sounds like first philosophy, but remember that, on Quine's view, the work of clarifying and reducing the ontological requirements of philosophy and mathematics is the same type of work that scientists in other fields do when they clarify and simplify theories in their home disciplines. The work of Frege, Russell, Quine, and others in unifying the foundations of mathematics using class theory is as much a piece of theoretical science as Von Neumann's reformulation of quantum mechanics using Hilbert spaces.

Some philosophers and mathematicians have argued that mathematics has no need for a reduced ontology. One might read such an argument into some of Hilbert's writings. On this reading, all that mathematicians need are clearly specified axioms for the branch of mathematics within which they are working. It is enough that they be assured that some things satisfy those axioms. Otherwise the nature of these things is of no concern.[14] I am not aware of Quine's responding explicitly to such reasoning, but a response is implicit in the passage quoted above. It would run as follows: Mathematics, even on Hilbert's' view of it, needs some ontology in order to be assured that its axioms are not vacuous. There are too few concrete physical objects to fill the bill. (Ditto for mental objects—if you are so rash as to countenance them.) Thus mathematics requires abstract objects. These are best provided through a unified ontology of classes.

Quine was aware of other proposals for ontological foundations for mathematics. For example, some mathematicians have urged that category theory is a much better vehicle for mathematics than set theory, and that it should take over that role. In *Set Theory and Its Logic* (1963), Quine does mention that category theory is useful for dealing with very large collections. But he does not defend sets and classes against the attacks from advocates of category theory in that book, and I am not aware of any place where he does.

As late as *Theories and Things* (1981c) Quine runs through his usual brief for an ontology of physical objects and sets of physical objects. But then he takes the argument a step further. In the name of simplicity we reduce physical objects to the space–time regions they occupy, and these in turn to sets of space–time coordinates relative to some fixed coordinate system. But space–time coordinates are ordered quadruples of real numbers, and these in turn reduce in familiar ways to sets built up from the empty set, leaving nothing but pure sets (Quine 1981a, 16–18).

[14] Hilbert expressed views similar to this in discussing his axiomatic work with Frege. See, for example, Resnik (1980).

Quine's discussion then turns from this reduced ontology to other ontological candidates generated by reinterpreting our theory of the world while preserving its observational consequences. Here "we... merely change or seem to change our objects without disturbing either the structure or the empirical support of a scientific theory in the slightest" (Quine 1981a, 19). To do this, it suffices that we be able to specify a one-one function—Quine calls it a "proxy function"—which maps each object in the original ontology to an object in the new ontology. Then, using a well-known technique from formal logic, we can reinterpret each predicate of our theory so that it is true of something in the new ontology if and only if it was true (as originally interpreted) of this things inverse under the proxy function.

> The apparent change is twofold and sweeping. The original objects have been supplanted and the general terms reinterpreted. There has been a revision of ontology on the one hand and of ideology, so to say, on the other; they go together. Yet verbal behavior proceeds undisturbed, warranted by the same observations as before and elicited by the same observations. Nothing really has changed. (Quine 1981a, 19)

Quine concludes from this that reference is inscrutable, that is, there is no saying absolutely whether our words refer to this or that, but what they refer to relative to a fixed interpretation. As we move from one ontology to another, our words change their reference, too; yet we have arranged for the truth-values of our sentences—the facts, so to speak—to stay the same. There is no fact of the matter concerning the references of our words—at least in the sense that we can vary their references at will while leaving the facts unvaried. Besides calling this the inscrutability of reference, Quine refers to it as ontological relativity: there is no fact as to what the ontology of a theory is absolutely, but only relative to a means of interpreting it in a language we take at face value (Quine 1981a, 19–20; see also Quine 1969a, 50–54 and 1990, 30–34).

Notice how nicely this fits with Quine's holism. According to holism, it is only relative to a local conception of evidence that we can say that a particular experience verifies or falsifies a particular sentence of some theory. Otherwise, experience bears upon the theory as a whole. Similarly, it is only relative to our parochial ontology and interpretation of our language that we can say that a particular word picks out a particular object of our experience. But doesn't something (e.g., our behavior or the causal relations between our words and the world) fix the references of our words? No. Just as there are many ways to reconcile a theory with a recalcitrant experience, so there many candidate references for our words, and no way to single out one. After applying a proxy function, "verbal behavior proceeds undisturbed, warranted by the same observations as before and elicited by the same observations. Nothing really has changed" (Quine 1981a, 19).

Of course, if we take our home language at face value, then we can simply single out an intended interpretation, using our own words to do so. Thus we simply state: the intended model of the Peano Axioms is the natural number sequence. So long as we don't question the reference of "the natural number sequence," we have no trouble distinguishing the standard model from, say, the sequence of even natural numbers or in ruling out an interpretation that construes the successor function, $s(x)$, as $x+2$. But once we question the references of our own words, nothing precludes the phrase "the natural number sequence" from referring to the even number sequence, the finite Von Neumann ordinals, or any progression.

These considerations led Quine to put an emphasis on the logical structure of a theory, since this is what remains invariant under applications of proxy functions. As Quine put it, "structure is what matters to a theory, and not the choice of its objects." The objects "serve merely as indices along the way, and we may permute or supplant them as we please as long as sentence-to-sentence structure is preserved" (Quine 1981a, 20; see also Quine 1990, 31).

Yet after presenting these ideas Quine was moved to reaffirm his realism with respect to ordinary bodies and the theoretical entities of science and mathematics. As a naturalist, he is committed to the ontological commitments of current science. As to his previous reflections that seem to belittle objects, Quine points out that these belong not to ontology but to "the methodology of ontology and thus to epistemology." The considerations showed that we could turn our backs upon external things and classes and "ride the proxy functions to something strange and different without doing violence to any evidence. But all ascription of reality must come rather from within one's theory of the world; it is incoherent otherwise" (Quine 1981a, 21).[15]

REFERENCES

Azzouni, Jody (1998). "On 'On what there is.'" *Pacific Philosophical Quarterly*, 79: 1–18.

Chihara, Charles (1990). *Constructibility and Mathematical Existence.* (Oxford: Oxford University Press).

Colyvan, Mark (2001). *The Indispensability of Mathematics.* (Oxford: Oxford University Press).

Duhem, Pierre (1954). *The Aim and Structure of Physical Theory.* (Princeton, N.J.: Princeton University Press), translation by Philip P. Wiener of *La Théorie physique, son object, son structure*, 2nd ed. (1914).

[15] Quine (1990) also emphasizes the primacy of structure but, interestingly, this book contains no brief for a mathematical ontology consisting of just sets. However, Quine (1995) does (cf. pp. 40–42). Furthermore, Quine (1992) reaffirms a mathematical ontology of classes while also emphasizing ontological relativity.

Field, Hartry (1984). "Is Mathematical Knowledge Just Logical Knowledge?" reprinted with additions in Field (1989): 79–124.
——— (1989). *Realism, Mathematics and Modality.* (Oxford: Blackwell).
——— (1998). "Epistemological Nonfactualism and the Aprioricity of Logic," *Philosophical Studies*, 92: 1–24.
Levi, Isaac (1980). *The Enterprise of Knowledge.* (Cambridge, Mass.: MIT Press).
Maddy, Penelope (1992). "Indispensability and Practice." *Journal of Philosophy*, 89: 275–289.
Parsons, Charles (1979–1980). "Mathematical Intuition," *Proceedings of the Aristotelian Society*, N.S. 80: 145–168.
——— (1986). "Quine on the Philosophy of Mathematics," in L.E. Hahn and P.A. Schilpp (eds.), *The Philosophy of W. V. Quine.* (La Salle, Ill.: Open Court).
Putnam, Hilary (1971). *Philosophy of Logic.* (New York: Harper).
Quine, W.V. (1948). "On What There Is," reprinted in Quine (1961).
——— (1951). "Two Dogmas of Empiricism," reprinted in Quine (1961).
——— (1960). *Word and Object.* (Cambridge, Mass.: MIT Press).
——— (1961). *From a Logical Point of View*, 2nd ed. (Cambridge, Mass.: Harvard University Press).
——— (1963). *Set Theory and Its Logic.* (Cambridge, Mass.: Harvard University Press).
——— (1969a). "Ontological Relativity," reprinted in Quine (1969d).
——— (1969b). "Epistemology Naturalized," reprinted in Quine (1969d).
——— (1969c). "Existence and Quantification," reprinted in Quine (1969d).
——— (1969d). *Ontological Relativity and Other Essays.* (New York: Columbia University Press).
——— (1981a). "Things and Their Place in Theories," in Quine (1981c).
——— (1981b). "Five Milestones of Empiricism," in Quine (1981c).
——— (1981c). *Theories and Things.* (Cambridge, Mass.: Harvard University Press).
——— (1986) "Reply to Charles Parsons," in L.E. Hahn and P.A. Schilpp (eds.), *The Philosophy of W. V. Quine.* (La Salle, Ill.: Open Court).
——— (1990). *Pursuit of Truth.* (Cambridge, Mass.: Harvard University Press).
——— (1992). "Structure and Nature," *Journal of Philosophy*, 59: 5–9.
——— (1995). *From Stimulus to Science.* (Cambridge, Mass.: Harvard University Press).
Quine, W.V., and Ullian, Joseph (1970). *The Web of Belief.* (New York: Random House).
Resnik, Michael D. (1980). *Frege and the Philosophy of Mathematics.* (Ithaca, N.Y.: Cornell University Press).
——— (1997). *Mathematics as a Science of Patterns.* (Oxford: Oxford University Press).
Sober, Elliott (1993). "Mathematics and Indispensability," *Philosophical Review*, 102: 35–57.
——— (2000). "Quine's Two Dogmas," *Proceedings of the Aristotelian Society*, supp. 74: 237–280.

CHAPTER 13

THREE FORMS OF NATURALISM

PENELOPE MADDY

MANY philosophers—from Hume[1] to the pre-Fregean German materialists,[2] from Reichenbach to Arthur Fine[3]—have been classified as "naturalists," in some sense or other of that elastic term, but the version most influential in contemporary philosophy of logic and mathematics undoubtedly comes to us from Quine. For him, naturalism is characterized as "the recognition that it is within science itself, and not in some prior philosophy, that reality is to be identified and described" (1981a, p. 21). Agreeing wholeheartedly with this sentiment and the spirit behind it, some post-Quinean naturalists, including John Burgess[4] and myself, occasionally find ourselves uncomfortably at odds with particular doctrines Quine

My thanks to John Burgess for all he has taught me over the years, as well as his thoughts on earlier drafts of this chapter. I am also grateful to Patricia Marino for helpful comments and discussions.

[1] Mounce (1999) and Stroud (1977) seem to attribute different forms of "naturalism" to Hume.

[2] See Sluga (1980), pp. 17–34. As Sluga notes (1980, p. 178), Stroud (1977, p. 222) sees these "scientific materialists," rather than the logical positivists, as the true descendants of Hume.

[3] I discuss these last two versions of "naturalism" in (2001a).

[4] As the central texts of Burgess's naturalism, I'll be using Burgess (1983, 1990, 1998) and Burgess and Rosen (1997). With apologies to Professor Rosen for downplaying his contributions, I will treat the "naturalism" of the co-authored work as an elaboration of the position in Burgess's earlier papers.

develops in his pursuit of philosophy naturalized, doctrines that seem to us less than completely true to his admirable naturalistic principles. Yet Burgess and I sometimes disagree on just which doctrines those are and on how to go about correcting the situation!

My plan here is to sketch the outlines of the Quinean point of departure, then to describe how Burgess and I differ from this, and from each other, especially on logic and mathematics. Though my discussion will touch on the work of only these three among the many recent "naturalists," the moral of the story must be that "naturalism," even restricted to its Quinean and post-Quinean incarnations, is a more complex position, with more subtle variants, than is sometimes supposed.[5]

1. Quinean roots

When Quine describes his naturalism as the "abandonment of the goal of a first philosophy" (1975, p. 72), he alludes to Descartes, who viewed his *Meditations on First Philosophy* as the only hope for "establish[ing] anything at all in the sciences that [is] stable and likely to last" (1641, p. 12). His approach, of course, was to doubt everything, including all of science and common sense, in order to uncover prescientific, first philosophical certainties that would then underpin our knowledge.[6] Few would suggest, at this late date, that Descartes succeeded in this, but Quine goes further, rejecting the project itself:

> I am of that large minority or small majority who repudiate the Cartesian dream of a foundation for scientific certainty firmer than scientific method itself. (Quine 1990, p. 19)

The simple idea is that no extrascientific method of justification could be more convincing than the methods of science, the best means we have.

The Quinean naturalist, then, "begins his reasoning within the inherited world theory as a going concern" (Quine 1975, p. 72). Alongside the familiar pursuits of physics, botany, biology, and astronomy, the naturalist asks how it is that human beings, as described by physiology, psychology, linguistics, and the rest, come to reliable knowledge of the world, as described by physics, chemistry,

[5] Obviously, I won't do justice to the details and subtleties of the three positions in this chapter. The interested reader is urged to consult the references for more careful and nuanced discussions.

[6] Broughton (2002) gives a fascinating and wonderfully readable account of how Descartes took his first philosophical method to work.

geology, and so on.[7] This is the task of epistemology naturalized, "the question how we human animals can have managed to arrive at science" (Quine 1975, p. 72). Ontology is also naturalized:

> Our ontology is determined once we have fixed upon the over-all conceptual scheme which is to accommodate science in the broadest sense... the considerations which determine a reasonable construction of any part of that conceptual scheme, for example, the biological or the physical part, are not different in kind from the considerations which determine a reasonable construction of the whole. (Quine 1948, pp. 16–17)
>
> Ontological questions... are on a par with questions of natural science. (Quine 1951, p. 45)

Insofar as traditional philosophical questions survive in the naturalistic context, they are undertaken "from the point of view of our own science, which is the only point of view I can offer" (Quine 1981c, p. 181).

One portion of this naturalistic undertaking will be a scientific study of science itself. Obviously, this intrascientific inquiry can deliver no higher degree of certainty than that of science. As Quine remarks, "Repudiation of the Cartesian dream is no minor deviation" (1990, p. 19). "Unlike the old epistemologists, we [naturalists] seek no firmer basis for science than science itself" (Quine 1995, p. 16). The naturalist

> sees natural science as an inquiry into reality, fallible and corrigible but not answerable to any supra-scientific tribunal, and not in need of any justification beyond observation and the hypothetico-deductive method. (Quine 1975, 72)

Science, on this picture, is open to neither criticism nor support from the outside.

But this leaves ample room for both vigorous criticism and rigorous support of particular scientific methods: "A normative domain within epistemology survives the conversion to naturalism, contrary to widespread belief..." (Quine 1995, p. 49).[8] What's changed is that the normative scrutiny comes not from an extrascientific perspective, but from within science: "Our speculations about the world remain subject to norms and caveats, but these issue from science itself as we acquire it" (Quine 1981c, p. 181). Here Quine returns to a favorite image:

> Neurath has likened science to a boat which, if we are to rebuild it, we must rebuild plank by plank while staying afloat in it. The philosopher and the scientist are in the same boat. (Quine 1960, p. 3)

[7] I depart here slightly from Quine (1969), where epistemology naturalized is said to take place inside psychology, but he gives more inclusive characterizations later (see II.1 below), so I take this as a friendly amendment.

[8] Cf.: "They are wrong in protesting that the normative element, so characteristic of epistemology, goes by the board" (Quine 1990, p. 19).

> He [the naturalist] tries to improve, clarify and understand the system [science] from within. He is the busy sailor adrift on Neurath's boat. (Quine 1975, p. 72)

This process is familiar: norms of confirmation and theory construction often arise in scientific practice, from simple canons of observation through elaborate guidelines for experimental design to highly developed maxims like mechanism.[9] As science progresses, these are put to the test, sometimes successfully and sometime not, and in this way their claim to a role in shaping future science is correspondingly strengthened or undermined. As Quine remarks, "We were once more chary of action at a distance than we have been since Sir Isaac Newton" (1981c, p. 181).

When Quine begins his naturalistic scientific study of science, he is struck by a simple but important observation of Duhem:

> The physicist can never subject an isolated hypothesis to experimental test, but only a whole group of hypotheses; when the experiment is in disagreement with his predictions, what he learns is that at least one of the hypotheses constituting this group is unacceptable and ought to be modified; but the experiment does not designate which one should be changed. (Duhem 1906, p. 187)

This phenomenon undermines the picture of a single scientific claim enjoying "empirical content" by itself, and leads Quine to holism and his famous "web of belief":

> Our statements about the external world face the tribunal of sense experience not individually but only as a corporate body. . . . The totality of our so-called knowledge or beliefs, from the most casual matters of geography and history to the profoundest laws of atomic physics . . . is a man-made fabric which impinges on experience only along the edges. . . . (Quine 1951, pp. 41–42)

Somewhat later, Quine tempers this holism to something more "moderate":

> It is an uninteresting legalism . . . to think of our scientific system of the world as involved *en bloc* in every prediction. More modest chunks suffice . . . (Quine 1975, p. 71)

But the moral—that particular scientific theories are tested and confirmed as wholes—remains intact.

Faced with a failed prediction, then, Quine notes that, strictly speaking:

> Any statement can be held true come what may, if we make drastic enough adjustments elsewhere in the system . . . . Conversely, by the same token, no statement is immune to revision. (Quine 1951, p. 43)

[9] I discuss the rise and fall of mechanism in (1997, pp. 111–116). This normative element is also present, for example, when a physiological, psychological theory of perception indicates why perceptual beliefs are largely reliable, and therefore reasonable, under certain conditions and largely unreliable, and therefore unreasonable, under others.

Practically speaking, we are guided by the "maxim of minimum mutilation" (Quine 1990, p. 14), "our natural tendency to disturb the total system as little as possible" (Quine 1951, p. 44), so we quite properly prefer to alter simple statements about observable physical objects—deciding that the swami only seems to levitate—rather than highly general laws (e.g., the law of gravity) if this is at all possible. In the image of the web, altering a statement closer to the experiential edges causes less widespread disturbance than revising a centrally located generality.

Granting that confirmation accrues holistically to scientific theories, on what sort of evidence is this confirmation based? On what grounds, for example, do we adopt atomic theory? Quine addresses this question as he continues his pursuit of a "scientific understanding of the scientific enterprise" (1955, p. 253).

> The benefits . . . credited to the molecular doctrine may be divided into five. One is simplicity. . . . Another is familiarity of principle. . . . A third is scope. . . . A fourth is fecundity. . . . The fifth goes without saying: such testable consequences of the theory as have been tested have turned out well, aside from such sparse exceptions as may in good conscience be chalked up to unexplained interferences. (Quine 1955, p. 247)

In another place ((1970), ch. V), Quine and Ullian give a slightly different list of theoretical virtues—conservatism, generality, simplicity, refutability, modesty, plus conformity with observation—and elsewhere (1990, p. 95), Quine lists economy and naturalness as examples, but the general flavor is the same throughout. Finally, as Quine notes, the various virtues can conflict; they must be balanced off against one another in particular cases.

Quine acknowledges that such a defense of atomic theory is indirect, and he considers the possibility that

> the benefits conferred by the molecular doctrine give the physicist good reason to prize it, but afford no evidence of its truth. . . . Might the molecular doctrine not be ever so useful in organizing and extending our knowledge of the behavior of observable things, and yet be factually false? (Quine 1955, p. 248)

Quine begins his response by pushing this skeptical line of thought even further, calling into question the tendency to "belittle molecules . . . leaving common-sense bodies supreme":

> What are given in sensation are variformed and varicolored visual patches, varitextured and varitemperatured tactual feels, and an assortment of tones, tastes, smells and other odds and ends; desks [and other common-sense bodies] are no more to be found among these data than molecules. (Quine 1955, p. 250)

This line of thought tempts us to conclude that

> In whatever sense the molecules in my desk are unreal and a figment of the imagination of the scientist, in that sense the desk itself is unreal and a figment of the imagination of the race. (Quine 1955, p. 250)

The upshot would be that only sense data are real, but this conclusion

> is a perverse one, for it ascribes full reality only to a domain of objects for which there is no autonomous system of discourse at all. . . . Not only is the conclusion bizarre; it vitiates the very considerations that lead to it. (Quine 1955, pp. 254, 251)

We can hardly see ourselves as positing objects to explain our pure sense data when those sense data can't even be described without reference to objects.

All this, Quine counts as a *reductio*: "Something went wrong with our standard of reality" (1955, p. 251). To correct the situation, he urges that we turn this tendency of thought on its head:

> We became doubtful of the reality of molecules because the physicist's statement that there are molecules took on the aspect of a mere technical convenience in smoothing the laws of physics. Next we noted that common-sense bodies are epistemically much on a par with the molecules, and inferred the unreality of common-sense objects themselves. (Quine 1955, p. 251)

But surely "the familiar objects around us" are real if anything is; "it smacks of a contradiction in terms to conclude otherwise." So,

> Having noted that man has no evidence for the existence of bodies beyond the fact that their assumption helps him organize experience, we should have done well, instead of disclaiming the evidence for the existence of bodies to conclude: such, then, at bottom, is what evidence is, both for ordinary bodies and for molecules. (Quine 1955, p. 251)

This, then, is Quine's conclusion: the enjoyment of the theoretical virtues is, at bottom, what supports all our knowledge of the world.

With this quick summary of Quine's views on science and the scientific study of science as backdrop, we can turn to his naturalist's position on logic and mathematics. He sees our knowledge of both as part of our web of belief, part of our best scientific theorizing about the world, confirmed with the rest by cooperative enjoyment of the theoretical virtues:

> A self-contained theory which we can check with experience includes, in point of fact, not only its various theoretical hypotheses of so-called natural science but also such portions of logic and mathematics as it makes use of. (Quine 1954, p. 121)

Confining our attention to logic for the moment, this means, for example, that the pursuit of the theoretical virtues might one day lead us to revise one or another of our current laws:

> Revision even of the logical law of the excluded middle has been proposed as a means of simplifying quantum mechanics; and what difference is there in principle between such a shift and the shift whereby Kepler superseded Ptolemy, or Einstein Newton, or Darwin Aristotle? (Quine 1951, p. 43)

The great weight of the maxim of minimum mutilation would stand against such a move, and Quine remarks, skeptically, that "the price is perhaps not quite prohibitive, but the returns had better be good" (1970, p. 86; 1986, p. 86).[10]

Readers familiar with this classical Quinean position on the revisability of logic are sometimes puzzled by later remarks to the effect that a deviant logician cannot disagree with the classical logician because his embrace of different logical laws shows that he actually means something different by the logical connectives. As Quine puts it: "Here, evidently, is the deviant logician's predicament: when he tries to deny the doctrine he only changes the subject" (1970, p. 81; 1986, p. 81). This is less jarring than it might seem, given that a change to new connectives can apparently be motivated by the same scientific reasons that were first imagined as motivating a change of logical laws:

> By the reasoning of a couple of pages back, [the deviant logician] changes the subject. This is not to say that he is wrong in doing so.... he may have his reasons. (Quine 1970, p.83; 1986, p. 83)

It is undoubtedly odd to hear Quine distinguishing change of meaning from change of theory,[11] but the central thesis of the revisability of logic on empirical grounds remains untouched.

However, a real departure on the revisability of logic can be found in a late discussion of the holism. Here Quine returns to the scientist facing a falsified prediction:

> We have before us some set *S* of purported truths that was found jointly to imply the false [prediction].... Now some one or more of the sentences in *S* are going to have to be rescinded. We exempt some members of *S* from this threat by determining that the fateful implication still holds without their help. Any purely logical truth is thus exempted.... (Quine 1990, p. 14)

This qualification of the revisability doctrine to rule out revision of logic is perhaps not unwelcome: one common objection to the original Quinean web has been that some laws of logic are needed for the simple manipulations of web maintenance[12] (the law of noncontradiction, for example, is what tells us we have

[10] In his (1981b), Quine discusses the costs of retaining the law of the excluded middle, but he reports, "My inclination is to adhere to it for the simplicity of theory it affords" (p. 32). Other proposed deviations from classical logic meet with even less enthusiasm.

[11] Cf. Quine (1951, pp. 36–37): "It is obvious that truth in general depends on both language and extra-linguistic fact.... Thus... it... seems reasonable that in some statements the factual component should be null; and these are the analytic statements. But, for all its a priori reasonableness, a boundary between analytic and synthetic statements simply has not been drawn. That there is such a distinction to be drawn at all is an unempirical dogma of empiricists, a metaphysical article of faith."

[12] For example, see Wright (1986) or Shapiro (2000).

to change something when we reach a falsified prediction[13]), so it is hard to see how these laws could be revised without crippling the scientific enterprise. But we are left with no replacement for the holistic justification of the assumption that our logic (with our meanings), as opposed to some deviant logic (with deviant meanings), is more suitable for our scientific theorizing about the world.

Finally, mathematics. First, we observe that our scientists typically make use of mathematics in their theorizing. This might be a mere manner of speaking—like saying "patience is a virtue" to mean that a patient person is to that extent virtuous—but strenuous efforts to reconstrue science in a mathematics-free idiom, from Quine and Goodman to Field, have all failed.[14] Second, as naturalized metaphysicians, we take science to be our best guide to what there is and how it operates. Third, as holists, we take a scientific theory to be confirmed as a whole, the mathematical along with the physical hypotheses.[15] To conclude, we apply Quine's criterion of ontological commitment (Quine 1948), which takes science to establish the existence of precisely those things that appear in its existential claims. Thus we arrive at Quine's mathematical realism by means of his "indispensability argument." Notice that the evidence for mathematical objects is the same as for molecules and for common-sense objects—participation in a theory with the theoretical virtues—and recall that "such . . . at bottom . . . is what evidence is" (Quine 1955, p. 251).

This famous argument only supports the existence of those mathematical entities that appear in our best scientific theory. But there is more to mathematics than this, as Quine recognizes:

> A word finally about the higher reaches of set theory itself and kindred domains where there is no thought or hope of applying in natural science. When

[13] Perhaps Quine has this case in mind when he writes: "On learning 'not' and 'and,' the child already internalizes a bit of logic; for to affirm a compound of the form 'p and not p' is just to have mislearned one or both particles" (1995, p. 23). But, of course, there are those who defend dialetheism (see Priest and Tanaka 2002).

[14] See Goodman and Quine (1947), Field (1980, 1989). This claim remains debatable, of course.

[15] In the late discussion of holism that "exempts" logic, mathematics seems to retain its original status: the maxim of minimum mutilation "constrains us . . . to safeguard any purely mathematical truth; for mathematics infiltrates all branches of our system of the world, and its disruption would reverberate intolerably. . . . Simplicity of the resulting theory is another guiding consideration, however, and if the scientist sees his way to a big gain in simplicity he is even prepared to rock the boat very considerably for the sake of it" (Quine 1990, p. 15). And in (1995), he stresses "the difference between logic, narrowly construed, and the rest of mathematics. . . . However, I am inclined to lighten somewhat the emphatic contrast usually drawn between mathematics and natural science. I already equated the roles of mathematical laws and laws of nature in implying [empirical predictions]" (pp. 52–53).

> I likened mathematical truths to empirical ones... I was disregarding these mathematical flights.... how should we view them? (Quine 1995, p. 56)

At one point, they're dismissed as meaningless:

> So much of mathematics as is wanted for use in empirical science is for me on a par with the rest of science. Transfinite ramifications are on the same footing insofar as they come of a simplificatory rounding out, but anything further is on a par rather with uninterpreted systems. (Quine 1984, p. 788)

Later on, Quine grudgingly relents:

> What of the higher reaches of set theory? We see them as meaningful because they are couched in the same grammar and vocabulary that generate the applied parts of mathematics. We are just sparing ourselves the unnatural gerrymandering of grammar that would be needed to exclude them. (Quine 1990, p. 94)

Having allowed them meaning, he also allows truth-value, but given their complete isolation from the data of experience, no evidence is available either way.

In the absence of holistic confirmation or disconfirmation from experience, Quine proposes we proceed simply by applying the remaining theoretical virtues. In particular, he suggests, simplicity or economy supports Gödel's axiom of constructibility, $V = L$, and opposes large cardinal axioms. This choice "inactivates the more gratuitous flights of higher set theory" (Quine 1990, p. 95), Quine remarks with approval. He insists this approach is "no threat to the starry-eyed set theorist for whom the sky is the limit," because the set theorist's statements are still meaningful and his theory of large cardinals "still makes proof-theoretic sense" (Quine 1995, p. 56). But this concession can't mask the fact that Quine's preference for $V = L$ contradicts the near-unanimous opinion of practicing set theorists.

II. Two Post-Quineans

Given this sketch of Quine's naturalism and its consequences, as he sees them, for logic and mathematics, let's turn to the views of our two post-Quinean naturalists, to illuminate both their departures from Quine himself and their disagreements with each other. To bring some order to this three-ringed circus, I'll break down this exercise in compare-and-contrast under a series of headings. Let's begin by considering the "science" in which "reality is to be identified and described."

II.1. Science

Quine plainly acknowledges that "I use 'science' broadly," including not only the "hard sciences" but also "softer sciences, from psychology and economics through sociology to history" (1995, p. 49). All these presumably display the markers of "observation and the hypothetico-deductive method" (1975, p. 72) and some attention to the theoretical virtues. And, conversely, any undertaking that shares these markers is likewise science, regardless of whether or not it falls squarely within some established branch. For Quine, the scientific study of science, a part of naturalized epistemology, is of this sort:

> The inquiry proceeds in disregard of disciplinary boundaries but with respect for the disciplines themselves and appetite for their input. (Quine 1995, p. 16)

We theorize about science, using the results and methods of science itself.

Burgess's "science" seems in some ways narrower than Quine's; for example, he speaks of the "scientific community ... whether understood narrowly, as including only specialist professionals, or broadly, as including also informed laypeople" (Burgess 1990, p. 5). Even the broad sense here seems limited to the established scientific disciplines, each a specialty unto itself, which may leave us wondering where the naturalist's scientific study of science is to find a home. And the semblance of a serious departure from Quine is encouraged by Burgess's insistence that the naturalist's project must be purely descriptive: "It seems that prescriptive methodology could not be a branch of science, though descriptive methodology is" (Burgess 1990, p. 6).[16]

In tone, at least, this tune diverges from Quine's. "The busy sailor adrift on Neurath's boat" is out to "improve, clarify and understand the system from within," not simply to describe the behaviors of the tribe of credentialed scientists. For Quine, the naturalistic study of science examines the methods of science with an eye to understanding how and why they are effective. Shoulder to shoulder with scientists, the naturalist strives to appreciate the reasons behind their design of experiments, their evaluation of evidence, and their preference for one theoretical elaboration over another; the ideal practicing scientist should be prepared to explain these things in terms of the general canons of scientific inquiry they both share. If the naturalist is a member of the scientific community, she should have the same grounds for scientific ratification and critique of scientific methods as are available to her fellow scientists. Science is a self-corrective process in which the naturalist participates.

At least on the score of normativity, I suspect that the semblance of disagreement here is largely illusory. Burgess (writing in collaboration with Gideon

[16] See also Burgess and Rosen (1997, pp. 208–209).

Rosen) clearly holds that the naturalist is a "citizen of the scientific community" (Burgess and Rosen (1997), p. 33).[17] And there is this important passage:

> science is not a closed guild with rigid criteria of membership. Philosophers professing naturalism often do contribute to debates in semantical theory or cognitive studies or other topics in the domain of linguistics or psychology, even though they are not officially affiliated with a university department in either of those fields. In principle nothing would bar such philosophers from participating in discussions on topics in the domain of chemistry or geology, though in practice they seldom do. The naturalists' commitment is at most to the comparatively modest proposition that when science speaks with a firm and unified voice, the philosopher is either obliged to accept its conclusions or *to offer what are recognizably scientific reasons for resisting them.* (Burgess and Rosen 1997, p. 65; added)

This leaves room for Quine's naturalistic justification and critique of scientific methods, for normative, prescriptive stands, since his "busy sailor" was never tempted to offer anything other than "recognizably scientific" grounds. Thus, it seems that Burgess's insistence that the naturalist's task is purely descriptive doesn't contradict Quine's insistence that normativity survives the move to naturalism because Quine is endorsing evaluations internal to science and Burgess is rejecting evaluations external to science.

But perhaps one more subtle difference remains. In Burgess's phrasing, the philosopher "should become a citizen of the scientific community... should become naturalized" (Burgess 1990, p. 5). In contrast, Quine's "busy sailor" would seem to be a native, not someone in need of conversion. Perhaps this is merely stylistic, but for my part, I much prefer the latter formulation. My naturalist "begins his reasoning within the inherited world theory as a going concern" (Quine 1975, p. 72). Such a naturalist, asked why she believes in, say, atoms, will react as an ordinary scientist, citing the usual scientific evidence (more on this below). Another sort of naturalist, the sort who started as a first philosopher and subsequently became a naturalized citizen of science, might be tempted to reply to the same question by citing the fact that her fellow scientists so believe and that she now believes as her fellow citizens do (i.e., "science says there are atoms and I, the naturalist, believe the utterances of science"). I have no reason to think that the Quinean or Burgessite naturalist would give the second answer, but my naturalist would certainly give the first.[18]

Where our three naturalists' views of "science" clearly diverge is on the status of mathematics. As we've seen, Quine includes mathematics as scientific, but only insofar as it takes part in empirical science, plus a bit more for "simplificatory roundings out" and a bit more again to avoid "unnatural gerrymandering of

[17] Burgess and Rosen are speaking of Quine here, but they voice no dissent of their own. See also Burgess (1990, p. 5).

[18] See Maddy (2001a) or (2002), I.

grammar." But evidence, for Quine, is always empirical evidence; proper methods are always those of natural science. Even in the grudgingly admitted higher reaches of set theory, it is the familiar theoretical virtues common to empirical theories that carry the day.

My naturalist sees things differently.[19] She begins, as Quine's does, within empirical science, and eventually turns, as Quine's does, to the scientific study of that science. She is struck by two phenomena: first, most of her best theories involve at least some mathematics, and many of her most prized and effective theories can only be stated in highly mathematical language; second, mathematics, as a practice, uses methods different from those she's turned up in her study of empirical science. She could, like the Quinean, ignore those distinctive methods and hold mathematics to the same standards as natural science, but this seems to her misguided. The methods responsible for the existence of the mathematics she now sees before her are distinctively mathematical methods; she feels her responsibility is to examine, understand, and evaluate those methods on their own terms; to investigate how the resulting mathematics does (and doesn't) work in its empirical applications; and to understand how and why it is that a body of statements generated in this way can (and can't) be applied as they are.

Burgess (with Rosen) goes further in his disagreement with Quine, sharply criticizing any naturalism according to which

> the mathematical sciences—so often considered by non-philosophers the very model of a progressive and brilliantly successful cognitive endeavor—must somehow be expelled from the circle of "sciences." (Burgess and Rosen 1997, p. 211)

To do this is to "mak[e] invidious distinctions . . . marginaliz[e] some sciences (the mathematical) and privileg[e] others (the empirical)" (Burgess and Rosen 1997, p. 211). Here it must be admitted that both Quine and I are guilty of so privileging natural science and of giving special attention to mathematics only because of its role in science.

But I do not follow Quine in the final move decried by Burgess (with Rosen), the one taken by those who "simply discard whatever of pure mathematics has not yet found application in the empirical sciences" (Burgess and Rosen 1997, p. 211). My naturalist, unlike Quine's, does not hold that those parts of mathematics that have been used in applications should be treated differently from the rest. She notes that branches of mathematics once thought to be far removed from applications have gone on to enjoy central roles in science[20] and, perhaps more important, that the methods that have led to the impressive practice she now observes, the practice so liberally applied in our current science, are the actual methods of mathematics, not the methods of natural science (as the Quinean naturalist would have it) nor

[19] See Maddy (1997, pp. 183–184).
[20] See Maddy (2001b).

some artificially gerrymandered subset of mathematical methods (as exclusive attention to the methods of applied mathematics, as distinct from pure mathematics, would require). She concludes that the entire practice of mathematics should be taken seriously, a practice including both applied and unapplied portions of contemporary pure mathematics, intricately intertwined.[21]

One more question arises for naturalists like Burgess and myself who venture beyond Quinean naturalism. Addressed to my version of naturalism, it takes this form: If the naturalist, engaged in her scientific study of science, discovers that one practice of human beings (namely, mathematics) is carried out using methods different from those of her natural science, why should she view this mathematical practice as different in kind from other practices with methods of their own, like astrology or theology? Addressed to Burgess, this becomes: Why should mathematics, but not astrology or theology, figure in the list of sciences?

My answer to this question, suggested above, has been that mathematics is used in science, so the naturalist's scientific study of science must include an account of how its methods work and how the theories so generated manage to contribute as they do to scientific knowledge. Astrology and theology are not used in science—indeed, in some versions they contradict science—so the naturalist needs only to approach them sociologically or psychologically.[22] Perhaps Burgess intends a similar answer to the analogous question for his view, for he writes (with Rosen):

> You cannot simply dismiss mathematics as if it were mythology on a par with the teachings of Mme Blavatsky or Dr. Velikovsky. A geologist interested in earthquake prediction or oil prospecting had better steer clear of Blavatsky's tales about the sinking of lost continents and Velikovsky's lore about the deposition of hydrocarbons by passing comets; but no philosopher will urge that the geologist should also renounce plate tectonics, on the grounds that it involves mythological entities like numbers and functions. (Burgess and Rosen 1997, p. 5)

Presumably, Burgess would also join my naturalist in holding that mathematics as a whole, not just its applied portion, is so-separated from Blavatsky tales and Velikowsky lore, perhaps for reasons not unlike those my naturalist gave a moment ago.[23]

[21] It seems to me that the two can't be separated without serious distortion. Tappenden (2001, p. 497) quite reasonably proposes that this claim be tested by a detailed study of the actual interactions between the various branches of mathematics.

[22] See Maddy (1997, pp. 203–205). Tappenden (2001, pp. 496–497) gives a fair outline of the debate on this score between my naturalist and her critics (in particular, Hale, Dieterle, Rosen, and Tennant; see Tappenden for references).

[23] Burgess rightly emphasizes that the "scientific" uses of mathematics include those of common sense and everyday belief, as in "stock indices, precipitation probabilities and batting averages" (personal communication, quoted with permission). See Burgess (1983, p. 94) on the slippery slope.

In sum, then, "science" for Quine and for me is natural science, while for Burgess it is a variety of natural and mathematical sciences,[24] but Quine and I differ on the status mathematics earns for itself in the course of our scientific study of natural science. I'll examine the differing results of this study for our three naturalists in more detail below, but first a brief look at logic.

II.2. Logic

The status of logic is not often the focus of Burgess's discussions, but passing comments (with Rosen) suggest that he would adopt something like the second Quinean position sketched above: "logical and analytic knowledge . . . is ultimately knowledge of language" (Burgess and Rosen 1997, p. 42). It isn't clear whether he would then follow the Quine who continues to count logic as empirically revisable or the Quine who exempts logic from holistic jeopardy (or takes some other stance entirely). As indicated above, I think both Quinean moves have their downsides.

My own rather speculative suggestion (see Maddy 2002) has been that humans are so constructed as to conceptualize the world in terms of some simple fundamental categories (e.g., as comprised of individual objects standing in various relations); that the world, to a large extent, is properly described as so structured (up to the point of quantum mechanics, at least); and that a rudimentary logic is implicit in these shared structures (e.g., fair versions of "or," "and," cruder counterparts to "not" and "if/then," and simple quantifications). This much logic, then, is obvious to us, as part of our most basic conceptual machinery, true of the world (for the most part), and, furthermore, can be known by us when we verify that the simple structures required for its support are present in the situations to which logic is applied. Beyond this rudimentary basis, we add idealizations of various sorts—bivalence, the truth-functional conditional, assumptions about our domains of quantification—whose wisdom must be judged as the wisdom of any scientific idealization is judged: by their appropriateness in particular contexts. Most cases for deviant logics can be seen largely as arguments that the relevant idealization is not advisable for one reason or another.

This brief sketch may well be too quick to be decipherable; I commend the interested reader to the longer sketch (Maddy 2002), and (I hope) to more detailed future elaborations. For present purposes, one point is worth noting: simple arithmetical claims like "$2+2=4$" can be expected to correspond to logical truths of the most rudimentary sort, and thus to robust truths about the world.

[24] Burgess writes, "I believe in the community, but not the unity, of science" (personal communication, quoted with permission).

II.3. The Scientific Study of Science

We've seen that Burgess disagrees with Quine on the scope of "science" and that I disagree on what the scientific study of science tells us about mathematics, but both Burgess and I also depart from Quine on fundamental aspects of the scientific analysis of natural scientific method. This in turn impinges on how we understand the notion of "best current scientific theory," the central notion of our naturalized metaphysics and epistemology.

Recall that Quine's analysis of the method of natural science comes down to holism and the theoretical virtues. Burgess's critique focuses on the virtue of economy, in particular the preference for theories that posit fewer things. According to Quine, this virtue implies that if we could do science without mathematical objects (and without seriously compromising other theoretical virtues), we should do so, for this would rid our theory of a vast ontology of abstracta, yielding a more economical, and thus better, scientific theory. Thus it is only the failure of his attempt to reconstruct science without abstracta that leads Quine to his indispensability argument for the existence of mathematical entities.

In contrast, Burgess doubts that scientific standards actually include a preference for theories with smaller ontologies of abstracta.[25] Most often, such economy is "a matter to which most working scientists attach no importance whatsoever" (Burgess 1983, p. 98). In most cases, "proposed changes in the mathematical apparatus of physics that have received ultimate acceptance have increased its power and freedom" (Burgess 1990, p. 11). On occasions when scientists have hesitated over adding new mathematical ontology (e.g., the introduction of analytic methods into geometry or of infinitesimals in the calculus), he argues that "decrease in rigor and/or a danger of inconsistency," not the new ontology, was the cause of concern, and points out that "rigor and consistency are already usually conceded . . . to be weighty scientific standards" (1990, pp. 11–12). Thus Burgess (with Rosen) proposes a modification of the list of theoretical virtues to reflect, among other things, a more limited version of economy (Burgess and Rosen 1997, p. 209).

On the main point here I completely agree: scientists feel free to adopt any mathematical apparatus that is convenient and effective, without concern for its abstract ontology.[26] But my own discomfort with the Quinean picture goes beyond the detail of the theoretical virtues, to the holistic model of confirmation itself. Based on a look at the historical case of atomic theory, I suggest that the theory enjoyed the Quinean virtues in abundance by 1860, when its successes in chemistry were crowned by the computation of stable atomic numbers, and even more so by 1900, after the rise of kinetic theory in physics. But scientists were

[25] See Burgess (1983, 1990), and Burgess and Rosen (1997, pp. 214–219).

[26] See Maddy (1997, pp. 154–157).

not satisfied until the direct detection of atoms by Perrin, verifying the crucial predictions of Einstein's 1905 calculations. This, I claim, means that the theoretic virtues are not enough; that enjoyment thereof is not what evidence is; that our best scientific theory is not confirmed as a whole; that some of its posits are properly regarded as fictional until further, more specific testing is possible.[27]

This anti-holistic point of view makes room for the wider range of things that scientists want to say—for example (in 1900), that atomic theory is one of our most highly accepted theories, but we don't yet know whether or not atoms exist, or (right now) that general relativity is one of our most highly accepted theories, but we don't know for sure if space-time is a continuous manifold.[28] The actual attitudes of practitioners toward their best theories are complex and nuanced, as when scientists worry over whether or not some aspect of their best theory is or isn't an artifact of their mathematical modeling, or when Einstein admits to using the continuum to formulate general relativity because "I have been unable to think of anything organic to take its place" ((1949), p. 686). Burgess (with Rosen) expresses some sympathy for this line of thought: it "might be that science itself makes invidious distinctions"; to show this would require "presenting studies of the distinctions and divisions observed within the community of working scientists" (Burgess and Rosen 1997, p. 213). This is what I've tried to do.

Finally, we post-Quineans owe one more item in our scientific study of science, namely, an account of mathematical methods. Burgess's general observations on this line have already been noted: rigor and consistency are important mathematical standards; economy of ontology is not. He also clearly regards the methods of mathematics as distinct from those of the natural sciences:

> Among sciences, mathematics is, owing to its distinctive methodology of deductive proof, a special case (Burgess 1992a, p. 437)
>
> . . . Rigorous proof is clearly distinguishable from systematic observation or controlled experiment. (Burgess 1992b, p. 10)

Beyond this, however, he sets aside the problem of axiom selection and, it seems, the related dynamics of concept formation, which have been my focus. Let me take a quick look at how this goes.

One moral of metaphysics naturalized is that natural science itself tells us a considerable amount about ontology. For example, medium-sized physical objects

[27] See Maddy (1997, pp. 135–143). Oddly enough, Burgess's naturalist might disagree with mine about the current status of atoms; he writes (with Rosen): "The naturalized epistemologist may largely accept the sceptic's description according to which our method of positing a physical system with parts we do not perceive is just the only effective way for us, with such cognitive capacities as we have, to cope with what we do perceive" (Burgess and Rosen 1997, p. 33; see also p. 212).

[28] See Maddy (1997, pp. 143–151).

exist in space and time, interact with one another causally, are as they are largely independently of our thought and knowledge of them; unobservable atoms also so exist and compose these more familiar objects; and so on. As my naturalist begins her study of the methods, justificatory procedures, and conclusions of pure mathematics, she might well ask an analogous question: What does mathematics tells us about its ontology? Some answers come quickly—there are numbers, sets, spaces, and so on—but very little is forthcoming about the nature of this existence. Further claims—that mathematical things exist in a non-spatiotemporal, acausal world; that mathematical things are mental constructions of an idealized mathematician; that mathematical things exist only as fictions—can be found in the literature, but a look at their role, especially in resolution of various historical debates, suggests that they are not integral parts of mathematical method, but extramathematical philosophizing.[29] If this is right, the upshot is that mathematics, in contrast to natural science, tells us nothing about the metaphysical nature of its objects beyond the bare claim that they exist.

In her analysis of mathematical methods, then, the naturalist should ignore such extramathematical metaphysical debates and attend to the explicit or implicit intramathematical reasons being offered for one course of action or another. The hope is that an understanding of the goals of a particular mathematical undertaking can be reached, and that alternative methods can then be evaluated in terms of their effectiveness as means toward those goals.[30] In (2001b), I sketch such analyses of the development of the concepts "group" and "topological space." In (1997, pp. 206–232), I argue on such grounds that the set theorist's rejection of $V = L$ is rational, given the goals of set-theoretic practice. This highlights the contrast between Quine's naturalist and mine: the Quinean endorses $V = L$, applying the methods of natural science, while my naturalist rejects it, applying the methods of the relevant branch of mathematics, that is, the methods of set theory.

Finally, let me emphasize that a purely internal, methodological study of mathematics is not the only investigation of the subject that my naturalist can undertake. Mathematics is a form of human activity, a distinctive linguistic practice, and as such it can be studied like any other such practice: by linguistics, psychology, and so on, as well as various natural scientific studies spanning the standard disciplines. Here the naturalist will face questions about the similarities and dissimilarities between mathematical and natural scientific language, and questions about how this practice manages to function so effectively in natural science. These inquires will raise naturalized versions of traditional philosophical questions about the ontology and epistemology of mathematics. This naturalized

[29] This is a difficult distinction to draw. See Maddy (1997, pp. 185–193).

[30] See Maddy (1997, pp. 193–200). Notice that this is the sort of intramathematical justification and critique that ought to be acceptable to Burgess.

philosophy of mathematics[31] is distinct from the naturalized methodology of mathematics discussed above but, given that we want a philosophical analysis of mathematics as it is, our naturalized philosopher must respect the practice; justification and criticism are internal to the practice, the province of internal methodology, not philosophy, even naturalized philosophy.

II.4. The Indispensability Argument

We can now assess the effects of these post-Quinean departures on the crucial indispensability argument. On Burgess's analysis, economy of abstract ontology is not among the theoretical virtues, so our "best scientific theory" will include mathematical entities regardless of the success or failure of strenuous nominalistic efforts to remove them.[32] There is still room here for a naturalistic argument to the existence of mathematical entities via holism and the criterion of ontological commitment,[33] but this is not the line Burgess takes. There are, no doubt, various more subtle reasons for this, but one straightforward motivation is clear:

> A thorough-going naturalist would take the fact that abstracta are customary and convenient for the mathematical (as well as other) science to be sufficient to warrant acquiescing in their existence. (Burgess and Rosen 1997, p. 212)

For Burgess (writing with Rosen), mathematics is a science in its own right, and fully capable of justifying its own existence claims.[34] Thus, Burgess embraces an ontology of mathematical entities, but on grounds quite different from those of the Quinean naturalist.

Since I share Quine's starting point in natural science, my own engagement with his indispensability argument focuses on our disagreement over holism. If cases like atomic theory show that proper scientific method does not regard the

[31] Critics of Maddy (1997) have sometimes complained that the position leaves no room for philosophy (as opposed to methodology) of mathematics, but naturalized philosophy of mathematics, as understood here, is described and endorsed on pp. 200–205.

[32] Burgess (1998) calls this "unconditional anti-nominalism." Of course, if the nominalists were to produce math-free scientific theories that improved on our current ones on other theoretical virtues, these would be preferred, but Burgess argues in some detail that they have not done this.

[33] Burgess (with Rosen) expresses some doubts about Quine's criterion (Burgess and Rosen 1997, pp. 225–232), but in the end, he seems happy to conclude that current science is committed to mathematical entities.

[34] He writes, "Numbers, even if real, aren't physical, and so the physicist can pass the buck to the mathematician . . . mathematical existence questions are questions for mathematicians and not for . . . empirical scientists" (personal communication, quoted with permission).

existence of all posits of our best scientific theory as confirmed, then we have to look more carefully at the status of its mathematical posits in particular. Obviously much mathematics occurs in explicitly idealized situations, when physical situations are mathematized in terms of simple geometrical structures, when large finite collections are treated as infinite, when discrete situations are treated as continuous, and so on; surely no simple ontological morals should be drawn from these appearances of mathematics.[35] In more fundamental theories, the most convenient and effective mathematics is used, seemingly without qualm, as Burgess suggests. But it is also true that the appearance of, say, a continuous manifold in our best description of space-time does not seem to be regarded as establishing the continuity of space-time; the microstructure of space-time remains an open question.[36]

These observations suggest, first, that holism is incorrect, that our best scientific theory is not simultaneously confirmed in all its parts, that at least in cases like atomic theory, some variety of "direct detection" is required. Recall Quine's claim that the theoretical virtues served to establish the existence of atoms *and* of medium-sized physical objects—that this is, in the end, what evidence is; applying this conclusion to mathematical objects yields the indispensability argument. But if the evidence for atoms is something more direct—and surely medium-sized physical objects are "directly detectable" if anything is—it follows that the "evidence" for mathematical objects is not the same as for the others. And, second, when we look more closely at the considerations that actually move scientists to include the mathematical posits they do, we find the likes of convenience and effectiveness. Perhaps unsurprisingly, when structures are posited on such a basis, the success of the overall theory in which they appear is not regarded as confirming their existence. On this analysis, the indispensability argument fails because ordinary scientific standards do not confirm all parts of our best theory, and because the mathematical posits are not among the posits that are confirmed.

There are passages where Burgess might be taken to agree with this conclusion:[37]

> An ontology of abstracta may be one feature of [our] current theories that is merely *conventional.* (Burgess 1983, p. 99)
>
> Our science [is] the way it is in part because the universe is the way it is and in part because we are the way we are . . . the presence of "Avogadro's number" in the language of science is not *caused* by the presence of Avogadro's number

[35] See Maddy (1997, pp. 143–146).

[36] We should distinguish between the purely mathematical existence assumptions involved in this application of mathematics—the existence of a continuous manifold—and the physical structural assumptions that accompany it—continuous space-time. Both seem to be added for convenience and effectiveness (we know of no other way to represent space-time), and neither seems to be regarded as confirmed. For more, see Maddy (1997, pp. 154–157).

[37] Though, in contrast, the passage quoted in note 27 seems more in sympathy with a holistic view like that of Quine (1955).

> in the universe. The relation between name and object is thus in one crucial respect unlike that between mirror image and object in the scene reflected. (Burgess 1990, p. 13)

And with Rosen:

> The naturalized epistemologist may largely accept the nominalist's description according to which our method of positing mathematical systems as models is just the most efficient way for us, with such cognitive capacities as we have, to cope with physical systems. (Burgess and Rosen 1997, p. 33; see also pp. 238–244)

If we can see, in the course of our scientific study of science, that certain parts of our theory are there by convention, that they don't reflect what's actually present in the physical situation, that we posit them merely because we have no better way of describing things, then it seems reasonable to conclude that these parts of our theory are not, in fact, confirmed by our scientific methods. But we've seen that, for Burgess, this is beside the point; the ground of mathematical ontology lies in mathematics, not natural science.

So, in short, Burgess sees the indispensability argument as flawed—it concedes too much to the nominalist in requiring that mathematics be indispensable—and beside the point—because mathematical science by itself gives us grounds for accepting the existence of mathematical entities. My own position is both more Quinean, in granting that natural science is the final arbiter of existence, and more firmly non-Quinean, in its explicit rejection of holism. The upshot, for my naturalist, is that the role of mathematics in natural science does not seem to support the claim that mathematical things exist.

II.5. Post-Quinean Philosophy of Mathematics

The outlines of Burgess's naturalism should now be clear, but there remains an open question for my version: Does pure mathematics provide support for the existence of mathematical objects? It surely implies that there are numbers, sets, functions, geometric and topological spaces, real numbers, continuous manifolds, Hilbert spaces, algebraic structures, complex numbers, and much more. Burgess would count this as the end of the story: the existence of these items is implied by one among our best sciences. But for my naturalist, natural science is the final arbiter of what there is, and it doesn't seem to support its mathematical ontology, so my story will have to be more complex.

Granting that mathematics itself offers no ontological guidance beyond the minimal "mathematical things exist," naturalized methodology of mathematics is of no further use; we can only turn to naturalized philosophy of mathematics. One chapter of this study is the aforementioned conclusion that the confirmation of natural scientific theories in which mathematics figures does not confirm its ontology, that the empirical confirmation does not transfer holistically to the mathematical existence claims. But perhaps support will come from other quarters of this study—for example, from the investigation of the semantics of mathematical language,[38] or from the analysis of human mathematical experience, or of how pure mathematics comes to be so effective in applications.[39]

My guess is that, in the end, the explanations and accounts of naturalized philosophy of mathematics will not involve a literal appeal to the objects of pure mathematics.[40] But even if this is so, it would not settle the case between, say, "set-theoretic claims are true, and sets just are the kind of things that are referred to and known about by set-theoretic methods" and "set-theoretic claims aren't true and sets don't exist, but it's perfectly rational and proper to say that they are and do while developing set theory in the pursuit of various mathematical goals." In fact, I suspect that a decision on these matters will have more to do with the theory of truth than with the methodological or naturalized philosophical facts about mathematics or natural science.[41]

But I leave this for another time, along with a large number of other worthy naturalistic undertakings: for example, a more complete account of set-theoretic methods, of the distinctive methods of the various other branches of mathematics, and of their interrelations;[42] a full discussion of how mathematics works in natural science, from the sorts of detailed analyses given by applied mathematicians (e.g., an explanation of why it's proper to regard fluids as continuous substances when we do fluid dynamics), to consideration of general questions

[38] As we've seen, Burgess admits that there are significant differences between mathematical and natural scientific terms: "the presence of 'Avogadro's number' in the language of science is not *caused* by the presence of Avogadro's number in the universe" (Burgess 1990, p. 13). I wouldn't have chosen this example—given the suggestion that claims like $2+2=4$ correspond to rudimentary and robust logical truths about the world—but I second the point about most of mathematics. Whether or not this is a fact about reference will depend on one's theory of reference (see below).

[39] Steiner (1998) draws strong "anti-naturalistic" conclusions from his analysis of the way mathematics works in application. He takes "naturalism" to be the view that "the human race is [not] in some way privileged, central to the scheme of things . . . that the universe is indifferent to the goals and values of humanity" (p. 55). This does seem to be the view of natural science, and hence of all three naturalisms considered here.

[40] That is, any appeal outside the context of some idealized or conventional mathematical modeling.

[41] Burgess (with Rosen) follows Quine in holding that the naturalist must embrace a disquotational theory of truth (Burgess and Rosen 1997, p. 33). I disagree, taking the decision between disquotational and more robust theories to be an open scientific question (see Maddy 2001a, III), but I don't think this disagreement is what's at issue here.

[42] As urged by Tappenden (2001, p. 497).

about the applicability of mathematics (the "miracle" of applied mathematics); a broad and careful study of how science can assess the ontological morals of its own best theories, beginning with a study of how we tell when an aspect of our best scientific theory is an artifact of the mathematical modeling; and so on. My more modest aim here has been to illuminate the outlines of three versions of naturalism, and perhaps to demonstrate that philosophizing naturalistically involves attention to matters of considerable subtlety and detail. My hope is that the reader may be inspired to further investigation of one or another of these, or some improved descendant thereof!

REFERENCES

Broughton, Janet (2002) *Descartes's Method of Doubt* (Princeton, NJ: Princeton University Press).
Burgess, John (1983) "Why I Am Not a Nominalist," *Notre Dame Journal of Formal Logic* 24, pp. 93–105.
Burgess, John (1990) "Epistemology and Nominalism," in A. D. Irvine, ed., *Physicalism in Mathematics* (Dordrecht: Kluwer Academic Publishers), pp. 1–15.
Burgess, John (1992a) "How Foundational Work in Mathematics Can Be Relevant to Philosophy of Science," *PSA 1992*, vol. II, (East Lansing, MI: Philosophy of Science Association), pp. 433–441.
Burgess, John (1992b) "Proofs About Proofs: A Defense of Classical Logic," in M. Detlefsen, ed., *Proof, Logic and Formalization* (London: Routledge), pp. 8–23.
Burgess, John (1998) "Occam's Razor and Scientific Method," in M. Schirn, ed., *Philosophy of Mathematics Today* (Oxford: Oxford University Press), pp. 195–214.
Burgess, John, and Rosen, Gideon (1997) *A Subject with No Object* (Oxford: Oxford University Press).
Descartes, René (1641) *Meditations on First Philosophy*, in *The Philosophical Writings of Descartes*, vol. II, J. Cottingham, R. Stoothoff, and D. Murdoch, trans. (Cambridge: Cambridge University Press, 1984), pp. 3–62.
Duhem, Pierre (1906) *The Aim and Structure of Physical Theory* (Princeton, NJ: Princeton University Press, 1954).
Einstein, Albert (1949) "Reply to Criticisms," in P. A. Schilpp, ed., *Albert Einstein: Philosopher-Scientist*, vol. 2 (La Salle, IL: Open Court Press), pp. 663–688.
Field, Hartry (1980) *Science Without Numbers* (Princeton, NJ: Princeton University Press).
Field, Hartry (1989) *Realism, Mathematics and Modality* (Oxford: Basil Blackwell).
Goodman, Nelson, and Quine, W.V.O. (1947) "Steps Towards a Constructive Nominalism," *Journal of Symbolic Logic* 12, pp. 105–122.
Maddy, Penelope (1997) *Naturalism in Mathematics* (Oxford: Clarendon Press).
Maddy, Penelope (2001a) "Naturalism: Friends and Foes," in J. Tomberlin, ed., *Metaphysics 2001, Philosophical Perspectives* 15 (Malden, MA: Blackwell), pp. 37–67.

Maddy, Penelope (2001b) "Some Naturalistic Reflections on Set Theoretic Method," *Topoi* 20, pp. 17–27.
Maddy, Penelope (2002) "A Naturalistic look at logic," in *Proceedings and Addresses of the APA* 76 (November), pp. 61–90.
Mounce, H.O. (1999) *Hume's Naturalism* (London: Routledge).
Priest, G., and Tanaka, Toji (2002) "Paraconsistent Logic," in E. Zlata, ed., *Stanford Encyclopedia of Philosophy*, Spring 2002 ed. http://plato.stanford.edu/archives/spr2002/entries/logic-paraconsistent/.
Quine, W.V.O. (1948) "On What There Is," reprinted in his (1980), pp. 1–19.
Quine, W.V.O. (1951) "Two Dogmas of Empiricism," reprinted in his (1980), pp. 20–46.
Quine, W.V.O. (1954) "Carnap and Logical Truth," reprinted in his (1976), pp. 107–132.
Quine, W.V.O. (1955) "Posits and Reality," reprinted in his (1976), pp. 246–254.
Quine, W.V.O. (1960) *Word and Object* (Cambridge, MA: MIT Press).
Quine, W.V.O. (1969) "Epistemology naturalized," in his *Ontological Relativity and Other Essays*, (New York: Columbia University Press), pp. 69–90.
Quine, W.V.O. (1970) *Philosophy of Logic* (Englewood Cliffs, NJ: Prentice-Hall).
Quine, W.V.O. (1975) "Five Milestones of Empiricism," reprinted in his (1981c), pp. 67–72.
Quine, W.V.O. (1976) *The Ways of Paradox*, rev. ed. (Cambridge, MA: Harvard University Press).
Quine, W.V.O. (1980) *From a Logical Point of View*, 2nd ed. (Cambridge, MA: Harvard University Press).
Quine, W.V.O. (1981a) "Things and Their Place in Theories," in his (1981c), pp. 1–23.
Quine, W.V.O. (1981b) "What Price Bivalence?," in his (1981c), pp. 31–37.
Quine, W.V.O. (1981c) *Theories and Things* (Cambridge, MA: Harvard University Press).
Quine, W.V.O. (1984) "Review of Charles Parsons' *Mathematics in Philosophy*," *Journal of Philosophy* 81, pp. 783–794.
Quine, W.V.O. (1986) *Philosophy of Logic*, 2nd ed. (Cambridge, MA: Harvard University Press).
Quine, W.V.O. (1990) *Pursuit of Truth* (Cambridge, MA: Harvard University Press).
Quine, W.V.O. (1995) *From Stimulus to Science* (Cambridge, MA: Harvard University Press).
Quine, W.V.O., and Ullian, J.S. (1970) *The Web of Belief* (New York: Random House).
Shapiro, Stewart (2000) "The Status of Logic," in P. Boghossian and C. Peacocke, eds., *New Essays on the A Priori* (Oxford: Oxford University Press), pp. 333–366.
Sluga, Hans (1980) *Gottlob Frege* (London: Routledge).
Steiner, Mark (1998) *The Applicability of Mathematics as a Philosophical Problem* (Cambridge, MA: Harvard University Press).
Stroud, Barry (1977) *Hume* (London: Routledge).
Tappenden, James (2001) "Recent Work in Philosophy of Mathematics," *Journal of Philosophy* 98, pp. 488–497.
Wright, Crispin (1986) "Inventing Logical Necessity," in J. Butterfield, ed., *Language, Mind and Logic* (Cambridge: Cambridge University Press), pp. 187–209.

CHAPTER 14

NATURALISM RECONSIDERED

ALAN WEIR

We're all naturalists now, aren't we? Well, not quite. It may be said, moreover, that the term "naturalism" means so many different things to different philosophers that calling someone a naturalist has very little substantive content; but this would also be an exaggeration. In what follows, I will not spend too much time on terminological matters; in my view, the phrase has a fairly definite content, in the context of discussions in the philosophy of mathematics and science such as this one, and picks out a program which attracts a great deal of support.

I will, though, make one terminological distinction: between *methodological naturalism* and *ontological naturalism*. The methodological naturalist assumes there is a fairly definite set of rules, maxims, or prescriptions at work in the "natural" sciences, such as physics, chemistry, and molecular biology, this constituting "scientific method." There is no algorithm which tells one in all cases how to apply this method; nonetheless, there is a body of workers—the scientific community—who generally agree on whether the method is applied correctly or not. Whatever the method is, exactly—such virtues as simplicity, elegance, familiarity, scope, and fecundity appear in many accounts—it centrally involves an appeal to observation and experiment. Correct applications of the method, the methodological naturalist goes on, have enormously increased our knowledge, understanding, and control of the world around us to an extent which would scarcely be imaginable to generations living prior to the age of modern science. The methodological naturalist therefore

My thanks to Penelope Maddy and Stewart Shapiro for very helpful comments on earlier drafts.

prescribes that one ought to follow scientific method, at a level of sophistication appropriate to the problem at hand, whenever attempting to find out the truth about anything and whenever attempting to understand any phenomenon.

The ontological naturalist holds first of all that the ontology of the natural sciences consists of physical objects and perhaps also physical properties and relations. Explaining just exactly what these are is a somewhat delicate matter, even in the case of physical objects. The most obvious account in the latter case is a Cartesian one: physical objects are extended in space and time; they occupy a region of space and an interval in time or, in post-Einsteinian terms, occupy a region of space-time. Even that is not entirely unproblematic, given interpretations of quantum mechanics in which, it is claimed, fundamental particles lack definite position. Explicating what it is for a property to be physical is more tricky still.[1] At any rate, let us concede to the ontological naturalist that we have a rough idea what is meant by a physical ontology. The ontological naturalist's position is then straightforward: everything which exists is physical, either a physical object or, if such entities are countenanced, a physical property or relation.

Further light on the nature of naturalism can be shed by contrasting it with paradigmatically anti-naturalist positions, in methodology or in ontology. In the former case, the paradigm of an anti-naturalist is someone who holds a view on the basis of an appeal to an authority whose credentials are not in turn based on experiment or observation: the authority of religious leaders and sacred texts is the standard example, since this does not typically seem to rest on observation or experiment in the usual sense. A related anti-naturalist method is appeal to internal "intuition" or some mystical form of revelation or apprehension, distinct from ordinary sense perception, in order to justify a belief or attitude whose object is supposed to be an external entity or phenomenon. A classic anti-naturalist ontological position is that of Cartesian dualism, with its commitment to disembodied souls existing in time but not extended in space, though mysteriously interacting with physical bodies and in particular human brains.

As noted at the outset, not all contemporary philosophers are naturalists in one or other of these senses; but it would be fair, I think, to say that a naturalistic attitude predominates among "analytic" philosophers. This is why philosophy of mathematics, quite apart from its intrinsic interest, is of such importance in contemporary philosophy. For, on the face of it, mathematics is an enormous Trojan Horse sitting firmly in the center of the citadel of naturalism. Modern natural science is mathematical through and through: it is impossible to do physics, chemistry, molecular biology, and so forth without a very thorough and quite extensive knowledge of modern mathematics (indeed, this is true to an increasing extent of social sciences such as psychology and economics). Yet, prima facie, mathematics provides a counterexample both to methodological and to ontological naturalism.

[1] Cf. Weir (2003) for some suggestions.

The central mathematical method of proof from axioms shares one feature with empirical science, namely, the key role of deduction of conclusions from premises; indeed, this role is far more central to mathematics than to science. But there is a crucial difference: where the ultimate premises in science are hypotheses and conjectures up for testing against experiment, in mathematics they are "axioms," traditionally held to be known a priori, in some accounts by virtue of a form of intuitive awareness. The epistemic role of the axioms in mathematics, then, seems uncomfortably close to that played by the insights of a mystic.

When we turn to ontology, matters are, if anything, worse: mathematical entities, as traditionally construed, do not even exist in time, never mind space. Difficult though it may be to see how Cartesian souls can interact with physical bodies, since they persist and change through time, such interaction does at least make some sort of sense, certainly on broadly Humean notions of causality. If, by contrast, mathematical entities are abstract and causally inert, how on earth do we have any knowledge of them?[2] This classic epistemological problem provides both a ground for ontological naturalism, for denying the existence of abstract entities, and a severe internal problem, since naturalism seems to presuppose mathematics, which seems in turn to presuppose abstract objects.

In this chapter I intend to look critically at some responses which naturalists have made to these difficulties. I will not examine each with the same detail, of course. Since many of the positions which are discussed elsewhere in this volume are motivated, in part, by a desire to deal with the questions raised above, this would be superfluous, not to mention impossible. Rather, I will attempt to give some sort of overview or map of the strategies of response, looking in more detail only at a select few.

I

First, then, I will look at three responses the methodological naturalist might make. I start with logicism. It is undeniable that logical inference is part of scientific method, whatever else is. But mathematics raises the additional problem of our knowledge of the axioms: the naturalist does not want to base this on some mysterious intuition or direct apprehension that the axioms characterize the appropriate mathematical domain. If, however, logicism is right and mathematics is just a branch of logic (plus definitions), then will these problems not simply disappear?

This might seem a very unpromising route for the naturalist. Was not logicism conclusively refuted both by the paradoxes flowing from Frege's Axiom V and by

[2] A highly influential exposition of this problem is found in Benacerraf (1973).

Gödel's incompleteness results? Moreover, there are clear differences between logic and mathematics; the most a sensible logicist could maintain is that mathematics has the same *status* as logic, as a priori, or analytic, or some such. But then the naturalistic logicist faces criticism from Quineans—supporters, in these matters, of the views of W. V. Quine—for they contend that such notions have been discredited.

The first two problems are not insuperable, according to the "neologicists" or "neo-Fregeans" such as Crispin Wright and Bob Hale, who seek to found mathematics not on the inconsistent Axiom V but on formally similar but consistent abstraction principles such as "Hume's Principle" for arithmetic and, for set theory, weakened forms of Axiom V such as George Boolos's "New V." The recent trend among neologicists has been to treat such abstraction principles as implicit definitions of the relevant concepts—"number," "set," and so forth. Even if not all mathematical truths are derivable (given an effectively decidable notion of theoremhood), the logicists' goals will be achieved if the main branches of contemporary mathematics, certainly of those areas utilized in modern science, can be founded on abstraction principles (i.e., if axiomatizations of each branch can be derived).[3]

As to the Quinean problem, it is worth remarking that the later Quine seems to step back somewhat from the very extreme empiricism on logic which appears to be entertained in Quine (1951), "Two Dogmas of Empiricism." Examples of his retreat start with §13 of *Word and Object* (Quine 1960), on translating the connectives. This retreat, in my view, is essential for the coherence of the Quinean position. The famous quotation "Any statement can be held true come what may, if we make drastic enough adjustments elsewhere in the system. . . . Conversely, by the same token, no statement is immune to revision" (Quine 1951, p. 43) is either trivially true or incoherent, at least if we extend the position from one concerning the holding of statements to one concerning the adoption of inference rules.[4] Certainly anyone who believes in determinacy of meaning for logical particles will

[3] The launch pad for contemporary neologicism is Wright (1983). See also Hale (1987). The most important articles on neologicism by Wright and Hale, jointly and severally, are to be found in Hale and Wright (2001). For a different, constructivistic strand of neologicism, see Tennant (1987, 1997). Much work in neologicism is in response to the detailed criticisms of (and technical elucidations of) the position to be found in the work of George Boolos, most of it collected in his (1998, §II). See chapter 6 of this volume.

[4] Puzzlingly, for one well acquainted with natural deduction systems, Quine seems wedded to the misconception of logic as a body of theorems rather than as the study of inference and the relation of consequence. Even by the stage of *The Roots of Reference* (1974), his qualification of radical empiricism in logic focuses on laws such as $P \rightarrow (P \vee Q)$ rather than inference rules such as $\vee$I (Quine 1974, p. 78). The abstraction principles of the neologicist can also be seen as principles of inference rather than as axioms, if we treat them as pairs of schematic inference rules in which, in the one case, all instances of the left side of the embedded biconditional are derivable from the other side, and vice versa for the other form of the rule.

find it trivial: any rule we accept could be abandoned if the particles involved in it take new meanings (whether of old particles or not); likewise, mutatis mutandis, for rules we reject. The nontrivial version of revisability says that any rule is revisable in a way which does not require us to impute meaning change (of course someone who rejects determinacy of meaning for logical particles will be committed to the nontrivial version, since we will never be required to impute meaning change, there being no meanings to change).

But now the proponent of the nontrivial thesis faces a dilemma. What are the (syntactic) consequences of theory *T*? Any sentence *A* derivable from *T* using some system of rules we could come to accept without changing the meaning of terms in *T* and *A*? But any sentence *A* is a consequence in this sense, according to the nontrivial revisability thesis, so every theory ends up having the same useless content, the set of all sentences of the language. Yet theories are individuated by their consequences—according to Quine, by their observational or empirical consequences—so if logical consequence is wholly indiscriminate in this way, the notion of a theory is, too. Thus critics of the radical Quinean position maintain that if there are no a priori principles, consequence is trivially undiscerning: everything entails anything.[5] If logic is nontrivially and unrestrictedly revisable, the connections which determine the web of belief dissolve.[6]

If not every sentence which could be derived by some rule or other is a consequence of *T*, what does determine *T*'s consequences, for the radical Quinean? The most obvious response would be to say that *T*'s consequences (or its consequences relativized to today) are all those sentences derivable by rules we accept, or are disposed to apply today (or perhaps some canonical subset of them). However, those inferential dispositions and practices could change; we might even be disposed to change them in systematic ways. Quine's theory of language, for instance, has a prominent place for higher-order dispositions to change lower-order ones (e.g., in the account of observation categoricals). On this horn of the dilemma we debar sentences derivable by means of the changed rules from counting as consequences of the theory, from determining the identity of the theory. But this position is just a traditional "logical connections are analytic" one. Change today's rules and you are no longer working with today's theory. It would be merely an uninteresting terminological maneuver to deny that the rules in question are "meaning-constitutive;" they are not revisable except trivially, and thus are presumably a priori.

The only other option available to a Quinean is a type of ultraholism. We take the networks which face the tribunal of "experience" to be pairs comprising a logic

[5] See, for example, Dummett (1981, pp. 596ff.) and, for a somewhat different argument along these lines, Wright (1986).

[6] Cf. Maddy's "one common objection to the original Quinean web has been that some laws of logic are needed for the simple manipulations of web maintenance" (Maddy 1997, ch. 13, §I).

L and a theory *T*. The right (or a right) notion of logical consequence is any such *L* which in combination with some *T* or other yields all true (pegged) observational sentences (or observation categoricals), and in that sense faces the tribunal of experience successfully. But this ultraholist view makes logical consequence unknowable, since we have no idea what the set of all true observation categoricals is; hence logical consequence becomes potentially divorced from even best inferential practice.

So Quine was wise to move on this issue of the analyticity of logic. Even so, this restriction of empiricism on logic does not necessitate a blanket welcome to all of logic in Quine's sense—namely, first-order classical predicate calculus with identity—back into the analytic fold. For the failure of a radical revisability of logic thesis does not entail that all the logic we uphold must be validated by rules which are meaning-constitutive or in some sense analytic (forced on us by the verdict matrices, say). A "moderate" Quinean might argue this is true only of some very basic rules, such as &E, and $\vee$I; other rules or principles—excluded middle, distributivity principles, various of the structural rules needed to generate a suitable proof architecture, and so forth—may be nontrivially revisable in the light of empirical experience, it might be urged. Logic, as usually construed, divides up into "analytic" and "synthetic" parts, on this view. And similarly, even if the considerations advanced above against total revisability are accepted, the Quinean might still claim that mathematics, as well as substantial portions of logic, is empirical and not a priori.

This point illustrates a serious problem for the neologicist. In order to generate standard mathematics from her favored abstraction principles—the Peano–Dedekind axioms from Hume's Principle, for example—the neologicist needs a powerful background logic: second-order logic with the full axiom scheme of comprehension and untrammeled by the restrictions on first-order and second-order $\forall$E and $\exists$I found in free logic systems. The anti-naturalist might well argue that these components of the powerful logic wielded by the neologicist pose the same epistemological problems as the axiom systems of standard mathematics which the neologicist is hoping to put on a sound footing by deriving them, in this potent logic, from abstraction principles.

Moreover, even if it is legitimate for the neologicist to exclude classically inconsistent abstraction principles such as Axiom V from functioning as analytic or as implicit definitions, this still leaves too many abstraction principles in play, since there are indefinitely many consistent but pairwise inconsistent abstraction principles. Setting out a non-ad-hoc criterion for winnowing out kosher from non-kosher principles is no easy task. Whether, then, the neologicist route is a promising one for the naturalist cannot be decided without further, more detailed debate.[7]

[7] For more, see the contributions on logicism in this volume (chapters 5–7) and Shapiro and Weir (1999, 2000); Weir (2003b).

II

A diametrically opposed strategy for naturalizing mathematics is to be found in Quine. The anti-naturalist argues that the methodology of mathematics goes beyond scientific methodology by laying down axioms as a priori starting points, in contrast with the provisional and contingent hypotheses of empirical science. But where the naturalistic neologicist attempts to show that it is naturalistically acceptable to take the axioms as a priori starting points, Quine adopts a radical empiricism. The axioms of mathematical systems are hypotheses just like the hypotheses of physics or chemistry, each confirmed or disconfirmed not directly, but indirectly by dint of contributing to a total theory of the world which directly confronts experimental and observational data. The difference between empirical science, on the one hand, and mathematics (and the synthetic parts of logic, if the considerations of the last section are accepted), on the other, is not a qualitative one. Mathematics and (synthetic) logic pervade all areas of science, with the result that alterations to the mathematics or logic have very far-reaching and disruptive effects; consequently we are more loath to make changes there than in a more localized discipline. But the difference is purely one of degree. Though we are more likely to revise an inflationary hypothesis in cosmology than the distributive laws of logic, in principle we may do the latter, too (as quantum logicians, in fact, have done); neither the cosmological nor the mathematical hypothesis is directly confirmed or disconfirmed by experiment, both can be confirmed or disconfirmed indirectly.

Though the Quinean position falls foul of pretty strong intuitions that some, at least, mathematical truths—Kant's $7+5=12$ and so forth—are necessary and a priori certain, this is a brilliantly bold naturalist response to the problem posed by mathematics. The main problem I will raise for it comes under the heading of "ontological naturalism." But a methodological worry can be raised by considering that old favorite in discussions of positivism, "The Absolute is lazy." Suppose T is a wonderfully successful total theory of the world. Add to T, "The Absolute is lazy" (abbreviate it by A) and we still get, at least as far as empirical adequacy is concerned, a successful total theory. Does this mean that "The Absolute is lazy" is confirmed just as much as the fundamental theorem of calculus (cf. Bergstrom 1993, p. 333)?

The obvious response is to say that T, A contains some unnecessary fat. Economy being a scientific virtue, we should hone it down to T. The problem is: Where does this honing stop? If E is the set of T's empirical consequences, why not treat everything in T minus E as blubber and slim our total theory down to the lean, mean E, excising theoretical "fat"—all of mathematics along with theoretical science? If our accepted theory T is recursively enumerable, our language is countable, and there is an effective way of deciding whether a sentence is observational or theoretical, then the set of observational consequences E of T is also recursively enumerable. Indeed, Craig's method shows how we can construct a decidable set "visibly equivalent," as Quine says, with it. Why not work simply with the program for enumerating E? To be sure, this will involve some mathematics, but only a weak amount, sufficient arithmetic for proof theory.

Quine's answer is that we have no realistic chance of getting at E except via deduction from T, or rather a formulation of T: "The theory formulation is a device for remote control and mass coverage" (1975, p. 324)—mass coverage of observation sentences, that is. Furthermore, if E turns out to contain falsehoods, revision to a more acceptable E' will be practically possible only by looking at ways of amending formulations of T. Quine's empiricism counsels, then, adopting the leanest, meanest theory which we can grasp and which is empirically adequate. True, Quine also admits scientific virtues other than empirical adequacy, scope, and coverage; he admits, for instance simplicity and conservatism (Quine 1955, p. 247, 1990 p. 95; see Maddy 1997, ch. 13, §I).[8] But his empiricist emphasis on economy and empirical adequacy leads him to jettison parts of mathematics which seem to play no role in scientific applications—for example, the higher flights of transfinite set theory. This rejection takes a number of forms. Penelope Maddy charts the different nuances (ch. 13, §I this volume). Sometimes Quine is tempted to write off the investigation of large cardinal axioms as a purely logical enterprise—the investigation of logical connections among essentially meaningless sentences (1984, p. 788). At other times (1990, p. 95) he advocates the adoption of the axiom of constructibility, $V = L$; this "thins" out the bloated hierarchy V of sets and commits him to the nonexistence of a vast range of sets which he views as empirically redundant.

However, this vast range of nonconstructible sets form the main topic of research for contemporary set theorists. They may be reluctant to accept that they are engaged merely in logic-chopping with empty formulas. Are they to be accused of delusions of transfinite grandeur which have led them to expend their energies arguing about nonentities? Must we urge them to get a life and join their colleagues across the corridor working in analysis or group theory and dealing with respectable denizens of reality? Quine's naturalism has led him into a radical revisionism with respect to pure mathematics, which might seem at odds with his insistence that there is no first philosophy and with his reluctance to let philosophy decide matters of objective existence and nonexistence.

[8] These additional methodological desiderata and others often adduced—elegance, explanatory power, and so on—pose problems for naturalism independently of any raised by mathematics. Prima facie, they pick out human-relative virtues of an "aesthetic" nature; naturalists thus have a job explaining why any such species-relative virtue yields a reliable guide to objective reality. Steiner (1998) makes similar points with respect not to science in general but to mathematics in particular.

III

Maddy, therefore, proposes a variant on Quinean naturalism which recognizes the autonomy of mathematics. She joins (ch. 13, §I) Quine in that "large minority or small majority who repudiate the Cartesian dream of a foundation for scientific certainty firmer than scientific method itself" (Quine 1990, p. 19).[9] Maddy adds: "The simple idea is that no extra-scientific method of justification could be more convincing than the methods of science, the best means we have."

However Maddy then goes on to urge that no extramathematical method of justification (including naturalistic philosophizing) could be more convincing than the methods of mathematics, which lead most set theorists, for instance, to affirm $V \neq L$. Philosophers should eschew any imperialistic ambitions to interfere in the internal affairs of the mathematical community.

> To judge mathematical methods from any vantage-point outside mathematics, say from the vantage-point of physics, seems to me to run counter to the fundamental spirit that underlies all naturalism: the conviction that a successful enterprise, be it science or mathematics, should be understood and evaluated on its own terms, that such an enterprise should not be subject to criticism from, and does not stand in need of support from some external, supposedly higher point of view. (Maddy 1997, p. 184)

Note that whereas Quine seemed to be congenitally hostile to boundaries (leaving aside those on maps, at any rate), seeking to replace them with differences of degree across spectra, Maddy's naturalism, taken neat, presupposes fairly determinate boundaries between philosophy, science, and mathematics. They need not be totally precise, nor need there be nontrivial necessary and sufficient conditions which characterize the distinctions, but if there is not even a vague boundary between philosophy and science or philosophy and mathematics then it does not make sense to say that "if our philosophical account of mathematics comes into conflict with successful mathematical practice, it is the philosophy that must give" (1997, p. 161).

[9] Note that one can reject the Cartesian dream without following Quine any further in rejecting any firm, relatively determinate, and principled distinction between (legitimate) philosophy and natural science. One could adhere to a traditional view of philosophy as a priori conceptual analysis, for example, while still being a fallibilist about the results of that analysis (a priori $\neq$ certain), a fallibilist in science, and an anti-foundationalist in epistemology. More plausibly, one could accept some looser version of the view of philosophy as discontinuous from science and as nonempirical, one which does not place much store on finding tight necessary and sufficient conceptual connections, while still being a fallibilist. Quineans sometimes move too quickly from rejection of Cartesian foundationalist epistemology to acceptance of a full Quinean package.

A treaty of noninterference across boundaries makes sense only where the boundaries exist. How can a Quinean class something as respectable philosophy and simultaneously deny it the right, so to speak, to criticize science? Maddy seems to accept this: "I do not assert that philosophy is irrelevant to the assessment of mathematical methodology; I could hardly do so without a principled distinction between philosophy and mathematics" (1997, p. 200). Yet she denies (1997, 197–198) that there is such a principled distinction. The upshot seems to be a less absolutist, more fallibilistic, prescription on interference: "What I do assert, on the basis of the historical analysis, is that certain types of typically philosophical considerations have turned out to be irrelevant in the past" (ibid.).

At any rate, if naturalistic philosophers should only describe, not prescribe, where the methods and results of mathematics are concerned, why does this not apply also to astrology or theology or "creation science" (cf. Dieterlie 1999)? Consider, for example, cosmologists who believe in a plurality of causally isolated mini universes, as a response to the "fine-tuning" problem; in this multiverse plurality all parametric values which have to be put in "by hand" in current physics are randomly distributed. Imagine, in addition, there is subgroup with specific views about what types of mini universe there are. In particular, one cosmological sect says there are exactly forty-two subuniverses in which there are Twin Earths where there has developed a parallel human or humanlike form of life, in exactly one of which John F. Kennedy (i.e., his parallel clone) is not assassinated; exactly one in which Scotland, not England, wins the (Association) Football World Cup in its equivalent of 1966 (our cosmologists call this mini universe Paradise); and so on. Perhaps these twists on (a substrain of) modern cosmology are based on scriptures they hold sacred or on interpretations of revelations of the sect leaders, grounds about which they debate apparently quite fruitfully and using methods they mostly seem to agree on. Since their position does not disagree at all with what standard science says about the universe we are all causally connected to, why is this not a successful enterprise with an extra flourish, a flourish comparable to the higher flights of transfinite set theory or some other branch of pure mathematics remote from any obvious possibility of empirical application?

Maddy answers that she is not advocating a quietist tolerance of all systems of thought and belief. She is a scientistic naturalist like Quine, but one who disagrees with Quine, *on scientific grounds* (for the Quinean considers epistemology and methodological prescriptions a subdiscipline of natural science), about scientific methodology. Science is less of a seamless whole than Quine suggests, even when he allows that science is "neither discontinuous nor monolithic. It is variously jointed, and loose in the joints in varying degrees" (Quine 1975, p. 314). Possession of the theoretical virtues adumbrated by Quine is insufficient to generate belief in a theory, in all circumstances and departments of science. Using a detailed investigation of the fate of the atomic theory in the nineteenth and early twentieth century, up to its general adoption after the experimental work of Perrin, Maddy concludes:

> The theory enjoyed the Quinean virtues in abundance by 1860, when its successes in chemistry were crowned by the computation of stable atomic numbers, and even more so by 1900, after the rise of kinetic theory in physics. But scientists were not satisfied until the direct detection of atoms by Perrin, verifying the crucial predictions of Einstein's 1905 calculations. This, I claim, means that the theoretic virtues are not enough, that enjoyment thereof is not what evidence is, that our best scientific theory is not confirmed as a whole, that some of its posits are properly regarded as fictional until further, more specific testing is possible. (Maddy 1997, ch. 13, §II.3)

Maddy's conclusion seems to be that science as a whole is rather more heterogeneous than Quine thinks, that some parts of our overall system we may not believe in, but nonetheless accept "as fictional." She even moots the possibility that the existential deliverances of mathematics may come under that score—"My guess is that, in the end, the explanations and accounts of naturalized philosophy of mathematics will not involve a literal appeal to the objects of pure mathematics" (Maddy 1997, ch. 13, §II.5)—with one way to cash that out being "mathematical claims aren't true and mathematical things don't exist, though it's rational and proper to assert that they are and do in the course of doing mathematics" (ibid.). Moreover, given the tight interconnections inside mathematics, pure and applied, the same ontological reading should apply to all such deliverances, from the existence of infinitely many primes to the existence of inaccessible cardinals.

To distinguish different types of ontological commitment in different types of existential assertions, to discern fundamental ambiguities in "exist" in this way, is to proceed on very un-Quinean lines (see Quine 1960, pp. 131, 241–242) which certainly need substantive justification. If, on the other hand, the claim is that mathematicians and nonatomistic chemists hold that atoms and prime numbers do not exist, in the only sense of "exist" which exists (!), but nonetheless find it a useful fiction to pretend that they do, then there certainly needs to be some work doing making that claim good, at least for the mathematical community. Moreover, one does not need to be a Quinean to ask what makes the atomic hypothesis part of the overall theory "held" by chemists such as Ostwald who treated (pre-Perrin) the hypothesis as a useful instrument. How is atomism part of their theory if they did not actually believe in atoms?[10] At any rate, even the second-class, nonliteral existence accorded by some to atoms in the nineteenth century, and

[10] Furthermore, Maddy's argument requires ruling out the possibility that the refusal of Ostwald and company to accept the existence of atoms was based on empiricist metaphysics, proto-verificationism perhaps, rather than on scientific reasons alone. Cf. Colyvan (2001, ch. 5).

which might perhaps be accorded to prime numbers in the twenty-first, Maddy certainly does not want to allow to the cosmological sect's forty-two Twin Earths.

Speaking as a scientific naturalist from within the perspective of current science, for she knows of no higher or better perspective, Maddy would reject the weird cosmology because there are no scientific reasons to believe in the forty-two Twin Earths. But there are overwhelming scientific reasons (the indispensability argument) for believing in, or at least "accepting," contemporary mathematics. Moreover, detailed investigation of current mathematics, Maddy conjectures, will show that, unlike science in general, it is, if not a seamless whole, at least highly interconnected. It may well prove very difficult to separate out those parts which are clearly "pure" mathematics, with no possible practical applications, from those which are applied or liable to be applicable, especially when one considers the highly esoteric branches of mathematics—non-Euclidean geometry, infinite dimensional Hilbert vector spaces—which later have turned out to be highly useful.

For Maddy, the direct reasons for believing that $V \neq L$ are mathematical reasons—not scientific ones but still worthy of respect because of the indispensability of mathematics as a whole in science; so in that sense, indirectly scientific. The Twin Earth cosmologist might claim that her scriptural, or whatever, reasons for believing in the forty-two Twin Earths have a similarly indirect scientific justification. The idiosyncratic parts of her cosmology cohere very well, she may claim, with the more standard parts of her cosmological and scientific views in such a way that it would spoil the overall elegance of the theory to rip the idiosyncratic, scriptural parts out and leave the rest. Insofar as the rest is indispensable in accounting for the phenomena, so are the details of the forty-two Twin Earths.

Such a cosmologist might claim this, but without any plausibility. True, much will depend on providing a philosophical justification for the claim that "scientific methodology," whatever it is, exactly, is more rational than the methods used by creationists and others to arrive at results contrary to science. But it would be obtuse to deny that the methods of set theory, "higher" and "lower," are much closer in nature to the mathematical methods utilized in, for example, quantum physics than the imagined disputations and dogmas of the forty-two Twin Earths sect. However, Quine himself allowed into his total theory some currently unapplied parts of pure mathematics for "simplificatory rounding out" (1984, p. 788) and to spare oneself "unnatural gerrymandering of grammar" (1990, p. 94). So maybe the difference here between Maddy and Quine is largely one of judgment as to how far one should go in "rounding out."

Furthermore, Maddy may still be vulnerable to less extreme theistic views than those considered or imagined above. After all, there have been, and continue to be, eminent scientists who have been or are theists, often theists of a rationalistic, nonfidiestic stamp. Many of them do not see their religious beliefs as sharply discontinuous in content or in their underlying justification from their scientific ones, but rather as complementing or even enhancing the coherence of their scientific worldview. Only a belief in a beneficent God, they might argue, renders reliance on inductive extrapolation reasonable, or explains adequately the "fine-tuning" of natural constants to values which permit stable complex molecules to exist or explains the "unreasonable effectiveness" of mathematics, some of it highly esoteric and without obvious physical application when introduced, in generating deep and plausible scientific theories with staggeringly good records of predictive accuracy.[11]

The Quinean may say that in order to make a sharp break with theistic scientists of this type, one should adhere only to total theories which meet Quine's empiricist theoretical virtues, and that will necessitate excluding not only theism but also much of transfinite set theory. Maddy could retort that the examples of esoteric pure mathematics turning out to have unexpected empirical applications provide a purely empiricist reason for accepting the methods and results of mathematicians as a whole, pure and applied. Alternatively, an atheistic naturalist could nonetheless allow that some theists provide naturalistic reasons, albeit ones the atheist does not accept, for their theism, in much the same way as some methodological naturalists are led to the prima facie anti-naturalist doctrine of Platonism in mathematics—that is belief in abstract, eternal, mind-independent numbers, functions, sets, and the rest.

IV

I turn, finally, to ontological naturalism. The ontological naturalist holds that we have a reasonably determinate conception of what it is to be physical and avers that everything is physical; in particular, all objects are physical, as are all properties and relations (trivially, for a nominalist in the medieval sense, who believes there are no properties or relations). Providing a characterization of physicality is not an entirely straightforward matter (cf. Burgess and Rosen 1991, §1.A.1), but I will assume for present purposes that it is not impossible. Certainly we seem to have a clear idea in particular cases: the sun is a physical object (albeit its boundaries are vague) and the mass of the sun a physical property, while Cartesian souls, if they exist, are nonphysical.

We can then divide naturalists in general into two camps with respect to ontological naturalism: the ontological naturalists and the revolutionary defeatists. The second group reject ontological naturalism but argue that this rejection is

[11] Mark Steiner (1998) has provided some thought-provoking examples of generalizations or extensions of physical theories on the basis of what seem to be purely mathematical, often even notational, similarities which have been highly successful empirically; he argues that this is hard to explain naturalistically (cf. the more general comment—note 8—on aesthetic considerations in scientific methodology).

compatible with—perhaps even implied by—their methodological naturalism, the strand of naturalism which is important to them. Logicists and Quineans typically belong to this grouping, taking mathematics pretty much at face value and arguing, if in Quine's case somewhat reluctantly, that this commits us to the existence of an infinite mathematical ontology of objects and (for the logicists) properties which are clearly nonphysical.

The ontological naturalist camp divides further into fictionalists such as Field, who deny the existence of mathematical entities but seek to account for the importance of mathematics via its instrumental utility, and less radical philosophers who seek to construe or reinterpret mathematics so that its theorems come out as true but with no commitment to abstract, nonphysical objects and properties. One strategy here is to take modality as primitive and read mathematics as dealing not with an actual infinity of abstract objects but with possibilities that infinite structures could exist or with endless possibilities for constructing concrete but nonetheless mathematical objects—tokens of mathematical formulas for example.[12] Examples include the modal structuralism of Geoffrey Hellman (1989), the modalized type theory of Charles Chihara (1990), and perhaps Philip Kitcher's empiricist constructionalism (1983). The main difficulties here, from a naturalistic perspective, are the legitimacy of taking modality as primitive or, more radically, the respectability of modality from a naturalistic perspective. Quine, of course, took a very dim view on this score.

A rather different strategy builds on the idea that the finite cardinals are properties of properties: two is a property shared by all doubly instantiated properties, and so on (Bigelow 1988 is an example of this strategy). This is quite an attractive strategy for anyone who believes in the objective existence not only of first-order physical properties but also of higher-order ones, arguably a reasonable position for a physicalist to take. If position properties exist, should we not see velocity as dependent on higher-order properties of position properties, acceleration as dependent on third-order properties of velocity properties, and so on? The classic difficulty with this position is that if there are in actual fact only finitely many first-order properties, then arithmetic, never mind analysis and so forth, will not get off the ground. There certainly seems to be no a priori reason for supposing there are infinitely many physical objects or properties. Logicism runs into similar difficulties, at least if the argument of Shapiro and Weir (2000) is right.[13]

[12] Intuitionism can also be seen as a "modalize mathematics" strategy for rejecting Platonism. But intuitionism, even if shorn of the highly Kantian elements in Brouwer's original formulation, is perhaps not very promising for naturalists, given its radically revisionary nature and the question mark, to say the least, as to whether it allows for sufficient mathematics to meet the needs of the sciences, today and tomorrow.

[13] Which it is. (Ed.).

Since these problems are treated more fully elsewhere, I want to focus finally on a problem which afflicts, I will urge, both Quine's account of mathematics and also Field's, despite their opposing stances on indispensability. We might call it the problem of the *conceptual* indispensability of mathematics for theoretical science as opposed to any deductive or methodological reliance on mathematics for churning out predictions from theories, or ontological dependence of science on mathematical objects as playing a role in causal explanation—Field's "heavy-duty platonism" (Field 1980).

According to scientific realists, many, at least, of our theoretical terms refer to objects way too small or large for us to come into contact with them through normal perceptual means. How do we manage to do this referring? Naturalists, who generally adopt a realist position[14] ought to want to explain this. True, the enormous problems which have beset all attempts at giving a naturalistic explanation of language understanding, at accounting for the normativity of language, and so on, are very daunting and it is not surprising that many philosophers have thrown in the towel and adopted a sort of Wittgensteinian quietism. Even so non-Wittgensteinian a philosopher as Tim Williamson writes:

> The epistemic theory of vagueness makes the connection between meaning and use no harder to understand than it already is. At worst, there may be no account to be had, beyond a few vague salutary remarks. Meaning may supervene on use in an unsurveyably chaotic way. (Williamson 1994, p. 209)

But it would be unnaturally defeatist for a naturalist to adopt this stance.[15]

Now what account does Quine have of the mechanism by which theoretical words link to theoretical objects? It is largely a negative account: operationalist, verificationist and other "reductionist" explanations of this link are hopelessly

[14] Perhaps some anti-realist philosophers would want to adopt the "naturalist" label. At least if one restricts to "methodological" as opposed to ontological naturalism, this does not seem incoherent. To such philosophers I would say that the following criticisms apply only to "realist naturalists," but that for convenience (and because I think this represents the usual naturalist position) I will restrict the term "naturalist" to apply only to realists.

[15] Perhaps there really is a plurality of mini universes with Twin Earths. Perhaps on some of them sophisticated culture and philosophy flourish, but only a stunted science and technology exists because anti-naturalism has gained an unshakable hold on the intellectuals. On one Twin Earth, Wittgenstein's twin, Ludwig Twittgenstein, say, declared that the wandering stars roam round the night skies in an unsurveyably chaotic way and not even Babylonian astronomy emerged. On another, L. Twittgenstein II persuaded everyone that the basic elements combine to form the vast array of substances around them in a similarly chaotic and nonsystematizable way and chemistry did not emerge. Elsewhere (metaphorically speaking) Twittgenstein III pointed persuasively to the incredibly chaotic nature of the brain and physiology was killed at birth. It is the mark of the naturalist, it seems to me, to soldier on regardless, looking for system amid the chaos.

inadequate: the holistic points of "Two Dogmas" alone show this. On the positive side, Quine develops an account of empirical meaning for observation sentences—the stimulus theory (albeit one which he held with less and less assurance and more and more qualification in his later work; see George 1997))—and argues that we grasp theoretical terms solely by grasping the interconnections among theoretical sentences and the empirical meaning of their empirical consequences. As we saw in section II, "consequences" can mean little more, for Quine, than the basic analytic logical connections. Constraints imposed by "synthetic logic," mathematics and nondeductive methodology are simply further nodes in the overall theory.

Suppose an interpreter selects a set of operators O of our language as subject to meaning-constitutive constraints.[16] Call a sentence nonlogical if no member of O is its dominant connective; and consider pairs of functions $\langle f, g \rangle$ where f maps the nonlogical sentences (including open sentences) onto sentences (logical or nonlogical) and g is a permutation of the members of O. Let an interpretation function I be a mapping of the language such that (1) if S is nonlogical, $I(S) = f(S)$; and (2) where $C \in O$, $I[C(S_1, \ldots, S_n)] = g[C](I(S_1), \ldots, I(S_n))$, where the S_i are the immediate constituents of the sentence $[C(S_1, \ldots, S_n)]$ (in which C is dominant). Finally, declare an interpretation to be Quinean if f preserves empirical content and g maps connectives onto connectives with the same meaning, according to our meaning-theory (e.g., the verdict matrix theory) for logical connectives.

If all Quinean interpretations are correct, if they all preserve objective meaning, then will there not be a plurality of widely diverging, indeed incompatible, interpretations? For the only constraints on interpretation are to preserve empirical meaning and the meaning of the logical connectives, as given by the verdict matrices, or something similar. Quine, of course, agrees. This is indeterminacy "from above" (Quine 1970), and the argument need not rely on any demand for behavioristic reduction of meaning: the main premises are that the above two constraints are the only ones, plus a modest Quine/Duhem holist denial that the typical theoretical sentence has any empirical meaning. What conclusions should then be drawn, with respect to realism? Compare Quine's indeterminacy from below, his argument for ontological relativity. I am not concerned here with the plausibility of the argument, merely with the conclusions he draws.

Quine claims that for some singular terms, "Bugs Gavagai" and "Bugs Gavagai's left ear," say, we can have "the objective meaning of 'Bugs Gavagai' = the objective meaning of 'Bugs Gavagai's left ear.'" But surely we also have the following principle concerning the dependence of reference on meaning (omitting, for simplicity, a uniform relativization to context). "If the objective meaning of singular term t = the objective meaning of singular term u then the referent of t = the referent of u."

[16] The following proposal generalises easily to interpretation of other languages.

Putting the above two together, we infer that the referent of "Bugs Gavagai" = the referent of "Bugs Gavagai's left ear" and then, disquoting with respect to reference, that Bugs Gavagai = Bugs Gavagai's left ear, which is absurd. But Quine does not conclude from all this that his thesis of indeterminacy from below is absurd. Rather,

> Reference *is* nonsense, except relative to a coordinate system.... What makes sense is to say not what the objects of theories are, absolutely speaking, but how one theory of objects is interpretable or reinterpretable in another. (Quine 1969, pp. 48, 50)

Now, however, consider the "argument from above," the interpretation of (late) Quinean holistic empiricism as saying that the only constraints on interpretation are preservation of empirical content and of the meaning (as given by the verdict matrices, say) of the logical connectives. Take as example Newton's inverse square law $\forall x \forall y N$ where N is

$$F_{xy} = \frac{g\, m_x m_y}{r^2},$$

stating that the gravitational force exerted on the line between massive objects x and y is proportional to the product of their masses and inversely proportional to the square of the distance r between them. The law has no empirical content on its own, independent of auxiliary hypotheses concerning the other forces in play and boundary conditions telling us what other objects there are in their world. Neither has the negation of Newton's law, $\sim\forall x \forall y N$, any empirical content, and the same is true of each instance of N and each of its negations $\sim N$. If O contains just the usual sentential logical operators, the function f, which is trivial except that it maps Newton's law to its negation, preserves empirical content. Thus by a blunt application of the Quinean "argument from above," Newton's law and its negation have the same objective meaning.[17] Add the corresponding principle of dependency of truth on meaning—"If the objective meaning of sentence p = the objective meaning of sentence q then the truth-value of p = the truth-value of q"—and the disquotational schema—"The truth-value of p = the truth-value of q iff $A \leftrightarrow B$"—whose instances are generated by substituting a name of the substituend

[17] We can add the quantifiers into O and retain the trivial g by letting f map the open sentence N to $\sim N$ and otherwise be the identity map. This generates an interpretation which maps Newton's Law to $\forall x \forall y \sim N$ so that the same principles yield $\forall x \forall y N \leftrightarrow \forall x \forall y \sim N$, again absurd (in a nonempty domain). If all of this is too blunt, consider subtler examples of empirically equivalent (for us) but genuinely incompatible (at least on a realist perspective) sentences, such as a formulation of a standard big bang theory together with one which posits our old favorite, a multiverse of big bang universes.

for A for p and similarly a name of the substituend for B for q. Now we are able to conclude, in parallel fashion to the ontological relativity argument,

$$\forall x \forall y N \leftrightarrow \sim \forall x \forall y N,$$

which is absurd, not only in classical logic but in many weaker ones, too.

How to avoid this contradiction? The parallel conclusion to draw is a deep relativism regarding truth:

> Truth *is* nonsense, except relative to a coordinate system. . . . What makes sense is to say not whether a theory is true or not, absolutely speaking, but how one theory is interpretable or reinterpretable in another, taken to be true.

Even Newton, then, had he been a Quinean(!), could not have said that the inverse square law was true absolutely. It is true relative to a "homophonic" interpretation and false relative to others. The only sentences true or false absolutely are observation sentences which wear their empirical meaning on their sleeve. The problem for the naturalist is that Quineanism leads to radical instrumentalism, for on the Quinean picture there is a radical asymmetry between truth for observation sentences (absolute) and truth for theories (relativistic), and it is this metaphysical asymmetry between theory and observation which is the essential feature of instrumentalism.[18]

Now Quine recoils from relativity of truth—partly, I think from a residual realism, partly as a result of the false notion that any substantive form of relativism is paradoxical, and partly because he thinks his rejection of first philosophy immures him to relativism: "there is no extra-theoretic truth, no higher truth than the truth we are claiming or aspiring to" (Quine 1975, p. 327). But this is to confuse epistemological with metaphysical issues. The instrumentalist can adopt Quine's methodology, espouse nothing not endorsed or at least compatible with best scientific practice, but give a metaphysical reading of theoretical sentences at variance with that given to observational. Quine is in fact that instrumentalist.

Thus the dilemma for a naturalist who adopts a Quinean empiricism on mathematics based on a general holism plus empiricism about meaning: how to buy into so very much of the Quinean package without swallowing all the relativistic

[18] Some use "instrumentalism" to mean the doctrine that theoretical sentences have no truth-value (cf. Quine 1975, p. 327). But this is, at least in current philosophy, a rather uninteresting doctrine. Instrumentalists are prepared to affirm current science, so they should be prepared to affirm its truth, at least on a disquotational account of truth. They differ on the *metaphysical* interpretation of theoretical sentences, finding a cleavage between the metaphysical grounding of observation sentences and that, parasitic on the first, of theoretical ones. Thus I do not think Quine can escape a charge of instrumentalistic anti-realism as he seeks to do (1975, p. 327) simply by endorsing disquotational truth.

instrumentalism, which of course will apply to mathematics through and through. The least prime number theorem and its negation will have the same meaning for Quine, both being devoid of empirical content and a reinterpretation of, say, the theory of rationals in the theory of the natural numbers, will be perfectly licit.[19]

There is, though, an obvious strategy of response here, but it is one which brings no comfort to the naturalist. Let O be a Quinean observation sentence and consider O, $O \& O$ and $O \vee O$. These three sentences, being logically equivalent, are empirically equivalent. Furthermore, the interpretation I based on $f(O) = O \& O$ and the trivial mapping g sending all sentential operators to themselves is a good Quinean one, as is I* based on $f^*(O) = O \vee O$ and the same g. Yet there are clear materials within the Quinean philosophy of language itself for distinguishing the meaning of O, $O \& O$ and $O \vee O$. The phrase structure tree associated with the first is very different from that for the other two, and although those two in turn have structurally identical phrase trees, the constituents on the corresponding nodes where they differ—& and $\vee$—differ in meaning according to purely Quinean conceptions of meaning. So even an empiricist can discern a notion of meaning narrower than sameness of empirical content, narrower than mutual interpretability according to Quinean interpretation functions. The notion is evidently close to Carnap's notion of intentional isomorphism—same grammatical structure, synonymous ultimate constituents—though for semantic purposes it would be sensible to widen somewhat: to allow, for example, that sentences transformable into one another by very simple means, such as active/passive transformations, are synonymous. Quine himself objects to this sort of use of the Carnapian notion to stymie indeterminacy, but the main reason he gives is question-begging: that if we use intentional isomorphism to characterize a more fine-grained notion of synonymy, then the indeterminacy thesis fails.[20]

However, merely appealing to the meanings of the logical connectives in a Carnap-style account of meaning will not of itself explicate how logically *incompatible* but empirically equivalent sentences can differ in meaning. To do that, we need to explain how ultimate constituents other than logical operators (and observation sentences) can differ in meaning. One way is to steal some leaves out of the logicist's book. If S differs from R only in that S has the numerical operator $\mathrm{N}x\,\varphi x$ where R has a class operator $\{x : \varphi x\}$, then they differ in meaning because the rule governing the meaning of the numerical operator—the rule form of Hume's

[19] Indeed, if we have a countable infinite set F of false scientific claims devoid of empirical content, none of which has one of our "meaning-constitutive" logical operators as the dominant operator, then there will be a good Quinean interpretation which maps all the nonlogical arithmetical truths into F (with g trivial). Arithmetic can perfectly well be interpreted as about phlogiston as about the numbers. Though Quine might be prepared to bite the bullet, I take this as a reductio.

[20] At least that is how I read Quine (1960, §42), especially the paragraph pp. 205–206.

Principle, perhaps—differs from that governing the class operator, perhaps a weakening of the rule form of Axiom V.

In what way could all this lay the basis for a naturalistic explanation of how our theoretical terms reach out to items utterly inaccessible to perception? Here's a very crude story. Imagine we grasp a four-place predicate *xy* Cong *zw* which we apply to observable rods (or rodlike segments of surfaces) and whose meaning is that the rod whose two end slices are *x* and *y* is congruent with that whose end slices are *zw*. We might manifest our grasp of this purely empirical predicate by, among other things, transporting rods and laying them end to end before issuing in a judgment. (How closely our grasp of such empirical predicates is tied to such verificatory behavior is, of course, a highly controversial matter.) As well as simple judgments that one segment is congruent (or not) with another, we will be able to grasp such ideas as that there is a point *t* between the end points *zw* of a segment such that *xy* is congruent with *zt* and with *tw*.

Now imagine that we also have an *independent* grasp of basic arithmetic. We might then be taught that ⌜there is a point *t* between the end points *zw* of a segment such that *xy* is congruent with *zt* and with *tw*.⌝ is synonymous[21] with ⌜*zw* is 2 times the length of *xy*⌝. And so on for three times, four times, and so on. If, in our system, *ab* is a standard unit, like the old Parisian standard meter, then for many values *n*, ⌜*zw* is *n* times the length of *ab*⌝ will be synonymous with an expression which language users utterly innocent of arithmetic can produce. But in other cases not: we might find ourselves saying that *zw* is 10^{45} times the size of *ab* or 10^{-45} the size of *ab*, and neither sentence may have any translation back into our purely empirical notion of congruence. Yet it may be suggested that by this means we are able to refer to lengths vastly greater than, or less than, anything perceptually discernible.

This suggestion may, of course, be strongly resisted. I am not intending to suggest here that this naturalistic strategy for explicating how we get beyond acquaintance solely with medium-sized dry objects is obviously going to succeed in the case of length (far less other physical magnitudes), or indeed that this is the only model for the introduction of theoretical terms; there is, of course, the Lewisian structuralist model in terms of Ramsey sentences (Lewis 1970), although, as Lewis himself seems to accept, this will not take us from observation to theory on its own. My claim is just that it is hard to see another route for the naturalist to follow in attempting to explain our grasp of the nonobservable. But now the whole strategy depends on our having an *independent* grasp of arithmetic, or more generally mathematics, that is an independent understanding of the sentences of these disciplines and some significant, albeit fallible, ability to differentiate *correct* from incorrect mathematical sentences. Granted that, and a disquotational notion of truth, we are committed to the existence of a body of truths grasp of which is

[21] Or at least paraphrasable in an interesting sense.

independent of our understanding of empirical theory—indeed, the latter presupposes it. Grasp of the highly theoretical sector of language T presupposes, first grasp of observational fragment O and also grasp of a fragment M of mathematics. So now the naturalist has the problem explaining our grasp of M, and Quinean holism cannot provide the answer.

Moreover, this problem is also Field's, despite his rejection of Quinean indispensability. Suppose Field can show us that mathematics is *ontologically dispensable* in science—that is, that mathematical objects play no role in scientific explanation—and also that it is deductively indispensable—we do not need mathematics, other than as a useful proof-shortening device, either in the deduction of predictions from scientific theory or in a scientific explanation of deduction (so that model theory and proof theory are dispensable). Even if Field can achieve this much, he still has to give us a naturalistic explanation of how we manage to grasp the meanings of theoretical terms which refer to items far removed from those we are acquainted with through our normal perceptual abilities, and he has to do this without appeal to an independent grasp of mathematics. I am suggesting that this cannot be done.

V

I said earlier that one of the strongest, if not the strongest, versions of naturalism is Quine's radical empiricism. If the conclusions of the last section are correct, then mathematics proves to be a fatal stumbling block for that project. But I do not conclude that naturalism is fatally wounded. The conclusion to draw is that accommodating mathematics continues to pose an extremely thorny and as yet unresolved problem for naturalism, and that the attempts to accommodate it (as also the attempts to show it cannot be accommodated) have generated some of the best and most illuminating philosophical work of recent times.

REFERENCES

Benacerraf, Paul (1973): "Mathematical Truth?," *Journal of Philosophy* 70, pp. 661–679.

Bergstrom, Lars (1993): "Quine and Skepticism," *Journal of Philosophy* 90, pp. 331–358.

Bigelow, John (1988): *The Reality of Numbers: A Physicalist's Philosophy of Mathematics* (Oxford: Oxford University Press).

Boolos, George (1998): *Logic, Logic and Logic* (Cambridge, Mass.: Harvard University Press).

Burgess, John, and Rosen, Gideon (1997): *A Subject with No Object: Strategies for Nominalistic Interpretation of Mathematics* (New York: Oxford University Press).
Colyvan, Mark (2001): *The Indispensability of Mathematics* (Oxford: Oxford University Press).
Chihara, Charles (1990): *Constructibility and Mathematical Existence* (Oxford: Oxford University Press).
Dieterlie, Jill (1999): "Mathematical, Astrological and Theological Naturalism," *Philosophia Mathematica* 7, pp. 129–135.
Dummett, Michael (1981): *Frege: Philosophy of Language* (Cambridge, Mass.: Harvard University Press).
Field, Hartry (1980): *Science Without Numbers* (Princeton, N.J.: Princeton University Press).
George, Alexander (1997): Review of W.V. Quine *From Stimulus to Science, Mind*, 106, pp. 195–201.
Hale, Bob (1987): *Abstract Objects* (Oxford: Basil Blackwell).
Hale, Bob, and Wright, Crispin (2001): *The Reason's Proper Study* (Oxford: Clarendon Press).
Hellman, Geoffrey (1989): *Mathematics Without Numbers* (Oxford: Oxford University Press).
Kitcher, Philip (1983): *The Nature of Mathematical Knowledge* (New York: Oxford University Press).
Lewis, David (1970): "How to Define Theoretical Terms," in his *Philosophical Papers*, vol. 1 (Oxford: Oxford University Press, 1983), 78–95. Originally published in *Journal of Philosophy* 67 (1970): pp. 427–446.
Maddy, Penelope (1997): *Naturalism in Mathematics* (Oxford: Clarendon Press).
Quine, W.V. (1951): "Two Dogmas of Empiricism," in Quine's *From a Logical Point of View*, 2nd ed. (Cambridge, Mass.: Harvard University Press, 1980), pp. 20–46.
Quine, W.V.O. (1955): "Posits and Reality," reprinted in his *Ways of Paradox*, rev. ed. (Cambridge, Mass.: Harvard University Press, 1976), pp. 246–254.
Quine, W.V.O. (1960): *Word and Object* (Cambridge, Mass.: MIT Press).
Quine, W.V.O. (1969): "Ontological Relativity," in his *Ontological Relativity and Other Essays* (New York: Columbia University Press), pp. 26–68.
Quine, W.V.O. (1970): "On the Reasons for the Indeterminacy of Translation," *Journal of Philosophy* 67, pp. 178–183.
Quine, W.V.O. (1974): *The Roots of Reference* (La Salle, Ill.: Open Court Press).
Quine, W.V.O. (1975): "On Empirically Equivalent Systems of the World," *Erkenntnis* 9, pp. 313–328.
Quine, W.V.O. (1984): Review of Charles Parsons *Mathematics in Philosophy, Journal of Philosophy* 81, *pp.* 783–794.
Quine, W.V.O. (1990): *Pursuit of Truth* (Cambridge, Mass.: Harvard University Press).
Quine, W.V.O. (1995): *From Stimulus to Science* (Cambridge, Mass.: Harvard University Press).
Shapiro, Stewart, and Weir, Alan (1999): "New V, ZF and Abstraction," *Philosophia Mathematica* Series III 7, pp. 293–321.
Shapiro, Stewart, and Weir, Alan (2000): "'Neo-logicist' Logic Is Not Epistemically Innocent," *Philosophia Mathematica* Series III 8, pp. 160–189.

Steiner, Mark (1998): *The Applicability of Mathematics as a Philosophical Problem* (Cambridge, Mass.: Harvard University Press).
Tennant, Neil (1987): *Anti-Realism and Logic* (Oxford: Clarendon Press).
Tennant, Neil (1997): "On the Necessary Existence of Numbers," *Noûs* 31, pp. 307–336.
Weir, Alan (2003a): "Objective Content," *Supplementary Proceedings of the Aristotelian Society* 77, pp. 47–72.
Weir, Alan (2003b): "Neo-Fregeanism: An Embarrassment of Riches," *Notre Dame Journal of Formal Logic* 44, pp. 13–48.
Williamson, Timothy (1994): *Vagueness* (London: Routledge).
Wright, Crispin (1983): *Frege's Conception of Numbers as Objects* (Aberdeen: Aberdeen University Press).
Wright, Crispin (1986): "Inventing Logical Necessity," in J. Butterfield (ed.), *Language, Mind and Logic* (Cambridge: Cambridge University Press), pp. 187–209.

CHAPTER 15

NOMINALISM

CHARLES CHIHARA

UNDOUBTEDLY, the most enlightening published work dedicated to giving knowledgeable readers an overview of the topic of nominalism in contemporary philosophy of mathematics is *A Subject with No Object* by John Burgess and Gideon Rosen.[1] So I shall begin with a brief description of that work, in order to provide readers of this chapter with a solidly researched account of nominalism with which my own account of nominalism can be usefully compared. Part I, then, will briefly present the Burgess–Rosen account. My contrasting account will be given in the longer part II.

I. The Burgess–Rosen Account

According to Burgess and Rosen, nominalism is the philosophical view that everything that exists is concrete and not abstract (Burgess and Rosen, 1997, p. 3). Thus, according to these authors, one can also characterize contemporary nominalism as the view that *there are no abstract entities.* They view the contemporary nominalist, then, as someone who "denies that abstract entities exist, as an atheist is one who denies that God exists" (Burgess and Rosen, 1997, p. 11).

These characterizations raise a number of tricky questions: (1) What makes something abstract and not concrete? (2) Why should one not believe in the

[1] The subtitle of the book is *Strategies for Nominalistic Interpretations of Mathematics.* In their book the authors write: "The aim of this book is to chart the main currents in the nominalist stream" (Burgess and Rosen 1997, p. 8).

existence of any abstract entities? (3) Why should philosophers who do not believe in abstract entities try to produce nominalistic interpretations or constructions of mathematics? That is, why should nominalists attempt to reconstrue or "reconstruct" mathematics in such a way that its assertions can be seen not to assert the existence of abstract entities?[2]

Burgess and Rosen devote a whole section to question (1), but for reasons that will become clear later, I shall not go over their worries about the distinction at issue. I turn instead to (2) and (3). The latter question needs some preliminary discussion. *What is a nominalistic interpretation or construction of mathematics?* The answer: the nominalist attempts to construct a system of mathematics and/or physics that is consistent with his/her skepticism about abstract entities.[3] Thus, it is an attempt to reproduce what ordinary mathematics accomplishes without presupposing the existence of such things as numbers and sets.

Let us now reconsider question (2): Why should one not believe in the existence of any abstract entities? Burgess (1990) quotes the nominalistic reconstructivists Daniel Bonevac, Dale Gottlieb, and Hartry Field, who cite an argument of Paul Benacerraf's published in his article "Mathematical Truth"[4] as providing motivation for undertaking their reconstructions. This argument, based upon "a causal account of knowledge," is supposed to show that "belief in an assertion or theory implying or presupposing that there are numbers or objects of some similar sort cannot be knowledge" (Burgess 1990, p. 1). Burgess then spends most of the article blunting this argument,[5] arriving in the end at the rather awkward double negative conclusion that "it is not known that it cannot be known that there are numbers" (p. 12).

Burgess and Rosen (1997) take up, with great attention to detail and to subtleties, not only Benacerraf's original "causal theory" argument, but also refinements of that argument which nominalists have put forward to motivate their reconstructive programs. In particular, they take up Hartry Field's argument based upon a "reliability thesis" (pp. 41–49), as well as an epistemological argument involving the theory of reference (pp. 49–60). Burgess and Rosen provide

[2] Burgess and Rosen (1997), p. 12.

[3] Burgess and Rosen have in mind such nominalists as Hartry Field, Geoffrey Hellman, and myself. The reader can find expositions and references of those nominalist's works in Burgess and Rosen (1997); Field (in chapters II.A and III.B.1.a); Hellman (in chapters II.C and III.A.1.a); and Chihara (in chapters II.B and III.B.2.a and b). I also give an exposition of the above three nominalistic accounts of mathematics in Chihara (2004, ch. 5, sec. 3). My own approach to reconstructing mathematics nominalistically will be explained in part II of this essay. I explain my doubts about the Hellman and Field accounts in (2004, ch. 11) for Field and in (2004, Appendix A) for Hellman.

[4] Benacerraf (1973).

[5] I do not agree with some of Burgess's arguments against the causal account of knowledge. In part II, I shall give a specific objection to one of Burgess's refutations of the causal account of knowledge.

assessments of these arguments concluding, in each case, that a kind of stalemate between the nominalist and the anti-nominalist (sometimes referred to as "realist" or "Platonist") results, where no decisive conclusion can be drawn. For reasons to be articulated shortly, that there is such a stalemate between the two antagonists in this dispute is taken by these authors to be very significant.

Burgess and Rosen are clearly most sympathetic with a position that may be called "moderate realism" or "minimal anti-nominalism."[6] Minimal anti-nominalists do not present arguments for their belief in mathematical objects as does the typical realist. Nor do they attempt to explain how set theorists have come to know that, say, the null set exists or that the pair set axiom is true. Instead, the minimalist anti-nominalist starts with a "fairly uncritical attitude towards, for instance, standard results of mathematics."

> Having studied Euclid's Theorem, we are prepared to say that there exist infinitely many prime numbers. Moreover, when we say so, we say so without conscious mental reservations or purpose of evasion. . . .
>
> For those of us for whom something like this is the starting-point, any form of nominalism will have to be revisionary, and any revision demands motivation.[7]

In other words, minimalists start with their "fairly uncritical attitude" about mathematical existence, and demand of the nominalist a proof or convincing argument that their starting point cannot be maintained. It is a strategy that consists in throwing the onus upon the nominalist to show that the minimalists' anti-nominalist position is untenable. Thus, having arrived at the important stalemates described above, *they can argue that no compelling argument justifying the revisions demanded by nominalism has been provided and hence that the minimalists are justified in maintaining their anti-nominalistic stance.*

Additional Reasons for Rejecting the Nominalist's Position

In the last section of their book, "Conclusion," Burgess and Rosen (1997) question the philosophical motivation for developing the kinds of nominalistic reconstructions of mathematics discussed above. "What good are these reconstructions?" they can be understood to be asking. The authors confine themselves to two possible rationales for such reconstructions: *hermeneutic* and *revolutionary.*

[6] Burgess (1983) characterized his own philosophical position as a "moderate version of realism": it was a moderate kind of realism insofar as it held merely that our current scientific theories seem to assert the existence of mathematical entities and that we do not have good reason to abandon these theories. (Burgess's moderate realism will be discussed in more detail in part II.) The position that I am describing here as "minimal anti-nominalism" is essentially Burgess's moderate kind of realism.

[7] Burgess and Rosen (1997), pp. 10–11.

On the one hand, the nominalist's reconstructions of mathematics might be part of an attempt to give analyses of the mathematical propositions being asserted by actual practicing mathematicians. In other words, the nominalist might be claiming that the propositions of his or her reconstructions are what practicing mathematicians are actually asserting. A philosopher who held such a view would be classified by these authors as a *hermeneutic nominalist*. On the other hand, the nominalist might be proposing a new kind of mathematics that scientists ought to accept in place of their current mathematical theories. On this view, there is no suggestion that our actual mathematical theories are nominalistic; instead, it is argued that we ought to replace our current mathematical theories with the nominalistic reconstructions being developed. A philosopher who took this route would be classified as a *revolutionary nominalist*.

To understand adequately the reasoning underlying the Burgess–Rosen evaluations, it is necessary to review a paper cited earlier in which Burgess attacks reconstructive nominalism and defends his "moderate realism." What follows is a sketch of the argument.

We first need some definitions that Burgess gives:

> An *instrumentalist* maintains that "science is just useful mythology, and no sort of approximation to or idealization of the truth."
>
> A *hermeneutic nominalist* holds that when the language of mathematics is properly analyzed, one will see that the scientist, in asserting mathematical propositions, is not really asserting the existence of any abstract mathematical objects.
>
> A *revolutionary nominalist* is a nominalist who proposes a new version of science—a nominalistic version—in which there are no assertions of the existence of abstract mathematical objects.

The argument proceeds from just one premise:

> [%] A nominalist is either an instrumentalist, or a hermeneutic nominalist, or a revolutionary nominalist.

Burgess's strategy is to argue that none of the above three positions is really tenable. So let's turn to the details of his reasoning.

Few philosophers are "instrumentalists" in the extreme form given by the definition above. And I agree with Burgess that this sort of instrumentalism is not plausible. Thus, if one accepts Burgess's premise, one is faced with choosing between being a *hermeneutic nominalist* or being a *revolutionary nominalist*. Hermeneutic nominalism is a thesis about the proper analysis of sentences of mathematics—and since ordinary mathematics is expressed in English or some such natural language, it is a thesis about the proper analysis of sentences of ordinary languages. Now Burgess can find no linguistic evidence supporting the hermeneutic nominalist's thesis. And no philosopher has adduced any convincing evidence supporting such a thesis (that is, evidence that a linguist would take seriously).

The revolutionary nominalist, on the other hand, is proposing a new version of science. So the question is: Why should we adopt the revolutionary version? What grounds are we given for accepting such a new version of science? There are, he claims, no good scientific grounds for such a revolution, and there are strong practical reasons for not proceeding with such revolution. For example:

> [A]ny major revolution involves transition costs: the rewriting of text books, redesign of programs of instructions and so forth. [I]t would involve reworking the physics curriculum [to allow the student to take courses to learn the logical or philosophical concepts required to understand the nominalist's reconstructions].[8]

Thus, Burgess concludes:

> Unless he is content to lapse into a mere instrumentalist or "as if" philosophy of science, the philosopher who wishes to argue for nominalism faces a dilemma. He must search for evidence for an implausible hypothesis in linguistics, or else for motivation for a costly revolution in physics. Neither horn seems very promising, and that is why I am not a nominalist. (Burgess, 1983, p. 101)

In a more recent work, Burgess argues that the hermeneutic and the revolutionary nominalists both make weighty claims the burdens of proof of which have "not yet been fully met" (1990, p. 7). Returning to the even more recent work by Burgess and Rosen (1997), they concentrate, for the most part, on the revolutionary nominalist's rationale for their reconstructions. Specifically, they question "the scientific merits of a nominalistic reconstruction as an alternative to or emendation of current physical or mathematical theory" (p. 205), and they argue that the ultimate judgment of the scientific merits of the nominalist's reconstructions should be made by the scientific community and not by the philosophical community. They go so far as to claim that "the true test would be to send in the nominalistic reconstruction to a mathematics or physics journal, and see whether it is published, and if so how it is received."[9] They then write:

[8] Burgess (1983), p. 98. Klaus Stelau has suggested to me, in response to the above argument against revolutionary nominalism, based upon such pragmatic considerations, that the position being attacked is not committed to requiring physicists to do mathematics in the reconstructive way being advocated: there could be a "division of labor" according to which physicists would continue to practice their physics in the way being done today, while knowing that the mathematics they make use of could be done (in principle by foundationalists) in the nominalistic way being advocated.

[9] Burgess and Rosen (1997), p. 206. In Chihara (2004), p. 163, I question the idea that a true test of the merits of a paper detailing a nominalist's reconstruction of mathematics is whether the editors of a physics journal would judge the paper worthy of publication in their journal. I see no reason to conclude from a negative decision to publish that the editors are passing on the scientific value of the paper (after all, they could reasonably judge that such a paper is simply not appropriate for the readership of their journal).

> If nominalistic reconstruals are not plausible as analysis of the ordinary meaning of scientific language, and if nominalistic reconstructions are not attractive by our scientific standards as alternatives to current physical or mathematical theories—if nominalism makes no contribution to linguistic science, nor to physical or mathematical science—then must the programme of nominalistic reconstrual be judged... an intellectual entertainment addressed to no serious purpose? (Burgess and Rosen 1997, p. 208)

This last question is addressed at the very end of their book, where they ask: "What is accomplished by producing a series of such distinct and *inferior* theories?" (Burgess and Rosen 1997, p. 238; italics added).

Notice that the nominalist's reconstructions are now judged to be "inferior." It is not clear to this writer why Burgess and Rosen feel justified in using the term "inferior" in their question.[10] My own view is that their judgment of inferiority is based upon a mistaken understanding of what the reconstructions were devised to accomplish. If one thinks tweezers are nail pullers, then one would undoubtedly regard tweezers as inferior tools indeed.

In any case, they answer their question (the one that ends the paragraph preceding the last one) with "No advancement of science proper, certainly; but perhaps a contribution to the philosophical understanding of the character of science." Here is how they imagine such a contribution being made:

> Devising alternatives distinct from and inferior by our standards to our actual theories, but in principle possible to use in their place, is a way of imagining what the science of alien intelligences might be like, and as such a way of advancing the philosophical understanding of the character of science.
>
> It is just such an advance, we want to suggest, that is accomplished by the various reconstructive nominalistic strategies surveyed in this book. (Burgess and Rosen 1997, p. 243)

Thus, the work of the nominalist is likened to the work of science fiction writers, who help us to imagine what the intellectual products of alien intelligences might be like—quite a damning simile.[11]

All in all, Burgess and Rosen have assembled quite a case against the nominalist. Judging from the pun in the title of their book, "A Subject with No Object,"[12] it would seem that the Burgess–Rosen criticisms were aimed at undermining the very

[10] They write: "Since anti-nominalists reject all hermeneutic and revolutionary claims, from their viewpoint the various reconstruals or reconstructions are all distinct from and inferior to current theories" (Burgess and Rosen 1997, p. 238). I can see why they infer that the reconstructions are distinct from current theories, but I fail to see why these authors conclude that such reconstructions must be inferior to current theories.

[11] See Burgess (1990), pp. 13–14, for amplifications of his view of how reconstructive nominalism might be regarded as suggesting what the science of aliens might be like.

[12] The pun suggests that there is no objet (goal) to the nominalistic reconstructions of mathematics.

object or goal of nominalism. Shapiro, for one, thinks that the authors of that book have succeeded admirably in raising "sharp and penetrating criticisms of the nominalistic projects and of the whole point of nominalism."[13]

Part II. A Nominalist's Response

I shall now give reasons for holding that Burgess and Rosen have not succeeded in undermining nominalism, at least when it is directed at the sort of nominalism I advocate. I shall first present a rationale for the nominalist's reconstructions of mathematics that is very different from the epistemological one described by Burgess and Rosen. To understand this rationale, one needs to appreciate the view of philosophy which motivates much of the work done by the nominalist of the kind I have in mind.

1. A Nominalist's View of Philosophy

The field of philosophy is divided into a number of specialties: philosophy of language, philosophy of mind, philosophy of science, philosophy of logic, philosophy of art, and so on. For practically any area *X* of intellectual study, there is a philosophy of *X*. As a general rule, one can say that the philosophy of *X* is aimed at achieving a kind of understanding of *X* that is unique to philosophy. One might call this sort of understanding "big picture understanding." What one seeks in philosophy is the really "Big Picture:" What, in general and in broad outlines, is the universe like? What, in general and in broad outlines, is our (i.e., humanity's) place in the universe? How, in general and in broad outlines, do we (humans) gain an understanding of the universe? And, more specifically, how, in general and in broad outlines, does *X* fit into this big picture?

Of course, this goal of producing such a big picture should not mislead one into thinking that subtle distinctions, careful and detailed examination of conceptual matters, and lengthy and intricate reasoning about minute points should not matter. Philosophers are concerned with very fundamental concepts, so that their analyses, even about apparently small matters, have very far-reaching consequences for the big picture being constructed.

[13] Shapiro (1998), p. 600. Burgess and Rosen suggest that the last chapter of their book should be titled "In Lieu of Conclusion" (instead of "Conclusion"), because their final remarks about reconstructive nominalism are neither "conclusions drawn from anything established in previous chapters" nor "conclusive" (Burgess and Rosen 1997, p. 205). Shapiro responded to this suggestion by claiming that these final remarks of Burgess and Rosen are "about as 'conclusive' as polite, professional philosophy gets nowadays" (1998, p. 600).

In this search for the big picture, *coherence* is an essential ingredient. We seek an understanding of *X* that is consistent with, and holds together with, the other views we accept about the universe and about us. Take the philosophy of language, for example. Here, we seek an understanding of the nature of language and our mastery of language that is consistent with our general scientific, epistemological, and metaphysical views, both about the universe we inhabit and also about us as organisms with the features attributed to us by science. An account of the nature of language that made our ability to learn a language into a complete mystery would be considered by most philosophers of language to be in serious trouble. We seek a coherent and comprehensive big picture, where all the different *X*s fit together. In general, one would not expect a contemporary philosopher's account of language to contradict any of our prevailing views of science and scientific knowledge without very compelling reasons.

Revolutions in Science

This conception of philosophy suggests an explanation of a striking feature of the history of philosophy. Following the development and acceptance of a *revolutionary scientific theory*—a theory that undermines fundamental and central beliefs of the well-educated elite—there tends to appear a heightened amount of philosophical theorizing. Think of the important (and radical) philosophical writings that appeared following the seventeenth-century scientific discoveries that undermined much of the medieval conception of the universe. Or consider the philosophical activity that arose from the publication of Darwin's work on evolution.[14] Other examples: the enormous number of philosophical works dealing with Freud's writings on mental illness and childhood sexuality, and the lively discussions in present-day philosophy of science dealing with the remarkable implications of relativity theory and quantum mechanics that conflict with so many fundamental beliefs underlying Newtonian physics.

The above-noted activity of philosophers is fitting, given the conception of philosophy I have been describing. When science undermines fundamental and central beliefs—fractures our big picture of the universe—then philosophers feel a pressing need to put the pieces together again, to develop a new and coherent big picture of the universe.

The Importance of Paradoxes

Another characteristic of philosophy is its great attention to, and serious interest in, uncovering and solving paradoxes or antinomies. A paradox is an argument

[14] A work that emphasized the philosophical responses to these two cases, the seventeenth-century scientific discoveries and the Darwinian theory of evolution—is Girvetz et al. (1966).

that starts with premises that seem to be incontestable, that proceeds according to rules of inference that are apparently incontrovertible, but that ends in a conclusion that appears to be obviously false. In many cases, a paradox ends in a conclusion that is downright self-contradictory. From the earliest beginnings of philosophy in classical Greece (think of Parmenides, Heraclitus, and Zeno), paradoxes have played an important role in motivating and developing philosophical theories.

Consider what is undoubtedly one of the most striking cases of philosophical fervor brought about by the discovery of paradoxes: the discovery of the various paradoxes of mathematics and set theory in the late nineteenth and early twentieth centuries. These paradoxes stimulated much research in the foundations of logic and mathematics.[15] They led Frege eventually to abandon his logicism.[16] They also stimulated Poincaré to come up with his vicious-circle principle.[17] The paradoxes figured in Zermelo's defense of his axiomatization of set theory.[18] Russell, who discovered the paradox that bears his name, was led to develop his Theory of Types and his "no-class" theory during the many years he spent searching for a solution to the paradoxes.[19] Hilbert motivated certain aspects of his formalist philosophy of mathematics by the need to make certain that such paradoxes would never again be produced in mathematics.[20]

Reasons for the importance philosophers attach to paradoxes are not hard to find, given the above view of philosophy. An antinomy starts from premises that appear to be obviously true and proceed according to principles of inference that seem to be clearly valid. These premises and principles may be fundamental to our belief system (some may belong to a body of beliefs classified as "folklore"). An antinomy may show that some such beliefs and/or principles clash with recent developments in science, mathematics, or logic, or are simply inconsistent.

Alfred Tarski once wrote: "The appearance of an antinomy is for me a symptom of disease" (1969, p. 66). What is diseased, evidently, is a body of beliefs or system of principles that had been largely unquestioned or taken for granted. In those cases, philosophers take it upon themselves to try to heal the body or system.

[15] Cf. Raymond Wilder's assessment that "symbolic logic itself had its beginnings long before the discovery of the contradictions.... There can be little doubt, however, of the great impetus given to its development by the logical contradictions" (1952, p. 56).

[16] Near the end of his life, Frege completely abandoned his logicist view and came to the conclusion that the source of our arithmetical knowledge is what he called "the Geometrical Source of Knowledge." See Frege (1979).

[17] See Chihara (1973, ch. 1, sec. 1) for a discussion of Poincaré's reasoning.

[18] See Bach (1998, pp. 21–22) for supporting arguments.

[19] Chihara (1973, ch. 1) contains a detailed discussion of Russell's attempt to solve the paradoxes and his development of the Theory of Types, as well as of his "no-class" theory.

[20] In Hilbert (1983, pp. 190–191).

In the case of the mathematical and set-theoretical paradoxes, it was thought that the basic principles of mathematics or logic were shown to be diseased.[21] It is no wonder that the discoveries of these paradoxes brought about so much unease and disquiet.[22] One can see why foundationalists—mathematicians, logicians, and philosophers—put so much effort into an attempt to put the mathematical house in order. Since a paradox is a symptom that our body of beliefs and principles is not coherent, the above account of one of the principal goals of philosophy allows us to see why philosophers feel such a need to try to refashion our beliefs and principles into a coherent big picture in which the paradoxes can no longer be constructed.

2. Why Nominalistic Reconstructions of Mathematics?

I shall now present a picture of what at least one type of nominalistic reconstructivist is attempting to accomplish—a picture that is very different from the one Burgess and Rosen have sketched in their work. First of all, the kind of nominalist I have in mind does not begin with the thesis that there are no abstract entities. I see the nominalist, rather, as an anti-realist. So to understand this kind of nominalism, one needs some acquaintance with realism in mathematics.

What Is Realism (Platonism)?

In the philosophy of mathematics, the realist maintains that mathematical objects exist and that the mathematician is attempting to provide us with information about these objects. Thus, realists base much of their view of mathematics on the hypothesis that such things as numbers, sets, functions, vectors, matrices, spaces, and such truly exist: they generally assert, for example, that the theorems of set theory are true statements that tell us what sets in fact exist and how these mathematical objects are related to one another by the membership relationship. Mathematical entities are not supposed to be things that can be seen, touched, heard, smelled, tasted, or even detected by our most advanced scientific instruments. So a problem for the realist is to explain how mathematicians have been able to gain knowledge of such things. Realists, however, have backed their belief in the existence of such mathematical entities with a variety of philosophical arguments. One such argument has been especially influential.

[21] Cf. Russell's attitude toward the paradoxes. He was convinced that logic itself needed to be reformed. See Chihara (1973, p. 1).

[22] Frege was dismayed because the very foundations of his system of arithmetic were shaken by Russell's paradox. Russell, at first, thought that some relatively trivial error was responsible for the paradoxes. It was only later that he came to think some radical changes in logic were necessary to resolve the paradoxes. See in this regard Chihara (1973, ch. 1).

Quine's Challenge

It was Willard Quine who advanced the well-known "indispensability argument" for the existence of mathematical objects, by making such declarations as

> Mathematics—not uninterpreted mathematics, but genuine set theory, logic, number theory, algebra of real and complex numbers, differential and integral calculus, and so on—is best looked upon as an integral part of science, on a par with the physics, economics, etc., in which mathematics is said to receive its applications. (Quine 1966, p. 231)

With such a view of mathematics, it is not surprising that Quine would maintain that the mathematical nominalist "*is going to have to accommodate his natural sciences unaided by mathematics.*" Earlier, he had advanced a *criterion of ontological commitment*, according to which a theory *T*, expressed in a first-order logical language, is ontologically committed to entities of kind *K* iff such entities would have to be in the range of the bound variables of *T* in order that *T* be true.[23] Thus, when he asserted that "Mathematics, except for some trivial portions such as very elementary arithmetic, is irredeemably committed to quantification over abstract objects,"[24] he was in effect claiming that science itself was ontologically committed to the existence of mathematical objects. If the truth of all but trivial portions of mathematics requires that the quantifiers of the mathematical language range over mathematical objects or logically similar abstract objects, then it is hard to see how a nominalist can accept the pronouncements of the natural sciences without implicitly accepting the existence of mathematical objects. It is no wonder, then, that Quine thought that the nominalist would have to "*accommodate his natural sciences unaided by mathematics*" (Quine 1960, p. 269).

In my *Ontology and the Vicious-Circle Principle* (Chihara 1973), I took Quine to be proposing not so much a direct argument for the existence of mathematical objects, as a kind of challenge to the nominalist to produce a nontrivial system of mathematics which would be adequate for the needs of the natural scientist, but which would not require its quantifiers to range over mathematical objects.[25] I regarded him are arguing roughly as follows:

> *Let's see you nominalists accommodate science without committing yourself to mathematical objects. Let's see you produce a system of mathematics which would be adequate for the needs of the natural scientist, but which would not require its quantifiers to range over abstract objects.*

If this challenge cannot be met, then it can be plausibly argued, by a sort of inference to the best explanation, that we have good grounds for believing in the

[23] See Chihara (1973, ch. 3, sec. 3) for a detailed discussion of Quine's criterion.

[24] Quine (1960, p. 269, italics added).

[25] I was willing to grant Quine a version of his "criterion of ontological commitment." See my reconstruction of his argument in (Chihara 1973, ch. 3, sec. 5).

existence of mathematical objects (i.e., for being mathematical Platonists).[26] Burgess and Rosen put the argument as claiming that "we should believe in abstract entities, but only because nominalistic alternatives to standard scientific theories cannot be developed" (1997, p. 64).

Of course, Quine was convinced that no such nominalistic system of mathematics will ever be devised—the reason being that he believed mathematical objects were *indispensable* to the practice of science—so he reluctantly adopted the view of mathematics for which he is now famous: "reluctant Platonism."[27]

Mark Steiner has suggested that what Quine was getting at with his indispensability argument can be expressed in the following way:

> [T]o describe the experience of diversity and change requires mathematical entities. Imagine defining *rate of change* without the resources of analysis. We cannot say what the world would be like without numbers, because describing any thinkable experience (except for utter emptiness) presupposes their existence. (Steiner 1978, pp. 19–20)

Steiner's formulation can also be regarded as a form of challenge to the nominalist: defenders of the argument can be understood to be saying, "Let's see you describe any thinkable experience without making reference to or quantifying over mathematical objects."

Naturalized Epistemology

Another Quinean view that has been extremely influential is his doctrine of naturalized epistemology. Quine rejected the traditional a priori way in which philosophers tended to practice epistemology and advocated instead taking epistemology to be a subarea of psychology. He suggested that epistemology should study a "natural phenomenon, viz. a physical human subject":

> This human subject is accorded a certain experimentally controlled input—certain patterns of irradiation in assorted frequencies, for instance—and in

[26] The reader should be warned that the term "the indispensability argument" has been used to refer to several quite different kinds of arguments for the existence of mathematical objects. In particular, many scholars apply the term to an argument based upon Quine's holism: an argument that supposedly shows that we have strong empirical evidence supporting the hypothesis that mathematical objects exist. For discussions of this version of the indispensability argument, see Maddy (1997), Sober (1993), Resnik (1997), and Vineberg (1998). An over all assessment of this version of the argument is given in Chihara (2004, ch. 5).

[27] It is possible that one reason Quine was so sure no nominalist could meet his challenge is that he also adopted a strong thesis about the kind of language that science can legitimately be expressed in. This is the thesis that the language of science ought to be a logical first-order extensional language—a thesis that severely restricts what can be an acceptable nominalistic version of mathematics. For a discussion and criticism of this thesis, see Chihara (1990, pp. 8–14).

> the fullness of time the subject delivers as output a description of the three-dimensional external world and its history. The relation between the meager input and the torrential output is a relation that we are prompted to study for somewhat the same reasons that always prompted epistemology; namely, in order to see how evidence relates to theory, and in what ways one's theory of nature transcends any available evidence. (Quine 1969, pp. 82–83)

Quine's naturalized view of epistemology was adopted by Burgess and Rosen. Since these authors regard the typical nominalist as attempting to justify his (or her) position on epistemological grounds, it was reasonable for them to pay special attention to the field of epistemology. Thus, they contrast "the traditional alienated conception of epistemology"—the advocates of which they describe with the words "the epistemologist remains a foreigner to the scientific community, seeking to evaluate its methods and standards"—with the Quinean "naturalized conception of epistemology," according to which the epistemologist "becomes a citizen of the scientific community, seeking only to describe its methods and standards, even while adhering to them." They argue that since the alienated epistemologist attempts to evaluate the methods and standards of the scientific community, he or she needs to use methods and standards of evaluation that are "outside and above and beyond those of science" (Burgess and Rosen 1997, p. 33).

It is clear where the sympathies of these two authors lie when they describe the "the pretensions of philosophy to judge common sense and science from some higher and better and further standpoint," and then go on to characterize the following words of David Lewis as giving an especially forceful expression of the rejection of such "pretensions":

> Renouncing classes means rejecting mathematics. That will not do. Mathematics is an established, going concern. Philosophy is as shaky as can be. To reject mathematics on philosophical grounds would be absurd.... I laugh to think how presumptuous it would be to reject mathematics for philosophical reasons. How would you like to go and tell the mathematician that they must change their ways, and abjure countless errors, now that philosophy has discovered that there are no classes?... Not me![28]

[28] Burgess and Rosen (1997, p. 34). The passage quoted can be found in Lewis (1991, p. 58). Burgess and Rosen do not tell the reader, nor note the irony, of the fact that this same philosopher is willing to proclaim to the whole world (and presumably to astronomers) that he has made the remarkable discovery that there are planets in existence, distinct from those in our solar system, on which there are intelligent beings who speak English, who philosophize as we do, who have developed mathematics and science identical to our own, and who have a political system that is indistinguishable from the American system, with a president who looks, talks, thinks, and acts exactly like G. W. Bush—all this on the basis of philosophical reasoning! For a discussion of how Lewis thinks he has achieved such a discovery, see Chihara (1998, ch. 3).

Why Some Nominalistic Reconstructivists Should Not Be Classified as Alien Epistemologists

The nominalistic reconstuctivists of the sort I have in mind do not attempt to judge common sense and science from some higher, better, and further standpoint. They seek to piece together their account of mathematics in a way that is compatible both with what science teaches us about how we humans obtain knowledge and also with what we already know about how humans learn and develop mathematical theories. Furthermore, these nominalists do not reject mathematics—a fortiori, they do not reject mathematics on the basis of "some higher and better and further standpoint." On the contrary, their goal is to understand the nature of mathematics in a way that is compatible with the other features of the "big picture" they are attempting to construct.

It is true that, as nominalists, they do not believe that sets, classes, or extensions of concepts actually exist. But their skepticism about the existence of such things is not based upon the conviction that they have some decisive knockdown, a priori argument showing that mathematical objects cannot (or do not) exist. Rather, it is based upon a lot of different considerations such as:

(a) That set theorists, sitting in their offices and merely thinking about sets, were able to discover somehow the truth of the axioms of set theory (taken to be assertions about mathematical objects) is something that nominalists typically find either utterly mysterious or unintelligible.[29]

(b) Nominalists have serious doubts about the arguments (such as the various versions of the indispensability argument discussed earlier) that Platonists have given to support their belief in such things. All such arguments have seemed to them to be highly questionable.[30]

(c) They find implausible and unscientific the philosophical theories that Platonists have advanced to account for their supposed knowledge of the things they postulate.[31]

[29] Maddy (1990, ch. 2) has attempted to give explanations, based upon the work of the neurophysiologist Donald Hebb, as well as her theory of perception, for how the truth of some of the axioms of set theory (understood realistically) have been discovered. Needless to say, I have not found her explanations believable. However, because Maddy has now abandoned her realism, I shall not give here my reasons for rejecting her explanations.

[30] There is an enormous literature on this topic. See, for example, Vineberg (1998) for an assessment and other references.

[31] For example, Gödel's appeal to mathematical intuition, which is supposed to be something like a faculty of perception. (See Chihara 1982 for details). Another example is Maddy's (1990) view that humans are able to literally see sets.

(d) The Platonist's position that empirical scientists need to discover complex and complicated relationships between entities that do not exist in the physical world[32] in order to develop their empirical theories has seemed bizarre and counterintuitive to nominalists. They question the explanations that Platonists have given of why knowledge of mathematical objects is required in order to obtain sophisticated scientific knowledge of the physical universe.[33]

(e) Their assessment of the many doctrines, advanced by philosophers over the ages, that postulate some type of nonphysical, undetectable substance or object in order to account for some feature of language or belief has taught them to be skeptical of all such doctrines.

(f) They are convinced that the descriptions Burgess and Rosen give of the nominalist ("the pretensions of philosophy to judge common sense and science"), as well as Lewis's diatribe against the presumptuousness of philosophers, are fundamentally misleading. First of all, contrary to what Lewis has suggested, there are many nominalists who maintain neither that our mathematical theories consist of false assertions nor that mathematicians have to change their ways because of what philosophers have discovered.[34] Second, one could easily get the impression from the above-mentioned Platonists that it is only the presumptuous philosopher (with pretensions of having some special means of judging the affirmations of science) who has doubts about the existence of mathematical objects. In fact, many (if not most) of the outstanding researchers in mathematics have serious doubts about the Platonist's account of mathematical theorems and knowledge.[35] And it has certainly been my experience that many empirical scientists find the Platonic view of mathematics quite fantastic. Also, anyone who has taught philosophy of mathematics (as I have for a great many years) will no doubt remember the look of disbelief, and sometimes of amazement, on the faces of some students on being told that highly respected philosophers and mathematicians at major universities believe that, in addition to planets, stars, galaxies, atoms, electrons, molecules, and photons, such things as numbers, sets, and

[32] Gödel explicitly asserted that the objects of transfinite set theory do not exist in the physical world. See Gödel (1964), p. 271.

[33] See, for example, Brown (1999, ch. 4, esp. pp. 47–49).

[34] See, in particular, Chihara (2004), where I advance an analysis of mathematics that implies neither of the positions being attributed to nominalists.

[35] Paul Cohen has opined that "probably most of the famous mathematicians who have expressed themselves on the question have in one form or another rejected the Realist position" (1971, p. 13). Soloman Feferman has expressed his anti-realist views quite forcefully: "Briefly, according to the Platonist philosophy, the objects of mathematics such as numbers, sets, functions, and spaces are supposed to exist independently of human thoughts and constructions, and statements concerning these abstract entities are supposed to have a truth value independent of our ability to determine them. Though this accords with the mental practice of the working mathematician, I find the viewpoint philosophically preposterous" (Feferman (1998, p. ix)).

> functions also exist. These reconstructivists feel that any account of mathematics should take account of the skeptical attitudes of *nonphilosophers* about the Platonic account. It suggests to them that the belief that Platonists have in mathematical objects is not at all like the belief that, essentially, everyone has in ordinary material objects like tables and chairs, despite what some Platonists have argued.[36]

Such considerations as the above lead these nominalists to doubt the appropriateness of the Platonist's conception of mathematical theories as descriptions of a realm of objects that do not exist in the physical world. These nominalists are skeptical of the Platonic doctrine that to achieve genuine mathematical knowledge, one needs to know that such objects truly exist and are related to one another as the theory in question affirms. As a result, these nominalists search for an alternative way of understanding the nature of mathematics, and *their reconstructions are seen as an aid to achieving their overall goal of arriving at such an understanding.*

I would like to emphasize that my own reconstructive works were never motivated by Benacerraf-type epistemological arguments. My first reconstructive book, *Ontology and the Vicious-Circle Principle*, was published in 1973—the year Benacerraf's paper "Mathematical Truth" was published—and almost all of the main reconstructive ideas were developed many years earlier. I certainly based nothing in my book on the Benacerraf paper. Indeed, I began the book with the quotation from Quine's *Word and Object* (1960), in which Quine declares that the nominalist would have to "accommodate his natural sciences unaided by mathematics," followed by these words:

> When I first read these words in *Word and Object* several years ago, I wrote in the margin: "This philosophical doctrine should be soundly refuted." It was only much later, while I was working on an essay on the vicious-circle principle, that an idea came to me as to how one might construct such a refutation.[37]

What I felt should be refuted was Quine's claim that the nominalist would be unable to develop a nontrivial system of mathematics for the natural sciences that would not be "ontologically committed" to mathematical objects. I had this conviction long before any talk of causal theories of knowledge or Benacerraf-type anti-realist arguments were heard in philosophical circles. The aim of my reconstruction, then, was to refute the Quinean claim by producing a nontrivial system of mathematics that does not require quantification over mathematical objects. But the nominalistic reconstruction was not supposed to be the end of the

[36] Penelope Maddy, at one time, argued that Gödel is not required to supply a theoretical justification for his belief in sets since we humans have perceived and believed in sets since prehistoric times, suggesting that our belief in sets is very much like our belief in material objects. See (Chihara, 1982), Part 2, especially p. 226.

[37] Chihara (1973, p. xiii). The essay mentioned was written in 1964–1965.

story: it was to be an aid toward gaining an understanding of the actual functioning of the mathematical theories mathematicians actually use and develop.[38]

3. A Nominalistic Reconstruction of Mathematics

In order to give you some idea of what these nominalistic reconstructions of mathematics are like, I shall here very briefly sketch the main ideas of the Constructibility Theory of my book *Constructibility and Mathematical Existence* (1990).[39]

A Brief Exposition of the Constructibility Theory

Set theory is a theory about sets. It tells us what sets exist and how these sets are related to one another by the membership relationship. The Constructibility Theory is a theory about open sentences: it tells what open sentences (of a certain sort) are constructible and how these constructible open sentences would be related to one another by the *satisfaction relation.* (Thus, in this theory, one open sentence can satisfy another.) The theory is formalized in what is basically a many-sorted, first-order logical language[40] that utilizes, in addition to the existential and universal quantifiers of standard first-order logic, *constructibility quantifiers.*

Constructibility quantifiers are sequences of primitive symbols: either (**C**—) or (**A**—), where "—" is to be replaced by a variable of the appropriate sort. Using "$\Psi\phi$" to be short for "ϕ satisfies Ψ," "$(\mathbf{C}\phi)\Psi\phi$" can be understood to say

> It is possible to construct an open sentence ϕ such that ϕ satisfies Ψ,

whereas "$(\mathbf{A}\phi)\Psi\phi$" can be understood to say

> Every open sentence ϕ that it is possible to construct is such that ϕ satisfies Ψ.

And just as "$(\exists x)Fx$" is equivalent to "$-(x)-Fx$," and "$(x)Fx$" is equivalent to "$-(\exists x)-Fx$," we have "$(\mathbf{C}\phi)\Psi\phi$" is equivalent to "$-(\mathbf{A}\phi)-\Psi\phi$," and "$(\mathbf{A}\phi)\Psi\phi$" is equivalent to "$-(\mathbf{C}\phi)-\Psi\phi$".

The assertions of this theory are the sentences of the theory that can be derived from a set of axioms, which yield what is basically the mathematics of

[38] How the Constructibility Theory is supposed to aid the nominalist in obtaining such an understanding is explained in detail in Chihara (2004).

[39] This account has been criticized in detail by Stewart Shapiro (1997, ch. 7, sec. 4) and Michael Resnik (1997, ch. 4, sec. 2). A detailed and extensive reply to their objections appears in Chihara (2004, ch. 4).

[40] For a clear and rigorous discussion of such languages, see Enderton (1972, sec. 4.3).

Principia Mathematica. Essentially, the mathematics is what foundationalists call "the simple theory of types."

It needs to be emphasized that what are said to be constructible by means of the constructibility quantifiers are open-sentence *tokens* as opposed to open-sentence types. Open-sentence types are classified by Charles Parsons as "quasi-concrete" objects (in contrast to purely mathematical objects), because "they are directly 'represented' or 'instantiated' in the concrete" (Parsons 1996, p. 273). Despite the fact that these objects are "quasi-concrete," it can be argued that they are as epistemologically disreputable as numbers and sets. Open-sentence tokens, however, are not open to the same objection: typically, an open-sentence token consists of particular marks on paper, which exist at a particular place in the universe and at a particular time. Furthermore, to say that an open sentence of a particular sort is constructible is not to imply or presuppose that any such open-sentence token actually exists or, indeed, that anything exists. Constructibility quantifiers do not carry ontological commitments as do the quantifiers of standard extensional logic.

What Does "Possible" Mean?

In the phrase "it is possible to construct," the term "possible" needs some explanation. There are, of course, many different kinds of possibility: logical, metaphysical, physical, epistemological, technological, to name just a few. Epistemological possibility is concerned with what is known. Thus, to say that ϕ is epistemologically possible for agent X is to say that ϕ is not logically precluded by what X knows—that is, ϕ is logically compatible with everything X knows. To say that ϕ is physically possible is to say something like "ϕ is logically compatible with all the physical laws of our universe." The possibility talked about in the Constructibility Theory is what is called "conceptual" or "broadly logical" possibility—a kind of metaphysical possibility, insofar as it is concerned with *how the world could have been.* Every purely logical truth is necessary in this sense, but the set of conceptual necessary truths will include much more. What are called "analytic truths" (such truths as "All bachelors are unmarried") are also held to be necessary in this broadly logical sense.[41] As a rough guide, Graeme Forbes suggests that "it is possible that P" should be taken to mean "There are ways things might have gone, no matter how improbable they may be, as a result of which it would have come about that P" (Forbes 1985, p. 2). Another way of expressing this sort of possibility is to take it to mean that the world could have been such that, had it been this way, P would have been the case. There are many different systems of modal logic, but the type of system that is generally believed

[41] See Plantinga (1974, ch. 1, sec. 1) and Forbes (1985, ch. 1, sec. 1) for more examples and discussion.

to correctly formalize the logical features of this broadly logical sense of necessity is S5.[42]

What "($\mathbf{C}\phi$)$\Psi\phi$" Does Not Mean

"($\mathbf{C}\phi$)$\Psi\phi$" does not mean that *one knows how* to construct such an open sentence or that one has a method for constructing such an open sentence. Hence, the constructibility quantifier is not at all like the Intuitionist's existential quantifier. Furthermore, it does not mean that one can always, or even for the most part, determine what particular objects would satisfy such an open sentence or how one would determine what objects would satisfy ϕ. Nor does it mean that one can determine if a series of marks, sounds, hand signals, or what have you is or is not such an open sentence.

The Many Levels of the Constructibility Theory

The Constructibility Theory is similar to Frege's theory of concepts: just as Frege's hierarchy of concepts is stratified into levels, so the open sentences that are talked about in the Constructibility Theory are of different levels. Thus, consider the following situation. On the desk in my office, there are two pieces of fruit, which I have named Tom and Sue. On the blackboard in my office, I write the open sentence

> x is a piece of fruit on the desk in my office.

Both Tom and Sue satisfy this open sentence. The desk does not. This open-sentence token is of level 1. Now suppose that I write the open sentence

> There is at least one object that satisfies F

in the bottom left corner of my blackboard, where "F" is being used as a variable of level 1. This open-sentence token is satisfied by the open-sentence token I constructed earlier—the level 1 open sentence I first wrote on the blackboard. The second open sentence I constructed is a second-level open sentence. Clearly, then, we can go on to construct open sentences of levels 3, 4, 5, and so on.

In the next section, I shall sketch the development of finite cardinality theory within the framework discussed above.

[42] Thus, Kit Fine writes, "S5 provides the correct logic for necessity in the broadly logical sense" (Fine 1978, p. 151). For an interesting discussion of the development of S5 modal logic, see Kneale and Kneale (1962, ch. 9, sec. 4). See Chihara (1998) for a discussion of a variety of S5-type systems of modal logic.

Notation I Shall Use to Refer to the Entities of Different Levels

Level 0: *Objects* x, y, z, ...
Level 1: *Properties* F, G, H, ...
Level 2: *Attributes* $\mathscr{F}$, $\mathscr{G}$, $\mathscr{H}$, ...
Level 3: *Qualities* **F**, **G**, **H**, ...

It needs to be emphasized that what I am calling "properties," "attributes," and "qualities" are just open sentences. Thus, the open sentences of level 1 that will be talked about in this theory are to be called "properties" and the uppercase letters F, G, H, and so on are to be used as variables to refer to these open sentences. Extrapolating from the level 1 case, one can see that the level 2 open sentences to be talked about will be called "attributes" and the script uppercase letters $\mathscr{F}$, $\mathscr{G}$, $\mathscr{H}$, and so on will be used as variables to refer to these second-level open sentences. Thus properties, attributes, and qualities are not being used to refer to universals or abstract entities of some sort, as is generally the case in philosophical works. *I use this terminology simply to facilitate our keeping distinct and ordered the open sentences of different levels I shall be talking about.*

Quantifiers

(a) Quantifiers containing occurrences of level 0 variables are just the standard quantifiers of first-order logic.

(b) Quantifiers containing occurrences of level 1 or higher variables are constructibility quantifiers.

Relations

All the open sentences to be talked about in this theory will be monadic open sentences (i.e., such open sentences as "x is a human" that contain occurrences of only one variable). The reason I restrict the theory to just monadic open sentences is simplicity: it makes the task of formalizing the theory so much easier.

In what follows, I shall define relations à la Kuratowski.

Couples

A couple $\{x, y\}$ is a property that is satisfied by only the objects x and y.

EXAMPLE: "$x = \text{Tom} \vee x = \text{Sue}$" is a couple {Tom, Sue}.

Notice that I have given here an example of a monadic open sentence: only one variable (i.e., "x") occurs in the open sentence—of course, there are two *occurrences* of this one variable.

Notice also that I say "a couple" (instead of "the couple"), because it is possible to construct indefinitely many different such couples. Thus, "x is the very same person as Tom or x is the very same person as Sue" is also a couple {Tom, Sue}.

Ordered Pairs

An ordered pair $\langle x, y\rangle$ is an attribute that is satisfiable by all and only couples $\{x, x\}$ and $\{x, y\}$ that could be constructed.

Note that an ordered pair is an open sentence satisfied by other open sentences—it is not satisfied by the objects x and y.

EXAMPLE: The open sentence

F is a couple {Tom, Tom} or F is a couple {Tom, Sue}

is an ordered pair ⟨Tom, Sue⟩.

Relations A relation **R** is a *quality* that is satisfiable only by ordered pairs such that if an ordered pair $\langle x, y\rangle$ could be constructed that satisfied **R**, then **R** is satisfiable by every ordered pair $\langle x, y\rangle$ that could be constructed.

EXAMPLE: "$\mathscr{H}$ is an ordered pair $\langle x, y\rangle$ such that x is married to y" is a quality which is a relation corresponding to the intuitive relation of *being married to.*

A Notational Definition If R is a relation, then

"$x\mathbf{R}y$" means$_{\text{def}}$ "an ordered pair $\langle x, y\rangle$ satisfies **R**."

Notice that, as the term "relation" has been defined, if an ordered pair $\langle x, y\rangle$ satisfies some relation **R**, so that $x\mathbf{R}y$, then every ordered pair $\langle x, y\rangle$ that is constructible would also satisfy **R**.

In what follows, many of the definitions will look exactly like those given in Frege's theory of cardinal numbers.

Relation **R** *correlates F to G*

iff for every x that satisfies F, there is a y that satisfies G such that

$$x\mathbf{R}y;$$

and for every y that satisfies G, there is an x that satisfies F such that

$$x\mathbf{R}y.$$

Relation **R** *is a one-one relation*

iff for every x, y, and z,
if $x\mathbf{R}y$ and $x\mathbf{R}z$, then $y = z$;
if $x\mathbf{R}y$ and $z\mathbf{R}y$, then $x = z$.

F is equinumerous with G

iff it is possible to construct a one-one relation that correlates *F* to *G*.

EXAMPLE: Let us suppose that the objects under consideration here are people. Then consider the following open sentences:

J: "x is the junior senator from some state"
S: "x is the senior senator from some state"

and let **R** be the open sentence:

$\mathscr{H}$ is an ordered pair $\langle x, y\rangle$ such that x is the junior senator from a state in which y is the senior senator.

It can be seen that **R** is a one-one relation correlating property *J* with property *S*. Hence, *J* is equinumerous with *S*.

The developments in the following section deviate somewhat from the Fregean developments, because we don't have objects to serve as the cardinal numbers.

Cardinal Number Attribute of a Property

An attribute $\mathscr{N}$ is a *cardinal number attribute of property F* iff $\mathscr{N}$ is satisfiable by all and only those properties equinumerous with *F*.

Example of a cardinal number attribute of a property:

G is equinumerous with "x is the senior senator from some state"

Cardinal Number Attribute

An attribute $\mathscr{N}$ is a *cardinal number attribute* iff it is possible to construct some property *F* such that $\mathscr{N}$ is a cardinal number attribute of property *F*.

The next step is to define the *natural number attributes*. Thus, I first define

A *zero attribute* is any cardinal number attribute that is satisfied by "$x \neq x$."

Then the following four definitions yield what is desired:

1. The immediate predecessor relation **P**
   $\mathcal{M}\mathbf{P}\mathcal{N}$ iff it is possible to construct a property F which is such that some object x satisfies it and $\mathcal{N}$ is a cardinality attribute of F and it is possible to construct a property G such that $\mathcal{M}$ is a cardinality attribute of G and G is satisfiable by all (and only those) objects different from x which satisfy F.
2. **P**-hereditary qualities
   A quality **Q** is **P**-hereditary iff, for all cardinality attributes $\mathcal{M}$, $\mathcal{N}$, if $\mathcal{M}$ satisfies **Q** and $\mathcal{M}\mathbf{P}\mathcal{N}$, then $\mathcal{N}$ satisfies **Q**.
3. **P**-descendants
   A cardinality attribute $\mathcal{N}$ is a **P**-descendant of cardinality attribute $\mathcal{M}$ iff $\mathcal{N}$ satisfies every **P**-hereditary quality that $\mathcal{M}$ satisfies.
4. Natural number attributes
   A natural number attribute is any **P**-*descendant of a zero attribute.*

Specific Natural Number Attributes

A natural number attribute is

a *one attribute* iff it is possible to construct a zero attribute which immediately precedes it;
a *two attribute* iff it is possible to construct a one attribute which immediately precedes it;
a *three attribute* iff it is possible to construct a two attribute which immediately precedes it;
and so on.

Because of space restrictions, I shall end my exposition of the Constructibility Theory here. One can find a more detailed and expansive discussion of the constructibility development of finite cardinality theory in Chihara (2004, ch. 7). I turn now to the Burgess–Rosen objections.

4. An Examination of the Burgess–Rosen Evaluation

According to Burgess's principal premise, the reconstructive nominalist must espouse one of three positions: scientific instrumentalism, hermeneutic nominalism, or revolutionary nominalism. Since I do not espouse scientific instrumentalism (certainly not in the form characterized by Burgess), Burgess's argument, directed at my nominalistic account, can be regarded as resting on the view that my choices are restricted to either hermeneutic nominalism or revolutionary nominalism.

Now the position I developed in my book (Chihara 1990) does not fit comfortably any of the above descriptions of the possible types of nominalism that Burgess provides. I was not espousing an instrumentalist view of science, since I in no way considered science to be a myth. My view was that a great many (if not most) of the assertions of scientists are, if not literally true, at least close to being true, when the proper conditions are expressed.

I also made it clear in my book that my constructibility theory was not meant to be an analysis of the mathematical statements asserted by practicing mathematicians. Here's what I wrote in the first chapter:

> If we are puzzled about certain aspects of classical mathematics, why not construct *another kind of mathematics* that will avoid those features of the original system that gave rise to the puzzles? A study of these alternative mathematical theories will give us a new perspective from which to view classical mathematics, which could prove to be extremely enlightening. (p. 23; italics added)

So was I proposing a revolutionary nominalism? Not at all. What I was proposing was an answer to the highly theoretical and deeply philosophical question posed by Quine: Can our contemporary scientific theories be reformulated or reconstructed in a way that will not require the assertion or the presupposition of abstract mathematical objects? My Constructibility Theory was meant to show how the mathematics needed for our contemporary scientific theories can be developed without requiring a commitment to quantification over abstract mathematical objects. I was not advocating that mathematics developable in my system be actually used by scientists and mathematicians. The system was devised to undermine a philosophical argument.

This point can be illustrated by taking up again Steiner's version of Quine's argument. Recall that he had suggested that one cannot describe "any [genuine nontrivial] thinkable experience" without presupposing numbers, since any such description would involve rate of change: "Imagine defining *rate of change* without the resources of analysis," he had challenged, confident that such a definition was impossible. In response, the nominalist can define rate of change using the Constructibility Theory, and this without presupposing numbers. Thus Steiner's challenge can be met.[43]

[43] My research assistant William Goodwin has responded to Steiner's argument by asking if Steiner thinks that before the nineteenth century arithmetization of analysis, and certainly before Newton and Leibniz came up with their versions of the calculus, no human was able to describe any genuine nontrivial thinkable experience. Do contemporary humans with no understanding of analysis lack the means of describing any nontrivial thinkable experience? The implications are staggering.

Does the Use of Classical Mathematics in Science Ontologically Commit One to the Existence of Mathematical Objects?

These developments also raise the question as to whether the use of even standard classical mathematics in science genuinely commits one to accepting the existence of abstract mathematical entities. If, as I have been arguing for many years, science can be done using the Constructibility Theory instead of standard versions of classical mathematics, without serious theoretical loss, this suggests that the role that mathematics is required to play by our contemporary scientific theories is not that of *referring to* and *providing information about* esoteric objects that do not exist in physical space (since the Constructibility Theory does not do these things). It encourages the thought that our contemporary scientific theories do not truly require belief in mathematical objects, as Quine and other Platonists have thought. Thus, we need to address the question: Does the use of even standard versions of classical mathematics in science logically commit one to the belief in abstract mathematical objects? But the answer to this question depends crucially on answering the following question:

> [#] *Does the use scientists make of standard versions of classical mathematics depend upon (or presuppose) taking the theorems of classical mathematics (literally and platonically construed) to be true?*

Most Platonists have simply assumed that [#] is true. For example, Michael Resnik has included the following principle as a premise of several of his arguments for the existence of mathematical objects:

> [*] We are justified in drawing conclusions from and within science only if we are justified in taking the mathematics used in science to be true.[44]

Resnik gives no justification for [*], perhaps because he thinks it is just obviously true. However, I have heard philosophers argue for [*] along the following lines:

> If the mathematics used in science and engineering were not true, how could we, in good conscience, use it in the way we do to build bridges and to design rockets? If the mathematical theorems upon which we base our scientific inferences and theories were false, would not our bridges collapse and our rockets go off target more often than not?

[44] See Resnik (1998, p. 234), where the quoted sentences expresses a premise of an argument that is formed by coupling his indispensability argument with a pragmatic argument. The quoted sentence also expresses a premise of his "indispensability argument for mathematical realism that separates questions of indispensability from questions of confirmation" given on p. 233. See chapter 12 of this volume.

5. How the Constructibility Theory Is Used in Analyzing Mathematical and Philosophical Reasoning

The Constructibility Theory plays a role in undermining the above indented argument, as well as the plausibility of [*] and [#]. Although I do not have the space here to explain my strategy in detail, I can at least indicate one kind of role that the Constructibility Theory plays in my refutation.[45] In this way, I can give additional grounds for doubting the Burgess–Rosen suggestion that the nominalist's reconstructions of mathematics are good only for showing how an alien mathematician might reason.

To show how the inferences drawn in science and engineering can be sound, even if the mathematical theorems used in drawing the inferences are not true, I set out to show two things:

1. The standard inferences involving the use of mathematics are indeed sound; and
2. These inferences, justified by appeal to mathematical theorems, can be legitimately drawn, without assuming that the mathematical theorems (literally and platonically construed) are true.

To justify claims (1) and (2), I needed to analyze mathematical reasoning and applications of mathematics in a way that does not presuppose the truth of the very mathematical theorems being analyzed. The Constructibility Theory is used in such an analysis, as is illustrated by the following examples.

Standard Arithmetical Inferences

Consider the case in which one concludes:

There are twelve coins on table A at time t

from the following premises:

[1] There are five dimes on table A at time t.
[2] There are seven quarters on table A at time t.
[3] A coin is on table A at time t iff it is either a dime or a quarter.
[4] Nothing is both a dime on table A at time t and a quarter on table A at time t.
[5] $5 + 7 = 12$.

The Constructibility Theory can be used to establish the soundness of this inference in the following way. For each of premises [1]–[4], as well as the

[45] The details of my response to the indented argument are given in Chihara (2004, ch. 9).

conclusion, there is a corresponding sentence of the Constructibility Theory (which I shall call the c-version of the sentence) such that the sentence of the natural language is true iff the c-version of the sentence is true. It can be shown, using the Constructibility Theory, that if the c-version of the premises are all true, then the c-version of the above conclusion must also be true. We can thus see that (1) if the premises are all true, the conclusion must also be true; and (2) nowhere in this justification is it assumed that [5] must be literally true. All of this is worked out in detail (Chihara 1990, pp. 89–92) to show that our intuitive reasoning using arithmetic is indeed sound.

The above example indicates the kind of role that the Constructibility Theory can play in the nominalist's analyses of reasoning involving mathematics: here, it functions in the nominalist's metatheory as a sort of tool, which helps the theoretician to understand (and to explain) the workings of real mathematics without assuming that theorems of mathematics, literally and platonically construed, are true.

In the above account, I am not claiming that the c-versions of the premises give the meaning of the premises. I am not a *hermeneutic nominalist.* However, as a native speaker of English, I am in a position to know that each premise is true iff the c-version of the premise is true, even if I cannot confidently give a semantic analysis of the premise. Taking a page from G. E. Moore's philosophy, I take the position that I can know, for example, that *this* (pointing at my right hand) is a hand iff *that* (pointing at my left hand) is a hand, even when I am not in a position to give any precise semantic analysis of statements I may make using the sentence "This is a hand."

The Use of the Constructibility Theory to Evaluate Philosophical Arguments

The Constructibility Theory can also be used in the assessment of philosophical arguments involving arithmetic and cardinality theory. Consider Burgess's attempted refutation of a version of the causal theory of knowledge. Burgess (1990) notes that the statement

Avogadro's number is greater than 6×10^{23}

has been judged by scientists not only to be true, but to be even known to be true (p. 6). He also points out that the statement seems to imply that there are numbers. Evidently, we have scientific grounds for asserting that it is known that there are numbers. Consequently, Burgess suggests, science provides us with grounds for rejecting the version of the "causal theory of knowledge" according to which any statement or theory implying that there are objects of a certain sort cannot be known to be true unless some objects of that sort causally interact directly or indirectly with us.

Let us evaluate this reasoning. Certainly, scientists do claim to know such things as

The number of planets whose orbits are smaller than that of the Earth is two.

The question is: Are these scientists claiming to know something that implies the actual existence of abstract mathematical objects? Suppose we put to these scientists the question

What justification do you have for making this claim?

I am confident that whatever scientific grounds these scientists would provide in response to our question would amount to no more than the grounds they have for asserting that the open sentence "x is a planet whose orbit is smaller than that of the Earth" satisfies a two-attribute. In other words, the grounds that the scientist would supply would be no more than the grounds we have for asserting the constructibility version of the cardinality statement.

Would any of the grounds that scientists supplied in answer to our question justify, over and above the constructibility cardinality statement, the proposition that there exist entities with which we are in no direct or even indirect causal relations? Not likely. I am confident that these scientists would provide a reasonably cautious person with no good reason for asserting, "Now I know that there exist in the actual world abstract entities from which we are completely and totally cut off causally." Hence, I am skeptical that science has provided us with grounds for rejecting the causal theory in the way Burgess has suggested.

Here again, the Constructibility Theory functions as a tool of the nominalist's metatheory, enabling the theoretician to more easily assess the soundness of certain philosophical arguments. These uses of the theory should not be lumped with science fiction.

The Vineberg defense

Against the earlier Burgess–Rosen objection to the reconstructive nominalist, Susan Vineberg has marshaled a case supporting the relevance of nominalistic reconstructions for deciding whether belief in mathematical objects is required by science.[46] Vineberg investigates various views of compelling evidence that do not require a causal connection between the scientist and what she has evidence for. Of the above sort, Vineberg finds three: the well-known Bayesian view; the eliminativist view, advocated by Philip Kitcher[47]; and the contrastive view, proposed by Elliot

[46] This was done at the Pacific Division Meeting of the American Philosophical Association in 2001.

[47] See Kitcher (1993, esp. pp. 237–247).

Sober and Larry Laudan.[48] Omitting here the details of her reasoning, Vineberg concludes that, regardless of which of the above theories of evidence one may choose, "substantial confirmation requires eliminating, or revealing as improbable, alternative theories." Not surprisingly, it is common scientific practice to look for alternate theories that fit or explain the relevant experimental data and observations that have been gathered, and to attempt to find evidence that allows us to eliminate all but one of these theories. Within this context of scientific theory evaluation, the nominalistic reconstruction of mathematics discussed above can be seen to be fitting and reasonable: in the absence of genuine evidence (as opposed to mere pragmatic considerations such as "familiarity with established theory") that allows the Platonist to eliminate this alternative, the nominalistic reconstruction provides us with rational grounds for being skeptical about the existence of mathematical objects.

Vineberg's point can be illustrated by the following: it is thought by Platonists that the existence of mathematical objects is required to solve certain problems or puzzles, and that this fact alone counts strongly in favor of belief in mathematical objects. For example, according to Steiner, Frege completely solved what Steiner calls the "metaphysical problem of applicability".[49] Since Frege's solution presupposes the existence of mathematical objects, it is thought that the very success of Frege's solution amounts to a reason for believing in the existence of mathematical objects.[50]

The Constructibility Theory, however, allows one to devise an alternative solution to the "metaphysical problem of applicability"[51]—a solution that does not presuppose the existence of mathematical objects—thus undercutting the idea that the postulation of mathematical objects is required to solve the problem. We can thus see how a nominalistic alternative can serve to undermine various claims of the Platonist that we have compelling scientific grounds for postulating the existence of mathematical objects.

[48] Sober's eliminativist view is discussed in Sober (1993). Laudan's view is presented in Laudan (1997), which contains the idea, relevant to the discussion above, that "The evaluation of a theory or hypothesis is relative to its *extant* rivals. To accept *H* is to hold that it is more reliable than its known rivals; to reject *H* implies that it is worse than at least one of its known rivals" (p. 314). As Vineberg has noted, James Hawthorne has analyzed Bayesian induction in such a way that it can be seen to be a probabalistic form of eliminative induction, i.e., "a method for finding the truth by using evidence to eliminate false competitors" (Hawthorne 1993, p. 99).

[49] See Steiner (1995, p. 132). Steiner attributes the formulation of this problem to Carl Posy.

[50] Frege's solution is supposed to consist in showing how mathematical entities relate, not directly to physical objects but rather through concepts. Commenting on this solution, Steiner writes: "That physical objects may fall under concepts and be members of sets is a problem only for those who do not believe in the existence of sets" (1995, p. 138).

[51] Such a solution is given in Chihara (2004, ch. 9).

I should like to emphasize that the nominalistic reconstruction of mathematics described here was not put forward either as an analysis of the systems of mathematics used by practicing mathematicians or as a theory to be used in place of the standard mathematical systems in use today all over the world. But this does not mean that the only use left for the reconstruction is as an imaginative exercise in how alien mathematicians might have reasoned.

6. Summation

It makes a big difference whether one sees realism in the *positive* way Gödel did—as part of a positive account of the nature of mathematical truth and of how mathematics is related to science—or in the purely *negative* ("anti-nominalist") way Burgess and Rosen do. Burgess and Rosen seem to have hit upon a brilliant strategy for defending realism. By characterizing their own version of realism in the negative way they did, they seem to have succeeded both in shifting the onus of justification onto the nominalist and at the same time absolving themselves of the enormous burden of explaining how they obtained their supposed knowledge of mathematical objects. However, I find their minimal anti-nominalism very unsatisfying. I am inclined to call the Burgess–Rosen brand of realism "aphilosophical realism" instead of "moderate realism" or "minimal anti-nominalism," since it does not subject the doctrine being advanced to philosophical analysis or development, being content merely to assert the existence of mathematical objects without attempting to make sense of the doctrine, to reconcile the doctrine with the various epistemological views that are widely accepted, or to explain a number of very puzzling questions that arise in connection with what mathematicians and scientists do and say. If risk aversion is one's chief concern, then minimal anti-nominalism may be a good strategy, but from the perspective of those sympathetic to the view of philosophy described in the beginning of part II, minimal anti-nominalism will simply be a cop-out.

REFERENCES

Bach, Craig (1998). "Philosophy and Mathematics: Zermelo's Axiomatization of Set Theory." *Taiwanese Journal of Philosophy & History of Science* 10: 5–31.

Benacerraf, Paul (1973). "Mathematical Truth." *Journal of Philosophy* 70: 661–679.

Brown, James R. (1999). *Philosophy of Mathematics: An Introduction to the World of Proofs and Pictures.* London, Routledge.

Burgess, John (1983). "Why I Am Not a Nominalist." *Notre Dame Journal of Formal Logic* 24: 93–105.

Burgess, John (1990). "Epistemology and Nominalism." In *Physicalism in Mathematics*, A.D. Irvine (ed.). Dordrecht, Kluwer Academic Publishers: 1–15.
Burgess, John, and Rosen, Gideon (1997). *A Subject with No Object: Strategies for Nominalistic Interpretation of Mathematics.* New York, Oxford University Press.
Chihara, Charles S. (1973). *Ontology and the Vicious-Circle Principle.* Ithaca, N.Y., Cornell University Press.
Chihara, Charles S. (1982). "A Godelian Thesis Regarding Mathematical Objects: Do They Exist? And Can We Perceive Them?" *Philosophical Review* 91: 211–227.
Chihara, Charles S. (1990). *Constructibility and Mathematical Existence.* Oxford, Oxford University Press.
Chihara, Charles S. (1998). *The Worlds of Possibility: Modal Realism and the Semantics of Modal Logic.* Oxford, Oxford University Press.
Chihara, Charles S. (2004). *A Structural Account of Mathematics.* Oxford, Oxford University Press.
Cohen, Paul (1971). "Comments on the Foundations of Set Theory." In *Axiomatic Set Theory*, D. Scott (ed.). Providence, R.I., American Mathematical Society: 9–15.
Enderton, Herbert (1972). *A Mathematical Introduction to Logic.* New York, Academic Press.
Feferman, Solomon (1998). *In the Light of Logic.* New York, Oxford University Press.
Fine, Kit (1978). "Model Theory for Modal Logic: Part I—The De Re/De Dicto Distinction." *Journal of Philosophical Logic* 7: 125–156.
Forbes, Graeme (1985). *The Metaphysics of Modality.* Oxford, Oxford University Press.
Frege, Gottlob (1979). "Sources of Knowledge of Mathematics and the Mathematical Natural Sciences." In *Gottlob Frege: Posthumous Writings*, H. Hermes, F. Kambartel, and F. Kaulbach (eds.). Chicago, University of Chicago Press: 267–274.
Gödel, Kurt (1964), "What is Cantor's Continuum Problem?," *Philosophy of Mathematics*, P. Benacerraf and H. Putnam (eds.). Englewood Cliffs, N.J., Prentice-Hall: 258–273.
Girvetz, Harry, Geiger, George, Hantz, Harold, and Morris, Bertram (1966). *Science, Folklore, and Philosophy.* New York, Harper & Row.
Hawthorne, James (1993). "Bayesian Induction Is Eliminative Induction." *Philosophical Topics* 21: 99–138.
Hilbert, David (1983). "On the Infinite." In *Philosophy of Mathematics: Selected Readings*, P. Benacerraf and H. Putnam (eds.). Cambridge, Cambridge University Press: 377–393.
Kitcher, Philip (1993). *The Advancement of Science.* New York, Oxford University Press.
Kneale, William, and Kneale, Martha (1962). *The Development of Logic.* Oxford, Oxford University Press.
Laudan, Larry (1997). "How About Bust? Factoring Explanatory Power Back into Theory Evaluation." *Philosophy of Science* 64: 306–316.
Lewis, David (1991). *Parts of Classes.* Oxford, Basil Blackwell.
Maddy, Penelope (1990). *Realism in Mathematics.* Oxford, Oxford University Press.
Maddy, Penelope (1997). *Naturalism in Mathematics.* Oxford, Clarendon Press.
Parsons, Charles (1996). "The Structuralist View of Mathematical Objects." In *The Philosophy of Mathematics*, W. D. Hart (ed.). Oxford, Oxford University Press: 272–309.

Plantinga, Alvin (1974). *The Nature of Necessity.* Oxford, Oxford University Press.

Quine, Willard (1960). *Word and Object.* Cambridge, Mass., MIT Press.

Quine, Willard (1966). "The Scope and Language of Science." In his *The Ways of Paradox and Other Essays.* New York, Random House: 215–232.

Quine, Willard (1969). "Epistemology Naturalized." In his *Ontological Relativity and Other Essays.* New York, Columbia University Press: 69–90.

Resnik, Michael (1997). *Mathematics as a Science of Patterns.* Oxford, Oxford University Press.

Resnik, Michael (1998). "Holistic Mathematics." In *The Philosophy of Mathematics Today*, M. Schirn (ed.). Oxford, Oxford University Press: 227–246.

Shapiro, Stewart (1997). *Philosophy of Mathematics: Structure and Ontology.* New York, Oxford University Press.

Shapiro, Stewart (1998). "Review of *A Subject with No Object.*" *Notre Dame Journal of Formal Logic* 39: 600–612.

Sober, Elliott (1993). "Mathematics and Indispensability." *Philosophical Review* 102: 35–57.

Steiner, Mark (1978). "Mathematics, Explanation, and Scientific Knowledge." *Nous* 12: 17–28.

Steiner, Mark (1995). "The Applicabilities of Mathematics." *Philosophia Mathematica* 3: 129–156.

Tarski, Alfred (1969). "Truth and Proof." *Scientific American* 220: 63–77.

Vineberg, Susan (1998). "Indispensability Arguments and Scientific Reasoning." *Taiwanese Journal for Philosophy & History of Science* 10: 117–140.

Wilder, Raymond (1952). *Introduction to the Foundations of Mathematics.* New York, John Wiley.

CHAPTER 16

NOMINALISM RECONSIDERED

GIDEON ROSEN

JOHN P. BURGESS

1. Nominalism and Its Varieties

Nominalism is usually formulated as the thesis that only concrete entities exist or that no abstract entities exist. But where, as here, the interest is primarily in philosophy of mathematics, one can bypass the tangled question of how, exactly, the general abstract/concrete distinction is to be understood by taking nominalism simply as the thesis that there are no distinctively *mathematical* objects: no numbers, sets, functions, groups, and so on. As to the nature of such objects (if there are any), we need only say that it has come to be fairly widely agreed, under the influence of Frege and others, that they are very different both from paradigmatically physical objects (bricks, stones) and from paradigmatically mental ones (minds, ideas)[Frege 1884]. Modern nominalism emerged in the 1930s as a response to the view of Frege and others that numbers, sets, functions, groups, and so on belong to a "third realm."[1]

[1] [Goodman and Quine 1947] is the first important manifesto of modern nominalism, but the view was widely discussed among the Warsaw logicians in the 1930s; Lesniewski and Kotarbinski were its main advocates. In his autobiography, Quine reports conversations with Lesniewski in 1933 in which "I would argue far into the night, trying to convince him that his system of logic did not avoid, as he supposed, the assuming of abstract objects [Quine 1985: 104]. See [Simons 1998] for a survey of Polish nominalism between the wars.

This modern nominalism was not intended as a reactionary movement back toward pre-Fregean views that would take mathematics to be directly concerned either with the contents of the physical universe or with those of the human mind. It did not say, "Yes, there are numbers, but they are somehow parts of nature or of the mind," but rather, "There are no numbers. Mathematics is not a descriptive science with a special subject matter of its own. Either it is a body of general truths with no distinctive subject matter, or it is not a body of truths at all." Our concern here will be with nominalism thus conceived.

One way to classify the varieties of nominalists is by considering their responses to the following methodological/epistemological argument for anti-nominalism. It has three premises that seem scarcely deniable.

(1) Standard mathematics, pure and applied, abounds in "existence theorems" that appear to assert the existence of mathematical objects, and to be true only if such objects exist; which is to say, to be true only if nominalism is false. Such, for instance, are

> There are infinitely many prime numbers.
> There are exactly two abstract groups of order four.
> Some solutions to the field equations of general relativity contain closed timelike curves.

(2) Well-informed scientists and mathematicians—the "experts"—accept these existence theorems in the sense both that they assent verbally to them without conscious silent reservations, and that they rely on them in both theoretical and practical contexts. They use them as premises in demonstrations intended to convince other experts of novel claims, and together with other assumptions as premises in arguments intended to persuade others to some course of action.

(3) The existence theorems are not merely *accepted* by mathematicians, but are *acceptable* by mathematical standards. They, or at any rate the great majority of them, are supplied with proofs; and while the mathematical disciplines recognize a range of grounds for criticizing purported proofs, and while it occasionally happens that a widely accepted proof is undermined by criticism on one or another of these grounds, nonetheless the proofs of the existence theorems, or at any rate the great majority of them, are not susceptible to this kind of internal mathematical criticism.

There are four other premises a nominalist might try to challenge:

(4) The existence theorems really do assert and imply just what they appear to: that there are such mathematical objects as prime numbers greater than 1000, abstract groups of various orders, solutions of various equations of mathematical physics with various properties, and so on.

(5) To accept a claim in the sense of assenting verbally to it without conscious silent reservations, of relying on it in theoretical demonstrations and practical deliberations, and so on, just *is* to believe what it says, to believe that it is true.

(6) The existence theorems are not merely acceptable by specifically *mathematical* standards, but are acceptable by more general *scientific* standards. Not only do empirical scientists in practice generally defer to the mathematicians on mathematical questions, existence questions included; they are by scientific standards right to do so. There is no empirical scientific argument against standard mathematical theorems, existence theorems included.

(7) There is no philosophical argument powerful enough to override or overrule mathematical and scientific standards of acceptability in the present instance.

From (1), (2), (4), and (5) there follows an intermediate conclusion:

(8) Competent mathematicians and scientists believe in prime numbers greater than 1000; abstract groups of various orders, solutions of various equations of mathematical physics with various properties, and so on. Hence, if nominalism is true, expert opinion is systematically mistaken.

From (8) together with (3), (6), and (7) there follows the ultimate anti-nominalist conclusion:

(9) We are justified in believing (to some high degree) in prime numbers greater than 1000, abstract groups of various orders, solutions of various equations of mathematical physics with various properties, and so on, which is to say we are justified in disbelieving (to the same high degree) nominalism.

Nominalists who concede (8) are *revolutionaries*: They concede that their philosophical position is at odds with established mathematical and scientific opinion. Revolutionaries must reject either (6) or (7). Those who reject (6) are *naturalized* revolutionaries. They seek to correct entrenched errors of established mathematics and science only by appeal to the sorts of argument that typically weight with the experts themselves. Those who concede (6) but reject (7) are *alienated* revolutionaries. They seek to correct entrenched errors of established mathematics and science by appeal to arguments that may have little or no weight with mathematicians or scientists, but might come to have such weight if the experts could be persuaded to take a more philosophical view of the matter.

Nominalists who wish to stop the argument before (8), by rejecting either (4) or (5), are *hermeneuticists*. Those who reject (4) are *content* hermeneuticists, and those who reject (5) are *attitude* hermeneuticists. The former reject the straightforward "face value" construal of what the existence theorems themselves say, while the latter reject the straightforward "face value" construal of what mathematicians are doing when they assent verbally to such theorems. If either variety of hermeneuticism is correct, nominalist philosophy is fully compatible with the "deliverances" of ordinary mathematics and science.

We accept the schematic argument, and so reject both revolutionary and hermeneutic nominalism in all their guises. Our aim here, however, is not to refute the nominalists once and for all, but simply to point out some of the obstacles a compelling case for nominalism would have to overcome. We content ourselves with only a rather brief review of revolutionary nominalism and related versions of content-hermeneutic nominalism. We have treated these positions at considerable length in our book *A Subject with No Object* [1997] (henceforth *SWNO*), and to our knowledge nothing substantially new has been added concerning these positions in subsequent years. Under the label "fictionalism," attitude-hermeneutic nominalism (which makes play with a distinction between two verbs, "accept" and "believe," that can be used interchangeably when discussing other forms of nomnalism) has become quite influential in recent years, and we turn to it next. Finally we consider the distinctive form of hermeneuticism advocated by Stephen Yablo.

2. Naturalized Revolutionary Nominalism

The revolutionary nominalist concedes that scientists and mathematicians routinely assert and believe that mathematical objects of various sorts exist, and that the great majority of their assertions and beliefs of this kind are prima facie acceptable, given the standards that are brought to bear within mathematics and the sciences. Perhaps the best evidence is that existence theorems are often published in journals with the most stringent peer review procedures, and the vast majority of these published theorems survive subsequent scrutiny on the part of competent practitioners. A *naturalized* revolutionary nominalist must claim these theorems are nonetheless *un*acceptable by mathematical or scientific standards.

As we have already said, calling claim (3) of the preceding section "scarcely deniable," the claim will almost surely have to be that the existence theorems are unacceptable by broader scientific standards, not by narrowly mathematical standards. Such a position—one that criticizes mathematical results by "extra-mathematical" standards, even scientific as opposed to distinctively philosophical ones—is one that already will be dismissed out of hand by some anti-nominalists, those Penelope Maddy calls the "mathematical naturalists." Their credo is that

> Mathematics is not answerable to any extra-mathematical tribunal and not in need of any justification beyond proof and the axiomatic method. Where Quine takes science to be independent of first philosophy, [the] naturalist takes mathematics to be independent of first philosophy and natural science—in short, from any external standard. [Maddy 1997:184]

By contrast, scientific as contrasted with mathematical naturalism (whose most famous and influential representative, W. V. Quine, is mentioned by Maddy in the preceding passage), will at least be prepared to listen to an appeal beyond narrowly mathematical to broadly scientific considerations. Though there is a good deal to be said for mathematical naturalism, we will not rehearse the case for it here, but rather will pass on to consider what arguments a nominalist might present to a scientific naturalist.

Of course there will be nothing like a *direct* empirical "disconfirmation" of, say, the claim that the number 1 has three complex cube roots, in the sense that there may be a direct empirical "disconfirmation" of the claim that spontaneous generation occurs. The nominalist argument will have to be less direct and more "theoretical." The most common form of the argument runs as follows:

> There is no a priori objection to standard mathematics, any more than to the phlogiston theory of combustion or the doctrine of metempsychosis. But mathematics is a body of theory, and as such is subject to correction as science progresses. One ground for rejecting a scientific theory that purports to describe a special domain of objects is a demonstration that we can account for the phenomena without invoking those objects, by constructing an alternative theory whose substantive commitments are in some sense weaker, or at least less problematic, but which is nonetheless capable of "doing the work" of the original. And it turns out that it *is* possible to construct nominalist alternatives to standard mathematically formulated theories.

This is the inverse of what is known as the "Quine–Putnam indispensability argument" *against* nominalism, and might be called the "dispensability argument" *for* nominalism.

Of course, the larger part of the task of arguing for nominalism along the lines indicated must be the construction of the alternative theories. This has been attempted in different ways by different writers in the literature, and we have surveyed many of these in *SWNO*. Limitations of space preclude undertaking any such survey here. Suffice it to say that each of the proposed alternatives, while avoiding mathematical objects, involves auxiliary apparatus that will not be to everyone's liking (points and regions of physical space-time considered as substantive objects on one approach, and on other approaches, considerations of things there aren't but could have been, and how, if they had been, they would have compared with things there are). And even waiving any objections to the apparatus invoked, these alternatives still have the feature that they are significantly less perspicuous, tractable, familiar, and (so far as one can judge) fruitful for the further development of science.

The nominalist advocates of such alternatives hold that these drawbacks are outweighed by the fact that they avoid "ontological commitment" to mathematical objects. But can we really say that this avoidance is a factor that by *scientific* standards counts more than perspicuity, tractability, familiarity, fruitfulness, and the like? It is unclear why it should. There is no clear evidence from the history of science or mathematics that these disciplines are concerned to minimize their inventory of mathematicalia. Maddy has argued that, so far as the internal standards of mathematics are concerned, there is, on the contrary, evidence that mathematics seeks to *maximize* the number and variety of the mathematical objects it can accommodate [Maddy 1997]; and it is also easy to understand why empirical scientists, who turn to mathematics for models and tools, would be happier to have more of them rather than fewer.

Thus there are grounds to suspect that, in valuing avoidance of mathematical objects as highly as they do, the revolutionary nominalists of the kind we have been discussing are in fact promoting a distinctively *philosophical* rather than scientific, let alone mathematical, standard of evaluation. That they are themselves to some degree conscious that this is what they are doing is suggested by the fact that they generally publish their proposals not in scientific but in philosophy journals. If the nominalist argument does involve such appeal over the head of science to a philosophy outside, above, and beyond it, then scientific as well as mathematical naturalists will reject it out of hand. But as with mathematical naturalism, so with scientific naturalism; we are not going to rehearse the case for it here but, rather, will pass on to consider naturalist arguments that reject it.

3. Alienated Revolutionary Nominalism

If one is prepared to appeal above and beyond science to some superior and ulterior philosophy, however, then there are much more direct ways to argue for nominalism than by citing some alternative to current scientific and mathematical theories that is superior to them by philosophical, though inferior by scientific, standards. Some nominalists, for instance, report "intuitions" (though few have propounded *arguments*) to the effect that the positing of abstract, mathematical objects is *meaningless.* Many more nominalists express *epistemological* qualms about numbers, sets, and the rest, and propound epistemological arguments for nominalism.

Such arguments do not seek to show that mathematical existence claims are false (let alone meaningless), but rather seek to show that while such claims may be justified by mathematical and scientific standards, they are nonetheless not really justified, all things considered. Naturally such arguments, according to which meeting mathematical and scientific standards just isn't enough, are always published in philosophy, not in cognitive science, journals and are distinctively philosophical. There can be no question of surveying here all the different distinctively philosophical arguments for nominalism that might be found in the literature, but the epistemological type, because of its popularity, does call for comment.

Behind all epistemological arguments lies a certain picture. One key respect in which mathematical objects differ from physical and mental objects is that they are causally inactive and impassive—or, as nominalists like to say, causally *isolated* from us. This fact leads to a picture of some great wall or gulf separating and isolating us on the one side from the mathematical objects on the other side, and the picture leads to the worry, "How can we possibly know what is going on over on the other side?" Those nominalists who take it upon themselves to go beyond a mere picture to an articulated argument would seem to need an analysis or theory of some key epistemic notion, such as rational opinion or justified belief or warranted assertion. But in assuming the burden of producing such a theory, these nominalists place themselves at a disadvantage.

For it is notoriously difficult to produce convincing analyses in this area, and no one proposal is likely to have even a majority of nominalist party members behind it. Though there can be no question of an extended survey of epistemological arguments, the survey in *SWNO* (I.A.2) shows that nominalist epistemological premises typically turn out to be either too weak or too strong. (For instance, the principle that one cannot justifiably believe in objects of a certain kind unless they are mentioned in the causal explanation of how one came to believe it is too weak to have the implications the nominalist wants, since mathematical objects are appealed to in the detailed scientific explanation of almost anything, cognitive processes included [Steiner 1975]. The principle that one cannot justifiably believe in objects unless they exert a causal influence on oneself, by contrast, is too strong and has consequences the nominalist does not want, such as the impossibility of knowledge of the future.)

And waiving all objections of detail, assuming one has somehow found an epistemological principle that is not too strong, not to weak, but "just right" for nominalist purposes, there is a further, general difficulty. The development of theories in epistemology is a quasi-inductive affair. Typically one begins with what one takes to be clear cases of justified belief and clear cases of the opposite, and looks for a formula that covers the data and rings true on reflection. A theory developed in this spirit can then be used to correct pretheoretical verdicts at the margins. But it would seem that it cannot be used to undermine a vast and significant class of normally uncontroversial verdicts about justification.

If the theory really has the implications the nominalist wants, implying that our apparent mathematical knowledge is error or delusion, the anti-nominalist can simply claim that the novel theory, however meritorious in other respects, stands refuted by the counterexample of mathematical knowledge. It would be a gross mistake to repudiate the central claims of Mesozoic paleontology or Byzantine historiography on the basis of a theory of justification that had been developed and tested on examples drawn exclusively from, say, particle physics. The nominalist who wields a theory of justification developed by reflection on cases of empirical belief as a club against the mathematicians can with considerable plausibility be charged with a similar mistake.

We should not overstate the case, however. Philosophical theories of justified belief are only "quasi"-inductive. We may begin by extracting principles to cover a list of paradigms and foils, and find that once extracted, the principles strike us as so obviously correct that they could motivate a significant redescription of part of the data. What can be said without overstatement, however, is that in the present case no one has yet produced an independently compelling general theory of justification that would support the view that scientists and mathematicians are systematically mistaken about what it takes to justify an existence claim in mathematics.

This being so, nominalists are likely to want to shift the burden to the other side, claiming it is not up to *them* to articulate a theory of justification showing that belief in mathematical existence is *not* justified, but rather is on their opponents to articulate a theory showing that it *is* justified. The epistemological challenge then might be put like this:

> Mathematicians (to whom empirical scientists defer in these matters) generally obtain their existence theorems by deduction from previously established results, which ultimately depend on existence axioms. But such a deduction provides a justification of the theorems only if the axioms are themselves justified. (This is not a "theory" of justification, but just a platitude.) Now has anyone shown that the kind of process by the axioms were arrived at is a reliable one, tending to lead to *true* axioms? Have the axioms been justified? Not by the main corps of mathematicians, who seldom if ever consider such questions. Only a small cadre of specialists, with one foot in philosophy, ever bother to consider the epistemic status of the axioms, and even they do not manage to agree among themselves as to what the justification for them is supposed to be.

And indeed this last observation is quite true: There is no consensus among anti-nominalists who take upon themselves the burden of articulating a theory of justification according to which belief in the axioms is justified.

Some claim that the axioms "force themselves upon us as true" [Gödel 1947], to which the nominalist will reply that this kind of subjective conviction is not accepted as sufficient for justified belief in any other area [Chihara 1973]. Some claim the axioms are "constitutive of the meaning" of mathematical terms [Wright 1983], to which nominalists will reply that no existential assertion can be analytic [Field 1984]. Some claim that the axioms are justified by their consequences, since the mathematical theories deduced from them are indispensable for scientific applications [Putnam 1971], to which the nominalist will reply that (even setting aside the question broached in the preceding section as to whether mathematical objects really *are* indispensable) it is unclear that indispensability in the relevant sense is a cogent reason for belief [Maddy 1997; Melia 1995]. Moreover, in the case of each specific anti-nominalist epistemological claim, a large segment of the *anti*nominalist party will sympathize with the nominalist objection to that specific claim.

But there is another kind of anti-nominalist response, which simply rejects the presupposition of the nominalist challenge that our basic mathematical assumptions require some sort of positive defense. To see how rejection of the challenge might be motivated, consider the following skeptical parallel to or parody of the nominalist challenge itself:

> Scientists generally derive their results from ordinary perceptual judgments (about their observational instruments and experimental apparatus). But such a derivation provides a justification only if ordinary perceptual judgments are themselves justified. (This is not a "theory" of justification, but just a platitude.) Now has anyone shown that the kind of process by ordinary perceptual judgments are arrived at is a reliable one, tending to lead to *true* judgments? Have ordinary perceptual judgments been justified? Not by the corps of scientists, who never consider such questions. Only a small cadre of specialists, with both feet in philosophy, ever bother to consider the epistemic status of perceptual judgments, and even they do not even manage to agree among themselves as to what the justification for them is supposed to be.

And indeed this last observation is quite true: There is no consensus among anti-skeptics who take upon themselves the burden of articulating a theory of justification according to which belief in ordinary perceptual judgments is justified.

But many anti-skeptics simply reject the presupposition of the skeptical challenge that our basic ways of forming judgments of perception require some sort of positive defense. They take a standard for the justification of beliefs to be something that emerges from our evolving practice of criticizing the beliefs people actually form, one central feature of which practice is that we take our spontaneous perceptual beliefs for granted until we are given some sort of positive reason to doubt them. Perceptual judgments are justified by default, innocent until proven guilty, on this view. It is beyond dispute that we unhestiatingly accord them just such a status, as the view in question implies, when engaged in serious inquiry about matters of real importance. Why should we accept the demand of the skeptics that we hesitate to accept the results of important inquiries until a substantive justification for perceptual judgment is provided that would satisfy *them*? That is the question raised by one line of thought in response to the skeptical challenge, and insofar as the nominalist and skeptical challenges are parallel, a similar response would seem to be available to the anti-nominalist.

4. Content-Hermeneutic Nominalism

So far we have been discussing the admirably straightforward nominalists who agree that scientists and mathematicians typically believe standard existence theorems, and that these beliefs are justified by internal scientific and mathematical standards. Nominalists of this sort are self-conscious revolutionaries. They may struggle in a losing cause, but at least we know what sort of struggle they're engaged in: the struggle to correct a mistake embedded in existing science.

We did not attempt in our discussion in the preceding section to show that theirs *is* clearly a losing cause, but only that it is not clearly a winning one. A main lesson to take away from our discussion is that it is very difficult to settle issues when one rejects scientific standards for settling them. Argumentation over epistemology (or almost anything else) from an alienated, extrascientific, as opposed to a naturalized, intrascientific, standpoint usually tends to be inconclusive and often tends to bog down in issues of burden of proof. So it was that we saw that whichever side is saddled with the onus of articulating a detailed theory of justification finds itself at a serious disadvantage.

According to hermeneutic nominalism, to which we next turn, there is no need to enter into such inconclusive debates, since there is no conflict between nominalism and science at all. Of course, the hermeneuticist must explain away as somehow unreal the apparent conflict between the mathematician who affirms, "There are prime numbers greater than a thousand" and the nominalist who insists, "There are no numbers at all." Such is the task confronting nominalists who are not so straightforward as to admit that they are engaged in the business of correcting the errors of scientists and mathematicians. They must defend some subtle interpretation, either of the content of the existence theorem, or of the attitude of the mathematician who verbally affirms it, so as to reveal that the apparent or "surface" conflict is not, "deep down," a real one.

One way a nominalist might attempt to go about this would be to claim that one or another of what in our discussion of naturalized revolutionary nominalism we called nominalist *alternatives* to standard mathematical and scientific theories is not in fact an alternative, but a revelation of what "deep down" the standard theory has really meant all along. For instance, it might be maintained that

(A) "There exist prime numbers greater than a thousand"

is innocent because all it really means is

(A*) There could exist a prime numeral greater than a thousand.

or something of the sort.

There are, however, two serious difficulties with such a view. For one thing, such a nominalistic translation seems to work too well. If (A) is nominalistically acceptable because "deep down" all it means is (A*), then it would seem that

(B) "There exist numbers"

must be acceptable, too, because all it means is

(B*) There could have been numerals.

But to concede (B) (and the corresponding statement about other kinds of mathematical objects) is to concede all the anti-nominalist maintains [Alston 1958].

For another thing, there is a total lack of scientific evidence in favor of any such nominalistic reconstrual as a theory of what ordinary mathematical assertions mean. Or at least, no nominalists favoring such a reconstrual have ever published their suggestions in a linguistics journal with evidence such as a linguist without ulterior ontological motives might accept.

But there is another hermeneutic claim, not dependant on associating with mathematical existence assertions like (A) above another form of words such as (A*), and presumably not subject to the same kind of objections as views that are dependent on such an association. On this other view, there is simply an ambiguity in the word "exists," between a strong and a weak sense, which we may write as "exists" and "exists." (A) is supposed to be quite true if by "exists" one means "exists," and to become false only if one takes "exists" to mean "exists." Adherents of such a view maintain that there is no conflict between their species of nominalism and science or mathematics, since scientists and mathematicians, when immersed in their practice, are only talking about existence. The nominalist's quarrel is only with those "Platonist" philosophers who misread scientific and mathematical practice, and claim existence for mathematical objects.

Any proposal along these lines faces a considerable explanatory burden. The first task, obviously, will be to explain the distinction between the two supposed senses of existence. The second task will be to make it plausible that philosophical claims involve one sense, and internal mathematical and scientific claims the other. Suppose it is said, for instance, that for a thing to exist is for it to part of the ultimate furniture of the universe. However this last phrase is interpreted, it seems quite plausible that large, composite objects like the Eiffel Tower do not exist in this sense. But an anti-nominalist may be perfectly willing to grant that the Euler function may not exist in this sense either. The most the anti-nominalist wishes to claim that the Euler function exists in the same sense that the Eiffel Tower does. To put the matter another way, it is supposed to be uncontroversial that mathematical objects exist, and controversial whether they exist. But the genuinely controversial question is whether or not numbers, functions, and the like exist in the sense in which the planet Venus does and the planet Vulcan doesn't. The positing of two senses, strong and weak, of the same verb does nothing to settle this controversy.

5. Attitude-Hermeneutic Nominalism

A different but not unrelated nominalist view, instead of distinguishing between two senses of "exists," and two corresponding contents that a statement like (A) might have, distinguishes two *attitudes* with which a statement like (A) might be put forward. Often the view in question is expressed using a distinction between "acceptance" of and "belief" in existence theorems: mathematicians and scientists, it is said, only "accept" the existence theorems; "Platonist" philosophers err by believing them.

The advocates of such a nominalist view may avoid the task of explaining the differences between two senses of "exist," but they still have some explaining to do. The nominalist can hardly deny that informed practitioners routinely put forward existence theorems in serious conversation without conscious qualms, rely on existence theorems in high-stakes deliberations, and so on. So "acceptance" must be compatible with all of this and yet still somehow fall short of belief.

Now one doesn't have to be a behaviorist to think that when a person understands a sentence *S*, confidently affirms it without qualification and without conscious insincerity, organizes serious activity just as would be done if *S* were believed, and so on, then we have a *powerful* case for attributing to that person a belief that *S*: the attitude-hemeneuticist posits an attitude called "acceptance" that is supposed to be distinct from belief but to have all the most salient behavioral and instrospectable consequences of belief. But attitude types cannot be brutely different. There must be *some* difference between a believer and a mere accepter, and if there is such a difference, surely it is reasonable to expect that its presence will be betrayed *somehow* in behavior [Horwich 1991].

It would be one thing if scientific neophytes were regularly taken aside at some stage in their training by their mentors and told, "Don't believe it, just accept it. No one around here really *believes* what he says." We can imagine a community in which this happened, in which scientists and mathematicians were explicitly encouraged to regard the mathematics they employ in roughly the same spirit in which the aeronautical engineer regards Newtonian physics—as a theoretical instrument to be used in many important contexts, but not to be believed [Rosen 2001]. But while we can imagine such a community, we don't live in one. It is not part of the practice of ordinary mathematics to take steps to disavow belief in the standard existence theorems, to warn students against believing, or the like. Thus we lack what would be the best kind of direct evidence that the practice of mathematics and science involves something less than belief in existence theorems.

There are some phenomena of the behavior of mathematicians for which attitude-hermeneuticism can give a plausible partial explanation, and which therefore constitute some kind of *indirect* evidence in favor of that nominalist position. Perhaps the most important is the fact that scientists and mathematicians are strikingly *lighthearted* when it comes to the introduction of novel mathematical entities. Once theorists have convinced themselves that some new mathematical theory is consistent (internally and with previously accepted mathematics), nothing further seems to be needed to persuade them to accept the new theory, existence claims and all, when it would suit to do so. The attitude-hermeneuticist has a partial explanation of this phenomenon: since the theorists' decision to accept the new existence claims is not a decision to believe them, it is not surprising that they are unconcerned about the possibility that the theory may not be true because the objects it posits do not exist.

Another phenomenon is that of the indifference of mathematicians to certain kinds of questions about the identity mathematical objects. Mathematicians speak of numbers and of sets, but never concern themselves with such as questions as which set (if any) is identical with, say, the number 2. Such an attitude is readily intelligible on the supposition that the mathematician regards number theory and set theory only as useful fictions. For there are no end of questions about fictional characters one is quite prepared to leave undecided [Field 1989].

Yet another phenomenon is the reaction—or, rather, the varied reactions—of mathematicians when pressed by philosophers with questions about the existence of the entities mentioned in their theorems. Some dismiss the question as of no interest, some freely concede any skeptical charge their philosophical inquisitor raises, some frankly do not know what to say. All this is readily intelligible if mathematicians do not regard themselves as committed to the *truth* of their theorems, which would presuppose the existence of the entities those theorems are about, in a way the mere *utility* of the theorems would not.

But considerations of these kinds lend only rather weak support to attitude-hermeneuticism, because there are many alternative explanations of the cited phenomena available. Thus the lightheartedness about new existence assumptions may be accounted for by the fact that assurance of consistency—and mathematicians are seldom quite so lighthearted as to accept new existence assumptions without such assurance—is generally provided by describing models consisting of objects already recognized. But then the new entities, being identifiable with those already recognized objects, may not really be "new" after all. Again, the indifference to questions of identity may be accounted for by the fact that pure mathematicians are generally interested only in properties that structures share with all isomorphic structures. If statements ostensibly about 2 are really *general* statements about the two-element in any structure of a certain kind, then there can be no meaningful question about which set 2 is identical with, because the two-elements of different structures may be identical with different sets. Many other alternative explanations—more than there is space for here—will occur to readers of the literature in philosophy of mathematics.

As for mathematicians becoming flustered when grilled by skeptical philosophers, don't ordinary people become flustered when pressed by skeptical philosophers to defend the statement "I am awake"? And does this fact in any way suggest they don't "believe" they're awake, but only "accept" it? The indirect evidence for attitude-hermeneuticism is thus weak, and seems insufficient to overcome the presumption against the view established by the lack of *direct* evidence—by the absence of self-conscious disavowal among working mathematicians and scientists.

Moreover, it strains credulity to suggest that the internal norms governing scientific inquiry somehow demand such disavowal nonetheless. We conclude that attitude-hermeneutic nominalism is untenable. A philosopher may *advocate the adoption* of some attitude toward mathematics weaker than belief, considering that only this weaker attitude is ever justified. But such a philosopher is an alienated revolutionary nominalist, and subject to the doubts raised in the section above on that type of position.

6. Yablo's Figuralism

The most developed hermeneutic position in recent years has been that of Stephen Yablo [1998, 2000a, 2000b, 2001, forthcoming]. Yablo distinguishes figurative from literal statements or uses of statements. His position will be treated here as a form of content-hermeneuticism, though different from those considered two sections back, in that it is a form according to which an existence theorem is ambiguous between a literal and a figurative sense, which latter sense *need not coincide with the literal sense of any form of words.* Arguably his position could also be considered a form of attitude-hermeneuticism, a form according to which a verbal affirmation of an existence theorem may be an act of putting it forward as a literal truth, or putting it forward as a useful pretense. But regardless of how it is classified, Yablo's position deserves a section to itself.

Yablo's claim is that existence theorems are false when taken literally but true when taken figuratively. He makes the hermeneutic claim that ordinary uses of existential idioms in mathematics are mostly figurative, whereas in philosophy we are concerned with the literal. Thus the nominalists' philosophical claim—mathematical objects do not literally exist—is supposed to be compatible with the immersed practitioner's commitment to their figurative existence.

Here is a rough sketch of Yablo's account, in the case of arithmetic. Suppose, to begin with, we speak a language in which number-words are not used as nouns. We can use them as adjectives and say, "There are two cats in the yard," which can be explained without number-words as meaning simply, "There is a cat and another cat distinct from it, and the former is in the yard and so is the latter." What we cannot say is "The number of cats in the yard is the number two."

Now suppose that we begin to notice a regularity: there were two cats in the yard on Monday, three cats in the yard on Tuesday, four cats in the yard on Wednesday, and so on. We lack the resources to describe the pattern. These resources are ultimately provided by indulgence in a certain game of make-believe or "let's pretend" which we next describe.

We pretend there are some objects called numbers—zero, one, two, and so on—and there are some operations on them: successor, addition, multiplication, and so on. We pretend these objects satisfy certain laws, including those known as the Peano postulates and, more important in the present context, the following laws:

> The number of *F*s is zero if and only if there are no *F*s.
> The number of *F*s is the successor of *n* if and only if for some *y*, *y* is an *F* and the number of *F*s other than *y* is *n*.

Having adopted the pretense just described, it becomes appropriate, given certain facts about the world, to assent to yet further statements about numbers beyond those listed. Notably, if there are two cats in the yard, it is appropriate to say, while playing the game, that the number of cats in the yard is two. But this is only a pretense kept up while playing the game: we aren't really committing ourselves to there being such things as numbers.

We have here an instance of the general phenomenon, described by Kendall Walton, that in general what is true under a pretense depends in part on the content of pretense and in part on the way the world is [Walton 1990]. If we are playing a game in which we pretend tree stumps are bears, and you run toward a tree stump, then it is "true in the game" that you are running toward a bear. But of course we aren't really committed to the existence of a bear that you are running toward.

As Walton observes in his discussion of "prop-oriented make-believe," it may be difficult or impossible to say in sober, literal terms just how the world has to be in order for a claim to be true in a given game of make-believe [Walton 1993]. Suppose someone asks where Woods Hole is located on Cape Cod. The easiest thing for me to do may be to say, "Pretend Cape Cod is a human arm bent like *this*. Woods Hole is at the elbow." In this process, I convey the requested geographical information by invoking a pretense. And it may well be that this is the *only* way I'm in a position to supply that information. Of course, in supplying the information in the way I do, I am in no way committing myself to Cape Cod's being really anyone's arm, Woods Hole's being really at anyone's elbow, or anything of the sort.

The case is similar with arithmetic, according to Yablo. In the case of our number-game, we *can* say that the figurative meaning of "The number of cats in the yard on Monday was two" is the same as the literal meaning of "There were two cats in the yard on Monday" (which we have seen how to express without using number-words at all). But with "The number of cats in the yard on the *n*th day is *n*," the fictional meaning does not coincide with the literal meaning of any statement in our original language. Indulging in the pretense extends our descriptive resources. Indeed, that is precisely the point of indulging in it.

In general, even statements normally taken figuratively do have a literal meaning which may exceptionally be the appropriate one to consider. Thus "She has butterflies in her stomach" is normally taken figuratively, as meaning that she is nervous. But a doctor reading an X-ray of a patient with a gastric complaint might say, "Incredible! She has *butterflies* in her stomach!" and literally mean that there are insects of a certain kind in a certain portion of her digestive tract.

Similarly, according to Yablo, number-talk is normally meant figuratively by mathematicians, but it does have a literal meaning, and that is the meaning at issue in philosophical discussions. When Yablo says there are no numbers, he means that the assertion of the existence of numbers is false taken literally; he does not mean it is false when made by mathematicians, for he claims that they mean it only figuratively. This is how he reconciles his nominalism with ordinary mathematical practice, and this is why we have classified his nominalism as being of the hermeneutic variety.

Some questions of detail can be raised about this hermeneutic view, but let us focus on the fundamental issue. Yablo's claim that number-talk is systematically ambiguous is presumably an empirical hypothesis in semantics. What evidence is there for it?

Yablo's main argument is a version of the argument from lightheartedness considered in the preceding section.[2] Mathematicians and scientists pass unhesitatingly from statements about ordinary things to statements in the mathematical idiom, and back again:

There are n *F*s	The number of *F*s is n
There are just as many *F*s as *G*s.	There is a one-one correspondence between the *F*s and the *G*s.
The argument from P to Q is invalid.	There is a model in which P is true but Q is false.
Line a is to line b as the circumference of a circle is to its diameter.	The ratio of the length of line a to the length of line b is the ratio of the circumference of a circle to its diameter.

Taken literally, the item in the right column in each case entails the existence of some mathematical object, while the corresponding item in the left column does not. But ordinary practitioners *treat the claims as equivalent.* Not only do they freely pass back and forth between the two columns, but as they do so, it may be with a palpable sense that *nothing more is at stake* with the one formulation than

[2] See [Yablo 2000a, §XVI] for a list of "suggestive similarities" between figurative utterances and mathematical utterances. For reasons of space, our discussion is restricted to three exemplary arguments for Yablo's hermeneutic hypothesis.

with the other. This is readily explicable if the ordinary practitioner is concerned only with what, according to Yablo, is figurative content of the mathematical claim. For the *figurative* content of each item in the right column, according to Yablo, is simply the literal content of the corresponding item on the left.

A subsidiary argument involves appeal to intuitions as to what statements "are about." When someone says the number of cats in the yard today is greater than the number of cats in the yard yesterday, the speaker feels the statement to be *about the cats*, and not about anything over and above. When someone says the number of starving people is enormous, the speaker feels the statement is about a matter of great moral importance, and that the same matter of great moral importance could be equally well spoken of by saying instead that there are enormously many starving people.

Another subsidiary argument turns on a fantasy. Suppose the Oracle of Philosophy tells you that everything is concrete and nothing is abstract, and suppose that you believe her.

> Impressed by what the Oracle has told you, you return to civilization to spread the concrete gospel. Your first stop is [your school here], where researchers are confidently reckoning validity by calculations on models, equinumerosity by calculations on 1-1 functions, and so on. You demand that the practice be stopped at once. Given that models and functions don't really exist, all theoretical reliance on them should cease. They of course tell you to bug off and am-scray. Which, come to think of it, is exactly what you yourself would do if the situation were reversed. [Yablo, forthcoming]

The suggestion is that since working mathematicians would be utterly unfazed by the philosophical discovery that, literally speaking, there are no mathematical objects, it must be that their ordinary uses of existential idiom do not involve a commitment to the literal truth of the existence theorems.

Needless to say, none of these arguments is decisive. We have already discussed the first of them, and said that the phenomenon of lightheartedness admits of other explanations. The specific case cited by Yablo could in fact by explained by *any* assumption implying that for *whatever* reason, mathematicians don't regard the existence of numbers as problematic. They treat "There are *n* *F*s" and "The number of *F*s is *n*" as equivalent because they are equivalent modulo a background theory whose truth is not for them in question.

The second argument can be met by observing that the intuitions of aboutness are unstable and context-dependant. If someone says "The number of cats in the yard is two," and you ask what his statement is about, he may indeed say "cats." But suppose you then point out to him that his statement implies that the number of cats in the yard is a prime number, and ask him what *this* claim is about. At least after helpful prompting—"We're talking about a number here, aren't we?"—it is immensely likely that he will say, "Yes, I suppose we are." Think how bizarre it would be for him to say, "Certainly not; we're still just talking about cats." Now return to the original claim with this dialogue fresh in mind and ask, "Are you sure your original claim wasn't also about a number?" Who knows what he will say? But it seems well within reason that he will now report that his original claim was in part about the number 2, and not just about the cats.

The third argument can be met in a number of ways. Perhaps the mathematicians don't share your faith in the Oracle of Philosophy, and put zero value on her pronouncements. Perhaps they will quite naturally *adjust* to the striking philosophical discovery that the objects they had thought of themselves as studying do not exist, given that this revelation comes accompanied by an assurance that the pretense that they do, may still be relied on in all practical deliberations. To use hackneyed examples, people adjusted to Copernican astronomy, but went right on speaking of sunrise and sunset. People adjusted to special relativity, but went right on speaking of events that are simultaneous within the limits of observation in the most salient local frame of reference as being simultaneous *sans phrase.*

Against these three far-from-conclusive indirect arguments for Yablo's position and the many others he adduces must be set arguments opposing it, of which the most important is, as in our earlier discussion of attitude-hermeneuticism, the lack of *direct* evidence for it where one would expect to find it.

We fully grant that someone speaking figuratively may lack the ability to express in literal language just what is meant by the figurative speech. Even "She has butterflies in her stomach," which we earlier parsed as "She is nervous," really means that she is experiencing a certain physiological accompaniment of nervousness for which there may be no better description than the metaphorical "butterflies in the stomach." Likewise, we fully concede that someone speaking figuratively need not be consciously aware of doing so. Again, "butterflies in the stomach" is such a cliché that most users of it probably do not have a recognition of its metaphorical character present to consciousness in the act of uttering it. But it seems that in all uncontroversial cases of nonliteral language, the competent speaker can promptly recognize that the language was meant nonliterally if the question arises. If a child or a foreigner responded to "She has butterflies in her stomach" by asking, "Are you sure they're not moths?" would not the explanation "Oh, it's just a figure of speech" be immediately forthcoming?

Or consider the following exchange:

> "Where's Woods Hole?"
> "Cape Cod is an arm extending eastward into the Atlantic Ocean, and bent like this. [Demonstrates.] Woods Hole is at the elbow."
> "It must be an enormous arm! Whose is it, King Kong's? Is it very hairy? Will I have trouble finding my way through the hair?"
> "Hold on; you've got me wrong. It's not *literally* an arm. That was just a figure of speech. Cape Cod is a spit of land that is roughly *shaped like* an arm."

We submit that whenever a bit of language is used nonliterally, it is possible for an interlocutor to misconstrue it by taking it literally, and for the competent speaker to recognize this misunderstanding and correct it by pointing out that the remark was not meant literally. Certainly in all *clear* cases of figurative language—and it is worth stressing that the boundary between figurative and literal is as fuzzy as can be—the nonliteral character of the linguistic performance will be *perfectly obvious* as soon as the speaker is forced to turn attention to the question of whether the remark was meant literally.

We further submit that mathematical discourse fails this test for nonliteralness. What would it be to misunderstand a mathematical remark like "There are two abstract groups of order four" by taking literally what is meant only figuratively? What would be the analogue of the risible misunderstanding of the directions to Woods Hole? The best we can do is the following:

> Mathematician: There are two abstract groups of order four.
> Interlocutor: Fascinating. Where are these groups? What are they like intrinsically? How do you know that they exist?

And what would be the analogue of the original speaker's reply to the obvious misconstrual? Yablo himself no doubt would say something like this:

> Hold on; you've got me wrong. I didn't mean to say literally that any abstract groups exist. I was just speaking figuratively. It's hard to say in strictly literal terms what I did mean, but I only meant to commit myself to the view that *assuming standard mathematics*, there are two abstract groups of order four.

But we submit that an actual mathematician would be much more likely to say something like this:

> Groups don't exist in some special *location*. They're abstract. Abstract groups are equivalence classes of ordinary groups under isomorphism. I know abstract groups exist because I can describe concrete instances of them—for instance, the integers modulo 4 (and because I can always define an equivalence class, given a suitable equivalence relation).

These answers may not be satisfactory—no nominalist will be satisfied with them, obviously—but the fact that the ordinary practitioner would be disposed to reach for answers of this sort suggests that the questions do not constitute a "literalistic" misconstrual of remarks meant figuratively. And if there can be no literalistic misconstrual, then the language was not figurative in the first place.

In *SWNO* we were concerned with ambitious programs for nominalistic reconstruction or reconstrual of mathematics and mathematically formulated scientific theories. We argued that these nominalist theories were neither clearly better than standard theories by recognizably *scientific* criteria, nor at all plausible as hypotheses about what the standard theories had "really meant all along." In the present work we have been especially concerned with nominalist positions turning on some distinction between literal and figurative belief, between belief and acceptance, or something of the sort, since that kind of nominalism is the one that has been most actively pursued in recent years. We have just been expressing doubts about whether the actual attitude of working mathematicians and scientists is weaker than literal belief, having earlier expressed doubts about nominalist epistemological arguments that, if cogent, would show that anyone who *does* have a literal belief in mathematics *ought to* retreat to some weaker attitude.

As matters stand, the recommended attitude seems to be one that is adopted only by certain philosophers, who on account of worries engendered by a certain picture—us on one side of a wall or gulf, mathematical objects on the other—feel impelled to take back in their philosophical moments what they say at other times. We have to admit, however, that the picture is a perennially powerful one, and that until it can be exorcised—something we do not claim to have done—critics of existing varieties of nominalism can hardly hope to accomplish more than to inspire the development of new ones. We therefore expect the topic of nominalism to remain a lively one for years to come.

REFERENCES

Alston, W. [1958] "Ontological Commitments," *Philosophical Studies* 9: 8–17.

Burgess, J.P., and Rosen, Gideon [1997] *A Subject with No Object*, New York, Oxford University Press.

Chihara, C. [1973] *Ontology and the Vicious Circle Principle*, Ithaca, N.Y., Cornell University Press.

Field, H. [1984] Review of Wright [1983], *Canadian Journal of Philosophy*, 14, 637–662.

Field, H. [1989] *Realism, Mathematics and Modality*, Oxford, Basil Blackwell.

Frege, G. [1884] *Die Grundlagen der Arithmetik*, trans. J. L. Austin as *The Foundations of Arithmetic*, Evanston, Ill., Northwestern University Press, 1968.

Gödel, K. [1947] "What Is Cantor's Continuum Problem?" *American Mathematical Monthly* 54: 515–525.

Goodman, N., and Quine, W.V. [1947] "Steps Towards a Constructive Nominalism," *Journal of Symbolic Logic* 12: 105–122.

Horwich, P. [1991] "On the Nature and Norms of Theoretical Commitment," *Philosophy of Science* 58: 1–14.

Maddy, P. [1997] *Naturalism in Mathematics*, Oxford, Clarendon Press.

Melia, J. [1995] "On What There's Not," *Analysis* 55: 223–229.

Putnam, H. [1971] *Philosophy of Logic*, New York, Harper Torchbooks.

Quine, W.V. [1985] *The Time of My Life*, Cambridge, Mass., MIT Press.

Rosen, G. [2001] "Nominalism, Naturalism, Epistemic Relativism," *Philosophical Perspectives* 15 (Metaphysics), 60–91.

Simons, P. [1998] "Nominalism in Poland," in J. Srzednicki and Z. Stachniak, eds., *Lesniewski's Systems: Protothetic*, Dordrecht, Kluwer.

Steiner, M. [1975] *Mathematical Knowledge*, Ithaca, N.Y., Cornell University Press.
Walton, K. [1990] *Mimesis as Make-Believe: On the Foundations of the Representational Arts*, Cambridge, Mass., Harvard University Press.
Walton, K. [1993] "Metaphors and Prop-Oriented Make-Believe," *European Journal of Philosophy* 1, 39–57.
Wright, C. [1983] *Frege's Conception of Numbers as Objects*, Aberdeen, Aberdeen University Press.
Yablo, S. [1998] "Does Ontology Rest on a Mistake?" *Proceedings of the Aristotelian Society*, supp. 72: 229–261.
Yablo, S. [2000a] "A Paradox of Existence," in T. Hofweber, ed., *Empty Names, Fiction and the Puzzles of Non-existence*, Stanford, Calif., CSLI.
Yablo, S. [2000b] "Apriority & Existence," in P. Boghossian and C. Peacocke, eds., *New Essays on the A Priori*, Oxford, Oxford University Press.
Yablo, S. [2001] "Go Figure: A Path through Fictionalism," *Midwest Studies in Philosophy* 25, 72–102.
Yablo, S. [forthcoming] "The Myth of the Seven," in M. Kalderon, ed., *Fictionalist Approaches to Metaphysics*, Oxford, Oxford University Press.

CHAPTER 17

STRUCTURALISM

GEOFFREY HELLMAN

1. Introduction

With the rise of multiple geometries in the nineteenth century, and the rise of abstract algebra in the twentieth century, the axiomatic method, the set-theoretic foundations of mathematics, and the influential work of the Bourbaki, certain views called "structuralist" have become commonplace. Mathematics is seen as the investigation, by more or less rigorous deductive means, of "abstract structures," systems of objects fulfilling certain structural relations among themselves and in relation to other systems, without regard to the particular nature of the objects themselves. Geometric spaces need not be made up of spatial or temporal points or other intrinsically geometric objects; as Hilbert famously put it, items of furniture suitably interrelated could satisfy all the relevant axiomatic conditions as far as pure mathematics is concerned. A *group*, for instance, can be any multiplicity of objects with operations fulfilling the basic requirements of the binary group operation; indeed, the very abstractness of the group concept allows for its remarkably wide applicability in pure and applied mathematics. Similar remarks can be made regarding other algebraic structures, and the many spaces of analysis, differential geometry, topology, and so on. Of course, mathematicians distinguish between "abstract structures" and "concrete ones" (e.g., made up of familiar, basic items such as real or complex numbers or functions of such, or rationals, or integers, etc.). (For example, the space L^2 of square-integrable functions from $\mathbb{R}$ (or $\mathbb{R}^n$) to $\mathbb{C}$, with inner product $(f, g) = \int f^* g \, d\mu$, where μ is Lebesgue measure, is a "concrete" example of a Hilbert space.) But it is characteristic of a thoroughgoing structuralism to treat even these systems as like the more "abstract" ones, in that the "objects" involved serve only to mark "positions" in a relational system; and the "axioms" governing these objects are thought of *not* as *asserting definite truths*, but as *defining* a type of structure of mathematical interest. In some sense to be clarified, the objects serve only as relata of key relations, and their "individual nature" is of no mathematical concern, if one can even speak of such a nature. Sometimes it is even said that such objects—"structural objects"—have only the properties and bear only the relations to other such objects as required by the relevant axioms or defining conditions laid down in the branch of mathematics in question.

Now in the course of one paragraph, we have gone from a commonplace of the modern mathematical point of view to some pretty deep-sounding issues that will require some sorting out. We have various notions of "abstract" and "concrete" to contend with. Moreover, we have the tantalizing suggestion that perhaps mathematical objects "have no nature" at all, beyond their "structural roles." Does this make sense? And we confront the question of whether structuralist tenets apply at the most fundamental levels at which we speak of mathematical objects: numbers, sets, functions, even relations themselves; and if they do apply, how so? It is one thing to study algebraic, geometric, or topological structures independently of any particular objects making them up, in the sense that none are preferred, except for special purposes in particular contexts. But are we to understand talk of *the natural numbers* in the same way? And what about *collections* and *operations*, or more generally *relations* (which can include collections as unary relations)? In short, *where does structuralism start*? And then, *where does it take us*? Does it provide novel insights, perhaps even answers to, or dissolutions of, any long-standing metaphysical and epistemological problems concerning mathematics, or does it just shift these with some new terminology?

As we shall see, there are some strikingly different ways of developing the informal, intuitive ideas associated with structuralism, and a large part of our task in this chapter will be to delineate and compare these alternatives along several important dimensions. As will emerge, there are a number of interesting tradeoffs, and it is surprisingly difficult—perhaps impossible—to formulate a version that combines all the advantages exhibited by some version or other while avoiding the pitfalls. At this stage, it will help guide our investigation to formulate a number of central questions which any developed version of structuralism ought to answer. Here are five such:

(1) What primitive notions are assumed and, in particular, what is the background logic? Does it go beyond first-order logic? If higher-order logic is presupposed, what is the status of relations and functions as objects? What limitation does this imply on the structuralist approach in question?

(2) Already we have hinted that the term "*axiom*" is ambiguous, meaning either "*defining condition*" on a type of structure, in which case nothing is being *asserted*, or meaning "basic or initial assumption," as an assertion that can be true or false, rationally credible to some degree or not, and so forth. It is characteristic

of a "structuralist" approach to a branch of mathematics to appeal to "axioms" in the former sense, and this is precisely what Dedekind did in his treatment of arithmetic [9], which justifies calling his approach "structuralist" at least in a minimal sense. (Similarly, this was Hilbert's conception in his famous correspondence with Frege [13].) But then what are the *assertory axioms* of the framework in question? (To modify a good saying, "Not by definition alone!")

(3) Is there a thoroughgoing *elimination* of structures-as-objects? If not, what is a *structure*? Moreover, what is a *mathematical* structure? [21]

(4) As a special case of (2), what assumptions are made governing the *mathematical existence* of structures, and how is this understood? In particular, how, if at all, is the *indefinite extendability* of the realm of mathematical structures accounted for?

(5) How is *reference* to structures understood, and what account can be provided of our *epistemic access* to structures?

The main types of mathematical structuralism that have been proposed and developed to the point of permitting systematic and instructive comparison are four: structuralism based on model theory, carried out formally in set theory (e.g., first- or second-order Zermelo–Fraenkel set theory), referred to as STS (for set-theoretic structuralism); the approach of philosophers such as Shapiro and Resnik of taking structures to be *sui generis* universals, patterns, or structures in an *ante rem* sense (to be explained below), referred to as SGS (for *sui generis* structuralism); an approach based on category and topos theory, proposed as an alternative to set theory as an overarching mathematical framework, referred to as CTS (for category-theoretic structuralism); and a kind of eliminative, quasi-nominalist structuralism employing modal logic, referred to as MS (for modal-structuralism).[1] Let us take these up in turn, guided by questions (1)–(5), with the aim of understanding their relative merits and the choices they present.

2. Structuralism in Set Theory

This approach has, of course, arisen within mathematics itself and is the standard way of articulating a kind of structuralism. Mathematical structures are certain sets, ordered tuples of a domain together with distinguished relations, functions, and possibly individuals (members of the domain). (We take as familiar the

[1] For further details on these types of structuralism and their comparison, see [19], [33], and [20]. For a somewhat different, but largely overlapping, way of dividing up the subject (less category theory), see [29].

interdefinability of functions and relations.) Structures can be *models* of certain theories, *satisfying* their theorems. Such theories would include all the ones normally encountered in mathematics. Furthermore, structures may be related by *isomorphisms*, relation-preserving bijections, or *embeddings*, *homomorphisms*, and so on, all defined in set theory in familiar ways. Various important properties of theories, such as consistency, completeness, categoricity, and so on, correspond to or are defined by conditions on their models, a central concern of model theory.

In order to answer questions (1)–(5), framed above, it is necessary to identify a particular set theory in which model theory is carried out. The standard choice is Zermelo–Fraenkel set theory, usually with the Axiom of Choice (ZFC). But various other choices present themselves (e.g., are there to be *Urelemente* (in addition to the empty set) and—of greater significance—is the background logic to be *first-order* logic, or is it (an axiomatic part of) *second-order* logic? If proper classes are admitted, are they done so *conservatively* with respect to set theory proper, as is the case in the (two-sorted, first-order) system NBG, or as an essential enrichment of set theory, as in the second-order system of Morse-Kelly (with impredicative class comprehension)? To some extent, the answers to questions (1)–(5) depend on these, and perhaps other, choices. But let us approach them with first-order ZFC in mind, as the central and most common case.

Concerning (1) and (2), then, the logic is first-order with equality; the sole nonlogical primitive is set membership, and the nonlogical axioms, in the assertory sense, are those of ZFC. (3), the nature of mathematical structures according to STS, has already been described. Structures-as-objects are just certain sets, and so they are not eliminated. There is, however, elimination of any non-set-theoretic structures (e.g., for the natural numbers, or the reals, etc., as *sui generis* objects, as Dedekind, e.g., viewed them). [10] On STS, one identifies such objects in a particular way, but this is for convenience (e.g., for a smooth development of ordinal number theory, etc.). But it is recognized that the branch of mathematics in question (number theory, real analysis, etc.) concerns any structures of the relevant isomorphism type.[2] On this view, a question such as "What are the natural numbers, really?" is dismissed as misguided. As to the question of what distinguishes "mathematical" from other structures, a natural answer is forthcoming: that, typically, there is vocabulary (predicates, functional expressions) underlying given structures and that the source of that vocabulary indicates the

[2] Equivalence in somewhat weaker senses than isomorphism may of course be recognized, allowing for definitional extensions of given structures (e.g., adding addition and multiplication to successor in second-order arithmetic). Note that model theory can treat second- and higher-order theories and their models as objects, even if the background logic for the set theory is first-order. Also, note that our talk of "isomorphism types" is only a manner of speaking. In first-order ZFC, the relevant types cannot exist, as they would be proper classes.

field of inquiry (e.g., space-time physics, as opposed to purely mathematical geometry, although in a field such as mathematical mechanics, the line may not be sharp, especially if the theory in question is being explored not as an actual description but as merely a theoretical possibility).

Turning to question (4), one virtue of STS is the clarity of its standard of mathematical existence of structures, as this just means existence as sets in ZFC. Of course, many questions are not answered by these axioms, but for the most part these concern *extraordinary* mathematics (e.g., large cardinals, not the structures and spaces of ordinary mathematics, for which a system such as ZFC is more than adequate). What about the issue of extendability? While there is a built-in indefinite extendability of structures in the strict, set-theoretic sense (since the ordinals "go on and on"), there is a certain crucial limitation when it comes to structures for set theory itself. Here we encounter a massive exception to the structuralist point of view, in that, on its face-value interpretation, set theory itself is *not* treated structurally: its axioms are not understood as defining conditions on structures of interest but are taken as assertions of truths in an absolute sense. One speaks of "*the* cumulative hierarchy," or even "the real world of sets." *In* first-order ZFC, of course, no such thing is officially recognized, although in NBG or in second-order ZFC, one has "the universe" as a proper class. Clearly these latter theories explicitly violate *extendability* as a general principle, but first-order set theory as practiced violates it as well, for it implicitly recognizes "the sets," if only in a plural sense, as the very subject matter of the theory. Why should such higher-type totalities not be a subject of mathematical investigation? Why should not higher-order "collections" be recognized (i.e., what *prevents* the "collectibility" of "all sets," in a logical sense inherent in set theory itself)? And why aren't such collections subject to operations analogous to those of set theory itself, including formation of singletons, power collections, and so on? (Of course, this latter question applies to theories of proper classes as well, which is one of the reasons they are an embarrassment to set theory.)

As to question (5), reference to structures is just a special case of reference to sets, and this is usually just taken for granted. Once given a starting point, the null set, we know what its singleton is, what the pair of it and its singleton is, what the finite von Neumann or Zermelo ordinals are, what the hereditarily finite sets built on the null set are, what the power set of this is, and so on. Whether all these things, as intended—and this includes full power sets (containing *all* subsets of given sets) and enough ordinal levels to satisfy the Replacement Axiom—actually occur in "the real world of sets," and how we can know, are esoteric questions that surely no mathematician qua mathematician would bother about.

As this brief overview indicates, STS, despite its clarity and definiteness, is not without its problems. Three in particular stand out:

(i) As already indicated, set theory itself is not treated structurally, and this despite the fact that there are multiple set theories around, all legitimate as avenues of mathematical investigation. Thus, STS is deprived of one of the most natural ways of "justifying axioms" from a mathematical point of view, that is, by appeal to our interests in a particular kind of structure. For example, consider well-foundedness (Axiom of Regularity): rather than "defend" this as *true* in some absolute sense, a structuralist would simply cite our interest in exploring well-founded sets, without denying that non-well-founded sets "exist" or are a fit mathematical subject. Similar remarks can be made about the axiom of Replacement. It need not be guaranteed by "the meaning of 'set' " or by some a priori insight into "set-theoretic reality;" it is sufficient as a coherent condition on *domains* or *universes* of sets, guaranteeing their *largeness* relative to any of the members. But such an answer is not available to STS (for why should the real world of sets be large?). In short, this view fails to block bad questions about set theory.

(ii) STS is saddled with a maximal totality (or plurality) of sets, hence of structures, in conflict with intuition and with mathematical practice, which always has the potential of transcending any domain proposed as a limit. It violates a truly general extendability principle.

(iii) It confronts mathematically esoteric and seemingly intractable ontological and epistemological questions (e.g., in connection with the null set, with "full power sets," with "the order type of the universe," and so forth).

Thus, despite the great successes of model theory, considerations such as these have motivated some mathematicians, logicians, and philosophers to seek alternative ways of articulating structuralist ideas.

3. Structures as *Sui Generis* Universals

Like STS, this approach conceives of structures as *absolute objects*, but not as abstract particulars—rather, as universals, *patterns* in Resnik's terminology, answering to "what all particular systems of a given type—whether made up of concreta or abstracta such as sets—have in common." Patterns are thus reified; they *are* the types. One thus speaks literally of "*the* natural number structure": its constituents, the numbers, are not sets but are conceived as mere "places" or "positions" defined by the structural relations, themselves determined by mathematical axioms or conditions. To be the number *two*, for instance, is just to be the third place in this natural number structure (if we begin, with Frege, with 0). Since places are here treated as objects in their own right, rather than as "offices" to be filled by particulars, Shapiro uses the term "*ante rem*" for these structure types, in contrast to the usual *in re* structures of set theory or concrete realizations.

Many of the latter are instances of a unique one of the former.[3] Thus, SGS is not eliminativist at all. So long as some mathematical conditions—normally given as statements in second-order logic—are *coherent*, there will be an *ante rem* structure which they automatically describe. The various number systems, for instance, can be instantiated multiply by sets, but—in answer to Benacerraf's puzzle—no such instantiation is the true *ante rem* number structure. It is to such a structure that our ordinary designators actually refer, and set-theoretic reductions are merely convenient representations. Since *ante rem* structures are conceived as abstract types standing "above" already abstract instances (e.g., of sets) and moreover as instantiating themselves (i.e., as made up of objects fulfilling the structural defining conditions), the term "hyperplatonist" seems apt for this view.

In addressing questions (1)–(5), we will follow Shapiro's presentation which outlines a formal framework.[4] Concerning (1) and (2), Shapiro assumes a second-order background language, assumed "to include a rudimentary theory of collections" and the machinery for speaking of functions and relations among *places* in *structures*. It is natural to understand by this an axiomatic second-order logic including the usual unrestricted (impredicative) comprehension scheme for collections and relations. A system of axioms—structure theory—governing the primitives, *structure* and *places*, is presented: these include axioms very much like those of second-order ZF, including axioms of *Infinity*, *Power Structure*, and *Replacement*. In addition, there is a *Coherence* Axiom: "If Φ is a coherent formula in a second-order language, then there is a structure that satisfies Φ." Here "coherent" is a further primitive, corresponding to model theory's notion of *satisfiability*. (As Shapiro points out, it is not reducible to a formal condition such as consistency or even ω-consistency.) *As a direct consequence of this axiom, the basic distinction introduced above between axioms-as-defining-conditions and axioms-as-assertions collapses: any conditions that might possibly be satisfied are in fact satisfied* (i.e., *are true of* an *ante rem* structure), but since the axioms are understood to be *about* such a structure, according to SGS, they are *true simpliciter*.

In addition to these axioms, a (second-order) Reflection Principle can be assumed, guaranteeing large structures (corresponding to strongly inaccessible cardinals).

Turning to (3), its main part on the conception of structures as objects has already been addressed. And presumably we have also addressed the subsidiary question of distinguishing *mathematical* structures from others: it is natural to take these to be structures specified by coherent second-order conditions in the language of structure theory. Of course, only a very special subcollection of these

[3] Parsons [28], citing Tait, has described the thought process of moving from particular *in re* realizations to the *ante rem* structure as "Dedekind abstraction."

[4] Alternative presentations, such as Resnik's, which is more informal, may not be entirely accurately represented by Shapiro's formulation.

would be of genuine mathematical interest, and it would surely be misguided to attempt any general characterization of these; but physical and other nonmathematical structures would presumably not be found, insofar as their description requires vocabulary beyond that of pure structure theory (second-order logic plus the primitives "structure" and "place").

As to (4), existence of structures is of course spelled out in the axioms of structure theory. What about extendability? Clearly SGS improves on STS in applying to set theories without having to recognize any one as "the one true one," and so it avoids any maximal plurality or collection of sets. Still, there is a maximal collection of *places in structures* as a consequence of the second-order comprehension principle which structure theory presupposes. And while there is no collection of all *structures*, since this would be a third-order object, still *there are*, speaking plurally, *all the structures* (i.e., an inextendable "universe of structures"), informally speaking, whether or not the formalism allows talk of such a collection.

Finally, regarding (5), as indicated, reference to *ante rem* structures and their places is just supposed to occur when we gain mastery over the relevant mathematical vocabulary. Just as in learning the English alphabet, we learn to refer to the letters as types, so we learn to refer to the natural numbers, and then, somewhat later, to the rationals, the reals, and maybe even the complexes.

How does SGS fare with respect to the problems (i)–(iii) affecting STS, described above? (i) is overcome in SGS, as it treats set theory (theories) structurally like other mathematical theories. Sets are not conceived of as abstract particulars, existing in their own right and forming the subject matter of mathematics. Rather, like numbers or elements of algebraic structures, they are conceived merely as places in an abstract structure. The multiplicity of set theories fits nicely into the view; for instance, well-founded sets are realized in one *ante rem* structure (actually in many such), but there are others realizing non-well-founded sets. The "membership relation" of the latter simply behaves differently. Likewise with regard to other structure-characterizing axioms such as Replacement. In sum, SGS succeeds in blocking a number of bad questions about set theory where STS only encouraged them. Similarly, the puzzles for STS under (iii) do not arise for SGS. There is no mystery about the null set, for example. This is just a structural "starting point" in an *ante rem* structure for set theory, an object to which no other place bears the relevant two-place relation (ordinarily called "membership"). Moreover, there is no question about "real-world power sets," for such are not distinguished. So long as "full power set" (of a given infinite set) is coherent, an *ante rem* structure will have a place for such (as places). Of course, there are also many structures with less-than-full power sets, but that does not matter. The question of coherence can certainly be raised (and typically is by predicativists), but the issue is not an ontological one, according to SGS. This seems right. Similarly, there is no problem about "the order type of the universe," since no "universe" of all sets need be recognized. Many universes are possible; hence actual as *ante rem* structures. If power sets are full and second-order Replacement holds, the order type of such a universe will be a strongly inaccessible cardinal, but, presumably, there are boundlessly many of these.

At first blush, it seems that SGS also handles problem (ii) quite deftly, for there simply is no maximal totality of sets forming an *ante rem* structure. As far as these structures for set theory go, the extendability principle is respected. However, as already pointed out in connection with point (4) above, SGS is still saddled with maximal totalities of its own, namely, the class of all places (explicitly delivered by second-order logical comprehension in structure theory) and the plurality of all *ante rem* structures, implicit in the overall system with the Coherence Axiom. In effect, SGS has merely traded "the totality of sets" for "the totality of structures," and so does not endorse the general extendability principle.

There are, moreover, further problems that have been raised against SGS, problems that do not confront STS. Two interrelated such problems should be mentioned here. (Our numbering continues that of the list of problems begun above.)

(iv) *Places* as objects in *ante rem* structures raise questions concerning their very identity.[5] It seems, from a structuralist perspective, that intrastructural relations alone should suffice to distinguish these objects, without appeal to external relations or individual constants. But then places in a structure should satisfy a Leibnizian principle of *identity of structural indiscernibles*: any items bearing exactly the same intrastructural relations to other items should be not many but one.[6] But this immediately implies that the structure in question must be *rigid* (i.e., admitting no nontrivial automorphisms). While that is true of some key mathematical structures (e.g., the natural numbers, the field of real numbers, segments of the cumulative hierarchy of sets, etc.), nonrigid structures nevertheless abound in mathematics (e.g., the complex numbers [interchanging i and $-i$], the additive group of integers [interchanging $+1$ and -1], geometric figures with reflectional symmetry, homogeneous Euclidean n-space, etc.). While it is true that all these nonrigid structures can be recovered inside rigid ones (via reduction to sets), that seems counter to the whole thrust of SGS. Another reply suggests simply not seeking any criterion of identity for *places*. The debate over this continues.[7]

[5] This objection has been raised independently by Keränen [22] and Burgess [8].

[6] Note that this is weaker than the claim, sometimes made, that places in *ante rem* structures have only the properties—or only the essential properties—required by the axioms defining the structure in question. These latter conditions seem impossible to realize: for example, surely the natural numbers (according to SGS) are abstract, nonspatiotemporal, nonphysical, and so on, even essentially so; yet the axioms are silent on such matters.

[7] See Shapiro's "Structure and Identity," Keränen's "The Identity Problem for Realist Structuralism II," and Shapiro's "The Governance of Identity" in [23].

(v) Are *purely structural relations* intelligible in the context of putatively structural objects—*places*—as relata? This objection challenges the very notion of *ante rem* structure, whether rigid or not. If we do not appeal to the relata of a structure as somehow independently given (e.g., invoking reference by singular terms in our background language as standard Platonism does), but as determined by structural relations—which is surely part and parcel of the SG view—what have we to go on in specifying structural relations other than the axioms (defining conditions) themselves?[8] But, as Russell pointed out in his criticism of Dedekind's early expression of SGS [9], [10], the axioms don't distinguish any particular realization from among the many systems that satisfy them.[9] What, for example, can it mean to speak of "*the ordering*" of "*the natural numbers*" as objects of an *ante rem* structure unless we already understand what these numbers are *apart from their mere position in that ordering*? Surely the notion of "*next*" makes no sense except *relative to an ordering or function or arrangement of some sort*, something Dedekind was careful to take into account when describing *simply infinite systems*, which always involve objects "set in order by transformation φ." (Clearly, anything whatever can be "next after" anything else in some system or other.) Thus the notion of an *ante rem* structure seems to involve a vicious circularity: such a structure is supposed to consist of *purely structural relations among purely structural objects, but understanding either of these requires already understanding the other.* Whereas the Keränen–Burgess objection granted the relations and raised questions about how these alone could determine the objects (unless the structures are rigid), this objection questions such talk of relations in the first place, and thereby the very notion of "Dedekind abstraction," which is supposed to lead to them.[10]

[8] As Parsons put it in assessing the idea of structural objects as "incomplete objects," this view is "itself incomplete for it neglects the fact that the *relations* of a structure are themselves given only by formal conditions..." ([28], 334).

[9] As Russell wrote, "It is impossible that the ordinals should be, as Dedekind suggests, nothing but the terms of such relations as constitute progressions. If they are to be anything at all, they must be intrinsically *something*; they must differ from other entities as points from instants, or colors from sounds.... Dedekind does not show us what it is that all progressions have in common, nor give any reason for supposing it to be the ordinal numbers, except that all progressions obey the same laws as ordinals do, which would prove equally that *any* assigned progression is what all progressions have in common..." ([31], p. 249).

[10] In correspondence, Shapiro has pointed to "finite cardinal structures"—strucures with finitely many distinct "places" but no relations at all (other than (non-)identity)—as showing that places should not be thought of as dependent on structural relations. Neither places nor relations are prior to the other. But such structures seem an ultimate offense against Leibnizian scruples. For what distinguishes one "place" from another? How can we even make sense of mapping the places to or from the many finite collections such a structure is supposed to exemplify (e.g., all pairs, or triples, or quadruples, etc.)? Does it even make sense to think of labeling these "things"? In the case of identical bosons of quantum mechanics (a famous case where labeling is problematic), we ultimately have the

In fact, SGS seems ultimately subject to the very objection of Benacerraf [4] that helped inspire recent structuralist approaches to number systems in the first place. Suppose we *had* the *ante rem* structure for the natural numbers, call it $\langle \mathbb{N}, \varphi, 1 \rangle$, where φ is the privileged successor function and 1 the initial place. Obviously, there are indefinitely many other progressions, explicitly definable in terms of this one, which qualify equally well as referents for our numerals and are just as "free from irrelevant features"; simply permute any (for simplicity, say finite) number of places, obtaining a system $\langle \mathbb{N}, \varphi', 1' \rangle$, made up of the same items but "set in order" by an adjusted transformation, φ'. Why should this not have been called "the archetypical *ante rem* progression," or "the result of Dedekind abstraction?" We cannot say, for instance, "because 1 is really *first*," since the very notion "first" is relative to an ordering; relative to φ', $1'$, not 1, is "first." Indeed, Benacerraf, in his original paper, generalized his argument that numbers cannot really *be* sets to the conclusion that they cannot really *be* objects at all, and here, with purported *ante rem* structures, we can see again why not, since multiple, equally valid identifications compete with one another as "uniquely correct." Hyperplatonist abstraction, far from transcending the problem, leads straight back to it.

Other problems can be raised for SGS, in particular problems concerning the Coherence Axiom and the primitive "coherent." It can be thought of as a post-Gödelian substitute for formal consistency: the axiom mimics Hilbert's idea that consistency suffices for mathematical existence. But of course it is not a formal notion and seems no clearer than a primitive notion of (second-order) logical possibility—indeed, perhaps less so, for do we have anything as developed as modal logic governing "coherent?" And if we identify these notions, then the Coherence Axiom appears even more problematic, for why should mere logical possibility suffice for existence? Indeed, why not just rest with the former, which, as we shall see below, enables avoidance of the very maximality problems that plague both STS and SGS?

For all these reasons, then, we are motivated to look at some alternative, nonabsolutist approaches to structuralism.

4. Structuralism in Category Theory

Like set theory, category theory (CT) has arisen within mathematics both as a branch of mathematics making its own special contributions (in the case of

option of dropping the notion of two particles, say, in favor of a "boson-pair system." But such a move in the case of finite cardinal structures would destroy their cardinality (at least for any $n > 1$)!

category theory, to algebraic topology and algebraic geometry) and also as a general framework or setting for vast amounts of mathematics. Mac Lane and others have taken the further step of proposing topos theory as providing an alternative foundation for mathematics, comparable to and in some ways superior to set theory, and, more recently, Awodey has explicitly suggested that category theory provides a natural framework for realizing structuralism, one that philosophy of mathematics should employ and pursue.[11]

The basic idea behind Awodey's suggestion is that category theory provides an effective way of codifying and illuminating mathematical structure through its focus on families of structure-preserving mappings between objects having the relevant kind of structure and through functorial relations between different categories, which exhibit important relationships among different kinds of structures, of special interest in advanced mathematics. Moreover, the CT approach to mathematical structure has advantages over the Bourbaki (set-theoretic) approach, in that descriptions via mappings (morphisms, also functors) abstract from the initial means by which structures of a given type may have been introduced. Topological structure, for example, is determined via continuous maps to and from other spaces regardless of the details of their original introduction, via open sets, limit points, closure operations, and so on. Indeed, it is remarkable how much of what are normally thought of as operations and relations *internal* to a structure can be recovered via morphisms to and from other "objects" (what would normally be called "structures," formally treated as pointlike in CT). It is even possible, for example, to get the effect of elementhood itself, via morphisms between a terminal object and a given object (where an object 1 is *terminal* just in case, for any object C in the category there is a unique morphism from C to 1). Moreover, CT constructions exhibit in a precise way how mathematical structure is significant only "up to isomorphism," and they bring out "shared structure" through novel methods of generalization (e.g., via universal mapping properties of, for instance, products, generalizing the notion of Cartesian product in set theory).

Clearly, CT provides important insights into various kinds of mathematical structures and into "mathematical structure" in a general sense. But does it provide a genuine structuralist framework for mathematics on a par with STS or SGS? In assessing this, some basic distinctions must be borne in mind. First, it is essential to distinguish between CT *as mathematics* in its own right and CT *as a foundational framework*. Texts on CT as mathematics typically make reference to a background set theory or universe of sets relative to which CT is to be understood and carried out.[12] If generally enforced, this would make CT dependent on set

[11] In this discussion, we will presuppose a familiarity with the basic notions of "category" and "topos." Awodey [1] presents an accessible overview, as do Mac Lane [24] and Bell [2]. See also McLarty [27].

[12] See, for example, Mac Lane and Moerdijk [25], Bell [3], and Freyd and Scedrov [14].

theory, and it could not be considered to provide an autonomous framework. Thus, the issue of autonomy vis-à-vis set theory must be addressed. In particular, Mac Lane has proposed that mathematics generally can be developed in a certain kind of topos (satisfying a condition called "well-pointedness," leading to extensional discrimination of morphisms as in classical mathematics), and that this is an alternative to set-theoretic foundations. But then it must be possible to make sense of such topoi without falling back on set theory, either conceptually or ontologically. Similarly, Bell has proposed that any topos with a natural number object can serve as a background universe for (ordinary) mathematics, and that this results in an interesting relativity of ordinary mathematical concepts (e.g., "real number," "continuous function on reals," etc.) to background topos (whose internal logic is in general intuitionistic, but which may be classical if stronger conditions, such as well-pointedness, are built in). Some core mathematics turns out to be invariant over the relevant totality of topoi, but many things are not. To sustain this local, relativist view, again it must be possible to carry out the mathematics and metamathematics of topoi without falling back on set theory.

A second important distinction concerns the term "category theory" itself. On the one hand, there are the "first-order axioms on categories," the standard conditions interrelating the domains and codomains of morphisms with the binary "composition" operator and governing identity morphisms. To these may be added a variety of further conditions defining various kinds of topoi: for "elementary topoi," the basic requirements generalizing Cartesian products, functional exponentiation, and set theory's apparatus for (Boolean) propositional functions ("subobject classification"); for foundational purposes, further conditions guaranteeing a natural number object; and stronger principles such as an axiom of choice, well-pointedness, and so on. It is clear, however, that these are not "theories" in the sense in which ZFC is a theory, for these "axioms" on categories and topoi are merely defining conditions, simply telling us what first-order conditions must be satisfied by anything to qualify as a category or a topos. Thus, the primitives ("domain," "codomain," "composition") need not have any particular meanings, and these axioms *assert nothing* by themselves. As Awodey himself puts it:

> A category is *anything* satisfying these axioms. The objects need not have "elements," nor need the morphisms be "functions," although this is the case in some motivating examples.... We do not really care what non-categorical properties the objects and morphisms of a given category may have.... ([1], 213)

Thus, categories and topoi are conceived in the manner of algebraic structures such as groups, rings, and so on, and the informal "theory of categories," which of course contains genuine assertions including foundational ones *about categories and topoi*, must not be identified with these first-order "axiom systems," but must somehow be rich enough to express substantive claims about structures satisfying these axioms. This is essential in assessing the status of CT vis-à-vis set theory. For example, just because notions of "collection" and "operation" are not used in the first-order defining conditions does not mean that they are really bypassed in theorizing about categories in the relevant sense.

This is an appropriate place to recall an early critique by Feferman of Mac Lane's claims on behalf of CT as providing a foundational alternative to set theory [11]. In essence, Feferman argued that category theory presupposes and uses, informally, notions of *collection* and *operation*, both in saying what a category (or topos) is, and in relating categories to one another through *homomorphisms* or *functors*. Moreover, a foundational framework for mathematics must provide some systematic account of these notions, something set theory does but category theory does not. It was explicitly recognized that alternatives to standard set theory might also provide this, so that the claim was not that CT depends on set theory per se, but rather that, as it stands, it is inadequate.

Now there is a temptation to respond to this by taking CT to be a theory of families of functions related by composition, and to claim that this in principle is no different from set theory as a theory of sets (and possibly individuals) related by membership. The notion of "function" is one that mathematicians use all the time, and even set theory can be derived from an analogous theory of functions, as was originally carried out by von Neumann [34]. CT could be viewed as an alternative, perhaps more interesting, systematic approach based on the familiar notion of function. What's wrong with that?

The problem is that this depends on a special, privileged interpretation of "composition," hence of "morphism," in diametric opposition to the algebraico-structuralist reading that category theorists apply to their own systems (cf. the quotation from Awodey, above). Moreover, the "axioms" defining categories and topoi are silent on the matter of mathematical existence of these structures, leaving unresolved basic questions concerning their sizes and their scope. (E.g., is there really a category of *all* categories? If so, how is Russell's paradox avoided?, and so forth.)

Thus we see that an algebraico-structuralist reading of the CT axioms actually underlay Feferman's critique;[13] moreover, it simply vitiates the above response, which in any case is inadequate, as just indicated.

Turning back to questions (1)–(5) we have been putting to each version of structuralism, we find ourselves now in the awkward position of being unable to

[13] As Feferman originally wrote, "when *explaining* the general notion of structure and of particular kinds of structures such as groups, rings, categories, etc., we implicitly *presume as understood* the ideas of *operation* and *collection*; e.g., we say that group consists of a collection of objects together with a binary operation satisfying such and such conditions" ([11], 150).

answer any of them with any definiteness, save perhaps (3), concerning elimination of structures as objects. (For future reference, let us number this problem of underspecification (vi).) Consider (1) on primitives and background logic. Of course we know what the primitives of the first-order CT axioms are; however, the question is not about the definition of "category," but rather about the primitives of the background (informal) substantive mathematical-foundational (meta-) theory, which, as Feferman observed, employs notions of *collection* and *operation* and *functor*. The fact that these are not among the primitives of the definitional CT or topos-theoretic axioms is irrelevant. We cannot even say that the background logic is to be first-order, nor that it is intuitionistic, like the *internal* logic of a topos. Almost certainly classical logic is required for some of the results of category theory as substantive mathematics. Bell's approach, for example, which treats topoi as models of "local set theories," requires some substantial metalogic. The languages of local set theories are type-theoretic, but with no cardinality restrictions on the type symbols that may occur. (Bell even acknowledges at the outset that for some purposes, a background set theory such as NBG may be needed.)

The situation regarding (2) is no better. We are simply not in a position to identify agreed-upon substantive (assertory) axioms explicitly or implicitly assumed in category and topos theory qua genuine mathematics or metamathematics, without falling back on set theory.

Presumably, structures-as-objects are not eliminated in CT structuralism, for they make up one of the sorts (the *objects*) of typical categories (e.g., of groups, of vector spaces, of topological spaces, etc.).[14] This is a short answer to the first part of (3), but what about the second and third parts, concerning the nature of structures and what distinguishes *mathematical* from other structures? As objects inside categories, structures are treated as "pointlike," and everything is expressed in terms of morphisms and functorial relations, leaving the nature of structures themselves up in the air. While it is true that CT has the resources to express its own account of various spaces (e.g., topological), one wonders whether such an account would be intelligible without prior acquaintance with ordinary set-theoretic constructions (e.g., via open sets).

The first part of (4), of course, as a special case of (2), remains unanswered. Questions of mathematical existence of structures, including categories and topoi, are usually deferred to an unspecified background set theory or model thereof, but, as already said, this is appropriate only when pursuing CT as pure mathematics, but surely not when proposing it as a foundation or structuralist framework. Concerning the second part of (4), extendability, this principle is certainly in the spirit

[14] Even this point, however, is somewhat debatable, since the objects of a category can actually be dispensed with in favor of special morphisms, identity morphisms. In any case, morphisms carry structure, and clearly *they* are not eliminated, so that in a broad sense, distinctively mathematical objects are recognized.

of CTS, as Mac Lane's remarks on the open-endedness of mathematics attest.[15] But we saw above that SGS, while critical of "the real world of sets," nevertheless runs into violations due to second-order logic, and it remains to be seen how CTS can avoid such problems. As to (5), how CTS contributes to our understanding reference and epistemic access to structures, we remain quite in the dark.

Even without enough development to answer these questions, CTS does appear to escape the problems (i)–(iii) affecting STS and (iv) and (v) affecting SGS, on the proviso that it can avoid commitment to maximal totalities. Clearly it treats set theories structurally along with other branches of mathematics. The null set, for example, is treated simply as an initial object of a category, not as a mysterious empty abstract container. And there need be no commitment to any unique set-theoretic reality. Nor is CTS plagued by the special objects and relations of *ante rem* structures.

On the positive side, category theory has much to offer by way of insights into mathematical structure and through its novel approach to generalization via morphisms and functors. Equally clearly, however, in light of the foregoing discussion, it is inadequate as a foundational framework as it stands. If it is not to fall prey to a dependence on set theory after all, some alternative way of developing it which is responsive to the above basic questions must be found. In the final section below, we will mention one such alternative.

5. Modal-Structuralism

A good way into this approach (developed in [17] and [18]) is via Russell, who wrote early on in [32]:

> It might be suggested that, instead of setting up "0," "number," and "successor" as terms of which we know the meaning..., we might let them stand for *any* three terms that verify Peano's five axioms. They will then no longer be terms

[15] Thus, he writes:

> Understanding Mathematical operations leads repeatedly to the formation of totalities: the collection of all prime numbers, the set of all points on an ellipse... the set of all subsets of a set..., or the category of all topological spaces. There are no upper limits; it is useful to consider the "universe" of all sets (as a class) or the category *Cat* of all small categories as well as CAT, the category of all big categories. After each careful delimitation, bigger totalities appear. No set theory and no category theory can encompass them all—and they are needed to grasp what Mathematics does. ([24], 390)

> which have a meaning that is definite though undefined: they will be "variables," terms concerning which we make certain hypotheses, namely, those stated in the five axioms, but which are otherwise undetermined.... our theorems... will concern all sets of terms having certain properties. (p. 10)

No sooner had Russell offered this suggestion than he retracted it for what we today would regard as quite spurious reasons (the *definition* did not provide for the *existence* of models—this from a critic of the ontological argument!—and you could not account for ordinary counting on the proposed interpretation),[16] and pursued the Fregean absolutist line (cardinal numbers as definite classes of equinumerous concepts or classes), only to encounter various problems and paradoxes leading to the abandonment of classes. In the end, we find this remarkable suggestion concerning logical and mathematical propositions generally:

> We may thus lay down, as a necessary (though not sufficient) characteristic of logical or mathematical propositions, that they are to be such as can be obtained from a proposition containing no variables... by turning every constituent into a variable and asserting that the result is always true or sometimes true.... logic (or mathematics) is concerned only with *forms*...." ([32], p. 199)

When we consider that *relations (propositional functions)* as well as individuals count as "constituents" of propositions, then we realize that this criterion is met by formulating mathematics in higher-order logic without constants. In fact, second-order logic suffices. In the case of number theory, for example, an ordinary statement A naturally goes over to a conditional of the form

$$\forall R\,[\mathrm{PA}^2 \rightarrow A]\,(\mathrm{S}/R),$$

in which "PA^2" stands for the conjunction of the (Dedekind–) Peano axioms and "S/R" indicates systematic replacement of the successor constant with the relation variable "R" throughout. (Here "0" can be dropped because it is definable from successor.) If A is logically implied by the axioms, this is a truth of second-order logic; if not, the result of replacing "A" with "$\neg A$" is, in light of Dedekind's

[16] On the eliminativist strategy suggested, counting would be understood, roughly speaking, as indicating, with numerals, or related symbols, a one-one correspondence between the enumerated items and an initial segment of any progression (or any that there *might be*, on a modalized version). One can even put it constructively: counting provides a means of building a one-one correspondence between the enumerated items and a segment of any given progression. On such an account, *standing in such correspondences* plays the role that *membership* plays on the Frege–Russell account (i.e., the enumerated class (or concept) *belongs to* a Frege–Russell number).

categoricity theorem.[17] Even better, taking the predicate "number" into account as suggested, and generalizing, we obtain

$$\forall X \forall R\,[\mathrm{PA}^2 \rightarrow A]^X (\mathrm{S}/R),$$

where the superscript indicates relativization of all quantifiers to domain X. Thus, Russell has come full circle (at least "up to negation!"), for this is just the general interpretation "through variables" that he had suggested and dismissed at the outset. (But no matter. Russell eventually got there.) It is already a kind of structuralist interpretation, expressing that truths of arithmetic are what hold in any progression whatever. Formulated as it stands, however, it is inadequate, for suppose there *are no* progressions; then all such conditionals are vacuously true, regardless of the content of A. A better plan is to construe the generalization "always true" modally (i.e., as meaning "in any progression there might be, logically speaking"), prefixing the last displayed formula with a necessity operator ("$\Box$"). Then, to avoid vacuity, we may categorically lay down

$$\Diamond \exists X \exists R\,[\mathrm{PA}^2]^X (\mathrm{S}/R),$$

affirming the logical possibility of a progression, which of course is compatible with the actual absence of any. The same plan works for the real number system, for the complexes, for cumulative hierarchies of sets (characterized by cardinality of *Urelemente* and ordinal height), and, indeed, for any mathematical structure categorically characterized by second-order axioms. (For noncategorical theories—as, for instance, in abstract algebra—the result is incompleteness, but that is as it should be. Of course, more specific types of structures—e.g., transitive permutation groups of a given order, and so on—can be treated.) So far we have the basic plan of modal-structural interpretations of mathematical theories (the conditionals forming the "*hypothetical component*," the possible existence claims forming the "*categorical component*").

The background logic is second-order, with quantified S-5 modal logic, without the Barcan formula.[18] However, care must be taken in formulating second-order logical comprehension. In a modal context, unrestricted comprehension leads to intensions, transworld classes and relations. For example, suppose

[17] The question may be raised here as to what background theory this categoricity theorem is proved in. It need not be anything so strong as impredicative set theory. Indeed, as shown in [12], essentially only a weak axiomatic fragment of weak second-order logic is really required, in which quantification over finite subsets of the domain is taken as given. More precisely, the theorem is recovered in "EFSC," an elementary theory of finite sets and classes of individuals, conservative over Peano Arithmetic.

[18] That is, the inference from $\Diamond \exists x \varphi$ to $\exists x \Diamond \varphi$ is to be avoided.

the predicate 'planet' is available; then we would generate a class not merely of existing planets, but also of those together with "any that *might have existed.*" That is, we would be recognizing *possibilia.* (Notice that this follows even if the class quantifier is understood as a plural quantifier.) In general, we would be quantifying over relations, not merely relating objects in the actual world or in any hypothetical one being entertained, but *across* worlds.[19] In particular, we would generate a universal class of all *possible* objects, and corresponding universal relations among possibilia, directly violating the extendability principle (modally understood, appropriately, as "Any totality there might be, might be extended"). Ordinary mathematical *abstracta* seem tame compared to such extravagances; indulging them would deprive MS of much of its interest as a distinctive program. To avoid such commitments, therefore, an extensional version of comprehension is chosen:

(Logical Comp) $\Box \exists R \forall x_1 \ldots \forall x_n (R(x_1 \ldots x_n) \leftrightarrow \Phi)$,

where Φ lacks free "R" and is also modal-free. Note that the universal quantifiers are not boxed. In effect, (n-tuples of) individuals form collections and relations only *within* a world, not across worlds. (Officially, of course, neither worlds nor possibilia are recognized.)

This leaves us, however, with a level of abstract classes and relations or Fregean concepts as second-order entities. This can be dispensed with, however, by appealing to a combination of mereology and plural quantification, provided (the possibility of) an infinity of individuals is assumed. This itself can be expressed using mereology and plural quantifiers, for example, by

(Ax ∞) There are some individuals one of which is an atom and each of which, combined with an atom not part of it, is also one of them,

where an atom is an individual without proper parts. Given this, one can get the effect of ordered pairing of arbitrary individuals,[20] and this effects a reduction of polyadic second-order quantification to monadic, which itself can be interpreted plurally. *Thus, the relation quantifier in the above second-order comprehension scheme can be replaced with a class quantifier.* Also, Φ can contain the part–whole relation of mereology.

[19] Note that it is *quantification* over transworld relations that is to be avoided. There is nothing to prevent us from applying *particular predicates* to entertain relations among given objects and "others that there might have been," as when we say innocous things such as "There might have been a horse larger than any existing one," and so on. This does not commit us to possible horses as objects nor, of course, to worlds containing such beasts.

[20] See [6].

Completing the core system is a "comprehension" scheme of mereology itself, guaranteeing the whole ("sum," or "fusion") of any individuals satisfying a given (non-null) predicate:

$$(\Sigma\ \text{Comp}) \quad \exists x \Psi(x) \rightarrow \exists y \forall z[y \circ z \leftrightarrow \exists u(z \circ u \ \&\ \Psi(u))],$$

where " $\circ$ " "*overlaps*" is defined via "*part of*", $<$, as $x \circ y \leftrightarrow \exists v(v < x \ \&\ v < y)$; and, in Ψ, monadic second-order variables are allowed, free or bound. Thus formulated, MS is ontologically neutral to this extent: any objects whatever may stand in relevant structural relationships so long as it makes sense to speak of wholes, or of pluralities, of them, which is all that this machinery requires. In particular, a physical or spatiotemporal interpretation of "part–whole" is not required.[21]

This framework turns out to be surprisingly powerful. On the assumption of (the possibility of) just a countable infinity of atoms, not only can the modal existence of progressions (or $\mathbb{N}$-structures) be derived, but by repeated use of plurals and mereology, full, polyadic classical third-order number theory, equivalently second-order analysis, can be recovered. If one postulates the possibility of a continuum of atoms, this can be pushed up to fourth-order number theory. Even second-order is already rich enough to represent vast amounts of ordinary mathematics, yet there is no use of set-membership or even an abstract ontology of classes and relations. Structures and structural relations are gotten at indirectly; set-theoretic and higher-order logical constructions can be encoded by talk of pluralities (invoking pairing for relations), but without actually reifying structures or relations. We have a "structuralism without structures."

We have now provided answers to the questions under (1) for MS in some detail. Concerning (2), assertory axioms, the initial Axiom of Infinity (Ax ∞) and the displayed comprehension schemata form the core system, to which can be added more specific modal-existence claims for particular kinds of structures. That for $\mathbb{N}$-structures is already derivable from (Ax ∞) and instances of comprehension, as can even the modal-existence of continua or $\mathbb{R}$-structures, obtained by following well-known classical constructions (e.g., via Dedekind cuts). For *atomic* $\mathbb{R}$-structures, with reals as atoms, however, further postulation is necessary. Similarly, further postulates are needed for structures of higher cardinality (e.g., for Zermelo set theory and beyond). It should also be mentioned that the *unicity* (uniqueness up to isomorphism) of $\mathbb{N}$- and $\mathbb{R}$-structures is also derivable in the core system.

[21] It is worth noting, for example, that Goodman [15], who helped promulgate mereology (which he called the calculus of individuals), applied it to sense qualia, themselves conceived as "abstract" in the sense of "multiply instantiable" and not themselves spatiotemporally bound.

As to (3), this version of structuralism is thoroughly eliminativist, as already described.[22] Concerning what distinguishes *mathematical* from other structures, the same second-order logical criterion mentioned in connection with SGS is available (appealing to translation via plurals).

Concerning (4), MS is uniquely explicit in distinguishing *mathematical existence* from ordinary existence. Zermelo spoke of the former as "ideal existence," an idea similar to the notion of logical possibility. By bringing this to the fore, however, MS permits explicit principles of extendability of (pluralities of) structures (which Zermelo [35] articulated for models of set theory), but without generating any universal classes of structures or structural objects (as arise in SGS and in a straightforward second-order logical formalization of Zermelo [35] as well), due to the natural limitations set by extensional second-order comprehension. It simply makes no sense to speak of a collection or plurality of all structures or items in structures *that there might be.* (This builds on the more mundane fact that, on an ordinary understanding of "collecting," you cannot actually collect anything which only might have existed.)

As to (5), MS is distinguished from other versions in eliminating the need for reference to mathematical structures, since it is only possibilities of structurally interrelated objects that are entertained, and these are given by general descriptions. Ordinary designators (e.g., numerals) occurring in everyday use can be accounted for in various ways, e.g., as indicating relevant places in structures; as convenient devices in counting, measuring, and computing; as introduced in mathematical reasoning modulo the assumption of (the possibility of) a structure of a given type (e.g., $\mathbb{N}$-structures), following the logical move of existential instantiation, and so forth. But the usual puzzles (and "bad questions") concerning Platonistic reference to abstracta do not arise.

The real challenge for MS lies in the last question, concerning epistemic access, not in this case to structures themselves—since these are eliminated on this interpretation—but with respect to the possibilities of structures. What sort of evidence can we have for the various modal-existence postulates arising in mathematics, as illustrated above? Of course, we may gain evidence of formal consistency of the associated axiomatic systems (including various strong forms of consistency, such as ω-consistency), even if we cannot have a finitistic proof in the central cases of interest. But the second-order machinery of MS is adopted so that *standard* models of theories (e.g., number theory, analysis, set theories) will be describable, and the possibility of these is not guaranteed by formal consistency claims. It seems that we must fall back on indirect evidence pertaining to our successful practice internally and in applications, and, perhaps, the intuitive pictures and ideas we have of various structures as supporting the coherence of

[22] Of course, the elimination is for pure mathematics, without prejudice to any actual instantiations of structures there may happen to be in the material world.

our concepts of them. Perhaps that is the best that any version of structuralism can hope for. We will return to this below, when we come to address some specific challenges that have been raised for MS.

Turning briefly to the problems (i)–(v) raised above, it should be clear that none of them affects MS. Set theories are interpreted structurally, and questions about "the real world of sets" do not arise. Multiple structural possibilities are allowed for, including full "power sets"; less-than-full-power sets; well-founded domains; non-well-founded; and so on. Extendability principles are explicitly part of the interpretation of set theory, leading to the "small" large cardinals (inaccessible, hyper-inaccessible of all orders, Mahlo, n-Mahlo, etc.). And, as explained, extensional comprehension does not permit recognition of maximal totalities such as that of "all possible structures." Finally, the whole thrust of MS is to avoid postulating special abstract objects, so the puzzles concerning "places" and "purely structural relations" do not arise. The MS route to "abstractness" consists not in attempting to introduce "featureless objects," but in simply not building into the descriptions of hypothetical structures anything beyond what is of mathematical interest. Benacerraf's puzzle is solved by accepting his conclusion: numbers as objects are officially eliminated, although number-words can be introduced as aids in computation and reasoning. Finally, regarding (vi), MS is as explicit as any version regarding its assumptions.

The main new problem for MS is reliance on primitive modality (call this (vii)), analogous to SGS's reliance on the primitive "coherent." One would like a formal criterion for these notions, but that is not to be hoped for, and the approach confronts the epistemological questions broached above.

Space permits us to consider, briefly, only the core Axiom of Infinity (Ax ∞), which, with the MS machinery (as already explained), suffices for the vast bulk of ordinary mathematics.

The possibility of a countable infinity of objects seems so entrenched in, and indispensable to, our scientific and mathematical thinking that it is difficult to argue against the skeptic. Intuitionists who insist on human mental constructions of course cannot be satisfied because, of course, we have only finite resources to work with and can work only so fast. From the classical MS perspective, the issue of supertasks is entirely beside the point. As Hale recognized [16], it is the ready conceivability of situations in which infinitely many mind-independent objects exist that we naturally appeal to if pressed. Of course, the Platonist can claim, in a variety of ways, that our present situation is such, since, for instance, a proposition p exists and, for every proposition q, the statement "*It is true that q*" expresses a proposition, distinct from q (Bolzano); or that some object o is given, and that for any object a, the singleton of a exists and is distinct from a (Zermelo), and so on.

But nominalists are not deprived of the possibility of infinities just for not going along with propositions, sets, and such. It is sufficient if, for example, it could be the case that there is a moment of time and that for every moment of time there is a later one (perhaps forming a convergent series), or—as Dummett concedes to be perfectly intelligible—that there be stars such that, for any one of them, another one exists some further distance away, and so on. Now, in examining the move from conceivability to possibility, Hale [16] explicitly distinguishes between requiring that the conceived situation be one *in which* it *could* be verified that there are infinitely many things—a condition he (rightly) regards as too strongly verificationist—and requiring rather that the conceived situation be one *of which* it *can* be recognized that, *were it to obtain*, there would indeed be an infinity of things. The latter can be satisfied if a sufficiently detailed description is available from which it can (here, in *our* world) be inferred that an infinity exists wherever the description holds, whereas the former might be quite impossible without supertasks. And it is the latter requirement that is to govern, according to Hale ([16], 137–138).

So what, then, is wrong with the appeal to moments of time, stars, and such, as above? Our descriptions straightforwardly entail the existence of infinitely many things in those situations, and Hale seems to grant the possibility that those descriptions could hold. We certainly *can* see the entailment of infinity, right here and now. It appears that, several pages later, however, there is a slide in Hale's discussion back to a verificationist requirement, for he writes that "the imagined situation would have to be given by a description of which we *could* tell . . . that, were it to be satisfied, this *would* mandate acceptance of a theory which entails the existence of a completed concrete ω-sequence" ([16], 145, my emphasis), and just prior to this he writes that only *ordinary* tasks are relevant in assessing an imagined situation "as being one of which we *could* recognize that, were it to obtain, a concrete ω-sequence would exist" (*ibid.*, my emphasis).[23] Then, not surprisingly, the MS appeal to situations in which there is always a later moment of time or another star further away, and so on, will be found wanting. It seems that, after all, we would have to be able to *determine* such things *in the imagined situations*, in a strong sense (e.g., have evidence that could not be explained on a strictly finitistic basis). It seems that the red herring of supertasks is again out of the jar.

[23] Lest it be thought that we are resolving some subtle ambiguity of modal usage in a biased way, we note that, in the ensuing discussion, Hale writes:

> [the structuralist] must supply a description—necessarily finite—of a possible situation, no empirically adequate theoretical account of which could avoid postulating the existence of a completed concrete ω-sequence. But any desription which could—in the present context—be reckoned unproblematic will perforce mention only finitely many observationally ascertainable facts . . . which an empirically adequate theory must explain" (p. 145).

This then rules out closure conditions involving quantification (e.g., "for any moment of time, there is a later one"), on the grounds that its satisfaction is not "observationally ascertainable." The slide back to a verificationist requirement is complete.

A further criticism sometimes leveled against MS concerns its use of second-order logic. There are two interrelated parts to this: first, second-order logic, in its intended sense, is not formalizable; and, second, this reflects the fact that a substantial amount of mathematics is thus presupposed, raising the specter of circularity. In reply, the first point is correct and is a corollary of Gödel's incompleteness theorems (in the context of Dedekind's categoricity of the Dedekind–Peano axioms). It is the price paid for the gain in expressive power along with the failure of logical compactness. *In seeking a systematic formulation of structuralism, however, one is not attempting to formalize all of mathematics.* The advantages of explicitness and clarity concerning one's assumptions speak for themselves, despite the unattainability of completeness. Indeed, the open-endedness and extendability of mathematics are reasons enough to forgo the latter aim. Moreover, concerning the second part of the criticisim, there is no need to insist on an absolute distinction between logic and mathematics, for MS *does not seek a reduction of mathematics to logic* in anything like the traditional sense(s) (e.g., to demonstrate the "analyticity" of mathematics). It should be granted that some core mathematical content must be built into one's primitive notions if structuralism is to be articulated at all. Indeed, the notions of "arbitrary plurality" of infinitely many objects and "arbitrary part" of an infinitude of atoms are inherently mathematical, as the work they can do in the detailed development of MS makes clear. Nevertheless, it should be emphasized that no *primitive* notion of "relation" or "function" is needed: as noted above, *monadic* plural quantification, combined with mereology, enables a reduction of *polyadic* second-order quantification (i.e., of a full theory of relations). This is a nontrivial gain vis-à-vis versions of structuralism which presuppose "set-membership" or "function" or "relation." The claim would be that while far-reaching in their mathematical import, the notion "some of these things," in the plural sense, and that of "part of a whole" of pairwise discrete things, are accessible to us in ordinary contexts and not special to mathematics. Moreover, surely plural quantifiers belong to *logic* in a general sense, even if mereology's status remains moot. Thus, MS could be said to establish a *partial logicism*, less ambitious than the full program, to be sure, but significant nonetheless.

Other criticisms of MS have been offered, especially in connection with applications of mathematics, an issue that space has not permitted us to deal with in this chapter.[24] In general, structuralists are well-positioned to treat applications because these are naturally understood in terms of full or partial instantiation of

[24] Resnik ([30], pp. 74–75), for example, has argued that MS cannot treat ordinary scientific applications of probability and statistics, because of the need for abstract objects such as "events" (e.g., possible outcomes of experiments) and numbers. While Field-style nominalism may be threatened by his objection, I believe it has no force against MS, which can readily invoke the possibility of rich enough structures to *represent* or *model* applications of probability and statistics. What matters is not the metaphysical category of

mathematical structures by material systems, or, in some cases, just via mappings between these. As an eliminativist version, MS does require some artful maneuvering to express relevant relationships between material systems and hypothetical objects that merely *might* form mathematical structures (e.g., $\mathbb{N}$-structures, $\mathbb{R}$-structures, the various spaces of analysis, etc.). But the methods worked out in ([17], ch. 3), together with improvements from [7], I believe, essentially solve this problem.

6. Summation

Our comparative investigation thus far can be summarized in the following table, most of which should now be self-explanatory (a check means that the objection applies; the goal is to "draw a blank"):

	STS	SGS	CTS	MS
(i) Sets exceptional	√	—	—	—
(ii) Maximal totalities	√	√	?	—
(iii) Possibility of gross error	√	—	—	—
(iv) "Places" as objects	—	√	—	—
(v) Purely structural relations?	—	√	—	—
(vi) Underspecification	—	—	√	—
(vii) Primitive modality	? (informal)	√	?	√

The reason for the "?" in the last row under STS has to do with various informal appeals to possibility in motivating certain points of the theory (e.g., starting with the null set; explaining power sets, via all possible ways of selecting; and motivating largeness conditions, "... the hierarchy go[es] on as long as possible").[25]

As anticipated, none of the approaches is free of problems. But the paucity of checks in the last two columns encourages us to seek some kind of synthesis of CTS and MS. This is in fact achievable, with the effect of removing the check under CTS (vi) while replacing the "?" under (ii) with a blank and replacing that under (vii)

objects in such a model but rather the (applied) mathematical information they carry, which depends on our stipulations and on structural roles. Numbers in an absolute sense are no more required as values of probability functions than they are for ordinary counting or measuring. (Cf. n. 16.)

[25] See, e.g. ([26], p. 141). For a fuller discussion of these points, see [19].

with a check. Instead of relativizing CT to background universes of sets, one can introduce hypothetical large domains (corresponding to inaccessible cardinalities) merely by employing the language of mereology and plurals, as in MS; CT can be carried out relative to such domains, any one of which will also support many topoi, incorporating the relativity that Bell has described as well (resulting in an overall double relativity).[26] At the same time, set theory itself can be developed structurally, relative to such domains. Extendability principles can readily be formulated and adopted applying to these domains, and the same considerations that ensure a blank under (ii) for MS carry over. Puzzles about proper classes in set theory and large categories in category theory are handled in parallel fashion, by relativization to large domains (recovering, for set theory, Zermelo's method, and for CT, the Grothendieck method of universes). If we are right, use of modal notions is the price we must pay if we are to have a well-specified structuralism which respects the indefinite extendability of universes of discourse for mathematics.

REFERENCES

[1] Awodey, S. "Structure in Mathematics and Logic: A Categorical Perspective," *Philosophia Mathematica* (3) 4 (1996): 209–237.

[2] Bell, J.L. "From Absolute to Local Mathematics," *Synthèse* 69 (1986): 409–426.

[3] Bell, J.L. *Toposes and Local Set Theories* (Oxford: Clarendon Press, 1988).

[4] Benacerraf, P. "What Numbers Could Not Be" (1965), reprinted in P. Benacerraf and H. Putnam, eds., *Philosophy of Mathematics*, 2nd ed. (Cambridge: Cambridge University Press, 1983), pp. 272–294.

[5] Bolzano, B. *Paradoxes of the Infinite*, D.A. Steele, trans. (London: Routledge & Kegan Paul, 1950).

[6] Burgess, J.P., A. Hazen, and D. Lewis. "Appendix on Pairing," in D. Lewis, *Parts of Classes* (Oxford: Blackwell, 1991), pp. 121–149.

[7] Burgess, J.P., and G. Rosen. *A Subject with No Object: Strategies for Nominalistic Interpretation of Mathematics* (New York: Oxford University Press, 1997).

[8] Burgess, J.P. Review of Stewart Shapiro [1997], *Notre Dame Journal of Formal Logic*, 40 (1999), pp. 283–291.

[9] Dedekind, R. *Was sind und was sollen die Zahlen?* (Brunswick: Vieweg, 1888), trans. as "The Nature and Meaning of Numbers" in W.W. Beman, ed., *Essays on the Theory of Numbers* (New York: Dover, 1963), pp. 31–115.

[10] Dedekind, R. Letter to Heinrich Weber, in R. Fricke, E. Noether, and O. Ore, eds., *Gesammelte mathematische Werke*, vol. 3 (Brunswick: Vieweg, 1932), pp. 489–490.

[11] Feferman, S. "Categorical Foundations and Foundations of Category Theory," in R.E. Butts and J. Hintikka, eds., *Logic, Foundations of Mathematics, and Computability Theory* (Dordrecht: Reidel, 1977), pp. 149–169.

[26] For details, see [20].

[12] Feferman, S., and G. Hellman. "Predicative Foundations of Arithmetic," *Journal of Philosophical Logic* 24 (1995): 1–17.

[13] Frege, G. *Wissenschaftlicher Briefwechsel*, G. Gabriel, H. Hermes, F. Kambartel, and C. Thiel, eds. (Hamburg: Felix Meiner, 1976), trans. as *Philosophical and Mathematical Correspondence* (Oxford: Blackwell, 1980).

[14] Freyd, P., and A. Scedrov. *Categories, Allegories* (Amsterdam: North Holland, 1990).

[15] Goodman, N. *The Structure of Appearance*, 3rd ed. (Dordrecht: Reidel, 1977).

[16] Hale, B. "Structuralism's Unpaid Epistemological Debts," *Philosophia Mathematica* (3) 4 (1996): 124–147.

[17] Hellman, G. *Mathematics Without Numbers: Towards a Modal-Structural Interpretation* (Oxford: Oxford University Press, 1989).

[18] Hellman, G. "Structuralism Without Structures," *Philosophia Mathematica* (3) 4 (1996): 100–123.

[19] Hellman, G. "Three Varieties of Mathematical Structuralism," *Philosophia Mathematica* (3)9 (2001): 184–211.

[20] Hellman, G. "Does Category Theory Provide a Framework for Mathematical Structuralism," *Philosophia Mathematica* (3)11 (2003), pp. 129–157.

[21] Hersh, R. *What Is Mathematics, Really?* (Oxford: Oxford University Press, 1997).

[22] Keränen, J. "The Identity Problem for Realist Structuralism," *Philosophia Mathematica* (3)9 (2001): 308–330.

[23] MacBride, F. ed., *Identity and Modality: New Essays in Metaphysics* (Oxford: Oxford University Press, forthcoming).

[24] Mac Lane, S. *Mathematics: Form and Function* (New York: Springer, 1986).

[25] Mac Lane, S., and Moerdijk, I. *Sheaves in Geometry and Logic: A First Introduction to Topos Theory* (New York: Springer, 1992).

[26] Maddy, P. *Realism in Mathematics* (Oxford: Oxford University Press, 1990).

[27] McLarty, C. "Numbers Can Be Just What They Have To," *Noûs* 27 (1993): 487–498.

[28] Parsons, C. "The Structuralist View of Mathematical Objects," *Synthèse* 84 (1990): 303–346.

[29] Reck, E.H., and M.P. Price. "Structures and Structuralism in Contemporary Philosophy of Mathematics," *Synthèse* 125 (2000): 341–387.

[30] Resnik, M.D. *Mathematics as a Science of Patterns* (Oxford: Oxford University Press, 1997).

[31] Russell, B. *The Principles of Mathematics* (London: Allen and Unwin, 1903).

[32] Russell, B. *Introduction to Mathematical Philosophy* (New York: Simon and Schuster, 1919).

[33] Shapiro, S. *Philosophy of Mathematics: Structure and Ontology* (New York: Oxford University Press, 1997).

[34] von Neumann, J. "An Axiomatization of Set Theory," in J. van Heijenoort, ed., *From Frege to Gödel* (Cambridge, MA: Harvard University Press, 1967), pp. 394–413, trans. of "Eine Axiomatisierung der Mengenlehre," *Journal für die reine und angewandte Mathematik* 154 (1925): 219–240.

[35] Zermelo, E. "Über Grenzzahlen und Mengenbereiche: Neue Untersuchungen über die Grundlagen der Mengenlehre," *Fundamenta Mathematicae* 16 (1930): 29–47.

CHAPTER 18

STRUCTURALISM RECONSIDERED

FRASER MACBRIDE

1. INTRODUCTION

THE properties and relations that perform a role in mathematical reasoning arise from the basic relations that obtain among mathematical objects. It is in terms of these basic relations that mathematicians identify the objects they intend to study. The way in which mathematicians identify these objects has led some philosophers to draw metaphysical conclusions about their nature. These philosophers have been led to claim that mathematical objects are positions in structures or akin to positions in patterns.

Let us retrace their route from (relatively uncontroversial) facts about the identification of mathematical objects to high metaphysical conclusions. Begin with the natural numbers. How are they identified? The mathematically significant properties and relations of natural numbers arise from the successor function that orders them; the natural numbers are identified simply as the objects that answer to this basic function. But the relations (or functions) that are used to identify a class of mathematical objects may often be defined over what appear to be different kinds

Thanks to audiences in Edinburgh, Geneva, Helsinki, London, and Oxford for helpful discussion of this and related material on structuralism more broadly conceived. I am also indebted to Patrick Greenough, Geoffrey Hellman, Keith Hossack, Mark Kalderon, Mike Martin, Alex Oliver, Crispin Wright, and, especially, Stewart Shapiro. I gratefully acknowledge the support of the Leverhulme Trust, whose award of a Philip Leverhulme Prize made the writing of this chapter possible.

of objects. The successor function, for example, may be defined over real numbers, Zermelo ordinals (etc.). It follows that the identification of mathematical objects in terms of their basic relations fails to settle whether mathematical objects of one kind (natural numbers) are identical to or distinct from objects of an apparently different kind (real numbers, Zermelo ordinals). The identification of mathematical objects in terms of their basic relations also fails to establish what, if any, intrinsic features mathematical objects possess. The natural number 2 is not picked out by any intrinsic features it may have. 2 is identified only by its relation to other natural numbers—the fact that it is the successor of 1 and succeeded by 3 (etc.).[1]

The fact that mathematical objects are identified in terms of their basic relations leaves the mathematician unable to answer certain questions: whether mathematical objects of one kind are identical to objects of a different kind, or what intrinsic features mathematical objects possess. This suggests that the mathematician's knowledge of mathematical objects is essentially incomplete, that there are aspects of mathematical reality—the identity of its objects, their intrinsic features—that lie ineluctably beyond the bounds of mathematical inquiry. But this is not the only way to interpret the mathematician's inability to answer certain questions. For perhaps it is not the mathematician's knowledge that is incomplete but the mathematical objects themselves: it is not ignorance that hinders the mathematician from settling whether 2 is $\{\{\varnothing\}\}$; it is the very absence of answers to such questions. It is to make sense of this idea—the idea that mathematical objects are incomplete *in re*—that some philosophers identify (or compare) these objects with positions in patterns or structures.

The most extended and influential elaborations of the claim that mathematical objects may be fruitfully compared or identified with positions in patterns or structures are to be found in Michael Resnik's *Mathematics as a Science of Patterns* and Stewart Shapiro's *Philosophy of Mathematics: Structure and Ontology* (hereafter *MSP* and *PMSO*, respectively).[2] For the purpose of assessing whether

[1] These remarks introduce one version of the much-discussed "Caesar problem." In connection with the issues raised in the present chapter, the interested reader may usefully consult Frege [1884], §66; Benacerraf [1965], [1998], pp. 45–57; C. Parsons [1965], [1971], pp. 154–157; Quine [1969], pp. 43–45; Kitcher [1978]; Wright [1983], pp. 123–127; Field [1989], pp. 20–25; and Maddy [1990], pp. 80–98. MacBride [forthcoming] distinguishes some of the different dimensions along which versions of the Caesar problem may vary.

[2] *MSP* and *PMSO* draw upon and develop the views presented by their authors in a number of earlier papers. The most important of these include Resnik [1975], [1981], [1982], [1988] and Shapiro [1983], [1989], [1993]. However, it should not be assumed that *MSP* or *PMSO* advocates views that are identical to or even cohere with the structuralist ideas articulated in these papers. See Chihara [1990], pp. 125–145, for an outline and critique of some of these earlier ideas. C. Parsons [1990] develops an important and independent position that embeds both Kantian and structuralist components. For reasons of space, an evaluation of Parsons's rich and subtle views will have to be postponed until another occasion.

such conceptions of mathematical objects may be sustained, the (incompatible) views of Resnik and Shapiro will be explored and evaluated in turn.

2. Resnik on Ontology and Incompleteness

According to Resnik, the incompleteness of mathematical objects is to be enthusiastically embraced, simply a predictable and intelligible consequence of the fact that mathematical objects are akin to positions in patterns. He offers two complementary considerations in favor of this view.

Here is the first. Take an equilateral triangle *ABC* (*MSP*, pp. 202–203). Relative to this triangle, the points *A*, *B*, and *C* may be distinguished. But now ask yourself, Do these points—conceived in isolation from the triangle—enjoy any features that distinguish them from one another or, indeed, from any other points? Resnik recommends the answer: no: our intuitions speak in favor of the view that geometrical positions lack any features other than those that relate them to the other positions in the patterns they inhabit. Now take the corner *A* of a right triangle and the corner A^* of a rectangle (*MSP*, pp. 210–211). Ask yourself, Is *A* identical to A^*? Again Resnik answers: no: our intuitions simply deliver no verdict on this question. He surmises that positions (points) in geometrical patterns are incomplete objects—they lack intrinsic features and there is no "fact of the matter" concerning whether they are identical to or distinct from positions drawn from different patterns (or, indeed, any other kind of object). Resnik then proposes that mathematical objects themselves be conceived as positions in patterns—the natural numbers, for example, as positions in the natural number sequence. From this perspective it then appears that there is no more mystery to mathematical objects being incomplete than there is a puzzle concerning the incompleteness of geometrical points:

> But what would strike one as a problematic oversight if one thought of mathematical objects along the lines of ordinary objects, seems quite natural when it comes to positions in patterns. For restricting identity in the same pattern goes hand in hand with their failure to have any identifying features independently of a pattern. (*MSP*, p. 211)

Here is the second consideration that Resnik offers in favor of the incompleteness of mathematical objects. Is there is any mathematical evidence that speaks for or against the identity of (e.g.) the natural number 1 and the real number *e*? According to Resnik, there is no (local) mathematical evidence that could be adduced in favor of identifying or distinguishing between objects drawn from different mathematical structures, no novel theorem that could finally settle whether the natural numbers are real numbers, whether the real numbers are sets, and so on (*MSP*, pp. 219, 246, 269–70). It is out of respect for this feature of mathematical practice that Resnik disavows facts of the matter concerning the identity or character of mathematical objects that transcend the capacity of mathematics to settle. Resnik describes his "epistemic turn" in the following terms:

> Epistemology enters the picture through motivating my disclaiming these facts: since nothing mathematics countenances would fix these facts if we countenanced them, I have opted for denying that there is anything to be decided, instead of enlarging the notion of evidence applicable to the putative facts. (*MSP*, p. 270)

Resnik, then, is unwilling to countenance the possibility that the truth-values of sentences that affirm the identity of objects drawn from different structures (or ascribe extrastructural features to them) are irresolvable and inscrutable, in principle inaccessible to the best techniques of mathematical research (*MSP*, pp. 89, 219, 244, 248). So to avoid this kind of transcendental realism Resnik denies such sentences are either true or false, and modifies his logic to embrace the possibility that these sentences may be neither true nor false. Any impression we may have to the contrary is generated, Resnik suggests, by the fact that other sentences of the same grammatical form that identify positions drawn from the same pattern are either true or false. The view that emerges from these reflections has it that *within* a mathematical theory, classical logic obtains, each sentence being either true or false. But if attempts are made to frame identities between objects that are described by different theories, or to employ predicates from one theory to describe objects drawn from another theory, then excluded middle fails (*MSP*, pp. 245–246, 257).[3]

Resnik endeavors in this way to place the incompleteness of mathematical objects in perspective, a perspective from which it no longer appears troublesome that these objects lack internal features or determinate identity conditions. It is tempting to conceive Resnik's project as an attempt to provide thereby a novel ontological foundation for mathematics, a foundation in positions and patterns. To feel the pull of this interpretation, just read the title of his book. Or consider the declaration "Mathematics is a science of patterns with mathematical objects being positions in patterns" (*MSP*, p. 199; see also p. 9). But it would be a mistake to fall prey to the temptation. Resnik resists any ontological foundationalist reading of his structuralism (*MSP*, pp. 222–223, 258, 272–273). He seeks only to

[3] Resnik makes two further suggestions: (i) the underlying logic without a given theory (that is, internally classical) may be intuitionistic; (ii) bivalence will also have to be dropped (*MSP*, p. 245, n.4; p. 270).

provide "another way of viewing numbers and number theory" that places the problems that confront the understanding of numbers and number theory "in a clearer light." The idea that mathematical objects are positions in patterns may even be "a ladder ultimately to be kicked away," part of an "analogy" that—for one reason or another—"must be given up" (*MSP*, p. 261).

It is critical to a proper appreciation of Resnik's philosophy of mathematics that these remarks not be dismissed as coy disclaimers. They reflect the deep character of Resnik's enterprise and his conception of mathematical practice. To make progress toward a proper appreciation of this somewhat surprising fact, consider again the geometrical examples that Resnik employs to convince us that there is nothing intrinsically problematical about incomplete objects (whether they be points or numbers). Your responses may well have been different from those that Resnik predicted. You might have responded to the questions raised that the geometrical examples given are underdescribed, that we have yet to be supplied with sufficient information to determine whether, for example, $A = A^*$ or not. You might have said, It's not indeterminate whether $A = A^*$, it's epistemologically underdetermined; we just don't know whether $A = A^*$ or not, so there's no reason to think that geometrical points are *ontologically* incomplete in some way that is credible or intelligible to the understanding. The way in which Resnik responds to this concern is highly revealing.

The counterintuition that the identity of A and A^* is left epistemologically underdetermined—rather than indeterminate—presupposes that geometry is the theory of a single space, a space where each point enjoys a unique location and consequently there is a determinate fact of the matter concerning the identity of each point (*MSP*, p. 210). But geometry might have developed differently without commitment to a single space that brings each point into a determinate relation with every other. Instead, geometry might have developed as a collection of theories of shape without commitment to a single space of points, but only to a collection of spaces, a space for the positions of each shape so introduced. Then it would be indeterminate whether the points from one space were identical to the points drawn from another space. In fact, Resnik claims, this is just the situation that arises in mathematics. Number theory and analysis introduce, respectively, the positions in the natural number structure and the positions in the real number structure. But neither number theory nor analysis evinces a commitment to a common universe of positions such that there is a determinate fact of the matter concerning whether, for example, the natural number 1 is the real number *e*.

From a cursory reading one might be forgiven for thinking that Resnik's view derives from an act of intuitive insight into the ontological character of geometrical points. But, as we've just seen, this would be wrong. It is not geometrical intuitions—an intuitive appreciation of what it is to be a position in a geometrical pattern—that ultimately underwrites Resnik's view that mathematical objects are incomplete. It is, rather, an interpretation of mathematical theories, an interpretation according to which mathematical theories are neither ontologically committed to a common domain of discourse nor ideologically committed to a universally applicable identity predicate. It is true that Resnik has avowed an epistemic turn. But it is now evident that he has also taken a "linguistic turn," a transposition of intellectual key that plays no less of a role in legitimating Resnik's conception of mathematical objects. And it is a way of thinking about mathematical objects that does not rely upon commitment to a favored ontology of positions or patterns; it results from an appreciation of the manner in which mathematical objects are introduced into discourse by the theories that denote them.

Unfortunately, the way in which Resnik conceives of mathematical theories threatens to conflict with mathematical practice. This is because the mathematical properties of a domain of objects d will often be graspable only when d is embedded within a larger domain d^* (Kreisel [1967], p. 166). Therefore mathematicians are very often obliged to establish isomorphisms or, more generally, embeddings between the objects described by different mathematics theories in order to establish results that would otherwise be unobtainable (consider, for example, the embedding of the integers in the complex plane). But as we have just seen, Resnik denies that different mathematical theories enjoy a common universe of discourse. He also endorses the usual assumption that there cannot be an embedding between objects unless they belong to a common domain (*MSP*, p. 211). Consequently Resnik cannot admit any embedding between the domains of different theories (d, it appears, cannot be embedded in d^*).

In the same way, Resnik cannot admit any significant set-theoretic or category-theoretic reduction of mathematical objects. In these reductions the pattern or structure of the objects characterized by one theory (e.g., the natural number structure) is usually identified with an object (e.g., ω) in another theory. But Resnik is not in a position to admit such identities. Not only does he deny the intelligibility of identities between the domains of different theories, he also denies that mathematical theories typically denote the structure their objects exhibit; typically they do not even "countenance" the existence of such structures (*MSP*, p. 211). For example, Peano Arithmetic refers to and quantifies over the natural numbers but does not name or quantify over the natural number structure itself. Similarly, set theory quantifies over sets but not over the set-theoretic hierarchy. Resnik therefore offers the following picture of existential commitment: whereas a mathematical theory is existentially committed to the positions (the numbers, the sets) that lie within a structure, it is not committed to the existence of the structure that prima facie encompasses these positions (the natural number structure, the set-theoretic hierarchy). But this appears to leave him unable to countenance the identification of the structure of the objects characterized by a mathematical theory with a set-theoretical object.

Resnik strives to overcome these apparent clashes with mathematical practice by employing what I will dub the "replacement strategy" (*MSP*, pp. 214–219, 222,

246–248). It can hardly be denied that mathematicians establish embeddings and isomorphisms between different collections of objects. So Resnik denies, instead, that mathematicians establish embeddings or isomorphisms between the domains of distinct theories. We are in the grip of a picture according to which the mathematician begins with two independent theories, t and t^*, and then establishes an embedding of the domain d of the former theory in the domain d^* of the latter. But Resnik wishes to reject this picture of mathematical practice, to loosen its grip upon us. According to Resnik, the embedding takes place not between d and d^*, but between the subdomains Δ and Δ^* of the larger domain of a novel theory θ. His idea is that the embedding of one collection of objects in another makes sense only because the mathematician replaces two theories between whose domains there is no intelligible mathematical connection with a new theory that subsumes collections between which such a connection can be made out. What is critical from Resnik's point of view is that there "would be no fact of the matter as to whether the old and new theories had the same ontology" (*MSP*, p. 219).[4] The mathematician has simply replaced old theories with a new one and there can be no embeddings or isomorphisms between the domains of distinct theories.

Turning to the set-theoretic case, it can hardly be denied that mathematicians identify some structures with sets. But it can be denied that the structures in question are the structures of collections of objects characterized by theories that are developed outside the aegis of set theory. There can be no fact of the matter concerning whether the natural number structure described by number theory is an object of another theory (e.g., the set ω). Nor, indeed, can there be a fact of the matter concerning the identity of the objects denoted by number theory with the objects that pass for "natural numbers" in set theory. But, applying the replacement strategy, there can be a fact of the matter in the set theory that replaces (or supersedes) number theory concerning whether the "natural numbers" in its domain form a particular set.

Resnik's conception of how the domains of different theories relate not only threatens to conflict with mathematical practice, it also threatens to conflict with his own philosophical theorizing (*MSP*, pp. 248–250). According to Resnik, there is no fact of the matter concerning whether the structure or pattern a mathematical theory describes is an object or an entity of any other kind. Nevertheless, Resnik continually treats mathematical patterns as objects. He refers to them, quantifies over them, and applies predicates to them throughout the course of his study. Consider, for example, his definition of the subpattern relation: "A pattern P is a *sub-pattern* of another pattern Q just in case P occurs within Q and every position of P is a position of Q" (*MSP*, p. 206).

[4] Resnik sometimes speak of "explication" of the old theory by the new. But it is explication in "the Carnap–Quine sense of explication as elimination," and therefore no different from what I have described as replacement (*MSP*, p. 248, n.6).

Resnik's response to this difficulty provides further insight into his denial that patterns provide any kind of ontological foundation for mathematics. Resnik admits that a mathematically rigorous theory might be developed that would refer to and quantify over patterns and their positions in very much the same way that set theory quantifies over sets and their members (*MSP*, pp. 249, 254–257). But the utility of developing a mathematical theory of patterns would be questionable. For one thing, it is doubtful whether such a theory would constitute any mathematical advance over set theory or category theory. More significantly, no pattern theory could be foundational in the sense of revealing the kinds of things that mathematical theories had been talking about all along. Why? Well, because a mathematical theory of patterns would be just one more theory among others. So there would be no fact of the matter—from Resnik's point of view—concerning whether the patterns and positions quantified over by the envisaged theory were identical to or distinct from the structures and objects characterized by mathematical theories already established. It is for this reason that Resnik cannot accept a foundationalist reading of his slogan that mathematical objects are positions in patterns. It is a consequence of his own denial—the denial that distinct theories ever share a common domain—that Resnik is constrained to claim no more than an "analogy" between mathematical objects and positions in patterns. He can claim no more than that mathematics "treats the numbers *as if* they were positions in a pattern" (*MSP*, p. 250; my italics).

What assessment, then, is to be made of Resnik's "structuralism"? At first glance it may appear that his position admits of *refutation.* According to Resnik, there is no fact of the matter concerning whether the natural number 2 is identical to or distinct from the real number 2 (since these numbers are introduced by distinct theories). So, according to Resnik, it is indeterminate whether $2_{\text{natural}} = 2_{\text{real}}$. But it appears self-contradictory to assert that there are objects such that it is indeterminate of *them* whether they are identical or distinct. Here's the reason. Suppose that it is indeterminate whether $2_{\text{natural}} = 2_{\text{real}}$. Then 2_{natural} and 2_{real} differ with respect to the property of being determinately identical to 2_{natural} (2_{natural} has that property, whereas 2_{real} lacks it). Since 2_{natural} and 2_{real} differ with respect to this property, it follows by contraposition on Leibniz's Law (if $x = y$, then x and y share all the same properties) that $2_{\text{natural}} \neq 2_{\text{real}}$.

This objection to Resnik's position should have a familiar ring to it. In fact it is a version of Evans's famous argument against vague objects. Chihara has also independently advanced a version of the argument targeted specifically against Resnik (Evans [1978]; Chihara [1990], p. 143). Unfortunately, it is an objection that cannot be logically decisive. Arguments of this kind presuppose the rules of classical logic. For example, the Evans/Chihara argument sketched here presupposes the rule of contraposition. But, as has become all too familiar from the vagueness debate, proponents of indeterminate identity all too often reject classical logic, in particular the rule of contraposition in contexts involving an indeterminacy operator.[5]

This does not mean that another version of this argument might not be given, one that does not rely upon the questionable use of Leibniz's Law. For example, the Evans/Chihara argument relies upon the principle that if a and b have different properties, then $a \neq b$. This principle—sometimes called the Diversity of the Dissimilar—is derived by contraposition from Leibniz's Law. If, however, we are willing to entertain the Diversity of the Dissimilar as a logical principle (or, at least, a highly compelling principle) in its own right, then the argument can easily be rewritten so as to obviate the questionable use of contraposition. Simply assume the Diversity of the Dissimilar as a premise and forget about deriving it from Leibniz's Law. But the package of views to which Resnik is committed—his denial that identity is everywhere determinate, his disavowal of any domain shared among theories—runs counter to the Diversity of the Dissimilar along with many other presuppositions of classical reasoning. It is correspondingly doubtful whether a *refutation* of Resnik's system of beliefs might be constructed that is compelling by the lights of both Resnik and his opponents.

In a way this should come as no surprise. As an advocate for classical logic might put the point: if someone refuses to employ reason, then, of course, it's going to be difficult to hit upon an argument that persuades him of much about anything. Nevertheless this suggests another means of taking the attack to Resnik, an attack that does not threaten to become bogged down in discussion of what counts as logic. Resnik cannot accept that certain rules of inference or certain "logical" principles are compelling because there are other things he refuses to say (for example, that $2_{\text{natural}} = 2_{\text{real}}$ or $2_{\text{natural}} \neq 2_{\text{real}}$). But it may be that this refusal prevents Resnik from sayings other things that really need to be said. So Resnik's position may admit of pragmatic confutation even if it cannot be logically refuted.

It is to avoid confutation of this kind that Resnik adopts the replacement strategy. He refuses to allow that there can be any embeddings between the domains of distinct theories. But mathematicians appear to say that there are such embeddings. To avoid this pragmatic clash, Resnik reinterprets what mathematicians are doing so that it coheres with his own account.

Unfortunately, Resnik's response to this pragmatic difficulty is hardly credible, certainly no more credible than Russell's valiant attempts to accommodate the restrictions imposed by the theory of types. If Resnik is correct, then mathematical theories are inherently incommensurable in the very sense that will disturb a mathematician most: there can be no comparison of the size or structure of their domains unless—behind our backs, as it were—they are merely passages of a larger theory. (Why? Well, because size and structure are established by laying down forbidden embeddings between domains.) As a consequence, each novel

[5] See T. Parsons [2000], pp. 16–30.

embedding between theories signals a Kuhnian paradigm shift, the dawn of a new theory that subsumes the embedding in question, and the history of mathematics is a succession of Kuhnian revolutions of this kind. But this cannot be right; this picture belies the underlying *rationality* of mathematical practice. What sense can be made of the fact that mathematicians choose to replace two or more old theories with a new one? The only answer that appears to be available is this: *because* of the embeddings that obtain between the domains of these different theories and the novel theory introduced. But Resnik cannot say this. From his perspective it appears that each shift in practice is nothing more than an *arational* transition between subjects of enquiry.

It is a further corollary that there can be no *intelligible* attempt—of the kind that Zermelo and Fraenkel undertook—to unify mathematical practice. From Resnik's point of view, any attempt to do so will inevitably miss its mark. Any new theory that attempts to comprise the subject matters of established theories will simply introduce a domain of its own such that there is no fact of the matter concerning whether the objects in this novel domain are identical to or distinct from objects drawn from the domains of already familiar theories. The attempt to unify will inevitably result only in further fragmentation, never in the unification of practice.

These are not the only pragmatic difficulties that threaten to confute Resnik's account. His refusal to admit embeddings between domains leaves Resnik unable to say how mathematics gains application in the physical world. There is no sterner criticism (short of contradiction) to be made of a philosophy of mathematics. As Frege insisted, it is the fact that mathematics is capable of gaining application that elevates it from the status of a game to the rank of a science (Frege [1893–1903], vol. 2, §91).

Resnik says very little about the applications of mathematics.[6] However, it seems, at least, that he is committed to something like the following. It is because physical objects instantiate mathematical patterns (or segments of them) that the study of mathematical objects—objects that are positions in the aforementioned patterns—is capable of illuminating the character of the physical realm. For Resnik, instantiation is a special case of an embedding (a congruence) between mathematical objects and the physical things that exhibit the same pattern, a case where "the objects 'occupying the positions' of a pattern have identifying features over and above those conferred by the arrangements to which they belong" (*MSP*, p. 204).

By now it will be clear what problem arises. Since, for example, number theory does not include physical objects in its domain, it follows that there cannot be any kind of congruence between the natural numbers and physical things.

[6] Resnik does say a good deal by way of defense of the Quine–Putnam thesis that mathematics is indispensable to science (*MSP*, pp. 40–81). However, determining that mathematics is indispensable to science in general does not settle, but rather leaves a mystery, how, in detail, mathematics gains application to the physical realm.

It follows that the natural numbers cannot be instantiated by the physical realm and are therefore intrinsically incapable of application. The same goes for other kinds of mathematical objects that are introduced independently of physical theory. It is Resnik's distinctive claim that mathematical objects are incomplete by nature, incomplete in the sense that there is no fact of the matter concerning whether they are identical to or distinct from objects of other kinds and do not even belong to a common domain of discourse. Insofar as Resnik is correct to say this, he cannot account for the applications of mathematics[7]

Is there no "way out" for Resnik? The difficulties confronted so far turn upon the assumption (among others) that there can be an embedding between collections of objects only if they belong to the same domain. Perhaps, however, this assumption can be put under pressure. Consider the usual definition of isomorphism: *F*s 1–1 *G*s iff $\exists R \forall x((Fx \rightarrow \exists! y(Gy \,\&\, Rxy)) \,\&\, (Gx \rightarrow \exists! y(Fy \,\&\, Ryx))$. This definition presupposes that the *F*s and *G*s that are 1–1 correlated occupy the same domain of discourse. This presupposition is revealed by the fact that the universal quantifier "$\forall x$" ranges over both *F*s and *G*s. However, it appears that an alternative definition might be offered where the *F*s and *G*s are ranged over by separate quantifiers, and no assumption is made that they belong to a common domain: *F*s 1–1 *G*s iff $\exists R \forall x ((Fx \rightarrow \exists! y(Gy \,\&\, Rxy)) \,\&\, \forall y(Gy \rightarrow \exists! x(Fx \,\&\, Ryx))$. If there is a difficulty that confronts this proposal, it is nothing to do with the absence of facts of the matter concerning the identity of the *F*s with the *G*s. Such facts are not required to determine an isomorphism. For this, facts of identity and distinctness are required among the *G*s and among the *F*s—so that it is determined that *each F* bears *R* to a *unique G* and *each G* bears *R* to a *unique F*—but not between the *F*s and the *G*s. If this modified definition is tenable, then perhaps there can be isomorphisms (or, more generally, embeddings) between distinct domains after all.

Is this a "way out" for Resnik? An answer to this question will depend (in part) on what is meant by "relation" and "domain." There is certainly a very permissive notion of domain according to which the mere obtaining of a relation provides a sufficient condition for its relata to belong to a common domain. But there are also more restricted notions of domain that require some privileged kind of relation to obtain between the elements of a domain. For example, it is built into the idea of a spatiotemporal domain that the elements that fall within a domain of this kind are spatiotemporally related. Any proponents of the "way out" we are considering therefore face a substantial challenge. If they wish to allow embeddings between distinct domains, then they cannot merely mean "domain" in

[7] Russell and Dummett insist that structuralism cannot account for applications for another reason, the reason that structuralism fails—in some suitable sense that remains to be properly clarified—to incorporate the principles governing the application of mathematics into its foundations (Russell [1919], p. 9; Dummett [1991], pp. 295–297). For defense of structuralism against this charge, see Quine [1969], pp. 44–45 and C. Parsons [1990], p. 309.

the permissive sense. But if they mean "domain" in some more restricted sense, then whatever types of relation in fact embed distinct domains, these relations had better not be privileged for the kind of domain in question. For example, if the relation *R* that in fact correlates the *F*s and *G*s is privileged for the kind of domain under consideration, then the embedding between the *F*s and *G*s will indicate that they share a common domain after all. Whether appropriate notions of domain and relation—that are neither too permissive nor too restrictive—can be made out remains to be seen. This is an area that is too little investigated. It is disappointing that Resnik does not subject his own assumptions—concerning what *he* means by relation and domain—to scrutiny.

Suppose, however, that Resnik manages somehow to resolve the pragmatic difficulties so far considered. Does it follow that he has succeeded in coming to terms with the incompleteness of mathematical objects? No. Here is a preliminary indication to the contrary. It is usually assumed that it is mathematical objects rather than, say, concrete objects that are incomplete. Resnik shares this assumption, maintaining that mathematical objects are unlike "ordinary objects," and thus fit to be considered as positions in patterns (*MSP*, p. 211). But there is a tension here. Resnik "accounts" for incompleteness by denying any facts of the matter concerning the identity of mathematical objects to other kinds of objects drawn from the domains of other theories. But this makes every kind of object incomplete. Just as there is no fact of the matter about whether $2_{\text{natural}} = 2_{\text{real}}$, so there is no fact of the matter about whether Julius Caesar $= 2_{\text{natural}}$ (indeterminate identity is symmetric). Hence ordinary objects turn out to be no less incomplete than mathematical objects themselves.[8]

Perhaps finessing the kind of incompleteness involved may assuage this concern. Even if it transpires that ordinary objects are incomplete in one sense—that of failing to be everywhere determinately identical or distinct—it does not follow that they are incomplete in another sense—namely, that of failing to possess any intrinsic nature. So long as ordinary objects do not turn out to be incomplete in this latter sense, there need be no deep objection to the position that says they are incomplete in the former sense.

Yet even if this response is acceptable, there are deeper methodological objections to the way in which Resnik treats of the phenomenon of incompleteness. He begins from the premise that mathematics recognizes no facts of the matter concerning the identity of mathematical objects outside of the domains of the theories that introduce them. He then develops a truth-value gap account to accommodate this phenomenon:

[8] It should be noted that Resnik does countenance the possibility that all objects are incomplete, but later refrains from endorsing it (*MSP*, pp. 92, 267–268). If I am right, then Resnik is in fact committed to the claim that all objects are actually incomplete.

> I would like to add that mathematical realists are not committed to claims about mathematical objects beyond those they hold by virtue of endorsing the claims of mathematics. Since mathematics recognises no facts of the matter in the puzzling cases, mathematical realists are free to develop solutions that do not recognise them either. (*MSP*, p. 92)

However, the failure of mathematics to register facts of the matter in the relevant class of cases may be symptomatic of a variety of underlying causes. It may indeed be the case that mathematics fails to register these facts because none of the relevant facts are out there waiting to be discerned. But mathematics may fail to register these facts for reasons other than their nonexistence. It may simply be that they are *epistemologically* inaccessible to mathematical techniques (see Steiner [1975], p. 91). After all, no mathematician—not even an ideal mathematician—is a god. So why suppose that every fact about a mathematical object must be accessible to inquiry by mathematicians? Alternatively, the failure of mathematics to register any of the relevant facts of the matter may arise from indeterminacy in the language of mathematics rather than the quirky incomplete natures of mathematical objects. In other words, the failure of mathematics to settle whether $2_{natural} = 2_{real}$ may result from the *semantic* fact that the expressions "$2_{natural}$" and "2_{real}" are systematically ambiguous. Consequently, there is no fact of the matter about whether they absolutely refer to the same mathematical object (see Field [1974], pp. 221–222 and McGee [1997], p. 39 for different versions of this proposal). Until these (and other) alternative explanations of incompleteness are ruled out, Resnik is hardly "free" to draw the *ontological* conclusion he favors from a purely epistemological premise.[9] Until then, his claim that mathematical objects are incomplete *in re* can hardly be justified.[10]

This is not the only difficulty that confronts Resnik's account of incompleteness. His account also rests on the assumption that the incompleteness

[9] It is important to note that the difficulty Resnik confronts here—that of circumscribing the character of the incompleteness exhibited by mathematical objects—is entirely general. It applies not only to structuralists but also to supervaluationists, neologicists, and so on. See MacBride [2003], pp. 133–134 and [forthcoming] for further discussion.

[10] Resnik does briefly offer some justification for his position. The justification offered is contained in the following remark: "My *dicta* concerning identities between various kinds of mathematical objects are based upon likening them to positions in patterns. Thus they derive from an insight concerning the nature of mathematical objects. They are, if you will, consequences of my philosophical premises" (*MSP*, p. 246). But this suggests we have gone round in a circle. To be convinced that mathematical objects are like positions in patterns, we have to be assured that there are no facts of the matter of the relevant kind (that the facts in questions are not merely inscrutable from our point of view). But it seems, by the lights of the remark given, that we can be assured of the absence of these facts only if we are already convinced that mathematical objects are akin to positions in patterns. Resnik endeavors to pull himself up by his bootstraps. It is difficult to see how any stable justification results for his dicta concerning identities.

of mathematical objects arises from the way in which they are introduced into discourse:

> Mathematical objects are incomplete in the sense that we have no answers within or without mathematics to questions of whether the objects one mathematical theory discusses are identical to those another treats; whether, for example, geometrical points are real numbers. This springs from the way mathematics defines its terms. (*MSP*, p. 90)

But what sense are we to make of this? How can the manner in which an object is introduced into discourse—the way in which a term that first denotes an object is defined—determine that the object introduced is thereby constrained to have no more of a nature than is given expression by whatever served as an initial characterization of it?

Of course, there is one class of entities for which, arguably, it is true that their initial characterization exhausts their nature. This is the class of fictional objects. One might think that there is no more to a fictional character than arises from the descriptions of the author who invented them: there is, for instance, no more to Nostromo—no hidden aspect of his nature—that is not already contained in Conrad's descriptions of the character; it is not open to a future author to discover Nostromo's true nature concealed beneath these (perhaps) deceptive characterizations. But this does not hold for objects about which we are realist. For instance, it does not follow from the fact that we are first introduced to tables in macroscopic terms that tables are not also swarms of microscopic particles. It is worth noting that Resnik is keen to emphasize the contrasts between mathematics and fiction. He denies, in particular, that mathematical objects are, in any sense, constructed or invented by us (*MSP*, pp. 188–189). But if we are also realist about numbers, it no more follows that where the natural numbers are introduced by (e.g.) the axioms of Peano Arithmetic, they are not also real numbers (where the real numbers are introduced independently by a different axiomatic theory). Consequently, Resnik fails to provide sufficient reason to doubt that relevant facts of the matter about the identity of mathematical objects obtain in such cases. As a result, Resnik fails to establish or shed light upon the doctrine that mathematical objects are incomplete *in re.*

3. Shapiro on Ontology and Incompleteness

Does the structuralism that Shapiro avows make any more sense of the idea that mathematical objects are incomplete? By contrast to Resnik, Shapiro does endeavor to provide an ontological foundation for mathematics. According to the "*ante rem* structuralism" he favors, structures are universals and mathematical objects are identified (literally) with positions within these structures (*PMSO*, pp. 10–11, 100). According to this view, the progression of natural numbers, for example, is itself a structural universal, the individual numbers being places or positions within the structure, their properties and relations determined solely by the structural role they perform there, and every property or relation of a number *n* consequent upon its being the *n*th position in the natural number structure. Mathematical structures and positions are, in this sense, said to be "freestanding."

Ante rem structuralism accounts for the instantiation of structures in the following terms. A structure is instantiated (realized) when objects occupy the positions of the structure. Objects occupy positions when they exhibit the relevant battery of relations to the other objects in a system that exhibits the structure to which these positions belong; in effect, positions are conceived as relational properties, the overarching structure they participate in being a complex relation that obtains when these properties are satisfied. The objects that occupy these positions may in turn be positions drawn from other structures. More radically, a structure Σ will instantiate itself and may embed substructures that are also instances of Σ. This is made possible by the fact that—according to the *ante rem* view—the notion of "place" admits of a significant relativity; pronouncements about place are understood either relative to the "places-are-offices" perspective or to the "places-are-objects" perspective.

To see this last point, consider once more the progression of natural numbers. From the places-are-offices perspective, the natural numbers are offices or roles (the 0-role, the 1-role, the 2-role . . .) waiting to be filled by objects drawn from some background ontology (for example, the sets $\varnothing$, $\{\varnothing\}$, $\{\{\varnothing\}\}$ . . .). But from the places-are-objects perspective, the numbers are objects in their own right, conceived independently of any objects—set-theoretic or otherwise—that may occupy these positions. From this perspective, the series of natural numbers instantiates the natural number structure (0 fills the 0-role, 1 fills the 1-role, 2 fills the 2-role . . .). Moreover, the natural numbers also embed indefinitely many series of numbers—for example, the even numbers, the odd numbers, and so on—that satisfy Peano's Axioms and instantiate the natural number structure. So, from the places-are-objects perspective, the natural numbers can also occupy different positions within the natural number structure (for example, in the even numbers series 0 fills the 0-role, 2 fills the 1-role, 4 fills the 2-role . . .).

Against this ontological backdrop Shapiro offers two contrasting accounts of incompleteness. According to the first account, identities between the elements of different structures are indeterminate:

> But it makes no sense to pursue the identity between a place in the natural number structure and some other object, expecting there to be a fact of the matter. Identity between natural numbers is determinate; identity between other sorts of objects is not, and neither is identity between numbers and the positions of other structures. (*PMSO*, p. 79)

This much Shapiro shares in common with Resnik. Yet unlike Resnik, he continues to espouse classical logic, insisting that "each well-formed, meaningful sentence has a determinate and non-vacuous truth-value, either truth or falsehood" (*PMSO*, p. 4). Shapiro does not explain how indeterminacy and classical logic can peacefully coexist. But he does hint that to ask after the identity of objects of different kinds is to commit a category mistake ("It is similar to asking whether 1 is braver than 4, or funnier" [*PMSO*, p. 79]). This suggests that the identity "statements" in question are actually meaningless. If such meaningless utterances lack a truth-value, this hardly constitutes a violation of classical logic (excluded middle).

Shapiro's account faces a now familiar difficulty. It threatens to conflict with mathematical practice, the fact that mathematicians routinely embed and even "identify" the elements of different structures. To accommodate these facts, Shapiro offers a threefold theory of the copula, distinguishing the "is" of identity, the "is" of prediction and the "is" of fiat. Even if it is indeterminate whether 2 $\text{is}_{\text{identity}}$ $\{\{\phi\}\}$, it may still be the case that 2 $\text{is}_{\text{predication}}$ $\{\{\phi\}\}$ in the sense that $\{\{\phi\}\}$ performs the "2"-role (occupies the "2" position) in the Zermelo system of arithmetic. In the same sense it is also true that 2 $\text{is}_{\text{predication}}$ $\{\phi, \{\phi\}\}$, for $\{\phi, \{\phi\}\}$ performs the "2"-role in the von Neumann system. To accommodate the practice of identifying elements of different structures, Shapiro also acknowledges a further use of the copula: "For example, when set theorists settle on the von Neumann account of arithmetic, and thereby declare that 2 is $\{\phi, \{\phi\}\}$, they invoke what might be called "'the is of identity by fiat'" (*PMSO*, p. 83).

There are (at least) three obstacles faced by this account. The first, perhaps the most significant, is shared with Resnik's treatment of incompleteness. It cannot simply be a matter for declaration that identity between the elements of different structures is everywhere indeterminate, that there might not (e.g.) be facts of the matter that fall beyond our epistemic capacity to register or the precision of our language to encode. Shapiro owes us an argument for the absence of any underlying facts of the matter, but he gives none. Second, Shapiro must provide some means of policing the boundary between genuine category mistakes and statements that are just plain obviously false. If any of the "statements" of identity at issue fall on the wrong side of this boundary, then—at least on some occasions—the identity between places in different structures will be determinate. But it is difficult to foresee how the boundary between grammatical nonsense and self-evident falsity could be reliably policed. Notoriously, our linguistic intuitions do not speak in unison on distinctions of this kind, and it is unclear what else there is to go on. Third obstacle: the notion of "identity by fiat" makes dubious sense. If we are realist about our subject matter—Shapiro staunchly advocates such "realism in ontology" (*PMSO*, p. 4)—then the facts of identity cannot be manufactured by us. Moreover, it is mysterious how utterances that are category mistakes in some contexts can turn out to be true in others.

It is the clash with realism in ontology that motivates Shapiro to develop an alternative account of incompleteness. This second account says that positions drawn from different structures are determinately false (rather than neither true nor false, as Shapiro's first account suggested) (Shapiro [forthcoming]). Mathematical objects of different kinds ($2_{natural}$, 2_{real}, $\{\phi, \{\phi\}\}$) belong to different structures. For example, $2_{natural}$ belongs to the natural number structure and therefore has an immediate successor ($3_{natural}$). By contrast, 2_{real} belongs to the real number structure and enjoys no such successor. Mathematical objects drawn from different structures therefore "enjoy different relations to different objects." Because these objects are dissimilar—because (e.g.) $2_{natural}$ enjoys relations that 2_{real} lacks—it follows that they are distinct. Therefore identity statements to the contrary are strictly false.

This reasoning relies upon two key assumptions: (i) the same object cannot belong to different structures; (ii) mathematical objects of different kinds belong to different structures (natural numbers belong to the natural number structure, whereas real numbers belong to the real number structure). (i) derives from Shapiro's structuralist slogan that "mathematical objects are tied to the structures that constitute them." It is, Shapiro claims, "part of the essence" of a mathematical object that it belongs to its parent structure. But even if this is granted, it does not follow that the same object cannot belong to different structures. Nothing has been established that precludes the possibility that a single object might belong essentially to different structures, constituted jointly by them, much as a child might—if Kripke is right—be essentially the offspring of distinct parents.[11]

(ii) appears no less precarious an assumption. It is certainly built into the concept of *structure* that the natural number structure and the real number structure are different. However, it does not follow that the former cannot be a substructure of the latter. If indeed the natural number structure is a substructure of the real number structure, then—contrary to Shapiro—it is not in general the case that natural numbers and real numbers enjoy different relations to different objects. For example, $2_{natural}$ will enjoy the relation of having an immediate *natural* successor, but this does not preclude $2_{natural}$ from lacking an immediate *real* successor. Consequently, Shapiro has said nothing to preclude the possibility that $2_{natural}$ is 2_{real}. More generally, if one is willing to multiply different types of relation (natural successor, real successor, and so on), there is no need to multiply kinds of objects (structures and the objects that inhabit them). In fact, there is strong motivation for opting to multiply relations rather than structures and their objects. For by doing so, we are able to provide a smooth and natural reading of mathematical practice.

To see this, consider an objection that Shapiro raises to Resnik's structuralism (Shapiro [forthcoming]). He observes that it may, to choose just one pertinent example, be an interesting mathematical question whether a given closed integral

[11] See Kripke [1980], pp. 110–115.

is a natural number. Shapiro complains that Resnik cannot make sense of questions of this kind. This is because, according to Resnik, there can be no fact of the matter—and so no interesting mathematical question—about whether objects drawn from different structures are the same or different. Resnik actually anticipates this complaint. He argues that even if mathematical practice sometimes allows that interesting questions may be raised concerning the identities among mathematical objects of different kinds, we may nevertheless make sense of this practice without supposing there to be any facts of the matter about the identity of objects drawn from different patterns or structures.[12] For, Resnik claims, there will be a fact of the matter about whether, say, a given integer is a natural number if we liken the historical development of the integer number system to "the step-by-step construction of a complicated pattern through adding positions to an initially simple one" (*MSP*, p. 215). If, in this way, the integer structure is conceived as an *extension* of the natural number structure, it will be true that some natural numbers are integers. Of course, in that case—as Shapiro points out and Resnik readily admits—there will be no fact of the matter concerning whether the natural numbers and integers so conceived are identical to or distinct from naturals and integers conceived as the objects posited by independent theories (*MSP*, p. 247; Shapiro [forthcoming]).

What is important for present purposes is that a version of the complaint that Shapiro raises against Resnik may be turned against his own brand of structuralism. Consider the question—that a mathematician may surely raise—of whether a given integer is a natural number. If the natural numbers and integers belong to the same structure, this question may receive a clear and straightforward answer. It will be true or false depending upon whether the integer in question is identical to a natural number. But if Shapiro's account is right, the answer to this question (if it is taken at face value) can only be "false," since, according to Shapiro, integers and natural numbers belong to different structures. Moreover, it can hardly be an interesting question—it only takes a metaphysician who knows precious little about mathematics to figure out that answer. So Shapiro is obliged to deny a face-value reading of questions of this kind and is forced to uncover instead a subterranean interest in establishing whether an embedding obtains between distinct structures. But, other things being equal, a face-value reading of mathematical practice is to be preferred. It cannot be credible to suppose that mathematicians persist in speaking falsely even though the vocabulary of embeddings is so readily to hand for expressing what—by Shapiro's lights—they must really mean.

[12] Resnik considers the following example: "equations of the form "$x^2 = a$" have complex roots for any choice of *a*, but their roots are *integers* only when *a* is the square of a *natural number*" (*MSP*, p. 214). Here integers and natural numbers are conceived as species of complex number.

In sum, neither account that Shapiro presents provides a convincing treatment of the "incompleteness" of mathematical objects. There remains, however, the further question of whether structures—conceived as *ante rem* universals—provide an otherwise credible ontology for mathematics. Hellman has suggested otherwise: that purely structural relations on the *ante rem* conception are not even intelligible. He argues the point in the following terms. The positions of a purely structural relation are identified merely as its terms (they are purely structural positions). But the relation cannot be identified simply as the structure that formally satisfies whatever axioms may be available for the purpose. This is because any system that exhibits the relation in question will answer to the axioms, too. In short, the relation will be indistinguishable from the systems that exhibit it unless we already have access to the relata of the relation and can identify the relation as the relation of those relata. This, Hellman maintains, is a "vicious circularity": "to understand the relata we must be given the relation, but to understand the relation, we must already have access to the relata" (Hellman [2001]; also ch. 17 in this volume).

This difficulty does not appear to be critical. It presupposes that we are able to understand a relation only if we are able to uniquely identify it. But there seems no reason to assume that reference to structures is or ought to be determinate. It is certainly true that many proponents of the view—Shapiro included—have accepted this assumption, presenting structuralism as an antidote to referential indeterminacy. But since indeterminacy appears to be an ineliminable aspect of reference to mathematical objects (structural or otherwise), there appears no reason to accept it. And once this assumption is discarded, Hellman's criticism appears to express only a concern about the impredicative dependencies that obtain between structures and positions, a concern that should hardly carry force once it is recognized that structures and positions are not *constructed* by us (cf. Gödel [1944], pp. 455–456).

Far more worrisome is the concern that Shapiro's notion of a structural position cannot be made to cohere with the existence of structures admitting nontrivial automorphisms. The concern arises from structuralist slogans like "The essence of a natural number is its relation to other natural numbers . . . there is no more to the individual numbers "in themselves" than the relations they bear to each other" (*PMSO*, pp. 72–73).[13] This makes it appear as if Shapiro is committed

[13] Slogans of this kind trace back to Benacerraf's articulation of the structuralist conception: "Mathematical objects have no properties other than those relating them to other 'elements' of the same structure" (Benacerraf [1965], p. 285). It is worth noting, however, that Dedekind—the philosopher-mathematician with whom the structuralist conception originates—did not commit himself to the claim that mathematical objects are entirely constituted by their relations to other elements of their parent structure. Instead, Dedekind offers the following claim: "If, in considering a simple infinite system N, ordered by a mapping ϕ, we entirely disregard the particular nature of its elements, *retaining only their discriminability from each other*, and having only regard to the relations to one another imposed by the mapping . . ." (Dedekind [1888], par. 73; my italics).

to a version of the Identity of Indiscernibles for mathematical objects: if x and y share just the same intrastructural relations to other items, then $x = y$. There are, however, systems of mathematical objects that contain structurally indiscernible elements (systems that admit nontrivial automorphisms): the complex numbers (i *and* $-i$), the additive integers ($+1$ and -1), points in the Euclidean plane, geometric figures with reflectional symmetry, and so on. This means that Shapiro is committed to identifying these indiscernible elements: i with $-i$, $+1$ with -1, and so on. But we know mathematically that these elements are distinct. It follows that *ante rem* structuralism must be rejected (see Burgess [1999], pp. 287–288; and Keränen [2001]).

Whether this objection ultimately carries any force turns upon whether a version of the Identity of Indiscernibles really does follow from the structuralist slogans. But these slogans may be interpreted in a way that does not evince a commitment to the Identity of Indiscernibles. However, interpreting them in this way leaves *ante rem* structuralism almost indiscernible from traditional Platonism. Thus a dilemma arises for Shapiro: either bad news ($i = -i$) or old news (*ante rem* structuralism = good old-fashioned Platonism).

To work toward an appreciation of this dilemma, let us assemble some of the considerations that speak in favor of a structuralist commitment to the Identity of Indiscernibles. Keränen argues that general considerations to do with the metaphysics of identity elicit this commitment (see his [2001], pp. 312–318). According to Keränen, whenever a type of object Ω is introduced into discourse, it must be possible to supply necessary and sufficient conditions for Ωs to be identical. Next, he remarks: "All extant theories of ontology maintain that the identity of objects is governed by their *properties*." Such an account of identity may appeal either to purely *general* properties or else to *haecceities*. Keränen now develops a dilemma of his own for the *ante rem* structuralist. Suppose the structuralist appeals to purely general properties to provide necessary and sufficient conditions for the identity of mathematical objects. Then the properties he has available for the purpose of fashioning an account of identity are all structural properties and relations. This means that mathematical objects can be distinct, from the structuralist perspective, only if they differ in their structural properties and relations. However, we know this is wrong. We know mathematically that there are structurally indiscernible but distinct mathematical objects. Alternatively, if the structuralist appeals to haecceities, then structurally indiscernible objects may be distinguished by their possession of nonstructural properties (i enjoys the property *being identical to* i, whereas $-i$ lacks it). But then the structural slogan—"There is no more to the individual numbers 'in themselves' than the relations they bear to each other"—will have been given up.

The most questionable feature of this argument is its most basic assumption, the thesis that necessary and sufficient conditions for the identity of objects can and should be stated in exclusively property-theoretic terms. The idea that we should be able—at least in principle—to uniquely characterize each object in our ontology provides one motivation for upholding this thesis. That such a motive is operative in Keränen's mind is suggested by the frequent shifts he makes between talking about "identity" and talking about "individuation." But, as Shapiro points out, it is highly questionable to suppose that we should be able to individuate—even in principle—the elements of an uncountable ontology (see Shapiro [forthcoming]). More generally, Keränen has articulated a metaphysical thesis about identity (reductionism) that seems no less dubious. Why should the relevant class of identity facts not turn out to be metaphysically basic rather than reducible? Why should it be supposed that there is invariably a supply of properties sufficient to provide a reductive base for the facts of identity? To these and other relevant questions Keränen supplies no answers.

There are, however, less global considerations that speak in favor of an *ante rem* structuralist commitment to the Identity of Indiscernibles. For the slogan "There is no more to the individual numbers 'in themselves' than the relations they bear to each other" may be interpreted as articulating a local reductionist thesis about mathematical objects. According to this interpretation, the structuralist is committed to a bundle theory of mathematical objects: there are no "bare particulars" hidden behind the façade of structural relations; there are only bundles of these relations. But if mathematical objects are reducible to bundles of structural relations, then mathematical objects that are structurally indiscernible must be numerically the same. Since there are structurally indiscernible but numerically distinct mathematical objects, it follows that the structuralist slogan must be rejected. Mathematical objects cannot be just bundles of structural relations.

It is far from evident, however, that the *ante rem* structuralist is committed to a bundle theory (object reductionism) of this kind. For there is a further reading available of the slogan "There is no more to the individual numbers 'in themselves' than the relations they bear to each other." Instead of offering a reductionist thesis about mathematical *objects*, the slogan may be interpreted as a reductionist thesis about mathematical *properties*. According to this reading, there is no more to the properties of individual numbers than arise from the relations to other positions in their parent structure. But the fact that the properties of mathematical objects are reducible to structural ones does not imply that mathematical objects are themselves capable of such reduction. So it is consistent with this reading of the structuralist slogan that there are structurally indiscernible but numerically distinct mathematical objects.

Unfortunately, a property reductionist conception of *ante rem* structuralism is beset by difficulties of its own. To start with, it is far from evident that all the properties mathematical objects exhibit are reducible to structural relations (Hellman [1999], [2001]). There are three classes of plausible counterexamples to this claim. First, there are categorical properties such as *being abstract*. Second, there are intentional properties like *being Stewart's favorite number.* Third, there are properties of application (e.g., *being the number of planets*).

Some of these examples may be less troublesome than others. It may be doubted whether there are any genuine categorical properties rather than mere formal predicates. For example, the predicate "*x* is abstract" may be taken to be true of objects that *lack* a certain property (*being spatiotemporally located*) rather than of objects that possess some definite feature of their own. In a similar way, it may be doubted whether there are genuine intentional properties. For example, Nostromo may be my favourite fictional character, but it doesn't follow that there is some genuine property possessed by him (how could there be a property possessed by him? he doesn't exist!). But if there aren't any categorical or intentional properties, then these nonstructural properties cannot serve as counterexamples to the thesis that the properties of mathematical objects are reducible to structural ones.

Of course, these arguments against categorical or intentional properties are far from decisive. But even if it is granted—for whatever reason—that there are no such properties, properties of application present a more troublesome case. Recall that it is the possession of these properties that makes mathematical objects worth believing in. Properties of application cannot be dismissed as mere lacks or shadows of thought. For what use are the cardinal numbers if they can't be employed to count? What merit is there in the real numbers if they cannot serve to measure? But while properties of application cannot be dismissed in this way, they cannot be reduced to the obtaining of structural relations either. The fact that 9 has the property of *being the number of planets* is not reducible to the structural relations (derived from the obtaining of the successor function) that 9 bears to the other natural numbers. Those relations might have obtained among the numbers even if (e.g.) Neptune had never existed. The structuralist slogan cannot be sustained as a general thesis.

Recall the dilemma posed for *ante rem* structuralism: either it's bad news or it's old news. We have seen how Shapiro may avoid the first horn of this dilemma—the horn that says structurally indiscernible objects must be numerically the same. This horn may be avoided by insisting upon a property-theoretic reading of structuralist slogans. But now Shapiro is impaled on the other horn of the dilemma—*ante rem* structuralism collapses into Platonism. Why? The first horn of the dilemma reveals that object reductionism fails. Mathematical objects cannot simply be bundles of structural relations; they are a separate, irreducible category of existent. So the structuralist must admit (at least) a two-category ontology of objects and relations. The failure of property reductionism indicates that mathematical objects are also the bearers of properties and relations that take them outside their parent structures; for example, to the items in the physical world they are used to count and measure. But then *ante rem* structuralism turns out to be just a kind of traditional Platonism that also posits a two-category ontology and conceives mathematical objects as the bearers of structure-independent properties and relations. How, then, has the *ante rem* structuralist enabled us to pass beyond this traditional picture? Where's the news?

Shapiro may seek to evade this dilemma by pointing to what is ontologically innovative in his view, something that traditional Platonists do not routinely countenance, namely the identification of mathematical objects with positions in structures. But can this identification legitimately be made? Structures are conceived by *ante rem* structuralism to be a species of universal; they are considered to be relations (complex relations in the mathematically interesting cases) that obtain among systems of objects. Positions in structures then turn out to be the argument slots that are filled by objects when the universal (relation) in question is instantiated. However, the identification of mathematical objects with positions in structures rests upon the prior credibility of the thesis that positions are objects in their own right. But this thesis—it will be seen—is more incredible than credible, and it is far better to conceive of positions in a nominalistic spirit. Note the consequences of this: if the notion of a "position" is not that of a kind of object, if "positions" do not exist, then clearly Shapiro's identification of mathematical objects with positions misfires.

If there is a reason to believe in positions (as objects), then it should surely be a reason of the following kind. It should be because the positing of them oils the wheels of a theory of differential application. What does this mean? It means that positing positions should enable us to explain why relations obtain one way rather than another (why aRb rather than bRa, where R is asymmetric). Now it is indeed familiar enough for philosophers to introduce the notion of a position in the following terms: positions are "slots" in the relation universal, and the relation applies differently (in a different order) when the terms of the relation fit different slots. Speaking in these terms, it is a fact that aRb rather than bRa because a fits into the x slot, and b fits into the y slot rather than the other way around. But does an explanation of this kind—perhaps unexceptional in itself—provide any grounds for supposing that positions or slots are *entities*?

The problem is that this explanation goes awry if positions are conceived as self-standing objects. The proposed explanation of the differential application of relations to objects goes via the assignment of objects to slots (a and b to x and y, in that order). However this explanation will be effective only if it is presupposed that some other objects—the slots—are related to each other in the right order: the right order to secure the result that the objects assigned to them are then related in a given order (so that aRb rather than bRa). The application of a to x and b to y will therefore suffice for aRb rather than bRa only if x lies in the further relation (call it R^*) to y such that if a fills x and b fills y, then aRb. This shows that the purported explanation of the differential application of R presupposes the further relational fact that xR^*y rather than yR^*x (this is not, of course, the only relational fact presupposed). A regress now arises. In order to explain the relational ordering of the slots x and y, the relation R^* will require further slots z and w into which x and y

must fit to explain the differential application of R^* to x and y. But then the further relational fact $wR^{**}z$ will have been presupposed, and so on.[14]

This regress may be avoided by refusing to posit further entities to explain the ordering of the slots. In that case it will just be a brute fact that the slots x and y *are so ordered.* But, if that is the case, the differential application of relations (R) might have been explained all along by appeal to the brute ideology of order. Then there will simply have been no reason to inflate our ontology to include positions. And if there is no reason to believe in positions (conceived as objects), then there is no reason to believe that positions are candidates eligible to be identified with mathematical objects.

Of course the *ante rem* structuralist may continue to insist that positions are objects (to be identified with mathematical objects) for reasons other than their potential contribution to an account of order. But it is important to keep in focus that it is the raison d'être of relations to account for order and that positions are essentially the positions *of* relations. Therefore, unless positions (conceived as objects) are assigned some significant role in the account of order, the insistence that "positions" are objects will not differ from simply insisting that there are objects (for no content will have been given to the notion that these objects are positions). So unless such a role for positions is made out, the *ante rem* structuralist will differ no whit from the traditional Platonist who simply posits mathematical objects outright.

The *ante rem* structuralist faces a challenge, a challenge to explain how order obtains in the world. Unless *ante rem* structuralists—Shapiro included—answer this challenge, they will be unable to justify the critical thesis for which an argument remains wanting in all their writings: the thesis that positions are objects.

4. Conclusion

What are the primary obstacles that confront any attempt to vindicate the claim that mathematical objects are incomplete *in re*, to be fruitfully compared or identified with positions in patterns or structures? A variety of criticisms have been made of the attempts by Resnik and Shapiro to convince us that mathematical objects should be conceived in such a spirit. Nevertheless, two general difficulties stand out. They arise not only for Resnik and Shapiro but also for anyone else who would claim that mathematical objects are incomplete *in re*. First difficulty: that

[14] This regress argument is akin to, but distinct from, Bradley's famous regress argument (see his [1893], pp. 31–33). For whereas Bradley's regress concerns instantiation per se, the regress that threatens Shapiro's conception of positions as objects concerns order.

the evidence adduced (so far) in favor of the incompleteness of mathematical objects underdetermines whether it is the objects themselves that are incomplete rather than our knowledge or descriptions of them. Second difficulty: that holding mathematical objects to be incomplete conflicts or fits uncomfortably with the practice of mathematics. These concerns are not independent of one another. The fact that the incompleteness of mathematical objects fits ill with mathematical practice places even greater pressure upon the assumption that the evidence adduced in favor of the incompleteness of mathematical objects is at all convincing.

These and other difficulties that the views of Resnik and Shapiro encounter point toward the contrary view, that it is our knowledge or our description of mathematical objects that is incomplete rather than the objects themselves. So rather than speaking completely about objects that are incomplete *in re*, the mathematician should be interpreted as speaking incompletely about objects that are complete in themselves. Just over a century ago Russell advanced this idea in an especially pure form. He claimed that the role of mathematics is not to speak about a special realm of privileged objects—the "mathematical" objects—but rather to speak conditionally about objects of any kind whatsoever:

> We start, in pure mathematics, from certain rules of inference, by which we can infer that *if* one proposition is true, then so is some other proposition. These rules of inference constitute the major part of the principles of formal logic. We then take any hypothesis that seems amusing, and deduce its consequences. *If* our hypothesis is about *anything*, and not about some one or more particular things, then our deductions constitute mathematics. Thus mathematics may be defined as the subject in which we never know what we are talking about, nor whether what we are saying is true. (Russell [1901], p. 75)

Russell went on to express this idea more formally in the opening sentence of *The Principle of Mathematics*:

> Pure mathematics is the class of all propositions of the form "p implies q," where p and q are propositions containing one or more variables, the same in the two propositions, and neither p nor q contains any constants except logical constants. (Russell [1903], p. 3)

The difficulties that confront the claim that mathematical objects are incomplete *in re* indicate that it is in this direction—the direction that Russell pointed—that structuralism can expect to receive its most fruitful development.

REFERENCES

Benacerraf, P. [1965]: "What Numbers Could Not Be," *Philosophical Review*, 74, pp. 47–73; reprinted in Benacerraf and Putnam [1983], pp. 272–294.

Benacerraf, P. [1998]: "What Mathematical Truth Could Not Be—I," in M. Schirn (ed.), *The Philosophy of Mathematics Today* (Oxford: Clarendon Press), pp. 33–75.

Benacerraf, P., and Putnam, H. (eds.) [1983]: *Philosophy of Mathematics*, 2nd ed. (Cambridge: Cambridge University Press).

Bradley, F.H. [1893]: *Appearance and Reality: A Metaphysical Essay* (London: George Allen and Unwin).

Burgess, J. [1999]: Review of Shapiro [1997], *Notre Dame Journal of Formal Logic*, 40, pp. 283–291.

Chihara, C. [1990]: *Constructibility and Mathematical Existence* (Oxford: Clarendon Press).

Dedekind, R. [1888]: *Was sind und was sollen die Zahlen?* (Brunswick: Vieweg); translated as "The Nature and Meaning of Numbers" in *Essays on the Theory of Numbers* (New York: Dover Press, 1963), pp. 31–115.

Dummett, M. [1991]: *Frege: Philosophy of Mathematics* (Cambridge, Mass.: Harvard University Press).

Evans, G. [1978]: "Can There Be Vague Objects?," *Analysis*, 38, p. 208.

Field, H. [1974]: "Quine and the Correspondence Theory," *Philosophical Review*, 83, pp. 200–228.

Field, H. [1989]: *Realism, Mathematics and Modality* (Oxford: Blackwell).

Frege, G. [1884]: *Die Grundlagen der Arithmetik* (Breslau: Koebner); translated by J.L. Austin as *The Foundations of Arithmetic* (Oxford: Blackwell, 1953).

Frege, G. [1893–1903]: *Grundgesetze der Arithmetik, begriffsschriftlich abgeleitet*, 2 vols. (Jena: H. Pohle); partially translated in P.T. Geach and M. Black (eds.), *Translations from the Philosophical Writings of Gottlob Frege* (Oxford: Basil Blackwell, 1952), pp. 117–224.

Gödel, K. [1944]: "Russell's Mathematical Logic," in P.A. Schilpp (ed.), *The Philosophy of Bertrand Russell* (Evanston, Ill.: Northwestern University Press, 1944), pp. 125–135; reprinted in Benacerraf and Putnam [1983], pp. 447–469.

Hellman, G. [1989]: *Mathematics Without Numbers* (Oxford: Oxford University Press).

Hellman, G. [1999]: Review of Shapiro [1997], *Journal of Symbolic Logic*, 64, pp. 923–926.

Hellman, G. [2001]: "Three Varieties of Mathematical Structuralism," *Philosophia Mathematica*, 9(3), pp. 148–211.

Keränen, J. [2001]: "The Identity Problem for Realist Structuralism," *Philosophia Mathematica*, 9(3), pp. 308–330.

Kitcher, P. [1978]: "The Plight of the Platonist," in *Noûs*, 12, pp. 119–136.

Kreisel, G. [1967]: "Informal Rigour and Completeness Proofs," in I. Lakatos (ed.), *Problems in the Philosophy of Mathematics* (Amsterdam, North-Holland), pp. 138–186.

Kripke, S. [1980]: *Naming and Necessity* (Oxford: Basil Blackwell).

MacBride, F. [2003]: "Speaking with Shadows: A Study of Neo-logicism," *British Journal for the Philosophy of Science*, 54, pp. 103–163.

MacBride, F. [2004]: "Can Structuralism Solve the 'Acess' Problem,?" *Analysis 64.4*, pp. 309–317.

MacBride, F. [forthcoming]: "More Problematic Than Ever: The Julius Caesar Objection," in MacBride (ed.) [forthcoming].

MacBride, F., (ed.). [forthcoming]: *Identity and Modality* (Oxford: Oxford University Press).

Maddy, P. [1990]: *Realism in Mathematics* (Oxford: Oxford University Press).

McGee, V. [1997]: "How We Learn Mathematical Language," *Philosophical Review*, 106, pp. 35–68.
Parsons, C. [1965]: "Frege's Theory of Number," in M. Black (ed.), *Philosophy in America* (London: Allen and Unwin, 1965), pp. 180–203.
Parsons, C. [1971]: "Ontology and Mathematics," *Philosophical Review*, 80, pp. 151–176.
Parsons, C. [1990]: "The Structuralist View of Mathematical Objects," *Synthèse*, 84, pp. 303–346.
Parsons, T. [2000]: *Indeterminate Identity* (Oxford: Oxford University Press).
Quine, W.V. [1969]: "Ontological Relativity," in his *Ontological Relativity and Other Essays* (New York: Columbia University Press), pp. 26–68.
Resnik, M. [1975]: "Mathematical Knowledge and Pattern Cognition," *Canadian Journal of Philosophy*, 5, pp. 25–39.
Resnik, M. [1981]: "Mathematics as a Science of Patterns: Ontology and Reference," *Noûs*, 15, pp. 95–105.
Resnik, M. [1982]: "Mathematics as a Science of Patterns: Epistemology," *Noûs*, 16, pp. 95–105.
Resnik, M. [1988]: "Mathematics from a Structuralist Point of View," *Revue International de Philosophie*, 42, pp. 400–424.
Resnik, M. [1997]: *Mathematics as a Science of Patterns* (Oxford: Oxford University Press).
Russell, B. [1901]: "Mathematics and the Metaphysician," *The International Monthly*; reprinted in Russell's *Mysticism and Logic* (London: Allen & Unwin, 1917), pp. 74–94.
Russell, B. [1903]: *The Principles of Mathematics* (London: Allen & Unwin).
Russell, B. [1919]: *Introduction to Mathematical Philosophy* (London: Allen & Unwin).
Shapiro, S. [1983]: "Mathematics and Reality," *Philosophy of Science*, 50, pp. 523–548.
Shapiro, S. [1989]: "Structure and Ontology," *Philosophical Topics*, 17, pp. 145–171.
Shapiro, S. [1993]: "Modality and Ontology," *Mind*, 102, pp. 455–481.
Shapiro, S. [1997]: *Philosophy of Mathematics: Structure and Ontology* (New York: Oxford University Press).
Shapiro, S. [forthcoming]: "Structure and Identity," in MacBride (ed.) [forthcoming].
Steiner, M. [1975]: *Mathematical Knowledge* (Ithaca, N.Y.: Cornell University Press).
Wright, C. [1983]: *Frege's Conception of Numbers as Objects* (Aberdeen: Aberdeen University Press).

CHAPTER 19

PREDICATIVITY

SOLOMON FEFERMAN

WHAT is predicativity? While the term suggests that there is a *single idea* involved, what the history will show is that there are a *number of ideas* of predicativity which may lead to different logical analyses. I shall uncover these only gradually.[1] A central question will then be what, if anything, unifies them. Though early discussions are often muddy on the concepts and their employment, in a number of important respects they set the stage for the further developments, and so I shall give them special attention. Note that, ahistorically, modern logical and set-theoretical notation will be used throughout, as long as it does not conflict with original intentions.

[1] The subject of predicativity is one that has been of great interest to me and has periodically commanded much of my attention over the last forty years. It involves substantial developments in logic and mathematics and is of significance for the philosophy of mathematics. However, it is still unsettled how best to assess these various aspects of predicativity. On March 28, 2002, for a joint meeting of the American Philosophical Association and the Association for Symbolic Logic, Jeremy Avigad, Geoffrey Hellman, and I participated in a symposium organized by Paolo Mancosu titled "Predicativity: Problems and Prospects." In my lecture I concentrated on the idea of predicativity in its historical development and particularly its logical analysis, which has led to new problems of current interest; the present chapter is based on that lecture. Complementarily, Avigad and Hellman dealt respectively with questions concerning the mathematical and philosophical significance of predicativity. As of my checking proof for this chapter, an article based on Hellman's lecture "Predicativism as a philosophical position" is to appear with a response by me in an issue of the *Revue Internationale de Philosophie*. Avigad has reported to me that preparation of his symposium lecture, "Methodological predicativity," is still in progress. I greatly appreciate the help that Paolo Mancosu has given with this chapter, especially the part having to do with early developments.

Predicativity Emerges: Russell and Poincaré

To begin with, the terms *predicative* and *nonpredicative* (later, *impredicative*) were introduced by Russell (1906) in his struggles dating from 1901 to carry out the logicist program in the face of the set-theoretical paradoxes. Russell called a propositional function $\varphi(x)$ predicative if it defines a class (i.e., if the class $\{x: \varphi(x)\}$ exists), and nonpredicative otherwise. Thus, for example, the propositional function $x \notin x$ figuring in Russell's paradox is impredicative. Since the admission of classes defined by arbitrary propositional functions in Frege's execution of his logicist program led to its demise as a result of this paradox, if the program were to be resurrected, it would somehow have to incorporate a criterion for distinguishing predicative from impredicative functions. Russell's first attempts to separate these were highly uncertain, and it was only through the engagement of Henri Poincaré in the problem, starting in his article (1906a) that progress began to be made.

Poincaré took several paradoxes as examples to try to elicit what was common to them: the Burali–Forti paradox of the largest ordinal number, König's paradox of the least nondefinable ordinal number, and the Richard paradox defining, by diagonalization, a real number different from all definable real numbers; it was this last that Poincaré took as a paradigm. Note that in doing so, Poincaré shifted attention away from purported definitions of set-theoretical objects involving only purely set-theoretical notions such as those of class, membership, ordinal number, and cardinal number, to purported definitions of mathematical objects more generally in which the notion of *definability* itself was an essential component. In doing so, he could be considered to be at cross-purposes with Russell. At any rate, Poincaré came up with two distinct diagnoses of the source of the paradoxes via what he regarded as "typical" examples. The first was that there is in each case a *vicious circle* in the purported definition. For example, in the case of Richard (1905), since each definition of a real number via its decimal expansion can be written out using a finite number of symbols, the set D of definable real numbers is countable. Then, by Cantor's diagonal construction, one can define a real number r which is distinct from each member of D; but since r is defined, it is a member of D, which is a contradiction. According to Poincaré, in this case the vicious circle lies in trying to produce the object r in D by reference to the supposed totality of objects in D; indirectly, then, r is defined in terms of itself, as one of the objects in D. Poincaré's second diagnosis is distinct in its emphasis, that the source of each paradox lies in the assumption of the "actual" or "completed" infinite. Again with reference to the Richard paradox, one cannot assume that there is a completed totality of *all* definable objects of a certain kind; rather, each one "comes into existence" through a definition in terms of previously defined objects. As we shall see, in his own mathematics

Poincaré did not hew to the injunction against the actual infinite. And that is related to the issue of impredicative definitions as they occur in mathematical practice, to which Poincaré was to return a few years later. We'll also that take up below.

Russell's Elaborations

In his article "Les Paradoxes de la logique" (1906b), Russell quickly took both the vicious circle diagnosis and, to an extent, the objection to the completed infinite as the point of departure for his further work on predicative versus impredicative definitions of classes:[2]

> I recognize . . . that the clue to the paradoxes is to be found in the vicious circle suggestion; I recognize further this element of truth in M. Poincaré's objection to totality, that whatever in any way concerns *all* or *any* or *some* of a class must not be itself one of the members of a class. (Russell 1973, p. 198)

Russell then went on to make the first of his several attempts to formulate the VCP (Vicious Circle Principle) in syntactic terms that would be appropriate for a formalism in which to redevelop logicism:

> In M. Peano's language, the principle I want to advocate may be stated: "Whatever involves an apparent variable must not be among the possible values of that variable." (ibid.)

Insofar as this form of the VCP proscribed certain formulas $\varphi(x)$ from defining classes, its effect would be to exclude from φ any bound variables whose intended range includes the class $\{x\colon \varphi(x)\}$ as one of its values. But also the bound variable "x" in the designation $\{x\colon \varphi(x)\}$ of that class must not be in that range. These restrictions were then the lead-in to the formalism proposed in Russell's article "Mathematical Logic as Based on the Theory of Types" (1908), and on whose plan *Principia Mathematica* was erected. As is well known, Russell formulated the VCP in several different ways, and their precise significance and relation to each other has been the subject of much scrutiny and critique by a number of scholars, including Kurt Gödel (1944), Charles Chihara (1973), and Philippe de Rouilhan (1996). The formulation that is tied most closely to Russell's theory of types was given in the 1908 article as the following sharpening of the above:

> [The VCP], in our technical language, becomes: "whatever contains an apparent variable must not be a possible value of that variable." Thus [it] must be of a

[2] The quotation is from the English translation in Russell's *Essays in Analysis* (1973). Many of the original texts on the antinomies and predicativity in the period 1906–1912 are conveniently assembled in Heinzmann (1986).

> different type from the possible values of that variable. Thus we will say that whatever contains an apparent variable must be of a different type from the possible values of that variable; we will say that it is of a *higher* type. (Russell 1908, in van Heijenoort 1967, p. 163)

Before going into the actual structure of types in Russell's setup, let me draw attention to an earlier section of the article, headed "All and Any" (ibid., pp. 156–159). Here, in contrast to the first quotation from Russell above, a distinction was made between the use of these two words. Roughly speaking, in logical terms, the statement that *all* objects x of a certain kind satisfy a certain condition $\varphi(x)$ is rendered by the universal quantification $(\forall x)\varphi(x)$ in which "x" now is a bound variable, while the statement that $\varphi(x)$ holds for *any* x is expressed by leaving "x" as a free variable. In modern terms, the logic of the latter is treated as a scheme to be coupled with a rule of substitution. The importance of this distinction for Russell has to do with the injunction against illegitimate totalities. In particular, with p a variable for propositions, he would admit "p is true or false, where p is any proposition" (i.e., the scheme $p \vee \neg p$, but not the statement $(\forall p)(p \vee \neg p)$ that "all propositions are true or false" (in both cases using truth of p to be equivalent to p). Similarly, for properties $P(x)$; significantly, Russell pointed out that the proposed definition of the natural numbers in the form "'n is a finite integer' means 'Whatever property φ may be, n has the property φ provided φ is possessed by 0 and by the successors of possessors'" (ibid., p. 159). That is, in symbols,

$$N(n) := (\forall\varphi)[\varphi(0) \wedge (\forall x)(\varphi(x) \rightarrow \varphi(x')) \rightarrow \varphi(n)]$$

cannot be replaced by dropping the universal quantifier over properties "$(\forall\varphi)$."

Though the simplest paradoxes, such as Russell's of the class of all non-self-membered classes or the heterologicality paradox, can be construed as involving arbitrary classes or properties in the form of apparent variables, they do not involve them in the form of quantified variables. Nevertheless, these considerations led Russell to ban the use of "all" in the form of unrestricted quantification over propositions and properties, among other things. He then faced the question of when it is legitimate to apply universal quantification over any given kind of object, and here he veered away from Poincaré's injunction against the "actual" infinite:

> It has often been suggested that what is required in order that it may be legitimate to speak of *all* of a collection is that the collection should be finite. Thus "all men are mortal" will be legitimate because men form a finite class. But that is not really the reason why we can speak of "all men." What is essential . . . is not finitude, but what may be called *logical homogeneity*. This property is to belong to any collection whose terms are all contained within the range of significance of some one function. It would always be obvious at a glance whether a collection possessed this property or not, if it were not for the concealed ambiguity in common logical terms such as *true* and *false*, which gives an appearance of being

> a single function to what is really a conglomeration of many functions with different ranges of significance. (ibid., p. 163)

Here a new idea became central:

> ...we can speak of *all* of a collection when and only when the collection forms part or the whole of the *range of significance* of some propositional function, the range of significance being defined as the collection of those arguments for which the function in question is significant, that is, has a value. (ibid.)

That is, in more modern terms, like all functions in ordinary mathematics, propositional functions are *partial* in the sense that they have a prescribed domain which is given in advance of the function, and outside of which they are undefined. In particular, the arguments x of a propositional function φ must somehow be "prior" to the function itself. This is achieved by the type distinction and an ordering of types, which makes the type of x lower than the type of φ. Without further restrictions, this would lead to the formalism of the *simple theory of types.* But the VCP in the form given above requires that each apparent (bound) variable in φ also have lower type than φ. Together these result in the formalism of the *ramified theory of types.* At the bottom are variables of type 0 ranging over an unspecified nonempty domain of "individuals." In the simple theory of types, types can be identified with natural numbers; classes of individuals are of type 1; classes of classes of individuals, of type 2; and so on. In the ramified theory of types, (e.g., as elucidated by Myhill 1974), *types* are finite descending sequences $t = \langle m_1, m_2, \ldots m_k \rangle$ of natural numbers $m_1 > m_2 > \cdots > m_k = 0$, of *order* or *level* m_1. If $\varphi(x)$ has x of type t and m is least such that $m > m_1$ and all bound variables of φ have order less than m, then φ and its extension $\{x\colon \varphi(x)\}$ are of type $\langle m \rangle ^\frown t = \langle m, m_1, m_2, \ldots m_k \rangle$, and variables of that type are interpreted as ranging over all such classes. The atomic formulas of this language are of the form $x = y$ where x, y are of the same type and $x \in y$ where x is of a type t and y is of a type $\langle m \rangle ^\frown t$. Russell says that the type of one object is lower (resp. higher) than that of another if its level is lower (resp. higher).

If the domain of individuals is finite, then under this interpretation, for each type t the domain of objects of type t is finite, and the ramified interpretation collapses to the simple interpretation. Thus, if one is to define the notion of natural numbers in purely logical terms, as would be required by the logicist program, it must be assumed that the domain of individuals is infinite. The assumption of this Axiom of Infinity was the first crack in Russell's attempt to continue the logicist program within the ramified theory of types. The second came with the definition of natural number itself in the form, as above:

$$N(a) := (\forall \varphi)[\varphi(0) \wedge (\forall x)(\varphi(x) \rightarrow \varphi(x')) \rightarrow \varphi(a)],$$

where 0 is the empty class at a given type t and the successor operation takes objects of type t to objects of type t, φ is of a type $\langle m \rangle ^\frown t$, and the variable "$a$" is also of type t. Here, with natural numbers interpreted as cardinal numbers in the sense of equivalence classes under the relation of one–one correspondence, t must be of a type of classes of classes.

There were two problems with this. First, the notion of natural number is relative to any such type t, and second, most usual proofs of induction can't be carried out. To show, for example, for any given natural number a that $(\forall x)\ [N(x) \rightarrow N(a+x)]$, where $+$ is suitably defined in cardinal-theoretic terms, one usually has to carry out an induction as follows: take $\psi(x) := N(x) \rightarrow N(a+x)$, and establish $(\forall x)\ \psi(x)$ by showing $\psi(0) \wedge (\forall x)(\psi(x) \rightarrow \psi(x'))$. But ψ is of a higher type (level) than that of the quantified variable φ in the definition of N, so we can't apply that definition to carry out the required induction. As Russell says, "[i]t is obvious that such a state of things renders elementary mathematics impossible" (ibid., p. 167). In order to get around this quite serious obstacle, Russell introduced his "oddly devious" (in Quine's words[3]) Axiom of Reducibility. This is formulated in terms of a notion of *predicative* function $\varphi!x$ which may be described as one in which the bound variables, if any, are all of the same or lower type than the type of x. (Note that here "predicative" is used in a very restricted sense.) Using variables $f, g, \ldots$ to range over predicative functions, the Axiom of Reducibility states that any propositional function is coextensive with some predicative function. Schematically:

$$(\exists f)(\forall x)[\varphi(x) \leftrightarrow f!x]$$

(ibid., p. 171). In particular, that allows one to carry out the blocked induction above by replacing ψ with an associated predicative function g.

Evidently, the Axiom of Reducibility completely vitiates the system of ramified types and makes it equivalent to simple type theory, for which, however, there is no predicative justification if one assumes the Axiom of Infinity. In the introduction to the second edition of *Principia Mathematica*, (presumably) Russell wrote:

> [The Axiom of Reducibility] has a purely pragmatic justification: it leads to the required results, and to no others. But clearly it is not the sort of axiom with which we can rest content. On this subject, however, it cannot be said that a satisfactory solution is as yet obtainable. (Whitehead and Russell 1925, p. xiv)

More specifically, at the end of that introduction Russell points out various places in the development of mathematics on the basis of his formalism, where the Axiom of Reducibility is required to carry out not only the usual inductions on the natural numbers, as well as their generalizations to transfinite ordinal numbers, but also in the foundations of the theory of real numbers. In conclusion, he writes:

> It might be possible to sacrifice infinite well-ordered series [i.e., well-ordering relations] to logical rigour, but the theory of real numbers is an integral part of

[3] In his introductory note to Russell (1908) in van Heijenoort (1967), p. 151.

> ordinary mathematics, and can hardly be the object of a reasonable doubt. We are therefore justified in supposing that some logical axiom which is true will justify it. The axiom required may be more restricted than the axiom of reducibility, but, if so, it remains to be discovered. (ibid., p. xlv)

Poincaré versus the Logicists and the Cantorians: From Paradoxes to Practice

In view of the lack of justification for the Axiom of Infinity and the Axiom of Reducibility, one must count as a failure Russell's attempt at a purely logical predicative foundation of mathematics beginning with the definition of the natural numbers. But even if it had been successful, it would have done nothing for Poincaré. In his excellent article, "Poincaré Against the Logicists," Warren Goldfarb writes:

> Although the great French mathematician Henri Poincaré wrote on topics in the philosophy of mathematics from as early as 1893, he did not come to consider the subject of modern logic until 1905. The attitude he then expressed toward the new logic was one of hostility. He . . . dismissed as specious both the tools devised by the early logicians and the foundational programs they urged. His attack was broad: Cantor, Peano, Russell, Zermelo, and Hilbert all figure among its objects. Indeed, his first writing on the subject is extremely polemical and is laced with ridicule and derogation. Poincaré's tone subsequently became more reasonable but his opposition to logic and its foundational claims remained constant. (Goldfarb 1988, p. 61)

In this article Goldfarb on the whole legitimately critiques various of Poincaré's objections to the indicated foundational programs, but the nature of those objections is relevant to a fundamental shift in the further development of the idea(s) of predicativity and its relation to mathematical practice, so we shall review them here. First, as Goldfarb explains, Poincaré's attack on the new logicians begins

> . . . with the avowed aim of showing that [they] have not eliminated the need for intuition in mathematics. By showing this, he says, he is vindicating Kant. . . . This avowal is misleading, for in Poincaré's hands the notion of intuition has little in common with the Kantian one. (ibid., p. 63)

Rather,

> For Poincaré, to assert that a mathematical truth is given to us by intuition amounts to nothing more than that we recognize its truth and do not need, or do

not feel the need, to argue for it. Intuition, in this sense, is a psychological term; it might just as well be called "immediate conviction." (ibid.)

In particular, for Poincaré the structure of natural numbers and the associated principle of induction are given in intuition and do not require a foundation; indeed, in his view, they are presupposed in any attempt at such a foundation. In the case of the logicist program, this is seen in the very description of the axiomatic system, with its inductive generation of formulas and proofs; hence that enterprise assertedly involves a petitio principii. Goldfarb argues that Poincaré is mistaken in ascribing a petitio in this case, as he is in other cases. In a useful review of Goldfarb's article, Michael Hallett (1990) defends Poincaré on this point, at least in part. What is important for us here is not which side is correct, but the content and influence of Poincaré's views.

Poincaré's objections to the Cantorians rests on their essential use of impredicative constructions violating the VCP. Curiously, he accepted the Axiom of Choice on the grounds that it "is a synthetic judgment a priori [note the Kantian terminology]; without it the theory of cardinals would be impossible, for finite numbers as well as for infinite ones."[4] But he criticized Zermelo's 1904 proof—from the Axiom of Choice—that every set can be well-ordered, at a point that made use of the very common set-theoretical operation of the union U of a set S of sets X, $U = \{x\colon (\exists X)(X \in S \wedge x \in X)\}$, on the grounds that what members U has, depends on which sets belong to S and what their members are. In particular, it depends on whether U belongs to S and what its members are; thus, Poincaré identified the operation of union as illegitimately impredicative. In his response to this criticism, Zermelo (1908, pp. 190–191) pointed out that "proofs that have this logical form are by no means confined to set theory; exactly the same kind can be found in analysis wherever the maximum or the minimum of a previously defined 'completed' set of numbers . . . is used for further inferences. This happens, for example, in the well-known Cauchy proof of the fundamental theorem of algebra, and up to now it has not occurred to anyone to regard this as something illogical."

The particular proof cited by Zermelo proceeds by forming the minimum of the set of values of $|p(x)|$ where p is a polynomial over the complex numbers. Relatedly, if $f(x)$ is a continuous function on a closed interval $[a,b]$ in the real numbers, then f has a minimum value (as well, of course, as a maximum value) in that interval. This seems to appeal to the general greatest lower bound (g.l.b.) principle for the real numbers R (i.e., any bounded subset S of R has a g.l.b. [and least upper bound, l.u.b.]). The existence of the g.l.b. can be recognized set-theoretically as the formation of a union in terms of the association of real numbers x with their upper Dedekind sections D_x in the rational numbers Q, where

$$D_x = \{r\colon r \in Q \wedge x < r\}.$$

[4] Poincaré (1906), p. 313.

Let $S^* = \{D_x : x \in S\}$. Then the *union* U of the sets in S^* is the Dedekind section D_m where m = g.l.b.(S); in this, one invokes the *Dedekind completeness* (or *continuity*) of R, according to which every Dedekind section determines a real number. Dually, the l.u.b.(S) is determined by the *intersection* of all the sets in S^*. (If one uses lower Dedekind sections, these operations are reversed.)

Poincaré responded to Zermelo's specific example by asserting that the minimum of $|p(x)|$ is equally described as the minimum of $\{|p(x)| : x \text{ is rational}\}$. More generally, for any continuous function $f(x)$ on a compact set S in the real or complex numbers, its minimum is the same as the minimum of the countable set $\{f(x): x \in S \wedge x \text{ is rational}\}$, and that does not require an impredicative step. Poincaré was correct in this response, though his argument was faulty. Simply, any bounded set S of rational numbers has a g.l.b. (and l.u.b.) by selecting a subsequence that converges to the same. In this case, one invokes *Cauchy completeness of* R, for which, in contrast to Dedekind completeness, there is a predicative argument, as will be explained below.

Though Poincaré rejected the actual infinite in his polemical writings (e.g., "There is no actual infinite; the Cantorians forgot that, and they fell into contradiction."—1906, p. 316), his acceptance of the structure of natural numbers and of the principle of induction on it, coupled with the implicit use of classical logic and its assumption of the law of excluded middle, at least takes quantification over the natural numbers to be definite, and that is a form of acceptance of the actual infinite. No doubt Poincaré assumed this and much more in his mathematical practice in analysis and topology, though he might have said that he could always account for his work predicatively by more careful considerations of the sort given in his response to Zermelo. There would be some historical interest in putting Poincaré's "philosophical" principles (to the extent that they can be made precise) up against the details of his practice, but that is neither here nor there for our present survey. What counts is what influence those principles had on the development of predicativity, not whether he abided by them himself. And what is most significant about Poincaré's dispute with Zermelo in 1906–1909 is that attention was shifted away from the role of purportedly circular definitions in the production of the paradoxes to their role at the very center of mathematical practice, namely, in the l.u.b. principle for the real numbers.

The other aspect of Poincaré's dispute with Cantorians lies in their extension of the actual infinite to the transfinite. Of course he was not alone in this. Other prominent critics were Leopold Kronecker in the nineteenth century and L. E. J. Brouwer in the twentieth century; but both were more radical than Poincaré, in rejecting it also at the level of the natural numbers. In Brouwer's case it lay in the identification of the law of excluded middle as the culprit in the Cantorian crimes, in its supposedly illegitimate employment when applied to infinite totalities. Since, as I have argued, Poincaré did not go that far, he can be considered to be laying out a middle ground between the constructivists and the set-theoretical Platonists: the position of being a realist with respect to the natural numbers and a definitionist in all else.

Weyl's Predicative Development of Analysis

Hermann Weyl was the first to give substance to this middle ground, in his famous 1918 monograph *Das Kontinuum*. An exegesis of that work is contained in the article "Weyl Vindicated" (Feferman 1988), and its significance has been discussed in several other articles reprinted in my collection *In the Light of Logic* (1998). There is thus no need to go over those here at any length. Rather, for the present purposes, a relatively brief summary is sufficient, and that is most simply accomplished by quoting from myself (1988, pp. 51ff.):

> In the introduction to [*Das Kontinuum*] Weyl criticized axiomatic set theory as a "house built on sand" (though the objects of, and reasons for, his criticism are not made explicit). He proposed to replace this with a solid foundation, but not for all that had come to be accepted from set theory; the rest he gave up willingly, not seeing any other alternative. Weyl's main aim in this work was to secure mathematical analysis through a theory of the real number system (the continuum) that would make no basic assumptions beyond that of the structure of natural numbers N.... Weyl did not attempt to reduce... reasoning about N to something supposedly more basic. In this respect Weyl agreed with Henri Poincaré that the natural number system and the associated principle of induction constitute an irreducible minimum of theoretical mathematics, and any effort to "justify" that would implicitly involve its assumption elsewhere... unlike Brouwer, Weyl accepted uncritically the use of classical logic at this stage (though at a later date he was to champion Brouwer's views).

In Weyl's redevelopment of analysis, the rational numbers are reduced to the natural numbers in a standard way going back to Kronecker. But to do analysis, one needs representations of real numbers either as sets or as sequences of rational numbers, and that reduces to the question of what sets or sequences of natural numbers may be asserted to exist. Having accepted the natural number system with its basic inductively defined operations such as addition and multiplication, Weyl accepted that each subset of N of the form $\{n \in N: A(n)\}$ exists, where A is an arithmetical formula (i.e., one that contains no quantifiers ranging over sets, only over natural numbers). Beyond those, what other kinds of sets may be asserted to exist on predicative grounds? Adapting ramified type theory to this specific second-order context, one could consider sets of level 1, 2, 3, and so on, where the sets of level 1 are the arithmetically definable ones, those of level 2 are defined by formulas in which all second-order quantifiers range over sets of level 1, those of level 3 allow second-order quantifiers to range only over sets of level 1 or sets of level 2, and so on. Carrying this over to analysis, one would correspondingly then have real numbers of different levels 1, 2, 3,.... But if S is a bounded set of real numbers of some level k, its g.l.b. and l.u.b. are defined by applying the operations of union (resp. intersection) to corresponding sets S^* of upper Dedekind sections of the rational numbers, and that requires quantification over sets of level k. Thus the g.l.b. (resp., l.u.b.) of S would be of level $k+1$.

Weyl concluded that a development of analysis in such ramified terms would be unworkable; at the same time, he rejected the Axiom of Reducibility as untenable on predicative grounds. His solution was to confine himself to arithmetical sets of natural numbers and the associated sets and sequences of rational numbers. With real numbers identified not as Dedekind sections but rather as Cauchy (or fundamental) sequences $\langle r_n \rangle_{n \in N}$ of rational numbers, sequences of real numbers can be treated as double sequences $r_{k,n}$ of rational numbers. Then one can show that if $s = \langle x_k \rangle_{k \in N}$ is a Cauchy sequence of real numbers given as $x_k - \langle r_{k,n} \rangle_{n \in N}$, then its limit $t = \langle q_n \rangle_{n \in N}$ exists arithmetically definable from the double sequence $\langle r_{k,n} \rangle_{k,n \in N}$. Thus, in this form, the Cauchy completeness of the real number system is justified by the arithmetical comprehension axiom.

With the real number system secured in this way, Weyl could move on to see which parts of analysis could be redeveloped in this restricted version of predicativity, given the natural numbers, which one might call *arithmetical analysis*.[5] His main achievement was to show that substantially all of nineteenth-century analysis of piecewise continuous functions could be accounted for in this way, since any continuous function f of real numbers is completely determined by its behavior at rational arguments, and hence can be represented at the second-order level. Subsequent research beginning in the 1970s has been able to extend Weyl's program much farther into twentieth-century analysis; more details about the reach of that will be described below. What is important for this part of our story is not that Weyl brought forth a new idea about predicativity, but rather that he showed mathematically the exceptional viability of the restricted part of predicativity, given the natural numbers, that is based unramifiedly on the second-order arithmetical comprehension axiom.

[5] The first formulation of arithmetical analysis in modern logical terms was given by Grzegorczyk (1955).

Predicativity Sidelined: 1920–1950

If predicativity was to progress following these early developments, it would need a leader and practitioners. As it turned out, neither were there, for a number of reasons. Its first champion, Poincaré, died in 1912, and in any case he had not engaged in the essential clarification or implementation of his ideas. As for Russell, predicative logicism in the form of ramified type theory was compromised by the Axiom of Reducibility. Then, following the exhausting task of producing the first edition of *Principia Mathematica* (1910–1913), Russell turned to general problems in philosophy from an analytic perspective, and during World War I was drawn aside into pacifist politics. Finally, Wittgenstein's critiques of Russell's logic and philosophy struck serious blows to his confidence in his ideas.

The case of Weyl is different: despite the relative success of his arithmetically predicativist program in *Das Kontinuum*, he fell under the spell of Brouwer as he became more familiar with the work of that crusading intuitionist. In 1921 he wrote, "I now abandon my own attempt and join Brouwer."[6] He then engaged in contributing to intuitionistic ideas and their relation to practice, and championed these in the following years, but eventually he became rather pessimistic about its prospects. Years later he wrote:

> Mathematics with Brouwer gains its highest intuitive clarity. He succeeds in developing the beginnings of analysis in a natural manner, all the time preserving the contact with intuition much more closely than had been done before. It cannot be denied, however, that in advancing to higher and more general theories, the inapplicability of the simple laws of classical logic eventually results in almost unbearable awkwardness. (Weyl 1949, p. 54)

The foundational program most prominently in competition with that of Brouwer was Hilbert's proof-theoretical consistency program. By restricting to finitist methods, it was more radical than Brouwer's, but it was much more ambitious in its aim to "secure" the practice of nonconstructive mathematics via consistency proofs of appropriate formal systems. Though that was dashed by Gödel's theorem on unprovability of consistency in 1931, the program took on new life with Gentzen's consistency proof for arithmetic and was transformed by the employment of limited transfinite methods.

What really pushed predicativity to the sidelines, though, was the success of axiomatic set theory—as developed by Zermelo, Skolem, and Fraenkel—in allaying fears about the paradoxes. Though not demonstrably consistent, intensive development of the subject without running into any difficulties gave comfort and confidence to its practitioners and gradually won the support of mathematicians at

[6] Weyl's intuitionistic excursion is fully described and exemplified with relevant articles in part II of Mancosu (1998); see p. 98 for the 1921 quote.

large. Nor did the impredicativity that Poincaré and Weyl had located in the l.u.b. principle in the real numbers generate any concern. For, in mathematical practice, the real numbers are regarded as a definite, completed totality independent of human constructions and definitions, not some vague collection of numbers "growing" under successive definitions. Then the l.u.b. of a bounded subset S of R merely serves to single out a specific element of R, not to "bring it into existence." In this respect, it is like the least number operator in the set of natural numbers, which serves to single out a specific number in N without our always being able to say *which* one it is in ordinary systems of representation. (For example, in the solution of Waring's problem, given k, there is a least n such that every natural number is a sum of at most n kth powers; for all but the first few k, the specific value of n is unknown.) Mathematicians were thus insensitive to impredicativity in practice.

A story recounted in Feferman 1998 is apropos here:

> . . . a famous wager was made in Zürich in 1918 between Weyl and George Pólya, concerning the future status of the following two propositions:
> (1) Each bounded set of real numbers has a precise upper bound.
> (2) Each infinite subset of real numbers has a countable subset.
> [The latter requires the Axiom of Choice.] Weyl predicted that within twenty years either Pólya himself or a majority of leading mathematicians would admit that the concepts of number, set and countability involved in (1) and (2) are completely vague, and that it is no use asking whether these propositions are true or false, though any reasonably clear interpretation would make them false. . . . the loser was to publish the conditions of the bet and the fact that he lost in the *Jahresberichten der Deutschen Mathematiker Vereinigung*. . . . (p. 57)

The wager was never settled as such, for obvious political reasons. According to Pólya (1972), "The outcome of the bet became a subject of discussion between Weyl and me a few years after the final date, around the end of 1940. Weyl thought he was 49% right and I, 51%; but he also asked me to waive the consequences specified in the bet, and I gladly agreed." Pólya showed the wager to many friends and colleagues, and, with one exception, all thought he had won.

In axiomatic Zermelo–Fraenkel (ZF) set theory, the fundamental source of impredicativity is in the Separation Axiom scheme, which asserts for each well-formed formula $\varphi(x)$ (possibly with parameters) of the language of ZF the existence of the set $\{x: x \in a \wedge \varphi(x)\}$ for any set a. Since the formula φ may contain quantifiers ranging over the supposed "totality" of all sets, this is impredicative according to the VCP. Mathematically, this is given teeth by the assumption of the Axiom of Infinity, guaranteeing the existence of the set ω of finite ordinals (or the natural numbers) and by the assumption of the Power Set Axiom, guaranteeing for any set a the existence of $\wp(a) = \{x: x \subseteq a\}$. Without the Axiom of Infinity, the impredicativity of the Separation Axiom becomes innocuous, since the system has a model in the hereditarily finite sets. With the Axiom of Infinity, power set and separation lead, among other things, to the existence of the real number system, represented, for instance, as the set of all upper Dedekind sections in the rational numbers, and thence to the l.u.b. principle for arbitrary bounded sets of real numbers.

Though predicativity went into hibernation until the 1950s, one important technical development in axiomatic set theory would prove to be significant for it. This was Gödel's model of the constructible sets (1939). The "standard interpretation" of ZF set theory, due to Zermelo in 1930, is in the *cumulative hierarchy* $\langle V_\alpha \rangle_{\alpha \in On}$, defined by transfinite recursion on the class *On* of (finite and transfinite) ordinal numbers α by transfinite iteration of the power set operation $\wp$ starting with the empty set $\varnothing$ as follows:

$$V_0 = \varnothing, V_{\alpha+1} = \wp(V_\alpha), \text{ and for limit } \alpha, V_\alpha = \bigcup\nolimits_{\beta<\alpha} V_\beta.$$

This is a *cumulative theory of types*, unlike simple type theory, since, as may be shown by induction on α, if $\beta < \alpha$, then $V_\beta \subseteq V_\alpha$. Gödel defined the *constructible hierarchy* $\langle L_\alpha \rangle_{\alpha \in On}$ by modifying the successor step, replacing the power set operation with an operation $\wp_{Def}(a)$ which, for any set a, consists of all sets of the form $\{x\colon x \in a \wedge \varphi^a(x)\}$ where the superscript "a" indicates that all quantifiers in φ are restricted to range over a, and where all parameters in φ are elements of a. Then L_α is defined recursively by

$$L_0 = \varnothing, L_{\alpha+1} = \wp_{Def}(L_\alpha), \text{ and for limit } \alpha, L_\alpha = \bigcup\nolimits_{\beta<\alpha} L_\beta.$$

The constructible hierarchy thus stands to the cumulative hierarchy as Russell's ramified theory of types stands to the simple theory of types. In fact, the constructible hierarchy may be considered to be entirely predicative except perhaps in its free use of arbitrary ordinals. Since ordinals are the order types of well-ordered sets and those are defined impredicatively by the condition that any nonempty subset has a least element, the constructible hierarchy is not on the face of it predicative. But it may be considered to be predicative in a modified sense, relative to the notion of arbitrary well-ordering or ordinal number. An interesting restriction of that notion is the constructible hierarchy taken up to the least uncountable ordinal ω_1, which comes into the next part of our story.

Predicativity in Transition, as a Chapter of Definability Theory

The logical analysis of predicativity reemerged in the 1950s as a chapter in the extension of recursion theory to various "higher" definability notions, especially for sets of natural numbers. A systematic study of hierarchies of definitions of sets of natural numbers was undertaken in the early 1950s by Kleene and, independently, Davis and Mostowski. In Kleene's hands, this took the following form, using the "Turing jump" operator which takes one from any set X to a set X' which is universal for sets that are Σ_1^0 in X. The sets which are obtained from N by finite iteration of the jump operation are, up to relative recursiveness, all the arithmetically definable sets. These thus take for granted the definiteness of quantification over N, and in that sense N as a completed totality. Kleene (1955) defined a transfinite extension of this hierarchy, using the set O of Church–Kleene notations for constructive ordinals. The details of the latter are not important in the following. Roughly speaking, there is a notation for 0 in O, and each $n \in O$ has a notation for its successor; finally, any effectively enumerated increasing sequence in O has a notation for its limit. For $n \in O$, we write $|n|$ for the ordinal α it denotes. As an analogue of the least uncountable ordinal ω_1, the least ordinal not constructive in the sense of Church–Kleene (i.e., the least α such that there is no $n \in O$ with $|n| = \alpha$), is denoted ω_1^{CK}. It was shown by Spector (1955) that the order types of recursive well-orderings of N are exactly the ordinals less than ω_1^{CK}. Kleene's extension of the finite jump hierarchy, and thence of the arithmetical sets, is defined inductively for each $n \in O$ yielding a subset H_n of N, as follows. $H_0 = \emptyset$ and the step from H_n to its successor set is by the jump operation; finally, at a limit notation, one takes the direct sum of the sets associated with the sequence approaching it. Then a set is called *hyperarithmetic* if it is recursive in H_n for some $n \in O$. The collection of all such sets is denoted HYP, which properly extends the class of arithmetical sets. One of the main results of Kleene (1955) is that

$$HYP = \Delta_1^1$$

where Δ_1^1 is by definition $\Pi_1^1 \cap \Sigma_1^1$. That is, it consists of all sets S such that for some arithmetical predicates $A(n, X)$ and $B(n, X)$, and all $n \in N$,

$$n \in S \leftrightarrow (\forall X) A(n, X)$$

and

$$n \in S \leftrightarrow (\exists X) B(n, X)$$

The connection with predicative definitions in the sense of the *ramified analytic hierarchy* was established as follows. The basic step in that hierarchy consists in passing from a collection D of subsets of N to a new collection D^* by putting a set S in D^* just in case there is a formula $\varphi(x)$ of second-order arithmetic such that for all n,

$$n \in S \leftrightarrow (\varphi(n))^D,$$

where the superscript "D" indicates that all second-order quantifiers in φ are relativized to range over D. Then we can define the collections R_α for *arbitrary* ordinals α by

$$R_0 = \emptyset, R_{\alpha+1} = (R_\alpha)^*, \text{ and for limit } \alpha, R_\alpha = \bigcup_{\beta<\alpha} R_\beta.$$

Thus the definition of the R_α proceeds exactly like the L_α in Gödel's constructible hierarchy, except that at each stage one is confined to collections of subsets of N. Though R_α is defined for all α, it is not hard to show that we get nothing new for uncountable α; in fact there is a countable α such that $R_\alpha = R_{\alpha+1}$. Kleene (1959) proved that

$$HYP = R_{\omega_1^{CK}}.$$

However, *HYP* does not exhaust the ramified hierarchy, as follows from results of Gandy (1960) and Spector (1960).

It was tentatively proposed by Kreisel (1960) to identify the predicatively definable sets, given the natural numbers, with the members of *HYP*, essentially for the following reasons. Call an ordinal α predicative(ly definable) if it is the order type of a predicatively definable well-ordering W of the natural numbers; then a set should be considered to be predicatively definable if it belongs to R_α for some predicative α. Clearly, then, on those grounds all recursive ordinals are predicative, and so by Kleene's equation above, each member of *HYP* ought to be accepted as predicative. For the converse, the predicative ordinals should be taken to be only those generated by the following "bootstrap" condition: if α is a predicative ordinal and W is a well-ordering relation in R_α and if β is the order type of W, then β is predicative. But then the predicative ordinals do not go beyond the recursive ordinals, using the result of Spector (1955) that every hyperarithmetic well-ordering has order type less than ω_1^{CK}: for if α is recursive and the well-ordering $W \in R_\alpha$, then $W \in HYP$, and hence the order type of W is recursive. This part of the argument shows that the predicative sets do not go beyond *HYP*.

Actually, analogous considerations leading to this identification had been proposed somewhat earlier by Wang (1954), who introduced ramified formal systems Σ_α, which have as a natural model the constructible sets to level $\omega + \alpha$ (since Wang starts with the hereditarily finite sets). The ordinals regarded to be predicative are again generated by a "bootstrap" condition: if α is predicative and β is defined by a well-ordering relation expressed in Σ_α, then β is predicative. The system Σ is taken to be the union of the Σ_α for α predicative. Spector (1957) established that the predicative ordinals in this sense are exactly those less than ω_1^{CK} and the sets of natural numbers definable in Σ are exactly the *HYP* sets. In a suitable sense, the sets definable in Σ are those which have *HYP* structure on their transitive closure.

Independently of Wang, Lorenzen (1955) dealt with systems S_α of a character similar to the Σ_α but did not impose precise conditions on the ordinals to be admitted. Lorenzen's main concern there is with what parts of mathematics can be developed predicatively. Incidentally, both Lorenzen and Wang claim that the l.u.b. principle holds for sets of reals defined at any limit level α, in particular for $\alpha = \omega$. In their systems, the members of a set at any level are all at a lower level. Hence, if a bounded set X of real numbers is defined in S_α for limit α, then it lies in S_β for some $\beta < \alpha$, and then l.u.b. (X) belongs to $S_{\alpha+1}$. However, the applicability of this form of the l.u.b. principle has to be questioned, since the sets of reals one deals with in practice, even if defined arithmetically, do not have members of a level restricted in advance.

Though the considerations leading to the identification of the predicative ordinals (resp. sets of natural numbers) with the recursive ordinals (resp. hyperarithmetic sets) have a certain plausibility, they ignored one crucial point if predicativity is only to take the natural numbers for granted as a completed totality: namely, that they involve in an essential way (both from above and below) the impredicative notion of being a well-ordering relation. A step away from that would be to talk of *predicatively provable well-orderings* in a way to be explained next.

Predicative Provability in the 1960s

The idea, to begin with, is that instead of dealing with the collections R_α, one deals (as in the systems Σ_α) with a transfinite progression of formal systems of ramified analysis RA_α, using variables $X^\beta, Y^\beta, Z^\beta, \ldots$ for each $\beta \leq \alpha$. Here ordinals are not to be considered set-theoretically but rather as notations chosen from O by some natural procedure. The formal systems RA_α incorporate as axioms comprehension principles expressing closure under the appropriate ramified definitions. These can provide closure conditions only on the sets at each level; the minimal model of RA_α is given by $\langle R_\beta \rangle_{\beta \leq \alpha}$, but larger collections can also satisfy the axioms of RA_α. In particular, we can interpret the variables $X^0, Y^0, Z^0, \ldots$ as satisfying the closure conditions of *any* larger R_β.

The crucial new point (compared to Wang's systems) is that the predicative ordinals not only are those that can be *defined* by (what happen to be) well-ordering relations in the given systems, but also must previously be *proved* to be such relations. The problem is how to meet this requirement without unrestricted second-order quantification; the answer comes from the provability condition as follows. Given W, a binary recursive relation in the natural numbers, let $WO^{\beta}(W)$ express that W is a linear ordering such that every nonempty subset X^{β} of the field of W has a W-least element. Then if one proves $WO^{0}(W)$, it follows that one can "lift" the proof to establish $WO^{\beta}(W)$ for each β that comes to be accepted. In this sense, the proof of the predicatively meaningful statement $WO^{0}(W)$ can ensure all predicatively meaningful consequences $WO^{\beta}(W)$ of the impredicative statement of well-ordering $WO(W)$; indeed, from the outside it ensures that W is a well-ordering. Now Kreisel's proposal (1958, 1960) can be formulated as follows. The predicative(ly provable) ordinals are generated from 0 by the bootstrap or *autonomy* condition: if α is predicative and RA_{α} proves $WO^{0}(W)$ for a given recursive W, and β is the order type of W, then β is predicative. The least nonpredicatively provable ordinal is then proposed to be the least ordinal which cannot be obtained in that way.

Kreisel called for an independent characterization of that limit of the predicatively provable ordinals in the sense just described. That problem was solved independently by Schütte (1965a, 1965b) and me ("Systems of Predicative Analysis," 1964), in terms of the Veblen hierarchy $\langle\chi_{\alpha}\rangle_{\alpha}$ of critical functions of ordinals defined for each ordinal α by $\chi_0(\xi) = \omega^{\xi}$, and when $\alpha \neq 0$, χ_{α} enumerates in order of size the set of common fixed points ξ of all χ_{β} for each $\beta < \alpha$; that is, $\{\xi$: for all $\beta < \alpha$, $\chi_{\beta}(\xi) = \xi\}$. Then the function of ξ given by $\chi_{\xi}(0)$ is normal; let Γ_{α} be the αth fixed point ξ of the equation $\chi_{\xi}(0) = \xi$. Then what Schütte and I showed is that the least nonpredicatively provable ordinal is Γ_0. Outlines of proofs are to be found in the cited references. A full exposition can be found in Schütte's book *Proof Theory* (1977), Chapter VIII.

Predicatively Reducible Systems

Taking the intrinsic interest of predicativity given the natural numbers for granted, after determining Γ_0 to be the least nonpredicatively provable ordinal according to the above proposal, one had to return to the question of how much mathematics could be developed predicatively. This was pursued both theoretically, via alternative formal systems, and by means of case studies, to be discussed below. On the theoretical side, since (as Weyl stressed) ramified theories are unsuitable as a framework for the development of analysis, the first question was to see which unramified systems could be justified on predicative grounds. A formal system T is said to be *(locally) predicatively reducible* if it is proof-theoretically reducible to the autonomous progression of ramified systems described in the previous section.[7] If T has

[7] See Feferman (1998), p. 193 for the notion of proof-theoretic reduction.

the same proof-theoretic strength as that progression, then its proof-theoretic ordinal is Γ_0. In that case, though the system T as a whole may not be justifiable predicatively, each theorem φ of T rests on predicative grounds, at least indirectly. In practice, more can be said: T is conservative over the autonomous ramified progression for arithmetic sentences (i.e., if φ is arithmetical and provable in T, then it is provable in that progression). For second-order T this can often be strengthened to conservativity for Π^1_1 sentences (i.e., for φ of the form $(\forall X)A$, where A is arithmetical), which on the ramified side is taken to be $(\forall X^0)A$. In particular, in that case, any provable well-ordering of T is also predicatively provable.

The first two examples of unramified second-order predicatively reducible systems were given in Feferman (1964). The first of these was obtained by replacing the progression of ramified theories RA_α with a progression HC_α of unramified second-order theories, based on the Hyperarithmetic (or Δ^1_1) Comprehension Rule:

$$(\Delta^1_1\text{-CR}) \quad \text{From } (\forall x)[P(x) \leftrightarrow Q(x)] \text{ infer } (\exists X)(\forall x)[x \in X \leftrightarrow P(x)],$$

where $P(x)$ is any Π^1_1 formula and $Q(x)$ is any Σ^1_1 formula (parameters allowed).

The motivation for Δ^1_1-CR is the recognizable absoluteness (or invariance) of provably Δ^1_1 definitions, in the following sense. At each stage one has recognized certain closure conditions on the "open" universe of sets, and the definitions $D(x)$ of sets introduced at the next stage should be independent of what further closure conditions may be accepted. In the words of Poincaré, the definitions used for objects in an incomplete totality should not be "disturbed by the introduction of new elements." Thus, if U represents a universe of sets (subsets of N) satisfying given closure conditions and is extended to U' (satisfying the same closure conditions and possibly further ones), one wants D to be *provably invariant* or *absolute* in the sense that $(\forall x)[D^U(x) \leftrightarrow D^{U'}(x)]$. This requirement is easily seen to hold for provably Δ^1_1 formulas in the sense that $(\forall x)[D(x) \leftrightarrow P(x)]$ has been proved, where P and Q satisfy the hypothesis of the rule Δ^1_1-CR.[8] The progression of theories HC_α is then obtained by suitable transfinite iteration of the rule HCR. Again, one can apply the notion of autonomy to such a progression: it was shown that the least nonautonomous ordinal of this progression is Γ_0; that the union of the HC_α for $\alpha < \Gamma_0$ is of the same proof-theoretical strength as the union of the autonomous ramified progression; and that one has conservativity for Π^1_1 formulas.

The second predicatively reducible system introduced in Feferman (1964) was obtained by replacing the autonomous progression of the HC_α with a single second-order system, denoted IR. This is axiomatized by the rule Δ^1_1-CR together

[8] It has been shown in Feferman (1968) that every provably invariant formula is equivalent to a provably Δ^1_1 formula, so the rule Δ^1_1-CR is fully general for this requirement.

with what is called the Bar Rule in the proof-theoretic literature, which allows one to infer the full scheme of transfinite induction on a recursive ordering W when $WO(W)$ has been established; correspondingly, one has a rule for inferring a scheme of transfinite recursion on W under the same hypothesis. (The "I" in "IR" is for Induction, and the "R" is for Recursion.) The main result concerning IR stated in Feferman (1964) is that it proves the same theorems as the union of the HC_α for $\alpha < \Gamma_0$, and thus it is predicatively reducible with conservativity for Π^1_1 formulas.

The system of what is often referred to as second-order arithmetic has the *full* (Π^∞_1) *comprehension axiom*,

$$(\Pi^\infty_1\text{-CA}) \quad (\exists X)(\forall x)\,[x \in X \leftrightarrow \varphi],$$

where φ is an arbitrary second-order (Π^∞_1) formula which does not contain "X" free. In 1976, Harvey Friedman introduced several subsystems of second-order arithmetic whose common feature is that induction is restricted to its second-order form, the *induction axiom*,

$$(\text{IA}) \quad (\forall X)\,[(0 \in X) \wedge (\forall x)(x \in X \rightarrow x' \in X) \rightarrow (\forall x)(x \in X)].$$

In the presence of the full comprehension axiom, one can infer from the induction axiom IA the *induction scheme*, considered as the collection of all formulas of the form

$$(\text{IS}) \quad \varphi(0) \wedge (\forall x)[\varphi(x) \rightarrow \varphi(x')] \rightarrow (\forall x)\,\varphi(x)$$

for arbitrary second-order formulas φ. But in the restricted systems considered by Friedman, in which the comprehension axiom is considerably weakened, that step is not possible. Following Friedman, these systems are indicated by a subscript 0. An obvious choice for such to consider with respect to the present subject is the system ACA_0, obtained by replacing the full comprehension axiom scheme with the subscheme in which only arithmetical formulas φ (no bound second-order variables) are admitted. It is a classical result of proof theory that the system ACA_0 is a conservative extension of the first-order system of Peano Arithmetic (PA). On the other hand, the system ACA, which is obtained by use of the full induction scheme (IS) in place of (IA), is stronger than PA; it is still predicatively reducible, but its proof-theoretical ordinal is far below Γ_0.

Another predicatively reducible system introduced in Friedman (1976) is denoted ATR_0. It has a certain similarity to IR, using axioms in place of rules, as follows. In place of the Δ^1_1-CR one has the Δ^1_1-CA, that is, the axiom scheme

$$(\forall x)[P(x) \leftrightarrow Q(x)] \rightarrow (\exists X)(\forall x)[x \in X \leftrightarrow P(x)],$$

where P is Π^1_1 and Q is Σ^1_1. In place of the transfinite recursion rule of IR, one has an axiom expressing that for all well-ordering relations Z, one can construct the Turing jump hierarchy along Z starting with any set X (i.e., the relative hyperarithmetical hierarchy along Z). The transfinite induction rule in IR is not replaced by the corresponding axiom, since that would be too strong; in place of it, the system ATR_0 uses only the induction axiom (IA) for natural numbers as above. Friedman's main result announced in 1976 was that ATR_0 is predicatively reducible with proof-theoretical ordinal Γ_0 and is conservative over IR for Π^1_1 sentences; a full proof of this with further interesting results about ATR_0 is given in Friedman, McAloon, and Simpson (1982).

Various predicatively reducible systems of higher type are surveyed in Avigad and Feferman (1998), sections 8.2 and 8.3. Predicatively reducible systems of set theory related to IR were treated in Feferman (1974), and ones related to ATR_0 have been dealt with in Simpson (1982). See also Simpson (2002),"Predicativity: The Outer Limits."

The Mathematical Reach of Predicativity: Positive Developments

Having established theoretical bounds to predicativity and workable unramified predicatively reducible systems, the next questions of interest are to see what parts of "everyday" mathematics can be carried out within those bounds, and what parts are essentially impredicative; the latter question is treated in the next section. It turns out in practice that if a known mathematical result can be established predicatively, it can already be done in a system conservative over Peano Arithmetic (PA). In other words, the predicative part of everyday mathematics is *robustly predicative.* There is no theorem which can establish this; one can only depend on various case studies for confirming evidence. One avenue is that pursued in the so-called Reverse Mathematics (RM) program due to Friedman and carried on most extensively by Simpson in his book *Subsystems of Second Order Arithmetic* (1999). The primarily relevant system for the positive work on predicative mathematics in that framework is ACA_0, described above, and conservative over PA. Another relevant system is much weaker, being based only on the weak König's Lemma (i.e., the tree lemma for infinite branching binary trees; it is shown in Simpson (1999) that the associated system WKL_0 is conservative over the system PRA of Primitive Recursive Arithmetic. A second avenue in which the positive reach of predicative mathematics has been studied is via a system W (for Weyl) of flexible finite types introduced in the final part of the article "Weyl Vindicated" (Feferman 1988). The main proof-theoretical result concerning W is that it is a conservative extension of PA (Feferman and Jäger 1993, 1996). Only some points of general character concerning both of these approaches will be indicated here.

In W, types are variable; for any two types X and Y, one has existence of their Cartesian product $X \times Y$, and the type $X \xrightarrow{\sim} Y$ of all partial functions from X to Y; finally, for any type X and bounded formula φ one has the subtype $\{x \in X: \varphi(x)\}$ of X determined by φ. Starting from the type N of natural numbers, one can introduce, as usual, the type Q of rational numbers, and then, by separation from $N \xrightarrow{\sim} Q$, the type R of real numbers, considered as Cauchy sequences of rational numbers identified under the usual equivalence relation. Then (partial) functions of real numbers can be treated as members of $R \xrightarrow{\sim} R$, and various spaces of such functions (e.g., continuous, measurable, etc.) form examples for Hilbert spaces and Banach spaces. Given any such space S, the functionals from S to R are then certain members of $S \xrightarrow{\sim} R$. In practice, only separable spaces can be treated predicatively, via a given countable dense subset.

As for the program initiated by Weyl in 1918, the full classical analysis of continuous functions can be carried out directly in W. Turning to more general classes of functions from twentieth-century analysis, such as come out of the Lebesgue theory of measure, one first notes that the general notion of outer measure can't be defined in W, since it makes essential use of the g.l.b. as applied to sets of reals, not sequences of reals. But the notion of measurable set can be treated via sequences of open covers whose measure is directly defined, and the notion of measurable function can be defined in terms of that or in terms of sequences of approximating step functions. One cannot prove the existence of Lebesgue nonmeasurable sets of reals; put in other terms, it is consistent with W that all sets of reals are Lebesgue measurable. Various parts of standard functional analysis have been verified in unpublished notes: the Riesz representation theorem, the Hahn–Banach theorem, the uniform boundedness theorem, and the open mapping theorem for separable Hilbert and Banach spaces. Finally, one can obtain the principal results of the spectral theory of bounded as well as unbounded self-adjoint linear operators on a separable Hilbert space. Subsequent to this work, it has been shown in the dissertation of Feng Ye (1999) that much of this work can already be carried out in a constructive subsystem of W, conservative over PRA.

By comparison, in the RM program, sharper results have been obtained, of the form that over a weak base system RCA_0, various of these results in analysis are actually equivalent to ACA_0 (if not already to WKL_0); these are detailed in Simpson (1999). However, since the systems considered in the RM program are all subsystems of second-order arithmetic, there is a cost involved, namely, that higher-type notions, such as those of functions of real numbers, function spaces, and functionals, cannot be dealt with directly but must somehow be coded in second-order terms. If one is not concerned with obtaining exact equivalences between mathematical results, and the second-order set and function existence principles on which they ultimately rest, the positive predicative development of analysis is carried out much more naturally in systems like W.

The Outer Mathematical Bounds of Predicativity

Turning now to mathematical results which cannot be carried out predicatively, the easiest proofs of independence are those which can be established by models in the hyperarithmetic sets. For example, independence of the existence of non-measurable sets of reals stems from the fact that every Δ^1_1 set of reals has *HYP* Lebesgue measure. Another example using *HYP*, was found by Kreisel (1959), is the impredicativity of the Cantor–Bendixson theorem, which asserts that every closed set of reals is the union of a perfect set and a countable (scattered) set. The *HYP* model can also be used to obtain examples of impredicative theorems in algebra. For example, in the Ulm structure theory of countable Abelian *p*-groups *G*, one starts by dividing out *G* by its largest divisible subgroup *H*, to yield a reduced group G/H. An example is given in Feferman (1975) of a recursive Abelian *p*-group *G* for which the union *H* of its *HYP* divisible subgroups is not *HYP*; thus the existence of a largest divisible subgroup of any countable Abelian *p*-group is not predicatively provable.

For independence results closer to the bounds of predicativity, one must fall back on proof-theoretic results. For example, any theorem equivalent to ATR_0 is at the exact limit Γ_0 of predicativity, and is thus impredicative. Simpson (2002) gives a number of examples of theorems from descriptive set theory that are equivalent to ATR_0 over RCA_0, including the following: (i) every disjoint pair of analytic sets can be separated by a Borel set; (ii) every uncountable closed (or analytic) set contains a perfect subset; (iii) clopen (or open) determinacy. Also, ATR_0 is equivalent to (iv) comparability of countable well-orderings, and (v) Ulm theory for countable reduced Abelian *p*-groups. Other independence results have come from combinatorics rather than analysis, set theory, or algebra. They are of interest because the statements are arithmetical, in fact, in Π^0_2 form. One group of these has to do with variants of the Ramsey coloring theorem, such as the Paris–Harrington (1977) version independent of PA. Friedman, McAloon, and Simpson (1982) gave a mathematically natural finite combinatorial theorem which is equivalent to ATR_0 over RCA_0. Friedman showed that certain simple Π^0_2 consequences of Kruskal's theorem about embeddings of finite trees are not even provable in ATR_0 (cf. Simpson 2002 for references and further developments).

PREDICATIVITY AND THE INDISPENSABILITY ARGUMENTS

The idea that natural science justifies a part of mathematics because of its indispensability to science—and that that is the *only* part of mathematics that is justified—is due to Willard Quine and Hilary Putnam, among others. Famously, Quine has written:

> So much of mathematics as is wanted for use in empirical science is for me on a par with the rest of science. Transfinite ramifications are on the same footing insofar as they come of a simplificatory rounding out, but anything further is on a par with uninterpreted systems. (Quine 1984, p. 788)

(See chapters 12 and 13 above.) Quine argued that we need the power set operation in set theory to establish the existence of the real numbers R and then its application once more to obtain such sets as that of all functions from R to R. That led him by a "simplificatory rounding out" to acceptance of finite iterations of the power set, but not its ωth iteration over the natural numbers. In other words, Quine was led thus to accept Zermelo set theory but nothing stronger, which he looked upon as "mathematical recreation and without ontological rights" (Quine 1986, p. 400). Penelope Maddy has characterized the indispensability arguments (for critical purposes) as follows:

> We have good reason to believe our best scientific theories, and mathematical entities are indispensable to those theories, so we have good reason to believe in mathematical entities. Mathematics is thus on an ontological par with natural science. Furthermore, the evidence that confirms scientific theories also confirms the required mathematics, so [that part of] mathematics and science are on an epistemological par as well. (Maddy 1992, p. 278)

In (Feferman 1993, reprinted 1998) I considered the significance for the indispensability arguments of the positive developments of parts of mathematics by predicative means described above in the following terms (1998, p. 284): "If one accepts the indispensability arguments [of course one might argue against them on philosophical grounds] there remain two critical questions:

> Q1. Just which mathematical entities are indispensable to current scientific theories?
> Q2. Just what principles concerning those entities are needed for the required mathematics?"

The positive developments described above are brought to bear on these questions as follows. On their basis, I had formulated *the working hypothesis that all of scientifically applicable analysis can be developed in the system* W, and argued that this has been verified in its core parts (cf. 1998, pp. 280–283, 293–294). Of course, there are results of theoretical analysis which cannot be carried out predicatively, either because they are essentially impredicative in their very formulation or because they are independent of predicative systems such as the examples given above. However, none of those affects the working hypothesis because they do not figure in the applicable mathematics. What is more to the point are examples closer to the margin scientifically (e.g., the proposed use of certain nonseparable spaces for a quantum-mechanical model involving infinitely many degrees of freedom in Emch [1972, p. 103]; contra that, one can appeal to the arguments on physical grounds by Streater and Wightman [1978, p. 87]).[9] Of course the working hypothesis may yet prove to be wrong by other examples, but as of now, all evidence is in its favor. If one accepts it, one can return to questions Q1 and Q2 as follows:

> By the fact of the proof-theoretical reduction of W to PA, the only ontology it commits one to is that which justifies acceptance of PA. But even there, the answer to Q1 and thence to Q2 is underdetermined. One view of PA is that it is about the natural numbers as independently existing abstract objects; that is . . . a platonistic view, albeit an extremely moderate one. . . . Or one can make use of the fact that PA is reducible to HA [Heyting Arithmetic] to justify it on the basis of a more constructive ontology. (Feferman 1998, p. 296)

From this and related arguments, I drew the conclusion that

> . . . if one accepts the indispensability arguments, practically nothing philosophically definitive can be said of the entities which are then supposed to have the same status—ontologically and epistemologically—as the entities of natural science. That being the case, what do the indispensability arguments amount to? As far as I'm concerned, they are completely vitiated. (ibid., p. 297)

Rethinking Predicativity II: 1970–1996

In Kreisel's "Principles of Proof and Ordinals Implicit in Given Concepts" (1970), he criticized existing proof theory for "*the lack of a clear and convincing analysis of the choice of methods of proof*," and took as his ultimate aim "*the discovery of objective criteria for such a choice*" (italics in the original). "What one is after is a (phenomenological) description of certain kinds **P** of mathematical reasoning; the *objective question* is then simply this: whether the proofs represented . . . by derivations of a given formal system **F** are in **P** (soundness of F); [and] whether all proofs in **P** are represented in **F** (completeness with respect to . . . provability in **P**)." The particular kinds of reasoning **P** considered with respect to this aim were

[9] See Feferman (1998), p. 282 for a full discussion.

described as an answer to the following: "What principles of proof do we recognize as valid once we have understood (or, as one sometimes says, "accepted") certain given concepts?" (Kreisel 1970, p. 489). As a further elaboration, "[t]he process of recognizing the validity of such principles (including the principles for *defining* new concepts, that is, formally, of extending a given language) is here conceived as a *process of reflection*.... Granted that we have to do with an area **P** which lends itself to the kind of analysis indicated, it is evident that *ordinals* play a basic role. They index the stages in the reflection process." (ibid.)

The two principal basic concepts considered by Kreisel (p. 490) are, in his terminology:

1. The concepts of ω-sequence and ω-iteration
2. The concepts of set of natural numbers and numerical quantification.

Kreisel pointed out that this is related to earlier work on autonomous progressions for *finitist mathematics* (in Kreisel 1958, 1965) and for *predicative mathematics* (in Feferman 1964).

This rethinking of predicativity and, relatedly, of finitism was persuasive to me except for the idea that one would expect the stages of reflection to be indexed by (possibly) transfinite ordinals, though such ordinals might well be used metamathematically in an evaluation of the proof-theoretical strength of the system **F** proposed to represent **P**. In my view, the **F** considered for a given **P** should not be taken to involve the notions of ordinal or well-ordering in any way that is not already contained in the basic concepts of **P**. The formulation of this went through several stages, marked by, among others, the publications Feferman (1979, 1991), arriving most recently at the general notion of unfolding, as first explained in "Gödel's program for new axioms: why, where, how, and what?" (Feferman 1996). That also pointed to the possible applicability of the notion to systems of set theory at the other extreme to finitism, in which certain axioms for "large cardinals" would be derived that one could argue would fulfill Gödel's view that the familiar systems such as ZFC "can be supplemented without arbitrariness by new axioms which are only the natural continuation of those set up so far" (Gödel 1947, p. 520). The idea of unfolding is outlined next.

Predicativity as Unfolding

It is of the essence of the notion of unfolding that we are dealing with schematically presented formal systems.[10] In the usual conception, *formal schemata* for

[10] This section is extracted from Feferman (1996).

axioms and rules of inference employ *free predicate variables* P, Q, ... of various numbers of arguments $n \geq 0$. An appropriate substitution for $P(x_1, \ldots, x_n)$ in such a scheme is a formula $\varphi(x_1, \ldots x_n)$, possibly with additional parameters. Familiar examples of *axiom schemata* in the propositional and predicate calculus are

$$\neg P \rightarrow (P \rightarrow Q) \quad \text{and} \quad (\forall x)P(x) \rightarrow P(t).$$

The induction axiom scheme in nonfinitist arithmetic is given by

$$P(0) \wedge (\forall x)[P(x) \rightarrow P(x')] \rightarrow (\forall x)P(x).$$

In set theory, familiar axiom schemes which can be represented similarly are those for the Separation and Replacement axioms.

Also, rules of inference may be represented schematically, such as modus ponens in the propositional calculus and universal generalization in the predicate calculus, given respectively by

$$P,\ P \rightarrow Q\,/\,Q \quad \text{and} \quad [P \rightarrow Q(x)]\ /\ [P \rightarrow (\forall x)Q(x)].$$

In finitist arithmetic, in which quantification over the natural numbers is not accepted, the basic principle of induction is given by the rule:

$$P(0),\ P(x) \rightarrow P(x')\,/\,P(x).$$

The informal philosophy behind the use of schemata in the concept of unfolding is their *open-endedness*. That is, they are not conceived of as applying to a specific language whose stock of basic symbols is fixed in advance, but rather as applicable to *any* language which one comes to recognize as embodying meaningful basic notions. Put in other terms, *implicit in the acceptance of given schemata is the acceptance of any meaningful substitution instances*. But *which ones* those substitution instances are, need not be determined in advance. Thus, for example, if one accepts the axioms and rules of the classical propositional calculus given in schematic form, one will accept all substitution instances of these schemata in any language which one comes to employ.

The question which the notion of unfolding is supposed to address is: *given a schematic system* S, *which operations and predicates—and which principles concerning them—ought to be accepted if one has accepted* S? The answer for operations is straightforward: *any operation from and to individuals is accepted which is determined explicitly or implicitly from the basic operations of* S. Moreover, the *principles* which are added concerning these operations are just those which are derived from the way that they are introduced. Ordinarily, we would confine ourselves to the *total operations* obtained in this way (i.e., those which have been proved to be defined for all values of their arguments), but it should not be excluded that the introduction might depend on prior *partial operations* (e.g., those introduced by recursive definitions of a general form). The question concerning predicates in the unfolding of S is treated in operational terms as well; that is, *which operations on and to predicates—and which principles concerning them—ought to be accepted if one has accepted* S? For this, it is necessary to tell at the outset *which logical operations on predicates are taken for granted in* S. For example, in the case of nonfinitist classical arithmetic, these would be (say) the operations $\neg$, $\wedge$, and $\forall$, while in the case of finitist arithmetic we would be limited to positive propositional connectives and (in one formulation) the $\exists$ operator.

As a general background theory to unfolding for arbitrary S, one assumes a kind of protomathematical theory of operations and predicates, which makes only those assumptions that appear in every mathematical theory. The theory of operations can be typed or untyped; the latter, which is formally simpler, is taken to be a form of partial combinatory algebra with pairing and projection operations. These provide for closure of both operations and predicates under explicit definition. In addition, the combinatory setup allows one to construct a *generalized recursion* or *fixed point operator* r, satisfying

$$r(f) = f(rf)$$

for any f. In other words, we allow implicit definitions of operations g via

$$g = f(g)$$

for any given f. One general (logic-independent) operation on predicates is given a distinguished role, since we consider the case when it is not used. Given an operation f on a domain M to n-ary predicates over M, $fx = P_x$ for each $x \in M$, form the *join* predicate $J(f) = P$, where

$$P(x, x_1, \ldots, x_n) \leftrightarrow P_x(x_1, \ldots, x_n).$$

The *operational unfolding* of a schematic system S, in symbols $U_0(S)$, makes use only of the background theory of operations over the given operations of S (i.e., it does not make use of any operations on or to predicates. The *full (operational and predicate) unfolding of* S, in symbols U(S), also admits the background theory of operations on and to predicates over the given logical operations of S, including the join operator J. The *intermediate (operational and predicate) unfolding of* S, in symbols $U_1(S)$, is the same without the join operator. Systems are unfolded by establishing the definedness and uniqueness of more and more operations of the various kinds. These serve to expand the language, and thence the S formulas φ which can be admitted to the

(Substitution Rule) $A(P)/\ A(\hat{x}.\varphi(x))$.

As an example, the schematic system NFA of nonfinitist arithmetic is given with basic operations on individuals of successor *Sc*, predecessor *Pd*, and the 0-ary operation (constant) 0, and a schematic predicate symbol $P(x)$. In the intermediate and full unfoldings, the basic logical operations assumed are $\neg$, $\wedge$, and $\forall$. The basic axioms of NFA are simply the following three, where we write x' for $Sc(x)$:

Ax 1. $x' \neq 0$.
Ax 2. $Pd(x') = x$.
Ax 3. $P(0) \wedge (\forall x)[P(x) \rightarrow P(x')] \rightarrow (\forall x)\ P(x)$.

In my paper with Thomas Strahm, "The Unfolding of Non-finitist Arithmetic" (2000), the following theorem is proved, where $\equiv$ is the relation of proof-theoretical equivalence, PA is the usual system of Peano Arithmetic, and $\mathrm{RA}_{<\alpha}$ is the union of the systems RA_β of ramified analysis to level β, for $\beta < \alpha$.

Theorem 1.
(i) $\mathrm{U}_0(\mathrm{NFA}) \equiv \mathrm{PA}$
(ii) $\mathrm{U}_1(\mathrm{NFA}) \equiv \mathrm{RA}_{<\omega}$
(iii) $\mathrm{U}(\mathrm{NFA}) \equiv \mathrm{RA}_{<\Gamma_0}$.

In other words, the full operational and predicate unfolding of NFA is proof-theoretically equivalent to predicative analysis as characterized via the autonomous progression of ramified theories. The first step in getting PA contained in $\mathrm{U}_0(\mathrm{NFA})$ is to establish successively the definedness of all primitive recursive functions.[11]

In an abstract with Strahm, "The Unfolding of Finitist Arithmetic" (2001) we announced the following result for a system FA of Finitist Arithmetic with the restricted operations on predicates indicated above, and with the induction axiom replaced by the quantifier-free induction rule. Here we obtained, in terms of the system PRA of Primitive Recursive Arithmetic:

Theorem 2. $\mathrm{U}_0(\mathrm{FA}) \equiv \mathrm{U}_1(\mathrm{FA}) \equiv \mathrm{U}(\mathrm{FA}) \equiv \mathrm{PRA}$.[12]

[11] Actually, in Feferman and Strahm (2000) we made use of a background theory of typed operations with general least fixed point operator, but we have also verified the results as stated here for the untyped background theory.

[12] Preparation of a full presentation of the system FA and the proof of theorem 2 is in progress.

This is in accord with Tait's argument (1981) that finitism is formally represented by PRA.

Kreisel's work on finitism in terms of certain autonomous progressions (1958, 1965) led to a system whose proof-theoretical strength is Peano Arithmetic (PA). It should be possible to expand the system FA to a system FA* whose unfolding is exactly PA in strength. The guess is that this would be obtained by adding a suitable form of the Bar Rule in the language of FA, informally expressed as follows: if W is a decidable ordering and it has been proved (with free variable f) that f is not descending in W, then one can apply transfinite induction to W.

CONJECTURE. $U(FA^*) \equiv PA$.

If this is right, what gives Kreisel's characterization of finitism its unexpected strength is his implicit use of a notion of finitist well-ordering.

What Is Predicativity? Summary

The idea of predicativity started in a negative frame of mind with the identification of a vicious circle in the use of definitions purported to single out an object D from a supposed totality V by essential reference to the entirety of V. Such a definition was considered to be problematic when V is in some sense essentially ill-defined, not "actual" or "complete." To begin with, the focus was on the supposed totalities of all sets or all ordinal or cardinal numbers in the framework of Cantorian set theory. The radical diagnosis of Kronecker, Poincaré, and Brouwer saw a threat to mathematics in *all* assumptions of the actual infinite, even down to the natural numbers. Hilbert, too, in his consistency program to secure infinitistic mathematics, was infected in this way and, in the words of Paul Bernays, "It became his goal to do battle with Kronecker with his own weapon of finiteness" by way of inoculation.

Less radical at first, Weyl accepted the set N of natural numbers as a completed totality, but not the real numbers or, what comes to the same thing foundationally, the totality of all subsets of N. Thus was born the concept of predicativity given the natural numbers, and the investigation of its reach in practice. That there is a fundamental difference between our understanding of the concept of natural numbers and our understanding of the set concept, even for sets of natural numbers, is undeniable. The study of predicativity given the natural numbers is thus of special foundational significance, and is the one that has received the most detailed study, as described above. This is *not* to say that *only* what is predicative in that sense is justified. What we are dealing with here are *questions of relative conceptual clarity and foundational status.* Parallel to the development of predicative mathematics was the pursuit of finitist mathematics in the hands, to begin with, of Skolem (1923) and, much later, Goodstein (1957). Both were overshadowed in the foundational wars by the intuitionists, the most radical of the radicals.

With the rise in the 1950s of metamathematics as a substantial discipline and broadly applicable tool, these directions became the subject of logical analysis at the hands of philosophically motivated but not necessarily ideologically committed logicians such as Kleene and Kreisel. Kleene was concerned, in effect, with what *definition processes* (recursive, hyperarithmetic, etc.) were implicit in accepting given notions. Kreisel shifted the attention to what *proof processes* were implicit in these. The notion of unfolding of schematic formal systems evolved from his 1970 paper "Principles of Proofs and Ordinals Implicit in Given Concepts." The aim was to put that in more general form by not assuming the notion of ordinal as part of the basic description. The first results (theorems 1 and 2 above) are gratifying in this respect: the full unfolding of the system NFA of nonfinitist arithmetic is equivalent to the characterization in terms of the autonomous progression of ramified systems of predicativity, given the natural numbers, while that of the system FA of finitist arithmetic is equivalent to primitive recursive arithmetic.

Are those the only two senses of predicativity to be considered? In one direction, stronger notions of predicativity than that, given the natural numbers, were suggested by work in the late 1950s of Paul Lorenzen, John Myhill and Hao Wang. In particular, Lorenzen and Myhill (1958) involves the acceptance in a certain constructive form of the countable ordinals and of inductive proof and definition on them. In metamathematical terms, their principles would legitimate a system of strength at least that of ID_1, the theory of one generalized inductive definition. In quite the opposite direction, Edward Nelson published a monograph, *Predicative Arithmetic* (1986), which perhaps more properly should have been titled *Strictly Finitist Arithmetic*, whose admissible principles place it within the systems of "feasible" or "poly-time" arithmetic that have been developed by Sam Buss and others, systems that form a very weak fragment of primitive recursive arithmetic (cf. Buss 1986). And somewhere between these are papers by Geoffrey Hellman and myself, "Predicative Foundations of Arithmetic" (1995) and "Challenges to Predicative Foundations of Arithmetic" (2000), where we argue in particular for a concept of predicativity given the notion of finite set, and in general for predicativity as a *relative* rather than an *absolute* concept. In addition, concepts of predicativity, given the notion of finitist ordinal, and even of predicativity, given the notion of the cumulative hierarchy of sets, have been indicated above. These open up a series of more or less specific problems to be tackled in terms of the concept of unfolding. To begin with, these problems call for the formulation of basic schematic systems as simple and natural as NFA and FA for the weaker or stronger notions indicated in the various mentioned developments, and then the determination of the reach of the corresponding systems of unfolding.

So if one accepts this more general standpoint, one answer to the question "What is predicativity?" is that it is a concept applicable to different foundational stances given by the rejection of the actual infinite for various domains, coupled with its possible limited acceptance for others. Then the logical problem in each case is to characterize exactly the limits of that particular stance. To the extent that such stances are restrictive, the potential positive value of this enterprise for mathematics is—to quote another Kreiselian slogan—to tell us what more we know if we know that something has been proved by limited means. And for philosophy, its value is to provide sharp, informative explications in terms of which arguments can more pointedly be mounted in favor of, or opposed to, one or another of those foundational stances.

REFERENCES

Avigad, J., and S. Feferman (1998), Gödel's functional ("Dialectica") interpretation, in *Handbook of Proof Theory* (S. Buss, ed.), Elsevier (Amsterdam), 337–405.

Buss, S. (1986), *Bounded Arithmetic*, Studies in Proof Theory Lecture Notes, Bibliopolis (Naples).

Chihara, C. (1973), *Ontology and the Vicious Circle Principle*, Cornell University Press (Ithaca, NY).

de Rouilhan, P. (1996), *Russell et le circle des paradoxes*, Presses Universitaires de France (Paris).

Emch, G.G. (1972), *Algebraic Methods in Statistical Mechanics and Quantum Field Theory*, Wiley (New York).

Feferman, S. (1964), Systems of predicative analysis, *Journal of Symbolic Logic* 29, 1–30.

——— (1968), Persistent and invariant formulas for outer extensions, *Compositio Mathematica* 20, 29–52.

——— (1974), Predicatively reducible systems of set theory, in *Axiomatic Set Theory*, part 2 (D. S. Scott and T. Jech, eds.), *Proceedings of Symposia in Pure Mathematics* XIII, American Mathematical Society (Providence, RI), 11–32.

——— (1975), Impredicativity of the existence of the largest divisible subgroup of an abelian p-group, in *Model Theory and Algebra. A Memorial Tribute to Abraham Robinson* (D. H. Saracino and V. B. Weispfenning, eds.), Lecture Notes in Mathematics 498, Springer-Verlag (Berlin), 117–130.

——— (1979), A more perspicuous system for predicativity, in *Konstruktionen vs. Positionen*, vol. II (K. Lorenz, ed.), Walter de Gruyter (Berlin), 68–93.

——— (1988), Weyl vindicated. *Das Kontinuum* 70 years later, in *Temi et prospettive della logica e della filosofia della scienza contemporanee*, vol. I, CLUEB (Bologna). Reprinted with a postscript in Feferman (1998), 249–283.

——— (1991), Reflecting on incompleteness, *Journal of Symbolic Logic* 56, 1–49.

——— (1993), Why a little bit goes a long way. Logical foundations of scientifically applicable mathematics, in *PSA 1992*, vol. 2, Philosophy of Science Association (East Lansing, MI), 442–455. Reprinted in Feferman (1998), 284–298.

——— (1996), Gödel's program for new axioms. Why, where, how and what?, in *Gödel '96* (P. Hajek, ed.), Lecture Notes in Logic 6, Association for Symbolic Logic, Springer-Verlag (New York), 3–22.

——— (1998), *In the Light of Logic*, Oxford University Press (New York).

Feferman, S., and G. Hellman (1995), Predicative foundations of arithmetic, *Journal of Philosophical Logic* 24, 1–17.

——— (2000), Challenges to predicative foundations of arithmetic, in *Between Logic and Intuition. Essays in Honor of Charles Parsons* (G. Sher and R. Tieszen, eds.), Cambridge University Press (Cambridge), 317–338.

Feferman, S., and G. Jäger (1993), Systems of explicit mathematics with non-constructive μ-operator, Part I, *Annals of Pure and Applied Logic* 65, 243–263.

——— (1996), Systems of explicit mathematics with non-constructive μ-operator, Part II, *Annals of Pure and Applied Logic* 79, 37–52.

Feferman, S., and T. Strahm (2000), The unfolding of non-finitist arithmetic, *Annals of Pure and Applied Logic* 104, 75–96.

——— (2001), The unfolding of finitist arithmetic (abstract), *Bulletin of Symbolic Logic* 7, 111–112.

Friedman, H. (1976), Systems of second order arithmetic with restricted induction, I, II (abstracts), *Journal of Symbolic Logic* 41, 557–559.

Friedman, H., K. McAloon, and S. Simpson (1982), A finite combinatorial principle which is equivalent to the 1-consistency of predicative arithmetic, in *Patras Logic Symposion* (G. Metakides, ed.), North-Holland (Amsterdam), 197–230.

Gandy, R.O. (1960), Proof of Mostowski's conjecture, *Bulletin de l'Académie Polonaise des Sciences*, série des sciences mathématiques, astronomiques et physiques 8, 571–575.

Gödel, K. (1939), Consistency proof for the generalized continuum hypothesis, *Proceedings of the National Academy of Sciences of the United States of America* 25, 220–224. Reprinted in Gödel (1990), 28–32.

——— (1944), Russell's mathematical logic, in *The Philosophy of Bertrand Russell* (P. A. Schilpp, ed.), Library of Living Philosophers 5, Northwestern University Press (Evanston, IL), 123–153. Reprinted in Gödel (1990), 119–141.

——— (1947), What is Cantor's continuum problem?, *American Mathematical Monthly* 54, 515–525; errata, 55, 151. Reprinted in Gödel (1990), 176–187.

——— (1990), *Collected Works*, vol. II, *Publications 1938–1974* (S. Feferman et al., eds.), Oxford University Press (New York).

Goldfarb, W. (1988), Poincaré against the logicists, in *History and Philosophy of Modern Mathematics*, (W. Aspray and P. Kitcher, eds.), Minnesota Studies in the Philosophy of Science 11, University of Minnesota Press (Minneapolis), 61–81.

Goodstein, R.L. (1957), *Recursive Number Theory*, North-Holland (Amsterdam).

Grzegorczyk, A. (1955), Elementary definable analysis, *Fundamenta Mathematicae* 41, 311–338.

Hallett, M. (1990), Review of Goldfarb (1988), *Journal of Symbolic Logic* 55, 1315–1319.

Heinzmann, G. (ed.) (1986), *Poincaré, Russell, Zermelo et Peano. Textes de la discussion (1906–1912) sur les fondements des mathématiques: Des antinomies à la prédicativité*, Albert Blanchard (Paris).

Kleene, S.C. (1955), Hierarchies of number-theoretic predicates, *Bulletin of the American Mathematical Society* 61, 193–213.

——— (1959), Quantification of number-theoretic functions, *Compositio Mathematica* 14, 23–40.

Kreisel, G. (1958), Ordinal logics and the characterization of informal concepts of proof, *Proceedings of the International Congress of Mathematicians, 14–21 August 1958*, Cambridge, Cambridge University Press, 1960, 289–299.

——— (1959), Analysis of the Cantor–Bendixson theorem by means of the analytic hierarchy, *Bulletin de l'Académie Polonaise des Sciences* 10, 621–626.

——— (1960), La predicativité, *Bulletin de la Société Mathématique de France* 88, 371–391.

——— (1965), Mathematical logic, in *Lectures on Modern Mathematics III* (T. L., Saaty, ed.), Wiley (New York), 95–195.

——— (1970), Principles of proof and ordinals implicit in given concepts, in *Intuitionism and Proof Theory* (A. Kino et al., eds.), North-Holland (Amsterdam).

Lorenzen, P. (1955), *Einführung in die operative Logik und Mathematik*, Springer-Verlag (Berlin).

Lorenzen, P., and J. Myhill (1959), Constructive definition of certain analytic sets of numbers, *Journal of Symbolic Logic* 24, 37–49.

Maddy, P. (1992), Indispensability and practice, *Journal of Philosophy* 89, 275–289.

Mancosu, P. (1998), *From Brouwer to Hilbert. The Debate on the Foundations of Mathematics in the 1920s*, Oxford University Press (New York).

Myhill, J. (1974), The undefinability of the set of natural numbers in the ramified *Principia*, in *Bertrand Russell's Philosophy* (G. Nakhnikian, ed.), Duckworth (London), 19–27.

Nelson, E. (1986), *Predicative Arithmetic*, Princeton University Press (Princeton, NJ).

Paris, J., and L. Harrington (1977), A mathematical incompleteness in Peano Arithmetic, in *Handbook of Mathematical Logic* (J. Barwise, ed.), North-Holland (Amsterdam), 1133–1142.

Poincaré, H. (1906), Les mathématiques et la logique, *Revue de métaphysique et de morale* 14, 294–317. Reprinted in Heinzmann (1986), 11–53.

Pólya, G. (1972), Eine Erinnerung an Hermann Weyl, *Mathematische Zeitschrift* 126, 296–298.

Quine, W. V. (1984), Review of Charles Parsons' *Mathematics in Philosophy, Journal of Philosophy* 81, 783–794.

——— (1986), Reply to Charles Parsons, in *The Philosophy of W. V. Quine*, vol. I (L. H. Hahn and P. A. Schilpp, eds.), Library of Living Philosophers 18, Open Court (Lasalle, IL), 396–403.

Richard, J. (1905), Les principes des mathématiques et le problème des ensembles, *Revue générale des sciences pures et appliquées* 16, 541. English translation in van Heijenoort (1967), 142–144.

Russell, B. (1906a), On some difficulties in the theory of transfinite numbers and order types, *Proceedings of the London Mathematical Society* 4, 29–53. Reprinted in Russell (1973), 135–164.

——— (1906b), Les paradoxes de la logique, *Revue de métaphysique et de morale* 14, 627–650. Reprinted in Heinzmann (1986), 121–144; English version under the title "On "insolubilia" and their solution by mathematical logic" in Russell (1973), 190–214.

——— (1908), Mathematical logic as based on the theory of types, *American Journal of Mathematics* 30, 222–262. Reprinted in van Heijenoort (1967), 150–182.

——— (1973), *Essays in Analysis* (D. Lackey, ed.), George Braziller (New York).

Schütte, K. (1965a), Predicative well-orderings, in *Formal Systems and Recursive Functions* (J. Crossley and M. Dummett, eds.), North-Holland (Amsterdam), 279–302.

——— (1965b), Eine Grenze für die Beweisbarkeit der transfiniten Induktion in der verzweigten Typenlogik, *Archiv für mathematischen Logik und Grundlagenforschung* 7, 45–60.

——— (1977), *Proof Theory*, Springer-Verlag (Berlin).

Simpson, S. (1982), Set theoretic aspects of ATR_0, in *Logic Colloquium '80* (D. van Dalen et al., eds.), North-Holland (Amsterdam), 255–271.

——— (1999), *Subsystems of Second Order Arithmetic*, Perspectives in Mathematical Logic, Springer-Verlag (Berlin).

——— (2002), Predicativity: The outer limits, in *Reflections on the Foundations of Mathematics. Essays in Honor of Solomon Feferman*, (W. Sieg, R. Sommer, and C. Talcott, eds.), Lecture Notes in Logic 15. Association for Symbolic Logic, (Natick, MA).

Skolem, T. (1923), Begründung der elementaren Arithmetik durch die rekurrierende Denkweise ohne Anwendung scheinbarer Veränderlichen mit unendlichem Ausdehnungsbereich, *Skrifter utgit av Videnskapsselskapet i Kristiania, I. Matematisk-naturvidenskabelig klasse* no. 6, 1–38. English translation in van Heijenoort (1967) 302–333.

Spector, C. (1955), Recursive well-orderings, *Journal of Symbolic Logic* 20, 151–163.

——— (1957), Recursive ordinals and predicative set theory, in *Summaries of Talks Presented at the Summer Institute for Symbolic Logic, Cornell University 1957*, facsimile in 1968 by microfilm-xerography, University Microfilms (Ann Arbor, MI), 377–382.

——— (1960), Hyperarithmetical quantifiers, *Fundamenta mathematicae* 48, 313–320.

Streater, R.F., and A.S. Wightman (1978), *PCT, Spin and Statistics, and All That*, 2nd ed., Benjamin/Cummings (New York).

Tait, W. (1981), Finitism, *Journal of Philosophy* 78, 524–546.

van Heijenoort, J. (*ed.*) (1967), *From Frege to Gödel. a Source Book in Mathematical Logic 1879–1931*, Harvard University Press (Cambridge, MA).

Wang, H. (1954), The formalization of mathematics, *Journal of Symbolic Logic* 19, 241–266.

Weyl, H. (1918), *Das Kontinuum. Kritischen Untersuchungen über die Grundlagen der Analysis*, Veit (Leipzig).

——— (1949), *Philosophy of Mathematics and Natural Science*, Princeton University Press (Princeton, NJ).

——— (1987), *The Continuum: A Critical Examination of the Foundation of Analysis*, English trans. of Weyl 1918 by S. Pollard and T. Bole, Thomas Jefferson Press (Kirksville, MO).

Whitehead, A.N., and B. Russell (1925), *Principia Mathematica*, vol. I, 2nd ed. Cambridge University Press (Cambridge).

Ye, F. (1999), *Strict Constructivism and the Philosophy of Mathematics*, Ph.D. dissertation, Dept. of Philosophy, Princeton University.

Zermelo, E. (1908), Neuer Beweis für die Möglichkeit einer Wohlordnung, *Mathematische Annalen* 65, 107–128. English translation in van Heijenoort (1967), 183–198.

CHAPTER 20

MATHEMATICS—APPLICATION AND APPLICABILITY

MARK STEINER

I.

AT the beginning of a book titled *Symplectic Techniques in Physics* (Guillemin and Sternberg 1990a), the authors (both mathematicians) state: "Not enough has been written about the philosophical problems involved in the application of mathematics, and particularly of group theory, to physics."[1] While I applaud this statement, and mean this chapter to address their concerns, if only partially (hoping to avoid replacing their "not enough" with "too much"), I must point out that the disregard by the philosophical community of issues of mathematical application is quite recent.

To an unappreciated degree, the history of Western philosophy is the history of attempts to understand why mathematics is applicable to Nature, despite apparently good reasons to believe that it should not be. A cursory look at the great books of philosophy bears this out.

[1] I would like to thank Professor Sternberg for valuable discussions concerning the role of symmetry in physics; Stanley Ocken for his insights into advanced aspects of elementary mathematics; and Mark Colyvan for his detailed suggestions. This research was supported by the Israel Science Foundation (Grant no. 949102), and I am very grateful for the support.

Plato's *Republic* invokes the theory of "participation" to explain why, for instance, geometry is applicable to ballistics and the practice of war, despite the Theory of Forms, which places mathematical entities in a different (higher) realm of being than that of empirical Nature. This argument is part of Plato's general claim that theoretical learning, in the end, is more useful than "practical" pursuits.

Descartes's *Meditations* invokes no less than God to explain why the ideas of "true and immutable essences" of mathematics (triangle, circle, etc.) that we grasp with our mind must represent existing entities in nature (meaning empirical space). Thus the applicability of mathematics is co-opted by the mind–body problem. And Spinoza's monism, such as it is, is intended to solve the same problem without invoking the explanatory, or other, power of the Deity.[2]

Berkeley's problem is the same as Plato's, except in reverse: for Berkeley, standard mathematics is an obfuscation, and is even incoherent; examples of incoherent ideas in mathematics are the dimensionless point, the infinitesimal, and numbers (understood as abstract objects). His problem, then, is: How does an inconsistent theory like Newtonian calculus give the right numbers? (Note that Hartry Field's explanation in the twentieth century would not help Berkeley, because it applies only to consistent mathematical theories.) His instrumentalist explanation is ingenious: the infinitesimal calculus gives the right predictions by a sort of canceling out of errors, which reminds me of our contemporary practice of renormalization in quantum field theory.

Kant's *Critique of Pure Reason* returns mathematics to its status of synthetic a priori truth without a return to Plato. Kant's transcendental idealism, according to which the mind itself imposes mathematical order on empirical reality, is an answer to the question of how a synthetic a priori truth can be applied to empirical reality, an answer which avoids the Theory of Forms, theology, and instrumentalism. It turns out, contrary to Plato, that the *only* synthetic a priori truths are about empirical reality.

John Stuart Mill's account of the applicability of mathematics to nature is unique: it is the only one of the major Western philosophies which denies the major premise upon which all the above accounts are based. Mill simply asserts that mathematics itself is empirical, so there is no problem to begin with.[3]

This short sample of Western philosophy illustrates that the central philosophical doctrines of these major philosophers were conceived in great measure

[2] I once heard this idea from Stuart Hampshire.

[3] Colyvan points out that I could have listed Quine as also holding that mathematics is empirical. This makes sense, but Quine rejects the empirical/a priori dualism in the first place, so it is far from clear what the doctrine as espoused by Quine comes to. Also, Quine believes that mathematics is the science of objects that have no empirical properties whatever. See also note 28.

to explain the applicability of mathematics to Nature. What is more, we conclude that (though they didn't use the word "apply") all the doctrines presupposed the same *concept* of application: they all assumed that application is a relation (some approximation to a homomorphism) between mathematical theorems and empirical facts, a relation which can be used to "read off" empirical facts from mathematical theorems. The question they ask is: Given the nature of mathematics, why should such a homomorphism exist? And their strategy is either to provide an explanation (participation, God, etc.) or to deny the existence of applicability in the first place—like Berkeley or Mill, but to explain only why it looks as though mathematics were applicable (what my teacher, Professor Sidney Morgenbesser, called "explaining away" the phenomenon).

We must now ask a question which is not often asked: What do we *mean* when we say that mathematics is "applicable"? And before this question, another: What are we doing when we "apply" mathematics to Nature?

II. Canonical and Noncanonical Empirical Applications

The sort of applications the classical philosophers puzzled over could be called "canonical empirical applications" of mathematics, because the available empirical applications of mathematics were either canonical or reducible to canonical explanations. The classical philosophers, like their counterparts today, mostly did not realize that some of the canonical applications of (even elementary) mathematics are not empirical, nor of course could they have predicted the rise of empirical applications that are not canonical. But let's define these terms.

I call an application of a mathematical theory *canonical* if the theory was developed in the first place to describe the application. For example, suppose we have an apparent empirical regularity R of some kind; mathematics M is developed in an attempt to describe this regularity; perhaps we should say that mathematics M is developed so that R should be, at least approximately, its model. Then R is a canonical empirical application of M. Our classical philosophers want to understand this procedure and how it could work—but they *don't* need an explanation why R, *rather than* some other regularity R^*, is an application of M, because M was introduced for this purpose.

An obvious case of canonical empirical application is the use of the differential calculus to describe accelerated motion. This was developed by Newton precisely for this purpose. Any philosophical problem concerning this that we find in the philosophers would be equally a problem concerning any canonical empirical application.[4]

Ironically, it is much harder to see the role of canonical empirical applications in elementary mathematics, because in these cases, typically, the mathematics evolved together with the applications over a long period of time, rather than being invented; and there is thus a tendency to confuse the applications of elementary mathematical theories with the theories themselves.

An example of what I speak is the "application" of addition to finding the size of collections of bodies. Many people, even mathematicians, find it hard to recognize that we are speaking of application of mathematics rather than mathematics itself, but let me try to get things clear.

Consider a set S of bodies. Consider the scattered physical object S^* which is the "mereological sum" of the elements of S. (The mereological sum of a set of objects is the smallest body of which each of the objects is a part.) A mereological sum is not a set and does not have members, but only parts; a set has both members and subsets.[5] Another difference is that the sum of a sum of bodies is no different from the sum itself; yet the set containing a set of objects is different from the latter.

Suppose, however, we think of bodies as maximal irregular polyhedrons (i.e., polyhedrons not part of larger polyhedrons). Then these bodies play the role, to some extent, of set members—since the parts of these bodies are not themselves bodies. There is, therefore, a sense in which a mereological sum of bodies is a model or image of a set of those same bodies. And since this is true because of the empirical properties of those bodies (stability, for example, as well as discreteness), we can say that mereological sums of bodies are empirical applications of the set concept. Let us call the mereological sum of bodies a "collection" of those bodies, so that collections are empirical applications of sets.[6]

But this works in the opposite direction as well: we can read some of the properties of the sets from the collections for reasons which are both mathematical and empirical.

[4] One caveat, nevertheless, is in order. Modern foundational achievements in analysis have turned the "calculus" almost into a logical device, analytic truths concerning accelerated motion. So it's not clear that we have here an *empirical* application, one which could be refuted by experience. In the seventeenth century, however, I believe it would be fair to say that Newton was "reading off" the characteristics of accelerated motion from his "calculus," which justifies the name. At the same time, the calculus was developed for just this purpose, justifying the label "canonical."

[5] There is a sense in which sets also have parts—we can view the subsets of S as parts of S. See Lewis (1991).

[6] The reader will find it instructive to compare the present approach with the thoughtful discussion of the relationship between sets and physical bodies in Maddy (1990).

Consider "counting." This is basically an empirical process in which we point to all the bodies in a collection while reciting numerals in order. The last *numeral* recited expresses the *number* or *size* of the set of those same bodies. This is true because, as we say, the collection is a physical model of the set. But it is also true for a mathematical reason: the finite numbers are both ordinal and cardinal numbers. (For infinite sets, it matters a great deal in what order we place them—the ordinal number of the natural numbers in the standard ordering is called ω, but if we "count" the numbers starting from 1 and put 0 after the rest, we get the ordinal $\omega + 1$.) This means, for example, that it matters not in what order we count the bodies in a collection—we will get the same result. The result, invariant over order, is thus the size of the set of those bodies. Once again reversing the story, if we know the cardinal number of a set of bodies, we can predict the result of counting the collection of the bodies, an empirical prediction which has a mathematical explanation.

Next consider addition. Thought of as an operation on cardinal numbers, m and n, which are the sizes of disjoint sets X and Y, $m + n$ is the size of the union of X and Y. Suppose X and Y are sets of bodies, as before, and X^* and Y^* the corresponding collections of those same bodies. Then the mereological sum of X^* and Y^* is an image or model of the union of X with Y. If we know the sum $m + n$ in canonical notation, we can then predict the result of counting the collection of X^* with Y^*. This is, of course, because of the above-mentioned identity between cardinal and ordinal finite numbers.

But the cardinal–ordinal equivalence has another implication, and a profound one. This is that cardinal addition is equivalent to ordinal addition; extensionally, they are the same operation. But ordinal addition is based on recursion (iteration), and recursion yields computational algorithms, such as the ones we learn in school. Many educators today decry the emphasis placed on these algorithms, and they have a point: these algorithms are for ordinal arithmetic, which has little to do with the more applicable cardinal arithmetic, which measures the sizes of various sets. On the other hand, the mathematicians have a point as well, since ordinal arithmetic can be immediately applied to cardinal arithmetic to make calculation a cinch. The mathematicians suffer from the lack of algorithms in infinite cardinal arithmetic, and as a result cannot make the simplest calculations, such as

$$2^{\aleph_0}.$$

Cardinal arithmetic, then, is of great application but does not allow calculation; ordinal arithmetic has no great application, but does allow algorithmic calculation. Together we have the following story: take two distinct collections—of bodies—and count them both. We then have the ordinal number, and thus the cardinal number, of the two sets corresponding to the collections. The cardinal number of the (disjoint) union of the sets corresponds to the sum of the two cardinal numbers. This can be calculated using ordinal arithmetic (i.e., algorithms). From this we can predict the result of counting the (mereological) sum of the two collections, the smallest collection of which the two original collections are parts.

It is a considerable intellectual strain, we see, to separate pure arithmetic from its canonical empirical applications. Small wonder that the distinction is seldom made.

To sum this all up (pardon the pun), the applicability of addition algorithms to counting physical collections is based on an amazingly complicated series of facts and mathematical "accidents": (a) the empirical stability of many objects; (b) the resulting model of sets of physical objects as collections of those objects; (c) the equivalence of cardinal and ordinal finite addition, which allows (1) finding the cardinal number of a set by counting the corresponding collection and (2) the use of powerful algorithms of ordinal arithmetic to solve problems in cardinal arithmetic.

A noncanonical, yet still empirical, application of addition would be to weights. Seven unit weights and another five give twelve unit weights. This is an empirical application, so empirical that it is even (slightly) wrong—according to Einstein's General Theory of Relativity, weight is not quite linear, just as velocity is not quite linear in Special Relativity. However, for ordinary purposes we can speak of the "linearity" of weight. In that case, we can use arithmetic to predict the weight of a collection of unit weights. (Of course, for a fuller story one would discuss the application of the theory of real numbers to weights which are incommensurable.)

This is not, strictly, a canonical application, insofar as the theory of arithmetic operations was not put together to do weights, but to count. However, weight is "near" canonical, in that (given the empirical properties of the magnitude "weight") we can reduce weighing to counting. (I again stress that for simplicity I am discussing only commensurable weights.)

A more advanced empirical, yet noncanonical, application is the application of the Apollonian theory of conic sections by Kepler to planetary orbits. Obviously the Greeks had not introduced this theory to describe the orbits of the planets, which in any case were thought of as strictly circular. Indeed, as far as I know, Kepler was the first to apply the theory in physics.

An even more advanced case is the application of non-Euclidean geometry to gravity, as in Einstein's General Theory of Relativity. What is noncanonical here is not necessarily the description of space as "curved" (it is plausible that both Gauss and Riemann had this in mind), but rather the application of Riemannian geometry to space-time.

This raises the question of whether all mathematics is applicable in physics. Of any given mathematical theory it is quite risky to say that it has no physical application.[7] Nevertheless, we can say that the beautiful theory of prime numbers, which so caught the imagination of the Pythagoreans, has to this day no certified applications in physics. Of course, arithmetic itself is applicable in physics, and arithmetic does treat prime numbers, but this is not the sense of "applicability" I have in mind, which demands that the concept "prime number" should actually appear in a physical theory. Suppose that the *n*th energy level of hydrogen had a certain property if and only if *n* is a prime number—that would be an example, but there aren't any like that.[8] If string theory is ever confirmed empirically, the role of the prime numbers in science will finally be established, because string theory actually does rest on some characteristic concepts of the theory of numbers.

The converse issue, raised by the celebrated physicist Eugene Wigner in a famous paper (Wigner 1967), is whether the level and type of applicability we do see in these noncanonical applications is "unreasonable." Wigner argued that it is, because:

(1) Mathematical concepts—at least in the last 150 years—are subject primarily to criteria internal to the mathematical community. Among these criteria are aesthetic ones; and most of the criteria are not such as to make it likely that mathematical concepts should be applicable at all. In some cases, the mathematical concepts have, and had, no *canonical* empirical applications.
(2) In physics, the reliance on mathematical concepts in formulating laws of nature has led to laws of unbelievable accuracy.

Skeptics argue that Wigner suppressed the failures in applying mathematics (there are hundreds monthly). Other skeptics argue the reverse, that we should expect mathematics to be applicable because (unlike what Wigner asserts) the real and ultimate source of mathematical concepts is experience; thus it is not at all unreasonable that mathematical concepts should return the favor.

I'm not particularly impressed by these objections (particularly because they contradict one another), since I think that each of the examples that Wigner gives is so extraordinary that it requires explanation.[9] For example, I feel that even

[7] Most physicists before 1930 would have said this of group theory—it is now, of course, the centerpiece of elementary particle physics. The famous mathematician G. H. Hardy argued that no truly mathematical theory could be applied to warfare, but this was wishful thinking, since one of his examples of a "truly mathematical theory" was $E = mc^2$.

[8] As Professor Sternberg pointed out to me, however, the concept of a prime number is crucial to cryptography, and of course he's right. This is another—rather whopping—counterexample to Hardy's wishful thinking (see note 7), since a lot of research in number theory today must be classified.

[9] I regret that Wigner, no doubt out of modesty, left out of his article the most striking examples of the unreasonable effectiveness of mathematics: his own contributions in the field of applying group theory to quantum mechanics.

Newton's law of gravitation, based on astronomical observations which were wildly inaccurate by today's standards, nevertheless has withstood the increase in the accuracy of our measurements to the extent that we can now say that the elementary particle known as the neutrino falls to earth just like the moon. There is nothing like the accuracy of the inverse square law, for example, in economic models, which retain roughly the experimental error of the data which suggested them in the first place. Kant once argued that the inverse square law is to be expected (if not a priori), because gravity spreads out spherically, and the area of the (surface of) the sphere is proportional to the square of its radius. But Charles Peirce already answered this by the telling rejoinder that what spreads out in space is the potential, not the force, and the gravitational potential goes as the inverse, not the inverse square, of the radius! Thus it's not clear that we have anything that would count as an explanation for how much more we "got out" of the inverse square law than we "put in."

My objection to Wigner's thesis is completely different: it's not a thesis at all. He gives persuasive examples of successes that cry out for explanation—but he doesn't prove that they add up to one phenomenon that cries out for explanation. Each success is a story in itself, which may or may not have an explanation. Wigner does not make a case that what is unreasonably effective is *mathematics*, even though the individual examples he gives are of concepts that happen to be mathematical. In other words, Wigner may give examples of a number of applications which are "unreasonably effective"—applications of concepts which happen to be mathematical. But he doesn't show that these successes have anything to do with the fact that the concepts are mathematical. Of course this is connected with Wigner's failure or inability to give a definition of "mathematical concept."

III. Canonical Nonempirical Applications

I would like now to speak, finally, of canonical applications of mathematics that are not empirical at all—these are applications in which one uses mathematics to "read off" results in another theory which itself is mathematical. One might think that these kinds of applications would have to be quite advanced, yet I was surprised to find that there is a very elementary example: multiplication.

Multiplication is often "defined" as repeated addition, but this, I hold, is a confusion between a definition and an application.

An intuitive way to see that something has gone wrong with the pseudo definition "multiplication is repeated addition" is to inquire concerning the commutativity of multiplication. Why is it always the case that adding x, y times, gives the same result as adding y, x times? To be sure, this fact can be proved by mathematical induction. But the proof is far from trivial, and my experience teaching this material from textbooks like Mendelson (1997) is that most college students are not capable of discovering the proof. Morever, proofs by mathematical induction usually prove only *that*, but not *why*, a theorem is true.[10]

The "definition" of multiplication as repeated addition has its source in ordinal arithmetic (i.e., recursion). We have already seen that ordinal arithmetic is dandy as algorithm, horrible for applications. Small wonder that children cannot understand why multiplication is commutative (the noted philosopher and logician Saul Kripke recalled—in a class I attended—being amazed as a child at the commutativity of multiplication). Nor can they understand why one can "add candies to candies" but not "multiply candies by candies." What we need is a *cardinal* definition of multiplication.

Now let's look at the matter from the logician's point of view. To see that "repeated addition" cannot be a definition of multiplication, consider the formula xy, having two free variables. The "repeated addition" definition (so called) would dictate that xy means

$$y+y+y+\cdots+y \text{ (x times).}$$

But the ellipsis here (...) is not defined mathematically; and the parenthetical comment (x times) means only that the letter y is repeated an unspecified number x of times. The most coherent interpretation of this "definition," then, is a schema containing infinitely many definitions of the following type:

$$1\times y=y$$
$$2\times y=y+y$$
$$3\times y=y+y+y$$
$$\cdots\cdots\cdots\cdots\cdots\cdots$$

The numbers 1, 2, 3, ... count the number of occurrences of the letter "y;" to put it another way, x is a metalinguistic, rather than a mathematical, variable, where y is a "true" variable.[11] To see what is wrong with this, take a specific example: $7\times 5=35$, which, according to the standard pseudo definition is to *mean* $5+5+5+5+5+5+5=35$. But standard rules of logic allow the inference

(E) There is x such that $x\times 5=35$

[10] See Steiner (1978a) for a discussion.

[11] Interestingly enough, Wittgenstein in RFM (Wittgenstein 1978) points this out.

from $7 \times 5 = 35$, but not from $5+5+5+5+5+5+5=35$. Hence the definition of multiplication in terms of repeated addition is wrong. The most we can say is that $5+5+5+5+5+5+5=35$, if we assume the axioms of Peano (which of course contain multiplication as a primitive, undefined notion), is equivalent to $7 \times 5 = 35$ (and therefore implies E), but the equivalence is not purely logical.

A good way to look at multiplication is, in fact, set-theoretic: xy is the (cardinal) number of x disjoint sets, each of which has the cardinal number y. The role of x here is different from that of y (though it is *not* metamathematical); suppose we are calculating the number of candies we need to give four children five candies each—this is the same as calculating the number of elements in four disjoint sets with five candies each. Five, then, tells us the number of candies, where four tells us the number of sets. In other words, the "second-order" concept of "set of sets" is inherent in multiplication. We have already seen that though collections, in a sense, can be considered physical models for sets, there is no such model for "sets of sets." This is the deeper reason why we can't "multiply candies by candies," though we can "add candies to candies": adding candies to candies is to model addition by mereological sum, forming a larger collection from two smaller ones; for multiplication there is no such model.

At the same time, we see from this definition that each multiplication is equivalent to (though not synonymous with) a repeated addition—we can "read off" the results of a repeated addition from a product of two numbers. In this sense, it could be said that multiplication is being "applied" to addition! But this is not an empirical application at all!

Consider, now, the following table.

	candy no. 1	candy no. 2	candy no. 3
child no. 1	**A**	**B**	**C**
child no. 2	**D**	**E**	**F**
child no. 3	**G**	**H**	**I**
child no. 4	**J**	**K**	**L**

Fig. 20.1.

We have labeled the candies with capital letters. If we have four children and have to give each child three candies, we need 4×3 candies, the cardinal number of a disjoint union of 4 disjoint sets with 3 members each. The rows of the table, marked out as above, symbolize the sets of candies. If we permute "candy" with "child" everywhere and switch rows for columns, we get:

	child no. 1	child no. 2	child no. 3
candy no. 1	A	E	I
candy no. 2	B	F	J
candy no. 3	C	G	K
candy no. 4	D	H	L

Fig. 20.2.

It is clear from this that multiplication is commutative; from the abstract mathematical point of view, what we have done here is set up a one–one map between

$$\bigcup_{i=1}^{|A|} B_i,$$

all the Bs equinumerous, and $A \times B$, the set of ordered pairs

$$\{\langle x, y\rangle : x \in A \wedge y \in B\},$$

B being any set eqinumerous to the Bs, and then noted that, obviously,

$$|A \times B| = |B \times A|.$$

In effect, we have two set-theoretic definitions of multiplication here which are equivalent: in terms of disjoint unions, or in terms of Cartesian products. The former definition is the one that leads to applications—but distinguishes between the operands m and n in that the latter numbers sets; the former numbers sets of sets. In the latter definition, both m and n number sets.

A final comment: I maintained that "repeated addition" should be regarded as a *nonempirical*, if canonical, application of multiplication; what of *areas*, as in geometry? Is this not a canonical *empirical* application? Didn't the Egyptians and the Greeks develop this application of multiplication?

Yes and no. Though it is true that area is certainly an application of multiplication, it originally was not an application of the operation we teach today in school. The Greeks regarded multiplication as an operation on magnitudes, not numbers; when we multiply linear magnitudes, we get square magnitudes. In other words, multiplication of magnitudes was not a closed operation.

Furthermore, the use of multiplication in areas is connected with the Euclidean structure of the plane. If space is not Euclidean, then the area of a square is not actually the product of the lengths of two sides. We will still need multiplication to find the area, which is still definable (as an integral) as long as space is still locally Euclidean (the smaller the space we take, the "flatter" it gets). However, I don't think that we have here anything like a canonical application anymore.

Let us leave elementary arithmetic, and look at some of the modern theories of mathematics. Indeed, let us heed the call of the mathematicians we began with, and discuss group theory.

Group theory is known today as the theory of symmetries—we define a symmetry as a property which is invariant under a group of transformations. The classical symmetries, which are visual symmetries, involve transformations of space, such as rotations, translations, and reflections. For example, a cube is invariant under rotations of multiples of 90 degrees around axes that go through, and are perpendicular to, its faces. And one might have thought that group theory developed in order to describe these symmetries, for example, in crystallography—and it is true that group theory can indeed be applied to "read off" results in that field. Once we know, for example, the symmetry group of a substance like salt, we can predict all the possible forms of the salt crystal, even the forms that occur "naturally" and are not regular polyhedra. We can even use group theory to explain why so few visual symmetries are seen in nature—crystals are made up of a lattice structure (atoms), which restricts the number of symmetry groups possible.[12]

Yet this is not how group theory came into existence. In fact, group theory was constructed to be applied not to empirical reality, but internally—in mathematics. Namely, group theory was introduced into mathematics by Galois, to be applied to algebra. The Galois group of an equation, for example, can be thought of as a group of certain permutations of the roots of that equation.[13] By studying the group of an equation, mathematicians are able to extract information about the equation itself—in fact, the application of group theory in mathematics has precisely this form: we find that a certain group characterizes a much more complicated structure, and we can get needed information about that structure by studying the properties of the group. (This is how group theory is used in topology, for example.) Thus what Galois did was to inaugurate modern mathematics, in which the main applications of mathematics are—to mathematics itself. It would be fair, then, if surprising, to state that the canonical applications of group theory are mathematical, not empirical. Ironically, the origins of group theory in pure mathematics not only did not prevent physicists from applying group theory (after a period of resistance), but made it possible.

Briefly, the reason is this (for a detailed account, cf. Sternberg 1991; Sternberg also graciously offered his help in cleaning up this entire section). Already in

[12] Perhaps the best treatment of this subject is Sternberg (1994, ch. 1).

[13] Given an equation E with coefficients in field K, one can construct an extension field M in which the equation "splits" (can be factored, has roots). The Galois group of the equation is the group of automorphisms of M which leave K fixed. Since the image of a root of the equation under any element of the Galois group is also a root, the Galois group can be thought of as a group of permutations of the roots.

classical mechanics it was discovered (by the mathematician Emmy Noether) that there is a mathematical connection between symmetry groups and *conservation laws*. For each continuous group G of symmetries[14] of the laws of nature, there is a magnitude P_G which is conserved. If the law has rotational symmetry, then angular momentum is conserved, for example.

In quantum mechanics, the state of a system is represented by a unit vector in a (complex) vector space V of infinite dimension. The group G must act on this vector space V, rather than directly on physical space.[15] This action is called a "representation" of the group, in which each element of the group is represented by a square matrix.[16] Suppose that a particle has a determinate value v for property

[14] Technically, a one-parameter group of symmetry transformations. In addition, these transformations must not change the "symplectic" geometry of the phase space (see note 15). In the case of quantum mechanics, to generate a conservation law the action of the group must not change the geometry of the Hilbert space—hence it must be represented by unitary matrices (these are invertible complex square matrices M such that the inverse is equal to the adjoint, $MM^* = I$). So in both classical and quantum mechanics, conserved quantities arise from the action of continuous symmetry groups that preserve both the laws of nature and the geometry of the abstract space (not necessarily space-time) which is the mathematical "arena" of these laws.

[15] Even in classical mechanics, symmetry groups *need* not act on physical space. The fact, for example, that the orbit of a planet does not precess (turn) is a conservation law which arises from the action of a group of rotations, but not rotations of three-dimensional physical space as we now describe:

A deep comparison of classical physics with quantum mechanics is through the medium of geometry. In classical mechanics, the state of a system is given as a point in "phase space." For a free particle (no constraints or forces) this will be a six-dimensional vector space with three coordinates of position and three of momentum. As soon as we introduce forces or constraints, as in a particle in a central gravitational field, however, the coordinates are no longer rectilinear (Cartesian) but curved. Only at each point (locally) can we say that phase space looks like a vector space. Phase space is thus what is called a "manifold," indeed, a symplectic manifold (the term "symplectic" refers to the particular geometry of the manifold). The rotation group, when acting on the space dimensions, yields conservation of angular momentum for the planet; when acting on another submanifold of this "phase space" (namely, the submanifold of the phase space which contains all the possibilities for planetary motion), it yields a different conservation law. For details of this, see Guillemin and Sternberg (1990b). In quantum mechanics, on the other hand, the state of a system is given by a unit vector in an infinite dimensional complex vector space, known as a Hilbert space. (The term "Hilbert space" expresses the particular geometry of the vector space.) The coordinates remain rigidly rectilinear (even if complex), and thus the restrictive theory of group representations by square matrices must apply.

[16] We are, again, interested primarily in representation by single parameter groups of unitary matrices because, as stated above, unitary transformations do not change the geometry of the Hilbert space. The geometry of a Hilbert space is given by the scalar product of two vectors, analogous to the scalar product of Euclidean space, and unitary matrices preserve the scalar product.

P_G, which is conserved under the action of G. Then,[17] as G acts, (a) the unit vector never moves out of a subspace W of V which is associated with that very value v and (b) there is no subspace Y of W which so confines the unit vector. W, taken together with the action of G, is called an "irreducible representation" of G. Suppose W is an n dimensional subspace of V. The physical meaning of this is that a particle with value v of property P_G can assume n different states. It should not be assumed without proof that G has such an irreducible representation for every dimension n.

Consider as an example the electron and let P_G be the property known as "spin," so called because the electron is a little magnet analogous to a spinning charged ball in classical physics. As one might expect, rotating an electron (or, equivalently, rotating the measuring device relative to the electron) does not change the value of its spin (which remains at ½ Planck's constant). Yet rotating it 360 degrees does not bring the electron back to its original *state* (the unit vector is multiplied by -1); this requires two full turns. What is most significant is that no matter how we rotate the electron (or the coordinates), the electron is observed in only one of two states: (a) with its north pole "up" or (b) with its north pole "down." (This is of course utterly unlike a classical spinning ball, which can be observed spinning in any direction.) Rotating the electron changes only the *probability* of observing the electron in an up or down position. Thus it has two states corresponding to its spin value.

This might suggest the following group-theoretical description of the electron (in fact, this picture was actually resisted by physicists for quite a while): the symmetry group of electronic spin is the one called SU(2). This is a continuous group which has a two-dimensional representation[18] just as required, and is also the "double covering group" of the rotation group in ordinary space (i.e., a continuous action of this group which brings a unit vector back to itself is homomorphic to two rotations, again as required).

So far the group-theoretical description does not seem to add anything to the physical facts, which is why physicists referred to Hermann Weyl and his colleagues in the mathematical community as the "Gruppenpest." Yet, as Sternberg points out, one can reverse this procedure, and go from the group description to the physical facts. For example, in 1932, Heisenberg argued that the neutron and proton are actually two different states of the same particle (today called the "nucleon"), with the value ½ of some property yet to be explored, and with the same symmetry group as the electronic spin. He called the property, again by analogy, "isotopic

[17] The success of group theoretical methods in physics has actually made this statement into a definition: what we mean by a "particle with a fixed value of a physical magnitude" is described by definition as an irreducible representation of the group associated with the conservation of that magnitude.

[18] The way SU(2) is usually defined, as a group of 2×2 "unitary" matrices, this statement is trivial, true by definition; but there are other ways to define the group.

spin," though this is nothing more than saying "the property conserved in the nucleus by the spin group SU(2)." Actually, even today, there is no known physical connection between the spin of the electron and the isotopic spin (known today as isospin) of the nucleus—nature merely utilizes the same symmetry in both places.

Having gone from physics to group theory, we now go back to physics: SU(2) also has a three-dimensional irreducible representation, which means that it is possible that there is a particle which can take on three different states (we can think of these particles as being "made out of," respectively, a neutron-neutron, a proton-proton, and a neutron-proton). The different states of the particle will look like three different particles in the laboratory, just as the nucleon looks like two different particles. Sure enough, the three *pions* were found: particles analogous to the neutron and proton in that they are *hadrons*—they exert the powerful nuclear force. The three "directions" of the pions are marked out by their charge: positive, neutral, and negative. By now the group-theoretical method began impressing physicists. These predictions were being made with no knowledge of the properties of hadrons, except for their symmetry groups.

The next step was even more startling. The neutron and proton themselves turned out to be part of a family of eight hadrons with comparable rest mass. Aside from them, there were nine more hadrons with comparable mass (though by no means the same), greater than that of the first family. This suggested to Yuval Ne'eman and Murray Gell-Mann (independently) that there was a higher symmetry than SU(2). Mathematical analogies suggested a group known as SU(3), of which SU(2) is a subgroup, which has representations of dimensions 3, 8, 10, . . . And, indeed, experimental evidence suggested that the neutron and proton belonged to a family of eight similar hadrons. Furthermore, there was a family of nine heavier hadrons. But if SU(3) were the right symmetry group of the hadrons, the family of nine particles would require a tenth. Hence, both Gell-Mann and Ne'eman predicted the tenth, calling it in advance the omega minus particle, which was later discovered. Gell-Mann then predicted the existence of the triplet from the three-dimension representation, *none* of which had been observed—and for good reason: mathematical considerations indicated that such a triplet, called by Gell-Mann "quarks" after Joyce, would have to have fractional charge. At the same time, the quark hypothesis (if one could come up with an excuse why nobody had ever detected these ubiquitous particles) could serve as a partial explanation of the SU(3) symmetry: one could use quarks to "construct" all the families of the hadrons (as we "constructed" the pions out of neucleons). Note, however, that this explanation was ex post facto, like most explanations: the success of SU(3) symmetry led to quarks, not the other way around.

One could push back the use of group-theoretic methods even earlier than I have done, by including in this story the saga of "identical particles." This is, in fact, what Steven French does, in an illuminating attempt to account for the success of the group-theoretical method in physics (cf. French 2000).

The principle of identical particles is that there is no physical difference between a state of two identical particles, *A* and *B* (e.g., two photons in which *A* is in state *X* and *B* in state *Y*, or one in which *A* is in state *Y* and *B* in state *X*. The most straightforward application of this principle is the situation where the state vector of the system of two particles simply does not change on permuting *A* with *B*. Let us also consider the case in which *A* and *B* must be either in state *X* or state *Y*. Then there are three cases: both particles are in state *X*, both in state *Y*, or they are in different states (not four cases where we distinguish *A* in state *X*, *B* in state *Y*, from the converse case with *A* in state *Y*, *B* in state *X*). This means that the probability of finding the particles in the same state is 2/3, rather than 1/2, which we might have thought. Another application of this principle is that permuting *A* with *B* brings the state vector to minus itself, since there is no observable difference in quantum mechanics between the vector ψ and the vector $-\psi$. These two group actions lead to an important classification of particles: those that change sign are called "fermions," and those that do not are called "bosons."

Now there is no question that we can see this is as an application of group theory—in hindsight, though, the historical players would mostly have rejected this characterization. They did not see the category of "groups" as a natural way to categorize phenomena; hence the "success" of applying one group could not, for them, suggest the application of a completely different one (such as SU(2)).[19] The fact that we today see the permutation group and SU(2) as examples of the "same thing" and as relevant to physics does not mean that the previous generations did. Our inclination to see the quantum mechanical treatment of "identical particles" as a triumph of group theory betrays our implicit belief that mathematical language is the deepest language of physics and that mathematical classification of structures is the ultimate physical classification, too. But this belief amounted to nothing less than a revolution in thinking about nature, and prominent in this revolution was Eugene Wigner, one of the greatest pioneers of the group-theoretical approach to physics. For him, it could be said, a physical object is not much more than the sum of all its symmetries, including the symmetries of the geometry of space-time.[20]

[19] Suppose we have a physical theory, like string theory, which postulates a 26-dimensional space. The number 26 happens to be the numeral value of the Divine Tetragrammaton in Hebrew. Should this encourage us to try other of the Hebrew Names of God?

[20] Let's for the record make precise what we mean by "geometry," at least for a two-dimensional vector space. If we consider two vectors (x_1, x_2) and (y_1, y_2), then the different quadratic expressions we can make from the two define different kinds of geometries: Euclidean geometry would be defined by $x_1y_1 + x_2y_2$; a (two-dimensional complex) Hilbert space would be defined by $x_1\bar{y}_1 + x_2\bar{y}_2$, thinking of the numbers here as complex numbers; Minkowsky geometry by $x_1y_1 - x_2y_2$; and symplectic geometry by $x_1y_2 - x_2y_1$. We also have the four corresponding symmetry transformations which leave each of these quadratic forms invariant.

We have, then, a new kind of application of mathematics. In the past, mathematics was used to get quantitative descriptions of phenomena that could also be described qualitatively. Today mathematics gives even the qualitative descriptions, because we often have no deeper language than mathematics. Related to this is the use of mathematics to make discoveries and "predictions." Our story here has shown that pure mathematics was the basis of analogies for which there was (at least at the time) no underlying physical basis. Scientists postulated symmetries without knowing what they were symmetries *of.* "Predictions" were made in a new way: in the past, mathematical calculations were made to show what necessarily had to be the case. In our story here, mathematics was used to show what is possible, the assumption being that what is mathematically possible is physically actual.

All of this suggests (and Ne'eman has recently written this explicitly) that modern physics has a Pythagorean streak to it. So far, at least the quality of *mass* has not been shown to be amenable to a group-theoretical treatment. If that one day happens, however, the world itself could turn out to be nothing but a mathematical structure: a reducible representation of the ultimate symmetry group,[21] a bizarre possibility we will nevertheless discuss in the concluding section of this essay.

IV. Logical Applications

Another nineteenth-century development that influences our subject is the development of modern logic and the foundations of mathematics. An attempt was made to put mathematical deductions on an a rigorous footing and, at the same time, to find the foundations of mathematics. This research program presupposed, or perhaps introduced, concepts of application of mathematics which could not have been formulated before. These are the logical concepts of application which we can now survey.

For example, one sense in which mathematics is applied is as an engine of deduction. Even if mathematics is not literally logic, as Frege and many others once thought, one of its functions is certainly akin to that of logic: as what some called a "juice extractor" in which we use mathematics to derive consequences of nonmathematical premises. Of course, any set of sentences can be used to derive consequences from others, but mathematical sentences seem "topic neutral" in a way other sentences are not.

But for this purpose, attention must be paid to the logical form both of pure mathematics and of "mixed" contexts, in which the same sentence contains both mathematical and nonmathematical references. An elementary example is that of

[21] This is a point made by Steven Weinberg (1986).

the use of arithmetic to shorten or eliminate counting, as when we conclude that there are twelve fruits on the table after observing that there are seven apples and five pears, and no other fruits on the table, from the pure mathematical proposition $7 + 5 = 12$.

The question arises: What does the pure mathematical proposition, which on its face is about "natural numbers," have to do with fruits? This question has two aspects, the logical and the metaphysical. The metaphysical question arises from the ontological gap, noted already by Plato, between mathematical and empirical objects. How can truths about mathematical objects be in any way relevant to the empirical world?

The logical question arises from the difference between the numerals "seven" and "five," used as adjectives, and the numerals 7 and 5, used as nouns. This ambiguity produces an equivocation which threatens to spoil the logical validity of our deduction. Obviously the two questions are related.

A strategy which suggests itself to many observers, past and present, is to argue that there is no such thing as pure mathematics, or at least that there are two kinds of mathematics, pure and applied. Applied mathematics (whether or not it is the only mathematics) is the empirical theory of certain properties, such as "seven" and "five." There are no mathematical entities, but only mathematical properties, and these are empirical properties like any other. The seven apples are a collection with the property "seven," for example. The arithmetical statement $7 + 5 = 12$, then, is a general empirical law about any such collections. This is the position, in fact, of the great nineteenth-century British philosopher J. S. Mill in his *A System of Logic.*

However, there are so many problems with this view that, strikingly, even the empiricists have by and large rejected it—with some notable exceptions. Most of the objections have been epistemological—the empiricist account of arithmetic does not seem to account for the various peculiarities of mathematical knowledge. I would like, however, to draw attention to a logical objection: What is the status of a "collection" of which a number is a property? If a collection is a physical object (as is probably intended) rather than, say a set, then there is no one number that characterizes it: A heap of socks could be characterized by the number of socks, the number of pairs of socks, or the number of molecules in the heap, or perhaps the number of atoms, and so on. Thus numbers cannot be regarded as properties of empirical bodies like collections. This objection was raised by Frege in his *Grundlagen*, which contains a scathing critique of "pebble and gingerbread arithmetic," referring to Mill. Frege does not seem to be aware, however, that basically the same objection was raised by Plato in the *Theaetetus*, against the more general view that "perception is knowledge."[22]

[22] "[Soc.] Very good; and now tell me what is the power which discerns, not only in sensible objects, but in all things, universal notions, such as those which are called being and not-being, and those others about which we were just asking—what organs will you

Instead, Frege, like Plato before him, took the opposite tack: he eliminated the adjectival numerals, and with them the numerical properties, entirely, leaving only the mathematical entities. Plato's solution, however, is not entirely coherent (as he himself pointed out in various places): he thought of the empirical world as *participating* in the mathematical world. Since, however, as he himself had argued, the empirical world can participate in the mathematical world in conflicting ways, he ended up arguing that the empirical world itself is subject to logical conflict (and thus is not the object of knowledge). Frege, though he gives no credit to Plato, in fact made two revisions in Plato's account: first, he replaced "participation" with something like "class membership." Second, he eliminated the logical incoherence by denying outright that arithmetic relates directly to the world of empirical objects. Rather, it is the empirical (and nonempirical) *concepts* which are members of the different numbers. Thus, the number two is the class of all *concepts* which are true of two objects.[23] The metaphysical problem (How can facts about the world of abstract entities be relevant to the empirical world? also is dissolved: in Frege's scheme, numbers are related (directly) only to concepts.

Frege's account of (deductive) applicability comes with a price, however. It is heavily committed to the existence of nonmaterial objects (and, of course, concepts), and thus heir to all the traditional attacks on Platonism. I cannot discuss those attacks here, except to point out that some philosophers have been influenced enough by these attacks to attempt to gain the benefits of Frege's account of logical applicability without paying the price. One of these strategies is "fictionalism," understood as the doctrine that the Platonist commitments of Frege's account are real, in the same way that Shakespeare's *The Merchant of Venice* is committed to Shylock as well as to Venice.

The version of fictionalism set forth in Field (1980) is actually a research project: given a theory *MT* of mathematical physics, to find a "nominalist" theory

assign for the perception of these notions? [Theaet.] You are thinking of being and not-being, likeness and unlikeness, sameness and difference, and also of unity and other numbers which are applied to objects of sense; and you mean to ask, through what bodily organ the soul perceives odd and even numbers and other arithmetical conceptions. [Soc.] You follow me excellently, Theaetetus; that is precisely what I am asking. [Theaet.] Indeed, Socrates, I cannot answer; my only notion is, that these, unlike objects of sense, have no separate organ, but that the mind, by a power of her own, contemplates the universals in all things."

[23] Though this seems circular, in fact it is not, when the whole definition is written out. The "class of all concepts true of two objects," on the other hand, is a very tricky notion, and when Frege tried to axiomatize it, he landed up in Russell's Paradox. This is not the place to discuss the (by now) hoary question, is Frege's definition of the numbers "correct"—and what, indeed, constitutes "correctness" in this context? Since his definition cannot even be given in standard set theory (ZF), mathematics students are usually deprived of the pleasure of studying Frege's interesting ideas.

T which has the same "nominalist" consequences as *MT*. Field claims to have done just that for the classical equation of the gravitational field. Success in such a replacement program shows that mathematics is theoretically not necessary and thus can be regarded as pure fiction. Mathematics, Field argues, has only an instrumental value in shortening proofs. Hence it need not be regarded as true, but only as useful.

Field's book occasioned a great deal of comment and criticism, much of it centering around his major example of classical gravity: whether his replacement for it was truly nominalist (he helps himself to the entire continuum of space-time points, whereas Aristotle had rejected even one space-time point as too Platonist for his liking); whether he had truly demonstrated that it was a replacement; whether he had demonstrated that mathematics is really not deductively necessary in physics;[24] whether he could reproduce his success in other areas of physics, particularly quantum mechanics, where, as we have already seen, the mathematical formalism is central in a way that it is not in classical mechanics. Many of these criticisms seem to me cogent; in fact, I made some of them myself,[25] although in Hebrew.[26]

I would like to point out something else in this connection that seems to have been missed: we have seen already that mathematics is not only a medium of proof; it is an engine of discovery. We have seen, that is, how the mathematical form of a theory can serve as a springboard for its development or even, paradoxically, its replacement. The great scientists relied on mathematical analogies to suggest replacements or at least generalizations of existing theories. The idea was to replace false theories by (hopefully) true ones, in which what was preserved was mathematical structure.

Consider the following example. Measuring devices, no matter how accurate, give results in rational numbers only. On the other hand, the solutions of differential equations are functions on the continuum. One could presumably replace current theories, involving the continuum, with physical theories invoking only rational numbers—with no observable difference. Differential equations, for example, can be replaced by "difference equations," which are used as algorithms

[24] In this category, we can cite an ingenious paper (Shapiro 1983), which exposes a rather subtle error in Field's argument that mathematics does not add any logical power to (nominalist) physics. Field has confused the notion of logical consequence with deductive consequence, so that even if it is true that physics with mathematics does not *imply* more facts than physics without mathematics, there are nevertheless facts, so implied, which cannot be *deductively proven* without mathematics. The distinction between logical consequence and deductive consequence arises in Field's book because he does not use first-order logic (for which there is no difference between logical and deductive consequences) as his underlying logical system.

[25] In the next section, however, I will point out a great virtue of Field's book. Hang on.

[26] Cf. Steiner (1982).

to calculate the solutions of differential equations. In so doing, however, we would destroy the mathematical form of the latter; and it is the mathematical form which was used in finding generalizations or corrections of present equations. Physicists insist on the form of equations even when their content is obscure. For example, Richard Feynman introduced mathematical notation for calculations in quantum electrodynamics which he himself suspected of inconsistency and which, in any case, lacks a consistent mathematical interpretation even today. The inconsistency of this notation, however, has not prevented scientists from using it to calculate the magnetic moment of the electron correctly to twelve decimal places!

Why is this relevant to Field's program? Field regards Platonism as refuted if, for each "Platonist" physical theory *A* with nominalist consequences *N*, we can replace it with a "nominalist" theory *B* with the same nominalist consequences *N*. Since, then, *B* "can do everything that *A* can," Ockham's razor suggests that we are committed only to what *B* is committed to. This being a pragmatist concept of ontology, one could argue that epistemic values should be factored into this equation: there is an epistemic sense in which *B cannot* do everything that *A* can.[27]

V. Speculative Concluding Remarks

What philosophical problems are presented by the applicability of mathematics?

Since we have seen there is more than one concept of applicability, there is more than one problem. In discussing them, I will begin with the issues raised by the *logical* application of mathematics.

We saw that Frege showed how to use pure arithmetic alone in making deductions about the number of physical bodies. He did this by standardizing the expressions of the form "There are *n Fs*" as "The number of *Fs* = *n*," an equation. The variable *F*, for Frege, takes concepts as values. Hence the number *n* is directly associated only with a concept when saying "The number of *Fs* is *n*." The number *n* is not related to the *Fs* themselves, thus eliminating the traditional puzzlement concerning the relevance of mathematical objects to the empirical world. In fact, they are not relevant, and need not be.

However, Frege's solution raises the specter of Platonism, since his reduction of applied to pure mathematics involves the commitment to mathematical objects at the most elementary level. One can, of course, say that Frege's accomplishment was to *reduce* the (logical) problem of applicability (of arithmetic, since his success with analysis is questionable) to that of Platonism. Of course, Platonists

[27] After writing this, I discovered that Mark Colyvan (2001) already made this point. See chapter 4 there, which contains some nice examples.

don't see any problem with Platonism; they feel entitled to say that Frege has solved the (logic) problem of applicability. What of the nominalists?

Earlier, we saw how Hartry Field attempts to have his cake and eat it, too: he accepts all the details of the Fregean solution, but whisks away the ontological commitment involved with the claim that one "could," in principle, dispense with numbers (i.e., "number-talk") entirely. Fregean or "Platonist" mathematics can be regarded instrumentally. Field thus claims that he can get all the benefits of Frege's program without paying any ontological price.

Although I have given some of my reasons for being skeptical about Field's "free lunch" solution, I would like to point out an underappreciated virtue of Field's "piecemeal" approach to eliminating Platonism. That Field aims to replace each physical theory individually with a nominalist counterpart is regarded as a serious flaw to his argument by writers such as Dummett (cf. Dummett 1991), who complain that what he really should do is replace all the theories simultaneously, or are suspicious that Field won't be able to carry through the program in other, more intractable cases (like quantum mechanics). Yet the very piecemeal character of the project makes it ideal for solving a completely different problem of applicability, Wigner's problem of the noncanonical empirical applications of mathematics. This is not a logical problem but a descriptive problem, involving the use of mathematics in describing nature—rather than in making deductions.

For, as I suggested above,[28] Wigner's problem is not one problem: every unreasonable success in using mathematics to describe any physical phenomenon is a separate problem. Field's reduction of classical gravitational field theory to a theory of spatial regions may not be a nominalist reduction, but it does give an account of the usefulness of analysis in physics without using analysis. The bugaboo of Platonism—and the inability of most philosophers, including perhaps Field himself—is to see that there is more than one problem of the applicability of mathematics. Wigner's problem is not Frege's. Wigner's problem, for each application of a mathematical theory M, is solved by articulating, without using M, the conditions under which M applies.

Wigner's problem—or really problems, as I believe—is to give an account of various noncanonical empirical applications. Wigner says this is difficult to do on account of the gap between the goals of mathematics and the goals of physics, so that it is unreasonable to expect a mathematical concept to describe an empirical phenomenon with as much precision as we find in physics, typically. Field's method gives an example of a solution.

We have seen, moreover, that there are also philosophical problems connected with *canonical* empirical applications. Here, of course, Wigner's problem does not arise—by definition. But the opposite problem arises—the close relationship

[28] In section 2. See Steiner (1998) for more argumentation on this point.

between the mathematics and its application makes it often quite difficult to distinguish between them. We saw an example of this in the application of addition to predict the result of empirical counting. Inability to distinguish between mathematics and its *canonical* applications leads to positions like that of J. S. Mill, according to which mathematics is a crudely empirical science, a position which most philosophers regard as severely flawed.[29]

Finally, we discussed the application of mathematics within mathematics itself. For example, we use one kind of mathematical structure in order to characterize another. A striking illustration of this is the concept of group representations, in which the elements of a single group G are homomorphically "represented" by square matrices (i.e., linear transformations of a vector space). The existence of irreducible representations of various dimensions characterizes G by its possible actions.

The reason that these internal applications are so important to our subject is the existence of mathematical formalisms in physics. These are mathematical structures which, though they do not directly describe physical systems, nevertheless contain an enormous amount of information about those systems, which can be extracted from the formalism by rules whose major justification is that they work (or at least, they worked). Of course the mathematical structure of quantum mechanics is such a formalism. In this formalism a particle "is" (Pythagoreans will remove the scare quotes) not much more than an irreducible representation of its most comprehensive symmetry group. In that case, the theory of group representations, and the categories it works with, turn into the fundamental classification of reality, deeper than the fire/air/earth/water of the ancients, and even deeper than the periodic table of the elements. And this classification, in turn, makes it possible to draw analogies even from false theories in the process of guessing the true ones, using the mathematical hierarchies as (what Nelson Goodman called) "projectible predicates," or what were called, in the Middle Ages, natural kinds.

My claim, then, is that a good deal of what passes for applications of mathematics in physics, is really the application of mathematics to itself.[30] The end result remains empirical—make no mistake—because there are rules connecting the mathematical formalism to empirical predictions which could, but don't, fail. Nevertheless, the procedure is deeply Pythagorean, because the classification of empirical phenomena is induced by the classification of mathematical structures by mathematicians.

[29] By "crudely" empirical, I mean that the arithmetical operations are identified with everyday operations such as gathering (for plus). Even Quine, who regards mathematical theories as part of our entire scientific doctrine, which does face the tribunal of experience, rejects such a crude interpretation of arithmetic. Arithmetical theory for him is more like the theory of quarks than the theory of pebbles and gingerbread (to borrow phrases from Frege).

[30] This claim is analogous to the one I made (Steiner 1978b) concerning mathematical *explanations* in science.

There is a great vindication here of Galileo's vision of the "Book of Nature," written in the language of circles and triangles, except of course that the circles and the triangles themselves are replaced by their symmetry groups. The vision is still empiricist, as long as we distinguish between the Book and Nature itself. It is Nature which gives validity to the Book, but we use the Book as a map of Nature.

Yet the boundary between the Book and Nature itself has recently been becoming more and more blurred. Good old space-time is now in some quarters regarded as a mere low-energy approximation to what there really is in the way of dimensions (I have heard lectures in which over fifty dimensions were postulated).[31] We can detect by observation only four dimensions, or at least it appears to us that we do—and these appearances remain the basis for accepting the various hypotheses offered. Other dimensions may be, for example, cylindrical, so tightly rolled up that we cannot detect them. We may fundamentally misrepresent what we do detect—for example, when we think that the four dimensions of space-time are continua (they may, for all we know, be lattices). Finally, mass itself, together with gravitation, may be amenable to the group-theoretic approach. This would be the ultimate irony: Pythagoras and Democritus might turn out ultimately to have been saying exactly the same thing—of course, the world is made only of matter, but look what matter is! Materialism as a doctrine might turn to be not so much wrong as pointless—in a world in which matter, energy, and space-time turn out to be mathematical structures. What of atomism? Well, what are atoms? When we say "protons are made of quarks," all we might mean is that a certain irreducible representation of the symmetry group SU(3) is "constructed" from its basic ("atomic"), three-dimensional, irreducible representation, by means of mathematical operations such as "tensor product."[32] (Pythagoras thought the world is made of numbers, but that is too simple: it's made of matrices of numbers, and the numbers are not real, but complex!) In such a world, by the way, causality itself would be a pointless concept,[33] valid only of the world as it appears to our poor receptors (as Bertrand Russell argued almost one hundred

[31] Charles Peirce (1999) foresaw this possibility.

[32] For the definition of the tensor product of representations, see Sternberg (1994, Appendix B, pp. 320ff.). The tensor product of two irreducible representations is again irreducible, which is the basis of our Pythagorean "atomism."

[33] I am referring to the pointlessness of causal *laws*. Causal *reasoning* will continue to play a role in physics, for example, in the labeling of certain solutions of a valid equation "nonphysical"—as in Einstein's controversial rejection of travel faster than the speed of light on causal grounds, in Special Relativity. Ironically enough, however, in *General* Relativity, recent studies of black holes seem to allow the kind of "time travel" that Einstein ruled out! (Thanks to Shlomo Sternberg for pointing out this work to me.) This just shows that causal reasoning (even of Einstein) is fallible, not that it doesn't exist.

years ago). For as the late Pythagoreans argued, Aristotle's Four Causes can themselves be reduced to mathematical properties and relations.[34]

Does this raise any philosophical problems? You bet. First of all, there is the question of whether Pythagoreanism is a *coherent* doctrine—I believe that it is. I am persuaded by John Locke's central insight that the world could turn out to be fundamentally unlike what it appears to be, and the framework of space and time is no exception. Of course, Locke's rigid separation of primary from secondary qualities makes it seem as though it is some kind of an a priori truth that the world has the geometric properties it seems to have. Yet some astute commentators have pointed out passages in the *Essay* that hint that Locke saw that even the primary qualities may not be the last word, and that if we had the ability to perceive what is going on at the atomic level, even the primary qualities might turn out to be appearances.

The ultimate problem, then, would be whether Pythagoreanism is an *acceptable* doctrine.[35] If it is, then we end this article with the greatest irony of all.

REFERENCES

Colyvan, Mark. 2001. *The indispensability of mathematics.* New York: Oxford University Press.

Dummett, Michael. 1991. *Frege: Philosophy of mathematics.* Cambridge, Mass.: Harvard University Press.

Field, Hartry. 1980. *Science without numbers: A defense of nominalism.* Princeton, N.J.: Princeton University Press.

French, Steven. 2000. The reasonable effectiveness of mathematics: Partial structures and the application of group theory to physics. *Synthèse* 125: 103–120.

Guillemin, Victor, and Shlomo Sternberg. 1990a. *Symplectic techniques in physics.* Cambridge: Cambridge University Press.

Guillemin, Victor, and Shlomo Sternberg. 1990b. *Variations on a theme by Kepler.* American Mathematical Society Colloquium Publications, 42. Providence, R.I.: American Mathematical Society.

[34] Cf. O'Meara (1989).

[35] Readers familiar with the writings of W. V. Quine will recall that Pythagoreanism (especially rampant Pythagoreanism) was a sign that something had gone seriously wrong. But the context there was the use of the "Skolem–Löwenheim Theorem" to reinterpret any consistent first-order theory (including the theory of the continuum) in a universe consisting solely of natural numbers. There, however, the *predicates* turn out to be true of arbitrary classes of numbers (i.e., the predicates are not arithmetical). Here, everything is interpreted as usual, except it turns out that every property of the physical world is mathematical.

My own view, that Pythagoreanism is ultimately an anthropocentric doctrine, is expounded in Steiner (1998).

Lewis, David. 1991. *Parts of classes.* Oxford: Basil Blackwell.

Maddy, Penelope. 1990. *Realism in mathematics.* Oxford: Oxford University Press.

Mendelson, Elliot. 1997. *Introduction to mathematical logic*, 4th ed. New York: Chapman & Hall.

O'Meara, Dominic J. 1989. *Pythagoras revived: Mathematics and philosophy in late antiquity. Appendix; The Excerpts from Iamblichus; On Pythagoreanism V–VII in Psellus. Text, Translation, and Notes.* Oxford: Clarendon Press.

Peirce, Charles. 1999. Note on the analytical representation of space as a section of a higher dimensional space. In *Writings of Charles S. Peirce: A chronological edition*, ed. Nathan Houser, vol. 6. *1886–1890*: 260–262. Indianapolis: Indiana University Press.

Shapiro, Stewart. 1983. Conservativeness and incompleteness. *Journal of Philosophy* 80: 521–531.

Steiner, Mark. 1978a. Mathematical explanation. *Philosophical Studies* 34: 135–151.

Steiner, Mark. 1978b. Mathematics, explanation, and scientific knowledge. *Noûs* 12: 17–28.

Steiner, Mark. 1982. Review (in Hebrew) of Hartry Field, *Science without numbers. Iyyun* 31: 211–217.

Steiner, Mark. 1998. *The applicability of mathematics as a philosophical problem.* Cambridge, Mass.: Harvard University Press.

Sternberg, Shlomo. 1994. *Group theory and physics.* Cambridge: Cambridge University Press.

Weinberg, Steven. 1986. Lecture on the applicability of mathematics. *Notices of the American Mathematical Society* 33: 725–728.

Wigner, Eugene. 1967. The unreasonable effectiveness of mathematics in the natural sciences. In his *Symmetries and Reflections*: 222–237. Bloomington: Indiana University Press.

Wittgenstein, Ludwig. 1978. *Remarks on the foundations of mathematics*, 3rd ed., rev., ed. G.H. von Wright, R. Rhees, and G.E.M. Anscombe. Cambridge, Mass.: MIT Press.

CHAPTER 21

LOGICAL CONSEQUENCE, PROOF THEORY, AND MODEL THEORY

STEWART SHAPIRO

1. Proof Theory and Model Theory

The first entry under "logic" in the *Oxford English Dictionary* is "the branch of philosophy that treats of the forms of thinking in general, and more especially of inference and of scientific method." The corresponding entry in Webster's *New Twentieth Century Unabridged Dictionary* is "the science of correct reasoning; the science that deals with the criteria of valid thought." Of course, it is not a good idea to let Oxford and Webster dictate our agenda, but they are correct that logic is, at root, a philosophical enterprise. Since at least the beginning of the twentieth century, however, logic has become a branch of mathematics as well as a branch of philosophy. As a youngster, I was informed of the four great fields of mathematical logic: proof theory, model theory, set theory, and computability. Our main question here concerns how that wonderful mathematics relates to the philosophical targets: correct reasoning, valid thought, inference. Model theory and proof theory each provide for a notion of logical consequence, but the notions employed by these two branches are quite different from each other, at least conceptually. Although set theory and computability are interesting and important in their own right, their connection to "correct reasoning" is more indirect. Set theory deals with the realm of models for model theory, and computability relates to the notion of a formal deductive system, in proof theory.

Textbooks and advanced treatments in mathematical logic work with formal languages, which are rigorously defined sets of strings on a fixed alphabet. The resulting mathematics is, of course, interesting and important, but we can query its philosophical ramifications and its relation to correct reasoning.

Typically, parts of a formal language roughly resemble parts of a natural language. Characters like "&," "$\vee$," "$\rightarrow$," "$\neg$," "$\forall$," and "$\exists$" correspond to the English expressions "and," "or," "if... then," "it is not the case that," "for every," and "there is," respectively. These are sometimes called the *logical terms* of the formal or natural language. Some formal languages include specific nonlogical terms, such as the sign for the "less-than" relation over the natural numbers, but it is more common to include a stock of schematic letters which stand for arbitrary, but unnamed, nonlogical names, predicates, and functions.

In this chapter, I shall use lower-case Greek letters for items from formal languages and upper-case Greek Letters for corresponding sentences from natural languages (or propositions, or whatever the medium of correct reasoning is). Let γ be a set of formulas and ϕ a single formula of a formal language. A typical logic text formulates two rigorous notions of consequence, two senses in which ϕ follows from γ. For one of them the author presents a *deductive system D*. The simplest of these consist of a list of axioms and rules of inference. An argument $\langle\gamma, \phi\rangle$ in the formal language is *deductively valid* (via *D*) if there is a sequence of formulas ending with ϕ, such that each member of the sequence is either a member of γ or an axiom of *D*, or follows from previous formulas in the sequence by one of the rules of inference of *D*. Natural deduction systems and sequent calculi are a bit more complex, but similar in spirit. In each case, there is a rigorous notion of deductive validity. If $\langle\gamma, \phi\rangle$ is deductively valid in a system *D*, we write $\gamma \vdash_D \phi$, or simply $\gamma \vdash \phi$ if it is safe to suppress mention of the deductive system.

The other notion of consequence invokes a realm of *models* or *interpretations* of the formal language. Typically, a model is a structure $M = \langle d, I\rangle$, where d is a set, the *domain* of M, and I is a function that assigns extensions to the nonlogical terminology. For example, if c is a constant, then Ic is a member of the domain d, and if R is a binary predicate, then IR is a set of ordered pairs on d. Then one defines a relation of *satisfaction* between interpretations M and formulas ϕ. An interpretation M satisfies ϕ, written $M \models \phi$, if ϕ is true under the interpretation M. Finally, one defines ϕ to be a *model-theoretic* consequence of γ if every interpretation that satisfies every member of γ also satisfies ϕ. In other words, ϕ is a model-theoretic consequence of γ if there is no interpretation that satisfies every member of γ and fails to satisfy ϕ. In this case, we write that the argument $\langle\gamma, \phi\rangle$ is model-theoretically valid, or $\gamma \models \phi$.

Recall that the deductive notion of consequence is defined relative to a deductive system. Although, to be consistent, one should define the model-theoretic notion relative to the background set theory, or the background class of models, this is not often done. It is usually assumed, perhaps tacitly, that either the background set theory is ZFC, or else that all of the reasonable, competing set theories deliver the same relation of logical consequence.

Since model-theoretic consequence (via a set theory) and deductive validity (via a deductive system *D*) are sharply defined notions on a given formal language, relations between them are mathematical matters. The system is *sound* if every deductively valid argument is also model-theoretically valid, and the system is *complete* if every model-theoretically valid argument is also deductively valid.

Typically, soundness is easily established. One just checks each axiom and rule of inference to make sure that it does not lead from true premises to a false conclusion in any interpretation. Virtually every system presented in treatments of logic is sound. When completeness obtains, it is a usually a deep and interesting mathematical result. Gödel's [1930] completeness theorem entails that first-order logic (with or without identity) is complete. It is a corollary of Gödel's *incompleteness* theorem that second-order logic is not complete (see Shapiro [1991, ch. 4]). Its consequence relation is not recursively enumerable, and so no effective, sound deductive system is complete for it.

Our main concern here is the notion of model-theoretic consequence. What does it have to do with correct reasoning? We will take on deductive consequence only by way of contrast. Do these two notions answer to different intuitive notions of consequence? Is one of them primary, and the other secondary? Or perhaps they are autonomous and independent. Maybe there are two distinct notions of correct reasoning, valid thought, and/or inference.

For what it is worth, treatments of mathematical logic usually presuppose that the model-theoretic notion is the primary one. For example, one says that a deductive system is sound or complete (or not) *for the semantics*—not the other way around. If a deductive system is not sound for a given semantics, then that alone disqualifies the deductive system. Why? Because the deductive system allows us to deduce a falsehood from truths in some interpretation of the language. But one could perhaps argue instead that it is the model theory that is at fault. Any counterexample to soundness—any "interpretation" in which we can deduce a false conclusion from true premises—is perhaps not a legitimate interpretation of the language. For better or worse, however, most mathematical logicians do not think that way.

To a lesser extent, it is the same for completeness. Suppose that a given deductive system is not complete for a given semantics. Again, a typical, but not universal, response is that the deductive system is at fault. Why? Because the deductive system does not have the resources to establish every valid argument. Again, one might argue instead that it is the semantics that is at fault. It does not have enough interpretations to provide counterexamples to every argument that is not deducible in the deductive system.

In this case, some logicians do put things this way. As noted, the usual deductive systems for second-order languages are not complete for the so-called standard semantics. It is not a matter of having left out one or two invalid deductions. *No* effective deductive system is sound and complete for second-order validity—assuming standard semantics. Among those who accept second-order logic, the most common conclusion is that no sound, effective deductive system captures the relevant notion of consequence. But some thinkers take a different line. The usual deductive systems for second-order languages are complete for Henkin semantics. This last, in effect, contains more interpretations of the language—so-called nonstandard interpretations (see Shapiro [1991, ch. 4] for details). Some philosophers and logicians argue that Henkin semantics (or something equivalent to it) is the appropriate model theory, not standard semantics.

We won't delve any deeper here into the specific issues concerning second-order logic (see chapters 25 and 26 of this volume), except occasionally for purposes of illustration. Present concern is with the general notion, or notions, of logical consequence, and how those relate to model theory.

2. Logical Consequence

Our next task is to explore the intuitive, or pretheoretic, notion of logical consequence. What are correct reasoning, valid thought, and inference—the supposed subject matter of logic? Answers to this question will lend perspective to the mathematical notions of model-theoretic consequence and deductive consequence. What follows is based loosely on Shapiro [1998] and [2002].

2.1. Modality

Aristotle provided the earliest systematic treatment of logical consequence (or at least the earliest surviving treatment):

> A syllogism is a discourse in which, certain things having been supposed, something different from the things supposed results of necessity because these things are so. By "because these things are so," I mean "resulting through them" and by "resulting through them," I mean "needing no further term from outside in order for the necessity to come about." (*Prior Analytics*, book I, ch. 2)

Aristotle thus holds that a given proposition Φ is a consequence of a set Γ of propositions only if Φ is different from any of the propositions in Γ. Nowadays, logicians reject this and allow that Φ trivially follows from Γ when Φ is a member of Γ. But this is not a substantial issue. Aristotle's phrase "because these things are so" might indicate that in order to have a consequence, or "syllogism," the premises in Γ must all be *true*. We will not enter into this exegetical issue. With the possible exception of Gottlob Frege, most modern conceptions of consequence allow instances of logical consequence in which the premises are false. For example, "Socrates is a puppy" follows from "All men are puppies" and "Socrates is a man."

With the phrase "results of necessity," Aristotle introduces a modal element into the definition. He glosses the phrase "because these things are so" as "resulting through them," and he glosses that as "needing no further term from outside in order for the necessity to come about." These clauses seem to indicate that in a valid argument the premises *alone* guarantee the conclusion, or that the premises are sufficient for the conclusion.

Our first conception of logical consequence is modeled on this reading of Aristotle's definition:

> **(M)** Φ is a logical consequence of Γ if it is not possible for the members of Γ to be true and Φ false.

Following the contemporary practice of phrasing modal notions in terms of possible worlds, the thesis (M) becomes

> **(PW)** Φ is a logical consequence of Γ if Φ is true in every possible world in which every member of Γ is true.

According to (M) and (PW), "Bill is heavier than Joe" follows from "Joe is lighter than Bill," since it is impossible for Joe to be lighter than Bill and for Bill to fail to be heavier than Joe. Surely, Bill is heavier than Joe in every possible world in which Joe is lighter than Bill. Or so one would think. To adapt an example from Bernard Bolzano, a religious person who accepts (M) or (PW) would say that "Caius has an immortal soul" follows from "Caius is a human" since (according to the person's theology) the premise cannot be true and the conclusion false.

On most contemporary accounts of logic, neither of these conclusions is a logical consequence of its corresponding premise. It is a routine exercise to formalize these arguments and show that the conclusions do not follow on either model-theoretic or deductive notions of consequence. Perhaps we can bring (M) and (PW) closer to the contemporary notions by invoking a special notion of *logical* possibility and necessity. It is perhaps physically or metaphysically or analytically impossible for Joe to be lighter than Bill and Bill to fail to be heaver than Joe, but it is logically possible for this to happen. Along similar lines, there are perhaps no theologically possible worlds in which Caius is human and fails to have an immortal soul, but perhaps our theologian will concede that there are logically possible worlds where this happens.

To pursue this tactic, we would have to articulate the distinctive notion of logical possibility. Aristotle's final clause, that a syllogism needs "no further term from outside in order for the necessity to come about," seems appropriate for this purpose. In the example about Bill and Joe, we need to invoke some "outside" fact about the relationship between heaviness and lightness in order for the "necessity to come about," or at least in order to understand the necessity. In the theological example, we need the fact (if it is a fact) that, necessarily, all humans have immortal souls. "Caius has an immortal soul" does follow logically from "Caius is a human" *together with* the relevant theological thesis that all humans have immortal souls.

On the other hand, I would think that there just is no possible world in which Bill is heavier than Joe without Joe being lighter than Bill. Joe being lighter than Bill is part of what it is for Bill to be heavier than Joe. And I presume that most theologians would insist that having an immortal soul is part of what it is to *be a human.* There is no possible world in which Caius is a man and yet fails to have an immortal soul. To adequately understand the formal notions of model-theoretic and deductive consequence, we must supplement the modal notions.

2.2. Semantics

According to Alberto Coffa [1991], a major concern of philosophers throughout the nineteenth century was to account for the necessity of mathematics without invoking Kantian intuition. Coffa proposed that the most successful line came from the "semantic tradition," running through the work of Bolzano, Frege, and Ludwig Wittgenstein, and culminating in the Vienna Circle. The idea is that the relevant necessity lies in the use of language, or *meaning.* Although this is a departure from Aristotle, we can invoke the semantic program to further articulate the notion of logical consequence:

(S) Φ is a logical consequence of Γ if the truth of the members of Γ guarantees the truth of Φ in virtue of the meanings of the terms in those sentences.

Thesis (S) takes care of our theologian. I presume that even the most religious linguist or philosopher of language does not take it to be part of the *meaning* of the word "human" that humans have immortal souls. If someone who denies that humans have immortal souls merits divine or earthly retribution, it will not be on account of a *linguistic* deficit. Indeed, outside the classroom, one does not usually get punished for failing to understand certain words. So everyone should agree that according to (S), "Caius has an immortal soul" does not follow from "Caius is human."

Nevertheless, our other troubling example is left intact. According to (S), "Bill is heavier than Joe" does indeed follow from "Joe is lighter than Bill," since the meanings of "heavier" and "lighter" determine that these relations are converses of each other. Someone who understands the meaning of the two sentences has all she needs to know that if the premises are true, then so is the conclusion.

The thesis (S) captures what is sometimes called "analytic consequence," which is often distinguished from logical consequence, due to examples like the one considered here. We now refine the semantic idea.

2.3. Form

There is a long-standing view that logical consequence is a matter of *form*. The idea is that an argument is valid if and only if every argument with the same (logical) form is valid. As far as I know, Aristotle does not explicitly say that validity is a matter of form, but his work in logic presupposes this. He sometimes presents "syllogisms" by just giving the forms of the propositions in them, not bothering with their content. More telling, perhaps, is that when Aristotle wants to show that a given conclusion does *not* follow from a given pair of premises (i.e., that a given sequence is not a syllogism), he typically gives an argument in the same form with true premises and false conclusion. Prima facie, this practice presupposes that if an argument is valid, then every argument in the same form is valid. The presented "counterargument" is surely invalid, since its premises are true and its conclusion is false. The original given argument has the same form as the presented counterargument. Thus, the original argument is not valid.

Consider a paradigm case of a valid argument:

All men are mortal; Socrates is a man; therefore, Socrates is mortal.

The validity of this argument does not turn on anything special about mortality and Socrates. *Any* argument in the following form is valid:

All *A* are *B*; *s* is an *A*; therefore *s* is a *B*.

That is, if one replaces the schematic letters *A*, *B* with any predicates or common nouns and *s* with any name or definite description that describes something, the result is a valid argument.

However, one might say similar things about our examples. Consider the following "forms":

s is lighter than *t*; therefore *t* is heavier than s.
s is human; therefore *s* has an immortal soul.

An argument in one of these forms has the same status, vis-à-vis (M), (PW), and even (S), as the argument it was taken from. So one can still think of the arguments about Bill, Joe, and Caius as valid in virtue of their forms.

Nevertheless, the prevailing view is that neither of these arguments has a valid form. Although the argument about weight does not turn on anything special about the denotations of "Bill" and "Joe," it does turn on specific facts about (the meaning of) "heavier" and "lighter," namely, that these relations are converse to each other. Similarly, the argument about immortal souls does not turn on anything about Caius, but it does turn on putative theological facts about humans. On the prevailing view, the requisite logical form of both of the above arguments is

s is *A*; therefore *s* is *B*.

It is, of course, straightforward to find an argument in *this* form with a true premise and a false conclusion. Consider: "Kansas City is in the United States, therefore Kansas City is in Ohio." Thus the original arguments are not valid in virtue of *this* form, and for the arguments at hand, this is the form that matters.

If someone wanted to be stubborn, he could point out that even though the paradigm valid argument does not turn on anything special about Socrates, humanity, or mortality, it does turn on the specific meaning of "all," "are," and "is." We can give the following "form" to the paradigm valid argument:

Π *A* is *B*; *s* is *A*; therefore, *s* is *B*,

and then give a counterargument in that "form":

Some people are senators; Jorge Posada is a person; therefore Jorge Posada is a senator. This has true premises and a false conclusion. So even the paradigm argument is not valid in virtue of the last-displayed "form."

Of course, the standard response is to claim that the last-displayed "form" is not a *logical* form of the paradigm argument. But this raises the next question. How are we to characterize logical form? One might say that a form is logical if the only terms it contains (besides the schematic letters) are *logical terms.*

We now face the task of characterizing the logical terms. How do we go about designating a term as logical? The logician, or philosopher of logic, has three options. One is to attempt a principled definition of "logical term," perhaps by focusing on some of the traditional goals and purposes of logic (see, e.g., Peacocke [1976], Hacking [1979], McCarthy [1981], Tarski [1986], Sher [1991]). The proposals and theories cover a wide range of criteria and desiderata, such as a priori knowledge, analyticity, formality, justification, and topic neutrality. It would take us too far afield to examine the proposals here. A second tactic, implicitly followed in most logic textbooks, is to merely provide a list of the logical terms and leave our task with this act of fiat. Typically, logical terms consist of truth-functional connectives ("not," "and," "or," "if... then"), quantifiers ("some," "all"), variables, and perhaps the sign for identity. This might leave the readers wondering what is going on, and of course it provides no insight into the choice of logical terms. A third option (following Bolzano [1837] and Tarski [1935]) is to make the notions of logical form and logical consequence relative. That is, one defines an argument to have a certain logical form *relative to* a given choice of logical terms. The same argument might be valid relative to one set of logical terms and invalid relative to another. Here we just adopt the easier, second course.

Our next notion of consequence combines the notion of logical form with a semantic conception like (S) above:

> **(FS)** Φ is a logical consequence of Γ if the truth of the members of Γ guarantees the truth of Φ in virtue of the meanings of the logical terminology.

2.4. Epistemic Matters

We still have not addressed the role of logical consequence in organizing and extending knowledge. As noted above, a common slogan is that logic is the study of correct *reasoning*. Truth be told, we have not said much about reasoning yet. It seems that one reasons from premises to conclusion via valid arguments. If we believe the premises, we must believe the conclusion, on pain of contradiction.

Let us propose another definition of consequence:

> **(R)** Φ is a logical consequence of Γ if it is irrational to maintain that every member of Γ is true and that Φ is false. The premises Γ alone *justify* the conclusion Φ.

This much also seems consonant with at least part of Aristotle's definition of syllogism.

Back to our examples. A theologian might admit that it is not irrational to hold that Caius is a human being without an immortal soul. Indeed, our theologian should concede that someone can know that Caius is a human without knowing that he has an immortal soul. Some wretched folks are simply ignorant of the relevant theology, or too stubborn to understand it. On the other hand, there does seem to be something irrational in maintaining that Bill is shorter than Al while denying that Al is taller than Bill—unless, of course, one does not know the meaning of "shorter" or "taller." But perhaps one can also rationally deny that Socrates is mortal while affirming that all men are mortal and Socrates is a man—if one pleads ignorance of the meaning of "all."

Surely, in attributing rationality and irrationality to our subjects, we must credit them with understanding of the meanings of the terms in their arguments. So (R) seems related somehow to the semantic notion (S) from section 2.2.

What is the penalty for being irrational? What, exactly, is the "pain" of contradiction? The idea is that one who affirms the premises and denies the conclusion of a valid argument has thereby said things which *cannot* all be true. This broaches modal notions, as in (M) and (PW) from section 2.1. I submit, however, that the pain of contradiction goes further than this. The charge is not merely that our subject has said something impossible, but that she *could have known better*, and indeed *should have known better*. In this sense, logical consequence is a *normative* notion. It concerns the epistemic burdens on those who reason.

The most common way to articulate the normativity here is in terms of *deduction*. A sentence Φ is a consequence of Γ in this sense if there is a process of inference taking one from members of Γ to Φ. The purpose of deduction is to provide a convincing, final case that someone who accepts the members of Γ is thereby committed to Φ. So we have:

(Ded) Φ is a logical consequence of Γ if there is a deduction of Φ from Γ by a chain of legitimate, gapfree (self-evident) rules of inference.

Arguably, this notion also has its pedigree with Aristotle. He presents a class of syllogisms as "perfectly" valid, and shows how to reduce other syllogisms to the perfectly valid ones by inference (see Corcoran [1974]).

2.5. Recapitulation

I do not claim that the foregoing survey includes every notion of logical consequence that has been seriously proposed and maintained. For example, there is a tradition, going back to antiquity and very much alive today, that maintains that Φ is not a logical consequence of Γ unless Γ is *relevant* to Φ (see chapter 23 of the present volume). But to keep the treatment from getting any more out of hand, we will stick with the above notions. Here they are:

(M) Φ is a logical consequence of Γ if it is not possible for the members of Γ to be true and Φ false.

(PW) Φ is a logical consequence of Γ if Φ is true in every possible world in which every member of Γ is true.

(S) Φ is a logical consequence of Γ if the truth of the members of Γ guarantees the truth of Φ in virtue of the meanings of the terms in those sentences.

(FS) Φ is a logical consequence of Γ if the truth of the members of Γ guarantees the truth of Φ in virtue of the meanings of the logical terminology.

(R) Φ is a logical consequence of Γ if it is irrational to maintain that every member of Γ is true and that Φ is false. The premises Γ alone justify the conclusion Φ.

(Ded) Φ is a logical consequence of Γ if there is a deduction of Φ from Γ by a chain of legitimate, gap-free (self-evident) rules of inference.

3. Proof Theory and Model Theory Again

Each of these ideas captures, or models, a different notion of logical consequence, and there are some issues concerning their interrelations. The framework lends some perspective to the differences between the formal, mathematical notions of deductive consequence and model-theoretic consequence.

Deductive consequence is a formal analogue, or model, of (Ded). Let us say that a deductive system is *correct* if—or to the extent that—each of its primitive rules of inference corresponds to a legitimate, gapfree move in reasoning. This is a measure of the extent to which a deductive system is an accurate account of reasoning in natural language, or whatever the medium of actual reasoning is. Unlike soundness and completeness, which are sharp mathematical notions, correctness is a vague matter, not an all-or-nothing affair. If a deductive system D is (more or less) correct, then each deduction in D (more or less) corresponds to a legitimate, or valid, derivation in ordinary reasoning. That is, $\gamma \vdash_D \phi$ only if (Ded) holds for any natural language argument in that form. And (Ded) entails (R). So if $\gamma \vdash_D \phi$, then for any natural language argument in that form, it is not rational to maintain the premises and deny the conclusion.

What of the converse of correctness? Say that a deductive system D is *adequate* if every chain of legitimate gapfree inferences in natural language (or every chain that turns on certain features of certain natural language terms, or on certain formal features) can be recaptured in D. That is, D is adequate if, for each correct chain of reasoning (of the right sort), there is a deduction in D that captures its form. Like correctness, adequacy is a vague matter, especially if we are limiting its scope to certain kinds of arguments.

Prima facie, we have intuitions concerning what is a legitimate, gapfree inference in ordinary language. Aristotle's perfect syllogisms are a perfect example (pun intended). If there are such intuitions, they can pronounce on the correctness of a given deductive system *D*. We just check each primitive rule of *D*, to make sure that it corresponds to a gapfree inference. I take it, without much in the way of further argument, that standard first- and higher-order deductive systems are correct.

The intuitions in question, plus a bit of mathematical work, might tell us that a given deductive system *D* is *not* adequate. This will happen if we recognize a gapfree inference in ordinary language that has no counterpart in *D*. However, said intuitions concerning instances of correct, gapfree inference give us little clue on how to establish, or even argue for, adequacy. Do we have intuitions concerning the full extent or range of legitimate inference patterns? The thesis that a given deductive system is adequate is the same sort of thing as Church's thesis. It can perhaps be argued for on quasi-empirical grounds, noting that no exceptions are known so far. And different deductive systems, constructed independently, might converge on the same consequence relation. Also, adequacy is subject to analysis. I give an argument for adequacy below, restricted to first-order languages which, I claim, is (and ought to be) convincing.

To place the model-theoretic consequence relation in the present framework, let us focus briefly on satisfaction, the relation that holds between a set-theoretic interpretation and a formula of the formal language that is true on that interpretation. I suggest that the recursive clauses in the definition of satisfaction represent the relationship between sentences in natural language and the world. In other words, satisfaction conditions are a formal analogue of (part of) the *truth conditions* of sentences in natural language. As Hodes [1984] once put it, truth in a model is a model of truth. This recapitulates the slogan that validity is truth-preserving.

Model-theoretic consequence does not correspond very well to the semantic notion (S) of analytic consequence. Any "meaning" assigned to nonlogical terms is quite irrelevant, since these terms get different extensions from interpretation to interpretation. The definition of validity has a bound variable ranging over the entire class of interpretations, and this includes all possible extensions for each nonlogical term, independent of the meaning of any term from natural language corresponding to it.

Not so the logical terminology. In the context of model-theoretic semantics, the logical terminology of the formal language derives its meaning from the recursive clauses in the definition of satisfaction. For example, a formula in the form $\phi \vee \psi$ is satisfied in an interpretation (under an assignment) just in case either ϕ is satisfied in that interpretation or ψ is satisfied in that interpretation (under the same assignment). This fixes the meaning of the connective "$\vee$": it is the same truth function in every interpretation. In other words, model-theoretic consequence does turn on the meaning of the logical terminology (as it is manifest in the recursive definition of satisfaction). Nevertheless, the model theory does not recapitulate the formal notion (FS), since the meaning of the logical terminology is not all that matters. A modal element remains.

From this perspective, an interpretation in the model theory represents a way the world might be—a possible world—and the relation of satisfaction tells us which formulas would be true, had the world been that way. In particular, the different domains of discourse represent various possible contents to the universe. But this is a rather attenuated use of modality, due to the range of extensions of the nonlogical terminology.

Two interpretations M_1, M_2 are *isomorphic* if there is a one-to-one function f from the domain of M_1 onto the domain of M_2 that preserves the relations. For example, if R is a binary relation in the language, then the pair $\langle a, b\rangle$ is in the extension of R in M_1 if and only $\langle fa, fb\rangle$ is in the extension of R in M_2. Interpretations are isomorphic if they have the same structure.

Any model-theoretic semantics worthy of the name has the *isomorphism property*: if M_1 and M_2 are isomorphic; then for any sentence ϕ in the formal language, M_1 satisfies ϕ if and only if M_2 satisfies ϕ. In other words, isomorphic interpretations are equivalent. The formal language does not distinguish between them. The isomorphism property is established by induction on the complexity of each formula.

It follows from the isomorphism property that the individual objects in a domain of discourse do not matter. Suppose that $M = \langle d, I\rangle$ is an interpretation, and let d' be any set that is the same size as d. Then we can define a function I' such that the interpretation $M' = \langle d', I'\rangle$ is isomorphic to M. Indeed, let f be a one-to-one function from d' onto d. Then, for example, if R is a binary relation symbol, let a pair $\langle a, b\rangle$ of elements of d' be in $I'R$ if and only if the pair $\langle fa, fb\rangle$ is in IR. As above, M' is indistinguishable from M, even if the objects in the two domains are radically different from each other.

What does matter for model-theoretic consequence is the *size* of each interpretation. The class of interpretations represents the range (or a range) of sizes the universe might be. I submit that this is the only modality that is registered in model-theoretic consequence.

Model-theoretic consequence thus has elements of (M), (PW), and (FS). Let us say that a sentence Φ (in natural language) is a consequence of a set Γ of sentences in a *blended* sense if it is not possible for every member of Γ to be true and Φ false, and this impossibility holds in virtue of the meaning of the logical terms. In the terminology of possible worlds, Φ is a logical consequence of Γ in this blended sense if Φ is true in every possible world under every reinterpretation of the nonlogical terminology in which every member of Γ is true. Perhaps it is not too much of a stretch to see this blended notion of consequence as in line with Aristotle's notion: "a discourse in which, certain things having been supposed, something different from the things supposed results of necessity because these things are so (see §2.1 above)." The blended notion also recapitulates the slogans that logical consequence is truth-preserving, and that consequence is a matter of logical form.

There is a direct analogue of the above treatment of deductive consequence. Define a model-theoretic semantics to be *correct* if—or to the extent that—for any model-theoretically valid argument in the formal language, each argument in natural language with that form is a logical consequence in the blended sense. It might be clearer if we focus on the contrapositive. The model-theoretic semantics is correct if it has enough interpretations to refute any argument that is not valid in the blended sense. Suppose that a given sentence in natural language fails to be a logical consequence, in the blended sense, of a given set of such sentences. Then there is a possible world and an interpretation of the nonlogical terminology that makes the premises true and the conclusion false. If the model-theoretic semantics is correct, then there is an interpretation in the model-theoretic semantics that satisfies the formal counterparts of the premises but does not satisfy the formal counterpart of the conclusion.

Notice that the correctness of a model-theoretic semantics does not entail that its realm of interpretations has a counterpart to every possible world. Consider, for example, ordinary set-theoretic model theory. It may be that in some possible world (like this one), the range of the variables is not a set, in which case there is no interpretation in the model theory that corresponds to it. All that correctness requires is that if an argument is not valid in the blended sense, then at least one item in the realm of interpretations in the model theory refutes (a formal counterpart to) it.

Turning to the converse, define a model-theoretic semantics to be *adequate* if each argument in natural language that is valid in the blended sense—or each argument whose validity in the blended sense turns on certain features of the language—corresponds to a model-theoretic consequence. Again, consider the contrapositive. Suppose that a given formula is not a model-theoretic consequence of a given set of formulas. Then there is an interpretation in the model theory that satisfies every premise but does not satisfy the conclusion. If the semantics is adequate, then there is also a possible world in which one can interpret the nonlogical terms to make the premises true and the conclusion false. In other words, the model-theoretic semantics is adequate if each interpretation in the semantics corresponds to a genuinely possible world. In light of the isomorphism property, all that adequacy requires is that each interpretation represent a possible *size* of the universe.

For model theory, adequacy is the easier property to establish, or at least support. We seem to have intuitions concerning what is a correct interpretation, or at least what is a possible *size* of the universe, and we have intuitions concerning what is a legitimate interpretation of the nonlogical terminology. I take it for granted that the interpretations of standard model theory are indeed legitimate possibilities for first- and higher-order languages, and so for these languages, at least, model-theoretic consequence is adequate. Correctness is another matter, supported indirectly or in a quasi-empirical manner.

Correctness and adequacy concern how the formal notions of deductive and model-theoretic consequence relate to their informal counterparts: (Ded) and the blended notion, respectively. We can also ask how the informal notions relate to each other. If there is a deduction of a sentence Φ in natural language from a set Γ of premises by a chain of legitimate, gapfree (self-evident) inferences, then is Φ a consequence of Γ in the blended sense? In other words, does (Ded) imply that the argument is a logical consequence in the blended sense? This is the counterpart to soundness. Conversely, if Φ is a consequence of Γ in the blended sense, then is there a deduction of Φ from Γ by a chain of legitimate, gapfree (self-evident) rules of inference? Does validity in the blended sense imply (Ded)? This is the counterpart to completeness.

The pretheoretic or intuitive version of soundness is a sine qua non of rationality. If there were a chain of legitimate, gapfree inferences that were not valid in the blended sense, then it would be possible to reason correctly from truth to falsehood. In other words, there would be a possible world and an interpretation of the nonlogical terminology in which the premises are true, and the chain of supposedly correct, gapfree reasoning would deliver a falsehood (in that world). We had better not have that—at least for deductive reasoning. Fortunately, the pretheoretic version of soundness is easily checked. We would not call an inference legitimate (let alone gapfree) unless it is evident that it will not lead from truth to falsehood.

The counterpart of completeness is less crucial. It might well be the case that there are arguments that are valid in the blended sense whose conclusions cannot be deduced from their premises via a chain of gapfree inferences.

Let us assume, for now, that the deductive and semantic behavior of the logical terms of first-order formal languages ($\vee$, &, $\forall$, etc.) corresponds to the deductive and semantic behavior of their natural language counterparts. An instance of Kreisel's [1967] "informal rigor" can be adapted to show, or at least make it plausible, that the intuitive counterpart of completeness holds for arguments whose validity turns on the first-order logical terminology. Suppose that a given natural language sentence Φ is a consequence of a set Γ of sentences in the blended sense, and that this turns on the (counterparts to) the logical terminology of first-order languages. Let ϕ be a formula in the form of Φ, and let γ be the set of formulas that correspond to the sentences in Γ. By the adequacy of first-order model-theoretic consequence, ϕ is a model-theoretic consequence of γ. As noted above, this is checked by noting that each model-theoretic interpretation does indeed correspond to a possibility (in the sense of being a possible size of the universe). By the completeness theorem, ϕ can be deduced from γ in a standard deductive system for first-order logic. From the correctness of first-order deductive systems (checked by noting that each rule of inference in the deductive system corresponds to a correct inference in natural language) it follows that there is a chain of legitimate, gapfree inferences from members of the original Γ to the original conclusion Φ. In other words, (Ded) holds. So we have the pretheoretic counterpart of completeness.

Essentially the same result also indicates that for first-order languages, the standard deductive systems are adequate, in the sense that for each correct chain of reasoning (that turns on first-order structure) there is a deduction in a standard deductive system that captures its form. Suppose that a natural language sentence Φ is deducible from a set Γ of premises by a chain of legitimate, gapfree inferences. Again, let ϕ be the formal counterpart of Φ and let γ be the set of formal counterparts of the premises. If ϕ were not deducible from γ in a standard deductive system, then, by completeness, there would be an interpretation $M = \langle d, I \rangle$ in which the members of γ are true and ϕ is false. By the adequacy of the model-theoretic semantics, M corresponds to a possibility (i.e., a possible size), and thus the original Φ does not follow from Γ in the blended sense. This undermines the pretheoretic notion of soundness.

Similarly, we have good reason to hold that for first-order languages, the model-theoretic semantics is correct, in the sense that for any model-theoretically valid argument in the formal language, each argument in natural language with that form is a logical consequence in the blended sense. Again, let Φ be a sentence in natural language, and ϕ a formal counterpart, and let Γ be a set of premises, with γ a set of formal counterparts. Suppose that $\gamma \models \phi$. Then, by completeness, ϕ can be deduced from γ in a standard, first-order deductive system. So, by the pretheoretic version of soundness, the original sentence Φ is a consequence of the original Γ in the blended sense.

In short, for first-order languages, we can safely assume that the deductive systems are correct, the model-theoretic consequence is adequate, and the pretheoretic version of soundness holds. In the case of first-order formal languages and their natural language counterparts, it follows from this, and the completeness theorem, that the pretheoretic version of completeness holds, that the deductive system is adequate, and that the model-theoretic consequence is correct. Not bad for a series of supposedly informal notions.

However, these Kreisel-style arguments turn on Gödel's completeness theorem, and so are limited to first-order languages. Since second-order languages are inherently incomplete, one would think that the pretheoretic version of completeness fails as well, but this depends on the extent of legitimate, gapfree inference patterns. As far as I know, we have no way to ascertain that. We do know that the collection of second-order logical truths is not recursively enumerable (see Shapiro [1991, ch. 4]). So if the pretheoretic version of completeness holds, then the totality of legitimate, gapfree inference patterns available to us humans is not recursively enumerable. We thus broach issues of mechanism, and I won't speculate further on this.

Nevertheless, we can maintain the correctness of standard deductive systems (every inference in the formal system corresponds to a correct inference) and we can maintain the adequacy of model-theoretic consequence (each argument in natural language that is valid in the blended sense corresponds to a model-theoretic consequence). These are verified by an intuitive check of the rules of inference (to make sure they do not lead from truth to falsehood) and of the model-theoretic semantics (to make sure that each set-theoretic interpretation corresponds to a possible size of the universe). In contrast, the adequacy of the deductive system and the correctness of the model-theoretic consequence are left open. In Shapiro [1987] (and Shapiro [1991, ch. 6]), I argue that the correctness of the model theory is independent of ordinary set theory, in that it implies the existence of so-called small large cardinals.

To summarize so far, I submit that model-theoretic consequence ($\models$) and deductive consequence ($\vdash_D$) each correspond to a different intuitive notion of logical consequence: the blended notion and (Ded), respectively. Both of the latter are legitimate notions, and they are conceptually independent of each other. So there is no dispute between advocates of model theory and advocates of deductive systems. However, I know of no arguments to support this eclectic attitude. How does one show that a given, intuitive notion of consequence is legitimate? And how does one show that one notion of consequence is conceptually independent of or dependent on another?

The formal notion of deductive consequence ($\vdash_D$) depends on the particular deductive system *D*. It is clear that not just any collection of axioms and rules has a claim to be *a* (let alone *the*) notion of logical consequence. A. N. Prior's "The Runabout Inference Ticket" [1960] introduces an extension of an ordinary deductive system with a strange connective "tonk." It has an elimination rule like "&" and an introduction rule like "$\vee$." In the runabout deductive system the following are legitimate inference forms:

from ϕ infer (ϕ tonk ψ)
from (ϕ tonk ψ) infer ψ.

The deductive system that results is not very interesting. *Every* formula is a runabout consequence of each set of formulas. The inference system, if one can call it that, does not represent any notion of logical consequence.

The existence of "systems" like Prior's raises the question of what makes a deductive system legitimate. What conditions are necessary (and perhaps sufficient) for a proposed deductive system to be a candidate formalization of a notion of consequence like (Ded)?

With the exception of intuitionists, relevance logicians, opponents of higher-order logic (see chapters 9, 10, 23, and 26 of this volume) and some others, everyone agrees that the common deductive systems for first- and second-order languages are legitimate. But what makes those deductive systems legitimate (if they are)? How are we to distinguish between deductive systems that represent logical consequence and those, like Prior's runabout system, that do not?

The most common response to questions like this is to cite the soundness theorems, and perhaps the completeness theorem for first-order logic, as above. The argument is that the deductive systems are acceptable and legitimate just because they are sound for the model theory. This way of thinking presupposes that model-theoretic consequence, and its pretheoretic counterpart, are the prior or more important notions. We use model theory to determine which deductive systems are legitimate. Soundness to an adequate model theory is a necessary condition for a deductive system to be correct.

Advocates of the deductive notion of consequence reject this claim of priority. They argue that a notion in the neighborhood of (Ded) is the primary notion of consequence. Accordingly, each logical term is characterized by its inference rules. We need not, and should not, refer to a separate model theory.

The most common claim by advocates of the deductive notion is that the introduction and elimination rules in natural deduction systems provide each logical term with its meaning. But the connective "tonk" has introduction and elimination rules, and the resulting deductive system is clearly illegitimate. Apparently, the connective simply has no meaning. In these terms, our question becomes: Which rules of inference confer legitimate meaning on a logical connective, in order for the deductive system to represent a legitimate notion of logical consequence?

The response from the advocates of deductive consequence is that a pair of rules successfully characterizes a logical term only if they are in harmony (see, for example, chapters 22 and 23 of this volume). We are to think of the logical terms as introduced into the vocabulary one at a time. The result of adding a proposed new term, like "$\vee$" or "tonk," must be a conservative extension of the old system. In other words, if an inference that does not involve the new term is deducible in the new system, then it should be deducible in the old one, before the new connective was added. The rules for "$\vee$" meet this requirement, but those for "tonk" fail it miserably (unless the system was already useless).

The thesis is thus that (harmonious) rules of inference give the meaning of the logical terms of formal languages. Model theory—if there is to be one—must respect this meaning. The logic must be complete. Soundness is a bonus.

To repeat, the other school, in which model theory is primary (or at least autonomous), holds that the meanings of the logical terms are given by their satisfaction conditions. That is, the truth conditions of a compound are given in terms of the truth conditions of the parts. A deductive system—if there is to be one—must respect this. The logic must be sound. Completeness is a bonus.

Frankly, I am not sure just what the dispute is, let alone how to adjudicate it. When it comes to *formal* languages, there should be no dispute at all. It is just a matter of terminology. I think we can agree that when we are doing model theory, the meaning of a formal connective like "$\vee$" is given by its satisfaction conditions. After all, that is how the connective is presented and studied in model-theoretic semantics. There is no "∨" in natural language. When the connective is introduced into logical theory, it has the meaning that the theorist confers on it. Similarly, we should agree that when we are doing proof theory, the meaning of "∨" is given by its inference rules. The "∨" of model theory thus has a very different (sort of) meaning of the "∨" of deductive systems. Can we go on to wonder about the real meaning of the formal connective "∨," independent of whether we are doing model theory or proof theory, or if we are doing both at once (e.g., proving soundness and completeness)? I don't think we can. The meaning of a formal connective depends on what one is doing with it, and this is different in model theory and in deductive systems.

Perhaps the serious issue turns on the meaning of the English word "or" (or the German word "oder", etc.) when that particle is used to connect sentences. If so, then our dispute is an empirical one, to be decided by a controlled, blind study of the correct use of the word in ordinary conversation and writing by ordinary, competent speakers of the language. This would be conducted by the standards of professional linguistics. Or should the issue be decided with a study, conducted by child psychologists, of how the word is learned as children acquire facility with language? As far as I know, studies like this are not cited in the literature on model-theoretic and deductive consequence. The issue seems to be one that can be argued from an armchair. If it is an issue concerning the meaning of words in natural language, the participants in the debate consider themselves sufficiently expert, and do not feel a need to consult empirical linguistics or developmental psychology.

Perhaps the issue is to be decided on some sort of holistic grounds, looking for the overall theory that gives the best account of logical consequence. Perhaps it is meant as a quasi-empirical claim concerning the English word "consequence" or "entails." Or perhaps it is a normative claim that this is what the phrase ought to mean.

Against this, I would urge the thesis of this chapter, that there are different, but closely related, notions of logical consequence. Model-theoretic consequence and deductive consequence provide formal accounts of two such notions, one in the neighborhood of the blended semantic notion above and the other in the neighborhood of (Ded). Both notions are legitimate, and epistemological insight is provided by studying them and their interrelations. So neither proof-theoretic consequence nor model-theoretic consequence is primary. Instead, they illuminate the various informal, pretheoretic notions of logical consequence, and they illuminate each other. However, I do not know how to further defend this package of views against the theses proposed by advocates of deductive consequence or by advocates of model-theoretic consequence. The rules of the holistic enterprise are rarely articulated, and it is notoriously difficult to agree on what counts as reflective equilibrium, independent of logical theory. So I close by referring the issue to the reader's intuitions.

REFERENCES

Bolzano, B. [1837], *Theory of science*, translated by R. George, Berkeley, University of California Press, 1972.

Coffa, A. [1991], *The semantic tradition from Kant to Carnap*, Cambridge, Cambridge University Press.

Corcoran, J. [1974], "Aristotle's natural deduction system," in *Ancient logic and its modern interpretations*, edited by J. Corcoran, Dordrecht, Reidel, 85–131.

Gödel, K. [1930], "Die Vollständigkeit der Axiome des logischen Funktionenkalkuls," *Montatshefte für Mathematik und Physik* 37, 349–360; translated as "The completeness of the axioms of the functional calculus of logic," in *From Frege to Gödel*, edited by J. van Heijenoort, Cambridge, Mass., Harvard University Press, 1967, 582–591.

Hacking, I. [1979], "What is logic?," *Journal of Philosophy* 76, 285–319.

Hodes, H. [1984], "Logicism and the ontological commitments of arithmetic," *Journal of Philosophy* 81 123–149.

Kreisel, G. [1967], "Informal rigour and completeness proofs," in *Problems in the philosophy of mathematics*, edited by I. Lakatos, Amsterdam, North-Holland, 138–186.

McCarthy, T. [1981], "The idea of a logical constant," *Journal of Philosophy* 78, 499–523.

Peacocke, C. [1976], "What is a logical constant?," *Journal of Philosophy* 73, 221–240.

Prior, A. N. [1960], "The runabout inference ticket," *Analysis* 21, 38–39.

Shapiro, S. [1987], "Principles of reflection and second-order logic," *Journal of Philosophical Logic* 16, 309–333.

Shapiro, S. [1991], *Foundations without foundationalism: A case for second-order logic*, New York, Oxford University Press.

Shapiro, S. [1998], "Logical consequence: Models and modality," in *The philosophy of mathematics today*, edited by M. Schirn, Oxford, Oxford University Press, 131–156.

Shapiro, S. [2002], "Necessity, meaning, and rationality: The notion of logical consequence," in *A companion to philosophical logic*, edited by Dale Jacquette, Oxford, Blackwell, 227–240.

Sher, G. [1991], *The bounds of logic*, Cambridge, Mass., MIT Press.

Tarski, A. [1935], "On the concept of logical consequence," in Tarski's *Logic, semantics and metamathematics*, Oxford, Clarendon Press, 1956, 417–429.

Tarski, A. [1986], "What are logical notions?," edited by John Corcoran, *History and Philosophy of Logic* 7, 143–154.

CHAPTER 22

LOGICAL CONSEQUENCE FROM A CONSTRUCTIVIST POINT OF VIEW

DAG PRAWITZ

In spite of the great advancement of logic in our time and the technical sophistication of disciplines such as model theory and proof theory, the concept of logical consequence—the most basic notion of logic—is still poorly understood. Basic intuitions are often in conflict with each other, and rather few attempts have been made to sort this out in a systematic fashion. Here I shall critically review some of the attempts that have been made to articulate intuitions about logical consequence,[1] but the review makes no claim to be comprehensive or to settle the issues. Most attention will be given to how the notion of logical consequence may be developed from a constructivist point of view.

The idea of logical consequence has been with us since the time of Plato and Aristotle, and since then has been central for our understanding of philosophy and science. Presumably it arose at least partly in reflections upon Greek mathematics. It must have been a challenge to Greek intellectuals to explain the noteworthy form

[1] This article was written after I had seen a draft of Stewart Shapiro's contribution to this volume, "Logical Consequence, Proof Theory, and Model Theory" (chapter 21), the aim of which is similar to mine. As the reader can see, our viewpoints have some common elements but our conclusions are very different.

that mathematics took in their time. How could it be that the Greeks could not only calculate such things as the length of one side of a right-angled triangle given the lengths of the other two—something that the Babylonians already knew perfectly well how to do—but also could *prove* general laws such as Pythagoras' theorem? Somehow the idea occurred that the new practice was to be seen as involving the drawing of logically valid inferences. The idea of such inferences is present in Plato's writing and is clearly formulated by Aristotle.

From this beginning there have been at least three basic intuitions about what it is for an inference to be logically valid—or, as we also say, for its conclusion to *follow logically* from its premisses, or to *be a logical consequence* of its premisses.[2] The most basic one, which goes almost without saying, is that a valid inference is truth-preserving: if the premisses are true, so is the conclusion. If the conditional here is understood as material implication, this means only that it is not the case that the premisses are true and the conclusion is false. Of course, the satisfaction of this condition is not enough to make the inference valid. Two further conditions, which occur more or less explicitly in Aristotle's writings, must also be satisfied:

(1) It is because of the logical form of the sentences involved, and not because of their specific content, that the inference is truth-preserving.
(2) It is impossible that the premisses are all true but the conclusion is false—or, in positive terms, it is necessary that if the premisses are true, then so is the conclusion.

It remains to be discussed to what extent these two conditions are independent of each other, but there has been a general consensus that both conditions are necessary. While the modal notion that occurs in (2) is notoriously difficult to explicate, the main idea of (1) is comparatively easy to develop into a more precise form. Let us therefore first turn to that.

Variations of Specific Contents

Condition (1) comes into play in a well-known manner, already employed by Aristotle, when giving counterexamples to the logical validity of an inference.

[2] Here I take the field of the consequence relation to consist of sentences and also take the premisses and conclusions of inferences to be sentences. Consequences and inferences are thus treated to be on a par in this respect. However, it may be said that an inference act is a transition from judgments to judgments while consequence is a relation between propositions. This distinction is made by Martin-Löf [1985] and Sundholm [1998], who discuss validity of inferences and consequence in partly the same direction as I, but plays no role in my discussion here.

When a conclusion B is inferred from premisses $A_1, A_2, \ldots, A_n$, one may object to the logical validity of the inference by presenting true sentences $A_1^*, A_2^*, \ldots, A_n^*$ and a false sentence B^*, of the same logical form as $A_1, A_2, \ldots, A_n$ and B, respectively. To make this more precise, one must specify the logical forms and the different categories of nonlogical terms. This is indeed something that has been done in modern logic for various languages, the first-order languages being one kind of example.

Given a language where these things have been specified, we may quantify over all variations of the content of nonlogical terms. Let A be the conclusion and Γ the set of premisses of an inference considered in this language, and let a *substitution* S be a pairing of the nonlogical terms in these sentences with other nonlogical terms, each pair consisting of terms of the same category. Let A^S be the result obtained from A by replacing the nonlogical terms in A with the ones they are paired to by S, and let Γ^S be the result of carrying out such a substitution in each sentence of Γ. A counterexample to the inference then consists of a set Γ^S of true sentences and a false sentence A^S. Since S replaces only nonlogical terms with other nonlogical terms, the counterexample has the same logical form as the original inference.

The idea of (1) may be expressed by saying that there is no counterexample to a logically valid inference or, in positive terms, an inference is logically valid only if all inferences of the same logical form as the given one are truth-preserving. Letting S vary over all substitutions of the kind described above, condition (1), or at least part of its content, may now be expressed by saying simply:

(1′) For all substitutions S, if all the sentences of Γ^S are true, then so is A^S.

This way of expressing condition (1) occurs already in medieval times and is taken up again by Bolzano, except that he considered propositions instead of sentences and concepts instead of terms. Disregarding some minor further conditions, which we need not consider here, Bolzano took (1′) also as a sufficient condition for A being a logical consequence of Γ. This last idea we shall discuss in the next section.

An obvious objection to (1′) (when one comes to think about it) is that (1′) may be satisfied simply because of a poverty of the language that limits the number of substitutions S. One should therefore consider extensions of the given language by introducing names of all existing things in the categories involved, and demand that (1′) hold for all such extensions of S. Or, alternatively (and better, not having to consider languages with uncountably many terms), one speaks of interpretations instead of substitutions. An *interpretation* of a given language is an assignment I to the nonlogical terms of the language which assigns denotations appropriate for the various categories of nonlogical terms. Speaking of truth under such assignments, as this is nowadays defined in elementary textbooks of logic, we can replace (1′) with

(1″) For all interpretations *I*, if all sentences of Γ are true under *I*, then *A* is true under *I*.

This is how Tarski [1936] developed condition (1). If we take quantifiers to be always relativized by explicit specification of the domains, the interpretations also take care of the variations of these domains. Sentences such as "everyone is mortal" and "some numbers are prime" may then rendered "($\forall x \in$ humans)(x is mortal)" and "($\exists x \in$ numbers)(x is prime)"; and an interpretation of these sentences then also assigns some extensions to the domain names "humans" and "numbers."

Like Bolzano, Tarski takes his variant of (1) not only as a necessary but also as a sufficient condition for logical consequence. This is what is done also in model theory, where (1″) is slightly modified by making the domains explicit. Restricting ourselves to first-order languages and letting all the quantifiers range over the same individual domains *D*, we say that a structure or a *possible model* for a first-order language is a pair $M = \langle D, I \rangle$ where *D* is a set and *I* is an interpretation in *D* of the nonlogical terms of the language. This means that *I* assigns set-theoretical objects that can be obtained from *D* (i.e., individual constants are assigned elements of *D*, first-order functional symbols are assigned functions from *D* to *D*, one-place first-order predicates are assigned subsets of *D*, and so on). Saying that a possible model $\langle D, I \rangle$ is a *model for* a sentence *A* when *A* is true in *D* under *I* (i.e., *A* is true when the domain of the quantifiers is taken to be *D* and the values of the nonlogical terms are taken to the ones assigned by *I*), and that it is a model for the set Γ of sentences when it is a model for all the sentences of Γ, we may replace (1″) by the simple and well-known condition

(1‴) All models of Γ are models of *A*.

I consider condition (1″) to be a reasonable explication of the idea expressed in (1), and condition (1‴) to be a way of expressing a slight variation of (1″) in set-theoretical terms. However, the idea of making a condition of this kind a defining one for the concept of logical consequence must be discussed.

In discussing this question it is sometimes convenient to consider the closely related notion of logical truth instead of logical consequence. Logical truth may be considered the special case of logical consequence arising when the set Γ is empty. All the variants of (1) thus give a necessary condition for a sentence to be a logical truth, and this condition is turned into a definition of logical truth by Tarski and the model theory inspired by him. Hence, according to these analyses, a sentence *A* is *logically true* if and only if *A* is true under all interpretations of *A*, or in all possible models of *A*, respectively.

Criticism of the Tarskian Analysis of Logical Consequence

Tarski agrees that there is a modal ingredient in our intuitive idea of logical consequence and that it can be formulated in the way of (2). He must thus claim that condition (1″), which he equates with A being a logical consequence of Γ, contains a modal element that takes care of the intuition expressed in (2). Notwithstanding that our modal intuitions may be obscure, there is reason to be suspicious of that claim.

The validity of inferences is not a small theme of philosophy. It is said that with the help of valid inferences, we justify our beliefs and acquire knowledge. The modal character of a valid inference is essential here, and is commonly articulated by saying that a valid inference *guarantees* the truth of the conclusion, given the truth of the premisses. It is because of this guarantee that a belief in the truth of the conclusion becomes justified when it has been inferred by the use of a valid inference from premisses known to be true. But if the validity of an inference is equated with (1) (or its variants), then in order to know that the inference is valid, we must *already* know, it seems, that the conclusion is true in case the premisses are true. After all, according to this analysis, the validity of the inference just means that the conclusion is true in case the premisses are, and that the same relation holds for all inferences of the same logical form as the given one. Hence, on this view, we cannot really say that we infer the truth of the conclusion by the use of a valid inference. It is, rather, the other way around: we can conclude that the inference is valid after having established for all inferences of the same form that the conclusion is true in all cases where the premisses are.[3]

To say that a valid inference guarantees the truth of the conclusion, given the truth of the premisses, is to give the modal character of an inference an epistemic ring, and it seems obvious that condition (1″) has no connection with such a modality. What, then, about a modality of a more ontic kind that is pruned from epistemic matters? Does the claim that (1″) contains a modal ingredient (and in that way explicates condition (2)) fare better for a modality of that kind?

[3] This is a point made essentially by Etchemendy [1990] (e.g., on p. 93). As for the question how we establish the validity of an inference (in the sense of (1″)), it is to be noted that this has to be done on the metalevel by the use of some inference. As is pointed out in Prawitz [1985], this is typically done by simply using the very same type of inference on the metalevel, provided the inference is not naturally broken down into simpler inferences. In the latter case, we may of course resort to these simpler inferences on the metalevel. But then the given inference could have been supported equally well on the object level by replacing it with a series of such simpler inferences; see Dummett [1991] (pp. 200–204) on this point.

A verdict on that question may most easily be arrived at by considering the Tarskian definition of logical truth. One may then note that the definition equates the logical truth of a sentence *A* with the actual truth of a closely related sentence, namely, the universal closure of the open sentence A^* obtained from *A* by replacing all its nonlogical terms with variables of the same category. This observation enforces the suspicion that there is no modal ingredient at all in (1″). Clearly, the *actual* truth of the universal closure of A^* in no way implies the *necessary* truth of its instance *A*.[4] It is therefore difficult to see how the Tarskian analysis of logical consequence, which proceeds entirely in terms of actual truth (although under different interpretations of the descriptive terms), touches any modality.

But perhaps there is another way to look at the Tarskian analysis, or at least at the closely related model theory which grew out of Tarski's semantics. The basic intuitions behind our conditions (1) and (2) may be summarized by saying that two kinds of invariance are required of a logically valid inference: it must be truth-preserving regardless of the meaning of the descriptive terms of the sentences involved, and also regardless of the facts of the world. Etchemendy [1990] coined the term *interpretational semantics* for a theory that describes how the truth-value of a sentence varies with what the nonlogical terms of the sentence mean, and the term *representational semantics* for a theory that describes how it varies with how the world is. If it now has to be admitted that the Tarskian semantics occurring in (1″) is formulated as an interpretational semantics, it may be thought that at least the model theory occurring in (1‴) can be viewed both as an interpretational and as a representational semantics. If so, there would be a modal ingredient in condition (1‴) after all.

This stand seems to be the one taken by Shapiro (in his contribution concerning logical consequence in this volume). He argues that in a possible model $\langle D, I \rangle$, the domain *D* gives the size of a possible world, and by assigning values (extensions) to the descriptive terms, *I* gives the facts of a possible world. The possible model may therefore be seen as representing a way the world might be. The idea is thus that a possible model can be viewed in two ways, either as depicting the meaning of descriptive terms or as representing a possible world. Viewed in the first way, (1‴) says that the inference is truth-preserving regardless of meaning, and viewed in the second way, it says that it is truth-preserving regardless of the world.

This standpoint is exactly one of the main targets of Etchemendy's criticism. Etchemendy argues that the standpoint is reasonable only if it can be made

[4] This criticism against the Tarskian definition was leveled by Prawitz [1985] and is a main topic of Etchemendy [1990]. That (1″) deals with actual relations between the truth-values of the involved sentences rather than with any necessary links between them is quite obvious but seems nevertheless often to be overlooked. Tarski [1936] is, however, close to making this observation when he observes that a sentence that contains no descriptive constants is logically true just in case it is true.

plausible that the semantic theory simultaneously represents both all possible meanings of the descriptive terms and all the ways the world might be. There are simple instances where this is the case, but for several reasons ordinary model theory is not such a case according to Etchemendy. One objection is that some models cannot coherently be looked upon as representing a possible world for example, ones in which analytically true sentences such as $2+2=4$ come out false. A second, equally serious objection is that there are possible worlds that are not represented by any model.

The question of whether the model-theoretical notion of consequence contains a modal ingredient is thus controversial. The view that it does, is, in my opinion, not very well developed, and has to meet a number of challenges raised by Etchemendy. In any case, there is certainly general agreement that if the Tarskian or model theoretic notion of logical consequence contains a trace of modality, then it is not of an epistemic kind. Let us therefore now turn to condition (2) stated in the introduction and ask how it can be explicated.

Necessity of Thought

Two ways to articulate condition (2) have already figured in the discussion of the previous section, one ontic and one more epistemic. The ontic alternative is to speak, with Leibniz, of possible worlds and to say simply:

(2a) A is true in all possible worlds in which all the sentences of Γ are true.

In a more epistemic formulation I spoke of the truth of the premisses *guaranteeing* the truth of the conclusion. Another way of bringing out an epistemic force of necessity more clearly is to say

(2b) The truth of A follows by necessity of thought from the truth of all the sentences of Γ.

In the same direction there are formulations such as one is committed to holding A true, having accepted the truth of the sentences of Γ; one is compelled to hold A true, given that one holds all the sentences of Γ true; on pain of irrationality, one must accept the truth of A, having accepted the truth of the sentences of Γ.

To develop the idea of a necessity of thought more clearly we must bring in reasoning or proofs in some way. It must be because of an awareness of a piece of reasoning or a proof that one gets compelled to hold A true given that one holds all the sentences of Γ true.

Let us call a verbalized piece of reasoning an *argument*, and let us speak of an argument *for A from* Γ, if Γ is the set of hypotheses on which the reasoning depends and A is the conclusion of the reasoning. By a *proof* of A from Γ we may understand either a valid argument for A from Γ or, more abstractly, what such an argument represents; I shall here reserve it for the latter use.

A direction in which we may try to explicate (2b) may then be put in one of the following two ways:

(2b′) There is a valid argument for A from Γ; there is a proof of A from Γ.

It may be said that condition (2b′) does not in itself involve any idea of necessity of thought, since the mere existence of a valid argument or of a proof does not compel us to anything. Condition (2b′) may nevertheless be said to bring in the idea of a necessity of thought by requiring the existence of something such that when we know of it, we are compelled to hold A true, given that we hold the sentences of Γ true.

On the face of it, (2a) and (2b) (or (2b′)) may seem very different. Both formulations have of course to be developed further in order to be something more than mere phrases or metaphors, and before this is done, one cannot say how the two ideas are related. If the notion of a possible world is thought of as something determined by set of sentences Γ that are consistent in the sense of there being no valid argument or proof from Γ ending in a contradiction, then the two kinds of modalities may go together in the end. Here I shall leave the question how (2a) should be further analyzed and shall concentrate on (2b′).

We have thus to account for what it is that makes something a proof or, alternatively, to account for the validity of arguments. It should be noted at this point that not infrequently the validity of an argument is defined in terms of logical consequence. The idea here is to reverse the order of explanation in an attempt to catch a modal element in logical consequence. In other words, the validity of an argument for A from Γ is to be taken as a more basic notion, and is to be analyzed so that it constitutes evidence for A when given Γ, i.e., something that by necessity of thought makes us conclude A, given Γ.

What is it that makes an argument valid and thus compels us, by necessity of thought, to hold the conclusion true, given the truth of the premisses? It is difficult to think of any answer that does not bring in the meaning of the sentences in question. In the end it must be because of the meaning of the expressions involved that we get committed to holding one sentence true, given the truth of some other sentences.

To get further, we should thus turn to the notion of meaning. Since we are here interested in logical consequence, we shall focus on how one is to understand that an argument is valid in virtue of the meaning of the logical constants occurring in the sentences of the argument.

The Meaning of the Logical Constants

One idea about meaning in general going back to Frege is that the meaning of a sentence is given by its truth condition. Its rationale is the notion that the question of whether a sentence is true or not depends both on the meaning of the sentence and how the world is, the meaning of the sentence being exactly that feature of the sentence which determines how the world has to be in order for the sentence to be true. Although far from being unanimously accepted, the idea that the meaning of a sentence is the same as its truth condition has been influential both in philosophy of language and in logic. Wittgenstein in the *Tractatus* took it up after Frege, and it is a cornerstone in, for instance, Davidson's philosophy of meaning.

As regards logic, the idea is expounded in Church's [1956] classical textbook and is nowadays often presented in teaching of elementary logic, where sentential operators are said to get their meaning by their truth tables. More generally, the meaning of a logical constant c is said to be determined by the uniform truth condition of the sentences with c as the main sign, given in the form of an equivalence such as:

> $\neg A$ is true if and only if A is not true.
> $A \wedge B$ is true if and only if both A and B are true.
> $A \vee B$ is true if and only if at least one of A and B is true.
> and so on.

For a quantified sentence $(\forall x \in D)A(x)$ the truth condition may be stated:

> $(\forall x \in D)A(x)$ is true if and only if $A(t)$ is true for all terms t that denote elements in D.

(This formulation requires that names for all elements of D are available in the language. Alternatively, if we do not want to rely on such an assumption, we speak of assignments of values to the variables and define the notion of truth (or satisfaction) under such assignments in a similar way.)

Questions about the meaning of the logical consequence thus seem to have a straightforward answer in terms of such truth conditions. However, the substance of this answer depends on what we take truth to be. One must here beware of a possible confusion, which Dummett (see, e.g., Dummett [1978], pp xx–xxi), especially, has drawn attention to. The truth conditions for compound sentences of different logical form coincide formally with recursive clauses that occur in the definition of truth as given by Tarski [1935/1936] or in the definition of truth relative to a possible model (as is standard in model theory), and one may therefore be led to think that truth is what is defined in this way. But obviously the truth conditions cannot simultaneously do service both in a definition of truth and in an explanation of the meaning of the sentences in question. In other words, the equivalences of the kind exemplified above cannot be taken as clauses in a recursive definition of truth, and at the same time can be taken as explaining the meaning of the logical constants exhibited—this would be like solving two unknowns, given only one equation. If we have defined a set S of sentences by saying that it is the least set of sentences containing certain atomic formulas and satisfying certain equivalences, such as

$A \wedge B$ belongs to S if and only if both A and B belong to S,

then obviously we get no information about the meaning of the logical constants by being told again that these equivalences hold. Similarly, a person who does not know what truth is, but is informed that it is a notion satisfying certain equivalences of the kind given above, does not get to know the meaning of the logical constants by then being told again that these equivalences hold.

We must conclude that truth conditions can serve as meaning explanations only if we already have a grasp of truth. Accordingly, we find that Frege and those who follow him in thinking that meaning is given by truth conditions are careful to note that truth is taken as an already known notion. In contrast, Tarski, who wants to define truth but is anxious that his definition be adequate, is careful to note that he is taking the meaning of the sentences for which truth is defined as already known. Tarski takes instances of the equivalences given above, so-called T-sentences, which of course follow from his definition of truth, as showing the material adequacy of his definition of truth. In order that they will serve in this way, showing us the correctness of the truth definition, we must of course know that the T-sentences themselves are correct, and this we can know only if we already know the meaning of the sentences that are mentioned in the T-sentences.[5]

An informative answer to the question of what the logical constants mean thus requires that we further ask what truth is.

[5] Elementary expositions of logic are often less clear on this point, and give, rather, the impression that the truth conditions both are clauses in the definition of truth (in a model) and serve to give the meaning of the logical constants.

Truth

The notion of truth comes up not only when we ask about the meaning of the logical constants and want to use the idea that it is given by truth conditions, but occurs prominently also at the outset of an analysis of logical consequence. It can therefore not be passed over in an analysis of logical consequence. The question of what a true sentence is true in virtue of, is particularly significant. There are at least two different ways of thinking about this that are relevant here because of how they connect with the notion of evidence.

What may be called a *realist* conception takes sentences to be true in virtue of facts of the world given independently of what it is for us to know them. According to an *epistemic* conception, truth is instead determined in terms of what it is for us to acquire knowledge, and sentences are true in virtue of the potential existence of evidence for them. If the facts of the world are what are described by true sentences, then on this conception the world is seen as constructed in terms of potential evidence, and in the end, in terms of what it is for us to acquire evidence. One may therefore also call it a *constructive* conception of truth.

To analyze the modal ingredient of logical consequence in terms of evidence seems a hopeful project only if truth is understood in this constructive way. From this point of view, evidence or what it is to acquire knowledge must be taken as a more fundamental concept than truth—truth may then be defined as the potential existence of evidence. This is the path that I shall follow in the sequel.

With this understanding of truth we may go back to the idea that meaning is determined by truth conditions. A more profound way of accounting for the meaning of a sentence is now opened up, namely, in terms of what counts as evidence for the sentence. Knowing what counts as evidence for the sentence, one also knows the truth condition of the sentence. Hence, there is no conflict between the present suggestion to account for meaning in terms of evidence and the idea that meaning is determined by truth conditions, provided that truth is understood constructively.

A vital theory of meaning should be able to connect linguistic meaning with how language is used and functions. The presence of evidence is the ground on which we assert sentences. Thus, the use of sentences in assertions and the meaning of them now become clearly linked to each other.[6] It remains, however, to discuss the notion of evidence.

For observation sentences, evidence typically takes the form of a relevant observation. In general, we may speak of conclusive evidence for the truth of

[6] Such a link between use and meaning may not be possible to establish when meaning is accounted for in terms of realistically understood truth conditions. Doubts in this direction concerning a theory of meaning based on a realistic conception of truth have forcefully been raised by Dummett in several writings (see, e.g., Dummett [1976]).

a sentence as *verification.* A verification may take the form of a combination of observations and arguments. In mathematics the former fall away, and evidence occurs in the form of valid arguments or proofs. We shall here restrict ourselves to mathematics, and therefore to the notions of valid argument and proof.

PROOFS

We turned to the meanings of the logical constants, thinking that these meanings are what may make an argument valid or something toward a proof. But now the question at issue has been turned around, and it is suggested that meaning be accounted for in terms of proofs. There seem to be two clashing intuitions at work here, which also occur in more general discussions concerning the relation between the meaning and the use of a term; as was noted above, what counts as a proof of a sentence is one feature of the use of the sentence.

Sometimes it seems reasonable to equate meaning and use in line with the slogan "meaning is use," inspired by the later Wittgenstein. If someone asks why $3+1=4$, a natural answer is that this is what "4" means, or that this how "4" is defined and used. Similarly, what can we answer someone who questions the drawing of the conclusion $A \to B$, given a proof of B from A, except that this is how $A \to B$ is used, it is a part of what $A \to B$ means? But a similar answer to the question why $2+2=4$ or why we infer $A \to B$ from $\neg B \to \neg A$ seems inadequate. That $2+2=4$ or that we infer $A \to B$ from $\neg B \to \neg A$ is not reasonably looked upon as a usage that can be equated with the meaning of the expressions involved, but rather is something that is to be justified in terms of what the expressions mean. To answer *all* doubts about a certain usage of language by saying that this is how the terms are used, or that this is a part of their meaning, would be a ludicrously conservative way of meeting demands for justification. But for *some* such doubts the reference to common usage is very reasonable and may be the only thing to resort to.

Leaving the question of how use and meaning are related to each other aside for a while, one may ask what kind of proofs one has in mind when it is suggested that it is in terms of them that the notion of logical consequence and the meanings of sentences are to be accounted for. Are we to think of proofs as formal proofs given by formal deductive systems?

A formal system such as the one for Peano Arithmetic is of course to be seen as an attempt to codify valid reasoning within a part of intuitive mathematics. But the notion of logical consequence cannot be analyzed in terms of formal systems. An arithmetical sentence is not a logical consequence of other arithmetic sentences because of the existence of a proof in Peano Arithmetic. Nor can the meaning of an arithmetic sentence A be accounted for in terms of what counts as a proof of A within Peano Arithmetic. In the case of arithmetic we know by Gödel's Incompleteness Theorem that such an analysis would be outright wrong. We even know that no formal system can be used to exhaustively describe the intuitive meaning of the arithmetic sentences or the relation of logical consequences for arithmetic sentences: indeed, because of the meaning of these sentences, for every consistent system there will be true sentences that cannot be proved in the system.

However, we really do not need to refer to Gödel's Incompleteness Theorem to see that ordinary proof theory has nothing to offer an analysis of logical consequence. A deductive system is, as already said, an attempt to codify proofs within a given language, but when setting up such a system, one does not ordinarily try to analyze what makes something a proof. Nor does proof theory ordinarily try to justify a deductive system except for trying to prove its consistency.

An ironic illustration of the need for justification, and of the fact that not every use of a language affords the involved sentences meaning, is given by Prior's example of a sentential connective "tonk" (also referred to by Shapiro in his Chap. 21, this volume). "Tonk" is governed by two inference rules, one allowing the inference of "*A* tonk *B*" from a premiss consisting of either A or B, and one allowing the inference from "*A* tonk *B*" either of A or of B. We cannot find a meaning for "tonk" that accords with such inference rules; as is easily seen, they allow the deduction of any sentence we like. One cannot, therefore, say that sentences mean whatever the rules that govern them make them mean—the rules that govern them may very well reduce the sentences to sheer nonsense.

The notion of proof in a formal deductive system is thus not relevant for our purpose of analyzing logical consequence. Instead, we need a notion of proof such that a proof of a sentence can reasonably be said to constitute evidence in view of the meaning of the sentence. Within intuitionism one has often tried to explain the meaning of the logical constants by resorting to a notion of proof. This may proceed by recursive clauses, not in principle unlike the ones occurring in Tarksi's definition of truth, as follows:

A proof of a conjunction $A \wedge B$ is a pair consisting of one proof of A and one of B.
A proof of $A \vee B$ is a proof of either A or B together with an indication of which of the two it is a proof of.
A proof of $A \rightarrow B$ consists of a function which is recognized to yield a proof of B when applied to a proof of A.
A proof of $\forall x A(x)$ consists of a function which is recognized to yield a proof of $A(t)$ when applied to a term t denoting an element in the domain of the quantifier.
A proof of $\exists x A(x)$ consists of a pair whose first member is a term t denoting an element in the domain of the quantifier and whose second member is a proof of $A(t)$.

A problem with this explanation is that the notion of proof used here cannot stand for whatever establishes the truth of sentences in a normal intuitive sense. For instance, a proof of a disjunction $A(t) \vee B(t)$ may very well proceed even intuitionistically by first proving $\forall x(A(x) \vee B(x))$ and then applying universal instantiation to infer $A(t) \vee B(t)$. Given such a proof, we do not know which of the two disjuncts holds. Hence, it is not correct to say that a proof of a disjunct needs to consist of a proof of one of the disjuncts together with indication of which disjunct is proved. In intuitionistic meaning explanations of the kind exemplified above, proof must thus be meant in a quite restrictive way.

It was clear at the beginning of this section, when discussing the relation between meaning and use, that the project of accounting for the meaning of a sentence in terms of what counts as a proof of it needs a more restrictive notion of proof than the usual one. Some particular use of a linguistic expression may very well be seen as constitutive of its meaning, and other uses must then be accounted for or justified in terms of this meaning. More specifically, it may be constitutive of the meaning of sentence A that certain ways of arguing for its truth are recognized as a proof of A. But it would be counterintuitive to identify the meaning of a sentence with all the ways in which it can be proved. This is so for the simple reason that we may very well understand a sentence without having any clear idea about all the ways in which it may be proved. Furthermore, the finding of new ways to prove a sentence does not automatically amount to a change of its meaning.

In a satisfactory approach to meaning via proofs, therefore, we cannot use proofs in general to account for meaning, but must instead single out something we may call *direct* or *canonical* proofs that are constitutive of meaning.[7] This possibility has now to be explored.

Canonical Proofs

The first to suggest that certain ways of proving a sentence could be seen as determining its meaning seems to have been Gentzen [1934]. After having presented his system of natural deduction, in which logical inferences are broken down into two kinds of basic steps, called *introductions* and *eliminations*, one of each kind for every logical constant, Gentzen remarks:

> The introductions constitute, as it were, the "definitions" of the symbols concerned, and the eliminations are, in the final analysis, only consequences of this.

[7] The need to distinguish between proofs in general and direct or canonical proofs has been argued for by Dummett and Prawitz in several papers (see, e.g., Dummett [1975] and Prawitz [1974]).

Since it is clear that the introductions are not literal definitions, a better way to express the idea is to say that the rules for introduction inferences determine the meanings of the logical constant concerned, while the rules for elimination inferences are justified by these meanings.

An introduction for a logical constant is a form of inference whose conclusion has the constant as its main sign. For each logical constant of the language of first-order predicate logic, Gentzen gives a rule for how to form its introductions, which may be represented by the following figures:

$$\wedge\mathrm{I})\ \frac{A \quad B}{A \wedge B} \qquad \vee\mathrm{I})\ \frac{A}{A \vee B} \qquad \frac{B}{A \vee B} \qquad \exists\mathrm{I})\ \frac{A(t)}{\exists x A(x)}$$

$$\rightarrow\mathrm{I})\ \frac{\begin{matrix}[A]\\ B\end{matrix}}{A \rightarrow B} \qquad \forall\mathrm{I})\ \frac{A(y)}{\forall x A(x)}$$

The sentence A shown within the square brackets in the introduction rule for implication ($\rightarrow$I) indicates that in such an inference, it is allowed to discharge occurrences of A standing as hypotheses.

Negation, which does not occur in the list of introductions, is here supposed to be defined with the help of $\rightarrow$ and the constant $\perp$ for absurdity (i.e., $\neg A$ is short for $A \rightarrow \perp$), which makes negation introduction (better known as the constructive form of reductio ad absurdum) a special case of implication introduction. The introduction rule for $\perp$ is empty (i.e., it is the rule that says there is no introduction whose conclusion is $\perp$).

The question is now in what way these introductions determine the meanings of the logical constants. A way to answer this question is in terms of the notion of a direct or a canonical proof, mentioned at the end of the preceding section. The general idea is that the meaning of a sentence A is given by what counts as direct evidence for it. Using the term "canonical proof" for what in mathematics constitutes direct evidence, we have to state for various sentences what forms canonical proofs of them are supposed to have. This will now be indicated for first-order sentences understood constructively. In the next section, I shall carry out the same thing on a more concrete linguistic level by specifying forms of verbal arguments that are to count as canonical.[8]

[8] When speaking about canonical proofs and related notions on an abstract level, I shall be following Martin-Löf [1985] in the main lines. (For the source of some differences, see note 2. It is also to be noted that in later publications Martin-Löf's approach is different in several respects, in particular when he distinguishes between proof objects and demonstrations. Only the latter are supposed to have epistemic significance.) When speaking about canonical arguments and related notions, I shall essentially be following Prawitz [1973]. (See also Prawitz [2005].)

The general form of a canonical proof of a compound sentence C with the logical constant c as main sign can be written $O_c(\Pi)$, where O_c is an operation that stands for the recognition that we have obtained direct evidence for C because of Π (e.g., in the first case below, $O_\wedge$ is the recognition of the fact that because of having evidence both for A and for B, one is in possession of evidence for $A \wedge B$). We have to specify what Π is in the different cases.

If C is a conjunction $A \wedge B$, then Π is to consist of canonical proofs of A and of B conjoined.

If C is a disjunction $A \vee B$, then Π is to consist of either a canonical proof of A and an indication of the fact that it is the first disjunct that one has obtained evidence for, or a canonical proof of B and an indication of the fact that it is the second disjunct that one has obtained evidence for.

If C is an existentially generalized sentence $\exists x A(x)$, then Π is to consist of a individual term t (in the domain of quantification) together with a canonical proof of $A(t)$.

When C is an implication or a universally generalized sentence, we need to rely on the notions of hypothetical proofs and general proofs in order to specify the forms of canonical proofs. To do this, we need first of all to say what a proof is—we may say *categorical* proof to distinguish it from other kinds of proof.

The explication of these notions was of course a main goal for the whole present approach to the modal ingredient of logical consequence. That they are needed already in the specification of the canonical proofs may seem circular, but means only that we must proceed by recursion. The specification of the canonical proofs of implications and universally generalized sentences requires the notions of categorical, hypothetical, and general proof of sentences of less complexity, which in turn require that we know what the canonical proofs of these less complex sentences are.

To elucidate the notion of categorical proof, we may note that one may rightly assert a sentence A without having a canonical proof of A. But the idea that one can single out some proofs as canonical involves the idea that a proof can in principle be given in canonical form. The assertion of a sentence is thus constructively to be understood as claiming the existence of a canonical proof of it. This existence is not to be understood in the sense that a canonical proof has already been constructed and is at hand, but only that there is such a proof and that it therefore could in principle be constructed. Consequently, we may rightly assert a sentence A as long as we know that there is a categorical proof of A, which, understood constructively, means that we have an effective method for finding one. Such a method is thus to count as a proof. By a *categorical* proof of A, we shall just mean such an effective method for finding a canonical proof of A.

The notion of a *hypothetical* proof, that is a proof *of* a sentence A *from* sentences $A_1, A_2, \ldots, A_n$, is in the same vein proposed to be analyzed as being an n-ary effective function which, applied to categorical proofs of $A_1, A_2, \ldots, A_n$, yields a categorical proof of A. Finally, we propose that a *general* proof of an open sentence $A(x_1, x_2, \ldots, x_n)$, relative to a set of individual terms T, is an effective function which, applied to any elements $t_1, t_2, \ldots, t_n$ of T, yields a proof of $A(t_1, t_2, \ldots, t_n)$.

We can now specify that a canonical proof of a sentence $A \rightarrow B$ is to be of the form $O_{\rightarrow}(\Pi)$ where Π is a hypothetical proof of B from A, while a canonical proof of a sentence $\forall x A(x)$ is to be of the form $O_{\forall}(\Pi)$ where Π is a general proof of $A(x)$ (relative to the domain T of quantification).[9] To this has to be added that nothing is a canonical proof of $\bot$.

Canonical Arguments

It may be instructive to see how a corresponding specification proceeds if we stay on the more concrete level of verbal arguments; Gentzen was obviously thinking of deductions composed of sentences or formulas. One need then lay down some conventions for how the sentences of the arguments are to be arranged. It is convenient to assume that they are arranged in tree form. A sentence in a tree that stands immediately below some other sentences is claimed to follow from these sentences, and such a part of the tree thus represents an inference. The occurrences of sentences at the top of the tree are either axioms, *assumptions* of the argument, or *hypotheses* that are made temporarily (for the sake of the argument, as one says). In the latter case the hypothesis is discharged or, as I shall say, *bound* by some inference in the course of the argument. Above we have seen one kind of inference, $\rightarrow$I, that may bind assumptions. An occurrence of a sentence is said to *depend* on the assumptions and hypotheses that stand above it in the tree and are not bound by some inference occurring above it.

An inference may also bind a variable in the argument, in which case it must occur free neither in any assumption or hypothesis that a premiss of the inference depends on, nor in the conclusion of the inference. Above we have seen one inference, namely, $\forall$I, that binds the variable indicated in the premiss.

An argument whose last sentence (the sentence at the bottom of the tree) is A and whose assumptions (sentences at the top of the tree that are neither axioms nor hypotheses bound in the course of the argument) are $A_1, A_2, \ldots, A_n$ is said to be a argument *for A depending on $A_1, A_2, \ldots, A_n$*. Something is an argument for A *from* Γ if it is an argument for A depending on no other sentences than those

[9] Since in the case of implications and universal sentences we cannot require that the subproofs of a canonical proof also be canonical, we may as well leave out this requirement in the case of conjunction, disjunction, and existence sentences, and only require that the subproofs be categorical.

belonging to Γ. An argument is said to be *closed* provided it has no assumptions and all the variables that occur free in some open sentence of the tree are bound in the course of the argument. An argument that is not closed is said to be *open.*

An open argument is to be understood as a schema that becomes an argument when we make appropriate substitutions for the variables and assumptions that are not bound. This is to say that an open argument

$$\Delta(x_1, x_2, \ldots, x_m; A_1, A_2, \ldots, A_n)$$

containing exactly variables $x_1, x_2, \ldots, x_m$ and assumptions $A_1, A_2, \ldots, A_n$ that are not bound in Δ is *valid* if and only if certain results of carrying out substitutions in Δ are valid, namely, all results

$$\Delta(t_1, t_2, \ldots, t_m; \Delta_1, \Delta_2, \ldots, \Delta_n)$$

obtained from Δ by first substituting any terms $t_1, t_2, \ldots, t_m$ denoting individuals in the domain of quantification for $x_1, x_2, \ldots, x_m$ and then any valid closed arguments $\Delta_1, \Delta_2, \ldots, \Delta_n$ for the assumptions $A_1{}^*, A_2{}^*, \ldots, A_n{}^*$, where $A_i{}^*$ is the result of carrying out the first substitution of terms for variables in A_i.

The *canonical arguments* for the various first-order sentences may now be specified by simply saying that something is a canonical argument for A if and only if it is an argument for A whose last inference is an introduction and whose immediate part(s) is (are) valid argument(s) for the premiss(es) of that introduction.

It is to be noted that in case the last inference is an $\rightarrow$I or an $\forall$I, the immediate part of the argument is an open argument for which we have already defined what it is to be *valid.* This definition depends, however, on the notion of valid argument for sentences of lower complexity than A, actually subsentences of A. There is thus a recursion involved in these definitions of the same kind as the one noted above in connection with the notion of canonical proof.

It remains to say when a closed argument in general, one that is not necessarily canonical, is valid. In the same way that we have to require of a categorical proof that it is either canonical or is a method for finding a canonical proof, a closed argument should be defined as valid if it either is canonical or provides a method for finding a canonical proof. However, it is not immediately obvious how an argument may provide such a method. At this point we should consider some examples.

An open argument of the form

$$\frac{A_1 \wedge A_2}{A_i},$$

which coincides with what Gentzen calls an $\wedge$-elimination, is valid according to the definition given above, if the result of replacing the assumption $A_1 \wedge A_2$ (which for simplicity we assume to be a closed sentence) with a valid closed argument Δ for $A_1 \wedge A_2$ is valid. The question is now what it is that makes the result

$$\dfrac{\begin{array}{c}\Delta\\ A_1 \wedge A_2\end{array}}{A_i}$$

valid. According to what was just said above, the validity of the closed argument Δ should provide a method for finding a canonical argument Δ^* for $A_1 \wedge A_2$. Recalling how the notion of canonical argument was defined, we see that by replacing Δ with this canonical argument Δ^*, we get

$$(1) \qquad \dfrac{\dfrac{\begin{array}{cc}\Delta_1 & \Delta_2\\ A_1 & A_2\end{array}}{A_1 \wedge A_2}}{A_i}$$

where the two immediate sub-arguments of Π^*

$$(2) \qquad \begin{array}{c}\Delta_i\\ A_i\end{array}$$

are valid. Extracting the left immediate subargument of Δ^* if i is 1 and the right immediate subargument of Δ^* if i is 2, we thus get a valid argument for A_i. The extraction of (2) from (1) is known as $\wedge$-reduction and is an operation used in the normalizion of deductions.[10] It is in view of this operation that we see that an argument consisting of a $\wedge$-elimination is valid, and this is also how Gentzen meant that $\wedge$- elimination is justified (or is a "consequence" of the $\wedge$-introduction, as he expressed it in the quote given in the beginning of the previous section).

In an argument each inference that is not an introduction should be supplied with such a justifying operation in view of which it can be seen that a valid argument for the conclusion can be obtained, given valid arguments for the premisses. We say that an argument Δ *reduces* to the argument Δ^* relative to such operations if Δ^* is obtained from Δ by replacing subarguments with what is

[10] The reductions (for the different logical constants) were introduced in Prawitz [1965] (see also Prawitz [1971]) in his proof of the normalization theorem for first-order predicate logic, and correspond to Gentzen's cut-eliminations in the proof of his *Hauptsatz*.

obtained by applying the operations to them in the way that has just been exemplified. It is in this way, when the arguments have been supplied with justifying operations, that they may provide methods for finding canonical arguments, as must be required for validity in accordance with what was said above. A closed argument supplied with justifying operations is thus to be defined as valid if it reduces, relative to suitable operations, to a canonical argument. I shall not here give the formal details of how this is to be defined, but shall show how it works by considering two more examples.[11]

We illustrate the idea by two slightly more complicated reductions. The validity of implication elimination, that is, an argument of the form

$$\frac{A \quad A \to B}{B},$$

is first of all due to the fact that given a closed valid argument Δ_1 for A and a closed valid argument Δ_2 for $A \to B$, we get a closed argument for B

$$\frac{\begin{matrix}\Delta_1 \\ A\end{matrix} \quad \begin{matrix}\Delta_2 \\ A \to B\end{matrix}}{B},$$

which because of the validity of Δ_2 reduces to

$$\frac{\begin{matrix}\Delta_1 \\ A\end{matrix} \quad \dfrac{\begin{matrix}[A] \\ \Delta \\ B\end{matrix}}{A \to B}}{B},$$

where Δ is a valid open argument for B from A. We now see that there is an operation, so-called $\to$-reduction, which, applied to this argument, yields

$$\begin{matrix}\Delta_1 \\ [A] \\ \Delta \\ B\end{matrix}$$

[11] For a more detailed exposition see Prawitz [1973] or Dummett [1991]. A more recent exposition is Prawitz [2005].

(i.e., the result of substituting the valid closed argument Δ_1 for the open assumption A in Δ). According to the definition of validity for the open argument Δ, this result is valid, and we thus get a closed valid argument for B, as is required of a justifying operation for $\rightarrow$-elimination.

To see the validity of $\exists$-elimination

$$\frac{\exists x A(x) \quad \begin{array}{c}[A(y)]\\ B\end{array}}{B},$$

where the variable y is bound by the inference (which means that it must not occur in B nor in any hypothesis that the conclusion depends on), we have to show that it preserves validity, which means that we must show that a closed argument Δ of the form

$$\frac{\begin{array}{c}\Delta_1\\ \exists x A(x)\end{array} \quad \begin{array}{c}[A(y)]\\ \Delta_2(y)\\ B\end{array}}{B}$$

is valid, assuming that its immediate subarguments Δ_1 and $\Delta_2(y)$ are valid. By the validity of the closed argument Δ_1 we know that it reduces (relative to the operations assigned to the inferences of Δ_1) to canonical form, and hence that Δ reduces to the form

$$\frac{\dfrac{\begin{array}{c}\Delta_1{}^*\\ A(t)\end{array}}{\exists x A(x)} \quad \begin{array}{c}[A(y)]\\ \Delta_2(y)\\ B\end{array}}{B}.$$

To see that this argument is valid, we make use of the operation $\exists$-reduction, which transforms the argument into

$$\begin{array}{c}\Delta_1{}^*\\ [A(t)]\\ \Delta_2(t)\\ B\end{array}$$

We want to see that when $\exists$-reduction is applied to an argument whose immediate subarguments are valid, it yields a valid argument. This now follows from the validity of the open argument $\Delta_2(y)$, which just means that the result of first substituting t for y in $\Delta_2(y)$, and then the valid argument Δ_1^* for the assumption $A(t)$ is valid.

Summing Up

To sum up our analysis of the modal ingredient in logical consequence, we recall that a main goal was to bring out the sense in which it can be said that a sentence follows by necessity of thought from other sentences. It was suggested that to this end we need to bring in something about which it can be said that it commits us or compels us to hold one sentence true, given that we hold some other sentences true. It was proposed that proofs or arguments may be given this role. The idea is thus first that the necessity of logical consequence is expressed by the condition that for A to be a logical consequence of Γ, there must exist a proof or valid argument for A from Γ, and second, that the awareness of such a proof or valid argument compels or commits us to hold A true, given that we hold the sentences of Γ true.

Another main starting point was the idea that the meaning of a sentence is given by what counts as direct evidence for it. In mathematics, evidence takes the form of proof or valid argument, and the meaning of a mathematical sentence, according to this idea, is thus to be given by what counts as a direct or, as we have called it, canonical proof or argument.

In the same vein, the assertion of a sentence is understood constructively as the claim that there is direct evidence for it, and is to be taken as true if such evidence exists. In mathematics the truth of a sentence thus becomes equated with the existence of a canonical proof or argument for it. (In general, the existence of evidence for a sentence also depends, of course, on empirical matters.)

Presupposing an understanding of the general notion of evidence, which is taken as a more basic notion than truth, the forms of the canonical proofs or arguments for sentences within first-order languages understood constructively have been specified. This has been done by recursive clauses making use of Gentzen's rules for introductions. (They have to be supplemented with specifications of the forms of canonical proofs and arguments for atomic sentences.) At the same time (actually by a simultaneous recursion) the notions of categorical, hypothetical, and general proof—or, alternatively, of closed and open valid argument, have been defined.

As a result of how all these concepts have been developed, we can say of certain things that they constitute proofs of a sentence or valid arguments for a sentence in virtue of the meaning of the sentence. A certain Π is a canonical proof of a sentence A or a certain Δ is a canonical argument for A just because of the way in which the meaning of A is given. Furthermore, some inferences—namely, the inferences by introduction—become valid by the very meaning of the conclusion of the inference. Because of this meaning, we are compelled to hold the conclusion true when holding the premisses true. For instance, an inference by $\exists$-introduction of $\exists x A(x)$ from $A(t)$ is compelling, because by applying $O_\exists$ to a categorical proof of $A(t)$ or applying $\exists$-introduction to a closed argument for $A(t)$, we get by the very meaning of $\exists x A(x)$ a canoncial proof of $\exists x A(x)$. Therefore, if we hold that $A(t)$ is true, and consequently that there is a canonical proof or argument for $A(t)$, we are also committed to holding $\exists x A(x)$ true, since from the canonical proof or argument for $A(t)$ we can form a canonical one for $\exists x A(x)$, which guarantees the truth of $\exists x A(x)$.

An inference in general is compelling when we know a hypothetical proof of its conclusion A from its set of premisses Γ or, alternatively, an open valid argument for A from Γ. This is because knowing such a proof is to be in possession of an effective method which, applied to categorical proofs of the sentences of Γ, yields a categorical proof of A. Similarly, knowing a valid open argument Δ for A from Γ, we get a valid argument for A by replacing the open assumptions in Δ with closed valid arguments for them. For the same general reasons as before, we are in both cases committed to holding the conclusion true if we hold the premisses true.

We have thus shown how knowledge of a (hypothetical) proof of A from Γ or, alternatively, a valid argument for A from Γ—as we have developed these notions—commits us to holding A true if we hold the sentences of Γ true. It seems reasonable, therefore, to say that the way the notions of proof and valid argument have been developed, make condition (2b') account for the sense of necessity of thought involved in logical consequence.

One might say that any satisfactory analysis of the notions of proof and arguments should have the result that by knowing a proof or valid argument, one gets committed in the way discussed above. The question is how to achieve this result.

The notions of proof and valid argument are often defined as composed of valid inferences. This puts the burden on the notion of valid inference. How is it to be analyzed? If we do not simply say that an inference is valid when its conclusion is a logical consequence of the premisses, which would bring us back to the beginning of this investigation, we have to try to develop some concept of gapfree inference. The "gapfree" inferences must then be shown to have a compelling force. It is far from clear how this could be done.

To show an inference to be valid by breaking it down into simpler inferences that are known to be valid is certainly a good strategy in itself. Although a valid argument has not been defined here as a composition of valid inferences, it is still true that an argument built up of valid inferences is valid. But to define validity in that way would of course have made the whole analysis circular, as already noted when condition (2′) was first introduced. Instead, our basic notion is that of canonical proof or argument (linked to the meaning of the sentences). As shown above, this amounts to making inferences by introduction valid—valid by definition, so to say. It should be noted, however, that not every valid inference can be broken down into a sequence of introductions. In other words, there are inferences other than introductions that are gapfree in the sense that they cannot reasonably be broken down into simpler inferences. An essential ingredient in the analysis proposed here is a way of demonstrating the validity of such inferences. We have illustrated how this can be done by some examples where $\wedge$-, $\rightarrow$-, and $\exists$-eliminations were shown to be valid by applying reductions that are seen to transform arguments into ones that are known to be valid.

Finaly, it should be noted that we have been talking only about the meaning of the logical constants, and how proofs and arguments can have a compelling force in virtue of this meaning. But the same compelling force can arise in virtue of the meaning of nonlogical terms. Condition (2b) or (2b′) should therefore not be taken as a defining condition of logical consequence—it explicates a broader notion of consequence. To narrow it down to logical consequence, we have to combine it with a way of formulating condition (1) so that it fits the framework in which (2b′) has been explicated. This can easily be done in the manner of condition (1′) by requiring that for all substitutions S of the kind explained in that context, there is a proof of A^S from Γ^S. When this requirement is understood constructively, it amounts to dcmanding the existence of a uniform procedure that for each S takes proofs of the sentences of Γ^S into a proof of A^S. When speaking about arguments, this uniformity comes out directly by requiring that there be an argument Δ such that for each substitution S, Δ^S is an argument for A^S from Γ^S. To meet the objection that the force of this condition depends on the richness of the language, we have either to extend the language or to introduce a way of speaking of something being an argument under an assignment of values to the nonlogical terms, and then vary over all such assignments, as done in condition (2″).

Acknowledgments I thank Professor Cesare Cozzo and Professor Dugald Murdoch for valuable suggestions on an earlier draft, and participants in seminars where I have presented some of the above ideas, for helpful comments.

REFERENCES

Church, Alonzo, [1956], *Introduction to Mathematical Logic*, Princeton. NJ: Princeton University Press.

Dummett, Michael, [1975], "The Logical Basis of Intuitionistic Logic," in *Logic Colloqium '73*, ed. H. E. Rose et al. Amsterdam: North-Holland. Reprinted in Dummett [1978].

———, [1976], "What Is a Theory of Meaning? (II)" in *Truth and Meaning*, ed. G. Evans and J. McDowell, Oxford: Clarendon Press.

———, [1978], *Truth and Other Enigmas*, London: Duckworth.

———, [1991], *The Logical Basis of Metaphysics*, London: Duckworth.

Etchemendy, John, [1990], *The Concept of Logical Consequence*, Cambridge, MA: Harvard University Press.

Gentzen Gerhard, [1934], "Untersuchungen über das logische Schliessen," *Mathematische Zeitschrift* 39, 176–210, 405–431. Translated in *The Collected Papers of Gerhard Gentzen*, Amsterdam: North-Holland, 1969.

Martin-Löf, Per, [1985], "On the Meanings of the Logical Constants and the Justifications of the Logical Laws," *Atti degli Incontri di Logica Matematica*, vol. 2, pp. 203–281. Siena: Scuola di Specializzazione in Logica Matematica, Dipartimento di Matematica, Università di Siena. Reprinted in *Nordic Journal of Philosophical Logic* 1, 11–60.

Prawitz, Dag, [1965], *Natural Deduction: A Proof-Theoretical Study*. Stockholm: Almqvist & Wiksell.

———, [1971], "Ideas and Results in Proof Theory," in *Proceedings of the Second Scandinavian Logic Symposium*, ed. J. E. Fenstad, pp 235–307. Amsterdam: North-Holland.

———, [1973], "Towards a Foundation of a General Proof Theory," in *Logic, Methodology and Philosophy of Science*, ed. P. Suppes et al., pp. 225–250, Amsterdam: North-Holland.

———, [1974], "On the Idea of a General Proof Theory," *Synthese* 27, 63–67.

———, [1985], "Remarks on Some Approaches to the Concept of Logical Consequence," *Synthese* 62, 153–171.

———, [2005], "Meaning Approached via Proofs," in *Proof-Theoretic Semantics*, ed. R. Kahle and P. Schroeder-Heister, special issue of *Synthese*.

Sundholm, Göran, [1998], "Inferences Versus Consequence," in *The LOGICA Yearbook*. Prague: Czech Academy of Sciences.

Tarski, Alfred, [1935/1936], "Der Wahrheitsbegriff in den formalisierten Sprachen," *Studia Philosophica* 1, 261–405. Translated into English as "The Concept of Truth in Formalized Languages," in Tarski [1956].

———, [1936], "Über den Begriff der logischen Folgerung," in *Actes du Congrès International de Philosophie Scientifique* 7, 9–11. Translated into English as "On the Concept of Logical Consequence," in Tarski [1956].

———, [1956], *Logic, Semantics, Metamathematics*. Oxford: Clarendon Press.

CHAPTER 23

RELEVANCE IN REASONING

NEIL TENNANT

1. Introduction

WE advance here an unabashedly partisan view of how best to "relevantize" a logic. The view is laid out as informally as possible, given the technical nature of the subject matter.

Here, we understand "relevantizing" as the project of formulating a decent system of logic that does not endorse Lewis's First Paradox:

$A, \neg A : B.$

Such a system will be *paraconsistent*, in that it will allow for distinct inconsistent theories (within a given language). But it will not be *dialetheist*. That is, it will not allow for true contradictions. Dialetheism does not follow from (though, in order to avoid trivialization, it requires) a refusal to infer whatever one pleases from a contradiction.

We can pose our question as follows:

> How best might one restrict the deducibility relation of a familiar system of logic (such as intuitionistic or classical logic) so as to avoid the First Lewis Paradox, but still provide all the proofs needed for mathematics and for the hypothetico-deductive method in natural science?

We are not taking sides in the debate between the classicist and the intuitionist. We seek to enable the intuitionistic [resp. classical] mathematician to do intuitionistic [resp. classical] mathematics with just the relevant fragment of intuitionistic [resp. classical] logic. The relevantized restriction of each system is to be judged by reference to the original system that has been restricted. Once we have found the right restrictions, the resulting relevant systems ought to be as smooth and as elegant and as powerful as possible. These are of course evaluative notions. But we believe they can be made to apply, even in the eye of the logical beholder.

2. Some Historical Background

The method of relevantizing that is to be explained in this chapter derives from work by the author beginning with [15] and [16], and further developed in [17], [18], [19], [20], [21], [22], [23], [24], and [25]. It stands outside the so-called Anderson–Belnap tradition (for which, see [1]).

It does not fall within the scope of this discussion to examine the details of the Anderson–Belnap tradition. (A few more details will be provided in a later section.) I do need, however, to explain why this chapter is devoted to an alternative approach. I take the very title assigned to me—"Relevance in Reasoning"—to be an invitation to discuss (what I judge, on balance, to be) a *viable* account of how mathematical and scientific reasoning can be "relevantized" when it is logically regimented.

I believe that Burgess—building on the earlier critical contributions of Copeland [6], [7]—prevailed in his critique, in the early 1980s, of Anderson–Belnap relevance logic. This critique was laid out in his well-known *Notre Dame* exchange with Mortensen and Read. (See Burgess [3], [4] and [5]; Mortensen [11], and Read [14].)

One must nevertheless hasten to point out that, however diverse and protean the Anderson–Belnap tradition might be (as remarked by Burgess [5], p. 217), Burgess's critique was directed *only* against this tradition. In [3], p. 104, he wrote:

> ... I have been concerned here solely with the original Anderson–Belnap account of "relevant" logic, and with their claim that their systems *E*, *R*, etc., are in better agreement with common-sense than is classical logic.

Again, in [4], p. 41, he wrote, "All relevantists agree in rejecting *disjunctive syllogism* ..." (emphasis in original).

While that might have been true of all relevantists known to Burgess at the time of writing, it is so no longer. As we shall see presently, one can be a relevantist yet preserve disjunctive syllogism as an essential derived rule of one's logic. Although many relevantists (such as Anderson and Belnap) object to Disjunctive Syllogism, I do not. It strikes me, intuitively, as being in perfect order. In this I side with Burgess.

Mathematics would be crippled without Disjunctive Syllogism. There is no metatheorem showing that classical (or intuitionistic) mathematics survives any Anderson–Belnap style of relevantizing proofs of mathematical theorems from mathematical axioms, especially if one imposes the reasonable requirement that the formal relevantized proofs bear a natural structural resemblance to their informal counterparts in ordinary mathematical practice. Burgess's critique of Anderson–Belnap relevant logic in the *Notre Dame* exchange focused clearly on this, its main shortcoming. In [5], pp. 222–223, Burgess observed that (then unpublished) work of R. K. Meyer

> ... demonstrates beyond doubt that standard mathematical arguments cannot be formalized relevantistically.... For half of the most famous theorems of elementary number theory (including the theorem that every integer is a sum of four squares) it seems that no relevantistically acceptable proof is known, for all Meyer's work. For the other half of the most famous theorems of elementary number theory (including the theorem that no cube is the sum of two cubes) relevantistic proofs are available, but they cannot be regarded as formalizations of standard proofs. They involved carrying along caveats of the form "unless $0 = 1$," which one would never have heard from the lips of Fermat or Euler or Lagrange or Gauss or Louisville, and eliminating these caveats at the end of the proof by various manipulations. Meyer's manipulations are undeniably clever, but the very need for such cleverness demonstrates that relevantism conflicts with standard mathematical practice.
>
> It seems that a relevantist [in the Anderson–Belnap sense—NT] *must* be a revisionist, at least as far as mathematics is concerned....

Exactly how the work of Meyer in question (presumably [10]) "demonstrate[d] beyond doubt that standard mathematical arguments cannot be formalized relevantistically" is not altogether clear from what Burgess says. Since the time of Burgess's critique, however, there has been a highly significant, dispositive result in this regard. Friedman has definitively proved (see Friedman and Meyer [9]) that there are indeed limitations to relevant arithmetic: there is a strictly positive theorem of Peano Arithmatic that cannot be proved in the relevant Peano Arithmetic $R^{\#}$. ($R^{\#}$ is the closure of the Peano axioms under the first-order version of the Anderson–Belnap relevance logic R.)

We shall return in a later section to a comparison between the Anderson–Belnap approach and the one to be commended here. At this point we must prepare the ground for the latter.

3. What Do We Require of a Logic?

We begin by asking a question that is seldom raised explicitly, even when departures from classical logic are being proposed. *What, exactly, do we require of a logic?* Logic is an instrument of reasoning, and is central to our intellectual investigations. But what are the exact uses that we need to be able to make of it?

These uses will concern three main logical properties or relations:

(i) "Genuine" logical consequences (where a claim that is falsifiable follows from a set of assumptions that is satisfiable)
(ii) Logical falsehoods
(iii) Logical truths.

We take our task to be that of *restricting* what can be proved, in either intuitionistic or classical logic. In each case, the relevant fragment will be contained in the nonrelevant logic concerned. Moreover, because the systems of intuitionistic and classical logic have their respective notions of validity, and their own completeness theorems, we shall be able to speak interchangeably of deducibility or logical consequence; of logical truth or theoremhood; and of unsatisfiability or inconsistency.

We shall be eschewing the *nonrelevant parts* of the deducibility (or consequence) relation in each of these logics. Fallacies of relevance will not be regarded as genuine logical consequences, even when they appear to count as logical consequences according to the standard, Boolean definition that involves preservation of truth under all interpretations. Because we use a sentential definition of object-linguistic logical consequence in the metalanguage—

> X logically implies φ if and only if every model that makes all of X true makes φ true also; i.e.,
> $X \models \varphi \leftrightarrow_{df} \forall M[(\forall\psi(\psi \in X \rightarrow M \models \psi) \rightarrow M \models \varphi]$—

it is no surprise that the First Lewis Paradox comes out as a logical consequence in the object-language. The metalinguistic claim

> Every interpretation that makes A and makes $\neg A$ true makes B true also; i.e.,
> $\forall M[(M \models A \wedge M \models \neg A) \rightarrow M \models B]$,

is indeed a metatheorem even in a *relevantized* metalanguage. The question of genuine logical consequence should, rather, be posed as follows: Is it possible, in the metalanguage, to deduce from the premises that A is true under interpretation M and that $\neg A$ is true under interpretation M, the conclusion that B is true under interpretation M? When the question is posed this way, the respective answers from the standard metalogician and the relevant metalogician will differ. The standard metalogician will say "yes"; the relevant metalogician will say "no."

We now devote a subsection of clarification to each of our requirements (i), (ii), and (iii) above.

3.1. Ad (i): Genuine Consequence

Our logic should enable us to draw those conclusions that follow from whatever satisfiable assumptions (or axioms) we might adopt. Two areas of inquiry immediately come to mind. In *mathematics*, we should be able to derive all consequences of our axiom systems (assuming that these have models); and in the *natural sciences*, we should be able to derive all the empirical predictions that would follow from our scientific hypotheses taken in (consistent) conjunction with statements of boundary conditions, initial conditions, and so on. In both these cases, we are considering (what we believe to be) satisfiable sets of assumptions, and are concerned to be able to deduce from them all logically falsifiable conclusions that follow from them. Let us devote the next two paragraphs to underscoring this last claim.

In the case of mathematics, we take our axioms to be consistent (hence, satisfiable); and we seek proofs of nontrivial *mathematical* (i.e., nonlogical) theorems from them. It is in this sense that a nontrivial mathematical theorem is "logically falsifiable": for any such theorem p, there will be *some* logically possible interpretation of the extralogical primitives (the specifically "mathematical" vocabulary) that will make p false. Of course, such an interpretation will not be one in which the mathematical axioms themselves are true.

In the case of natural science, we take our scientific hypotheses to be consistent with the statements of such boundary conditions and initial conditions as might be involved in an experiment to test the predictions derivable from those combined assumptions. Call the set of such statements Δ. Note that Δ is consistent. It is only when observations and measurements are made, yielding further statements (say, Γ) treated as extra assumptions, that a contradiction might follow: $\Delta, \Gamma \vdash \bot$. Then, but only then, will the evidence Γ potentially falsify the hypotheses in Δ. (The famous so-called Duhem–Quine problem, of course, is to decide which statements in $(\Delta \cup \Gamma)$ to reject.) Note again that the predictions derived from Δ will be logically falsifiable statements. No natural scientist would ever wittingly produce a logical truth as a supposedly empirically "testable" prediction of a scientific theory!

3.2. Ad (ii): Logical Falsehoods

If in either case (mathematical or natural-scientific) the assumptions in question are *not* satisfiable, then we require only that our logic be able to reveal the fact: that is, that it furnish a proof of their inconsistency. Our logic should be able to tell us when a mathematical investigation would be fruitless to pursue, and when a scientific theory has run foul of the data.

3.3. Ad (iii): Logical Truths

Thus far we have focused on assumptions. But what about conclusions? We are interested in conclusions that follow from our assumptions (assumed to be satisfiable). But certain conclusions, called logical truths, have a nasty way of "following" (in standard systems) from *any* set of assumptions. So there is nothing special about deducing them "from" any particular satisfiable set of assumptions (let alone any particular *un*satisfiable one!). What is special about logical truths is that one needs no assumptions at all in order to establish them. So let us require of our logic no more than that. Let us insist, with regard to logical truths as potential conclusions, only that our logic should furnish proofs of them *from no assumptions at all.*

3.4. Some Definitions

The set X of assumptions is unsatisfiable just in case the sequent $X:\emptyset$ is valid. And the sentence φ is logically true just in case the sequent $\emptyset:\varphi$ is valid. (We write φ for the succedent when the latter is the singleton $\{\varphi\}$.) We shall also define the useful notion of *perfect* validity: a sequent is perfectly valid just in case it is valid and has no valid proper subsequents. That is, every sentence on the left and every sentence on the right is needed for the validity of the sequent concerned.

3.5. Summary of Methodological Requirements

We can summarize our three methodological requirements thus far as follows:

1. If $X:\emptyset$ is valid, then there should be a proof of $X:\emptyset$.
2. If $\emptyset:\varphi$ is valid, then there should be a proof of $\emptyset:\varphi$.
3. If $X:\emptyset$ is invalid and $\emptyset:\varphi$ is invalid, but $X:\varphi$ is valid, then there should be a proof of $X:\varphi$.

Note that requirements (1) and (2) take care of the two extremes, logical falsity and logical truth, respectively. And requirement (3) takes care of what we are here calling "genuine" logical consequence.

Moreover, (1) is all that is needed for the hypothetico-deductive method of testing empirical theories. Insofar as *deductive* logic features in scientific reasoning, it does so in the context of inferring predictions, or refuting theories modulo observational evidence (once certain predictions have been inferred). For the refutations, (1) suffices (with the observational evidence reckoned to X, along with the theory being refuted). For the derivation of predictions, (3) suffices, as noted above (since in such a case X does not contain any observational evidence that might refute it).

Finally, in the case of mathematical reasoning, one is interested only in two things: (a) being able to reveal that one's axioms are unsatisfiable, if indeed they are (a very rare occurrence); and (b) being able to deduce all interesting consequences of one's axioms, if indeed those axioms are satisfiable. Note that (1) takes care of (a). And in connection with (b), as already noted, no mathematician is going to find a *logical truth* to be an "interesting consequence" of her axioms. She will be interested only in such consequences of her axioms as are not, themselves, logically true. Hence (3) takes care of (b). So, for the purposes to be served by a system of logic within mathematics, (1) and (3) suffice.

4. Cumulative Deductive Progress

There is another aspect of the deductive enterprise, however, which is of special importance. We need to be able to make *cumulative deductive progress.* In mathematics, for example, we have the practice of proving interesting lemmas as "halfway houses" on the way to those deeper results that we dignify with the label "theorem." (From a logical point of view, of course, these psychological classifications are irrelevant.) We need, therefore, to be able to do the following:

1. Prove a lemma ψ, say, from axioms X
2. Then prove a theorem φ, say, from the lemma ψ along with further axioms Y
3. Finally, conclude to φ on the basis only of $X \cup Y$.

In a natural-deduction setting, where proof-trees have their assumptions at leaf-nodes, and have their conclusions at their root-nodes, this two-step process can be pictured as follows:

$$\begin{array}{c} X \\ \Pi \\ \underbrace{Y, (\psi)} \\ \Sigma \\ \varphi \end{array}$$

The lemma ψ stands as the conclusion of the proof Π and as a premise of the proof Σ. It is an "accumulation point" of the overall proof-tree. Proofs are defined in such a way that accumulations of proof-trees are themselves proof-trees. This is because, in the standard nonrelevant systems, proofs are allowed *not* to be in normal form. Here, for example, the sentence ψ is likely to stand as the conclusion (within the natural deduction Π) of an introduction rule, and as the major premise (within the natural deduction Σ) of the corresponding elimination rule. So the "grafting points" ψ could in general be maximal sentence-occurrences, and the overall proof of φ from $X \cup Y$ accordingly not in normal form.

In a sequent setting, where the nodes of proof-trees are labeled with the whole sequent that has been established thus far, the same cumulative effect is achieved by applying the rule of Cut:

$$\frac{\begin{matrix}\Pi & \Sigma \\ X : \psi & Y, \psi : \varphi\end{matrix}}{X, Y : \varphi}.$$

In general, dividing one's deductive work in this way into two stages—the first stage, Π, toward the lemma ψ, the second stage, Σ, away from it—enables one to reduce quite dramatically the overall length of proof. The combined length of the two proofs Π and Σ is in general much less than the length of whatever proof might codify an argument proceeding *directly* from $X \cup Y$ to φ without "going through" the lemma ψ. Abnormality of the overall proof is the price one pays for the reduction in length of proof that results from interpolating the lemma ψ between one's axioms $X \cup Y$ and the sought theorem φ.[1]

That is all very well, but considerations of feasibility or tractability can serve at best to *explain* unjustifiable deductive procedures rather than to justify them. Here is the explanation why the procedure of "proceeding via lemma ψ" works, according to the pattern just given, in mathematics:

> Mathematics is (presumed to be) *satisfiable.* So any collection of our mathematical axioms is satisfiable. Moreover, φ is a logical consequence of $X \cup Y$ (and, presumably, an interesting one, i.e., one that is not a logical truth in its own right). Hence one should be able to deduce φ from axioms drawn from $X \cup Y$.

[1] In this connection, Harvey Friedman has achieved, in an as yet unpublished work, some spectacular results about just how bad proof-length blowup can be upon elimination of Cuts when formalizing everyday mathematical proofs of some very accessible mathematical results. Georg Kreisel and Friedman, among others, knew in the 1960s about the *existence* (i.e., the possibility-in-principle) of such blowups; but it is only in Friedman's more recent work that these blowups have been shown to arise within ordinary mathematics, involving informal proofs of feasible length (but containing Cuts). For another striking example, albeit one that was carefully contrived to serve this specific theoretical purpose, see Boolos [2].

> Indeed, given that one has already deduced ψ from X, and deduced φ from Y and ψ, further deductive work would be superfluous: we already know, by virtue of these two proofs, and on the presumption that the overall set of premises $X \cup Y$ is satisfiable, that $X, Y : \varphi$ is valid.

We cannot, however, conclude that $X, Y : \varphi$ is *perfectly* valid, even if both $X : \psi$ and $Y, \psi : \varphi$ are perfectly valid. Courtesy of Cut, therefore, we push back the frontiers of deductively generated mathematical knowledge. But we are using a rather blunt instrument in doing so. For consider: *if* (admittedly, a big "if") our axioms $X \cup Y$ are inconsistent, might we not, in applying Cut, miss this fact altogether in charging ahead and inferring φ? Moreover, we do not need to consider possible inconsistencies in our axioms in order to see that we might still miss out on opportunities to be more economical with them. We might not realize, for example, that the grain of truth in ψ that is needed, along with our choice Y of axioms, to yield φ, might be considerably less weighty than the bushel that is ψ itself, inferred from the axioms X. In taking ourselves (courtesy of Cut) to have inferred φ "from" $X \cup Y$, we might be missing the fact that φ follows from some *proper subset* of $X \cup Y$ (even if neither from X itself nor from Y itself).

Furthermore, it is not enough for the natural-deduction theorist[2] to respond by suggesting that one should simply normalize the overall proof of φ from $X \cup Y$ in order to prune away such assumptions as might not be needed for the resulting proof in normal form. If the system within which such normalization is carried out tolerates irrelevancies such as the First Lewis Paradox, then potential "reductions in premises" can *still* escape detection!

5. Avoiding the First Lewis Paradox: Banning Dilution

If we wish to avoid the First Lewis Paradox, we have to disable its simplest proofs. Here is one, in the sequent calculus:

$$\dfrac{\dfrac{A : A}{A, \neg A : }\neg L}{A, \neg A : B}\text{Dilution}$$

[2] An analogous remark would of course apply to the sequent-proof theorist.

There cannot be anything wrong with the rule of initial sequents $A:A$. Nor can there be anything wrong with the rule

$$\neg L \quad \frac{X:A}{X, \neg A:}$$

for introducing a negation sign on the left of the colon.

There remains only one diagnostic possibility: the culprit here must be the final step of Dilution (on the right).[3] So, in the interests of relevance, Dilution (on the right) must be banned.

So, too, must Dilution on the left. Otherwise, we shall be able to prove the Lewis-type sequent $A, \neg A:\neg B$, which is just as objectionable, on relevantist grounds, as the sequent $A, \neg A:B$. If the arbitrary proposition B bears no connection to the premises involving A, then neither does its negation $\neg B$. With Dilution on the left, however, we can form the following otherwise unobjectionable proof:

$$\cfrac{\cfrac{\cfrac{A:A}{A, B:A}}{A, \neg A, B:}}{A, \neg A:\neg B}$$

In the context of the little example involving the First Lewis Paradox, the ban on Dilution means that the first proof above would get only as far as

$$\frac{A:A}{A, \neg A:} \neg L.$$

And isn't that just as it should be? It is, after all, a more informative result than $A, \neg A:B$. (This is an intuition to be explicated presently.) In general, given any sequent of the form $X:B$, it might be quite difficult to tell that B "follows" from X by dint of X's inconsistency rather than by dint of any genuine deductive connection between X and B. We can lose sight of this general point when looking at the Lewis sequent $A, \neg A:B$, because its antecedent is so obviously contradictory. But with generality we lose obviousness. And, I contend, it is better to know that X is inconsistent than to know only that X logically implies B. The sequent $X:\emptyset$ represents *epistemic gain* over the sequent $X:B$.

[3] Dilution is often called weakening.

6. Epistemic Gain in Logic

What, exactly, is epistemic gain? It is a matter of learning that a *tighter* logical result holds than one had previously thought. Examples of epistemic gains made on the deducibility statement $A, B \vdash C$ might be

- One finds a proof of the sequent $A : C$, thereby learning that B is unnecessary as a premise.
- One finds a proof of the sequent $A, B : \emptyset$, thereby learning that one's premises are inconsistent.
- One finds a proof of the sequent $\emptyset : C$, thereby learning that the conclusion one was trying to prove "from" the premises A and B is a logical truth.

A proof of any of these *proper subsequents* of the original sequent $A, B : C$ would represent epistemic gain. Knowing that $A \vdash C$ is better than knowing that $A, B \vdash C$; and likewise for knowing that $A, B \vdash \emptyset$ or knowing that $\vdash C$. The more one "subsets down," on the left and/or on the right, while still achieving the turnstile, the stronger the logical result one has learned. Of course, one cannot subset down all the way, since there is no proof of the empty sequent $\emptyset : \emptyset$. So one will always achieve a local optimum in any course of "subsetting down." In an obvious sense, it will be a "(locally) strongest possible result." It will be a valid sequent that has no valid proper subsequent. Remember that we are calling such sequents *perfectly valid.*

An easy corollary of a theorem to be stated below, in conjunction with the usual completeness theorems, is that every perfectly valid sequent in classical (resp. intuitionistic) logic has a classical (resp. intuitionistic) *relevant* proof.

7. Avoiding the First Lewis Paradox: Banning Cut

We have seen that Dilution is a source of irrelevance. But what about Cut? Cut is a structural rule in a sequent system that registers the (assumed) *unrestricted transitivity of deduction.* The orthodox statement of the unrestricted transitivity of deduction (in single-conclusion systems) has the form

If $X \vdash B$ and $Y, B \vdash C$ then $X, Y \vdash C$.

Correspondingly, the structural rule of Cut allows one to infer from two premise-sequents to a conclusion-sequent as follows:

$$\frac{X : B \quad Y, B : C}{X, Y : C}$$

If we can get rid of Cuts (as Gentzen's *Hauptsatz* says we can), why should we be worried about having the rule of Cut in our deductive system? The simple answer is that Cut, too, can be a source of potentially undetected irrelevance. For, consider the following two perfectly good proofs, of $\vee$-Introduction and of Disjunctive Syllogism respectively:

$$\frac{A : A}{A : A \vee B} \qquad \frac{\dfrac{A : A}{A, \neg A :} \qquad B : B}{A \vee B, \neg A : B}$$

Now, if we apply Cut to the sentence $A \vee B$, we obtain the unwanted Lewis sequent $A, \neg A : B$.

So, if we wish to avoid the First Lewis Paradox, we must either ban Disjunctive Syllogism or ban Cut. We wish to avoid the First Lewis Paradox. Therefore, we must either ban Disjunctive Syllogism or ban Cut. But we cannot ban Disjunctive Syllogism without unacceptable consequences for the codification of mathematical practice. My own methodological conclusion, therefore, is that *Cut* must be banned *as a rule within the system.* And I note the virtuous irony in the fact that I am arguing for this methodological conclusion by means of (a deontic version of) Disjunctive Syllogism:

We must either ban Disjunctive Syllogism or ban Cut.
We should not ban Disjunctive Syllogism.
Ergo, We should ban Cut.

7.1. On How Best to "Ban" Cut

To say that Cut must be banned is *not* to say that we shall thereby forfeit transitivity of deduction altogether. Our conclusion "We should ban Cut" should really be written: *We should ban Cut in those situations where its use can allow irrelevancies to creep into our reasoning. But we expect the vast preponderance of Cuts in actual mathematical and scientific reasoning to be innocuous in this regard.* Making good on this informally expressed expectation is the main task before us.

Recall our earlier contrast between the statement of transitivity (which involved turnstile-claims) and the rule of Cut as a means of inferring one sequent from two others. As far as the transitivity of deduction is concerned, our relevant systems will still enjoy it—*where it counts.* For the following metatheorem holds for the relevant systems:

If $X \vdash B$ and $Y, B \vdash C$ then $[X, Y] \vdash [C]$.

(Remember that $[X, Y]$ ranges over subsets of $X \cup Y$, and $[C]$ over subsets of $\{C\}$. We shall also use "$X \vdash \bot$" as a synonym for "$X \vdash \emptyset$," and talk of being able to deduce absurdity ($\bot$) from X.) Another way of stating this metatheorem would be to say that Cut, though no longer applicable unrestrictedly (as it was in the non-relevant systems *C* and *I*), will nevertheless be "gainfully admissible" in our respective relevant systems *CR* and *IR*.

This statement of *restricted* transitivity of deduction is all that one needs for the serious scientific purposes broached above. Those purposes, as we have already seen, are (1) to prove the inconsistency of any unsatisfiable set of assumptions, (2) to prove all logical truths from no assumptions, and (3) to derive all falsifiable consequences of any satisfiable sets of assumptions. There is nothing for a canon of deductive reasoning to accomplish over and above these three aims. By seeking to provide more, the classical Tarskian notion of consequence or deducibility is guilty of a misleading *overprovision*, which can blind one to epistemic gains that stand to be made by the logically more wary. *Without a rule of Cut*, one can still be assured of transitivity of deduction wherever it matters. Why, then, spoil that austere provision by adopting Cut as a rule? The only answer is: in order to achieve speedup in the search for "proofs." But that is a merely pragmatic matter of tactics, of strategy, and of expenditure of time and energy. It has nothing to do with normative issues of justification.

There is no rule of Cut, then, within our relevant systems. And Cut in its orthodox unrestricted form is not even an *admissible* rule for either of the systems *IR* and *CR*. For in each of these systems we have

$P \vdash P \vee Q$ and $\neg P, P \vee Q \vdash Q$ but $P, \neg P \nvdash Q$.

The orthodox Deduction Theorem for logical calculi states that

(i) $X \vdash B$ only if $X \setminus \{A\} \vdash A \rightarrow B$.
(ii) $X \vdash A \rightarrow B$ only if $X, A \vdash B$.

In *IR* and *CR*, the implication (i) holds. But the converse implication (ii) fails, for in these relevant systems we have $\neg A \vdash A \rightarrow B$, but not $\neg A, A \vdash B$.

Notice that the two little proofs that were put together above for a final step of Cut are so degenerate that the inconsistency of the antecedent of the final sequent is obvious. But it is worth stressing once again that in general this will not be the case. One might have proved $X: B$ (with X consistent) and $Y, B: C$ (with Y, B consistent), while yet the final sequent $X, Y: C$ obtained by Cut has an *inconsistent* antecedent X, Y. The latter inconsistency, however, might go undetected; the Cut-applier might be unaware of it. Accumulating proofs for deductive progress by means of Cut brings with it the risk that the overall result is *relevantly unsound.* It might be that the conclusion C "follows from" the set of premises $X \cup Y$ "only because" $X \cup Y$ is inconsistent.

Our method of relevantizing can be thought of as a measure for logical quality control. It is always in search of epistemic gain. It seeks to ensure that the results we can prove "relevantly" are those whose conclusions do not follow from their premises only because of the joint inconsistency of those premises. Viewed another way, it seeks to ensure that the deductive progress afforded by the transitivity of deduction is *genuine* deductive progress. Pooling of assumptions must not get us consequences on the cheap. For such consequences are worth no more than the assumptions that they depend upon. Better to know that the pooled assumptions are inconsistent than to continue in ignorance of this fact, and rely on any conclusion that has been "deduced" from them.

The fact that Cut is not a rule in our relevant systems does not mean that someone possessed of relevant proofs

$$\begin{array}{cc} X & \underbrace{Y, \psi} \\ \Pi & \Sigma \\ \psi & \varphi \end{array}$$

would be at a loss to find a relevant proof of some subsequent of $X, Y: \varphi$. In complaining that the Rule of Cut is too crude an instrument of cumulative deductive progress, I do not intend to leave the reasoner unable to achieve the benefits of accumulation. In situations such as these, the sought relevant proof can be obtained as follows. (For perspicuousness of display, we switch here to natural-deduction mode.)

1. Graft Π on top of Σ as one would normally do. Call the resulting proof-tree Θ:

$$\begin{array}{c} X \\ \Pi \\ \underbrace{Y, (\psi)} \\ \Sigma \\ \varphi \end{array}$$

2. Normalize Θ:

$$\Theta \mapsto \begin{array}{c} Z \\ \Xi \\ \varphi \end{array}$$

   The resulting normal proof Ξ has conclusion φ, and its set Z of undischarged assumptions is a subset of $X \cup Y$.
3. Extract from this normal form Ξ a *relevant* proof, of either $\perp$ or φ, from some subset W of Z.

That is to say, our metatheorem stated earlier—

If $X \vdash B$ and $Y, B \vdash C$, then $[X, Y] \vdash [C]$—

has a constructive proof. (W will be the subset of $X \cup Y$ that makes the consequent hold.)

To be sure, step (2) can result in exponential blowup. That is to say, the length of the normal proof produced as output in step (2) is bounded below by an exponential function in the length n of the nonnormal, input proof—in the sense that, for any n, some input proof (not in normal form) of length $k > n$ produces an output of length $> 2^k$.[4] Step (3), however, can be completed in polynomial time. That is to say, there is some polynomial function $p(n)$ such that for all m, for all input proofs (in normal form) of length $< m$, the relevantized proof (of course, still in normal form) is produced in fewer than $p(m)$ units of time. Normalizing carries a heavy computational price, whether or not one is a relevantist. But at least going one step further and relevantizing adds negligibly to the computational costs involved. The potential epistemic gains come cheap at this price.

8. The Metalogical Upshot

Our motivation can be summarized, but now in the sequent calculus setting, as the requirement that we should be able to prove a Dilution Elimination Theorem

[4] Indeed, Orevkov has demonstrated *hyper*exponential blowup, by exhibiting a hyperexponential function $f(k)$ and a sequence of sentences C_k each of which has a non-normal proof of length linear in k, but no normal proof of length $< f(k)$. See Orevkov [12].

in addition to the well-known Cut Elimination Theorem. That means that we must not rely on having Dilution in the system in order to show that the system admits of Cut-elimination.

Gentzen's famous Hauptsatz, or Cut-elimination theorem, is that applications of Cut can be eliminated from any proof. Gentzen happened to formulate his logical rules in such a way that in eliminating cuts from proofs, one would need to be able to take steps of Dilution in order to ensure that newly required applications of the logical rules would be formally correct. Perhaps that is why Gentzen himself never raised the question of whether one could eliminate Dilutions from proofs, along with Cuts. Given his chosen way of stating the logical rules, the answer would have appeared to be an obvious negative. Since, however, we are avoiding Dilutions as well as Cuts—because these are sources of irrelvance—we shall need to formulate the logical rules in a subtly different way from Gentzen.

The motivating considerations behind our approach to capturing relevance determine not only the form of the rules for the logical operators but also some very general restrictions on their applications in proofs. These restrictions are imposed in order to avoid Dilutions. Of course, if we can get rid of all Dilutions in a proof, what we end up proving is some *subsequent* of the original sequent proved. But that is precisely what I call epistemic gain.

The relevantizing of both intuitionistic logic *I* and classical logic *C* is accomplished uniformly, by means of the same techniques. Thus considerations of relevance are orthogonal to the considerations of constructivity that lead one from classical logic *C* to its intuitionistic subsystem *I*.

We relevantize both *C* and *I*, then, by modifying the usual Tarskian assumptions about what is desired of a deducibility (or consequence) relation. If one needs a slogan to help with orientation, ours is a method of relevantizing "at the level of the turnstile." As a result, there will be some tweaking of the rules for the logical operators (in the natural deduction or sequent setting); but these tweakings are not such as to change their established meanings.

Before I state the rules of the system *IR*, it is worth mentioning two of its proven virtues. It suffices for intuitionistic mathematics[5] and for the hypothetico-deductive method in science.[6] So for all serious epistemic purposes (from an intuitionistic point of view), *IR* is a system that relevantizes without loss. Exactly analogous claims hold in the classical case. The relevant system *CR* suffices for classical mathematics and for the hypothetico-deductive method in science. So for all serious epistemic purposes (from a classical point of view), *CR* is a system that relevantizes without loss.

[5] Cf. [22].
[6] Cf. [18].

9. The System *IR*

9.1. Natural Deduction

We now set out the system of natural deduction for *IR*. We begin with three explanatory remarks, and an explanation of some new notational conventions.

(i) In all applications of the elimination rules stated below, the major premise *stands proud*; that is, it is not the conclusion of any rule.
(ii) No applications of the absurdity rule are permitted.
(iii) All elimination rules are in "parallelized" or "general" form, including those for $\rightarrow$ and $\wedge$.

We also need to explain the perhaps unusual notational conventions used below, involving boxes and diamonds appended to discharge strokes over assumptions used for the sake of argument. The *permissive construal* with a discharge-rule is that one is permitted to discharge any assumption of the indicated form, *if* one has used it. The permissive construal allows one to apply a "discharge-rule" even in cases where no assumption of the indicated form has been used. In such a case one could speak of "vacuous discharge." In our statement of the rules below, this permissive construal applies to the rules $(\neg I)$ and $(\rightarrow I)$. Here, permissiveness is indicated by a diamond appended to the discharge-stroke. For all the other discharge-rules, we adopt instead a construal according to which discharge is *obligatory*. Obligatoriness is indicated by a box appended to the discharge-stroke. Here, the affected rules may be applied *only if* at least one assumption of the indicated form(s) has indeed been used to obtain the sub-conclusion concerned. (For the parallelized version of the rule of $(\wedge E)$ given below, one must have used either A or B as an assumption, but not necessarily both.)

The graphic presentations of the following rules do not explicitly reveal an important structural fact. Construed as definitional clauses for the inductive formation of proofs, inference rules with more than one premise allow their immediate subproofs, in general, to have *distinct* sets of undischarged assumptions. For example, the rule of $\wedge$-Introduction, construed as a clause in the inductive definition of the notion "Π is a proof of A from the set Δ of undischarged assumptions," would read as follows:

> If Π is a proof of A from the set Δ of undischarged assumptions and if Σ is a proof of B from the set Γ of undischarged assumptions, then $\dfrac{\Pi \quad \Sigma}{A \wedge B}$ is a proof of $A \wedge B$ from the set $(\Delta \cup \Gamma)$ of undischarged assumptions.

Such a clause allows the formation of the proof of $A \wedge B$ even if Δ is distinct from Γ. We adopt here the convention that distinct sets of assumptions will be presumed to be allowed, unless there is explicit mention to the contrary. And we shall see below that constraints to the contrary are what characterize the more intractable notion of proof in the Anderson–Belnap system *R*.

	INTRODUCTION	ELIMINATION
$\neg$	$\overset{\square\text{---}(i)}{A}$ $\vdots$ $\dfrac{\bot}{\neg A}\,(i)$	$\dfrac{\neg A \quad A}{\bot}$
$\wedge$	$\dfrac{A \quad B}{A \wedge B}$	$(i)\text{---}\square\text{---}(i)$ $\underbrace{A\,,B}$ $\vdots$ $\dfrac{A \wedge B \quad C}{C}\,(i)$
$\vee$	$\dfrac{A}{A \vee B} \quad \dfrac{B}{A \vee B}$	$\overset{\square\text{---}(i)}{A} \quad \overset{\square\text{---}(i)}{B}$ $\vdots \quad \vdots$ $\dfrac{A \vee B \quad \bot/C \quad \bot/C}{\bot/C}\,(i)$ where neither *A* nor *B* remains undischarged
$\rightarrow$	$\overset{\diamond\text{---}(i)}{A}$ $\vdots$ $\dfrac{B}{A \rightarrow B}\,(i)$ $\overset{\square\text{---}(i)}{A}$ $\vdots$ $\dfrac{\bot}{A \rightarrow B}\,(i)$	$\overset{\square\text{---}(i)}{B}$ $\vdots$ $\dfrac{A \rightarrow B \quad A \quad C}{C}\,(i)$ where *B* does not remain undischarged

Some further explanatory comments are now in order.

9.1.1. *Preventing Dilutions from Creeping In*

In the rule of $\vee$-Elimination (proof by cases), the restriction means that the first case-assumption cannot be an undischarged assumption in the second case-proof, and the second case-assumption cannot be an undischarged assumption in the first case-proof. In the rule of $\rightarrow$-Elimination, the restriction means that the consequent B of the major premiss $A \rightarrow B$ cannot be an undischarged assumption in the "minor" proof of A.

9.1.2. *Liberalized Proof by Cases*

The rule of proof by cases, or ($\vee E$), looks unusual. By stating it graphically as we have, we are providing for the possibility that one of the case assumptions might lead to absurdity ($\perp$). We are then permitted to bring down as the main conclusion whatever is concluded from the *other* case assumption. Thus the rule as stated is shorthand for the following possibilities regarding the sequence *Conclusion of first case-proof, Conclusion of second case-proof, ergo Main conclusion*:

C, C, *ergo* C (including, as a special case, $\perp$, $\perp$, *ergo* $\perp$)
C, $\perp$, *ergo* C
$\perp$, C, *ergo* C.

Liberalizing proof by cases in this way is entirely natural, given how we reason informally. Suppose one is told that $A \vee B$ holds, along with certain other assumptions X, and one is required to prove that C follows from the combined assumptions X, $A \vee B$. If one assumes A and discovers that it is inconsistent with X, one simply stops one's investigation of that case, and turns to the case B. If C follows in the latter case, one concludes C as required. One does *not* go back to the conclusion of absurdity in the first case, and artificially dress it up with an application of the absurdity rule so as to make it also "yield" the conclusion C.

9.1.3. *Vacuous vs. Nonvacuous Discharge of Assumptions*

Note that in the rules for *IR* above we make some unusual demands on the *discharges of assumptions* allowed by certain of these rules. By contrast with minimal logic, we retain the permissive construal only for (the first half of) the rule of ($\rightarrow I$). Neither minimal nor intuitionistic logic insists on nonvacuous discharge when applying the rule ($\neg I$). The intuitionist logician has a reason of sorts for this omission: any application of ($\neg I$) to infer $\neg A$ *without* having used A to obtain the preceding absurdity could simply be regarded instead as an application of the absurdity rule. The latter rule, however, is conspicuously absent from our relevant system *IR*, and for good reason: it leads to the First Lewis Paradox $A, \neg A : B$. Just as disagreeable is the negated-conclusion version of the paradox: $A, \neg A : \neg B$. Thus we have good reason to insist that $\neg I$ be applied only with nonvacuous discharge.

9.2. Sequent Calculus

We follow the convention, when stating sequent rules for intuitionistic systems, whereby succedents are at most singletons. But we write the sequents as though their succedents, when nonempty, are single sentences. Thus $X:\{C\}$ is rendered as $X:C$. Another useful notation is $X:[C]$, which is ambiguous between $X:C$ and $X:\emptyset$. Likewise, $[X]$ will range over subsets of X.

The only structural rule is the rule of initial sequents

$A:A.$

All other rules are logical rules, for introducing a dominant occurrence of the operator concerned on the *right* or on the *left* of the sequent.

RIGHT	LEFT
$\dfrac{X,A:}{X:\neg A}$ A not in X	$\dfrac{X:A}{X,\neg A:}$
$\dfrac{X:A \quad Y:B}{X,Y:A\wedge B}$	$\dfrac{X:C}{X\backslash\{A,B\},A\wedge B:C}$ $X\cap\{A,B\}$ nonempty
$\dfrac{X:A}{X:A\vee B} \qquad \dfrac{X:B}{X:A\vee B}$	$\dfrac{X,A:[C] \quad Y,B:[C]}{X,Y,A\vee B:[C]}$ A not in $X\cup Y$, B not in $X\cup Y$; conclusion-succedent empty iff both premise-succedents are
$\dfrac{X,[A]:[B]}{X:A\rightarrow B}$ A not in X and either A or B occurs in the premise sequent	$\dfrac{X:A \quad Y,B:C}{X,Y,A\rightarrow B:C}$ B not in Y

10. The system *CR*

10.1. Natural Deduction

The natural deduction system for *CR* is obtained by adding to *IR* the rule of classical reductio ad absurdum, subject, of course, to the requirement that the assumption for reductio really be used:

$$\begin{array}{c} \square\overline{\neg A}^{(i)} \\ \vdots \\ \dfrac{\bot}{A}(i) \end{array}$$

We still insist, as we did with *IR*, that major premises of eliminations stand proud. Thus no conclusion of classical reductio ad absurdum can be the major premise of an elimination.

The decision problem for theoremhood in *IR* (equivalently, deducibility from finite sets of premises in *IR*) is PSPACE-complete, just as it is for the parent system *I* of intuitionistic logic. The decision problem for theoremhood in *CR* is co-NP-complete, just as it is for the parent system of *C* classical logic.

10.2. Sequent Calculus

The sequent calculus for the classical system of relevant logic is obtained in the usual way, by allowing succedents with more than one member.

The Sequent Rules for *CR*

Right	Left
$\dfrac{X, A : Y}{X : \neg A, Y}$	$\dfrac{X : A, Y}{X, \neg A : Y}$
A not in X	A not in Y
$\dfrac{X : A, Z \quad Y : B, W}{X, Y : Z, W, A \wedge B}$	$\dfrac{X : Y}{X \backslash \{A, B\}, A \wedge B : Y}$
A not in Z, B not in W	$X \cap \{A, B\}$ nonempty

$$\frac{X : Y, A}{X : Y, A \vee B} \qquad \frac{X : Y, B}{X : Y, A \vee B} \qquad \frac{X, A : Z \quad Y, B : W}{X, Y, A \vee B : Z, W}$$

A not in Y $\quad$ B not in Y $\qquad$ A not in $X \cup Y$, B not in $X \cup Y$

$$\frac{X, [A] : Y, [B]}{X : Y, A \to B} \qquad \frac{X : A, W \quad Y, B : Z}{X, Y, A \to B : W, Z}$$

A not in X, B not in Y and either A or B occurs in the premisesequent $\qquad$ A not in W, B not in Y

Unrestricted Cut fails to be admissible for *CR*, as it failed for *IR*. But the same compensating metatheorem holds for *CR* as for *IR*:

If $X \vdash B, Z$ and $Y, B \vdash W$, then $[X,Y] \vdash [Z,W]$

where we now allow for more than one sentence in succedents.

11. Contrast with the Anderson–Belnap Approach to Relevantizing

IR and *CR* are to be contrasted with competing systems in the Anderson–Belnap tradition. These other systems (such as the relevant logic *R*) differ from standard logic by having unusual, "intensional" connectives, subject to more restrictive axioms (in a Hilbert-style proof system) and rules of inference (in a natural deduction or sequent setting) than the standard connectives. But on the Anderson–Belnap approach, the *deductive structure* of the relevantized logical system is in one important regard the same as in the nonrelevant case. The Anderson–Belnap relation of deducibility admits of *unrestricted Cut.*

The Anderson–Belnap system *R*, like the systems of classical and intuitionistic logic, was originally presented as a Hilbert-style proof-system based on many axioms and one or two rules of inference. Since we have been using natural deduction in this chapter, we present the system *R* here in that format, in order to facilitate comparison.

	INTRODUCTION	ELIMINATION
$\neg$	$\overline{A}^{(i)}$ ⋮ $\dfrac{\bot}{\neg A}(i)$	$\dfrac{\neg A \quad A}{\bot}$

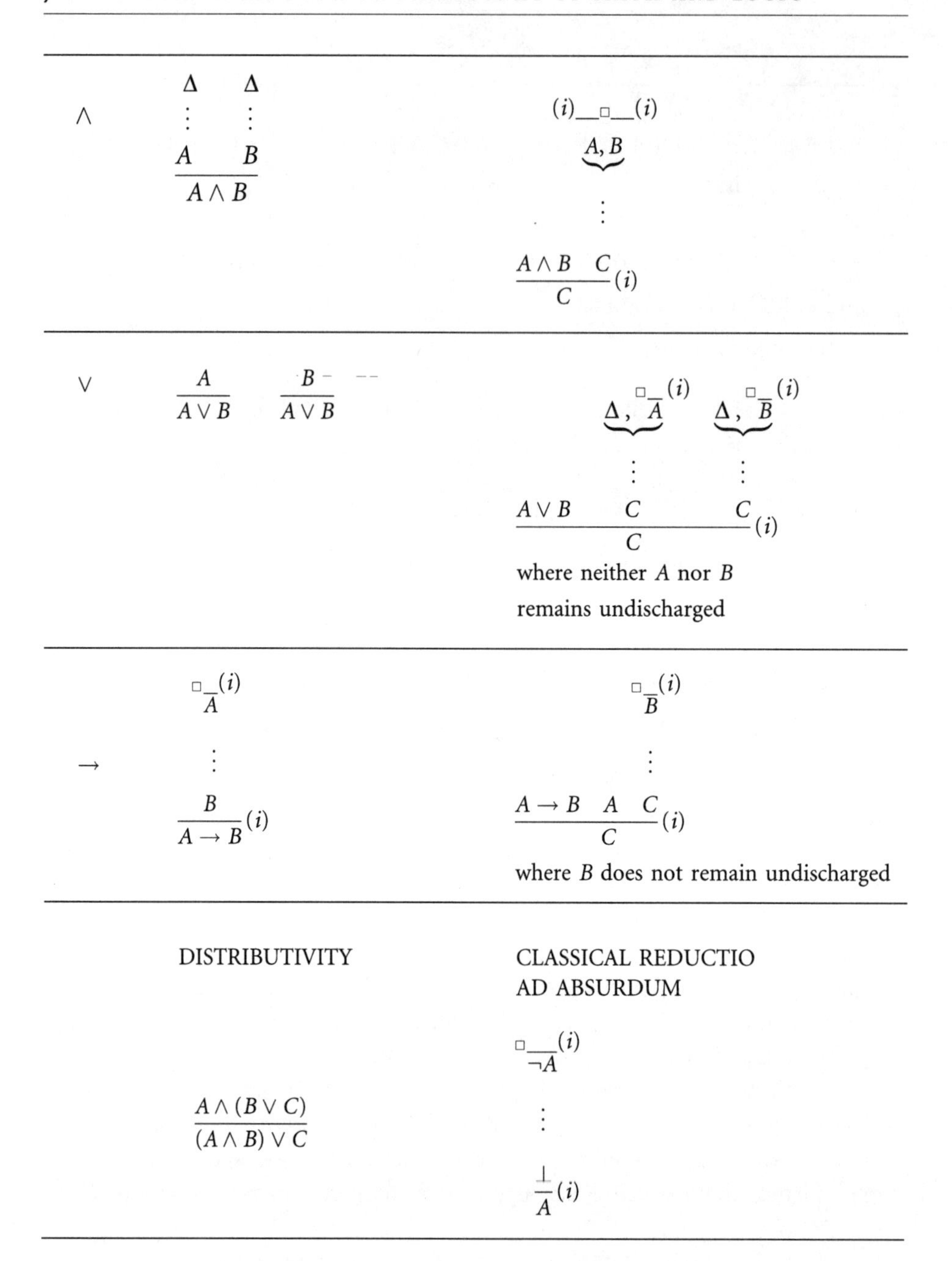

We stress again that major premises for eliminations must stand proud, which ensures that no conclusion of classical reductio ad absurdum can be a major premise for an elimination.

We have indicated, in $\wedge$-Introduction and $\vee$-Elimination, that these rules can be applied only when the same set Δ of undischarged assumptions is employed in the two places indicated. Hence the need for the independent postulation of Distributivity—for the usual proof of Distributivity would involve *distinct* sets of undischarged assumptions when applying the usual versions of $\wedge$-Introduction and $\vee$-Elimination.

We are also requiring that the discharge operations indicated with boxes are to discharge at least one occurrence, and indeed all occurrences, of the assumption of the indicated form. For $\wedge$-Elimination, we require use either of the assumption A or of the assumption B, and permit use of both. All such assumption-occurrences, however, must be discharged by the application of the rule. Thus the system above can be described as having the *universal discharge requirement.* It does not allow "partial discharge," that is, discharge of only some, but not necessarily of all, of the available occurrences of an assumption. In this regard our system differs from that of Prawitz [13].

Major premises of eliminations have to stand proud, with no proof-work above them. This ensures that every proof is in normal form.

Definition.

(i) If there is a proof of φ whose undischarged assumptions form the set Δ, then $\Delta \Rightarrow \varphi$.

(ii) If $\Delta \Rightarrow \varphi \in \Gamma$ and $\Gamma \Rightarrow \psi$, then $\Delta \cup \{\Gamma \setminus \{\varphi\}\} \Rightarrow \psi$

(iii) $\Delta \Rightarrow \varphi$ only if this can be shown by means of (i) and (ii).

Our system of proof requires proofs to be in normal form. So we can speak of "normal-form deducibility" in this system. Intuitively, $\Rightarrow$ is just the "deductive closure" of the (possibly more limited) relation of normal-form deducibility. (We could achieve $\Rightarrow$ directly as the deducibility relation if we allowed partial discharge with the rules of $\neg$-Intro and $\rightarrow$-Intro. This is essentially what Prawitz does.) We can talk of $\Rightarrow$ as representing "extended deducibility" in our ND-system.

Anderson and Belnap ([1], pp. 340–341) provide two slightly different sets of axioms for R, for the generation of Hilbert proofs using only the rules

($\wedge$I) From A, B infer $A \wedge B$

($\rightarrow E$) From A, $A \rightarrow B$ infer B.

Every line in such a proof is a theorem; whence the rules ($\wedge$I) and ($\rightarrow$E) apply only to theorems. Clearly, ($\wedge$I) is derivable in our ND-system for R, since we can take $\Delta = \emptyset$. Likewise, we can derive ($\rightarrow$E); whence its applications in Hilbert proofs can be mimicked by extended deducibility in our ND-system. Finally, one can show that every one of the Anderson–Belnap axioms ([1], pp. 340–341) has a *normal* proof in our ND-system (with *universal* discharge!). Hence, every theorem of R is extendedly deducible in our ND-system.

Conversely, one can show that every extended deducibility in our ND-system for R can be captured as a theorem of the Hilbert-system for R in the following sense:

(i) if $\varphi_1, \ldots, \varphi_n \Rightarrow \psi$, then there is a Hilbert-proof of $\varphi_1 \rightarrow (\varphi_2 \rightarrow \ldots (\varphi_n \rightarrow \psi) \ldots)$

(ii) if $\varphi_1, \ldots, \varphi_n \Rightarrow \perp$, then there is a Hilbert-proof of $\varphi_1 \rightarrow (\varphi_2 \rightarrow \ldots (\varphi_{n+1} \rightarrow \neg\varphi_n) \ldots)$.

Because of its restrictions on applications of rules, theoremhood in propositional R is undecidable (Urquhart [26]). The fragment obtained by dropping Distributivity is decidable, but the decision problem is at least ESPACE-hard.

11.1. Discussion

The practitioner of a system such as R might acquire the impression that it will be "business as usual" on the deductive front. One will be able to prove lemmas from one's axioms, and then prove theorems from one's lemmas (plus perhaps other axioms), and put one's proofs together so as to get the theorems directly from the axioms used. The usual strategies of "breaking down" a logical problem into a sequence of more manageable steps will therefore apply. One just has to find the right lemmas—the right interpolants—and the reasoning within one's favorite mathematical theory will surely be relevantizable.

Alas, this is mistaken. The effects of relevantizing by "going intensional" on the connectives (and quantifiers) are rather disastrous. They leave the relevantized theory in a relatively much less complete state than the unrelevantized one. One loses not just proofs embodying fallacies of relevance, but also, in many cases, the very results established by such proofs—even when, from our point of view, those results do admit of proofs free of any fallacies of relevance. In the systems that place these intensionalist demands on relevance it will often happen that familiar proofs in standard logic cannot be reconstructed so as to obey the relevantist strictures. Just as the intuitionist loses certain lovely classical proofs, and with them even the theorems that they establish,[7] so too the relevantist in the Anderson–Belnap tradition loses proofs and, with them, theorems. Yet no mathematician proving such "lost" theorems in the original system would brook the accusation that she has committed some fallacy of relevance that, once recognized, deprives her of her entitlement to assert the theorem. Such a suggestion from the reforming relevantist logician would strike her

[7] Example: Every polynomial of several variables and integer coefficients assumes a least absolute value. The classical proof exploits the least-number principle. But any constructive proof would yield an algorithm for determining that least absolute value, thus violating the negative solution to Hilbert's tenth problem. (There is no algorithm for determining whether a given polynomial expression has a root.) See Friedman [8].

as quite outrageous. Moreover, the relevantist (on the approach to relevantization which is being recommended here) can agree. The relevantist's aim should be so to regiment the mathematician's reasoning that its result is shown to be valid even by relevantist lights. Relevantizing is a matter of regimenting reasoning without epistemic loss—indeed, at times with epistemic gain.

Note that this is unlike the situation that is involved when disagreement arises between a classical mathematician and an intuitionist over the validity of a piece of classical mathematical reasoning. The classicist will be convinced by a particular proof that employs strictly classical forms of reasoning (such as excluded middle), while the intuitionist (in the absence of any intuitionistic proof of the same result) will refuse to assert the "result" that the proof (according to the classicist) establishes. Here a genuine doctrinal or epistemic difference arises over the issue of constructivity, which cannot be resolved. It is doubtful, however, that an analogous disagreement on the issue of *relevance*—between a mathematician using standard logic (embodying fallacies of relevance) and a mathematician committed to using only a relevant logic—will ever result in the former being prepared to make an assertion based on proof, and the latter being unprepared to endorse the same. This is because (or so I contend) the following explanatory conjecture is true:

> All intuitively convincing mathematical reasoning can be regimented without loss in the kind of relevance logic advocated here.

Thus what may look like a use of a fallacy of relevance in the course of a piece of mathematical reasoning is, upon closer analysis, no such thing. Intuitively, relevance is already "built into" our deductive sensibilities, in a way that a predilection for constructivity, for example, is not. If our conjecture is true, it follows that the standard systems of logic, which embody fallacies of relevance, are guilty of "overprovision." Fallacies of relevance are never of any help in mathematical reasoning, and ought to be expunged from the systems of logic that are designed to regiment such reasoning. It is an interesting question why (from Frege onward) our systematizations of deductive reasoning have allowed fallacies of relevance to count as formally correct and/or as semantically valid. I believe that it was because a certain kind of reflective equilibrium was struck, in which an overly succinct set of constraints was imposed on the relations of deducibility and consequence, at just that juncture when the object-language conditional was being teased apart from the metalinguistic turnstile. Within this reflective equilibrium, the price paid for unrestricted transitivity and time-saving accumulation of proofs was the admission of the fallacies of relevance. The moral of my story here is that this reflective equilibrium can be refashioned, and relevance regained. By being a little more attentive to what is really going on (and what we want to have going on) when we link proofs together, we can find a better way to characterize the structure of the turnstile relation uncolored by considerations of efficiency. We can fashion a system of proof that allows even more faithful regimentations of actual patterns of mathematical reasoning, even though mathematicians make profligate use of Cuts.

The deep reason why the Anderson–Belnap relevantist loses proofs and, with them, theorems, is that their systems do not contain *Disjunctive Syllogism*:

$$A \vee B, \neg A : B.$$

This impedes their logical investigation of the various cases yielded by such disjunctive principles as the axiom of trichotomy:

$$x < y \vee y < x \vee x = y.$$

Such principles are indispensable in our axiomatizations of various branches of mathematics. The mathematician needs, like Chrysippus's dog, to be able to chase the truth down a different fork in the road as soon as it is discovered that a chosen fork is a dead end. Without Disjunctive Syllogism, one cannot in general do that.

By contrast with the Anderson–Belnap "intensional–connective" systems, the systems *IR* and *CR*, which I advocate as the respective *properly relevant* versions of intuitionistic and classical logic, preserve the whole stock of mathematical theorems—and indeed, in each case, from the very same axioms—in the respective kinds of "nonrelevant" mathematics. Nonrelevant classical mathematics is the mathematics that can be derived from one's axioms by means of the full resources of nonrelevant classical logic. That means that one can use the rule that allows one to infer any statement at all from a contradiction; one can use Disjunctive Syllogism; one can use the "paradoxes of implication"; and so on. If a classical mathematician produces a "nonrelevant" proof of, say, Fermat's last theorem in some interesting arithmetical theory such as exponential function arithmetic (EFA), then my *relevantizing* classical mathematician will be able to prove Fermat's last theorem *relevantly*, using only axioms of EFA that had been used in the nonrelevant proof. (Mutatis mutandis with "intuitionistic" in place of "classical.")

12. The Maxim of Narrow Analysis

Relevantizing brings benefits for automated deduction. In automated deduction, the system is given a queried sequent $X?\text{-}Y$, and seeks to construct a normal or Cut-free proof of that sequent, without human intervention. The deductive problem is broken down into more manageable subproblems by foreshadowed applications of logical rules; one does not seek, in automated deduction, to program the system to find interpolants. That is something that only human reasoners do.

We noted earlier that every perfectly valid sequent in classical (resp. intuitionistic) logic has a classical (resp. intuitionistic) *relevant* proof. This observation prompts the following maxim:

> MAXIM OF NARROW ANALYSIS
> When given a deductive problem $X?\text{-}Y$ that admits of proof, always seek a proof of a minimal subsequent of $X: Y$ that admits of proof.[8]

Suppose one has a proof Π of the sequent $X: Y$. Suppose that one then discovers a new proof Π^* of some proper subsequent $X^*: Y^*$. Then one should be able to substitute Π^* for the old proof Π of the original sequent $X: Y$, without ever forfeiting (the results of) any proofs involving Π as a subproof. Making a subproof prove a tighter logical result should never result in having to forfeit larger proofs containing that subproof. This sounds obvious, and one feels it ought to be true.

In some systems of logic, however, one can be thwarted when trying to follow the maxim of narrow analysis as vigorously as one would like. In these systems of logic, if one substitutes in all one's proofs Σ, the better subproof Π^* (of the stronger result), for the worse subproof Π (of the weaker result), then one does not in general preserve proofhood. Nor is one in general able to recover or restore the situation by producing new proofs in the light of that substitution.

Some relevant systems (such as *R*) can be unreasonably finicky about the use one may make of assumptions for the sake of argument, especially with a rule like the rule of conditional proof. In a new would-be subordinate proof Π^* of a stronger result, some of those assumptions may be eschewed, or the subordinate conclusion may be of the wrong form ($\perp$ instead of the "sought" conclusion B). When such Π^* is substituted to replace the subproof Π (of the weaker result) within some larger proof Σ, the result $\Sigma(\Pi/\Pi^*)$ may fail to be a proof, because the "lost assumptions" or the changed form of the subordinate conclusion makes some rule-application in the wider proof-context Σ illicit. And there might not be any way, in general, of capitalizing on the epistemic gain represented by the new subproof Π^*, and distributing its potential returns within the wider proof Σ.

13. Nonforfeiture of Epistemic Gain

This complaint cannot be leveled, however, against *IR* and *CR*. These systems, unlike other systems of relevance logic, are devoted to the nonforfeiture and potentially

[8] The proof-finding algorithms in *Autologic* [20] are designed with this in mind.

wider distribution of epistemic gains. *IR*, for example, satisfies the following principle:

> Principle of Nonforfeiture of Epistemic Gain
> Let Π be a proof of the sequent $X : A$, occurring as a subproof of the proof Σ of the sequent $Y : B$. Let Π^* be a proof of some proper subsequent of $X : A$. Then $\Sigma(\Pi/\Pi^*)$—the result of substituting Π^* for Π in Σ—can be effectively transformed into a proof of some subsequent of $Y : B$. Indeed, the transformation can be effected in polynomial time.

Note that I am not saying that a strict gain via Π^* with respect to some subproof Π will always turn into a strict gain with respect to the overall proof Σ. In many cases it will; but in some cases the result established by (the transform of) $\Sigma(\Pi/\Pi^*)$ will be the original result $Y : B$.

Still, nonforfeiture of epistemic gain would be very good news. For we would never lose any of our erstwhile deductive knowledge by forgetting Π and remembering Π^* instead. And this cannot be said for many a rival system of relevance logic.

14. Summary

We have tried to show how to relevantize the two main logics—intuitionist and classical—without methodological loss. In the sequent setting, we ban Dilutions and Cuts. And we state the logical rules in a slightly more relaxed form, so that we can get by without the structural rules of Dilution or Cut. In the natural-deduction setting, we ban the absurdity rule and abnormalities, and keep exigent but well-motivated checks on how we have used assumptions for discharge. We also liberalize slightly certain logical rules, so that the loss of the absurdity rule is not crippling.

Elsewhere, I have shown how this process of relevantizing (in the intuitionistic case) brings the system of natural deduction and the corresponding system of sequent calculus into "deep isomorphism."[9] The relevant systems are also ideally suited for computational logic, since they allow one never to forfeit any epistemic gains that one might achieve in the course of solving deductive subproblems. This approach to relevantizing salvages those "local" forms of reasoning, such as disjunctive syllogism, that seem, intuitively, so indispensable for ordinary mathematical reasoning. It locates the issue of relevance at the level of the turnstile, and

[9] N. Tennant, "Ultimate Normal Forms for Parallelized Natural Deductions, with Applications to Relevance and the Deep Isomorphism Between Natural Deductions and Sequent Proofs," *Logic Journal of the IGPL* 10, no. 3 (May 2002): 1–39.

challenges certain orthodoxies as to what structural features of the deducibility relation it is methodologically necessary to preserve. The upshot is that the student of logic who is dismayed by the First Lewis Paradox can be assured that it is irrelevant to the deductive needs served by one's logic, whether one is an intuitionist or a classicist. We can make do without it, and stand only to gain by eschewing it. Logical rules suffice for our logical reasoning. We do not even need any "structural" rules—except, of course, the rule of initial sequents $A:A$. That, however, is the case only in a sequent system. In natural deduction, the counterpart to this is that one is allowed to write down an assumption A for the sake of argument. It might be left standing as an undischarged assumption by the end of the proof, or it might be discharged on the way to the conclusion. Without such allowance, no proof would ever get started. So I don't propose to revoke *that* licence. I have banned a great deal, but with no untoward effect at all. Our canon of deductive reasoning has been relevantized; but our powers of reasoning have been left intact.

Perhaps we should call the resulting position "compassionate relevantism."

REFERENCES

[1] Anderson, Alan Ross, and Nuel D. Belnap, Jr. *Entailment: The Logic of Relevance and Necessity*. Vol. I. Princeton University Press, Princeton, NJ, 1975.

[2] Boolos, George. Don't eliminate cut! *Journal of Philosophical Logic*, 13: 373–378, 1984.

[3] Burgess, John P. Relevance: A fallacy? *Notre Dame Journal of Formal Logic*, 22: 97–104, 1981.

[4] Burgess, John P. Common sense and "relevance." *Notre Dame Journal of Formal Logic*, 24: 41–53, 1983.

[5] Burgess, John P. Read of relevance: A rejoinder. *Notre Dame Journal of Formal Logic*, 25: 217–223, 1985.

[6] Copeland, B.J. On when a semantics is not a semantics: Some reasons for disliking the Routley–Meyer semantics for relevance logic. *Journal of Philosophical Logic*, 8: 399–413, 1979.

[7] Copeland, B.J. The trouble Anderson and Belnap have with relevance. *Philosophical Studies*, 37: 325–334, 1980.

[8] Friedman, Harvey. Demonstrably necessary uses of abstraction. Lecture 1. *www.mathpreprints.com*, 2002.

[9] Friedman, Harvey, and Robert K. Meyer. Whither relevant arithmetic? *Journal of Symbolic Logic*, 7: 824–831, 1992.

[10] Meyer, Robert K. Arithmetic formulated relevantly. 1979.

[11] Mortensen, Chris. The validity of disjunctive syllogism is not so easily proved. *Notre Dame Journal of Formal Logic*, 24: 35–40, 1983.

[12] Orevkov, V.P. Lower bounds for the lengthening of proofs after cut-elimination. *Journal of Soviet Mathematics*, 20: 2337–2350, 1982.

[13] Prawitz, Dag. *Natural Deduction: A Proof-Theoretical Study*. Almqvist & Wiksell, Stockholm, 1965.

[14] Read, Stephen. Burgess on relevance: A fallacy indeed. *Notre Dame Journal of Formal Logic*, 24: 473–481, 1983.

[15] Tennant, Neil. Entailment and proofs. *Proceedings of the Aristotelian Society*, 79: 167–189, 1979.

[16] Tennant, Neil. A proof-theoretic approach to entailment. *Journal of Philosophical Logic*, 9: 185–209, 1980.

[17] Tennant, Neil. Perfect validity, entailment and paraconsistency. *Studia Logica*, 43: 179–198, 1984.

[18] Tennant, Neil. *Anti-Realism and Logic: Truth as Eternal*. Clarendon Library of Logic and Philosophy. Clarendon Press, Oxford, 1987.

[19] Tennant, Neil. Natural deduction and sequent calculus for intuitionistic relevant logic. *Journal of Symbolic Logic*, 52: 665–680, 1987.

[20] Tennant, Neil. *Autologic*. Edinburgh University Press, Edinburgh, 1992.

[21] Tennant, Neil. Automated deduction and artificial intelligence. In R. Casati, B. Smith, and G. White, eds., *Philosophy and the Cognitive Sciences: Proceedings of the 16th International Wittgenstein Colloquium*, pp. 273–286. Hölder-Pichler-Tempsky, Vienna, 1994.

[22] Tennant, Neil. Intuitionistic mathematics does not need *ex falso quodlibet*. *Topoi*, 127–133, 1994.

[23] Tennant, Neil. Transmission of truth and transitivity of proof. In Dov Gabbay, ed., *What Is a Logical System?*, pp. 167–177. Oxford University Press, New York, 1994.

[24] Tennant, Neil. Delicate proof theory. In *Logic and Reality: Essays on the Legacy of Arthur Prior*, Jack Copeland, ed., pp. 351–385. Oxford University Press, New York, 1996.

[25] Tennant, Neil. *The Taming of the True*. Oxford University Press, Oxford, 1997.

[26] Urquhart, Alasdair. The undecidability of entailment and relevant implication. *Journal of Symbolic Logic*, 49: 1059–1073, 1984.

CHAPTER 24

NO REQUIREMENT OF RELEVANCE

JOHN P. BURGESS

1. Relevantism: Rejecting the "Paradoxes" of Classical Logic

Classical geometry defines a conic section to be the intersection of a plane with a (double) cone. This definition obliges the geometer to recognize such degenerate conics as the hyperbola consisting of two crossed lines (the intersection of the cone with a vertical plane through its apex) or the circle consisting of a single point (the intersection of the cone with a horizontal plane through its apex).

Classic logic defines entailment to hold between a premise (or set of premises) and a conclusion if and only if their logical form guarantees that the either the premise (or at least one element of the set of premises) is false, or the conclusion is true. The definition obliges the logician to recognize certain degenerate entailments. A premise (or set of premises) that is contradictory in the sense that its logical form guarantees that it is false (or that at least one element of the set of is false) entails any conclusion: *ex falso quodlibet.* And a conclusion that is tautologous in the sense that its logical form guarantees that it is true is entailed by any premise (or set of premises): *ex quolibet verum.*

The commitment of classical logic to these principles has frequently been attacked by indignant critics who denounce the degenerate cases of entailment as "paradoxes." Among these critics there have been many with little use for any sort of formal logic; but the concern of the present survey will be with schools of anticlassical logicians who have proposed rival formal logics to the classical, avoiding the so-called paradoxes.

Since the most common single complaint of such critics is that in the degenerate cases of entailment recognized by classical logic the premise is (or the premises are) "irrelevant" to the conclusion, the critics in question will be called "relevantists." Note, however, that as they are used in the literature, "relevance logic" and "relevant logic" are trademarks—both for the same school of logic,[1] everything about which is highly contentious, beginning with whether "relevance" or "relevant" is the right label for it—and are not to be equated with "relevantism" as here defined. There have been other kinds of relevantists than "relevance" or "relevant" logicians; and there have been important contributions to "relevance" or "relevant" logic by logicians (Kit Fine, Harvey Friedman, Saul Kripke, Alasdair Urquhart) whose endorsement of relevantist criticism of classical logic has been qualified or nonexistent.

Now there would be no room for contention between classicism and relevantism if "entailment" were simply being introduced as a term previously without meaning (outside probate law), which was now simply being *stipulated* by the classicist to mean "the relation that obtains between premise and conclusion when logical form alone guarantees that either the former is false or the latter is true." There is room for disagreement only if both sides claim to be analyzing a notion of entailment that already exists, though perhaps not under that label.

And indeed classical logic's theory of entailment was largely developed in order to provide an analysis of what is meant by rigorous proof or rigorous provability in orthodox mathematics: classical logic can be viewed as an attempt to describe explicitly the implicit norms of orthodox mathematicians, and it is largely the practice of mathematicians that will be invoked in its defense. (Classical logicians do generally regard the applicability of classical logic as extending beyond mathematics as well, but are at least open to the possibility that it may require modification for extramathematical applications, since mathematics has so many features peculiar to itself: mathematicians insist that definitions should be fully precise, whereas ordinary language is full of vagueness; mathematicians insist that before a term is introduced, the existence and uniqueness of the item it is to denote must be proved, whereas in ordinary language such presuppositions often fail; and so on.)

At this point it is necessary to note a contrast between two different kinds of criticism of classical logic. It is agreed on all sides that in some sense logic is concerned with norms (of valid argument). But there is a great difference—all the difference between *is* and *ought*—between describing what the norms operative in

[1] The school in question is that represented by Anderson and Belnap (1975) and Anderson, Belnap, and Dunn (1992). These works have extensive bibliographies covering "relevantism" in all senses of the term.

a given community are, and prescribing what they ought to be. In linguistics this distinction (the distinction between descriptive and prescriptive grammar, between Chomsky and Fowler) is universally recognized. It needs to be recognized in logic also. There is a difference between *prescriptive* criticism, alleging that classical logic correctly describes unacceptable practices in the community of orthodox mathematicians, and *descriptive* criticism, alleging that classical logic endorses incorrect principles that are *not* operative in the practice of that community.

The paradigm of prescriptive criticism is the intuitionist attack on classical logic. Neither Brouwer nor any of his disciples ever suggested classical logicians were wrong in *describing* what pass for proofs in the orthodox mathematical community as often involving the inference from the double negation of a proposition to the proposition itself. The intutionists' quarrel was not with the small body of classical logicians, but with the much larger body of orthodox mathematicians, whose practices, they insisted, were in need of revision. (Classical logicians were liable to censure only if they went beyond describing orthodox mathematicians as appealing double negation elimination to *endorsing* the practice of doing so.)

The prescriptive character of the intuitionist critique is evident from the mathematical practice of intuitionist logicians themselves, in proving metatheorems about intuitionistic logic. A genuine intuitionist proving a metatheorem will take care never to appeal to double negation elimination or, more generally, to make any argumentative move that cannot be represented in *intuitionistic* logic. Or if the intuitionist logician does sometimes reason like the orthodox, it is only because there has previously been established some metatheorem to the effect that in the particular kind of case in question, a classical proof can always be replaced by an intuitionistic one. And (it ought to be needless to say) "established" here can only mean established *by intuitionistically acceptable proofs.*

By contrast with all this, relevantists have historically virtually *never* troubled themselves to check through their proofs of metatheorems to make sure that no principle they reject has been appealed to. The most natural explanation for this insouciance is that their criticism has implicitly been meant as descriptive rather than prescriptive, and that the fault of the classical logicians, according to them, has been that of endorsing unacceptable principles *that are not in practice used by orthodox mathematicians.* Also, the rhetoric attending the historically most important statements of the relevantist position—their pose as indignantly championing common sense against the monstrous "paradoxes" propounded by classical logicians—is a more explicit indication that the criticism has been intended as descriptive.

If the descriptive criticism fails—and in particular, if it turns out that relevantists themselves constantly argue in what purport to be proofs of metatheorems about their own formal systems by appeal to principles rejected in those systems—the historically earliest form of the dispute between relevantism and classicism will have been decided in the classicist's favor. The options remaining to the relevantist will then be two.

On the one hand, a relevantist might adopt a *compatibilist* position, presenting relevantist logic not as an alternative intended to supplant classical logic but, at most, as an adjunct intended to supplement it somehow. This would be to give up criticism of classical logic altogether.

On the other hand, a relevantist might become a presciptive critic, adopting a *revisionist* position, demanding, like the intuitionist, a reform of orthodox mathematical practice (though not the *same* reform the intuitionist demands). This would require in the first place a reform of their own mathematical practice, appealing henceforth only to laws of relevantist logic when proving mathematical metatheorems about their own formal systems.

2. The Relevantist's Options

So much, for the moment, about the general character of the dispute between classicists and relevantists. As for the particulars on which relevantists differ from classicists, and among themselves, the most crucial issues can be illustrated at the level of sentential logic, and in connection with entailments where the only logical operators involved in premise(s) and conclusion are conjunction, disjunction, and negation. Thus the forms of premise(s) and conclusion alike can be represented by formulas built up from *atoms* or atomic formulas $p, q, r, \ldots$ by means of the symbols caret and wedge and tilde ($\wedge$ and $\vee$ and $\sim$) for the three logical operations just mentioned. We call these *elementary* formulas.

Most relevantists (in contrast to intuitionists, who reject the law of the excluded middle, for example) have been willing to agree, at least for the sake of argument, with the classical position as to which elementary formulas represent contradictions and which tautologies. The disagreement that will be examined below is over which elementary formulas entail which.

We write $\vdash$ for entailment. As already indicated, the primary target of criticism is the following principle:

(A) If $\alpha_1, \ldots, \alpha_m$ are contradictory, then $\alpha_1, \ldots, \alpha_m \vdash \beta$.

And the secondary target is this further principle:

(B) If β is tautologous, then $\alpha_1, \ldots, \alpha_m \vdash \beta$.

The cases most often discussed are

(A1) $p \wedge \sim p \vdash q$
(A2) $p, \sim p \vdash q$

and to a lesser extent

(B1) $p \vdash q \vee \sim q$.

But while in practice it is these cases that are primarily discussed, in principle the critics reject (A) and (B) quite generally.

Now how does the rejection of (A) and (B) square with orthodox mathematical practice? That is the issue to be considered in evaluating relevantism as a *descriptive* criticism of classical logic. On this issue, relevantists have been quick to remark that generally neither mathematicians nor other people set about trying to deduce conclusions from premises they recognize to be contradictory; nor do people generally spend much time looking for premises from which they might deduce a conclusion they recognize to be tautologous. This observation is clearly correct; what is not so clear is why it should be thought to motivate a nonclassical logic in which there is some *interesting distinction* between those conclusions that are and those that are not entailed by a given contradiction, and between those premises that do and those that don't entail a given tautology.

More important, despite the foregoing observation, it does often happen that a mathematician engaged in lengthy deductions may through some slip (say, leaving out a minus sign, or writing an inequality symbol backward) arrive at a contradiction or other absurdity without noticing the fact. And when this happens, it does often happen that the mathematician may then go on to deduce, by what appears to be quite ordinary mathematical reasoning, whatever conclusions the mathematician in question may have been seeking. This type of occurrence may be called the *Poincaré phenomenon*, after Henri Poincaré, who observed:

> The candidate often takes an immense amount of trouble to find the first false equation; but as soon as he has obtained it, it is no more than child's play for him to accumulate the most surprising results, some of which may actually be true.[2]

There is also a dual type of occurrence, where carrying out a long deduction, by quite ordinary mathematical reasoning (such as simple calculations involving rearranging terms in equations, or performing similar operations on both sides), results in obtaining what turns out at the last moment to be a tautology or other triviality. This may be called the *Quine phenomenon*, after W. V. Quine, who observed:

[2] "The New Logics," in Poincaré (1960), p. 161.

> ...the beginner in algebra works in danger of finding that his solution-in-progress reduces to "$0=0$."[3]

What the Poincaré and Quine phenomena suggest is that if one rejects (A) or (B), one is going to have to reject some quite ordinary pattern of deduction, such as is used by mathematicians and others who go in for long trains of arguments. And indeed, if one wishes to reject (A), the options are quite limited, as can be seen from the following derivation, establishing (A2).

(1)	p	premise
(2)	$\sim p$	premise
(3)	$p \vee q$	from (1)
(4)	q	from (2), (3)

The principle that gives (3) from (1) is the that a disjunction is entailed by either disjunct.

(C1) $p \vdash p \vee q$

(C2) $q \vdash p \vee q$

A slightly more elaborate variant of the first of these is the following:

(C3) $p \wedge r \vdash (p \vee q) \wedge r.$

Since those who reject any one of the principles just listed tend to reject the others as well, all of them, as well as other minor variants, will be referred to below collectively as (C) or *disjunctive weakening.*

The prinicple that gives (4) from (2) and (3) is

(D1) $p \vee q, \sim p \vdash q.$

This two-premise version has a one-premise variant.

(D2) $(p \vee q) \wedge \sim p \vdash q.$

Both of these, and other minor variants, will be referred to collectively below as (D) or *disjunctive syllogism.*

The derivation (1)–(4) is usually called the *Lewis argument* after C. I. Lewis, who first brought it to the general attention of logicians in modern times, after medieval antecedents had been largely forgotten.[4] The Lewis argument seems to show that one

[3] "Carnap and Logical Truth," in Quine (1966), p. 105.

[4] For the history of the argument, see Anderson and Belnap (1975), §16.1, pp. 163ff.

who rejects (A) must reject either disjunctive weakening (C) or disjunctive syllogism (D). But actually there is a *third* alternative—namely, to reject the *transitivity* of entailment, the principle that if certain initial premises entail each of certain intermediate steps as conclusions, and these in turn, taken as premises, entail a certain ultimate conclusion, then the initial premises entail the ultimate conclusion.

The issue here is most clearly seen in connection with the following variant of the Lewis argument.

(C4) $p \wedge {\sim}p \vdash (p \vee q) \wedge {\sim}p$
(D2) $(p \vee q) \wedge {\sim}p \vdash q$
(A1) $p \wedge {\sim}p \vdash q.$

Here the particular version (C4) of (C) used is an instance of the version (C3) cited earlier. The (meta-)principle needed to get from (C4) and (D2) to (A) is the simplest case of transitivity.

(E) If $\alpha \vdash \beta$ and $\beta \vdash \gamma$, then $\alpha \vdash \gamma$.

All three options have been taken up by critics in the literature: the least popular option has been to reject disjunctive weakening; the most popular has been to reject disjunctive syllogism; rejection of transitivity has been somewhere in between. The three options will be considered separately below, in order of increasing popularity.

One sometimes also meets in the literature with what at first glance appears to be a fourth option, namely, to claim there is a fallacy of equivocation in the Lewis argument. The equivocation is supposed to be between two senses of disjunction, a weaker *extensional* and a stronger *intensional* sense. It is claimed that disjunctive weakening holds for the former but not the latter, while disjunctive syllogism holds for the latter but not the former.

Usually extensional disjunction is written $\vee$, and classical logic is held to give a correct account of which formulas involving it represent contradictons and which tautologies, while intensional disjunction is written $+$, and the enunciation of its laws is taken to be one of the tasks for a new, nonclassical logic. But so understood, the apparent fourth option is really just a suboption under the second option: it amounts to rejecting (D) but softening the rejection by adding a further claim to the effect that there is an acceptable principle resembling (D), namely, the principle that results from it on replacing $\vee$ with $+$.

The option of rejecting disjunctive weakening has been favored by so few relevantists that it will be given only very brief attention.[5] The only principle that

[5] This option is involved in the "analytic implication" of Parry and the "connexive implication" of McCall, for which see Anderson and Belnap (1972), §§29.6–29.8, pp. 429ff.

has been advanced in support of such a rejection is the principle that in an entailment, the subject matter of the conclusion must be contained in the subject matter of the premise. Since formulas involving distinct atoms p and q will presumably have instances in which the subject matter of what is substituted for one letter p has nothing to do with that of what is substituted for q, the formal representation of the "inclusion of subject matter" requirement is that no atom should appear in the conclusion that does not appear in the premise, which of course immediately leads to the rejection of (C).

This has seemed unacceptable even to the great majority of relevantists. Yet one may wonder *why* anyone would wish to infer a disjunction $p \vee q$ from a disjunct p, since the former in an obvious sense contains less information than the latter. One reason is a desire to subsume a particular case under a generalization. Thus in mathematics one may want to prove $(\forall n)(P(n) \vee Q(n))$, and one way to do this may be by induction, proving $P(0) \vee Q(0)$ and $P(1) \vee Q(1)$ and then proving that if the theorem holds for n and $n+1$, it holds for $n+2$. And one way to prove $P(0) \vee Q(0)$ and $P(1) \vee Q(1)$ may be to deduce the former from $P(0)$ and the latter from $Q(1)$, by disjunctive weakening. Reasoning of this kind is in fact common in mathematical proofs, and insofar as the question before us is whether classical logic gives a correct description of orthodox mathematical standards of proof, this fact allows the relevantist position rejecting disjunctive weakening to be dismissed.

3. The Option of Rejecting Transitivity: Perfectionism

The great majority of relevantists do not wish to insist that the subject matter of the conclusion must be contained in that of the premise, though they do wish to insist that the two should overlap. This overlap is a bare minimum to require in the way of "relevance" between premise and conclusion. Now this bare minimum of relevance in fact obtains in all classical entailments where the premise is not contradictory and the conclusion not tautologous, as a consequence of the Craig Interpolation Theorem. That theorem says that, except in the degenerate cases just excluded, if $\alpha \vdash \gamma$, then there is a β, called an interpolant, such that $\alpha \vdash \beta$ and $\beta \vdash \gamma$, and every atom appearing in β appears both in α and in γ.

Actually, something still stronger is true, which it will be useful for future purposes to note now. We may distinguish, among the appearances of an atom p in an elementary formula α, the *positive* from the *negative*. When α is just the atomic formula p itself, the one and only appearance of p is positive. The positive appearances of p in $\alpha \wedge \beta$ and in $\alpha \vee \beta$ are just the positive appearances in α together with the positive appearances in β; and similarly for negative appearances. Finally, the positive appearances in $\sim\alpha$ are the negative appearances in α and vice versa. Thus in the premise $(p \vee q) \wedge \sim p$ of disjunctive syllogism, the first appearance of p is positive and the second negative, while the only occurrence of q is positive. The Lyndon Interpolation Theorem tells us that in the Craig Interpolation Theorem the interpolant β may actually be so chosen that the only atoms occurring positively in β are ones appearing positively both in α and in γ, and similarly for negative appearances.

Thus classical entailment does guarantee a bare minimum and more than a bare minimum of "relevance," except in the two degenerate cases. Relevantists will not allow those exceptions, and wish to insist in *all* cases on premise and conclusion having one or more atoms in common. Atoms being also called "sentential" (or "propositional") "variables" (or "letters"), the requirment of atoms in common is often referred to as the *variable-sharing* requirement.

The classical criterion of validity—that by virtue of logical form alone, if the premise is (or the premises all are) true, then so is the conclusion—is often call *truth-preservation.* Relevantists have sometimes written in a way that might suggest to the reader that they hold a "two-factor" theory of entailment according to which what is required for a genuine entailment is just the presence of both these features, truth-preservation and variable-sharing; but in fact relevantists generally hold no such a theory. Variable-sharing all by itself (as the sole addition to truth-preservation) is not enough to satisfy relevantists, who generally reject, for instance, the following version of (A):

(A4) $(p \wedge \sim p) \vee r \vdash q \vee r.$

Here the addition of the disjunct r to both sides of (A1) results in the variable-sharing condition being fulfilled, but fundamentally we still have truth-preservation here only because the first disjunct in the premise is contradictory. What relevantists really want is not just that variable-sharing should be present alongside truth-preservation, but that it should be variable-sharing that *underlies* truth-preservation, so to speak.

One case where variable-sharing must be what underlies truth-preservation is of course where the premise is *not* a contradiction and the conclusion *not* a tautology. One cannot, however, accept as correct laws of logic only those classical entailments $\alpha \vdash \beta$ where α is noncontradictory and β nontautologous. This is because formulas and entailments involving formulas are supposed to hold in *all* instances, no matter what is put in for p and q and r and so on. On the formal level, this understanding is represented by the rule of substitution, to the effect that if

$\alpha \vdash \beta$ and α' and β' are obtained from α and β by substituting formulas $\pi_1, \ldots, \pi_k$ for atoms $p_1, \ldots, p_k$, then $\alpha' \vdash \beta'$. Thus, from the entailment $\alpha \wedge \beta \vdash \alpha$, which even relevantists accept, it follows there follows the entailment $\alpha \wedge \sim\alpha \vdash \alpha$, despite the fact that the premise in this case is contradictory. Similary, $\beta \vdash \beta \vee \sim\beta$.

One school of relevantists, however, takes examples of the kind we have just been discussing to be the *only* exceptions to the rule that the premise must not be contradictory and the conclusion must not be tautologous. To state the position more precisely, call an entailment $\alpha \vdash \beta$ *perfect*—or "as good as an argument can be on formal grounds"—if it holds classically and α is not contradictory and β is not tautologous; and call it *perfectible* if it is obtainable by substitution from a perfect entailment. The school of relevantists we have in mind take the acceptable entailments to be precisely the perfectible ones. This position involves rejection of transitivity of entailment, since in the second version of the Lewis argument above (C4) and (D2) are easily seen to be perfectible, while (A1) is not.

There is some question of how this perfectionist criterion should be extended to arguments with multiple premises, and in the literature arguments with multiple conclusions have also been considered. Classically, a *sequent*

$$\alpha_1, \ldots, \alpha_m \vdash \beta_1, \ldots, \beta_n$$

is taken to hold if logical form guarantees that either at least one of the premises α_i is false or at least one of the conclusions β_i is true. Note that the multiple premises are being taken conjointly, and the conclusions alternatively. That is, classically the sequent holds if the simple one-premise, one-conclusion entailment

$$\alpha_1 \wedge \cdots \wedge \alpha_m \vdash \beta_1 \vee \cdots \vee \beta_n$$

holds.

These definitions are to be understood in such a way that, as a special case, if the set of conclusions β is *empty*, then

$$\alpha_1 \wedge \cdots \wedge \alpha_m \vdash \emptyset$$

holds if and only if the α_i are contradictory taken conjointly, and in particular

$$\alpha \vdash \emptyset$$

holds if α is contradictory. Similarly,

$$\emptyset \vdash \beta$$

holds if β is tautologous, and when

$$\varnothing \vdash \beta_1 \vee \cdots \vee \beta_n,$$

we may say that the β_j are tautologous (taken alternatively).

Now one way to extend the one-premise, one-conclusion version of the perfectionist criterion would be to take

$$\alpha_1, \ldots, \alpha_m \vdash \beta_1, \ldots, \beta_n$$

to be perfectible and acceptable if and only if

$$\alpha_1 \wedge \ldots \wedge \alpha_m \vdash \beta_1 \vee \cdots \vee \beta_n$$

is. But there is a different, and in some respects more natural, way to carry out the extension. If we look again at the above definitions, we see that

$$\alpha \vdash \beta$$

is perfect if and only if it holds classically while

$$\alpha \vdash \varnothing$$

and

$$\varnothing \vdash \beta$$

do not. Generalizing, we may take

$$\alpha_1, \cdots, \alpha_m \vdash \beta_1, \cdots, \beta_n$$

to be perfect if it holds classically, but no longer does so if the left-hand side is replaced by any proper subset of the α_i or the right-hand side by any proper subset of the β_j. We may then define the sequent to be perfectible if it is obtainable by substitution from a sequent that is perfect in the sense just indicated. This is Tennant's perfectionist logic, at the sentential level (see chapter 23 above).

Tennant's logic is decidable, at the sentential level. For note, first, that if a formula involves n sentence and connective and punctuation signs altogether, then any formula of which it is a substitution instance involves no more than n such signs. Note further that if a formula involves no more than m such signs and no more than k sentence letters, the formula is a substitution instance of a formula involving no more than *m* such signs and involving no sentence letters beyond the first *k*. Note further that there are only finitely many formulas having no more than *m* signs and no sentence letters beyond the first *k*. Note finally that this observation generalizes to sets of formulas. So for given sets of α_i and β_j, in seeking to determine whether the sequent having the former as premises and the latter as conclusions holds only finitely many pairs of other sets of sentences need to be evaluated classically (which may be done by truth tables or otherwise).

The main advantage of working with sequents is that though transitivity fails for perfectionist logic in the sense that

$$\alpha_1, \ldots, \alpha_m \vdash \beta_1, \ldots, \beta_n$$

and

$$\beta_1, \ldots, \beta_n \vdash \gamma_1, \ldots, \gamma_k$$

may be perfectible while

$$\alpha_1, \ldots, \alpha_m \vdash \gamma_1, \ldots, \gamma_k$$

is not, nonetheless one can develop (meta-)rules for passing from *sequents* taken as "premises" to a *sequent* taken as "conclusion," and the operation of these (meta-)rules is transitive. One has transitivity "one level up," so to speak. This makes it possible to represent perfectionist logic as a deductive system comparable to the deductive system for classical logic when the latter is developed in "sequent calculus" or Gentzen style.

Now in every branch of mathematics, axioms are assumed, various combinations of them are used to obtain lemmas, various combinations of axioms and lemmas are used to obtain theorems, various combinations of axioms and lemmas and theorems are used to obtain corollaries, and all the lemmas and theorems and corollaries are taken to follow from the axiom set initially assumed, as well as from any larger axiom set subsequently assumed. These are the facts that show that mathematicians treat entailment as transitive. And if we are concerned only with the question of whether classical logic correctly *describes* orthodox mathematical practice, any position rejecting transitivity would have to be rejected for this reason alone.

Nonetheless, a revisionist version of perfectionism, according to which, though mathematicians in practice *do* give "proofs" relying on transitivity, they in principle *ought not* to do so, might be thought to have certain attractions in view of some formal results of Tennant's. For Tennant seems to show that though orthodox mathematicians make moves in "proofs" not directly representable by sequent-calculus derivations in his system, if one is concerned only with prov*ability* and not directly with *proof*, there is no great loss.[6] It is not quite true that every classical "proof" can be replaced by a perfectionist "proof," leaving the class of conclusions provable from given premises unchanged; but something quite close to this, and quite attractive, does hold.

Let us try to describe Tennant's result more precisely. For a given set

$$A = \alpha_1, \ldots, \alpha_m$$

of premises and a given conclusion β, a mathematician's proof of a theorem whose logical form is represented by β from a set of premises whose logical form is represented by A can fairly routinely be represented by a derivation of $A \vdash \beta$ in a sequent-calculus version of classical logic with the *Cut* rule (a version of transitivity). By the Cut Elimination Theorem of Gentzen, the uses of Cut can in principle be eliminated. Then Tennant shows how to transform a Cut-free classical derivation of $A \vdash \beta$ into a derivation in his perfectionist system.

The result this two-step transformation is not always a derivation in the perfectionist system showing that $A \vdash \beta$ holds perfectionistically. This could or can fail in two situations. If A were inconsistent, $A \vdash \beta$ might fail perfectionistically, and if it did, the result of the two-step transformation would be a derivation showing that $A \vdash$. If the original hypotheses are not all needed, $A \vdash \beta$ may fail perfectionistically, and if it does, the result of the two-step transformation will be a derivation showing that $A' \vdash \beta$ for some proper subset A' of A. Since ultimately the mathematician is interested in proving results from consistent axioms and from the fewest of those possible, Tennant's result seems to say that nothing, or next to nothing, is lost by switching from classical logic to perfectionist logic: where a classical entailment is rejected, perfectionism supplies an improvement.

There is, however, a difficulty here. Transitivity, as noted above, is involved whenever a lemma is deduced from axioms and a theorem then deduced from the lemma, and claimed to follow from the axioms, and is thus ubiquitous in ordinary mathematical proofs. When these proofs are formalized in sequent calculus, instances of transitivity become instances of Cut. Obtaining perfectionist replacements for the proofs would require first eliminating Cut. But in general, Cut can be eliminated only at the cost of making proofs infeasibly long.[7] So even though *in principle* almost anything classically provable will be perfectionistically provable or else something even better will be, *in practice* the proofs mathematicians actually give not only *do* fail to adhere to the restrictions of perfectionism, but *must* do so if they are to be kept to a humanly comprehensible length.

[6] The interpretation of Tennant's results, and the question of whether they merely seem to show that there is no great loss or really do so, are unsurprisingly not uncontroversial. My evaluation will appear below.

[7] See "Don't Eliminate Cut," in Boolos (1998), pp. 365–369.

The perfectionist might reply that in practice mathematicians could go on working as they do now, since Tennant's work shows that in principle everything, or almost everything, they are doing could be justified from a perfectionist standpoint. The trouble with this response is that it relies on a theorem—Tennant's result described above—for which only a classically and not a perfectionistically acceptable "proof" has been given. How far can a logician who *professes* to hold that perfectionism is the correct criterion of valid argument, but who freely accepts and offers standard mathematical proofs, in particular for theorems about perfectionist logic itself, be regarded as *sincere* or *serious* in objecting to classical logic? This question will be left to the reader to ponder, as we turn to consideration of the one remaining relevantist option, rejection of disjunctive syllogism.

4. The Option of Rejecting Disjunctive Syllogism: Stringency

The very simplest elementary entailments are perhaps those of the form

$$(\sim)p_1 \wedge \ldots \wedge (\sim)p_m \vdash (\sim)q_1 \vee \ldots \vee (\sim)q_n,$$

where the premise is a conjunction, and the conclusion a disjunction, of plain or negated atomic formulas or sentence letters. These may be called *proto-elementary*. Classically, a proto-elementary entailment holds in three cases:

(I) Some atom occurs with the same sign in both premise and conclusion.
(II) Some atom occurs both plain and negated in the premise.
(III) Some atom occurs both plain and negated in the conclusion.

Here (I) means that either some atom occurs positively or plain in both, or that some atom appears negatively or negated in both. Perfectionist logic does not recognize cases (II) and (III), and takes (I) to be a necessary and sufficient condition for the entailment to hold. The major school rejecting Disjunctive Syllogism, "relevance" or "relevant" logic, adheres to the same criterion for proto-elementary entailments, but arrives at different results about more general elementary entailments, including Disjunctive Syllogism.

Since the roots of "relevance" or "relevant" logic are traced back by the logicians of that school itself to Wilhelm Ackermann,[8] we may sidestep the "relevance"/"relevant" terminological problem by adopting (a suitable English translation of) Ackermann's label for the kind of logic in question. Ackermann, writing in German,

[8] See the dedication to Anderson and Belnap (1972).

called his system *strenge Implikation.* Now though this was intended as a translation of Lewis's label "strict implication," Ackermann's system is very different from any of those considered by Lewis. So in translating his German back into English, we may use a different word than "strict," namely, "stringent."

The first (historically earliest) formulation of the stringency criterion for general elementary entailments was based on an alternative to truth tables as a method for testing classical elementary entailment. The classical method has four steps:

(i) Repeatedly apply the double negation and De Morgan laws of classical sentential logic, which stringent logicians accept, to replace subformulas

$$\sim\sim\alpha \text{ by } \alpha$$
$$\sim(\alpha \wedge \beta) \text{ by } \sim\alpha \vee \sim\beta$$
$$\sim(\alpha \vee \beta) \text{ by } \sim\alpha \wedge \sim\beta$$

so as to reduce premise and conclusion to formulas built up from plain and negated atoms by conjunction and disjunction.

(ii) Similarly apply the distributive laws of classical sentential logic, which stringent logicians also accept, to replace subformulas

$$\alpha \wedge (\beta \vee \gamma) \text{ by } (\alpha \wedge \beta) \vee (\alpha \wedge \gamma)$$
$$(\alpha \vee \beta) \wedge \gamma \text{ by } (\alpha \wedge \gamma) \vee (\beta \wedge \gamma)$$

in the premise and subformulas

$$\alpha \vee (\beta \wedge \gamma) \text{ by } (\alpha \vee \beta) \wedge (\alpha \vee \gamma)$$
$$(\alpha \wedge \beta) \vee \gamma \text{ by } (\alpha \vee \gamma) \wedge (\beta \vee \gamma)$$

in the conclusion, so as to reduce the premise to the form

$$\alpha_1 \vee \alpha_2 \vee \ldots$$

and the conclusion to the form

$$\beta_1 \wedge \beta_2 \wedge \ldots$$

where each α_i is a conjunction of, and each β_j is a disjunction of plain and negated atomic formulas.

(iii) Apply the laws of disjunction elimination and conjunction introduction of classical sentential logic, which stringent logicians also accept, to reduce the question of whether the premise entails the conclusion to the question of whether each α_i entails each β_j, a question of proto-elementary entailment.

(iv) Apply the classical criteria (I), (II), (III) to answer these questions of proto-elementary entailment.

Stringent logic's departure from classicism occurs only at the last step here, where instead of (iv) one would have

(iv*) Apply the criterion (I) to answer the questions of proto-elementary entailment.

To illustrate the process, consider Disjunctive Syllogism

(D2) $(p \vee q) \wedge \sim p \vdash q.$

At step (i) there is nothing to do. At step (ii) we get

$$(p \wedge \sim p) \vee (q \wedge \sim p) \vdash q$$

At step (iii) we get two entailments to consider:

$$p \wedge \sim p \vdash q$$
$$q \wedge \sim p \vdash q.$$

The latter is acceptable by (I), but the former, though acceptable classically by (II), is rejected by stringent logicians; so (D2) is rejected.

For another example, consider the dual to Disjunctive Syllogism

(D2*) $\sim(p \wedge q) \wedge p \vdash \sim q.$

At step (i) we get

$$(\sim p \vee \sim q) \wedge p \vdash \sim q.$$

At step (ii) we get

$$(\sim p \wedge p) \vee (\sim q \wedge p) \vdash \sim q.$$

At step (iii) we get

$$\sim p \wedge p \vdash \sim q$$
$$\sim q \wedge p \vdash \sim q.$$

And again the latter holds stringently, but the former holds only classically, not stringently; so (D2*) is rejected.

Now it seems that Disjunctive Syllogism (D2), and for that matter its dual (D2*), are used ubiquitously in mathematics—and not just in mathematics, either.[9] If this is so, then stringency fails as a *descriptive* criticism of classical logic. However, at this point we must take note of an option alluded to earlier, that of claiming that the apparent instances of (D2) in mathematical proofs are really instances of something else:

(D2†) $(p + q) \wedge \sim p \vdash q$,

where $+$ is an *intensional* disjunction, stronger than mere $\vee$.

The claim that "or" often means something stronger than $\vee$ has been advanced many times on many different grounds. And certainly, whenever a mathematician or whoever seriously wants to argue by Disjunctive Syllogism from "p or q" and "not p" to q, the reasoner will be in a position to say *something* stronger than just "$p \vee q$." Namely, something like, "$p \vee q$ and my present grounds for this assertion are not just the knowledge or assumption that p or the knowledge or assumption that q." For on the one hand, if the reasoner already knows q, there is no need to infer q by Disjunctive Syllogism or in any other way; while on the other hand, if the reasoner is assuming p to get $p \vee q$, presumably the same reasoner will not also at the same time want to be assuming $\sim p$, the other premise needed for an application of Disjunctive Syllogism.

But note that the "something extra" over and above $p \vee q$ is *purely subjective* and may consist of nothing but a *state of ignorance* on the part of the mathematician or whomever. The reasoner may have established $p \vee q$ yesterday by inferring it from q, and yet today recall only that $p \vee q$ has been established, and not how. Such a reasoner can still say, "My *present* grounds for $p \vee q$ are not simply knowledge that p or knowledge that q," for the reasoner's grounds are, rather, an imperfect memory of having *somehow* established $p \vee q$.

Indeed, the reasoner may even remember that it was *either* by inference from p *or* by inference from q, without recalling which. If the reasoner is now satisfied that "not p," then the inference by Disjunctive Syllogism may be a way of *recovering* the knowledge that has been forgotten, since then the inference must have been from q and not from p.

In mathematics, a later mathematician may start from the case $P(0) \vee Q(0)$ of an earlier mathematician's theorem that $\forall n(P(n) \vee Q(n))$ and establish $\sim P(0)$, and so establish $Q(0)$. And the later mathematician may do so without remembering (and, indeed, without ever having known) what the earlier mathematician's

[9] For a compendium of examples, due to E. M. Curley and others, see Burgess (1981).

proof of $\forall n(P(n) \vee Q(n))$ was. (Indeed, that proof may even have been an induction starting by inferring $P(0) \vee Q(0)$ from $Q(0)$. In that case, the later mathematician's inference that $Q(0)$ is not new to mathematics, but only to him- or herself; but that does not make the inference invalid.)

This issue has been extensively discussed in the literature, and the "intensional disjunction" option for defending stringency as a descriptive criticism of classical logic has, it seems, been pretty well given up.[10] The basic implausibility of this strategy ought to have been apparent from the fact that those who use it are committed not only to rejecting Disjunctive Syllogism (D2) but also to rejecting the dual version (D2*); and to defend the latter rejection along the same lines as the former, one would need to make the dual claim that there are both an extensional $\wedge$ and an intensional $\cdot$ conjunction, with the latter *weaker* than the former. But whatever plausibility may attach to the suggestion that there is a sense of "or" stronger than $\vee$, in which p does not entail "p or q," none at all attaches to the dual suggestion that there is a sense of "and" weaker than $\wedge$, in which "p and q" does not entail p.

5. "By Faith Unfaithful Kept Falsely True"

There remain to be considered some historically more recent rationales for stringency, based on certain equivalent formulations of the stringency criterion that are technically more convenient to work with than the original version given in the preceding section. Suppose the formulas of concern involve atoms $p_1, p_2, p_3, \ldots$. Let us then introduce new atoms $q_1, q_2, q_3, \ldots$ and $r_1, r_2, r_3, \ldots$. For elementary α involving only p atoms, let α^* (respectively $^*\alpha$) be the result of replacing each positive occurrence of p_i with the corresponding q_i (respectively r_i) and each negative occurrence with the corresponding r_i (respectively q_i). The second formulation of the stringency criterion is: *To test if elementary α stringently entails elementary β, test if α^* classically entails β^** (or equivalently, *if $^*\alpha$ classically entails $^*\beta$*; or, equivalently and more symmetrically, *if both*).

The second formulation of the criterion can be obtained from the first by looking at matters from from a classical point of view. From that point of view, ignoring (II) and (III) when testing for proto-elementary entailment amounts to

[10] For the later views of the originators of the option, see Anderson, Belnap, and Dunn (1992), §80.4, pp. 502ff.

forgetting that p and $\sim p$ are an atom and the negation of the *same* atom, and treating them as if they were just one atom q and the negation $\sim r$ of some *other* atom. More formally, performing step (iv*) in the first criterion amounts to first performing the *-substitution of the second formulation and then performing the step (iv) of evaluating the proto-elementary entailment resulting by the *classical* criterion (because the *-substitution prevents cases (II) and (III), which distinguish (iv) from (iv*) from arising).

But now it is readily checked that each of the laws used at steps (i), (ii), and (iii) is such that the result of first applying the law and then making the *-substitution or first making the *-substitution and then applying the law is the same. For instance, for double negation one has $(\sim\sim\alpha)^* = \sim\sim(\alpha^*)$. So instead of performing (i), (ii), (iii), *, (iv), one could perform (i), (ii), *, (iii), (iv) or (i), *, (ii), (iii), (iv) or *, (i), (ii), (iii), (iv). Since any other test for classical entailment would yield the same result as the test consisting in performing steps (i), (ii), (iii), (iv), the whole amounts simply to performing the *-substitution and testing *by one method or another* for classical entailment, which is the second formulation of the criterion. (This remains so whether "the *-substitution" means substituting q-atoms for positive and r-atoms for negative occurrences, or the reverse. Hence the parenthetical clause at the end of the second formulation.)

The second formulation shows that if α stringently entails β, then α perfectibly entails β. For if α stringently entails β, then α^* classically entails β^*; and since a formula like α^* or β^*, in which no atom occurs both positively or negatively, cannot be unsatisfiable or tautologous, because it can always be valued true (respectively, false) by so valuing the atoms that occur positively (respectively, negatively) and valuing oppositely the other atoms, it follows that α^* perfectly entails β^*; and since α and β are obtainable by substitution, namely of each p_i for both q_i and r_i, from α^* and β^*, it follows that α perfectibly entails β.

The second formulation shows that substitution holds for stringent entailment. For suppose α stringently entails β, so that α^* classically entails β^*, and consider the result of substituting a formula π for an atom p in α and β to obtain γ and δ, respectively. A little thought shows that the result of performing the *-substition to obtain γ^* and δ^* is the same as the result of substituting π^* for q and ${}^*\pi$ for r in α^* and β^*, and hence γ^* classically entails δ^*, and γ stringently entails δ.

The second formulation also shows that transitivity of entailment or hypothetical syllogism holds stringently. For if α stringently entails β, which stringently entails γ, then α^* classically entails β^*, which classically entails γ^*; thus α^* classically entails γ^* and α stringently entails γ.

At this point it is possible to make various "cosmetic" changes in the presentation of the second stringency criterion. A *Routley–Routley valuation* or *Sylvan–Plumwood valuation* or *double valuation* is a pair $(V^*, {}^*V)$ of assignments to atoms of truth-values, 1 for true and 0 for false, extended to compound formulas by the following recursion equations.

$$(V^*(\sim\alpha), {}^*V(\sim\alpha)) = (1 - {}^*V(\alpha),\ 1 - V^*(\alpha))$$

$$(V^*(\alpha \wedge \beta), {}^*V(\alpha \wedge \beta)) = (\min(V^*(\alpha),\ V^*(\beta)),\ \min({}^*V(\alpha),\ {}^*V(\beta)))$$

$$(V^*(\alpha \vee \beta), {}^*V(\alpha \vee \beta)) = (\max(V^*(\alpha),\ V^*(\beta)),\ \max({}^*V(\alpha),\ {}^*V(\beta)))$$

Then β is a (*right*) *double consequence* of α if $V^*(\alpha) \leq V^*(\beta)$ in all double valuations (and a *left double consequence* if ${}^*V(\alpha) \leq {}^*V(\beta)$, and a *symmetrical double consequence* if both). A third formulation of the stringency criterion is: *To test if elementary α stringently entails elementary β, test if β is a (right) double consequence of α* (or equivalently, *a left double consequence*, or equivalently *a symmetrical double consequence*).

The third formulation of the criterion can be obtained from the second by camouflaging the role of the q-atoms and r-atoms. The second formulation says to test whether for all valuations V of the q-atoms and r-atoms, extended classically to compound formulas, it is the case that $V(\alpha^*) \leq V(\beta^*)$. Take a valuation V of q-atoms and r-atoms and a pair $(V^*, {}^*V)$ of valuations of p-atoms to *correspond* if for all p_i one has $V^*(p_i) = V(p_i^*) = V(q_i)$ and ${}^*V(p_i) = V({}^*p_i) = V(r_i)$. Observe, first, that every V corresponds to exactly one $(V^*, {}^*V)$ and vice versa. Observe, second, that one will have $V^*(\alpha) = V(\alpha^*)$ and ${}^*V(\alpha) = V({}^*\alpha)$ not just for atoms but for all elementary α if one extends the pair $(V^*, {}^*V)$ to molecules by the above recursion equations for $\sim$, $\wedge$, and $\vee$, which are nonclassical for $\sim$ because it reverses positive and negative occurrences, and classical for $\wedge$ and $\vee$ because they preserve positive and negative occurrences. These observations together establish the equivalence of the second and third formulations of the criterion (with the parenthetical clause at the end of the second corresponding to the parenthetical clause at the end of the third).

A fourth formulation can be obtained by sensationalizing the third. A *dialectical valuation* is an assignment V to atoms of a subset of the set $\{0,1\}$ of classical truth-values, which may be $\{1\}$ or *true and unfalse*, or may be $\{0\}$ or *false and untrue*, or may be $\varnothing$ or *untrue and unfalse*, or may be $\{0,1\}$ or *true and false*, extended to molecules by the following recursion conditions:[11]

$0 \in V(\sim\alpha)$ iff $1 \in V(\alpha)$
$1 \in V(\sim\alpha)$ iff $0 \in V(\alpha)$
$0 \in V(\alpha \wedge \beta)$ iff $0 \in V(\alpha)$ or $0 \in V(\beta)$
$1 \in V(\alpha \wedge \beta)$ iff $1 \in V(\alpha)$ and $1 \in V(\beta)$

[11] The particular labels for the subsets given here represent only one of several options or "plans." One could, for instance, call the four values *true*, *false*, *gap*, and *glut*.

$$0 \in V(\alpha \vee \beta) \text{ iff } 0 \in V(\alpha) \text{ and } 0 \in V(\beta)$$
$$1 \in V(\alpha \vee \beta) \text{ iff } 1 \in V(\alpha) \text{ or } 1 \in V(\beta)$$

Then β is a (*right*) *dialectical consequence* of α if for any dialectical valuation, if 1 is among the values of α, then 1 is among the values of β (and a *left dialectical consequence* if whenever 0 is among the values of β, then 0 is among the values of α; and a *symmetrical dialectical consequence* if both). A fourth formulation of the stringency criterion is: *To test if elementary α stringently entails elementary β, test if β is a (right) dialectical consequence of α* (or equivalently, *a left dialectical consequence*, or equivalently *a symmetrical dialectical consequence*).

The fourth formulation can be related to the third as follows. Take a dialectical valuation V and a double valuation $(V^*, {}^*V)$ to *correspond* if and only if for all atomic α one has $1 \in V(\alpha)$ if and only if $V^*(\alpha) = 1$ and $0 \notin V(\alpha)$ if and only if ${}^*V(\alpha) = 1$, so that V^* says whether something is true or untrue, and ${}^*V$ whether something is unfalse or false. Then it can be observed that every dialectical valuation corresponds to exactly one double valuation and vice versa. And, comparing the two sets of recursion equations, it can be observed that the condition on atomic α defining correspondence actually holds for *all* α. This suffices to establish the equivalence of the third and fourth formulations (with the parenthetical clause at the end of third corresponding to the parenthetical clause at the end of the fourth).

Further ringing of changes is possible, but with the dialectical valuations we have come as far as we need to for present purposes. Note that with the last formulation, the notion that there is some subject-matter overlap requirement *additional to* truth-preservation has wholly dropped out, and entailment has become just truth-preservation for the enlarged set of four rather than the classical set of two truth-values. (By contrast, it can be shown that perfectionist logic is not a finite-valued logic.) This is one reason why some feel it inappropriate to call stringent logic "relevance logic" or "the logical of relevance." On the other hand, attractive as it may be to relevantists to call their logic "relevant logic" and classical logic an "irrelevant logic," this terminology is objectionable in the same way as would be a proposal to call Brouwer's logic "intuitive logic" and classical logic "counterintuitive logic".

Various rationales for stringency based on the alternate formulations have been advanced by various logicians with varying degrees of commitment, all based on the notion that something can be both true and false. This notion is understood by moderates in a *figurative* or imaginary sense, as "epistemic" truth according to a database or information corpus or belief system or whatnot; it is understood by radicals in a *literal* or real sense as "metaphysical" truth-in-the-world. In a terminology introduced earlier, to rely on the moderate's notion as a rationale would in effect amount to adopting a compatibilist position (though in actual fact the primary exposition of the moderate position was not accompanied by a retraction of the authors' previous criticisms of classical logic); while to rely on the radical's notion as a rationale would in effect amount to adopting a revisionist position (though in actual fact, enunciations of this view have not been followed by much work on revising "proofs" of metatheorems about "relevance" or "relevant" logics that rely on Disjunctive Syllogism).

The moderate idea is the basis for a suggestion about the potential utility of relevantism. The background is as follows. In ancient history, logic arose in a context of public debate in which argument was used at least as much for purposes of refutation as for purposes of demonstration: The recognition that a previously accepted premise entails some previously not accepted conclusion was taken to be grounds for ceasing to accept the premise at least as often as it was taken to be grounds for beginning to accept the conclusion. (This fact doubtless facilitated the recognition that logical validity of an argument is not simply a matter of truth or falsehood of premise or conclusion.) The critical faculty of being able to use the recognition of logical relationships thus in the reverse direction, like the more basic critical faculty of being able to reject some of what one is told, is an essential feature of natural intelligence. But it is a feature that "artificial intelligence" may not find it technically feasible to imitate in the foreseeable future.

In designing a computer to answer questions by inference from a database of things it has been told, allowance may have to be made for the fact that perhaps the computer cannot in the present state of the art be designed to reject any of its database (even when same item has been *both* said to be true *and* said to be false), or to use recognition of classical logical entailment relations between an accepted premise and an unaccepted conclusion as grounds for rejecting the premise (even when the conclusion has no overlap of subject matter with the premises, so that by the classical interpolation theorem, unless the conclusion is tautologous—in which case it is deducible even without the premise—the premise must be inconsistent). The computer, so to speak, may not be able to exhibit anything worthy of being called artificial intelligence, but only a kind of artificial bureaucratic stupidity. If so, it may well be appropriate *not* to allow such a computer to recognize all classical entailments, and it may even be appropriate to allow such a computer to recognize all and only stringent entailments.[12]

The classical response to this suggestion has a positive aspect, the acknowledgment that thus nonliterally interpreted stringent logic may well be useful in the manner described: that is for specialists in computer science to say. It also has a negative aspect, the insistence that the entailment relations *it would be appropriate to allow a computer with limitations of the kind described to recognize* are not the

[12] See Anderson, Belnap, and Dunn (1992), §81, pp. 506ff., though it is a lot easier to see why we would not want the computer to be able to infer anything whatsoever from a contradiction than why we would not want the computer to be able to infer a tautology from anything whatsoever.

only entailment relations that hold, and that the possible usefulness of stringent logic for the kind of application in question in no way sustains any relevantist criticism of classical logic.

The radical idea is that classical logic fails for the simplest reason, namely, that it is not truth-preserving, because there are cases where a premise of form $\alpha \wedge \sim\alpha$ is true, while not every conclusion β is true.[13] Tradition is scoured for examples of truths of the form $\alpha \wedge \sim\alpha$: Epimenides ("What I am now saying is false"), Heraclitus ("The way up and the way down are the same"), Zeno ("The arrow is both in motion and at rest"), Athanasius ("God is both one and three"), Leibniz ("Infinitesimals are both distinct from zero and less than anything greater than zero"), Kant ("Space and time are infinite in extent and divisibility, and space and time are finite in extent and divisibility"), Hegel-Marx-Engels-Lenin-Stalin-Mao ("There is contradiction in all things"), Meinong ("The round square is as surely round as it is square"), Russell ("The set of non-self-elements is a self-element and a non-self-element"), Bohr ("Waves are particles, and vice versa"), and others are canvassed.

The classical response, when there has been one—for some principles are so fundamental that there can hardly be non-question-begging argument about them, and many would hold that the law of noncontradiction is one of these[14]—has most often been that of Wellington ("If you can believe that, you can believe anything").

[13] Proponents of this idea—by some called "dialethism"—almost invariably claim that while adoption of a logic in which a contradiciton does not imply everything—by many called "paraconsistent"—enables one to deal with these inconsistent situations, somehow adoption of such a logic does *not* prevent one from using classical logic in consistent situations, so that adopting such a logic involves a pure benefit, without the enormous costs that would appear to be involved in giving up the ubiquitous Disjunctive Syllogism. The incoherence of this claim has been pointed out by Anderson, Belnap, and Dunn (1992, §80.4.2, p. 503):

> One might think as follows. The point of relevantism is to take seriously the threat of contradiction. But there is in this vicinity... no real such threat. So here it is OK to use the d.s.... That *sounds* OK; but is it? After all, we suppose that "here there is no threat of contradiction" is to be construed as an added premiss. But a little thought shows that *no* such added premiss should permit the relvantist to use the d.s., for a very simple reason: as we said, avoidance of the d.s. was bound up with the threat of contradiction, and one thing that is clear is that *adding* premisses cannot possibly *reduce* that threat.

The context of these remarks is a discussion of what relevantists should make of the fact (urgently brought to their attention by Saul Kripke) that the proof of a major metatheorem about their systems employs the forbidden disjunctive syllogism.

[14] Thus Lewis, "Logic for Equivocators," in (1998), p. 101, writes:

> The reason we should reject this proposal is simple. No truth does have, and no truth could have, a true negation. Nothing is, and nothing could

REFERENCES

Anderson, Alan Ross, and Nuel D. Belnap, Jr. (1975), *Entailment: The Logic of Relevance and Necessity*, vol. 1, Princeton, NJ: Princeton University Press.

Anderson, Alan Ross, Nuel D. Belnap, Jr., and J. Michael Dunn (1992), *Entailment: The Logic of Relevance and Necessity*, vol. 2, Princeton, NJ: Princeton University Press.

Boolos, George S. (1998), *Logic, Logic and Logic*, Cambridge, MA: Harvard University Press.

Burgess, John P. (1981), "Common Sense and Relevance," *Notre Dame Journal of Formal Logic*, 24, pp. 41–53.

Lewis, D.K. (1998), *Papers in Philosophical Logic*, Cambridge: Cambridge University Press.

Poincaré, Henri (1960), *Science and Method*, translated by F. Maitland, New York: Dover.

Quine, W.V. (1966), *The Ways of Paradox*, New York: Random House.

---

be literally true and false. This we know for certain, and *a priori*, and without any exception for especially perplexing subject matters. The radical case for relevance should be dismissed just because the hypothesis it requires us to entertain is inconsistent.

That may seem dogmatic. And it is: I am affirming the very thesis that [Richard Sylvan né] Routley and [Graham] Priest have called into question and—contrary to the rules of debate—I decline to defend it. Further, I concede that it is indefensible against their challenge. They have called so much into question that I have no foothold on undisputed ground. So much the worse for the demand that philosophers always must be ready to defend their theses under the rules of debate.

CHAPTER 25

HIGHER-ORDER LOGIC

STEWART SHAPIRO

> ... our definition of the [standard second-order] consequences of a system of postulates ... can be seen to be not essentially different from [that] required for the ... treatment of classical mathematics.... It is true that the non-effective notion of consequence, as we have introduced it ... presupposes a certain absolute notion of ALL propositional functions of individuals. But this is presupposed also in classical mathematics, especially classical analysis....
>
> —Church [1956, 326n]

FORMAL languages typically have variables that (purport to) range over some objects. These are called *first-order variables*, and the collection of objects they range over is called the "domain of discourse." *Second-order variables* range over properties, classes, relations, or functions of the items in the domain of discourse. *Third-order variables* range over properties, or relations, or functions on the items in the range of the second-order variables. And so on. A formal language is *first-order* if it contains first-order variables, and no others. A language is *second-order* if it contains first-order and second-order variables, and no others. And on it goes. A language is *higher-order* if it is at least second-order. Second-order logic is the logic of second-order languages, and higher-order logic is the logic of higher-order languages. To make a bad joke, if we could agree on what logic is, we would then be about done with our topic of higher-order logic. Stay tuned.

It is not an exaggeration to say that first-order logic is the paradigm of contemporary logical theory. The vast majority of the works in both philosophical and mathematical logic concern first-order languages exclusively. Most textbooks do not mention higher-order logic at all, and most of the rest give it scant treatment. In contrast, just about all of the central work that launched contemporary logic around the turn of the twentieth century concerns higher-order formal languages. Examples include Frege [1879], Peano [1889], and Whitehead and Russell [1910]. First-order logic appeared as a distinctive study only when some authors, beginning with Löwenheim [1915], separated out first-order languages as *subsystems* for special treatment. Early twentieth-century logicians often referred to the deductive system of first-order logic as the "*restricted* functional calculus." For more on the historical emergence of first-order logic, see Moore [1980], [1988], Shapiro [1991, ch. 7], and Eklund [1996].

The philosophical literature contains numerous claims on behalf of and numerous claims against higher-order logic. Virtually all of the issues apply to second-order logic (vis-à-vis first-order logic), so we will focus on that here. In this chapter, I develop the syntax of second-order languages and present typical deductive systems and model-theoretic semantics for them. This will help to explain the role of higher-order logic in the philosophy of mathematics. I assume that the reader has at least a passing familiarity with the theory and metatheory of first-order logic.

1. Formal Languages

Let K be a set of nonlogical terminology. We will consider formal languages built on such sets. In arithmetic, for example, the relevant set would be $\{0, s, +, \cdot\}$, the symbols for zero, successor, addition, and multiplication.

To establish notation, let us begin with the first-order language L1K. First-order variables are lower-case letters near the end of the alphabet, with or without numerical subscripts. The connectives are negation $\neg$, conjunction &, disjunction $\vee$, material implication $\rightarrow$, and material biconditional $\equiv$. The language has a universal quantifier $\forall$ and an existential quantifier $\exists$. A *sentence* is a formula without free variables, and a *theory* is a set of formulas.

The first-order language with identity L1K= is obtained from L1K by adding a binary relation symbol =. The identity symbol is regarded as logical and so is not in K. If t and u are terms, then we abbreviate $\neg t = u$ as $t \neq u$.

The language L2K is obtained from L1K (not L1K=) by adding a stock of (second-order) *relation variables* and *function variables.* Relation variables are uppercase Roman letters from the end of the alphabet, with or without numerical subscripts. Function variables are letters like *f*, *g*, and *h*, with or without numerical subscripts. To be pedantic, we should use a superscript to indicate the *degree*, or number of places, of each second-order variable: X^1 is a monadic predicate variable; f^2 is a binary function variable, and so on. If the context determines the degree of a variable, we usually omit the superscript.

There are four new formation rules:[1]

> If f^n is an *n*-place *function variable* and $t_1, \ldots, t_n$ are terms, then $f^n t_1, \ldots, t_n$ is a term.
> If R^n is an *n*-place *relation variable* and $t_1, \ldots, t_n$ are terms, then $R^n t_1, \ldots, t_n$ is an atomic formula.
> If Φ is a formula and V a relation variable, then $\forall V\Phi$ and $\exists V\Phi$ are a formulas.
> If Φ is a formula and f a function variable, then $\forall f\Phi$ and $\exists f\Phi$ are formulas.

Thus, for example, $\exists X \forall x \neg Xx$ asserts the existence of a property that applies to no objects.

The symbol for identity between (first-order) objects is introduced as an abbreviation,

$$t = u\text{: } \forall X(Xt \equiv Xu),$$

in which t and u are terms. This is not meant as a deep philosophical thesis concerning the nature of Identity. If she wishes, the reader is free to add a primitive, logical identity symbol, as in L1*K*=.

Incidentally, it is not necessary to include separate variables for functions, since a given *n*-place function can be represented by an $n+1$-place relation. For example, instead of a monadic function *f*, we can use a binary relation variable R^2, adding the clause $\forall x \exists y \forall z(R^2 xz \equiv y = z)$. We include variables for functions only as a convenience.

I do not include a symbol for identity between second-order items like relations and functions, mostly because I do not wish to engage issues concerning the nature and individuation of these items—or at least I wish to postpone such issues. For the time being, one can think of the higher-order items as intensional entities like properties or propositional functions, or one can think of them as extensional entities like classes or sets. I will use words like "property," "class," and "set" interchangeably, although we will briefly return to this issue in section 5 below, when we get to more philosophical matters.

Whitehead and Russell [1910] argued that intensional entities must be defined or constructed in *levels*, so that properties defined at a given level become available

[1] For readability, I do not mark distinctions like that between variables in the object language and metavariables that range over object language variables. Context will indicate which is meant.

for use in definitions at later levels. Say that a relation is of level 0 if it can be defined without referring to relations. For each natural number n, a relation is of level $n+1$ if it is not of level n, but can be defined with reference to level n relations only. To develop this *ramified type theory*, each second-order variable is marked somehow to indicate its level. We do not pursue this here, except by way of brief comparison (see Hazen [1983] for a readable development).[2]

We can expand the set K to include so-called higher-order nonlogical constants, terms that stand for such items as properties (or sets) of properties, or functions on properties. An example would be a property TWO of properties such that TWO(P) holds just in case P applies to exactly two things. The logicist and neologicist programs (see chapters 5 and 6 of this volume) invoke constants that denote functions from properties to objects. Each abstraction principle includes at least one such constant.

A third-order language L3K is constructed from L2K by adding *third-order variables* for relations on relations, functions of predicates, functions of functions, and so on. Then one could add nonlogical constants (to K) for relations on functions of predicates, and the like. Then one could add *fourth-order* variables ranging over such things, thus producing a fourth-order language, and so on. As noted above, no new conceptual issues arise, and we do not directly pursue such languages here.

2. Deductive System

I assume that the reader is familiar with deductive systems for first-order languages, and I will present an extension of one such system to our second-order languages L2K. It is straightforward to adapt the usual axioms or rules of inference for the quantifiers. In a natural deduction system (which allows formulas with free variables in deductions), the introduction and elimination rules for the higher-order quantifiers are:

∀E From $\forall X^n \Phi(X^n)$ infer $\Phi(T)$, where T is either an n-place relation variable free for X^n in Φ, or an n-place relation letter in the set K of nonlogical terminology. The conclusion $\Phi(T)$ depends on whatever premises $\forall X^n \Phi(X^n)$ depends on.

From $\forall f^n \Phi(f^n)$ infer $\Phi(g)$, where g is either an n-place function variable free for f^n in Φ, or an n-place function letter in the set K of

[2] Note that the above definition of identity is not unambiguous in the ramified system, since the variable "X" would have to be marked for level. The best route is to make the identity sign primitive, as it usually is in first-order logic.

nonlogical terminology. The conclusion $\Phi(g)$ depends on whatever premises $\forall f^n \Phi(f^n)$ depends on.

∀I From Φ infer $\forall X\Phi$, provided that X does not occur in any premise on which Φ depends. The conclusion $\forall X\Phi$ depends on whatever premises Φ depends on.

From Φ infer $\forall f\Phi$, provided that f does not occur in any premise on which Φ depends. The conclusion $\forall f\Phi$ depends on whatever premises Φ depends on.

∃E If one can infer a conclusion Φ from an assumption Ψ, then one can infer Φ from $\exists X\Psi$, provided that X does not occur free in Φ or in any premise on which Φ depends. The final conclusion depends on whatever premises $\exists X\Psi$ and Φ depend on, except Ψ.

If one can infer a conclusion Φ from an assumption Ψ, then one can infer Φ from $\exists f\Psi$, provided that f does not occur free in Φ or in any premise on which Φ depends. The final conclusion depends on whatever premises $\exists X\Psi$ and Φ depend on, except Ψ.

∃I From $\Phi(T)$ infer $\exists X^n\Phi$, where T is either an n-place relation variable free for X^n in Φ, or an n-place relation letter in the set K of nonlogical terminology. The conclusion $\exists X^n\Phi$ depends on whatever premises $\Phi(T)$ depends on.

From $\Phi(g)$ infer $\exists f^n\Phi$, where g is either an n-place function variable free for f^n in Φ, or an n-place function letter in the set K of nonlogical terminology. The conclusion $\exists f^n\Phi$ depends on whatever premises $\Phi(g)$ depends on.

The next item is the *comprehension scheme*. For each formula Φ in L2K and each relation variable X^n, the following is an axiom:

$$\exists X^n \forall x_1 \ldots \forall x_n (X^n x_1 \ldots x_n \equiv \Phi),$$

provided that X^n does not occur free in Φ. Taken together, the instances of the comprehension scheme register the thesis that every formula determines a relation or, more precisely, for every formula there is a relation with the same extension. In a deductive system for ramified type theory, the embedded formula Φ should not contain any bound variables whose level is the same as or greater than the level of the introduced variable X^n.

The final item is a form of the axiom of choice:

$$\forall X^{n+1}(\forall x_1 \ldots \forall x_n \exists y X^{n+1} x_1 \ldots x_n y \rightarrow \exists f^n \forall x_1 \ldots \forall x_n X^{n+1} x_1 \ldots x_n f x_1 \ldots x_n).$$

The antecedent of the embedded conditional asserts that for each sequence $x_1, \ldots, x_n$ there is at least one y such that the sequence $x_1, \ldots, x_n, y$ satisfies X^{n+1}. The consequent asserts the existence of a function that "picks out" one such y for each $x_1, \ldots, x_n$.

The reader who has qualms about calling the axiom of choice a principle of logic can drop it from the system. As above, one alternative is to eliminate variables from the system altogether. If our gentle reader who demurs from the axiom of choice wishes to maintain functions, she will probably wish to add a comprehension principle that relates functions to appropriate relations:

$$\forall X^{n+1}((\forall x_1 \ldots \forall x_n \exists y \forall z(X^{n+1}x_1 \ldots x_n z \equiv y = z)) \rightarrow \exists f^n \forall x_1 \ldots \forall x_n X^{n+1} x_1 \ldots x_n f x_1 \ldots x_n).$$

The antecedent of the embedded conditional asserts that for each sequence $x_1, \ldots, x_n$ there is *exactly* one y such that the sequence $x_1, \ldots, x_n, y$ satisfies X^{n+1}. The consequent asserts the existence of a corresponding function. This sentence follows from the above axiom of choice.

Call this deductive system D2. Recall that in the language L2*K*, a formula in the form $x = y$ is an abbreviation of

$$\forall X(Xx \equiv Xy).$$

As an exercise, we can check that the translations of the usual identity axioms are theorems of D2. The introduction rule is $\forall x(x = x)$, which translates as $\forall x \forall X(Xx \equiv Xx)$. It is immediate that this is derivable. The elimination rule is

from $t = u$ and $\Phi(t)$, infer $\Phi(u)$,

where t and u are terms, provided that any variables in u are free for t in $\Phi(t)$. This comes to

from $\forall X(Xt \equiv Xu)$ and $\Phi(t)$ infer $\Phi(u)$,

with the same proviso. The validity of this inference is a routine exercise in the natural deduction system D2. Take $\forall X(Xt \equiv Xu)$ and $\Phi(t)$ as premises. Assume $\forall x(Yx \equiv \Phi(x))$. So $Yt \equiv \Phi(t)$, and thus Yt. So Yu, and thus $\Phi(u)$. So, by $\exists$E above, we infer $\Phi(u)$ from the instance of comprehension $\exists X \forall x(Xx \equiv \Phi(x))$. This justifies the definition of identity.

It is straightforward, but perhaps tedious, to establish an indiscernibility principle for relation variables. That is, for each formula Φ such that Q is free for P, the following formula

$$\forall x_1 \ldots \forall x_n(Px_1 \ldots x_n \equiv Qx_1 \ldots x_n) \rightarrow (\Phi(P) \rightarrow \Phi(Q))$$

is derivable in D2. The proof of this proceeds by induction on the complexity of the formula Φ. This partly justifies an extensional orientation toward the higher-order terminology.

Deductive systems for the further extensions of L2*K* to third- and higher-order languages are straightforward extensions of those considered here. I shall not present details.

3. Model Theory

This section presents four (count 'em) model-theoretic semantics for the second-order languages L2*K*. Once again, I assume familiarity with the usual model-theoretic semantics for first-order languages L1*K* and L1*K*=. I pause only to establish the notation.

Each *model* or interpretation of the first-order L1*K* or L1*K*= is a structure $M = \langle d, I \rangle$, in which d is a nonempty set, the *domain* of the model, and I is an *interpretation function* that assigns appropriate items constructed from d to the items in K, the nonlogical terminology. For example, if R is an n-place relation symbol, then $I(R)$ is a set of n-tuples of members of d. A *variable-assignment* s is a function from the variables of L1*K* to d. We will have occasion to consider "interpretations" in which the domain is not a set. For example, in the intended interpretation of set theory, the variables range over the entire iterative hierarchy—which of course is not a set.

For each model and assignment, there is a *denotation function* that assigns a member of d to each term. The relation of *satisfaction* between models, assignments, and formulas is then defined in the usual manner. We write $M, s \models \Phi$ to indicate that the model M satisfies the formula Φ under the assignment s. If $M, s \models \Psi$ for every assignment s and every formula Ψ in a set Γ, then we say that M is a *model of* Γ. A formula Φ is a *semantic consequence* of Γ if for every model M and assignment s on M, if $M, s \models \Psi$ for every Ψ in Γ, then $M, s \models \Phi$. This is sometimes written $\Gamma \models \Phi$.

All four of our semantics for the second-order L2*K* build on the semantics for its first-order counterpart L1*K*, in the sense that each model has, as components, a domain d and an interpretation function I, as above. For three of the four semantics, what we add is a range for the relation and function variables, and, with that, an extension of the denotation function and the satisfaction relation to the new terms and the new atomic formulas, respectively. The rest is straightforward.

3.1. Standard Semantics

Although the name "standard" here is more or less common among writers on this topic, the emotive or normative connotations also reflect my own preferences (for what that is worth). Critics sometimes put the word "standard" in scare quotes, or use phrases like "so-called 'standard' semantics." In a sense to be made clear below, it is only standard semantics that makes the logic distinctively second-order.

A *standard model* of L2*K* is the same as a model of the first-order L1*K* and L1*K*=, namely, a structure $\langle d, I \rangle$. The range of the *n*-place relation variables is the entire power set of d^n, and the range of the *n*-place function variables is the set of all functions from d^n to *d*. Thus a *variable-assignment* is a function that assigns a member of *d* to each first-order variable, a subset of d^n to each *n*-place relation variable, and a function from d^n to *d* to each *n*-place function variable.

To belabor the obvious, function variables are assigned to *functions*. Thus the denotation function for the terms of L2*K* is a straightforward extension of the denotation function from L1*K*. The new clause is

> Let $M = \langle d, I \rangle$ be a model and *s* an assignment on *M*. If *f* is an *n*-place function variable and $t_1, \ldots, t_n$ are terms, the denotation of $ft_1 \ldots t_n$ is the result of applying the function *sf* assigned to *f* to the denotations of the terms $t_1, \ldots, t_n$.

Similarly, relation variables are assigned to relations (or sets of *n*-tuples). The clause for the satisfaction of the new atomic formulas is

> Let $M = \langle d, I \rangle$ be a model and *s* an assignment on *M*. If *X* is an *n*-place relation variable and $t_1, \ldots, t_n$ are terms, then $M, s \models Xt_1 \ldots t_n$ if and only if the sequence consisting of the denotations of the terms $t_1, \ldots, t_n$ is in *sX*.

The quantifiers are then interpreted as in the first-order case:

> $M, s \models \forall X \Phi$ if $M, s' \models \Phi$, for every assignment s' that agrees with *s* at every variable except possibly *X*.
> $M, s \models \exists X \Phi$ if $M, s' \models \Phi$, for some assignment s' that agrees with *s* at every variable except possibly *X*.
> $M, s \models \forall f \Phi$ if $M, s' \models \Phi$, for every assignment s' that agrees with *s* at every variable except possibly *f*.
> $M, s \models \exists f \Phi$ if $M, s' \models \Phi$, for some assignment s' that agrees with *s* at every variable except possibly *f*.

If Φ is a formula of L2*K*, then Φ is *standardly valid* or is a *standard logical truth*, if $M, s \models \Phi$ for every model *M* and assignment *s*; Ψ is *standardly satisfiable* if $M, s \models \Psi$ for some *M*,*s*. A set Γ of formulas of L2*K* is *standardly satisfiable* if there is an *M*, *s* such that $M, s \models \Psi$ for every Ψ in Γ. Finally, Φ is a standard consequence of Γ if for every model *M* and assignment *s*, if $M, s \models \Psi$ for every Ψ in Γ, then

$M, s \models \Phi$. To mark my preferences, the word "valid" is sometimes used for "standardly valid," "satisfiable" for "standardly satisfiable," and so on.

It is worth emphasizing that a standard model for L2*K* is the same as a model of its first-order counterpart L1*K*, namely, a domain and an interpretation of the nonlogical terminology. That is, in standard semantics, by fixing a domain one thereby fixes the range of both the first-order variables and the second-order variables. There is no further "interpreting" to be done. This is not the case with the next two semantics. In each case, one separately determines a range for the first-order variables and a range for the second-order variables.

3.2. Henkin Semantics

The central feature of Henkin semantics is that in a given model, the relation variables range over a *fixed collection* of relations on the domain, which may not include all of the relations; and the function variables range over a fixed collection of functions on the domain, which may not contain all of the functions. A *Henkin model* of L2*K* is a structure $M^{\mathrm{H}} = \langle d,D,F,I \rangle$, in which d is a domain and I an interpretation function for the terminology in *K*, as above. The new items are sequences. For each n, $D(n)$ is a nonempty subset of the power set of d^n and $F(n)$ is a nonempty set of functions from d^n to d. The idea is that $D(n)$ is the range of the n-place relation variables and $F(n)$ is the range of the n-place function variables in M^{H}. A *variable-assignment* on M^{H} is a function that assigns a member of d to each first-order variable, a member of $D(n)$ to each n-place relation variable, and a member of $F(n)$ to each n-place function variable.

The rest of the presentation of this semantics is the same as that of standard semantics, word for word:

> Let $M^{\mathrm{H}} = \langle d,D,F,I \rangle$ be a Henkin model and s an assignment on M^{H}. If f is an n-place function variable and $t_1, \ldots, t_n$ are terms, the denotation of $ft_1 \ldots t_n$ is the result of applying the function sf to the denotations of the terms $t_1, \ldots, t_n$.
>
> Let $M^{\mathrm{H}} = \langle d,D,F,I \rangle$ be a Henkin model and s an assignment on M^{H}. If X is an n-place relation variable and $t_1, \ldots, t_n$ are terms, then $M^{\mathrm{H}}, s \models Xt_1 \ldots t_n$ if and only if the sequence consisting of the denotations of the terms $t_1, \ldots, t_n$ is in sX.
>
> $M^{\mathrm{H}}, s \models \forall X\Phi$ if $M^{\mathrm{H}}, s' \models \Phi$, for every assignment s' that agrees with s at every variable except possibly X.
>
> $M^{\mathrm{H}}, s \models \exists X\Phi$ if $M^{\mathrm{H}}, s' \models \Phi$, for some assignment s' that agrees with s at every variable except possibly X.
>
> $M^{\mathrm{H}}, s \models \forall f\Phi$ if $M^{\mathrm{H}}, s' \models \Phi$, for every assignment s' that agrees with s at every variable except possibly f.

$M^H, s \models \exists f\Phi$ if $M^H, s' \models \Phi$, for some assignment s' that agrees with s at every variable except possibly f.

What distinguishes standard from Henkin semantics is the range of the quantifiers "every assignment" and "some assignment" in these last four clauses. In a Henkin model, we consider only the functions that assign members of the various $D(n)$ and $F(n)$ to the higher-order variables, while in standard semantics, we consider every such function.

The notions of *Henkin-validity*, *Henkin-satisfaction*, and *Henkin-consequence* are defined in the straightforward manner: Φ is Henkin-valid if $M^H, s \models \Phi$ for every Henkin model M^H and assignment s; Ψ is Henkin-satisfiable if $M^H, s \models \Psi$ for some Henkin model M^H and assignment s; Γ is Henkin-satisfiable if there is a Henkin model M^H and an assignment s such that $M^H, s \models \Psi$ for each $\Psi \in \Gamma$; and Φ is a Henkin-consequence of Γ if for every Henkin model M^H and assignment s, if $M^H, s \models \Psi$ for every Ψ in Γ, then $M^H, s \models \Phi$.

It is immediate that a standard model of L2K is equivalent to the Henkin model in which for each n, $D(n)$ is the entire power set of d^n, and $F(n)$ is the collection of all functions from d^n to d. Such Henkin models are sometimes called *full models*. In particular, let M be a standard model and let M^F be the corresponding full model. Then for each assignment s and each formula Φ, $M, s \models \Phi$ if and only if $M^F, s \models \Phi$. In effect, standard semantics just is the restriction of Henkin semantics to full models.

It follows that if a formula Φ is Henkin-valid, then Φ is standardly valid; and if Φ is a Henkin-consequence of a set Γ, then Φ is a standard consequence of Γ. And if Φ is standardly satisfiable, then Φ is Henkin-satisfiable. It will soon become clear that the converses of these conditionals fail.

On both of these semantical systems, the items in the range of higher-order variables are extensional entities—either sets or functions. Recall, however, that there is no symbol for "higher-order identity" in the language. As a result, one is free to maintain an intensional understanding of the higher-order entities, and think of them as attributes, properties, or propositional functions. For the purposes of model-theoretic semantics, sets can serve as *surrogates* for the relevant intensional items. If an advocate of intensional items believes that for every arbitrary collection S of n-tuples on the domain, there is an attribute whose extension is S (and similarly for functions-in-intension), then she will favor standard semantics. Otherwise, I presume that the attraction would be to Henkin semantics (see Cocchiarella [1988]).

3.3. Multisorted, First-order Semantics

As noted, on both standard semantics and Henkin semantics the range of the relation variables is a set of relations (or sets) and the range of function variables is a set of functions. With *first-order semantics*, this assumption is dropped. In effect, L2K is interpreted as a multisorted first-order language. The predication relation between the objects in the (first-order) domain and the items in the range of the predicate variables is non-logical, as is the application function from items in the range of the function variables and the objects in the domain.

A *first-order model* of L2*K* is a structure $M^1 = \langle d, d_1, d_2, \langle I, p, a \rangle \rangle$, in which d is a nonempty set and I an interpretation function assigning items constructed from d to the items in *K*, as usual. For each natural number n, $d_1(n)$ and $d_2(n)$ are nonempty sets. They are the ranges of the n-place relation variables and the n-place function variables, respectively. For each n, $p(n)$ is a subset of $d^n \times d_1(n)$. This represents the interpretation of the "predication" relation in M^1, between an n-place "relation" R in M^1 and n-tuples from the domain. For example, suppose that $d_1(1)$ has an object D which is supposed to represent the property of being a dog (or the set of dogs). Then $\langle c, D \rangle$ would be a member of $p(1)$ if and only if c is a dog. In like manner, for each natural number n, $a(n)$ is a function from $d^n \times d_2(n)$ to d. Here, $a(n)$ is the interpretation of the n-place "application" function in M^1, from the collection of sequences and the "functions" in M^1 to the first domain.

The rest is fairly straightforward, if a bit tedious. A *variable-assignment* on M^1 is a function that assigns a member of d to each first-order variable, a member of $d_1(n)$ to each n-place relation variable, and a member of $d_2(n)$ to each n-place function variable.

The denotation function is adapted from that of L1*K*. The new clause is:

> Let M^1 be a first-order model and s an assignment on M^1. If f is an n-place function variable and $t_1, \ldots, t_n$ are terms, then the denotation of $ft_1 \ldots t_n$ is the result of applying $a(n)$ to the denotations of $t_1, \ldots, t_n$ and sf.

In other words, the denotation of $ft_1 \ldots t_n$ is determined by the "application" function $a(n)$, evaluated at the denotations of the items in $t_1, \ldots, t_n$ and the item in $d_2(n)$ assigned to f by s.

The clause for the satisfaction of atomic formulas is similar:

> Let M^1 be a first-order model and s an assignment on M^1. If X is an n-place relation variable and $t_1, \ldots, t_n$ are terms, then $M^1, s \models Xt_1 \ldots t_n$ if the sequence consisting of the denotations of $t_1, \ldots, t_n$ and sX is a member of $a(n)$.

The clauses for the second-order quantifiers carry over:

> $M^1, s \models \forall X \Phi$ if $M^1, s' \models \Phi$, for every assignment s' that agrees with s at every variable except possibly X.
> $M^1, s \models \exists X \Phi$ if $M^1, s' \models \Phi$, for some assignment s' that agrees with s at every variable except possibly X.
> $M^1, s \models \forall f \Phi$ if $M^1, s' \models \Phi$, for every assignment s' that agrees with s at every variable except possibly f.

$M^1, s \models \exists f \Phi$ if $M^1, s' \models \Phi$, for some assignment s' that agrees with s at every variable except possibly f.

As above, Φ is *first-order-valid* if $M^1, s \models \Phi$ for every first-order model M^1 and assignment s; Ψ is *first-order-satisfiable* if $M^1, s \models \Psi$ for some first-order model M^1 and assignment s; Γ is *first-order-satisfiable* if there is a first-order model M^1 and an assignment s such that $M^1, s \models \Psi$ for each $\Psi \in \Gamma$; and Φ is a *first-order-consequence* of Γ if for every first-order model M^1 and assignment s, if $M^1, s \models \Psi$ for every Ψ in Γ, then $M^1, s \models \Phi$.

Notice that a given Henkin model $\langle d,D,F,I \rangle$ is equivalent to the first-order model $\langle d,d_1,d_2,\langle I,p,a \rangle\rangle$ such that d_1 is D, d_2 is F, p recapitulates the "real" predication (or membership) relation between d and the various $D(n)$, and a recapitulates the "real" application function from d^n and the various $F(n)$ to d. That is, for each n, $\langle u,v \rangle$ is in $p(n)$ if and only if $u \in v$; and $a(n)(u,w) = w(u)$. Thus, for every Henkin model M^H there is a first-order model M^1 such that for every assignment s on M^H there is an assignment s^1 on M^1, such that for every formula Φ of L2K, $M^H, s \models \Phi$ if and only if $M^1, s^1 \models \Phi$.

The converse holds as well. Let $M^1 = \langle d,d_1,d_2,\langle I,p,a \rangle\rangle$ be a first-order model of L2K. Then there is a Henkin model $M^H = \langle d,D,F,I \rangle$ such that for every variable-assignment s on M^1 there is an assignment s^H on M^H, such that for every formula Φ of L2K, $M^1, s \models \Phi$ if and only if $M^H, s^H \models \Phi$ (see Shapiro [1991, ch. 3] for details). The idea is that we replace an item c of, say, $d_1(2)$ with the set of pairs of elements of d that c "holds of," according to $p(2)$. To take a frivolous example, suppose that in M^1, d is a set of apples and $d_1(2)$ is a set of oranges. Then in the corresponding M^H, each orange is "replaced" by the set of pairs of apples that bear the M^1-interpretation of the predication relation to it.[3]

It follows that for each formula Φ of L2K, Φ is Henkin-valid if and only if Φ is first-order-valid, and Φ is a Henkin-satisfiable if and only if Φ is first-order-satisfiable. For each set Γ of formulas, Γ is Henkin-satisfiable if and only if Γ is first-order-satisfiable; and Φ is a Henkin-consequence of Γ if and only if Φ is a first-order-consequence of Γ. In short, Henkin semantics and first-order semantics are pretty much the same.

3.4. Plural Quantification

All three of the above model-theoretic semantics follow the prevailing custom of assigning each variable to a distinctive *range* in each model. On standard and

[3] See Shapiro [1991, ch. 3] and Gilmore [1957]. Our decision not to introduce a (primitive) symbol for identity-between-relations simplifies the present treatment somewhat.

Henkin semantics, the range of a given higher-order variable is a set of sets (of *n*-tuples) or a set of functions. As we just noted, even on first-order semantics, the range of a given higher-order variable is *equivalent to* a set of sets or a set of functions. This feature invites the charge that second-order logic is just a branch of set theory. Moreover, all three semantics seem to preclude an interpretation of a language like L2*K* in which the first-order variables do not range over a set. Thus we cannot give an account of second-order set theory in which the intended first-order variables range over every member of the iterative hierarchy. In that case, what do the higher-order variables range over?

In response to issues like these, George Boolos ([1984], [1985]) proposed a different way to understand at least monadic, second-order relation variables. According to both standard and Henkin semantics, an existential quantifier $\exists X$ should be read "there is a set X" or "there is a property X"—in which case, of course, the locution invokes sets or properties. Against this, Boolos suggests that a monadic, second-order quantifier be considered a counterpart of a *plural quantifier*, "there are (objects)" in natural language like English.

The sentence

Some critics admire only one another.

has a (more or less) straightforward second-order rendering, taking the class of critics to be the domain of discourse:

$$\exists X(\exists xXx \,\&\, \forall x\forall y((Xx(x \neq y \,\&\, Xy))).$$

According to standard or Henkin semantics, the formula would correspond to "there is a nonempty *set* (or property) C (of critics) such that for any x in C and any y, if x admires y, then $x \neq y$ and y is in C." However, this reading implies the existence of a set (or property), while the original sentence, "Some critics admire only one another," does not—or so it seems.

There is no doubt that native speakers of ordinary natural languages have no trouble with sentences that contain plural quantifiers. Boolos argues that we logicians can thus *use* this construction in the metalanguage in which we develop formal, model-theoretic semantics. The construction interprets the monadic, second-order quantifier.

So construed, a monadic second-order language has no ontological commitments beyond those of its first-order counterpart. Accordingly, second variables do not have a distinctive range. In a sense, the range of monadic, second-order variables is the same as the range of the first-order variables. It is just that the quantification is plural.

In set theory, the "Russell sentence," $\exists X\forall x(Xx \equiv x \notin x)$, is a consequence of the comprehension scheme. According to standard or Henkin semantics, it entails that there is a *set* that is not coextensive with any set in the domain. Admittedly, this does not sound right, even though we convince ourselves that it is correct—perhaps with talk of proper classes.[4] On Boolos's interpretation, the Russell sentence reads, "There are some sets such that any set is one of them just in case it is not a member of itself," a harmless truism. We forgo details of Boolos's [1985] rigorous, model-theoretic semantics for second-order languages with monadic relation variables (see Rayo and Yablo [2001] for an extension of the plural construction to multiplace relations and beyond).

It seems that the Boolos interpretation for monadic, second-order languages is intended to be equivalent to standard semantics, at least for interpretations in which the domain constitutes a set (so that no questions are begged). I have some doubts as to whether our independent or pretheoretic grasp of plural quantifiers is sufficiently determinate for this.

Consider a statement of second-order real analysis of the form

$$\forall X \exists Y \Phi(X,Y).$$

The opening second-order quantifiers can be given both a plural and an ordinary, standard reading. It had better be the case that if we read the quantifiers as plurals, we will get exactly the same truth-value, in general, as we would if we understand the quantifiers as ranging over sets of real numbers. In effect, there needs to be a "plurality" (so to speak) for each set of real numbers. Does the English plural construction have that determinate a meaning? To be sure, an advocate of the Boolos plural interpretation can *stipulate* that she intends the quantifier to have such a meaning in cases like that of real analysis. Notice, however, that one needs some set theory to do this stipulating. This consideration might sustain Michael Resnik's [1988] and my [1993] complaint that the sophisticated understanding of the plural construction used in justifying second-order logic is mediated by set theory.

4. Metatheory: Expressive Resources

This section provides brief sketches of some of the main metatheoretic properties of second-order languages. For more detail, see Shapiro [1991, chs. 4, 5].

[4] Shapiro [1991, ch. 2] invokes a distinction between logical sets and iterative sets for this purpose. For a retraction (of sorts), see Shapiro [1999].

4.1. Henkin and First-order Semantics

Since, as noted in the previous section, Henkin semantics is equivalent to first-order semantics for the languages L2*K*, we need only present the relevant results for whichever of these is most convenient. Anything said about one of them applies directly to the other.

First, the deductive system D2 is *not* sound for Henkin and first-order semantics. It is straightforward to verify that some Henkin models do not satisfy the comprehension scheme and some do not satisfy the axiom of choice. Consider, for example, a structure $M^{H} = \langle d,D,F,I \rangle$ in which the first-order domain d has two distinct members a,b; $D(2)$ has a single member, the relation $\{\langle a,a \rangle,\langle b,a \rangle\}$. Then M^{H} does not satisfy the following instance of the comprehension scheme:

$$\exists X \forall x \forall y (Xxy \equiv x = y).$$

In effect, this axiom asserts the existence of the identity relation, but M^{H} does contain such a relation. For similar reasons, the quantifier rules are not sound for Henkin and first-order semantics.

Define a Henkin model to be *faithful to* D2, or simply *faithful*, if it satisfies every instance of the comprehension scheme (and the axiom of choice, if that is included in the deductive system). That is, a Henkin model is faithful if it contains every relation definable via the comprehension scheme (and the functions promised by the axiom of choice). All subsequent discussion is restricted to faithful models. Of course, we now have soundness-by-fiat. We just restrict ourselves to Henkin and first-order models that satisfy the axioms and rules of D2.

As might be expected, languages like L2*K* under Henkin semantics and first-order semantics are pretty much like first-order languages. In a sense, they are first-order languages. First, the deductive system D2 is complete for Henkin semantics: if Γ is a set of sentences and Φ is a sentence of L2*K*, then Φ is true in every (faithful) Henkin model of Γ only if Φ can be deduced from Γ in D2. The downward Löwenheim–Skolem theorem holds: if a set Γ of formulas of L2*K* has a (faithful) Henkin model whose domain is infinite, then Γ has a (faithful) Henkin model whose domain is countable (or the cardinality of *K*, whichever is larger). Every Henkin model for L2*K* has an equivalent submodel whose domain is at most countable (or the cardinality of *K*, whichever is larger). And the upward Löwenheim–Skolem theorem holds: if for each natural number n, a set Γ has a model whose domain has at least n members, then for each infinite cardinal κ, Γ has a model whose domain has at least κ-many members. The proofs of these theorems are straightforward adaptions of the usual constructions for first-order logic (see Shapiro [1991, 89–95]).[5]

As usual for sound deductive systems, compactness is a corollary of completeness: let Γ be a set of formulas of L2*K*. If every finite subset of Γ is satisfiable in a (faithful) Henkin model, then Γ itself is satisfiable in a (faithful) Henkin model. So, as in the first-order case, no theory with an infinite Henkin-model is what may be called "Henkin-categorical." Thus, second-order languages with Henkin semantics are not adequate to characterize infinite structures up to isomorphism.

4.2. Standard Semantics

Recall that standard semantics is equivalent to the restriction of Henkin semantics to full models, structures in which the n-place predicate variables range over the collection of *all* n-tuples of elements from the domain, and the n-place functions range over the collection of all n-place functions on the domain. Since full models contain every relation and function, they certainly contain the functions and relations promised by the instances of the comprehension scheme and the axiom of choice (assuming separation and choice in the metatheory). So full models are faithful, and the deductive system D2 is sound for standard semantics.

The crucial feature of standard semantics is the existence of categorical axiomatizations of the natural numbers and the real numbers. Corollaries of these features include the refutation of compactness and the Löwenheim–Skolem theorems—and, in light of Gödel's Incompleteness Theorem for arithmetic, the refutation of completeness.

The *language of arithmetic* has $A = \{0, s, +, \cdot\}$ as its set of nonlogical terminology. The theory has first-order axioms stating that the successor function is one-to-one, that zero is not the successor of anything, and the usual recursive definitions of addition and multiplication:

$$\forall x \forall y(sx = sy \rightarrow x = y) \ \& \ \forall x(sx \neq 0)$$
$$\forall x(x + 0 = x) \ \& \ \forall x \forall y(x + sy = s(x + y))$$
$$\forall x(x \cdot 0 = 0) \ \& \ \forall x \forall y(x \cdot sy = x \cdot y + x).$$

[5] As a historical note, the now-common Henkin construction was first discovered (and reported in Henkin [1950]) in the context of higher-order logic under (what is here called) Henkin semantics. The same construction was later adapted to the proof of completeness for first-order languages. Gödel's original [1930] proof was different.

Then there is the *induction axiom*, a proper second-order statement:

$$\forall X[(X0 \,\&\, \forall x(Xx \rightarrow Xsx)) \rightarrow \forall x Xx].$$

Let AR (for "arithmetic") be the conjunction of these four axioms.

The intended interpretation of AR is the model N of L2A, whose domain is the set of natural numbers and which assigns zero to 0, and assigns the successor function, the addition function, and the multiplication function to s, $+$, and $\cdot$, respectively. It is immediate that N satisfies AR. Moreover, any two models of AR are isomorphic (see Shapiro [1991, 82–83] or the original Dedekind [1888] for a proof). It follows that if M is any (standard) model of AR, then the domain of M is denumerably infinite. This entails that the upward Löwenheim–Skolem theorem fails for second-order logic with standard semantics.

If follows from the categoricity of arithmetic that a sentence Φ of L2A is true of the natural numbers (i.e., $N \models \Phi$) if and only if AR $\rightarrow \Phi$ is a standard logical truth. That is, arithmetic truth is reducible to, or definable in terms of, standard second-order logical truth.

Let D be any effective deductive system that is sound for the language L2A (under standard semantics). Consider the set $\Gamma = \{\Phi | \vdash_D (\text{AR} \rightarrow \Phi)\}$ of sentences of L2A. Since D is effective, the set Γ is recursively enumerable, and as just noted, every member of Γ is true of the natural numbers. Gödel's [1931] incompleteness theorem entails that the collection of true first-order sentences of arithmetic is not recursively enumerable. So let Ψ be a true sentence of first-order arithmetic that is not in Γ. Then, the sentence (AR $\rightarrow \Psi$) is a (standard) logical truth, but it is not derivable in D. Thus, there is no effective, sound deductive system that is complete for standard semantics. Second-order languages with standard semantics are inherently incomplete.

Compactness also fails. Let c be a constant symbol and consider the following set Γ of sentences:

$$\{\text{AR},\ c \neq 0, c \neq s0, c \neq ss0, c \neq sss0, \ldots\}.$$

Each finite subset of Γ is satisfiable in the natural numbers. But the entire set Γ is not satisfiable, since the denotation of c would have to be different from the denotata of 0, $s0$, and so on. But, by the induction axiom and the categoricity result, the denotata of 0, $s0$, and so on exhaust the domain of each model of AR.

The general notion of finitude cannot be captured in any language which satisfies the upward Löwenheim–Skolem theorem. If a theory in such a language has a model of each finite cardinality, then it has an infinite model. However, the notion can be expressed with a second-order language with no nonlogical terminology (assuming standard semantics). The following purely logical sentence,

$$\text{(FIN)} \quad \forall f \neg(\forall x \forall y(fx = fy \rightarrow x = y) \ \&\ \exists x \forall y(fy \neq x)),$$

asserts that there is no one-to-one function from the domain to a proper subset of the domain. So (FIN) is satisfied by a model M if and only if the domain of M is Dedekind-finite.

Finitude is one of a number of related notions that have adequate characterizations in second-order languages with standard semantics (usually involving no nonlogical terminology) but cannot be characterized in any compact language, including second-order languages with Henkin or first-order semantics. Examples include countability, well-orderedness, well-foundedness, minimal closure, and the ancestral. Any consistent attempt to formulate a characterization any of these notions in a first-order or Henkin second-order theory will have unintended models that miss the mark (see Shapiro [1991, ch. 5, §5.1]).

All that remains is to refute the downward Löwenheim–Skolem theorem. In *real analysis*, the nonlogical terminology is $B=\{0,1,+,\cdot,\leq\}$. The axioms are those of an ordered field, all of which are first-order,[6] plus a second-order statement of *completeness*, asserting that every nonempty, bounded set (or property) has a least upper bound:

$$\forall X\{(\exists yXy \,\&\, \exists x\forall y(Xy \rightarrow y\leq x)) \rightarrow \exists x[\forall y(Xy \rightarrow y\leq x) \,\&\, \forall z(\forall y(Xy \rightarrow y\leq z) \rightarrow x\leq z)]\}.$$

Let AN (for "analysis") be the conjunction of the axioms of real analysis. The real number structure constitutes the intended model, and AN is categorical (see Shapiro [1991, 84]). Since AN has an uncountable model and no countable models, the downward Löwenheim–Skolem theorem fails for second-order languages with standard semantics.

In first-order arithmetic, the (second-order) induction principle is replaced by a scheme. If Φ is a formula in the language L1A= *of first-order arithmetic*, then

$$(\Phi(0) \,\&\, \forall x(\Phi(x) \rightarrow \Phi(sx))) \rightarrow \forall x\Phi(x),$$

is an axiom of first-order arithmetic. The theory thus has infinitely many axioms. Similarly, first-order real analysis is obtained by replacing the single completeness axiom with the *completeness scheme*,

$$(\exists x\Phi(x) \,\&\, \exists x\forall y(\Phi(y) \rightarrow y\leq x)) \rightarrow \exists x[\forall y(\Phi(y) \rightarrow y\leq x) \,\&\, \forall z(\forall y(\Phi(y) \rightarrow y\leq z) \rightarrow x\leq z)],$$

[6] The axioms for an ordered field are that addition and multiplication are associative and commutative, multiplication is distributive over addition, 0 is the additive identity, 1 is the multiplicative identity, $0\neq 1$, every element has an additive inverse, every element but 0 has a multiplicative inverse, $\leq$ is a linear order, $0\leq 1$, and the elements greater than or equal to 0 are closed under addition and multiplication.

one instance for each formula Φ of the language L1B= of real analysis that contains neither x nor z free.

The difference between, say, second-order real analysis and its first-order counterpart is that in the latter, one cannot directly state that every non-empty bounded set (or property) has a least upper bound. The closest one can come is a separate statement for each such set which is definable by a formula in the language of first-order analysis.

The same, or almost the same, goes for arithmetic and real analysis as formulated in a second-order language with Henkin semantics. To be sure, in those languages, the induction principle and the completeness principle are single sentences, with a second-order variable ranging over sets or properties. However, the only sets or properties that are guaranteed to exist in a given model are the definable ones—and we have those only thanks to the explicit restriction to faithful models, interpretations that satisfy each instance of the comprehension scheme. Without the restriction to faithful models, we are not guaranteed the existence of *any* particular sets or properties. One who works in first-order arithmetic or analysis, or one invokes Henkin semantics on the respective second-order language (restricted to faithful models), cannot apply the induction or completeness principle to a set or property until she shows that it is definable in the relevant language. The only difference between first-order arithmetic and second-order arithmetic with Henkin semantics is that a few more properties are definable in the second-order language than the first-order language.

The first-order theory and the Henkin theory have models which are not isomorphic to the real numbers. These are sometimes called *non-standard models.* Indeed, the Löwenheim–Skolem theorems indicate that for every infinite cardinal κ, there are models of first-order arithmetic and models of first-order analysis whose domain has cardinality κ. The same goes for Henkin semantics. The study of nonstandard models has proven fruitful in illuminating the original informal theories.

Per Lindström [1969] showed that, in a sense, first-order logic is *characterized* by the metatheoretic properties that distinguish it from second-order languages with standard semantics. Let L be any logic that is compact and has the property of the downward Löwenheim–Skolem theorem: if a theory is satisfiable, then it has a model whose domain is at most countably infinite. Then the language L cannot make any distinction among models that cannot be made with the corresponding first-order language (see Shapiro [1991, ch. 6, §6.5]).

For better or worse, then, standard semantics is what makes second-order logic distinctive. The categoricity results and the concomitant failure of the limitative properties are the source of both the expressive strength and the main philosophical and technical shortcoming of second-order logic.

5. Logic Choice: What Are We to Make of All This?

W. V. O. Quine is a long-standing and persistent critic of second-order logic. An early paper, Quine [1941] is critical of systems, like that of Whitehead and Russell [1910], in which the second-order variables range over intensional entities like properties, propositional functions, or attributes. As is well-known, Quine is skeptical of such entities. There is no consensus on which properties exist, nor on conditions under which two properties are identical or distinct. According to Quine's slogan "no entity without identity," one is not entitled to speak of a kind of item unless there is a clear and determinate criterion of identity on the items. So a Quinean would frown upon the present decision to forgo an identity relation on the items in the range of second-order variables, claiming that this just postpones the problem. At some point, we have to say *something* about what the variables range over, and this would include giving an identity relation.

Quine [1941] went on to suggest that variables ranging over properties be replaced with variables ranging over respectable extensional entities like sets. However, he later argued that by invoking sets in this way, we have crossed the boundary out of logic and moved into mathematics proper (Quine [1986, ch. 5]). He refers to (so-called) second-order logic as set theory in disguise. One might think that nothing turns on this. Unlike political borders, we can set and move the boundaries between disciplines at will. The more substantial issue concerns the nature of logic itself. We return to the bad joke mentioned at the beginning of this chapter.

The variables of formal languages, of any order, can range over any type of entity, so long as there are coherent things to say about them. Readers who share Quine's qualms about intensional entities are free to follow his lead and interpret higher-order variables as ranging over extensional things, like sets. To be sure, some philosophers doubt the existence of abstract objects like sets, but Quine is not among them (once he went beyond Goodman and Quine [1947]). Such nominalists will not have much use for higher-order languages as interpreted by either standard semantics or Henkin semantics, since the higher-order variables are interpreted as ranging over sets. Some nominalists attempt to gain the benefit of second-order languages, with standard semantics, by invoking the plural interpretation indicated in section 3.4 above (see Lewis [1993]). So understood, the higher-order variables do not have a distinctive range at all. But even this model theory, as well as that of ordinary first-order languages, is formulated in set theory. Models are themselves set-theoretic constructions. Thus our nominalist must either eschew model theory altogether or come up with an acceptable alternative. I will not address these issues further here, and simply declare a lack of interest in making higher-order logic safe for nominalism. I will freely invoke set theory, at least in the model-theoretic metatheory. Whatever the items in the range of higher-order variables are taken to be, they are represented by sets of n-tuples in the various models in the model theory.

As shown above, second-order languages with Henkin semantics are essentially the same as multisorted first-order languages. Quine may be right that there is something misleading about this, but so long as the theorist keeps the proper perspective, there is no serious objection to these semantics. However, with Henkin semantics (restricted to faithful models), the only relations/sets that are assumed to exist are those relations/sets *definable* in the language, and perhaps some choice functions. The distinctive expressive resources of higher-order logic come with standard semantics, where we assume that the second-order variables range over *every* relation/set and every function on the domain. This is the real target of contemporary criticisms of higher-order logic, including Quine's and I. Jané's companion article in the present volume (chapter 26).

Quine [1986, 68] wrote that "[s]et theory's staggering existential assumptions are... hidden... in the tacit shift from schematic predicate letter to quantifiable set variable." This, however, is an exaggeration. An advocate of second-order logic is not thereby committed to the entire iterative hierarchy. Suppose that we are considering a first-order theory whose intended model has an infinite domain of size κ. In the corresponding second-order theory, construed with standard semantics, each predicate variable ranges over 2^{κ}-many items. This, of course, is larger than κ, but it is hardly "staggering." For example, first-order arithmetic is already committed to a countably infinite universe. The total "ontology" of second-order arithmetic, on standard semantics, is the size of the continuum.

To be sure, the *metatheory* for second-order logic is a rich set theory, with axioms of infinity and power set. The presupposition of standard semantics is that each set x has a sufficiently determinate power set, in which we interpret the range of higher-order variables when x is the domain of discourse. Critics of second-order logic can and do challenge this assumption. The underlying issue concerns the extent to which this staggering metatheory compromises the status of second-order logic itself.

I might add that a lot of substantial mathematics is bound up with second-order logic with standard semantics. In section 4.2 above, for example, we saw that arithmetic truth is reducible to standard logical truth, in the sense that for any sentence Φ in the language of arithmetic, Φ is true (of the natural numbers) if and only if $AR \rightarrow \Phi$ is a standard logical truth, where AR is a categorical characterization of arithmetic. Something similar goes for any theory that has a categorical axiomatization. Truth-in-real-analysis, truth-in-functional-analysis, and truth-in-the-first-inaccessible-rank are each reducible to standard second-order logical truth.

Moreover, a lot of set theory can be expressed directly in a second-order language, with no nonlogical terminology. For example, there is a sentence C that is a standard logical truth if and only if the continuum hypothesis holds, and there is another sentence D that is a logical truth if and only if the generalized continuum hypothesis holds (see Shapiro [1991, ch. 5, §5.1.2]). We saw above (section 2) that there is a sentence that expresses the axiom of choice, at least within a given domain of discourse. It is perhaps counterintuitive to hold that such principles are "logical."

It would be ironic for a Quinean to dismiss higher-order logic on the grounds that it recapitulates a fair amount of mathematics. Quine is famous for championing the thesis that there no sharp borders between disciplines. He argues that the sum total of our beliefs forms a seamless web, with mathematics thoroughly enmeshed. There is no sharp distinction—no difference in kind—between mathematics and any respectable science, such as chemistry. As a naturalist, Quine holds that epistemology should be a chapter of psychology. Logic, presumably, is a branch of epistemology. So why should logic, especially the logic *of mathematics*, be an exception to the seamlessness of the web of belief? Given the Quinean themes, why should one expect logic to be free of substantial mathematical resources and mathematical ontology? No other science is.

The dispute surely concerns the nature and purpose of logic. I submit, however, that there is no single purpose for logic to serve. One traditional goal of logical study is to present a *calculus*, a deductive system that represents the canons of correct inference. The idea is that a given proposition Φ is a consequence of a set Γ only if it is possible to justify Φ on the basis of Γ alone. This conception of logic is behind the Fregean goal of giving proofs of various mathematical propositions that are free of gaps (see chapters 5–6 of the present volume). No intuition and no empirical, or otherwise substantial, knowledge should be required in proving theorems on the basis of premises or axioms.

A famous and influential logician once told me that second-order logic presupposes that we know just about every truth of mathematics. He took this as a reductio ad absurdum against the project. We saw above (section 4.2) that every truth of arithmetic is a model-theoretic consequence of the second-order axiomatization of arithmetic. Similarly, every truth of real analysis is a model-theoretic consequence of the axioms of the theory. The same goes for just about any mathematical theory with an intended interpretation, short of set theory itself. It seems that whenever mathematicians speak informally of a theory (short of set theory) that has a single intended model, up to isomorphism, then that intended structure has a categorical characterization in a second-order language with standard semantics.

So if the model-theoretic notion of second-order validity did meet the traditional epistemic goal of capturing gap-free chains of inference, idealized versions of ourselves would indeed know every arithmetic truth, every truth of real analysis, and so on. This does seem to be an unreasonable assumption. Even if it is true that "we" can know every truth, it is not wise to have one's logical theory presuppose it.

The fact is that the model-theoretic consequence relation of second- or higher-order logic with standard semantics does not meet this traditional goal of providing a calculus. For one thing, standard-validity is defined in terms of models, not inference patterns. Of course, that much applies to any model-theoretic system, including that of first-order languages and second-order languages with Henkin semantics. In those cases, however, a completeness theorem links the model-theoretic consequence relation with the deductive one, and this provides a connection with the epistemic goal of codifying correct inference patterns. In contrast, the incompleteness of second-order logic shows that the consequence relation of second-order logic, with standard semantics, is highly noneffective (section 4.2 above). Any effective deductive system is either unsound (in that it sanctions an invalid argument) or fails to codify at least one valid argument.

The study of canons of inference does not exhaust the traditional scope of logic. It is widely (but not universally) held that deductive systems must themselves adhere to a prior notion of logical consequence, often defined in modal terms. Near the beginning of the *Prior Analytics*, Aristotle defines a "syllogism" to be "a discourse in which, certain things having been supposed, something different from the things supposed results *of necessity*. . . ." Prior to reflection, there is no reason to expect that the pretheoretic, modal notion of logical consequence is effective, and that there is a complete deductive system for it.

I submit that model-theoretic consequence is an attempt to capture this pretheoretic, modal notion of logical consequence. The realm of models represents the possible interpretations, or meanings, of the nonlogical terminology. To say that a sentence Φ is a consequence of a set Γ is to say that Φ comes out true under every interpretation of the nonlogical terminology on which every member of Γ is true.[7] This is consonant with the slogan that logical consequence is a matter of "form." What must be determined, of course, is what counts as a nonlogical term, and what the realm of interpretations is. From this perspective, the issue of standard semantics is whether the membership (or predication) relation and bound variables ranging over sets (or relations) is logical. Alfred Tarski once wrote:

> . . . sometimes it seems . . . convenient to include mathematical terms, like the [membership] relation, in the class of logical ones, and sometimes I prefer to restrict myself to terms of "elementary [i.e., first-order] logic." Is any problem involved here? (Tarski [1987, 29])

From this eclectic perspective, the question now concerns the theoretical benefits of standard semantics. What is to be gained by adopting this highly noneffective consequence relation? I submit that standard semantics captures important

[7] See chapter 21 of this volume or Shapiro [1998] for an elaboration of this conception of logical consequence.

aspects of mathematical practice, aspects that are relevant to the study of correct inference, since they register the intended meanings of mathematical terms.

It is widely agreed that mathematicians succeed in describing and communicating information about various notions and structures. For example, most thinkers hold that when mathematicians refer to "the natural numbers," "the real numbers," and so on, they are talking about the same objects, or at least the same structures up to isomorphism. Along the same lines, there is not much doubt that notions like finitude and well-foundedness are clear and unequivocal, or that the minimal closure and ancestral relations are determinate. In short, there is almost a consensus that the informal language of mathematics has expressive resources sufficient for the ordinary description and communication of these basic structures and notions.

Part of the purpose of model theory is to delineate structures that are possible interpretations of various sentences containing nonlogical parameters. This is what it is to capture the *semantics* of informal mathematical languages. Using whatever linguistic resources are at their disposal, mathematicians (supposedly) manage to describe the natural number structure up to isomorphism, and they describe and communicate concepts like finitude and well-foundedness. So the expressive resources of the formal languages should match those of the mathematical discourse it models. As noted, first-order languages are inadequate to this task. As Jon Barwise [1985, 5] put it:

> As logicians, we do our subject a disservice by convincing others that logic is first-order and then convincing them that almost none of the concepts of modern mathematics can really be captured in first-order logic.

And Hao Wang [1974, 154]:

> When we are interested in set theory or classical analysis, the Löwenheim–Skolem theorem is usually taken as a sort of defect (often thought to be inevitable) of the first-order logic.... [W]hat is established [by Lindström's theorems] is not that first-order logic is the only possible logic but rather that it is the only possible logic when we in a sense deny reality to the concept of uncountable....

See also Montague [1965], Corcoran [1980], Isaacson [1985], and Shapiro [1991, ch. 5].

To be sure, the aforementioned near consensus is not a consensus. Some prominent thinkers hold that there are no unambiguous notions of "finite," "countable," "natural number," and so on (e.g., Skolem [1922], Field [2001, chs. 11–12]). Such skeptical views are sometimes supported by the Löwenheim–Skolem theorems, and for that reason are dubbed "Skolemite relativism." On such views, the model theory of first-order logic (or Henkin semantics for higher-order languages) reflects the situation of the mathematician. Standard semantics for second-order logic is deceptive, since it assumes that there is an unequivocal understanding of locutions like "all properties" or "all subsets," and it is this that gives rise to the equally misleading categoricity results.

Be this as it may, the majority of the mathematicians, logicians, and philosophers (including Quine) who reject higher-order logic also reject this Skolemite skepticism. Such thinkers, it seems, hold that there is a crucial gap between the informal language in which mathematics is done and the formal semantics that underlies this language (see Myhill [1951]). Logic must fail where informal language succeeds.

Jodi Azzouni [1994, ch. 1, sec. 3] claims that advocates of second-order logic, such as myself, claim that standard semantics *solves* the age-old problem of reference to mathematical objects (at least up to isomorphism) and that it somehow *explains* how mathematicians communicate with each other. Azzouni then points out that standard semantics cannot solve either problem (see also Wagner [1987] and Weston [1976]). Nonstandard, unintended interpretations are ruled out of the system by fiat—nonstandard models are simply not part of the semantics. Azzouni concludes that "second-order logic with standard semantics is *treacherous*: Its sirenlike notation can lull philosophers into an inadequate appreciation of how it gains its expressive power" (1994, p. 18).

I will leave it to the reader to determine who among the defenders of second-order logic are in fact lulled into a trance in this way. Most, I believe, are not blind to the connection with mathematics proper, and most advocates of higher-order logic do not make unreasonable claims on what is involved in understanding second-order languages. Alonzo Church, for example, was not lulled into Azzouni's trance, as indicated by the epigraph to this chapter:

> ... our definition of the [standard second-order] consequences of a system of postulates ... can be seen to be not essentially different from [that] required for the ... treatment of classical mathematics.... It is true that the non-effective notion of consequence, as we have introduced it ... presupposes a certain absolute notion of ALL propositional functions of individuals. But this is presupposed also in classical mathematics, especially classical analysis.... (Church [1956, 326n])

The key term here is "presupposed." The thesis that we understand second-order languages with standard semantics is of-a-piece with the thesis that we understand ordinary mathematical discourse. It is no less—and no more—problematic. The substantial claim of advocates of higher-order logic is that standard semantics has an important role in foundational studies, but, again, this role is not that of explaining or justifying mathematical practice or of explaining how successful reference and communication works. Church added that "... logic and mathematics should be characterized, not as different subjects, but as elementary and advanced parts of the same subject."

I close with some epistemological features of the second-order axiomatizations of mathematical theories (assuming standard semantics) noted by Georg Kreisel [1967]. Recall that first-order axiomatizations typically replace a single, second-order axiom with an axiom scheme. The scheme itself has no special status, since it does not occur in the object-language. It is only used to describe the theory in a finite number of words (in the metalanguage, or perhaps the meta-metalanguage). In the theory itself, as formulated in the object-language, each instance of the scheme is a (separate) axiom. Kreisel argued that this is an unnatural way to codify a mathematical theory like that of arithmetic or real analysis. Suppose, for example, that someone is asked why he believes that each instance of the completeness scheme of first-order real analysis is true of the real numbers. It is, of course, out of the question to give a separate justification for each of the axioms. Nor can one claim that the scheme characterizes the real numbers since, as we have seen, no first-order axiomatization can characterize any infinite structure. Kreisel argued that the reason mathematicians believe the instances of the axiom scheme is that each instance follows from the single second-order completeness *axiom* (via the corresponding instance of comprehension).

A related issue is that each first-order scheme or, to be precise, the collection of instances of each scheme, is tied to the ingredients of the particular first-order language in use at the time. As noted, this is to restrict the induction and completeness principles to those properties or sets definable in the given language. Mathematicians, however, are quick to apply the induction and completeness principles to sets regardless of whether they are definable in the given first-order language. Indeed, they usually do not check for definability in this or that language. Kreisel [1967, 148]) wrote:

> The choice of the first-order schema is not uniquely determined by the second-order axiom! Thus, Peano's own axioms mention explicitly only the constant 0 and the successor function, ... not addition nor multiplication. The first-order schema built up from 0 and [the successor function] is a weak, ... decidable, subsystem of classical first-order arithmetic ... and quite inadequate for formulating current informal arithmetic.

The same point applies to second-order languages understood with Henkin or first-order semantics. Although such theories do have a single induction or completeness axiom, the opening higher-order quantifiers are implicitly restricted to those properties or sets that are present in a given model of the theory. And the only properties or sets present in every such model are those definable in the respective language in the model (and we have those properties or sets only with the restriction to faithful models).

The issue concerning the interpretation of the induction and completeness principles is manifest in the practice of embedding structures in each other. Kreisel wrote:

> ... very often the mathematical properties of a domain D become only graspable when one embeds D in a larger domain D'. Examples: (1) D integers, D' complex plane; use of analytic number theory. (2) D integers, D' p-adic numbers; use of p-adic analysis. (3) D surface of a sphere, D' 3-dimensional space; use of 3-dimensional geometry. Non-standard analysis [also applies] here .... (Kreisel [1967, 166])

To follow the first example (modified slightly), when one sees that there is a structure isomorphic to the natural numbers in the complex plane, one can use complex analysis to shed light on the natural numbers. This works by applying the induction principle of arithmetic to sets of natural numbers definable in complex analysis, whether or not such sets can be defined in the language of arithmetic alone. One cannot tell in advance what resources are needed to shed light on a mathematical structure—even after the structure has been adequately characterized. The incompleteness theorem indicates that some theorems about the natural numbers proved in a rich theory, like complex analysis, are not provable in arithmetic alone.

In the case at hand, the induction principle is applicable simply because the set in question is a set of natural numbers (or is isomorphic to such a set). And the result is a theorem about the natural numbers. It would be a mistake to refuse to apply induction, simply because one has not defined the property in the language of arithmetic alone. This would run against the very nature of the natural numbers. As Shaughan Lavine [1994, 231 n. 24] put it:

> Part of what it is to define a property of natural numbers is to be willing to extend mathematical induction to it. To fail to do so is to violate our rules for extending and further specifying our arithmetical usage.

Although, like Lavine, Dummett is no friend of higher-order logic, he echoes the same idea:

> It is part of the concept of natural number, *as we now understand it*, that induction with respect to any well-defined property is a ground for asserting all natural numbers to have that property. (Dummett [1994, 337]; emphasis added)

The second-order induction axiom understood with standard semantics—and not the first-order induction scheme and not Henkin or first-order semantics of second-order languages—captures this feature of arithmetic truth. The same goes for the second-order completeness principle in analysis and just about every rich mathematical theory.

Acknowledgments I thank Ignacio Jané for reading an earlier draft of this chapter and providing insightful comments. I appreciate the spirit of collegiality.

REFERENCES AND FURTHER READING

Azzouni, J. [1994], *Metaphysical myths, mathematical practice*, Cambridge, Cambridge University Press.

Barwise, J. [1985], "Model-theoretic logics: Background and aims," in Barwise and Feferman [1985], 3–23.

Barwise, J., and S. Feferman (editors) [1985], *Model-theoretic logics*, New York, Springer-Verlag. (Extensive treatment of a lot of logics, many of which are intermediate between first-order and second-order. Quite technical, for the most part. Extensive bibliography.)

Boolos, G. [1984], "To be is to be a value of a variable (or to be some values of some variables)," *Journal of Philosophy* 81, 430–450; reprinted in Shapiro [1996], 331–350. (Initial account of plural quantification.)

Boolos, G. [1985], "Nominalist Platonism," *Philosophical Review* 94, 327–344; reprinted in Shapiro [1996], 351–368. (Another defense of plural quantification.)

Burgess, J. [1993], Review of Shapiro [1991], *Journal of Symbolic Logic* 58, 363–365.

Church, A. [1956], *Introduction to mathematical logic*, Princeton, NJ: Princeton University Press. (Still an excellent introduction to both first-order logic and second-order logic.)

Cocchiarella, N. [1988], "Predication versus membership in the distinction between logic as language and logic as calculus," *Synthèse* 77, 37–72; reprinted in Shapiro [1996], 473–508.

Corcoran, J. [1980], "Categoricity," *History and Philosophy of Logic* 1, 187–207; reprinted in Shapiro [1996], 241–261.

Dedekind, R. [1888], *Was sind und was sollen die Zahlen?*, Brunswick, Vieweg; translated as *The nature and meaning of numbers*, in *Essays on the theory of numbers*, edited by W. W. Beman, New York, Dover Press, 1963, 31–115.

Dummett, M. [1994], "Reply to Wright," in *The philosophy of Michael Dummett*, edited by B. McGuinness and G. Oliveri, Dordrecht, Kluwer Academic Publishers, 329–338.

Eklund, M. [1996], "How logic became first-order," *Nordic Journal of Philosophical Logic* 1, 147–167.

Field, H. [2001], *Truth and the absence of fact*, Oxford, Oxford University Press.

Frege, G. [1879], *Begriffsschrift, eine der arithmetischen nachgebildete Formelsprache des reinen Denkens*, Halle, Louis Nebert; translated in van Heijenoort [1967], 1–82.

Gilmore, P. [1957], "The monadic theory of types in the lower-predicate calculus," in *Summaries of talks presented at the Summer Institute of Symbolic Logic at Cornell*, Washington, DC, Institute for Defense Analyses, 309–312.

Gödel, K. [1930], "Die Vollständigkeit der Axiome des logischen Funktionenkalkuls," *Montatshefte für Mathematik und Physik* 37, 349–360; translated as "The completeness of the axioms of the functional calculus of logic," in van Heijenoort [1967], 582–591.

Gödel, K. [1931], "Über formal unentscheidbare Sätze der *Principia Mathematica* und verwandter Systeme I," *Montatshefte für Mathematik und Physik* 38, 173–198; translated as "On formally undecidable propositions of the *Principia Mathematica*," in van Heijenoort [1967], 596–616.

Goodman, Nelson, and W.V.O. Quine [1947], "Steps towards a constructive nominalism," *Journal of Symbolic Logic* 12, 105–122.

Hazen, A. [1983], "Predicative logics," in *Handbook of philosophical logic*, vol. 1, edited by D. Gabbay and F. Guenthner, Dordrecht, D. Reidel, 331–407. (Readable account and defense of ramified type theory.)

Henkin, L. [1950], "Completeness in the theory of types," *Journal of Symbolic Logic* 15, 81–91.

Hilbert, D., and W. Ackermann [1928], *Grundzüge der theoretischen Logik*, Berlin, Springer; translated as *Principles of mathematical logic* by L. Hammond, G. Leckie, and F. Steinhardt, New York, Chelsea, 1950. (Historical source. Still quite readable.)

Isaacson, D. [1985], "Arithmetical truth and hidden higher-order concepts," in *Logic Colloquium* 85, edited by The Paris Logic Group, Amsterdam, North-Holland, 1987, 147–169.

Jané, I. [1993], "A critical appraisal of second-order logic," *History and Philosophy of Logic 14*, 67–86; reprinted in Shapiro [1996], 177–196.

Kreisel, G. [1967], "Informal rigour and completeness proofs," in *Problems in the philosophy of mathematics*, edited by I. Lakatos, Amsterdam, North-Holland, 138–186. (Among other things, a multifaceted philosophical defense of the use of higher-order languages in foundational studies. Moderate difficulty.)

Lavine, S. [1994], *Understanding the infinite*, Cambridge, MA, Harvard University Press.

Lewis, D. [1993], "Mathematics is megethology," *Philosophia Mathematica*, 3rd ser., 1, 3–23.

Lindström, P. [1969], "On extensions of elementary logic," *Theoria* 35, 1–11.

Löwenheim, L. [1915], "Über Möglichkeiten im Relativkalkül," *Mathematische Annalen* 76, 447–479; translated in van Heijenoort [1967], 228–251.

Montague, R. [1965], "Set theory and higher-order logic," in *Formal systems and recursive functions*, edited by J. Crossley and M. Dummett, Amsterdam, North-Holland, 131–148. (Higher-order logic, put to linguistic use. The start of an extensive research program.)

Moore, G.H. [1980], "Beyond first-order logic: The historical interplay between logic and set theory," *History and Philosophy of Logic* 1, 95–137; reprinted in Shapiro [1996], 3–45.

Moore, G.H. [1988], "The emergence of first-order logic," in *History and philosophy of modern mathematics*, edited by W. Aspray and P. Kitcher, Minnesota Studies in the Philosophy of Science, 11, Minneapolis, University of Minnesota Press, 95–135.

Myhill, J. [1951], "On the ontological significance of the Löwenheim–Skolem theorem," in *Academic freedom, logic and religion*, edited by M. White, Philadelphia, American Philosophical Society, 57–70; also in *Contemporary readings in logical theory*, edited by I. Copi and J. Gould, New York, Macmillan, 1967, 40–54.

Peano, G. [1889], *Arithmetices principia, nova methodo exposita*, Turin, Bucca, translated in van Heijenoort [1967], 85–97.

Quine, W.V.O. [1941], "Whitehead and the rise of modern logic," in *The philosophy of Alfred North Whitehead*, edited by P. A. Schilpp, New York, Tudor, 127–163.

Quine, W.V.O. [1986], *Philosophy of logic*, 2nd ed., Cambridge, MA, Harvard University Press. (Argument against second-order logic, especially ch. 5.)

Rayo, A., and S. Yablo [2001], "Nominalism through de-Nominalization," *Noûs* 35, 74–92.

Resnik, M. [1988], "Second-order logic still wild," *Journal of Philosophy* 85, 75–87.

Shapiro, S. [1991], *Foundations without foundationalism: A case for second-order logic*, New York, Oxford University Press. (Extensive development of and justification for higher-order logic. Philosophical, mathematical, and historical treatment. Most of the book requires some background in logic. Extensive bibliography.)

Shapiro, S. [1993], "Modality and ontology," *Mind* 102, 455–481.

Shapiro, S. (editor) [1996], *The limits of logic: higher-order logic and the Löwenheim–Skolem theorem*, Aldershot, UK, Dartmouth Publishing Company. (Reprints of previously published articles.)

Shapiro, S. [1998], "Logical consequence: Models and modality," in *The philosophy of mathematics today*, edited by M. Schirn, Oxford, Oxford University Press, 131–156.

Shapiro, S. [1999], "Do not claim too much: Second-order logic and first-order logic," *Philosophia Mathematica*, 3rd ser., 7, 42–64. (Reply to some of the critics of Shapiro [1991].)

Shapiro, S. [2001], "Systems between first-order logic and second-order logics," in *Handbook of philosophical logic*, vol. 1, 2nd ed., edited by D.M. Gabbay and F. Geunthner, Dordrecht, Kluwer Academic Publishers, 131–187.

Skolem, T. [1922], "Einige Bemerkungen zur axiomatischen Begründung der Mengenlehre," in *Matematikerkongressen i Helsingfors den 4–7 Juli 1922*, Helsinki, Akademiska Bokhandeln, 217–232; translated as "Some remarks on axiomatized set theory" in van Heijenoort [1967], 291–301.

Tarski, A. [1987], "A philosophical letter of Alfred Tarski," edited by Morton White, *Journal of Philosophy* 84, 28–32.

Tharp, L. [1975], "Which logic is the right logic?," *Synthèse* 31, 1–31; reprinted in Shapiro [1996], 47–67.

Van Heijenoort, J. (editor) [1967], *From Frege to Gödel*, Cambridge, MA, Harvard University Press. (Collection of many original papers in logic, translated into English. Includes works by Frege, Peano, Dedekind, Cantor, Hilbert, Russell, Zermelo, Löwenheim, Skolem, Weyl, Ackermann, von Neumann, Bernays, and Gödel.)

Wagner, S. [1987], "The rationalist conception of logic," *Notre Dame Journal of Formal Logic* 28, 3–35.

Wang, H. [1974], *From mathematics to philosophy*, London, Routledge and Kegan Paul.

Weston, T. [1976], "Kreisel, the continuum hypothesis and second-order set theory," *Journal of Philosophical Logic* 5, 281–298; reprinted in Shapiro [1996], 385–402.

Whitehead, A.N., and B. Russell [1910], *Principia mathematica*, vol. 1, Cambridge, Cambridge University Press.

CHAPTER 26

HIGHER-ORDER LOGIC RECONSIDERED

IGNACIO JANÉ

In this chapter we discuss canonical (i.e., full, or standard) second-order consequence and argue against it being a case of logical consequence. Our discussion is divided into three parts.

The first part comprises the first three sections. After stating the problem in section 1, we devote sections 2 and 3 to examining the role that the consequence relation is expected to play in axiomatic theories. This leads us to put forward two requirements on logical consequence, which we call "formality" and "noninterference." It is this last requirement that canonical second-order consequence violates, as we set out to substantiate.

This we do in the second part, which consists of sections 4 to 6. In section 4 we argue that canonical second-order logic is inadequate for axiomatizing set theory, on the grounds that it codes a significant amount of set-theoretical content. Objections to our criticism are considered, in particular that we conflate semantical with deductive consequence, that even first-order logic is set-theoretically laden, and that we overlook the distinction between logical and iterative sets. The last two objections are met in sections 5 and 6, respectively.

The third part, which runs from section 7 to the end, deals with the issue of whether canonical second-order consequence is a determinate relation. In section 7 we emphasize the claim that it rests on the assumption that many undecided set-theoretical questions have a definite answer. From this we argue that unless we view canonical second-order consequence as an applied branch of set theory, its justified use requires embracing some strong form of philosophical realism. In section 8 we maintain that, in a relevant way, the use of canonical second-order consequence requires a stronger ontological commitment than set theory, and we sketch a non-Platonist account of the power set operation which, we submit, is suitable for set theory but not for the autonomous use of second-order consequence. As we point out in section 9, this account casts doubt on the import of the categoricity results for second-order languages. After a discussion of Boolos's plural interpretation of second-order quantification in section 10, the chapter ends with the suggestion that the existence of a complete calculus is an essential requirement of any fully determined, self-sufficient consequence relation, and thus of logical consequence—whatever it may reasonably be.

1. The Question

We deal with first- and second-order formal languages, mainly from the point of view of their use in the formalization of actual mathematical theories. Each language is supposed to be endowed with a definite consequence relation which we take to be an integral part of the language. This means that in order to identify a language, it is not enough to list its symbols and give the rules for generating its terms and formulas. It must be determined as well which sentences follow from which sentences and from which sets of sentences; in other words, it must be determined what particular relation the consequence relation is. In the case of first-order languages we have two equivalent ways to describe the consequence relation: by means of a deductive calculus and through Tarski's definition. For second-order languages canonically understood, the first option is not open.

We give a sketchy account of second-order languages, intended only to fix the setting and any terminological peculiarity. For a brief and clear exposition, refer to Enderton [11, ch. IV]; and for a thorough presentation to Shapiro [27, part II].

Second-order languages are extensions of first-order ones. Syntactically, the extension consists in the addition of n-ary predicate variables for each positive integer n. No new terms are added, but the set of formulas is expanded: for each n-ary variable X and n terms $t_1, \ldots, t_n$, the expression $Xt_1 \ldots t_n$ is a new atomic formula. From the atomic formulas, the rest are generated by the same rules as in first-order languages, taking into account that the new predicate variables are also quantifiable.

An interpretation for a second-order language has two components: (1) a structure $\mathcal{A}$ for the similarity type of the language, which is the same as its first-order restriction, and (2) for each positive integer n, a nonempty set D_n of n-ary relations on the universe of the structure, over which the n-ary predicate variables range (a 1-ary relation is a subset of the universe). This being enough to define satisfaction and truth, we shall speak freely of a second-order sentence being true in a given interpretation. We call an interpretation whose first component is the structure $\mathcal{A}$ an *interpretation over* $\mathcal{A}$. A *model* of a set of sentences is an interpretation in which all the sentences in the set are true.

Some conditions can be imposed on the interpretations over a structure $\mathcal{A}$ with universe A, thereby decreeing which of them are *admissible.* In the *full* or *canonical* interpretation over $\mathcal{A}$, each D_n contains all n-ary relations on A; in particular D_1 is the power set of A. In a *Henkin interpretation* one requires that the D_n be closed under impredicative comprehension, that is, that all the comprehension axioms be true in the interpretation.[1] One may also demand that some further conditions be met, provided that they are statable by second-order formulas or schemas. The canonical interpretation is also a Henkin interpretation, but if A is infinite, there are other Henkin interpretations over $\mathcal{A}$.[2] When dealing with the canonical interpretation there is no need to mention the D_n, because they are determined by A. Accordingly, we simply say that a sentence is true, or false, in the structure $\mathcal{A}$.

Consequence for second-order languages is defined in the Tarskian way. A sentence φ is a second-order consequence of a set of sentences Σ if and only if φ is true in all admissible models of Σ. What consequence relation this defines depends on which interpretations are admissible. Accordingly, we speak of the *canonical second-order consequence* relation, and of the various *Henkin consequence* relations (they vary according to which formulas the admissible interpretations are required to satisfy in addition to the comprehension axioms).

Many central mathematical structures and concepts can be characterized in second-order languages with the canonical semantics.[3] We can characterize, each with finitely many formulas, the closure of a set under a finite number of given operations—thus the structure of the natural numbers with the successor operation—the concept of well-ordering, the field of the real numbers, and so on. In contrast to this strong expressive power, canonical second-order logic (the logic of second-order languages with the canonical semantics) lacks those properties that have made work on first-order logic so fruitful. It lacks the Löwenheim–Skolem property, it is not compact, it admits no complete proof procedure, not even for universally valid sentences.

On the other hand, the logic of second-order languages endowed with Henkin semantics has all these properties in common with first-order logic. For it, both

[1] An n-ary comprehension axiom is the universal closure of a formula of the form

$$\exists X \forall x_1 \ldots x_n (X x_1 \ldots x_n \leftrightarrow \varphi),$$

where φ is a formula, $n \geq 1$, and X is a n-ary predicate variable not occurring free in φ.

[2] There are other natural possibilities for admissibility (e.g., that each D_n consists of all the finite, or all countable, or all first-order definable n-ary relations on A). But we won't deal with any of these save in examples, since our central concern is the canonical interpretation.

[3] See Shapiro [26] and [27, pp. 97–109].

the upward and downward Löwenheim–Skolem theorems hold, it is compact, and it has a complete proof procedure. Indeed, Henkin second-order consequence can be reduced to first-order consequence.[4]

We deal with the question of whether the canonical consequence relation of second-order languages is a case of logical consequence. We will argue that it is not. Of course, in order to do that, we must have some idea of what counts as *logical* consequence, but it is worth remarking that we don't have to bring in a fully precise notion of logical consequence in order to be able to answer the question in the negative. For this, it is enough to agree on one or more features that the consequence relation of a language should possess if it is to be considered logical and that second-order consequence lacks.

We don't mean to look for some alleged true nature of logic or of logical consequence in order to find these features. Instead, we follow what we may call a methodological approach by turning our attention to a context where it matters what we take logical consequence to be. Such a context is axiomatization. In discussing axiomatic theories, the notion of consequence is brought to the forefront, since it is needed to identify the theory we are dealing with. An axiomatic theory is given by a list of axioms in a certain language, but usually what matters about it are less the particular axioms than their consequences, as is obvious when we speak of different axiomatizations of the same theory. The theory consists of the consequences of the axioms, which means that the consequence relation at work is no less fundamental for the identification of the theory than the axioms.[5]

One advantage of looking at axiomatic theories for the purpose of assessing the role of the consequence relation is that in many of them the distance from the

[4] More precisely, the Henkin second-order consequences of a set of premises in a language of a given similarity type are just the first-order consequences of an enlarged set of premises in a language in some expanded similarity type. Let's restrict to monadic second-order languages (i.e., let's assume that all predicate variables are unary). Syntactically, a second-order monadic language can be viewed as a two-sorted first-order language with one hidden binary relation symbol, say E, represented by concatenation. We may make it explicit by rewriting each atomic formula Xt as EtX. For each second-order formula φ, let φ^* be the corresponding explicit two-sorted formula thus obtained, and for a set Σ of second-order formulas, let Σ^* the set of all formulas φ^*, with φ in Σ. It is obvious that for any structure $\mathcal{A}$ and any nonempty collection D of subsets of A, φ is true in $\langle \mathcal{A}, D \rangle$ if and only if $\langle \mathcal{A}, D, \in \rangle \models \varphi^*$ (in particular, φ is canonically true in the structure $\mathcal{A}$ if and only if $\langle \mathcal{A}, \mathcal{P}(A), \in \rangle \models \varphi^*$). The reduction of Henkin second-order consequence to first-order is this: if Δ is the set of sentences characterizing admissibility for interpretations, then for any set Σ of second-order sentences, φ is a second-order consequence of Σ if and only if φ^* is a first-order consequence of $\Sigma^* \cup \Delta^*$. For details, see Shapiro [27, sec. 4.3] or Enderton [11, pp. 277–289] (or chapter 25 above).

[5] To prevent any misunderstanding we want to emphasize that nowhere in this chapter do we assume that all the consequences of the axioms can be *proved* from them in some deductive calculus. Such an assumption would beg the question against canonical second-order consequence.

informal presentation to its formalized version is relatively short. This is so partly because even in informal axiomatizations the concepts that the theory is to deal with are duly specified, and this limits the freedom of formalization. This relative closeness may allow us to view the consequence relation of the formal language from a working perspective, thus helping us to gauge its adequacy to perhaps differing aims and conferring some valuable insight to our decisions.

2. Axiomatic Theories

When building an axiomatic theory, one distinguishes between primitive and defined terms or notions. Not all primitive terms receive the same treatment. Among them, the *specific* ones are singled out as standing for the entities (concepts, relations, operations) the theory is intended to be about. In informal presentations only the specific terms are usually explicitly mentioned, but formalization requires a definite list of all primitive terms, among which we normally find the equality sign, the propositional connectives, and the first-order quantifiers.

It is common to refer to the nonspecific terms as "logical," and to the specific ones as "nonlogical." In a clear sense this is right. Since the logic of a language rests on its consequence relation, and since the consequence relation depends on the behavior of the nonspecific terms, these may be called "logical." Nevertheless, if our task is to ponder whether the consequence relation of some particular languages is a case of logical consequence, it is advisable to keep to a more neutral terminology.

As a matter of fact, it is not hard to find axiomatic theories some of whose primitive nonspecific terms we would hardly count as logical. A simple and interesting example would be an axiomatization of Euclidean geometry presupposing the ordering of the real numbers. By this we mean an axiomatization like Hilbert's, but with a completeness axiom stating essentially that the order of any line is isomorphic to the order of the real numbers. In such an axiomatization, the terms standing for the set of real numbers and for its ordering relation would be nonspecific.

An essential, although imprecise requirement on axiomatization is that everything assumed about the specific (but not about all primitive) entities is to be explicitly stated in the axioms.[6] This implies treating the specific terms *formally* (i.e., as parameters, as mere place-holders for entities of the appropriate category). And treating them thus

[6] Thus Tarski: "Our knowledge of the things denoted by the [specific] terms . . . is, so to speak, our private concern and does not exert the least influence on the construction of our theory. In particular, in deriving theorems from the axioms, we make no use whatsoever of this knowledge, and behave as though we did not understand the content of the concepts involved in our considerations, as if we knew nothing about them that had not been expressly asserted in the axioms" (Tarski [31, pp. 121–122]).

has the effect that the consequences of the axioms hold not only for the intended specific entities, but for any entities for which the axioms hold. This is so fundamental to our conception of an axiomatic theory that many such theories are built with no specific entities intended. Any entities satisfying the axioms are as good as any other. Notice that the specific terms being treated formally is a property of the consequence relation of the theory, and thus of the language in which the theory is framed.

In informal axiomatizations, there always lurks the question of how well determined the consequence relation is. Often the consequence relation (and thus the theory itself) is left indeterminate in that, among other things, assumptions must be made about the entities for which the nonspecific terms stand, in order not only to draw certain consequences from the axioms but also for it to be a fact that these are indeed consequences, and thus belong to the theory. For instance, the axiomatization of Euclidean geometry hinted at two paragraphs above contains no axioms about the order of the real numbers, but the particular properties of this order are involved in determining what follows from the axioms, and hence in fixing the consequence relation.

In practice, this lack of determinacy is bearable, since one is concerned only with those theorems that have been proved, and from a given proof we can see what has been assumed. But such indeterminacy is not satisfactory in a foundational context, where we are not interested in developing the theory but rather in the question of what the theory is (i.e., of which sentences it consists).

In order to make the theory definite, the assumptions about the primitive nonspecific entities must be first singled out and made precise, and then incorporated into the theory, either through the axioms or through the consequence relation. If the former, we provide the original theory with new specific terms and axioms involving them, thus taking as specific some previously nonspecific terms (in our example, we should take the symbols for the set of real numbers and for its order as specific, and add the relevant axioms for them). The reformed theory has a wider subject matter than the original one, since it must deal with the new entities as well. If we choose the second way, we keep the assumptions as components of the consequence relation, either by building an auxiliary theory about them to be used in characterizing the consequence relation of our original axiomatic theory, or by some other means, as illustrated in the next section.

3. Beyond Formality

The formal treatment of the specific terms, which amounts to the acknowledgment that the theorems hold not only for the intended specific entities, but for any entities satisfying the axioms, is a basic requirement of an axiomatic theory. We shall refer to it as the requirement of *formality* of the specific terms.

The fulfillment of this requirement, which is a requirement on the consequence relation regarding the specific terms, is not enough to guarantee the fulfillment of the more basic, even if deeply imprecise, requirement that motivates it, namely, that everything assumed about the specific entities is to be explicitly stated in the axioms. For in some cases, even when formality is present, the consequence relation can code significant assumptions about the specific entities, which thereby need not be explicitly stated. Particularly clear examples can be readily found with the use of Lindström quantifiers, which allow for cheap characterizations of structures or classes of structures with single sentences.[7] As an example, we add to a first-order language the two-place quantifier Q, and extend the definition of formulas with the new clause that $Qxy\varphi$ is a formula whenever φ is. The satisfaction relation is extended thus: for any structure $\mathcal{A}$ with domain A and any assignment s of members of A to the variables,

$$\mathcal{A} \models Qxy\ \varphi[s] \quad \text{iff} \quad \langle A, \varphi^{\mathcal{A}}[s]\rangle \cong \langle R, < \rangle,$$

where $\varphi^{\mathcal{A}}[s] = \{\langle a,\ b\rangle \in A^2 : \mathcal{A} \models \varphi[s^{x\ y}_{a\ b}]\}$, and $\langle \mathbb{R}, < \rangle$ is the real-number order. It is obvious that the sentence

$$(1) \quad Qxy\, x < y$$

is a characterization of $\langle \mathbb{R}, < \rangle$, that is, all its models are isomorphic to $\langle \mathbb{R}, < \rangle$. Consequently, every first-order sentence true in $\langle \mathbb{R}, < \rangle$ is a consequence of (1). However, it is clear that (1) does not explicitly state anything about the order of the real numbers; rather, it presupposes it all.

As a means for characterizing the order of the real numbers, Q is an uninteresting quantifier, but it can be very useful in some other natural contexts. Thus we can use it (or a variant thereof) in the above-mentioned axiomatization of Euclidean geometry to express that the order of the points of a line is isomorphic to that of the real numbers.

We bring into consideration a second example, more akin to canonical second-order consequence. Syntactically, the formulas are those of a monadic second-order language, but the uppercase variables range over all countable subsets of the universe of the structure over which the language is interpreted. Every model of the following set Σ of axioms

1. $\forall xy\, (\forall z\, (zEx \leftrightarrow zEy) \rightarrow x = y)$
2. $\forall X\, (\exists yXy \rightarrow \exists y\, (Xy \wedge \forall z\, (zEy \rightarrow \neg Xz)))$

[7] See Ebbinghaus [10, sec. 4.1].

3. $\forall X \exists y \forall z (zEy \leftrightarrow Xz)$
4. $\forall y \exists X \forall z (zEy \leftrightarrow Xz)$

is isomorphic to the structure $\langle \mathrm{HC}, \in \rangle$ of the hereditarily countable sets with the membership relation.[8] But the fundamental notion about this structure, namely, countability, is not spelled out in the axioms, yet it is implicitly assumed and essentially used, being coded in the quantification. To expose it, we need only to fix any regular cardinal κ and let the uppercase variables of the language range over subsets of cardinality less than κ. If we do so, the same set of formulas characterizes the structure $\langle \mathrm{H}_{\kappa}, \in \rangle$ of sets hereditarily of cardinality less than κ.[9]

As these examples suggest, it is rather easy to build languages whose consequence relation stores large amounts of information about any desired class of mathematical entities or concepts. Building languages of this kind can be helpful for certain purposes, namely, when one wants to use or to apply facts about these concepts or entities in some other field without being distracted with lying down the details of what is being assumed. But it seems clear that a language with a consequence relation laden with characteristic assumptions about certain concepts or entities is unsuitable for axiomatizing a theory about those very concepts or entities. For in such a situation, the general principle that what is assumed about the specific entities should be explicitly stated in the axioms is clearly violated. In axiomatizations of this kind, the content of some theorems comes from the consequence relation rather than from the axioms. Thus, knowing that the above set of axioms characterizes $\langle \mathrm{HC}, \in \rangle$, we conclude that the set-theoretical axioms of infinity, union, and replacement are consequences of Σ. But if we want to see why this is so, we must be involved quite a bit with countability—about which the axioms say nothing, but presuppose everything.

These considerations help to enhance by contrast a general desideratum of logical consequence, namely, its universality of application. A language whose consequence relation is an instance of logical consequence could be used in principle to axiomatize theories about any entities or concepts whatsoever (provided only that it has the right stock of specific terms). Let's call this the requirement of

[8] Here is a sketch of a proof in ZFC. Since a relation R is well-founded if and only if every countable nonempty subset has an R-minimal element, by axioms 1 and 2 every model of Σ is extensional and well-founded; hence, by the Mostowski collapsing theorem, isomorphic to a transitive set M with the membership relation. By axiom 3, every countable subset of M is a member of M, which implies that $\mathrm{HC} \subseteq M$. That $M \subseteq \mathrm{HC}$ follows from axiom 4, which says that every member of M is countable. Everything needed to put flesh on this sketch can be found in Kunen [18, chs. III and IV].

[9] But these formulas do not characterize the structure $\langle \mathrm{HF}, \in \rangle$ of the hereditarily finite sets if the variables are taken to range over all finite subsets, the reason being that the second axiom does not then ensure that the E-relation is well-founded. To characterize HF we may add a fifth axiom asserting that each set is a member of some transitive set: $\forall x \exists y (xEy \wedge \forall zu ((zEu \wedge uEy) \rightarrow zEy))$.

noninterference. It is hard to make it precise, but as the examples above show, there are cases where it is clearly violated. Our first argument against canonical second-order consequence being logical consequence will be that it violates this requirement in a strikingly clear way.

4. Second-order Consequence

A valued strength of canonical second-order languages is their ability to characterize structures and classes of structures. This may be a ground for the use of canonical second-order consequence, but not, of course, for its logical character. First what this strength depends on must be examined. After all, Lindström quantifiers are fully adequate for characterizing structures.

Second-order consequence is defined in terms of sets. Introducing second-order quantification over a structure $\mathcal{A}$ amounts to being given, in addition to $\mathcal{A}$, the power set of its domain A (as well as that of the n-fold product A^n, for each positive integer n). There is, however, a noticeable difference between using Lindström quantifiers and second-order means to characterize a structure: Lindström quantifiers are ad hoc, while second-order quantification is uniformly defined. Nonetheless, one could suspect that this uniformity and ensuing generality are due only to the fact that all mathematical structures considered are set-theoretical entities, and set theory provides all that is needed for implementing canonical second-order quantification over them. In other words, one might suspect that the strength and versatility of canonical second-order consequence comes from its incorporating a significant amount of set theory.

In order to assess the dependence on set theory of canonical second-order consequence, it may prove useful to examine ZF2, the second-order version of Zermelo–Fraenkel ZF with the canonical consequence relation. Although they are redundant, we take as its axioms those of ZF but with separation, replacement, and foundation in second-order form. For a more succinct equivalent version see Shapiro [27, p. 85].

The only specific term of ZF2 is the binary predicate E for the membership relation (we reserve "$\in$" for the metalanguage). In order to avoid circumlocutions and limit the use of formulas when discussing ZF2, we shall often refer to values of first-order variables as "sets" and to values of second-order variables as "classes" (either "classes" *tout court* or "relational classes," among which are the "functional classes"). Thus, we may render the second-order axioms of separation, replacement, and foundation, respectively as (1) every subclass of a set is a set; (2) the image of a set by a functional class is a set; and (3) every nonempty class has a E-minimal member.

A remarkable property of ZF2, first established in Zermelo [34], is its quasi categoricity: every model of ZF2 is isomorphic to $\langle V_\kappa, \in \rangle$, for some inaccessible cardinal κ.[10] Since whenever κ is inaccessible, $\langle V_\kappa, \in \rangle$ is a model of ZF2, it follows that if we add to ZF2 a sentence asserting that there are no inaccessible cardinals, the resulting theory is categorical. It also follows that every set-theoretical question involving only sets of accessible rank is answerable in ZF2 in the sense that its correct answer is a consequence of the axioms. In particular, all propositions of set theory about sets of reals which are independent of ZFC are decided in ZF2, although we don't know in which way. The most famous of these is Cantor's Continuum Hypothesis (CH), first conjectured by Cantor in 1878 and the object of much research nowadays.

ZF2 decides CH in the sense that either CH or its negation is a canonical consequence of the axioms of ZF2. Since CH is independent of ZFC and the axioms of ZF2 are the second-order formulations of those of ZF, one might suspect that the answer to CH provided by ZF2 is contributed by the consequence relation rather than by the axioms. Admittedly, the distinction between what depends on the consequence relation and what on the axioms is not clear in general, but it turns out that the content of both CH and ¬CH can be expressed in a natural way without any appeal to the axioms, or even to the language of set theory. More precisely, there are sentences σ_1 and σ_2 of a pure second-order language such that (it can be shown in ZFC that) σ_1 is universally valid if and only if CH holds, and σ_2 is universally valid if and only if CH does not hold. Moreover, these sentences express in a strikingly direct way the content of CH and of ¬CH, respectively. This is worth emphasizing because the mere fact that for any set-theoretical sentence φ there is a pure second-order sentence σ which is universally valid if and only if φ is a consequence of ZF2 is not by itself indicative of the set-theoretical commitment of canonical second-order consequence. We can always find such a sentence from the finite axiomatizability of ZF2. For if ϑ is the conjunction of the axioms of ZF2, φ is a consequence of ZF2 if and only if the sentence $\vartheta \rightarrow \varphi$ is universally valid,

[10] It is worthwhile to pause and reflect on this short proof (carried out in first-order ZFC) of the quasi categoricity of ZF2, which is more an exercise in set theory than it has to do with logic. Since two of the axioms of ZF2 are extensionality and second-order foundation, every model of the theory is isomorphic to one of the form $\langle M, \in \rangle$, where M is a transitive set. By second-order separation, every subset of a set in M belongs to M, and thus, by the power set axiom, (i) the power set of every set in M is a member of M. On the other hand, by the union axiom and second-order replacement, (ii) whenever a belongs to M and $f{:}a \rightarrow M$, the set $\bigcup f[x]$ is a member of M. But one can show that any transitive set M satisfying (i) and (ii) is of the form H_κ, for $\kappa = \omega$ or κ inaccessible. By the axiom of infinity, the former is excluded. Thus, since for κ inaccessible, $\mathrm{H}_\kappa = V_\kappa$, $M = V_\kappa$ for some inaccessible cardinal κ. For details, refer to Kunen [18, chs. III and IV].

which it is just in case the pure sentence $\forall U \forall R (\vartheta^{U,R} \rightarrow \varphi^{U,R})$ is universally valid.[11] We can thus take this last sentence as σ. The formulas that we presently exhibit concerning CH and ¬CH are not of this form.

Cantor's first formulation of CH was that every set of real numbers is either countable or bijectable with the set all real numbers. Equivalently, if a is a countably infinite set and b is the power set of a, then every subset c of b is either injectable in a ($c \preceq a$) or bijectable with b ($c \sim b$). By writing down the definitions of the concepts involved, we can produce formulas φ_1, φ_2, φ_3, and φ_4 such that for any structure $\mathcal{M}$ (endowed with the canonical interpretation),

$\mathcal{M} \models \varphi_1(X)[A]$ iff A is a countably infinite subset of M
$\mathcal{M} \models \varphi_2(X, Y)[A, B]$ iff B is a faithful copy of the power set of A
$\mathcal{M} \models \varphi_3(Y, Z)[B, C]$ iff $C \subseteq B$,
$\mathcal{M} \models \varphi_4(X, Y, Z)[A, B, C]$ iff $C \preceq A$ or $C \sim B$.

Then σ_1 is the sentence

$$\forall X \forall Y \forall Z((\varphi_1(X) \wedge \varphi_2(X, Y) \wedge \varphi_3(Y, Z)) \rightarrow \varphi_4(X, Y, Z)),$$

while σ_2 is

$$\forall X \forall Y \exists Z((\varphi_1(X) \wedge \varphi_2(X, Y)) \rightarrow (\varphi_3(Y, Z) \wedge \neg\varphi_4(X, Y, Z))).$$

The formulas φ_1, φ_3, and φ_4 can be taken to be mere transcriptions of obvious set-theoretical formulas.[12] φ_2 is the formula saying that there is a relation R such that

1. $\forall xy (Rxy \rightarrow (Xx \wedge Yy))$
2. $\forall W (\forall x (Wx \rightarrow Xx) \rightarrow \exists! y (Yy \wedge \forall x (Rxy \leftrightarrow Wx)))$.

This state of affairs has drastic consequences for the use of second-order languages (with the canonical interpretation) in axiomatizing set theory. CH is an important open problem of set theory, and in order to investigate it, we could consider adding CH as a tentative assumption to ZF2. But this maneuver is either

[11] For any formula ψ of the second-order language of set theory, $\psi^{U,R}$ is the formula obtained from ψ by substituting the binary variable R for the membership relation E and suitably restricting the quantifiers to the unary variable U. It is assumed that neither U nor R occurs in ψ.

[12] See Shapiro [27, sec. 5.1.2] for details and for other pure second-order formulations of CH and ¬CH.

useless or fatal, because if CH is a consequence of ZF2, its addition is redundant, while if it isn't, its addition yields a contradictory theory.

One could object that this conclusion depends on conflating two distinct roles of the axioms, namely, as means for characterizing structures and as premises for deriving theorems.[13] That CH is semantically decided by ZF2 follows from the quasi categoricity of ZF2, but in order to find what the consequences of ZF2 are, we must set up some proof procedure. Indeed, if we limit ourselves to the usual deductive calculi, then we can be sure that (provided ZF2 is consistent in the calculus) neither CH nor $\neg$CH will be deducible from ZF2. These calculi, then, are suitable tools for axiomatizing second-order set theory with the purpose of deriving theorems.

But this objection amounts to not taking second-order consequence seriously.[14] As soon as a second-order language is given the canonical interpretation, this (presumably) determines the consequence relation, regardless of any deductive calculus. It is this consequence relation that we are concerned with. And with respect to it, either ZF2 + CH or ZF2 + $\neg$CH will be contradictory (i.e., will have a contradiction as a semantic consequence). That no contradiction be derived in some incomplete calculus is irrelevant to this.

A look at the axioms of the basic deductive calculus for second-order consequence makes it evident how, when using a second-order language, part of the content of the explicit set-theoretical axioms as formulated in a first-order language is transferred from the axioms of the theory to the consequence relation. Consider the separation schema of ZF: for each first-order formula $\varphi(x)$ with no free occurrences of b,

$$(2) \quad \forall a \exists b \forall x (x \in b \leftrightarrow x \in a \wedge \varphi(x)),$$

meaning that any condition on the elements of a set a determines a subset of a. In ZF2, the schema becomes the axiom

$$(3) \quad \forall X \forall a \exists b \forall x (x \in b \leftrightarrow x \in a \wedge Xx),$$

which can be rendered fairly faithfully as: "every subclass (i.e., every *true* subset) of a set a is a subset of a."

The axiom (3) is presumably stronger than the schema (2), but its strength depends on the implicit assumption that any condition determines a class.[15] In order for this strength to be available in a deduction from the axiom, a calculus for second-order consequence will include the comprehension schema:

[13] See Corcoran [8] on this distinction.

[14] Compare this argumentation with Shapiro [29, pp. 48–49, 56–57].

[15] Or, at least, that for every first-order formula $\varphi(x)$ the class $\{x: \varphi(x)\}$ is a value of the variable X in (3).

(4) $\exists X \forall x(Xx \leftrightarrow \varphi(x))$.

From (3) and (4) (and other rules that we may ignore here), we get (2).

We take the presence of such a schema in sound deductive calculi as evidence that canonical second-order consequence carries set-theoretical content relevant enough to be explicitly expressed in any first-order axiomatization, thus suggesting that the requirement of noninterference for logical consequence is being violated.

Comprehension axioms are not the only axioms in a deductive calculus that make it evident that the content of canonical second-order consequence overlaps widely with set theory. Other such axioms are forms of the set-theoretical axiom of choice. One of them is[16]

(5) $\forall X[\forall x \exists y Xxy \rightarrow \exists Y(\text{Func}(Y) \wedge \forall xy(Yxy \rightarrow Xxy) \wedge \forall x \exists y Yxy)]$.

Being among the axioms of a sound calculus, (5) is supposed to unfold second-order consequence. Now, from (5) and the rest of the calculus we can derive in ZF2 the usual axiom of choice in set theory, which, in one of its many equivalent forms, says that for every relation r there is a function f with the same domain such that for every $x \in \text{dom}(r)$, $\langle x, f(x)\rangle \in r$.

Thus, if the usual calculus with (5) is sound for second-order consequence, then the axiom of choice (AC) is a consequence of the axioms of ZF2, so that ZF2 $+ \neg$AC is contradictory. Accordingly, canonical second-order consequence is unsuitable not just for investigating but even for axiomatizing set theory without choice.

One could dismiss this complaint by replying that the second-order logic which is adequate for dealing with choiceless set theory is a logic without choice. But this reply amounts again to not taking canonical second-order consequence seriously. If canonical second-order consequence is a determinate relation, then either ZF2 has AC as a consequence or it doesn't. If it does (thus, if (5) is sound), then a second-order consequence without choice is some other consequence relation.

The insistence on the determinacy of second-order consequence is crucial. If second-order canonical consequence is not a determinate relation, then theories formulated in a language with canonical consequence are not determined by their axioms, and the alleged categoricity results for second-order axiomatizations become questionable.

[16] Here X and Y are binary predicate variables and Func(Y) is the usual formula expressing that Y is a functional class.

5. On the Existence of Structures

One might object that our complaint that canonical second-order consequence is too involved with set-theoretical matters is overstretched, adducing that, albeit to a lesser degree, it applies to first-order consequence as well. The objection to our complaint could be put thus. If, regardless of the language considered, we define the consequence relation according to Tarski (that is, if we declare that a sentence φ is a consequence of a set of sentences Σ just in case φ is true in all models of Σ), then what sentences turn out to be consequences of a given set will depend on what structures there are, and this is a set-theoretical matter. This dependence can easily be exemplified in the case of first-order consequence. For example, we can find a first-order sentence which is universally valid if and only if there are no sets of five or more members; more interestingly, we can produce a first-order sentence which is universally valid if and only if there are no infinite structures.[17] Thus, since the question of whether there exist infinite sets is certainly mathematical, first-order consequence includes or codes some definite mathematical content. Moreover, paraphrasing what has been said about canonical second-order consequence, if there are no infinite sets, then any first-order theory which has no finite models will be not only false but also contradictory.

We must be careful here. Whether or not we accept that there are infinite sets, we do accept that there are infinitely many sets or, at any rate, infinitely many objects, even if there is no set containing them all. If we are willing to discuss ordinary formal languages, we certainly accept a potentially infinite class of formulas, deductions of arbitrary length, and so on. Thus we reason about infinitely many things, and so our logic (that is, the logic we use when we reason informally about these mathematical objects) applies to them. Accordingly, whether or not we are willing to accept infinite sets, our logic does not condone the inference from "$\forall xy(fx=fy \rightarrow x=y)$" to "$\forall y \exists x fx=y$," or from "*R is a transitive and irreflexive relation*" to "*there is a maximal element with respect to R*," which are suitable for reasoning only about finite structures. According to the informal logic we reason by, these inferences are not valid, obvious counterexamples to them being provided by the successor operation on the natural numbers and the order relation among them, respectively. We reason about natural numbers, whether or not they make a set.

If we turn to set theory in order to build a theory of consequence for first-order languages agreeing with our logical practice in ordinary mathematical reasoning, our set theory must admit infinite sets, to serve as counterparts of the infinite multiplicities of numbers, formulas, and so on about which we reason.

[17] Take, for instance, a sentence with a binary predicate R expressing that *if* R is a linear order, *then* there is a maximal element with respect to R.

Otherwise, the logical theory we obtain will be manifestly inadequate. We should note that, strictly speaking, our set theory need not have an axiom of infinity, for the infinite counterparts of infinite multiplicities could be dealt with as proper classes. What must be asked of our set theory is that it be strong enough to allow one to define the syntactic and semantic notions of first-order languages and, perhaps, to prove the completeness theorem for first-order logic. This does not require much in the way of existence of sets or classes, since the model of a consistent set of sentences Σ obtained from the standard, Henkin's proof is arithmetical in Σ. Strictly speaking, only this is indispensable, namely, that the mathematical reconstruction of first-order consequence that we carry out in our set theory be in accordance with the informal version (at least as far as can be discerned). If our set theory is unsuitable for this purpose (perhaps because it is too weak), then we are not interested in its version of first-order consequence.

In short, if our mathematical reconstruction of first-order consequence is such that it declares a sentence σ to be a consequence of a set of sentences Σ if and only if σ is true in all finite models of Σ, then we declare it simply wrong. The crucial thing here is that we have a fairly good idea of what the relation of consequence is for first-order languages; we know at least that the usual rules of inference are correct, and this is enough (as the completeness theorem tells us) to accept the set-theoretical reconstruction as adequate.

But this, we contend, is not the situation as regards canonical second-order consequence, of which we have only the set-theoretical definition. Now, this contention might be objected to out of hand by appealing to our understanding of second-order reasoning previous to set theory, to the use of second-order inferences before the advent of set theory—but such an objection would miss the point. All that our non-set-theoretical practice of second-order languages requires is already embodied in Henkin consequence,[18] against which we have now nothing to say. It is precisely the requirement that the uppercase (unary) variables range exactly over all subsets of the universe of the structure that distinguishes canonical consequence from Henkin's, and what endows it with its impressive strength. Indeed, if the consequence relation of ZF2 were Henkin's, then neither CH nor its negation would be a theorem; more to the point, if σ_1 and σ_2 are the sentences construed above as expressing the continuum hypothesis and its negation, respectively, then neither σ_1 nor σ_2 is universally valid as regards Henkin consequence. This fact highlights an important difference between first-order and canonical second-order consequence. In the case of first-order logic, our inferential practice directed us toward a set theory allowing for infinite sets; here it does not direct to a set theory either with CH or with $\neg$CH. The set-theoretical, canonical version of second-order consequence goes significantly further than our logical practice, and cannot be justified by it.

[18] This is argued at length by Väänänen in [32].

6. Logical Versus Iterative Sets

One can object that our criticism of canonical second-order consequence as coding strong set-theoretical content is misguided, on the grounds that it overlooks a significant distinction, that between the logical and the iterative notions of set. Only the former, it is claimed, is needed to account for canonical second-order consequence, while the latter is the object of set theory proper. A set in the logical sense is said to be a subset of some domain, while an iterative set is an object in the cumulative hierarchy of sets.[19] Now, put this way, there seems to be no difference to account for, since we can view the cumulative hierarchy as an ordinal indexed sequence of increasing domains in such a way that the (logical) subsets of each domain are the members of the next. Since an iterative set is simply a subset (and a member) of any of these domains, it follows that all iterative sets are logical sets—or at least objects standing for them.

In order to find a sound distinction, we are asked to look more closely into what is needed for second-order quantification. It appears that we don't need to account for the whole power set operation, which assigns to each set its power set; when defining truth in a structure for second-order formulas, we need to focus only "on the subsets of a *fixed* universe or domain. That is, the context of the theory determines, or presupposes, a range of the first-order variables. A set is a subset of this universe (only).... Thus, in arithmetic, a logical set is a collection of numbers; in geometry, a logical set is a collection of points (or regions); and so on. There are no logical sets *simpliciter*, only logical sets within a given context".[20] Accordingly, for the canonical use of a second-order language, only the power set of the domain of objects that the language is about is needed, but by no means the whole power set operation.

Now, this limited scope may suffice when using a language for merely descriptive purposes, but it is not enough to support the logical functions which make second-order languages with the canonical interpretation strong; in particular it cannot account for their capacity of structure characterization and for second-order consequence. Consider a second-order characterization of the order $\langle \mathbb{R}, < \rangle$ of the real numbers by the sentence σ expressing that $<$ is a conditionally complete linear order without end points with a countable dense subset. In order to see that σ is categorical, it is not enough to evaluate σ in $\langle \mathbb{R}, < \rangle$. σ must also be evaluated in all relational structures, or at least in all dense linear orders without end points, in order to exclude those that are not conditionally complete or separable. The reason why some such order $\langle A, \prec \rangle$ is not a model of σ is to be

[19] See Shapiro [26, p. 721] and [27, pp. 18–22, 184–185].

[20] Shapiro [27, p. 18]. Later, Shapiro partly rejected his strong distinction between logical and iterative sets. See [29, pp. 60–61].

found in the contents of $\mathcal{P}(A)$. Thus, the power set of A needs to be taken into account to guarantee the categoricity of σ. The point of this remark is not that all the structures (of the relevant similarity type) have to be considered in order to exclude them as models of σ, but rather that, in addition to the structures, the whole power set of each of their domains must be surveyed as well.

The same happens when we turn to canonical consequence. In order for it to be determinate whether a sentence φ is a canonical consequence of a sentence ψ, it is not enough to be determinate what structures there are, but also what is the content of the power set of each domain. Since the power set of the domain of a structure is not given along with the structure, but is secured from it by the semantics itself, we must conclude that the full power set operation is needed for the proper account of canonical second-order consequence.

There is, however, a notion of set which is different from the one discussed so far and which can rather naturally be regarded as logical. It is the notion of a *conceptual* set (or class) given as the collection of all objects falling under some definite concept. This is to be opposed to the notion of a *combinatorial* set, an arbitrary collection of objects of a domain, regardless of there being any concept under which all and only its members fall. The combinatorial notion of set is the one that set theory is about, and thus the one needed for canonical second-order consequence. The conceptual notion can be attributed to Frege, who dismissed the idea behind combinatorial sets[21] and its relevance to logic.[22] Only if second-order consequence rested on a notion of set like the conceptual one could it be argued, with Frege, that second-order consequence is a case of logical consequence. But the ensuing consequence relation would not be the canonical one, but rather a variant of Henkin's.[23]

7. On the Determinacy of Canonical Second-order Consequence

When we say that canonical second-order consequence carries strong set-theoretical content, we are assuming that this is a determinate relation, at least for each fixed similarity type. It has to be a determinate relation if a list of axioms is

[21] "I do, in fact, maintain that the concept is logically prior to its extension; and I regard as futile the attempt to take the extension of a concept as a class, and make it rest, not on the concept, but on single things" (Frege [13, p. 228]).

[22] "Only because classes are determined by the properties that individuals in them are to have... it becomes possible to express thoughts in general by stating relations between classes; only so do we get logic" (Frege [13, p. 226]).

[23] See Jané [19] for a more detailed account of the distinction.

to determine a single theory as the totality of its consequences. Thus, when we argued that there is a pure second-order sentence which is universally valid if and only if the continuum hypothesis is true, we implicitly assumed that it is a fact whether such a sentence is valid (and that it is also a fact whether CH is true).

We have argued that the strong set-theoretical content coded by canonical second-order consequence makes it unsuitable for axiomatizing set theory. We want to emphasize that our complaints had nothing to do with the lack of a complete proof procedure (we have not been concerned at all with proofs). We complained about the specific set-theoretical content that the consequence relation contributes to the theory beyond the axioms. This becomes especially evident when we think of the many pure second-order sentences whose canonical validity is equivalent to some distinctly set-theoretical claim—which quite often, as in the CH case, can be seen to be expressed in a natural way by the sentence in question.

The equivalence of the validity of particular pure second-order sentences with set-theoretical assertions suggests a skeptical view of the determinacy of canonical second-order consequence. Namely, it suggests that whenever we speak of canonical second-order consequence, we are alluding only to a highly underdetermined relation, and that difficult set-theoretical decisions have to be made in order to *turn* it into a determinate one. And many of such decisions are open issues in set theory.

It also suggests that in order to admit that second-order consequence *is* a determinate relation, one has to acknowledge that a significant number of set-theoretical questions have a definite answer; that among the possibly conflicting ways we have of extending current set theory, exactly one is right. In other words, claiming that canonical second-order consequence is determinate requires taking a strong realist view of set theory. This counts against second-order consequence as being usable as logical consequence, since the use of logic as such should not depend on adopting a disputed philosophical position.[24]

It may be thought that we are going too far in involving philosophical realism in this issue; that mere methodological, or working, realism is enough; that we don't have to bring into consideration a notion of truth stronger than that implicit in mathematical practice, whatever that is.[25] When set theorists speak of measurable cardinals, we don't have to attribute to them any reference to a Platonic world in which these cardinals, if they exist, dwell. Why, then, assume such a thing with regard to second-order consequence?

We claim that we have to assume something like this if we want to use canonical second-order consequence as the self-sufficient consequence relation of axiomatic theories. For if the axioms must determine the theory completely, then

[24] Otherwise, philosophically nonrealist classical mathematicians would not be justified in using some part of logic.

[25] See Shapiro [28, pp. 38–44], for a description of the various degrees of working realism and its opposition to philosophical realism.

everything required to settle which particular sentences follow from the axioms, and which particular sentences don't follow from them, is to be embodied in the consequence relation. This should be the case if canonical second-order consequence were an instance of logical consequence.[26]

As a matter of fact, the standard use of canonical second-order consequence is not as logical consequence. In ordinary mathematical practice, one deals with second-order logic as applied set theory. Second-order languages with the canonical interpretation are, rather, set-theoretical tools which are handled in a set-theoretical context. The set-theoretical setting may be implicit, but it makes itself felt as soon as difficulties arise and we turn to a set theory book for help. In such a setting, no question about the determinacy of second-order consequence arises, since no question arises about the determinacy of the power set of any set we happen to be dealing with; all questions are internal to set theory, and all internal questions are treated as if they had a definite answer. This is just an aspect of classical reasoning, as can be easily shown: whatever proposition p is, $p \vee \neg p$ is a tautology and, as such, it is taken as true. Now according to the common use of the term "true," a disjunction is true just in case at least one of the disjuncts is, and a negation is true just in case the negated proposition is false. From this it follows, taking any conjecture (as the continuum hypothesis) as p, that either it is true or it is false.

Nevertheless, this piece of reasoning works only inside the set-theoretical setting (but logic, as such, should be usable outside the setting, namely, to systematize the setting itself). If any argument is needed, even one applying to set theory, it is that all we said in the previous paragraph is sensible when we take the setting to be some informal counterpart of first-order ZFC with a truth predicate, and it stays sensible even if we believe that all there is to sets is what this theory can prove.

We want to add a remark about the ontological commitment of canonical second-order logic. In order to refute Quine's assertion that second-order logic has "staggering existential assumptions,"[27] Boolos points to the fact that the sentence $\exists X \exists x \exists y (Xx \wedge Xy \wedge x \neq y)$ is not universally valid to conclude that "despite its affinities with set theory and its vast commitments, second-order logic is not committed to the existence of even a two-membered set" (Boolos [4, p. 40]). But this, we submit, is wrong. Boolos' line of reasoning would apply only if second-order logic were taken to be a set of theses, for only then would it be appropriate to judge its ontology by looking at its validities. But it is out of place if we view logic as dealing with the consequence relation of a language and if, as in

[26] Here Frege's maxim's is apt: "About what is foreign to it, logic knows only what occurs in the premises; about what is proper to it, it knows all" (Frege [14, p. 338]). One could cite as well Tarski's dictum that "logic itself does not presuppose any preceding discipline" (Tarski [31, p. 119]).

[27] "Set theory's staggering existential assumptions are cunningly hidden now in the tacit shift from schematic predicate letter to quantifiable variable" (Quine [24, p. 68]).

our case, consequence is defined in terms of structures. To assess the ontological commitment in this case, we have to look at what structures are required to exist in order that the consequence relation be a determinate relation. And this we find not by looking at the logical validities, but rather at the sentences that are not universally valid. That a sentence σ is not universally valid means that there is a structure failing to satisfy it. Hence, a logic that does not declare σ universally valid is committed to the existence of some such structure.

8. Subsets

It is a notorious circumstance that we don't know how to explain what an arbitrary subset of an infinite set is, let alone to describe the exact content of its power set. This wouldn't be much of a drawback if we had a procedure for generating all subsets of an infinite set from its members, but there is none. If all we know about a particular set is that it is infinite, then, even if allowed to use a name for each one of its members, we won't be able to specify a single infinite subset with infinite complement. So, how do we know that there is any? Unless we assume some form of the axiom of choice, it is consistent with our assumptions about sets as codified in ZF that there are infinite sets all of whose subsets are either finite or cofinite.[28] Even when, as in the case of the set of the natural numbers, we know how to specify infinitely many infinite subsets of the given set, we also know that every enumerating rule will leave out some of them.

Both set theory and second-order logic have to give some account of power sets, but not with the same urgency. Sets (and their power sets) are the subject matter of set theory, which thus can be loosely described as a theory pursued to gain knowledge of sets. Lacking a clear explanation of the power set operation is not a defect in set theory, since getting such an explanation is, rather, one of the aims it has to reach—not a starting point. Second-order logic, for its part, makes use of the power set operation, and an essential use at that, since on it rests the determinacy of the consequence relation. As we remarked above, any divergence in accounting for the content of power sets is bound to disturb canonical second-order consequence; in other words, different accounts of their content will give rise to different consequence relations. Accordingly, the autonomous use of second-order consequence (as opposed to its internal use in set theory, that is, to what we have described as its use in a set-theoretical setting) requires an explanation of (1) what is a set of objects of a domain and, most important, of (2) what is the exact content of the power set of a set—in terms of the content of the set.

[28] See Jech [20, exercise 21.6].

Let us fix some infinite set. For definiteness and simplicity we take the set $\mathbb{N}$ of the natural numbers. Whatever a set of natural numbers may be, we may regard it as the collection of its elements. This may be insufficient to account for full set theory, where sets are treated as objects, but it seems to be sufficiently adequate for our present purposes. We don't want to assume, at least not for the time being, that collections are objects; they may be mere pluralities or classes as many in Russell's sense (Russell [25, pp. 68, 76]).

There is a rather clear general notion of a collection of natural numbers, namely, that of a plurality specifiable with certain definite means. This general notion encompasses a variety of particular notions, obtained by fixing the means allowed for specification—for example, that of a collection first-order definable in the structure $\langle \mathbb{N}, +, \cdot \rangle$. Each particular choice of means yields a definite notion of subset with a definite extent, but none of them can deliver a power set rich enough to serve as the range of the unary variables of canonical second-order languages. No restriction of means is congenial with the intended absolute character of canonical second-order consequence.

However, placing no restrictions on the means allowed for specification, that is, taking a set to be (or to correspond to) a collection specifiable with any means whatever, will not do, because such a notion of a collection would lack definite extent: since it is not determinate what a possible means of specification is, it would hardly be determinate what collections there are.

The customary way of overcoming all limitations is to get rid of specifications altogether and to introduce the idea of an arbitrary, or combinatorial, set, that is, of a collection obtained "as the result of infinitely many independent acts deciding for each number whether it should be included or excluded" (Bernays [3, p. 260]). Of course, this is only a metaphor, since no agent is assumed to carry out the infinitely many steps of such a selection, not even in principle.

This proposal is apparently less effective than the previous one. By allowing arbitrary means of specification, our notion of collection was open-ended (there is no such thing as the totality of all arbitrarily specifiable sets of natural numbers), but at least we knew what an individual such set is. With combinatorial sets we have nothing but a metaphor which becomes even obscurer when we leave the domain of the natural numbers and consider in the abstract any (unstructured) infinite set. As Hermann Weyl put it:

> The notion that an infinite set is a "gathering" brought together by infinitely many individual arbitrary acts of selection... is nonsensical.... I contrast the [predicative] concept of set... with the *completely vague* concept of function which has become canonical in analysis since Dirichlet and, together with it, the prevailing concept of set. (Weyl [33], 23)

There is no doubt, however, that the combinatorial notion of set has been fruitful as a motivation for axioms in set theory. With its help we can justify impredicative separation and the axiom of choice. But it doesn't explain what a subset of a given infinite set is or what subsets an infinite set has. Moreover, the combinatorial notion of set is sensible only in a Platonist setting.[29] From a Platonist perspective, there is no need to be precise on what combinatorial sets are in order to succeed in referring to them. Indeed, if the combinatorial sets all exist, then, even if our hints about their nature fail to characterize them with any accuracy, they can nonetheless be good enough to single them out by separating them from other entities. But as an account of "combinatorial set," the Platonist way is helpless.[30]

We want to sketch a non-Platonist account of the power set operation which, albeit insufficient for second-order logic, is good enough for set theory. In order to overcome our inability to find a suitable notion of subset with a definite extent, we invert the priority relation between subsets and power set. Thus, we don't introduce the power set $\mathcal{P}(a)$ of a set a as the totality of all subsets of a; rather we define a subset of a to be a member of $\mathcal{P}(a)$—the latter being envisioned as some *maximal closed totality of subsets of a*. The idea is that all we need to assume about the power set of a is that it is a definite, closed totality (as opposed to an open-ended, increasable multiplicity) and that it is maximal. Surely these are not meant to be mathematical conditions, but neither is the requirement of being a combinatorial set. Nevertheless, they should be of help in motivating our choice of mathematical axioms. Maximality is understood as implying that no matter what collection of elements of a we would ever specify, $\mathcal{P}(a)$ will contain a set corresponding to it. In particular, all subsets of a which we know how to specify in any given context should be in $\mathcal{P}(a)$, just like those which are specifiable in terms of other members of $\mathcal{P}(a)$, which is thus envisaged as closed under various operations, some of them inspired by the metaphor of combinatorial sets. Of course, we don't want to add any axiom asserting or implying that any totality closed under such and such operations is the power set of a, but we may eventually add new axioms saying or implying that $\mathcal{P}(a)$ is closed under some new operations. By so doing, we won't enter into conflict with previous results, but we will possibly reject some previous interpretations as deficient.[31] This need not be understood as discovering new facts about sets, but rather as further determining our notion of set (and of subset). As to this further determination, it can be attributed to a deeper analysis of our original concept of set, but also to a refining of this concept. Nothing mathematical

[29] According to Gödel, a combinatorial subset "is conceived as something *which exists in itself* no matter whether we can define it in a finite number of words" (Gödel [16, p. 259, n. 14]; emphasis added).

[30] Gödel again: The notion of a combinatorial subset "cannot be defined satisfactorily (at least in the present state of knowledge), but can only be paraphrased by other expressions involving again the concept of set" (Gödel [16, p. 259, n. 14]).

[31] Thus we won't add $V = L$ as an axiom, but we may consider adding the axiom of measurable cardinals, which implies that the power set of ω is closed under sharps. See Kanamori [21, p. 110] and Maddy [23, p. 76 and passim].

depends on which stand we take. The important thing is that set theory does not have to be grounded on a fixed, absolute power set operation.

The set-theoretical cumulative hierarchy need not be seen as a single, well-determined universe of sets. This hierarchy, which can be thought of as generated by the iterated application of the power set operation along the ordinals, depends on these two parameters: the extent of power sets and the length of the ordinal sequence. According to set theory, they are both maximal, each in its own way: *all* ordinals are considered, and the power set of a set *a* contains *all* subsets of *a*. The mathematical import of these two *alls* is articulated in the theory, whose axioms are motivated, at least partly, by insights or ideas which are hard to make precise. The key idea behind the maximality of the ordinal sequence is that it is absolutely infinite in Cantor's sense. Of course, this notion of absolute infinity is not embodied in the theory, and even when engaged in philosophical reflection, we do not want to insist, let alone presuppose, that absolutely all ordinals (if that makes actual sense) are taken into account. Nevertheless, the idea that the ordinal sequence is absolutely maximal in its way is not idle, since it helps to motivate the choice of certain axioms of infinity. Something analogous is the case with respect to the intended maximality of power sets. Not, of course, that the way power sets are maximal is analogous to the way the ordinal sequence is. What is analogous is that the idea of the maximality of power sets works mainly as a motivating guide which cannot be fully expressed as a mathematical condition. It motivates the rejection of putative axioms that, like Gödel's axiom of constructibility, restrict the riches of power sets. Moreover, both in the case of the ordinal sequence and in that of the power set operation, there is no reason why maximality should be realized in an absolute sense.

If this outline of the power set operation is sound, canonical second-order logic is either an internal development of set theory, and thus not properly logic, or else it rests on a myth. But then, Henkin's emerges as the right semantics for the consequence relation of second-order languages. It does not require that to each domain an absolutely maximal power set corresponds, but only approximations to this intended, inaccessible ideal which are closed under whatever conditions we can state. Thus, all expressible assumptions we need to make about the existence of second-order entities will be fulfilled in Henkin interpretations, and will be taken into account in the shaping of the consequence relation—which now will be a determinate relation (as we argue in section 11).

9. A Remark on Categoricity

Our account of the power set operation impinges on the import of categoricity in canonical second-order languages. The meaning of a categoricity result is clear in a set-theoretical setting (i.e., for second-order logic as an application of set theory). In this setting (thus under the implicit assumption of a fixed assignment of a power set to every set), the import of categoricty is clear: a categorical set of sentences has a unique model up to isomorphism. However, from a higher, or external, standpoint, from which we perceive that no assignment has been fixed, categoricity is, at most, a relative matter—relative to just one of the ineffable and perhaps differing ways of complying with the ideal of maximality of power sets. From such a standpoint (the one we took in the preceding section) we see that despite categoricity, no single structure may have been absolutely characterized.[32]

10. The Plural Interpretation

George Boolos ([5], [6]) has proposed a plural interpretation of second-order quantification according to which second-order monadic variables do not range over the power set of the domain of the structure the language is interpreted in, but range over the domain itself, albeit plurally. Thus, the formula

$$\forall X(\exists x Xx \rightarrow \exists x(Xx \wedge \forall y(Xy \rightarrow x \leq y))),$$

which we usually read as "Every nonempty set (of whatever objects we are considering) has a least element," can be read plurally as "whenever there are some objects, one of them is the least." Similarly, the comprehension axiom $\exists X \forall x (Xx \leftrightarrow \varphi(x))$ can be read as "there are some objects which are precisely those satisfying the condition φ" or, more leisurely, as "there are some objects such that each of them satisfies the condition φ, and each object that satisfies the condition φ is one of them." Since there are conditions that no object satisfies, this example shows that the reading of "$\exists X \ldots$" needs some correction in order to take into account would-be empty pluralities. But we can ignore this point.[33]

A second-order sentence (with only monadic quantified second-order variables) is assumed to be true in a structure when given the plural interpretation just in case it is true in the canonical sense. More to the point, if A is the domain

[32] This applies in general, although in some special cases the situation may be better. Thus, in order to guarantee that the second-order Dedekind–Peano axioms characterize the structure $\langle \mathbb{N}, 0, S \rangle$ of the natural numbers, it is enough to admit that for any set A, any operation g on A, and any $a \in A$, the closure of $\{a\}$ under g exists (and is a value of the unary second-order variables). If we admit that, then whatever other subsets power sets may contain, the Dedekind–Peano axioms will single out $\mathbb{N}$ up to isomorphism.

[33] It turns out that, as Shapiro and Weir show in [30], this point is not as innocent as it may look.

of the structure under consideration, it is assumed that for each $B \subseteq A$ there are some objects in A which are precisely the elements of B. Thus, the various pluralities called up by the existential quantifier coincide with the collections corresponding to all subsets of A. That is to say, there are as many ways of referring plurally to some objects in A as there are subsets of A.

This is fundamental, since the plural version of the second-order consequence relation is meant to coincide with the canonical one. The only difference between the two accounts lies in the ontology required to make each work: the objects of the domain and all sets thereof in the canonical version, only the objects of the domain in the plural one.

We can very simply characterize the power set operation with the help of second-order quantification, whether plurally or canonically conceived. If A is a set and D is any collection of subsets of A, then $D = \mathcal{P}(A)$ if and only if the structure $\langle A \cup D, A, D, \in \rangle$ is a model of the sentences

1. $\forall xy(Exy \rightarrow Px \wedge Qy)$
2. $\forall xy(Qx \wedge Qy \wedge \forall z(Ezx \leftrightarrow Ezy) \rightarrow x = y)$
3. $\forall X(\forall x(Xx \rightarrow Px) \rightarrow \exists y(Qy \wedge \forall z(Ezy \leftrightarrow Xz)))$,

where P, Q and E are interpreted, respectively, as A, D, and the membership relation between objects in A and sets in D.

Let A be an infinite set and suppose that D is a rich but incomplete collection of subsets of A. In particular, D is to contain all subsets of A we know how to specify. It is clear that the structure $\langle A \cup D, A, D, \in \rangle$ satisfies the first two formulas above, but since D falls short of being the power set of A, the third one will be false in it. Thus, it will satisfy

$$\exists X(\forall x(Xx \rightarrow Px) \wedge \forall y(Qy \rightarrow \exists z(Ezy \leftrightarrow \neg Xz))),$$

that is, in plural parlance, there are some objects in A which coincide with the members of no set in D. Which objects these are we cannot say, for if we could, then the set whose members they are would be in D, contrary to our assumption. This makes it plain that understanding the ways of plural quantification needed to account for the full strength of canonical second-order consequence is tantamount to understanding the idea of a combinatorial set.[34] Indeed it is harder, since in the case of sets, one could at least appeal to their independent existence in order to account for them, but with plural quantification there seems to be nothing on which to base the assumption that some objects can be summoned that do not coincide with the elements of any set in D. Of course, we can base this

[34] In Shapiro's words: "Epistemic qualms about second-order variables become epistemic qualms about plural quantifiers" (Shapiro [28, p. 234]).

assumption on the existence of sets: it is the existence of a set *a* not in *D* which guarantees that there are some objects (namely, the members of *a*) which coincide with no set in *D*. If we do this, then, although sets are not used in evaluating second-order formulas in a structure, they are nevertheless needed to ensure that plural quantification works properly. This is not an important gain (if it is a gain at all), and moreover it is not adequate for the principal use to which plural second-order quantification is meant to be put, namely, to allow second-order quantification on the universe of sets without commitment to the existence of proper classes (Boolos [5, pp. 65–66]). What is required is to account for what we could describe as *all the ways of there being some objects* without presupposing that a set corresponds to each of these ways.

One side effect of this requirement is that plural second-order logic cannot be salvaged as an application of set theory. On the other hand, the objection that we have leveled against the legitimacy of a truly maximal totality of collections of (say) natural numbers applies also to the plural interpretation. There nothing was supposed about the nature of collections. They could be treated as objects, but also as mere pluralities. The argument we gave there against canonical second-order quantification is not that it requires strong ontological resources, but rather that the notion of an arbitrary (combinatorial) collection (even as a mere plurality) of objects of a domain *A* cannot be accounted for from below, that is, by appealing only to objects in *A*. What makes obscure the idea of the totality of all subsets of *A* is not worries (nominalistic or otherwise) about the individual nature of sets, but the utter arbitrariness of the collections corresponding to them—that is, of the pluralities of objects selected by no agent with no statable rule, pluralities that we are supposed to muster when we say that there some objects that so and so.

The main reason alleged for the legitimacy of plural quantification is that we understand it well enough in ordinary language.[35] However, assuming that we understand it well enough for everyday purposes is not a ground for believing that it can support canonical second-order consequence. The ways of singular

[35] Thus Boolos: "[T]here is a coherent and intelligible way of interpreting such second-order formulas.... The interpretation is given by translating them into the language we speak.... It cannot be seriously maintained that we don't *understand* these statements... or that any lack of clarity that attaches to them has anything to do with the plural forms found in the sentences expressing them. The language in which we think and speak provides the constructions and turns of phrase by means of which the meanings of these formulas may be expressed in a completely intelligible way" (Boolos [5, p. 69]). And Lewis: "Besides the elementary logical apparatus of truth functions, identity, and ordinary singular quantification, our framework also shall be equipped with apparatus of plural quantification.... This apparatus is not common in formal languages, but we know it well as masters of ordinary English" (Lewis [22, p. 62]).

and plural quantification are not equally smooth. Suppose we are given a definite domain *A* with all its elements. To explain singular quantification on *A*, we have to take into account no more than what we are given, whereas to explain plural quantification, we have to bring into consideration all possible ways of sorting elements of *A*. In view of the parallel complexities of the power set operation, it is more than doubtful that competence in one's mother tongue can accomplish that.

11. Completeness and Determinacy

We have raised two main charges against canonical second-order consequence. We have argued that unless a strong realist position is taken, canonical second-order consequence cannot be assumed to be a determinate relation. We have also argued that, even assuming that it is a determinate relation, it is not a case of logical consequence, since it cannot be used as a noninterfering consequence relation for axiomatic set theory. By far, the first charge is the stronger.

In view of the importance that a consequence relation be determinate, it would be desirable to have a warrant for determinacy. The existence of a sound and complete deductive calculus is such a warrant, for if the consequences of any recursively given list of axioms can be recursively generated, then the question of whether a definite sentence is among them will have a definite answer—whether or not we have an algorithm for it. To be sure, the claim of determinacy of the deducibility (thus of the consequence) relation rests on the assumption that the natural number sequence is a definite one. But this is already presupposed in the mere description of a formal language—for instance, in the definiteness of the notion of a formula as a finite sequence of symbols generated according to certain given rules. Without such an assumption one cannot even entertain a rigorous notion of an axiomatic theory. Thus, we don't have to argue for it now.

Besides guaranteeing its determinacy, the existence of a suitably chosen deductive calculus can be of help in deciding whether a given consequence relation should be taken as logical.[36] More profitably, by examining the explicit list of the axioms and rules of a given complete calculus, we can assess the assumptions behind the consequence relation, thus allowing us to gauge to what extent it can be used as the noninterfering consequence relation of some particular axiomatic theory. As an example, the presence of the comprehension axioms in a complete

[36] Section 3 of Cutler [9] can be read as an argument for this point.

calculus for Henkin's second-order consequence shows that this consequence relation is not suitable for axiomatizing set theory, although it can be fruitfully used in other theories, among them number theory.

As this example suggests, the explicitness of the axioms and rules of a complete deductive calculus may allow us to isolate part of the interfering content of a given consequence relation and to transfer it to the axioms of the theory being axiomatized. In set theory, this would correspond to passing from the second-order Henkin version of ZF to first-order Morse–Kelly.[37]

By doing this we can perhaps convince ourselves that the ensuing consequence relation is free of any content concerning the specific entities of the theory we are axiomatizing, and thus the noninterference requirement is met—but only as regards this particular theory. For it is clear that if the trimmed consequence relation is not altogether trivial (that is, if a set of sentences implies some sentence not in the set), then it must code some content. Every nontrivial consequence relation has to embody some implicit assumptions about the primitive nonspecific entities of the theory under consideration. These are the assumptions on which rests the validity of some particular inferences and the invalidity of some others.[38]

The question, then, is how to render these unstated assumptions innocuous. One obvious danger which has to be avoided is that of disagreement among practitioners. If the assumptions remain unstated, how do we know that they are definite and that we are dealing with one fixed consequence relation?[39] It is at this point that the existence of a deductive calculus settles the matter, since whatever those assumptions are, the determinacy of the consequence relation is guaranteed. In a deductive calculus the hidden assumptions about the workings of the primitive nonspecific terms are not stated, but their effect in fixing the consequence relation can be fully ascertained. It is a safe stopping point in the process of trimming a consequence relation. What we get may still not be a case of logical consequence, but it will be closer to it.

If these considerations are right, then admitting a deductive calculus is an essential feature of any self-sufficient consequence relation (in particular, of any sort of logical consequence)—not because a calculus is an instrument for proving theorems, but because the existence of a complete calculus is the surest guarantee that any given list of axioms determines a definite theory. Canonical second-order consequence is very far from this.

[37] Also called "Quine–Morse." See Fraenkel, Bar-Hillel, and Levy [12, pp. 138–141]).

[38] Thus, ordinary first-order consequence has implicit assumptions about the behavior of the connectives and the quantifiers.

[39] This is not a merely hypothetical possibility, since it is realized in second-order canonical consequence.

REFERENCES

[1] Barwise, J., and Feferman, S. (eds.) (1985). *Model-theoretic Logics.* Springer-Verlag, Berlin, Heildelberg, New York.

[2] Benacerraf, P., and Putnam, H. (eds.) (1983). *Philosophy of Mathematics,* 2nd ed. Cambridge University Press, Cambridge.

[3] Bernays, P. (1935). Sur le platonisme dans les mathématiques. *L'Enseignement mathématique,* 34, 52–69. English translation, Platonism in mathematics, in Benacerraf and Putnam [2], pp. 258–271.

[4] Boolos, G. (1975). On second-order logic. In Boolos [7], 37–53.

[5] Boolos, G. (1984). To be is to be a value of a variable (or to be some values of some variables). In Boolos [7], 54–72.

[6] Boolos, G. (1985). Nominalist Platonism. In Boolos [7], 73–87.

[7] Boolos, G. (1991). *Logic, Logic, and Logic.* Harvard University Press, Cambridge, Mass.

[8] Corcoran, J. (1980). Categoricity. *History and Philosophy of Logic,* 1, 187–207.

[9] Cutler, D. (1997). Review of Shapiro [27]. *Philosophia Mathematica,* 5, 71–91.

[10] Ebbinghaus, H.D. (1985). Extended logics: The general framework. In Barwise and Feferman [1], 25–76.

[11] Enderton, H. (1972). *A Mathematical Introduction to Logic.* Academic Press, New York.

[12] Fraenkel, A., Bar-Hillel, Y., and Levy, A. (1973). *Foundations of Set Theory.* North-Holland, Amsterdam.

[13] Frege, G. (1895). A critical elucidation of some points in E. Schröder, *Vorlesungen über die Algebra der Logik.* In Frege [15], 210–228.

[14] Frege, G. (1906). On the foundations of geometry: Second series. In Frege [15], 293–340.

[15] Frege, G. (1984). *Collected Papers in Mathematics, Logic, and Philosophy,* edited by B. McGuinness. Basil Blackwell, Oxford.

[16] Gödel, K. (1964). What is Cantor's continuum problem? In Gödel [17], 254–270.

[17] Gödel, K. (1990). *Collected Works,* vol. II, edited by S. Feferman et al. Oxford University Press, New York.

[18] Kunen, K. (1980). *Set Theory: An Introduction to Independence Proofs.* North-Holland, Amsterdam.

[19] Jané, I. (1993). A critical appraisal of second-order logic. *History and Philosophy of Logic,* 14, 67–86.

[20] Jech, T. (1978). *Set Theory.* Academic Press, New York.

[21] Kanamori, A. (1994). *The Higher Infinite.* Springer-Verlag, Berlin, Heidelberg, New York.

[22] Lewis, D. (1998). *Parts of Classes.* Blackwell, Oxford.

[23] Maddy, P. (1997). *Naturalism in Mathematics.* Clarendon Press, Oxford.

[24] Quine, W.V. (1970). *Philosophy of Logic.* Prentice-Hall, Englewood Cliffs, N.J.

[25] Russell, B. (1903). *The Principles of Mathematics.* George Allen and Unwin, London.

[26] Shapiro, S. (1985). Second-order languages and mathematical practice. *Journal of Symbolic Logic,* 50, 714–742.

[27] Shapiro, S. (1991). *Foundations Without Foundationalism.* Oxford University Press, New York.

[28] Shapiro, S. (1997). *Philosophy of Mathematics: Structure and Ontology.* Oxford University Press, New York.

[29] Shapiro, S. (1999) Do not claim too much: Second-oder logic and first-order logic. *Philosophia Mathematica (3)*, 7, 42–64.

[30] Shapiro, S., and Weir, A. (2000). "Neo-logicist" logic is not epistemically innocent. *Philosophia Mathematica (3)*, 8, 160–189.

[31] Tarski, A. (1946). *Introduction to Logic and to the Methodology of Deductive Sciences.* Oxford University Press, New York.

[32] Väänänen, J. (2001). Second-order logic and foundations of mathematics. *Bulletin of Symbolic Logic*, 7, 504–520.

[33] Weyl, H. (1917). Das Kontinuum. English translation by S. Pollard and T. Bole, *The Continuum.* Dover, New York, 1987.

[34] Zermelo, E. (1930). Über Grenzzahlen und Mengenbereiche. *Fundamenta Mathematicae*, 16, 29–47.

Index